NEET

National Eligibility cum Entrance Test

Objective CHEMISTRY

Part-I

NEET

National Eligibility cum Entrance Test

Objective CHEMISTRY

Part-I

Dr. Manu Kaushal

G K Publications (P) Ltd

Title : NEET 2022 : Objective Chemistry Part I - Guide

Language : English

Author's Name : Dr. Manu Kaushal

Typeset & Published by :

Career Launcher Infrastructure (P) Ltd.

A-45, Mohan Cooperative Industrial Area, Near Mohan Estate Metro Station, New Delhi - 110044

Marketed by :

G.K. Publications (P) Ltd.

Plot No. 9A, Sector-27A, Mathura Road, Faridabad, Haryana-121003

ISBN : **978-93-89310-53-5**

Printer's Details : Printed in India, New Delhi.

For product information :

Visit ***www.gkpublications.com*** or email to ***gkp@gkpublications.com***

ABOUT THE AUTHOR

Dr. Manu Kaushal is a renowned faculty of Chemistry and has teaching experience of more than 15 years. He has worked as a senior faculty with institutes like Aakash Institute (New Delhi) and Helix Institute (Chandigarh). Through this book, he wants to share his vast experience with each and every student in India and help them prepare the best for their most important exam.

PREFACE

Students appearing for NEET cannot rely on conventional books and measures to crack the entrance examination. Keeping in mind the cut-throat competition, it becomes mandatory for all the aspirants to have a thorough exposure of all types of problems, especially in Chemistry (Part I).

The objective of this book is to help students get rid of any phobia about Chemistry.

The questions have skillfully been set along with solutions to help students develop an all round problem solving methodology.

I am thankful to my parents, son 'Yuvraj', daughter 'Neyhal' and my wife who encouraged me to write this book.

Finally, I would like to thank GK Publications (P) Ltd. for giving me an opportunity to write this book.

Regards

Dr. MANU KAUSHAL

CONTENTS

CHAPTER 1

Some Basic Concepts in Chemistry

Questions Based on Calculation of Mole, Molecules, Atomic Mass& Laws of Chemical Combination

1. Match the following

	Column – I		Column – II
(a)	0.55 mole	(i)	4.48 L of SO_2 at STP
(b)	0.1 mole	(ii)	12.32 L of H_2 at STP
(c)	0.2 mole	(iii)	22 g of CO_2
(d)	0.5 mole	(iv)	2.24 L of NH_3 at STP

(a) a - i, b - ii, c - iii, d - iv

(b) a - i, b - iii, c - iv, d - ii

(c) a - i, b - iv, c - i, d - iii

(d) a - iii, b - iv, c - ii, d - i

2. How many moles of an electron weight one kilogram?

(a) 6×10^{23}

(b) $\frac{1}{9.1} \times 10^{31}$

(c) $\frac{6.02}{9.1} \times 10^{54}$

(d) $\frac{1 \times 10^{8}}{9.1 \times 6.02}$

3. If water sample are taken from rivers or lake, then hydrogen & oxygen will be in approximate ratio1:8 (by wt) . This statement is in accordance with which of the following laws?

(a) Law of reciprocal proportion

(b) Law of constant composition

(c) Law of multiple proportion

(d) Both 1 & 3

4. Hydrogen and oxygen combine to form H_2O & H_2O containing 5.93% & 11.2% hydrogen (by wt) respectively. This date illustrates–

(a) Law of multiple proportions

(b) Law of constant composition

(c) Law of reciprocal proportion

(d) Both 1 & 2

5. Which of the following combinations illustrate the law of reciprocal proportions?

(a) CO, CO_2, C_3O_2 (b) SO_2, SO_3, H_2O

(c) H_2S, SO_2, H_2O (d) Both (a) & (b)

6. At STP, 5.6 litres of a gas weighs 60 g. The vapour density of the gas is -

(a) 30 (b) 60

(c) 120 (d) 240

7. Let atomic mass of an element be 'A' grams. Then the mass of 10 atoms of element 'A' in amu is –

(a) $\frac{A}{10}$ (b) $\frac{A}{6.02 \times 10^{23}}$

(c) 10 A (d) $\frac{10A}{6.02 \times 10^{23}}$

8. Arrange the following samples in correct order of no. of molecules –

(1) 100 ml of di-oxygen at STP

(2) 100 ml of di-hydrogen at STP

(3) 100 ml of H_2O (liquid)

(4) 100 ml of Ozone at STP

(a) 1 = 2 = 3 = 4 (b) 2 > 3 > 1 > 4

(c) 3 > 4 > 1 > 2 (d) 3 > 4 = 1 = 2

9. Which of the following is correct ?

(a) 1gm atom of two different elements weighs 1gm each.

(b) 1gm molecule of two different molecules may have same weight.

(c) 1amu is reciprocal of Avogadro's No.

(a) Both a & b

(b) Both b & c

(c) a, b & c

(d) Both a & c

10. xgm of element P (at. wt = 15) contains y atoms. How many atoms are contained in 2*x* g of element Q (at. wt = 30)

(a) 2y (b) 4y

(c) $\frac{y}{2}$ (d) y

11. 10 moles of CO_2 do not contain

(a) 120g of C

(b) 6.02×10^{24} atoms of O

(c) $10N_A$ molecules of CO_2

(d) 20gm atoms of O

12. The number of molecules in 16g of CH_4 is–

(a) 3×10^{23} (b) 6×10^{23}

7 (c) $\frac{16}{N_A}(6.02 \times 10^{23})$ (d) $\frac{16}{3} \times 10^{23}$

13. The total number of atoms in 0.1 mole of a triatomic gas is ($N_A = 6.02 \times 10^{23}$)

(a) 1.8×10^{22} (b) 6×10^{23}

(c) 1.8×10^{23} (d) 3.66×10^{23}

14. If 10^{21} molecules are removed form 200 mg of CO_2, then the number of moles of CO_2 left are–

(a) 2.88×10^{-3} (b) 28.8×10^{-3}

(c) 0.28×10^{-3} (d) 1.66×10^{-2}

15. Calculate the number of the gram atoms of an element in 40 kg, if weight of 1 atom of an element is 6.644×10^{-23} g.

(a) 10^2 g atoms (b) 10 g atom

(c) 10^3 g atoms (d) None of these

16. The number of atoms in 558.5 g of Fe (at. wt = 55.85 g) is –

(a) Twice that in 60 g C

(b) 6.02×10^{22}

(c) Half in 8 g He

(d) $558.5 \times N_A$

17. Which of the following contains the greatest number of atoms?

(a) 1 g of C_4H_{10} (b) 1g of N_2

(c) 1g of Ag (d) 1g of H_2O

18. The number of neutrons in 1.8g of H_2O will approximately be –

(a) 4.216×10^{23} (b) 8.4×10^{23}

(c) 4.8×10^{24} (d) None of these

19. The numbers of atoms present in 52 amu of He are–

(a) 13 (b) 13 N_A

(c) N_A (d) $\frac{N_A}{6}$

20. Identify the correct statement among the following–

(a) 0.5gm atom of oxygen contains $\frac{N_A}{4}$ oxygen molecules.

(b) 0.5gm molecules of oxygen contains $\frac{N_A}{4}$ oxygen molecules.

(c) Total electrons present in 0.5g atom of oxygen are 4 N_A.

(d) Total electrons present in 0.5g atom of O_2^+ is $\frac{7N_A}{4}$.

(a) a & b (b) a & d

(c) b & d (d) a & c

21. Which one of the following is correct for 2 moles of N_3^-?

(a) The mass of 2 moles is 84g.

(b) It contains 32 N_A valence electrons.

(c) It contains 21 N_A total electrons.

(d) It contains 42 N_A total protons.

(a) a & c (b) b, c, d

(c) a, b & d (d) All of these

22. 11.2L of a gas at STP weighs 14g. The gas could be –

(a) N_2O (b) NO_2

(c) N_2 (d) NH_3

23. Which of the following weighs most?

(a) 50g of Fe (b) 5g atoms of nitrogen

(c) 0.1g atom of Ag (d) 10^{23} atoms of C

24. The number of gram molecules of oxygen in CO molecules is –

(a) 10g molecules (b) 5 g molecules

(c) 1 g molecule (d) 0.5 g molecules

25. What mass of H_2 will contain same number of atoms as that of 1.6g of O_2?

(a) 2gm (b) 0.2g

(c) 0.1gm (d) None of these

26. What is the mass of 1 molecule of oxygen?

(a) 5.314×10^{-22} g (b) 3.5×10^{-23} g

(c) 53.14×10^{-23} g (d) 5.314×10^{-23} g

27. The number of molecules of H_2S contained in 0.40 mole of H_2S are–

(a) 2.408×10^{23} (b) 24.08×10^{23}

(c) 2.4×10^{20} (d) 24.8×10^{22}

28. An organic compound on analysis was found to contain 0.032%. If its molecule contains 2 atoms of S, molecular mass of the compound is-

(a) 200 amu (b) 2000 amu

(c) 20000 amu (d) 200000 amu

29. Which one of the following is not correct for 1 mole of CH_4?

(a) It contains 6.02×10^{23} molecules

(b) It occupies 22.4L at STP

(c) It weighs 16 gms

(d) It weighs 16 amu

30. The percentage of Se in peroxide anhydrous enzyme is 0.5% by weight (at. wt 73.4). Then what is the minimum molecular weight of peroxide anhydrous enzyme?

(a) 14680 amu (b) 29360 amu

(c) 58720 amu (d) None of these

31. The volume at STP occupied by 14g of N_2, 1.5 moles of CO_2 & 10^{21} molecules of O_2 gas respectively is–

(a) 11.2 L, 33.6 L, 0.037 L

(b) 33.6 L, 44.8 L, 11.2 L

(c) 44.8 L, 44.8 L, 11.2 L

(d) None of these

32. The ratio of oxygen atoms in 1g O, 1g O_2 & 1g O_3 is–

(a) 1: 2 : 3 (b) 2: 3 : 1

(c) 3: 2 : 1 (d) 1: 1 : 1

33. The abundances of isotopes ^{54}Fe, ^{56}Fe & ^{57}Fe are 5%, 90% & 5% respectively. Then, the at. wt. of Fe is –

(a) 58 (b) 55.95

(c) 68 (d) 59.21

34. How many moles of $Mg_3(PO_4)_2$ will contain 0.25 mole of oxygen atoms?

(a) 2.5×10^{-2} (b) 0.02

(c) 3.125×10^{-2} (d) 1.25×10^{-2}

35. The masses of oxygen, which combine with fixed mass of hydrogen to form H_2O & H_2O_2 respectively, bears the simple ratio 1:2.

The above statement illustrates the –

(a) law of constant composition

(b) law of multiple proportion

(c) law of reciprocal proportion

(d) law of conservation of energy

36. 2g of an oxide of M contains 1g of M, 8g of another oxide of M contain 3.2g of M. This date illustrates the law of –

(a) Reciprocal proportions

(b) Constant proportion

(c) Conservation of energy

(d) multiple proportion

37. The total number of electrons in 180mg of glucose are –

(a) 9.6×10^{-2} N_A (b) 96 N_A

(c) 9.6×10^{-3} N_A (d) 96×10^{-19} N_A

38. P & Q are two elements which form P_2Q_3 & PQ_2 molecules. If 0.15 mole of P_2Q_3 & PQ_2 weighs 15.9g & 9.3g, respectively, the atomic masses of P & Q are, respectively

(a) 26 & 18

(b) 20 & 32

(c) 28 & 56

(d) 56 & 28

Questions based on limiting Reagent & Stoichiometry

39. If 1.5 moles of oxygen combine with Al to form Al_2O_3, then the mass of Al (at mass = 27) used in the reaction is-

(a) 45 gm (b) 54 gm

(c) 35 gm (d) 22 gm

40. Assuming fully decomposed, the volume of CO_2 released at STP on heating 9.85 g of $BaCO_3$ (Ba = 137, C = 12, O = 16) will be-

(a) 0.84 L (b) 2.24 L

(c) 4.06 L (d) 1.12 L

41. The mass of C anode consumed (giving only CO_2) in the production of 270 kg of aluminium metal from bauxite by Hall's process is–

$$2Al_2O_3 + 3C \longrightarrow 4Al + 3CO_2\uparrow$$

(at mass of Al = 27 gm)

(a) 90 kg (b) 540 kg

(c) 180 kg (d) 270 kg

42. When a mixture of 10 mole of SO_2, 15 mole of O_2 was passed over a catalyst, 8 mole of SO_3 was formed. How many moles of SO_2 & O_2 did not enter into combination?

(a) 2 & 11 (b) 3 & 14

(c) 8 & 12 (d) 2 & 5

43. A mixture of N_2 and H_2 is allowed to react in a closed container to form NH_3.The reaction ceases before either reactant has been totally consumed. At this stage, 4 moles each of N_2, H_2 and NH_3 are present.

Which of the following statements are correct regarding above experiment?

(1) Initial moles of N_2 were 6.

(2) Initial moles of H_2 were 10.

(3) Initial moles of H_2 were 8

(4) For complete reaction to occur, H_2 gas would have been the limiting reagent.

(a) 1, 2 & 3 (b) 2, 3 & 4

(c) 1, 2 & 4 (d) 1, 3 & 4

44. Cortisone is a molecular substance containing 21 atoms of C per molecule. The mass percentage of carbon in cortisone is 69.98%. The molecular mass of cortisone (in amu) is–

(a) 176.5 amu (b) 252.2 amu

(c) 287.6 amu (d) 360.1 amu

45. Air contains nearly 20% oxygen by volume. The volume of air needed for complete combustion of 100 cc of acetylene will be-

(a) 1.25 cc (b) 12.5 cc

(c) 125 cc (d) 1250 cc

46. If 0.5 mole of $BaCl_2$ is mixed with 0.2 moles of Na_3PO_4, the maximum number of moles of $Ba_3(PO_4)_2$ that can be formed are–

(a) 0.70 (b) 0.50

(c) 0.30 (d) 0.10

47. The volume (in litres) of CO_2 liberated at STP when 10 g of 90% pure lime stone ($CaCO_3$) is heated completely is-

(a) 2.016

(b) 20.16

(c) 2.24

(d) 22.4

48. A certain compound has the molecular formula X_4O_6. If 10 amu of compound contains 6.06 amu of X, the atomic mass of X is –

(a) 32 amu (b) 37 amu

(c) 42 amu (d) 48 amu

49. 146 g of HCl is neutralized by x gms of NH_3. The weight of ammonium sulphate required for the production of x gms of NH_3 is–

(a) 264 g

(b) 403 g

(c) 528 g

(d) 1056 g

50. Consider the reactions–

$$Mg_3N_2 + 6H_2O \rightarrow 3Mg(OH)_2 + 2NH_3$$

$$2\,NH_3 + \frac{5}{2}O_2 \xrightarrow{Pt} 2NO + 3H_2O$$

What amount of Mg_3N_2 must be used to produce sufficient NH_3 so that it may give 400 g of NO?

(a) $\frac{400 \times 10}{6}$ g

(b) $\frac{400 \times 3}{10}$ g

(c) $\frac{100 \times 3}{400}$ g

(d) $\frac{400}{100 \times 3}$ g

51. An element 'A' reacts with the compound BO_3 to produce A_3O_4 & B_2O_3. The no. of moles of A_3O_4 produced if 1 mole each of A & BO_3 are allowed to react, is–

(a) 3 (b) 1

(c) $\frac{1}{3}$ (d) $\frac{2}{3}$

52. When xg C is burnt with y gm oxygen in a closed vessel, no residue is left. Which of the following is correct regarding the relative amounts of oxygen & carbon?

(a) $\frac{y}{x}$ must be less than 1.33

(b) $\frac{y}{x}$ must be greater than 1.33

(c) $\frac{y}{x}$ must be greater than 2.67

(d) $\frac{y}{x}$ must be between 1.33 & 2.67

53. What volume of 20% by mass of HCl solution is required to obtain 213g of chlorine gas from the following reaction (density of HCl solution $= 1.20\frac{gm}{ml}$)

$$2KMnO_4 + 16HCl \longrightarrow 2KCl + 2MnCl_2 + 5Cl_2 + 8H_2O$$

(a) 2.22 litre (b) 3.75 litre

(c) 1.46 litre (d) None of these

54. Consider air as a 4 : 1 mixture of N_2 & O_2 by volume. The mass of O_2 in a hall with dimension $5m \times 5m \times 5m$ at STP

(a) 142.8 kg

(b) 160 kg

(c) 35.7 kg

(d) None of these

55. A compound contains 28% N_2 & 72% of metal by weight. Three atoms of metal combine with two atoms of N. The atomic weight of metal is

(a) 24 (b) 12

(c) 36 (d) 26

56. On heating 4.9 gm of impure $KClO_3$, its mass is reduced by 0.384 gm & converted into a gas. The percentage of original $KClO_3$ that has been decomposed is

(At. wt. of K = 39, Cl = 35.5, O = 16)

(a) 20% (b) 30%

(c) 14.7% (d) 12.25%

57. 90 ml of pure dry O_2 is subjected to Silent electric discharge. If only 10% of it is converted to O_3, the volume of O_2 & O_3 respectively in the reaction mixture is

(a) 84 ml & 78 ml (b) 81 ml & 6 ml

(c) 71 ml & 9 ml (d) 90 ml & 6 ml

58. $2SO_{2(g)} + O_{2(g)} \longrightarrow 2SO_{3(g)}$

Above reaction is carried out by taking 10 moles of SO_2 & 15 moles of O_2 to give 8 moles of SO_3. Then, Choose the correct option

(a) O_2 is excess reagent

(b) SO_2 is limiting reagent

(c) Percentage yield of reaction is 80%

(d) All are correct.

59. NaBr, used to produce AgBr for use in photography can be self prepared as follows

$Fe + Br_2 \longrightarrow FeBr_2$ ....(1)

$3FeBr_2 + Br_2 \longrightarrow Fe_3Br_8$ ....(2)

$Fe_3Br_8 + 4Na_2CO_3 \longrightarrow 8NaBr + 4CO_2 + Fe_3O_4$ ...(3)

If the yield of 2nd reactions 60% & 3rd reaction is 70% then mass of Fe required to produce 2.06×10^3 kg NaBr is

(a) 10^5 kg

(b) 10^7 kg

(c) 10^3 kg

(d) None of these

60. A mixture of formic acid & oxalic acid is heated with concentrated H_2SO_4. The gas produced is collected & on its treatment with KOH solution, the volume of the gas decreases by one sixth. Then, the molar ratio of two acids in the original mixture is

(a) $HCOOH : H_2C_2O_4 = 4:1$

(b) $H_2C_2O_4 : HCOOH = 1:4$

(c) $HCOOH : H_2C_2O_4 = 3:1$

(d) $H_2C_2O_4 : HCOOH = 1:3$

61. An oleum is labeled as 109%. The mass % age of free SO_3 & H_2SO_4 are respectively

(a) 60 & 40 (b) 40 & 60

(c) 30 & 70 (d) 70 & 30

62. Consider air as a 4 : 1 mixture of N_2 & O_2 by volume. The mass of O_2 in a hall with dimension 5m × 5m × 5m at STP is –

(a) 142.8 kg (b) 160 kg

(c) 35.7 kg (d) None of these

63. Reaction $N_2 + 3H_2 \longrightarrow 2NH_3$ is used to produce ammonia. When 450g of H_2 is reacted with (excess) N_2, then 1575g of NH_3 is produced. What is the percentage yield of this reaction?

(a) 41.5% (b) 30.8%

(c) 61.8% (d) 20.7%

64. An ore contains 1.34% of Ag_2S by mass. How much gram of this ore would have to be processed in order to obtained 1gm Ag?

(at. Mass of Ag = 108 & S = 32)

(a) 74.6 gm

(b) 85.7 gm

(c) 108 gm

(d) 1.14 gm

65. A metal oxide has formula M_2O_3. It can be reduced by hydrogen to give free metal and water. If 0.16 g of metal oxide requires 6 mg of hydrogen for complete reduction, then the atomic wt. of metal is –

(a) 27.90

(b) 159

(c) 79.86

(d) 56

66. 1 Mole of N_2 & 4 mole of H_2 are allowed to react in a vessel and 50% of expected moles of NH_3 are formed. Further, H_2O is added to this reaction mixture that dissolves all NH_3 produced. What is the mole fraction of H_2 in the gaseous mixture after reaction?

(a) $\frac{1}{6}$ (b) $\frac{5}{6}$

(c) $\frac{1}{3}$ (d) $\frac{2}{3}$

67. $CaCO_3 \xrightarrow{\Delta} CaO + CO_2 \uparrow$

$2KOH + CO_2 \longrightarrow K_2CO_3 + H_2O$

The decomposition of a certain mass of $CaCO_3$ gave 11.2 dm^3 of CO_2 gas at STP. The mass of KOH required to completely neutralize the gas is – (K = 39, Ca = 40)

(a) 28g (b) 20g

(c) 56g (d) 42g

68. What mass of phosphorous pentoxide will be obtained on heating 1.33 g phosphorous and 5.07 g of O_2?

(a) 3.05 g (b) 7 g

(c) 4 g (d) 9 g

69. A bangle made up of Cu – Zn alloy weighs 10g. It is accidently dropped into a beaker containing sulphuric acid & 224 ml of H_2 gas is evolved at STP. The percentage of zinc (by weight) in the bangle, if only zinc reacts with , is –

$Zn + H_2SO_4 \longrightarrow ZnSO_4 + H_2$

(At. wt. of Zn = 65, At. wt. of Cu =63.5)

(a) 22.4%

(b) 6.5%

(c) 65%

(d) None of these

70. $CaSO_4 \cdot xH_2O$ looses 6.2% of it by weight upon heating. The value of x will be–

(M. wt. of $CaSO_4$ = 136)

(a) 4

(b) $\frac{1}{2}$

(c) 2

(d) 3

71. Hydrated salt $Na_2CO_3 \cdot xH_2O$ undergoes 63% loss in mass on heating and becomes anhydrous. The value of x is –

(Na_2CO_3 = 106)

(a) 10

(b) 12

(c) 8

(d) 18

72. Potassium chlorate decomposed on heating as–

$2KClO_3 \xrightarrow{\Delta} 2KCl + 3O_2 \uparrow$

How much will mass of 80% pure $KClO_3$ produce 48g of oxygen? (at. mass of K = 39g)

(a) 98g sample (b) 153.12g sample

(c) 245g sample (d) 122.5g sample

Equivalent Weight & Molecular Mass

73. The equivalent weight of H_3PO_4 in the following reaction is–

$H_3PO_4 + Ca(OH)_2 \longrightarrow CaHPO_4 + 2H_2O$

(a) 98 (b) 49

(c) 32.66 (d) 40

74. An oxide of metal has 20% oxygen (by wt.), the equivalent wt. of metal and metal oxide are, respectively

(a) 32, 40

(b) 30, 46

(c) 38, 42

(d) 38, 38

75. When a metal is burnt, its weight is increased by 24%. Calculate the equivalent wt. of the metal.

(a) 67.7

(b) 33.3

(c) 12.25

(d) 69.91

76. Equivalent weight of a bivalent metal is 37.2. The molecular weight of its chloride is-

(a) 15.87 (b) 145.4

(c) 22.35 (d) 120.7

77. The equivalent weight of an element is 4. Its chloride has a vapour density 59.25. Then the valency of the element is –

(a) 4 (b) 3

(c) 2 (d) 1

78. 0.16 g of dibasic acid required 25 ml of decinormal NaOH solution for complete neutralization. The molecular weight of the acid will be –

(a) 32

(b) 64

(c) 128

(d) 256

79. Two oxides of a metal M contain 40% & 50% of the metal respectively. If the formula of the first oxide is M_2O_3, then the formula of the second oxide is–

(a) MO (b) MO_2

(c) M_3O_4 (d) M_2O_5

80. The chloride of a metal has 52.8% chlorine by mass. The equivalent wt. of metal is -

(a) 31.7 (b) 35.2

(c) 63.5 (d) 20

Questions Based on Concentration Terms

81. What is normality of 1M solution of H_3PO_4?

(a) 3N (b) 2N

(c) 4N (d) 1N

82. A mixture of ethanol & water contains 54% water by mass. Calculate the mole fraction of alcohol in this solution.

(a) 0.25 (b) 0.95

(c) 1.65 (d) 3.25

83. 214.2g of (sucrose) sugar syrup contains 34.2g of sugar. The molality of the solution is - (Formula of sucrose is $C_{12}H_{22}O_{11}$)

(a) 0.55 (b) 1.5

(c) 6.6 (d) 4.72

84. How much H_2O should be added to 200 c.c. of semi-normal solution of NaOH to make is exactly decinormal ?

(a) 200 cc (b) 400 cc

(c) 600 cc (d) 800 cc

85. 25 ml of 3M HNO_3 is mixed with 75 ml of 4M HNO_3. Then, the molarity of the final mixture is–

(a) 3.75M (b) 2.62M

(c) 4.31M (d) 5M

86. Molarity of 20% of H_2SO_4 by mass whose density is 1.02g /cc is –

(a) 8.4M (b) 2.08M

(c) 7.28M (d) 1.7M

87. What is the mass % of C in CO_2?

(a) 0.034% (b) 27.27%

(c) 3.4% (d) 28.7%

88. The density of 3 molal solution of NaOH is 1.110 g/ml. The molarity is

(a) 2.97 M (b) 5 M

(c) 6 M (d) 9 M

89. Which of the following terms are unitless?

(1) molality (2) molarity

(3) mole fraction (4) mass percent

(a) 1 & 2 (b) 3 & 4

(c) 1 & 4 (d) 2 & 3

90. If 250 ml of a solution contains 24.5g H_2SO_4, calculate the molarity & normality of the solution.

(a) 1M, 2N (b) 2M, 1N

(c) 3M, 3N (d) 3M, 1N

91. Find the density of an aqueous H_3PO_4 solution whose weight % age is 9.8% & molarity is 1.25.

(a) 1.25 g/ml (b) 2.5 g/ml

(c) 1.5 g/ml (d) 2 g/ml

92. When 644g of $Na_2SO_4 \times H_2O$ is dissolved in 4000 ml of aqueous solution, the molarity of solution was found to be 0.5 M. The value of x is –

(a) 5 (b) 10

(c) 8 (d) 2

93. Which of the following graphs is NOT correctly represented?

(a)

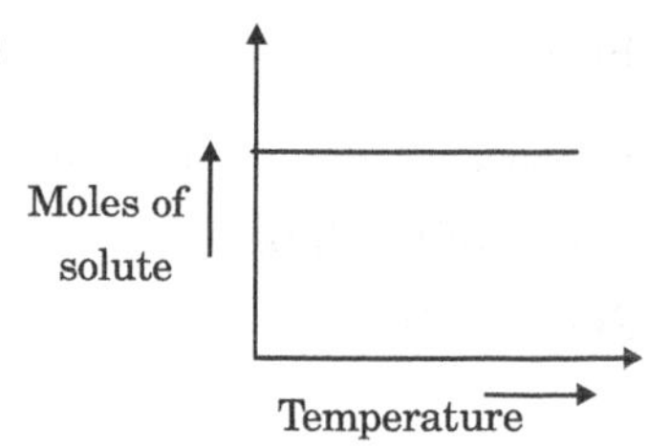

(b)

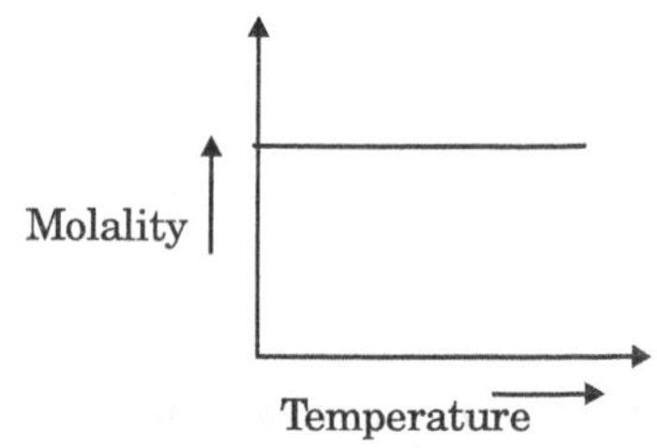

(c)

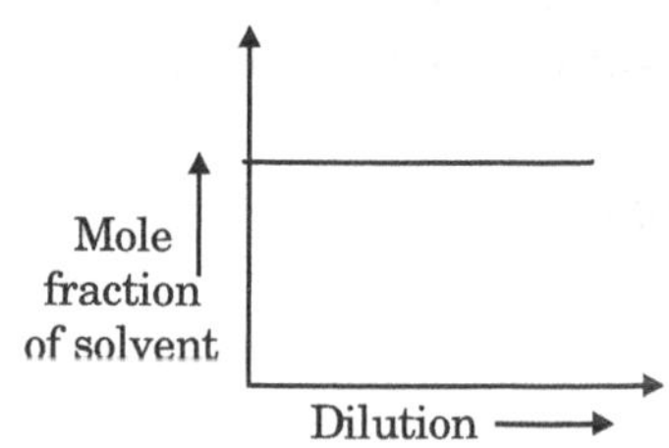

(d) All are correct

94. The mole fraction of a solute in a solution is 0.1. At 298 K, molarity of the solution is same as its molality. Density of this solution is 2g cm^{-3}. The ratio of molecular weights of solute & solvent is-

(a) 5 : 1

(b) 7 : 2

(c) 9 : 1

(d) None of these

95. What is the volume required of a 20% HCl solution of density 1.20 g/ml to prepare 363 g of $AsCl_3$, according to the following equations–

(As =75, Cl = 3.55)

$$2\,KMnO_4 + 16HCl \longrightarrow 2KCl + 2MnCl_2 + 5Cl_2 + 8H_2O$$

$$2As + 3Cl_2 \longrightarrow 2\,AsCl_3$$

(a) 2.56 L

(b) 0.73 L

(c) 1.46 L

(d) 2.9 L

96. In 1200g solution, 12g urea is present. If Density of solution = 1.2 g/ml, then molarity of solution of

(a) 0.2 M (b) 10 M

(c) 0.167 M (d) 12 M

97. Upon heating a litre of $\frac{M}{2}$ HCl solution, 2.675g of HCl is lost due to evaporation & the volume of the solution shrinks to 750 ml. Then, the molarity of resulting solution is X Molar & the number of millimoles of HCl in 100 ml of final solution are y. Then X &Y are, respectively

(a) 0.569, 56.9

(b) 0.769, 76.9

(c) 0.70 & 7

(d) 0.23 & 23

98. How many grams of Conc. HNO_3 solution should be used to prepare 250 ml of 2 M HNO_3?

Conc. Acid is 70% (w/w) HNO_3

(a) 90g (b) 70g

(c) 54g (d) 45g

99. What is the concentration of nitrate ions if equal volumes of 0.1M $AgNO_3$ & 0.1M NaCl are mixed together?

(a) 0.1M

(b) 0.2M

(c) 0.05M

(d) 0.25M

100. If the molality of an aqueous solution is 0.111, then the mole fraction of solute will be -

(a) 0.02 (b) 0.002

(c) 0.2 (d) 0.1

101. A compound contains 28% N_2 & 72% of a metal by weight. 3 atoms of metal combined with 2 atoms of N. The atomic weight of metal is -

(a) 24

(b) 12

(c) 36

(d) 48

Empirical Formula & Molecular Formula

102. Empirical formula for CH_3COOH is same as that for –

(a) $C_5H_5O_5$

(b) $C_{12}H_{22}O_{11}$

(c) $C_6H_{12}O_6$

(d) None of these

103. A compound contains C = 40%, H = 6.67% & rest is oxygen. The empirical formula of the compound is –

(a) $C_{12}H_{22}O_{11}$ (b) CH_2O

(c) C_2H_2O (d) CH_3O

104. The empirical formula of an organic compound containing carbon & hydrogen is CH_2. The mass of 1 litre of this organic gas is exactly equal to that of 1 litre of N_2 at STP. So, molecular formula of this gas –

(a) C_2H_4 (b) C_3H_6

(c) C_6H_{12} (d) C_4H_8

105. A gaseous hydrocarbon gives upon combustion 0.72 g of water & 3.08 g of CO_2. The empirical formula of the compound is

(a) C_3H_4 (b) C_6H_5

(c) C_9H_8 (d) C_2H_4

106. 0.2012g of an organic compound containing carbon, hydrogen & oxygen, gave on complete combustion, 0.4431g CO_2 & 0.1462g of H_2O. The molecular weight of the compound is 100g. Then, the molecular formula of the compound is–

(a) $CaCO_3$

(b) $C_5H_8O_2$

(c) $C_4H_{20}O_2$

(d) None of these

Miscellaneous

107. Consider the given reaction

$$H_4P_2O_7 + 2NaOH \longrightarrow Na_2H_2P_2O_7 + 2H_2O$$

If 534 gm of $H_4P_2O_7$ is reacted with 3×10^{23} molecules of NaOH, then total number of molecules produced in the product are

(a) $2.5\times N_A$

(b) $5\times N_A$

(c) $7.5\times N_A$

(d) $75\times N_A$

108. Equivalent weight of a metal is 20. If the metal forms a tripositive ion, its atomic weight will be

(a) 20

(b) 40

(c) 60

(d) None of these

109. 2g of a metal carbonate on decomposition gave 1.5 g of metallic oxide. The equivalent mass of metal is

(a) 58 (b) 29

(c) 5.8 (d) 2.9

110. The vapour density of a tetrabasic acid is x. The equivalent mass of the acid is

(a) $\frac{x}{2}$

(b) $\frac{x}{3}$

(c) $\frac{2x}{3}$

(d) $\frac{3x}{2}$

111. 0.36 gm of a metal was dissolved in 100 ml of 1N H_2SO_4 & the volume of solution was made upto 1000 ml by adding water. 40 ml of this solution was neutralized completely by 28 ml of 0.1 N NaOH. The equivalent mass of the metal is

(a) 3 (b) 6

(c) 12 (d) 24

112. V_1 ml of NaOH of molarity X & V_2 ml of of $Ba(OH)_2$ molarity Y are mixed together. The mixture is completely neutralized by 100 ml of 0.1 N HCl. If $\frac{V_1}{V_2}=\frac{1}{4}$ & $\frac{X}{Y}=4$. Then, what fraction of acid is neutralized by $Ba(OH)_2$?

(a) 0.5

(b) 0.25

(c) 0.33

(d) 0.67

113. What volume of 75% H_2SO_4 by mass is required to prepare 1.5 litres of 0.2 M H_2SO_4? (Density of the sample is 1.8 g/ml)

(a) 21.7 ml (b) 33 ml

(c) 92 ml (d) 56 ml

114. 'X' grams of $CaCO_3$ was completely burnt in air. The weight of solid residue formed is 28g. What is the value of 'X'? (in grams)

(a) 50 (b) 75

(c) 60 (d) 95

115. 12.5 ml of a solution containing 6 grams of dibasic acid in one litre was found to be neutralized by 10 ml of decinormal solution of NaOH, then the molecular mass of acid is

(a) 150 g (b) 257 g

(c) 267 g (d) 277 g

116. When N_A is avogadro's number then the number of oxygen atoms in one gm. equivalent of oxygen is

(a) N_A

(b) $\frac{N_A}{2}$

(c) $\frac{N_A}{2}$

(d) $2\times N_A$

117. What is the collective wt. (in gms) of 6.02×10^{23} molecules of N_2 & 3.10×10^{23} molecules of SO_2?

(a) 30g

(b) 32g

(c) 34g

(d) 60g

118. 'X' litres of Carbon monoxide is present at STP. It is completely oxidized to CO_2. The volume of CO_2 formed is litres of STP. What is the value of 'X' in litres?

(a) 11.207

(b) 34.4

(c) 19.9

(d) 27.7

119. The density of a solution containing x % by mass H_2SO_4 is y (g/ml). Then its normality is

(a) $\frac{xy \times 10}{98}$

(b) $\frac{20 \times xy}{98}$

(c) $\frac{100 \times xy}{98} \times \frac{1}{98}$

(d) Both (b) & (c)

120. 0.45 N and 0.6 N NaOH solution are mixed in 2 : 1 by volume. The amount of solute present in 1 litre of this solution is

(a) 20g (b) 75g

(c) 45g (d) 40 g

121. The density of 3M aq. $Na_2S_2O_3$ solution is 1.25 g/ml. Then, identify the correct statements

(a) % age by wt. of $Na_2S_2O_3$ is 37.92

(b) The moles fraction of $Na_2S_2O_3$ is 0.065

(c) Molarity of Na^+ is 6 & that of $S_2O_3^{2-}$ is 3

(d) All of these

122. A metal M of equivalent mass E forms an oxide of molecular formula M_xO_y. The atomic mas of metal is given by

(a) $2E\left(\frac{y}{x}\right)$ (b) $\frac{xy}{E}$

(c) $\frac{y}{E}$ (d) None of these

123. 0.1 mole of a carbohydrate with empirical formula CH_2O has 1g of hydrogen. What is its molecular formula?

(a) $C_5H_{10}O_5$ (b) $C_6H_{12}O_6$

(c) $C_{12}H_{22}O_{11}$ (d) None of these

Answers

1. (*c*)	2. (*d*)	3. (*d*)	4. (*a*)	5. (*c*)	6. (*c*)	7. (*c*)	8. (*d*)	9. (*b*)	10. (*d*)
11. (*b*)	12. (*b*)	13. (*c*)	14. (*a*)	15. (*c*)	16. (*a*)	17. (*a*)*	18. (*d*)	19. (*a*)	20. (*d*)
21. (*c*)	22. (*c*)	23. (*b*)*	24. (*b*)	25. (*c*)	26. (*d*)	27. (*a*)	28. (*d*)	29. (*d*)	30. (*a*)
31. (*a*)	32. (*d*)	33. (*b*)	34. (*c*)	35. (*b*)	36. (*d*)	37. (*a*)	38. (*a*)	39. (*b*)	40. (*d*)
41. (*a*)	42. (*a*)	43. (*c*)	44. (*d*)	45. (*d*)	46. (*d*)	47. (*a*)	48. (*b*)	49. (*a*)	50. (*a*)
51. (*c*)	52. (*b*)	53. (*c*)	54. (*c*)	55. (*a*)	56. (*a*)	57. (*b*)	58. (*)	59. (*c*)	60. (*a*)
61. (*b*)	62. (*c*)	63. (*c*)	64. (*b*)	65. (*d*)	66. (*b*)	67. (*c*)	68. (*a*)	69. (*b*)	70. (*b*)
71. (*a*)	72. (*b*)	73. (*b*)	74. (*a*)	75. (*b*)	76. (*b*)	77. (*b*)	78. (*c*)	79. (*a*)	80. (*a*)
81. (*a*)	82. (*a*)	83. (*a*)	84. (*d*)	85. (*a*)	86. (*b*)	87. (*b*)	88. (*a*)	89. (*b*)	90. (*a*)*
91. (*a*)	92. (*b*)	93. (*c*)	94. (*b*)	95. (*c*)	96. (*a*)	97. (*a*)	98. (*d*)	99. (*c*)	100. (*b*)
101. (*a*)	102. (*c*)	103. (*b*)	104. (*a*)	105. (*c*)	106. (*b*)	107. (*c*)	108. (*c*)	109. (*a*)	110. (*a*)
111. (*c*)	112. (*d*)	113. (*a*)	114. (*a*)	115. (*a*)	116. (*b*)	117. (*d*)	118. (*a*)	119. (*d*)	120. (*a*)
121. (*d*)	122. (*a*)	123. (*a*)							

Explanations

1. (c) Moles of $SO_2 = \frac{4\cdot48}{22\cdot4} = 0\cdot2$

Moles of $H_2 = \frac{12\cdot32}{22.4} = 0.55$

Moles of $CO_2 = \frac{22}{44} = 0.5$

Moles of $NH_3 = \frac{2.24}{22.4} = 0.1$

2. (d) 9.1×10^{-31}kg $\rightarrow$ weight of 1 electron

$1\text{ kg} \rightarrow \frac{1}{9.1\times10^{-31}}$ electrons

$\frac{1}{9.1\times10^{-31}}$ electron means $\frac{\frac{1}{9.1\times10^{-31}}}{N_A}$ moles

$= \frac{1}{9.1\times10^{-31}} \times \frac{1}{6\times10^{23}}$

$= \frac{1}{9.1} \times 6 \times 10^{-8}$

$= \frac{1}{9.1} \times \frac{10^8}{6.02}$

3. (d)

4. (a) H_2O_2 has 5.93% hydrogen by wt. So, 5.93 gm of hydrogen is present in 100 gm H_2O_2.

$\therefore$ Oxygen present in 100 gm $H_2O_2 = 94\cdot07$g.

5.93g H combined with O = 94.07

$\frac{1\text{g}}{(\text{fixed wt. of H})}$ H combined with O $= \frac{94\cdot07}{5\cdot93}$ ... (i)

H_2O has 11.2% hydrogen by wt. So, 11.2 gm hydrogen is present in 100 gm H_2O

So, 11.2 gm hydrogen is present in 100 gm H_2O.

$\therefore$ Oxygen is present in 100 gm = 100 – 11.2

= 88.8 g

H		O
$11\cdot2$g	$\longrightarrow$	$88\cdot8$g
1g	$\longrightarrow$	$\frac{88\cdot8}{11\cdot2}$g

....(ii)

(Fixed wt. of H)

Ratio of weights of oxygen $= \frac{\frac{94\cdot07}{5\cdot93}}{\frac{88\cdot8}{11\cdot2}}$

$= \frac{94\cdot07}{5\cdot93} \times \frac{11\cdot2}{88\cdot8} = 2:1$

5. (c)

6. (c) At STP

5.6 L of gas weighs = 60g

22.4 L of gas weighs $= \frac{60}{\cancel{5.6}} \times \overset{4}{\cancel{22.4}}$

= 240g

$\therefore$ M. mass of gas = 240g

Now, $2 \times$ V.D. = M. mass

$\text{V.D} = \frac{240}{2} = 120$

7. (c) 6.02×10^{23} atoms of element weigh = A gms

1 atoms of element weigh $= \frac{A}{6.02\times10^{23}}$ gms

10 atoms of element weigh

$= \frac{A}{6.02\times10^{23}} \times 10$ gms

Now, 1 amu $= \frac{1}{N_A}$ gms

So, mass of 10 elements

$= \frac{A}{6.02\times10^{23}} \times 10 \times 6.02 \times 10^{23}$ amu = 10 A amu

8. (d) 100 ml O_2 (at STP)

$\Rightarrow$ No. of moles $= \frac{100}{22400}$

$\Rightarrow$ Molecules $= \frac{N_A}{224}$

100 ml O_3 (at STP)

$\Rightarrow$ No. of moles $= \frac{100}{22400}$

$\Rightarrow$ Molecules $= \frac{N_A}{224}$

100 ml H_2 (at STP)

$\Rightarrow$ No. of moles $= \frac{100}{22400}$

$\Rightarrow$ Molecules $= \frac{N_A}{224}$

H_2O

ρ of liquid H_2O = 1g/ml

$1\,ml \rightarrow 1g$

$100\,ml \rightarrow 100g$

Now,

$18g\ H_2O \rightarrow N_A$ molecules

$100g\ H_2O \rightarrow \frac{N_A}{18} \times 100 = 5.55N_A$

9. (b) N_2O = 44gm, CO_2 = 44gm

1gm molecule of both these gases have same weight.

10. (d) Moles of P $= \frac{x}{15}$

Moles of Q $= \frac{2x}{30} = \frac{x}{15}$

So, both will contain same no. of atoms.

11. (b) CO_2

1 mole of CO_2 has C = 12g

10 mole of CO_2 has C $= 12 \times 10 = 120$g

CO_2

1 mole of CO_2 has O atoms = 2 moles

10 mole of CO_2 has O atoms $= 2 \times 10 = 20$ moles

$= 20 \times 6.20 \times 10^{23}$

$= 12 \times 10^{24}$

CO_2

1 mole CO_2 has molecules = NA

10 mole CO_2 has molecules $= 10 \times N_A$

Also, 1 mole CO_2 has O = 2 mole atoms of O

10 mole CO_2 has O = 20 mole atoms of O

= 20gm atoms of O

12. (b) 16gm (molar mass of CH_4) has molecules

$= 6 \times 10^{23}$

13. (c) 1 mole of a triatomic gas, contains molecules

$= N_A$

0.1 mole of a triatomic gas, contains molecules

$= N_A \times 0.1$

0.1 mole of a triatomic gas, contains atoms

$= N_A \times 0.1 \times 3$

$= 0.3 \times 6 \times 10^{23}$

$= 1.8 \times 10^{23}$

14. (a) Converting 200 mg CO_2 into moles

$$\text{Numbers of moles} = \frac{W}{M} = \frac{\overset{100}{\cancel{200}} \times 10^{-3}}{\underset{22}{\cancel{44}}}$$

$= \frac{10^{-1}}{22} = 0.045 \times 10^{-1}$

Converting 10^{21} molecules into moles

$n = \frac{10^{21}}{6 \times 10^{23}} = 0.166 \times 10^{-2}$

No, of moles left $= 0.045 \times 10^{-1} - 0.166 \times 10^{-2}$

$= 10^{-1}(0.045 - 0.166 \times 10^{-1})$

$= 10^{-1}(0.045 - 0.0166)$

$= 0.028 \times 10^{-1}$

$= \frac{28}{1000} \times 10^{-1} = 2.8 \times 10^{-3}$

15. (c) Weight of 1 atom of an element

$= 6.644 \times 10^{-23}$ g

Weight of N_A atom of an element

$= 6.644 \times 10^{-23} \times 6 \times 10^{23}$ g

$= 40 \times 10^{-23} \times 10^{23}$ g

= 40 g

$\therefore$ atomic weight = 40 g

No. of gm atoms (No. of moles of atoms)

$= \frac{40000}{40} = 10^3$ g atoms

16. (a) No. of atoms in 55.85 g Fe = N_A

No. of atoms in 558.5 g Fe $= \frac{N_A}{55.85} \times 558.5$

$= N_A \times 10$

$= 6.02 \times 10^{24}$

Atoms in 60 g C

$12\text{g C} \longrightarrow N_A$ atoms

$60\text{g C} \longrightarrow \frac{N_A}{\cancel{12}} \times \overset{5}{\cancel{60}} = 5 \times N_A$

Now, twice of $5 \times N_A = 6.02 \times 10^{24}$

17. (a) C_4H_{10}

Molar mass = 58g

58g C_4H_{10}, contains molecules = N_A

58g C_4H_{10}, contains atoms = $N_A \times 14$

1g C_4H_{10}, contains atoms

$= \frac{N_A \times 14}{58} = 0.241 \times N_A$

(b) N_2

Molar mass = 28g

28g N_2, contains molecules = N_A

28g, N_2, contains atoms = $N_A \times 2$

1g, N_2, contains atoms $= \frac{N_A \times 2}{28}$

$= \frac{N_A}{14} = 0.0714 \times N_A$

(c) Ag

M. mass = 108g

108g Ag, contains atoms = N_A

1g Ag, contains atoms $= \frac{N_A}{108} = 0.009 \times N_A$

(d) H_2O

M. mass = 18g

18g H_2O, contains molecules = N_A

18g H_2O, contains atoms = $N_A \times 3$

1g H_2O, contains atoms $= \frac{N_A \times 3}{18} = 0.166 \times N_A$

18. (d) 18g H_2O has molecules = N_A

1.8g H_2O has molecules $= \frac{N_A}{18} \times 1.8$

$= 6.02 \times 10^{22}$ molecules

Now,

1 H_2O molecule has neutrons = 8

6.02×10^{22} H_2O molecule has neutrons

$= 8 \times 6.02 \times 10^{22}$

$= 48 \times 10^{22}$

19. (a) 4 amu $\longrightarrow$ 1 atom of He

52 amu $\longrightarrow \frac{1}{\cancel{4}} \times \cancel{52}$ 13 atoms of He

20. (d) 0.5g atom of oxygen means 0.5 moles of atoms of oxygen.

1 atom of oxygen $\longrightarrow \frac{1}{2}$ molecules O_2

$0.5 \times N_A$ atom of oxygen $\longrightarrow \frac{1}{2} \times 0.5 \times N_A$

$= \frac{N_A}{4}$ molecules of O_2

0.5gm atom (= $0.5 \times N_A$ atoms)

1 atom of O $\longrightarrow$ 8 electrons

$0.5 \times N_A$ atom of O $\longrightarrow 8 \times 0.5 \times N_A$

$= 4 \times N_A$ electrons

21. (d) 1 mole of N_3^- weighs = 42g

2 mole of N_3^- weighs = $42 \times 2 = 84g$

1 molecule N_3^-, contains valencee e^- = 16

$\underset{(2\text{mole})}{2 \times N_A}$ molecule N_3^-, contains valence e^-

$= 16 \times 2 \times N_A = 32 \times N_A$

1 molecule N_3^-, contains total e^- = 22

$2 \times N_A$ molecule N_3^-, contains total e^-

$= 22 \times 2 \times N_A = 44 \times N_A$

1 molecule N_3^-, contains total protons = 21

$2 \times N_A$ molecule N_3^-, contains total protons

$= 21 \times 2 \times N_A$

$= 42 \times N_A$

22. (c) 11.2 L gas weighs = 14g

22.4 L gas weighs $= \frac{14}{\cancel{11.2}} \times \overset{2}{\cancel{22.4}} = 28g$

So, molar mass of gas is 28g

So, it is N_2.

23. (b) (a) 50g of Fe

(b) 1g atom (1 mole of atoms) of nitrogen weighs = 14g

5g atom (5 mole of atoms) of nitrogen weighs $= 14 \times 5 = 70g$

(c) 1g atom (1 mole of atom) of Ag weighs = 108g

0.1g atom (0.1 mole of atom) of Ag weighs $= \frac{108}{1} \times 0.1 = 10.8g$

(d) 6×10^{23} atoms of C, weigh = 12g

10^{23} atoms of C, weigh $= \frac{12}{6 \times 10^{23}} \times 10^{23} = 2g$

24. (b) 1 CO molecule has O = 1 atom

1 CO molecule has oxygen molecule $= \frac{1}{2}$

6.02×10^{24} CO molecule has oxygen molecule

$= \frac{1}{2} \times 6.02 \times 10^{24}$ molecule

To form gram molecules

$$\text{No. of moles of molecules} = \frac{\frac{6.02 \times 10^{24}}{2}}{\frac{6.02 \times 10^{23}}{1}}$$

= 5g molecules

25. (c) ATOMS IN 1.6gm of O_2

32gm O_2 = N_A molecules of O_2

32gm O_2 = $2 \times N_A$ atoms of oxygen

1.6gm O_2 $= \frac{\cancel{2} \times N_A}{\underset{2}{\cancel{32}}} \times \frac{\cancel{1.6}}{10}$

$= 6.02 \times 10^{22}$ atoms of oxygen

$2 \times N_A$ atoms are in H_2 = 2g

6.02×10^{22} atoms are in H_2

$= \frac{\cancel{2}}{\cancel{2} \times \cancel{6.02} \times 10^{23}} \times \cancel{6.02} \times 10^{22} = 0.1g$

26. (d) N_A molecules of O_2, weigh = 32g

1 molecules of O_2, weigh $= \frac{32}{6 \times 10^{23}} g$

27. (a) 1 moles of H_2S = N_A molecules of H_2S

0.40 mole of H_2S = $N_A \times 0.4$ molecules of H_2S

$= 6 \times 102^3 \times 0.4$ molecules of H_2S

$= 2.4 \times 10^{23}$

28. (d) 0.032 amu S is present in molecule = 100 amu

2×32 amu S is present in molecule

$= \frac{100}{0.032} \times 2 \times 32 \times 1000$

= 200000 amu

29. (d) 1 mole of CH_4 weighs 16 gms

1 molecule of CH_4 weighs 16 amu

30. (a) Minimum molecular weight of enzyme will be recorded when the amount of Se is minimum in it i.e. 1 atom of Se.

0.5 amu of Se is present in enzyme = 100 amu

73.4 amu of Se is present in enzyme

$$= \frac{100}{0.5} \times 73.4\,\text{amu}$$

31. (a) 14g N_2 means moles $= \frac{14}{28} = 0.5$ mole

1 mole occupied volume at STP = 22.4 L

0.5 mole occupied volume at STP $= 22.4 \times 0.5$

= 11.2 L

1.5 moles of CO_2

1 mole of CO_2 occupies vol. (at STP) = 22.4 L

1.5 mole of CO_2 occupies vol. (at STP) $= 22.4 \times 1.5$

= 33.6 L

10^{21} Molecules of O_2

No. of moles of O_2

$$= \frac{10^{21}}{6 \times 10^{23}} = 0.166 \times 10^{-2} = 0.00166$$

1 mole of O_2 = Vol. 22.4 L

0.00166 mole of O_2

$= 22.4 \times 0.00166 = 0.037$ litres

32. (d) 16g O has atoms = N_A

1g O has atoms $= \frac{N_A}{16}$ atoms

32 gm O_2 has molecules = N_A

32 gm O_2 has atoms $= 2 \times N_A$ O atoms

1 gm O_2 has atoms $= \frac{\not{2} \times N_A}{\not{32}\ 16} = \frac{N_A}{16}$ O atoms

48g O_3 has molecules of $O_3 = N_A$

48g O_3 has atoms of $O_3 = 3 \times N_A$

1g O_3 has atoms of $O_3 = \frac{3 \times N_A}{48}$

$= \frac{N_A}{16}$ O atoms

So, ratio = 1: 1 : 1

33. (b) Av. Atomic mass $= \frac{A_1x_1 + A_2x_2 + A_3x_3}{100}$

$$= \frac{54 \times 5 + 56 \times 90 + 57 \times 5}{100}$$

$$= \frac{270 + 5040 + 285}{100}$$

34. (c) 8 oxygen atom $\longrightarrow$ are in 1 molecule of $Mg_3(PO_4)_2$

$0 \cdot 25 \times N_A$ Oxygen atom are in

$\longrightarrow \frac{1}{8} \times 0 \cdot 25 \times N_A$ molecules

moles of $Mg_3(PO_4)_2 = \dfrac{\frac{0 \cdot 25 \times N_A}{8}}{N_A}$

$$= \frac{0 \cdot 25 \times N_A}{8} \times \frac{1}{N_A}$$

$$= \frac{0 \cdot 25}{8}$$

$$= 0 \cdot 03125 = 3 \cdot 125 \times 10^{-2}$$

35. (b)

36. (d)

37. (a) Glucose has the formula $C_6H_{12}O_6$

Now, each C has e^- = 6

each H has e^- = 1

each O has e^- = 8

Total e^- in one molecule $C_6H_{12}O_6$

= 36 + 12 + 48 = 96

Total e^- in 180g glucose $= 96 \times N_A$

Total $^-$in 180×10^{-3}g glucose

$$= \frac{96 \times N_A}{180} \times 180 \times 10^{-3}$$

$$= 0.096 \times N_A$$

$$= 9.6 \times 10^{-2}\ N_A$$

38. (a) 0.15 moles of P_2Q_3 weight = 15.9 gm

1 moles of P_2Q_3 weight $= \frac{15 \cdot 9}{0 \cdot 15} = 106 \text{ gm}$

$\boxed{\text{So, M. Mass of } P_2Q_3 = 106 \text{ gm}}$

0.15 moles of PQ_2 weight = 9.3 gm

1 moles of P_2Q_3 weight $= \frac{9.3}{0.15} = 62 \text{ gm}$

$\boxed{\text{So, M. Mass of } PQ_2 = 62 \text{ gm}}$

Now,

2×At.wt. of P + 3 × At. wt. of Q = 106

At.wt. of P + 2 × At. wt. of Q = 62

Let at .wt. of P = x

& at.wt. of Q = y

$2x + 3y = 106$ ....(1)

$x + 2y = 62$ ....(2)

From 2, $x = 62 - 2y$

Putting value in 1 –

$2(62 - 2y) + 3y = 106$

$124 - 4y + 3y = 106$

$124 - y = 106$

$\boxed{18 = y}$

$\boxed{x = 26}$

39. (b) $2Al + \frac{3}{2}O_2 \longrightarrow Al_2O_3$

$\underline{O_2}$ $\underline{Al}$

1.5 moles of O_2 combine with Al = 2 moles

$= 2 \times 27 = \underline{54 \text{ gm}}$

40. (d) $BaCO_3 \longrightarrow BaO + CO_2 \uparrow$

$BaCO_3$, M.Mass = 197 gm

197 gm $BaCO_3$ liberates CO_2 = 1 mole (= 22.4 L at STP)

9.85 gm$BaCO_3$ liberates $CO_2 = \frac{22.4}{197} \times 9.85 \text{ L}$

= 1.12 L

41. (a) $2Al_2O_3 + 3C \longrightarrow 4Al + 3CO_2 \uparrow$

To produce 4 mole Al, C consumed = 3 moles

To produce 4 × 27 gm Al, C consumed

$= 3 \times 12$ gm

To produce 270000 gm Al, C consumed

$= \frac{\cancel{3} \times \cancel{12}^{4}}{\cancel{4} \times \cancel{27}_{\cancel{9}_{3}}} \times \frac{90000 \text{ gm}}{\cancel{270000}}$

= 90 kg

42. (a) $\underset{(g)}{2SO_2} + \underset{(g)}{O_2} \longrightarrow \underset{(g)}{2SO_3}$

At t = 0	10	15	0
At t = t	$10 - x$	$15 - \frac{x}{2}$	x

$x = 8 \longrightarrow$ Moles of SO_3 formed

No. of moles of SO_2 which did not react

$= 10 - x = 10 - 8 = 2$ moles

No. of moles of O_2 which did not react

$= 15 - \frac{x}{2}$

$= 15 - \frac{\cancel{8}^{4}}{\cancel{2}} = 11 \text{ moles}$

43. (c) $\underset{(g)}{N_2} + \underset{(g)}{3H_2} \longrightarrow \underset{(g)}{2NH_3}$

At t = 0	a	b	0
At t = t	$a - x$	$b - 3x$	$2x$

Moles of N_2 at time, t = $a - x = 4$_____ (1)

Moles of H_2 at time, t = $b - 3x = 4$_____ (2)

Moles of NH_3 at time, t = $2x = 4$_____ (3)

From (3),

$2x = 4$

$x = 2$

Putting value of x in (1)

$a - x = 4$

$a - 2 = 4$

$a = 6$

Putting value of x in equation (2)

$b - 3x = 4$

$b - 6 = 4$

$b = 10$

44. (d) 69.98 amu of C is present in cortisone

= 100 amu

21×12 amu of C is present in cortisone

$= \frac{100}{69.98} \times 21 \times 12$

= 360.1 amu

45. (d) $\underset{\text{(Acetylene)}}{\underset{(g)}{C_2H_2}} + \underset{(g)}{\frac{5}{2}O_2} \longrightarrow \underset{(g)}{2CO_2} + \underset{(l)}{H_2O}$

To burn 1 vol. acetylene, O_2 required = 2.5 vol.

OR, To burn 1 cc acetylene, O_2 required =2.5 vol.

To burn (100 vol.) acetylene, O_2 required

$= 2.5 \times 100$

= 250 vol

= 250 ml

(or 250 cc)

Now,

20 vol. O_2 comes from air = 100 vol.

250 vol. O_2 comes from air

$= \frac{\overset{5}{\cancel{100}}}{\cancel{20}} \times 250 = 1250$ vol.

46. (d) $3BaCl_2 + 2Na_3PO_4 \longrightarrow Ba_3(PO_4)_2$

Na_3PO_4	$BaCl_2$
2 moles react with 3 moles of $BaCl_2$	

0.2 moles react with

So, $BaCl_2$ left = 0.5 – 0.3 = 0.2 moles

$\therefore$ $BaCl_2$ is excess reagent

Na_3PO_4 is LIMITING REAGENT

Now,

Na_3PO_4	$Ba_3(PO_4)_2$
2 moles react to form $\longrightarrow$	1 mole
0.2 moles react to form $\longrightarrow$	$\frac{1}{2} \times 0.2$

= 0.1 mole of $Ba_3(PO_4)_2$ is formed.

47. (a) $\underset{(s)}{CaCO_3} \longrightarrow \underset{(s)}{CaO} + \underset{(g)}{CO_2}$

$CaCO_3$ taken = 10g

But its percentage purity = 90%

$\therefore$ Pure $CaCO_3 = 1\cancel{0} \times \frac{9\cancel{0}}{1\cancel{00}} = 9$gm

$\therefore$ moles of pure $CaCO_3 = \frac{9}{100} = 0.09$ mole

So, moles of CO2 liberated = 0.09

Volume of CO_2 liberated (at STP)

$= 0.09 \times 22.4 = 2.016$ L

48. (b) Let atomic mass of X = y amu

$(4y + 96)$amu X_4O_6 has X = 4y amu

10amu X_4O_6 has X $= \frac{4y \times 10}{4y + 96}$

$\frac{4y \times 10}{4y + 96} = 6.06$

$40y = 6.06(4y + 96)$

$40y = 24.24y + 581.76$

$40y - 24.24y = 581.76$

$15.76y = 581.76$

$y = \frac{581.76}{15.76} = 36.91 \approx 37$amu

49. (a) $HCl + NH_3 \longrightarrow NH_4^+Cl^-$

moles $= \frac{146}{36 \cdot 5} = 4$

So, moles of NH_3 reqd. = 4

wt of NH3 reqd. $= 4 \times 17 = 68g$

$(NH_4)_2SO_4 \longrightarrow 2NH_3 + H_2SO_4$

Now, $2 \times 17g$ NH_3 is produced by $(NH_4)_2SO_4$

= 132 g

68g NH_3 is produced by

$(NH_4)_2SO_4 = \frac{132}{34} \times 68 = 264$ g

50. (a) $2NH_3 + \frac{5}{2}O_2 \xrightarrow{Pt} 2NO + 3H_2O$

n (moles) of NO $= \frac{400}{30}$

So, NH_3 reqd. $= \frac{40}{3}$ moles

Now, $Mg_3N_2 + 6H_2O \longrightarrow 3Mg(OH)_2 + 2NH_3 \uparrow$

2 mole NH_3 are produce from Mg_3N_2 = 1 mole

$\frac{40}{3}$ mole $NH_3 \longrightarrow \frac{1}{2} \times \frac{40}{3} = \frac{20}{3}$ moles

$Mg_3N_2 = 24 \times 3 + 28$

= 72 + 80 = 100g

wt. of Mg_3N_2 reqd. $= \frac{20}{3} \times 100$

51. (c) $\overset{(1)}{9A} + \overset{(1)}{8BO_3} \longrightarrow 3A_3O_4 + 4B_2O_3$

$\underline{A}$ $\underline{BO_3}$

9 moles $\longrightarrow$ 8 moles

1 moles $\longrightarrow \frac{8}{9}$ moles $= 0 \cdot 88$ moles

'A' is the limiting Reagent.

Now,

9 moles of A gives A_3O_4 moles = 3

1 moles of A gives $A_3O_4 = \frac{3}{9}$ moles $= \frac{1}{3}$ moles

52. (b) $\underset{x\,gm}{C} + O_2 \longrightarrow \underset{y\,gm}{CO_2}$

$\underset{12gm}{C} \longrightarrow \underset{32gm}{O_2}$

$xgm \longrightarrow \frac{32}{12} \times x = 2 \cdot 63\,gm$

$y = 2.67\,x$

$\frac{y}{x} > 1 \cdot 33$

53. (c) Now,

$2KMnO_4 + 16HCl \longrightarrow 2KCl + 2MnCl_2 + 5Cl_2 + 8H_2O$

Cl_2 HCl

$5 \times 71gm \longrightarrow 16 \times 36 \cdot 5g$

$213gm \longrightarrow \frac{16 \times 36 \cdot 5}{5 \times 71} \times 213 = 350 \cdot 4g$

HCl available = 20% (w/w)

20gm HCl is obtained from solution = 100 gm

350.4 gmHCl is obtained from solution

$= \frac{100}{20} \times 350 \cdot 4$

wt. of soln. = 1752 gm

reqd.

Now, ρ of soln. = 1.20 g/ml

1.20g soln. has vol. = 1 ml

1752 g soln. has vol. $= \frac{1}{1 \cdot 20} \times 1752$

= 1.46 litres

54. (c) Let volume occupied by O_2 = n

Volume occupied by N_2 = 4n

Total volume of the hall = 125 m^3

Now,

n + 4n = 125 m^3

5n = 125 m^3

$\boxed{n = 25\ m^3}$

$10^{-3}\ m^3 = 1L$

$$25\,m^3 = \frac{1}{10^{-3}} \times 25 = 25000L$$

At STP,

$$22.4L \longrightarrow 32g\ O_2$$

$$25000\,L \longrightarrow \frac{32}{22.4} \times 25{,}000\,g$$

$= 35714$ g

$= 35.71$ kg

55. (a) Formula of metal nitride will be M_3N_2.

Let atomic wt. of metal be yamu.

$$\text{So, \% age by wt. of metal} = \frac{3y}{3y+28} \times 100 = 72$$

$$300y = 216y + 72 \times 28$$

$$y = \frac{72 \times 28}{84} = 24\,amu$$

56. (a) $2\ KClO_3 \xrightarrow{\Delta} 2KCl + 3O_2 \uparrow$

Whatever KCI & O_2 is generated, it will come from pure $KClO_3$.

Mass of $KClO_3$ converted into $O_2 = 0.384\,gm$

$$\text{Moles of } O_2 \text{ formed} = \frac{0.384}{32} = 0.012 \text{ moles}$$

According to Stoichiometry,

3 moles O_2 are formed from $KClO_3$ = 2 moles

0.012 moles O_2 are formed from

$$KClO_3 = \frac{2}{3} \times 0.012$$

$$\therefore \text{ moles of pure } KClO_3 = \frac{2}{3} \times 0.012$$

$$\text{Wt. of pure } KClO_3 = \frac{2}{3} \times 0.012 \times 122.5$$

$= 0.98$ g

$$\text{\% age of pure } KClO_3 = \frac{0.98}{4.9} \times 100 = 20\%$$

57. (b) $3O_{2(g)} \rightleftharpoons 2O_{3(g)}$

At t = 0   90 vol   0

$$\text{At t = t} \quad 90 - \frac{10}{100} \times 90 \quad \frac{2}{3} \times \frac{10}{100} \times 90$$

$$\text{Vol. of } O_2 = 90 - \frac{10}{100} \times 90 = 90 - 9 = 81 \text{ ml}$$

$$\text{Vol. of } O_3 = \frac{2}{3} \times \frac{10}{100} \times 90 = 6 \text{ ml}$$

58. $2SO_{2(g)} + O_{2(g)} \longrightarrow 2SO_{3(g)}$

10 moles   15 moles

According to Stoichiometry, if 2 moles of SO_2 react then moles of O_2 reacted = 1

If 10 moles of SO_2 react then moles of

$$O_2 = \frac{1}{\cancel{2}} \times \overset{5}{\cancel{10}} \text{ reacted} = 5$$

O_2 left = 10 moles

O_2 is excess regent

SO_2 is limiting reagent

If 2 moles SO_2 react, moles of SO_3 formed = 2

If 10 moles SO_2 react, moles of SO_3 formed = 10

$\therefore$ Maximum moles of SO_3 formed = 10

According to question, moles of SO_3 formed = 8

$$\text{So \% age yield} = \frac{8}{10} \times 100 = 80\%$$

59. (c) $\text{Moles of NaBr reqd.} = \dfrac{2.06 \times 10^3 \times 10^3}{104}$

$$= \frac{206 \times 10^4}{104} = 1.98 \times 10^4 \text{ moles}$$

Fe_3Br_9 & NaBr are in the ratio 1 : 8 (by moles)

To form $1{\cdot}98 \times 10^4$ moles of NaBr, moles of

$$Fe_3Br_8 \text{ reqd} = \frac{1}{8} \times 1{\cdot}98 \times 10^4$$

$= 0{\cdot}247 \times 10^4$ moles

But yield of reaction is 70%, So moles of Fe_3Br_8 reqd. will be more than $0 \cdot 247 \times 10^4$.

Let x moles of Fe_3Br_8 are reqd.

$$x \times \frac{70}{100} = 0 \cdot 247 \times 10^4$$

$$x = \frac{0.247 \times 10^4 \times 100}{70}$$

$$= \frac{0 \cdot 247}{70} \times 10^6$$

$$= 0 \cdot 0035 \times 10^6 \text{ moles}$$

Now, from reaction 2 –

To form $0 \cdot 0035 \times 10^6$ moles of Fe_3Br_8 moles of $FeBr_2$ required

$$= 3 \times 0 \cdot 0036 \times 10^6$$

$$= 0 \cdot 0106 \times 10^6 \text{ moles}$$

But its yield is 60%, so actual $FeBr_2$ moles required will be y –

$$y \times \frac{60}{100} = 0 \cdot 0106 \times 10^6$$

$$y = \frac{0 \cdot 0106 \times 10^6}{60} \times 100$$

$$= 0 \cdot 000176 \times 10^8 \text{ moles}$$

If $FeBr_2$ required is $0 \cdot 000176 \times 10^8$ moles , Fe required Will be $0 \cdot 000176 \times 10^8$ moles

So, Fe reqd. $= 0 \cdot 000176 \times 10^8$ moles

wt. of Fe reqd. $= 0 \cdot 000176 \times 10^8 \times 56$

$$= \frac{99}{10000} \times 10^8 \text{ g}$$

$$= 99 \times 10^4 \text{ g}$$

$$= 100 \times 10^4 \text{ g}$$

$$= 10^6 \text{ g}$$

$$= \frac{106}{1000} = 10^3 \text{ kg}$$

60. (a) Let moles of HCOOH & $H_2C_2O_4$ are a & b respectively

$$HCOOH \xrightarrow{H_2SO_4} H_2O + CO$$

a	0	0
0	a	a

$$H_2C_2O_4 \xrightarrow{H_2SO_4} H_2O + CO + CO_2$$

b	0	0	0
0	b	b	b

H_2O is absorbed by H_2SO_4 & CO_2 is absorbed by kOH

$$\text{So, } \frac{\text{vol. of } CO_2}{\text{Total vol. of } (CO + CO_2)}$$

$$= \frac{1}{6} = \frac{b}{a + b + b}$$

$$\frac{a}{b} = 4$$

Molar ratio HCOOH & $H_2C_2O_4 = 4:1$

61. (b) If oleum is labeled as 109%.

It means 9gm water is added in 100g oleum. Let us calculate how much SO_3 can be converted into H_2SO_4 by 9gm of water

$$SO_3 + H_2O \longrightarrow H_2SO_4$$

80g 18g 98g

i.e., 9g H_2O can dissolve 40 g SO_3 to form 49 g H_2SO_4.

$\therefore$ mass of SO_3 in 100g oleum = 40g

% by mass of SO_3 = 40%

% by mass of H_2SO_4 = 60%

62. (c) Let volume occupied by $O_2 = x$

Volume occupied by $N_2 = 4x$

Total volume of the hall = 125m³

Now, $x + 4x = 125m^3$

$5x = 125n^3$

$x = 25m^3$

$$10^{-3} m^3 \longrightarrow 1L$$

$$25m^3 \longrightarrow \frac{1}{10^{-3}} \times 25 = 25000L$$

At STP,

$22.4L \longrightarrow 32g\, O_2$

$25000L \longrightarrow \frac{32}{22.4} \times 25000\, g$

$= 35714$ g

$= 35.71$ Kg

63. (c) $\underset{(g)}{N_2} + \underset{(g)}{3H_2} \longrightarrow \underset{(g)}{2NH_3}$

No. of moles of $H_2 = \frac{450}{2} = 225$ mole

No. of moles of NH_3 formed

$= \frac{1575}{17} = 92.64$ moles

$N_2 + 3H_2 \longrightarrow 2NH_3$

3 moles of H_2, produced NH_3 = 2 moles

225 moles of H_2, produced NH_3

$= \frac{2}{3} \times 225 = \frac{450}{3} = 150$

So, actually 150 moles of NH_3 should have been produced

Percentage yield $= \frac{92.64}{150} \times 100 = 61.8\%$

64. (b) Ag_2S = 1.34% by mass

1.34 gm of Ag_2S is present in 100 gm ore.

$Ag_2S \xrightarrow{\Delta} 2Ag + S$

2×108 gm Ag is obtained from Ag_2S = 248 g

1 gm Ag is obtained from $Ag_2S = \frac{248}{2 \times 108}$ gm

1.34 gm Ag_2S is present in ore = 100 gm

$\frac{248}{216}$ gm Ag_2S is present in ore $= \frac{100}{1.34} \times \frac{248}{216}$

$= \frac{24800}{289.44} = 85.7$g

65. (d) $M_2O_3 + 3H_2 \longrightarrow 2M + 3H_2O$

Let atomic weight of metal = y gms

Molar mass of M_2O_3 = $(2y + 48)$ gms

From balanced equation –

$(2y + 48)$ gms M_2O_3, requires H_2 = 6 gm

0.16 gms M_2O_3, requires $H_2 = \frac{6}{(2y+48)} \times 0.16$

$\frac{6}{(2y+48)} \times 0.16 = 6 \times 10^{-3}$g

$\frac{0.96}{2y+48} = 6 \times 10^{-3}$

$0.96 = 0.006 \times 2y + 0.006 \times 48$

$0.96 = 0.012y + 0.288$

$y = 56$gm

66. (b) $N_2 + 3H_2 \rightleftharpoons 2NH_3$

At $t = 0$ 1 4 0

Here N_2 will be limiting reagent. 1 mole of N_2 will combine with 3 moles of H_2 & 1 msole H_2 will be left.

So, 2 mole NH_3 is expected product.

But since 50% of expected product is formed

It means

NH_3 formed = 1 mole

N_2 reacted = 0.5 mole

H_2 reacted = 1.5 moles

Now, H_2O is added and NH_3 will be soluble in H_2O.

Mole fraction of $H_2 = \frac{nH_2}{\text{Total gaseous moles}}$

No. of moles of H_2 left unreacted = 4 – 1.5 = 2.5

No. of moles of N_2 left unreacted = 1 – 0.5 = 0.5

$\underset{(\text{Mole fraction of } H_2)}{H_2} = \frac{2.5}{2.5+0.5} = \frac{2.5}{3} = \frac{\cancel{25}^{\,5}}{\cancel{30}_{\,6}} = \frac{5}{6}$

67. (c) CO_2 produced = 11.2L

No. of moles of CO_2 produced $= \frac{11.2}{22.4} = 0.5$

$$2KOH + CO_2 \longrightarrow K_2CO_3 + H_2O$$

1 mole CO_2, requires KOH = 2 mole

0.5 mole CO_2, requires KOH = 2×0.5

= 1 mole

1 mole KOH mass = 39 + 16 + 1 = 56g

69. (b) Let wt. of zinc in alloy = $x$gm

$$Zn + H_2SO_4 \longrightarrow ZnSO_4 + H_2 \uparrow$$

H_2 evolved = 224 ml

No. of moles of H_2 evolved $= \frac{\cancel{224}}{\cancel{22400}_{100}} = 0.01$

So, from equation, the No. of moles of zinc which would have entered into reaction = 0.01

Moles of zinc reacted = 0.01

$$0.01 = \frac{x}{65}$$

x (given wt. of zinc in alloy) = 0.65g

Percentage by weight of zinc $= \frac{0.65}{10} \times 100 = 6.5\%$

70. (b) $CaSO_4 \cdot xH_2O \xrightarrow{\Delta} CaSO_4 + xH_2O$

Molar Mass of $CaSO_4 \cdot xH_2O$ = 136 + 18x

Now, $\frac{18x}{136 + 18x} \times 100 = 6.2$

$$1800x = 6.2 \times 136 + 6.2 \times 18x$$

$$1800x = 6.2 \times 136 + 111.6x$$

$$1688.4x = 6.2 \times 136$$

$$x = \frac{6.2 \times 136}{1688.4} = 0.5$$

71. (a) $\frac{Na_2CO_3 \cdot xH_2O}{\text{Molar mass} = (106 + 18x)}$

$$Na_2CO_3 \cdot xH_2O \xrightarrow{\Delta} Na_2CO_3 + xH_2O$$

Now, $\frac{18x}{106 + 18x} \times 100 = 63$

$$1800x = 63 \times 106 + 63 \times 18x$$

$$1800x - 1134x = 63 \times 106$$

$$666x = 63 \times 106$$

$$x = \frac{63 \times 106}{666} = 10$$

72. (b) $2KClO_3 \longrightarrow 2KCl + 3O_2 \uparrow$

3×32gm O_2 is produced by $KClO_3 = 2 \times 122.5$g

48gm O_2 is produced by $KClO_3 = \frac{2 \times 122.5}{3 \times 32} \times 48$g

Let actual $KClO_3$ reqd. = x

$$x \times \frac{80}{100} = \frac{2 \times 122.5 \times 48}{3 \times 32}$$

$$x = \frac{\cancel{2} \times 122.5 \times \cancel{48}^{3}}{\cancel{3} \times \cancel{32}_{16}} \times \frac{10\cancel{0}}{8\cancel{0}} = 153.12 \text{ sample}$$

73. (b) Here, n factor for $H_3PO_4 = 2$

$\therefore$ equivalent wt. of $H_3PO_4 = \frac{98}{2} = 49$

74. (a)

Metal + Oxygen	$\longrightarrow$	Metal Oxide
80g	20g	100g

No. of gmeqvts. of metal reacted = No. of gmeqvts. of metal oxide formed

$$\frac{W_{Metal}}{\text{Eq. wt of metal}} = \frac{W_{Metal\ oxide}}{\text{Eq. wt of metal oxide}}$$

$$\frac{80}{E_M} = \frac{100}{E_M + E_{O^{2-}}}$$

$$\frac{80}{E_M} = \frac{100}{E_M + 8} \qquad \left[\text{Eq. wt. of } O^{2-} = \frac{16}{2}\right]$$

$$80 E_M + 640 = 100 E_M$$

$$20 E_M = 640$$

$$E_M = \frac{\cancel{640}}{\cancel{20}} = 32$$

Eq. wt. of metal = 32

Eq. wt. of O^{2-} = 8

∴ Eq. wt. of metal oxide = 32 + 8 =40

75. (b) Metal + Oxygen ⟶ Metal Oxide

$$W + \frac{24}{100}W = \frac{124}{100}W = 1.24W$$

No. of gmeqvts. of metal reacted = No. of gmeqvts. of metal oxide formed

$$\frac{W_{Metal}}{\text{Eq. wt metal}} = \frac{W_{Metal\ oxide}}{\text{Eq. wt of metal oxide}}$$

$$\frac{W}{E_M} = \frac{1.24\,W}{E_M + E_{O^{2-}}}$$

$$\frac{W}{E_M} = \frac{1.24\,W}{E_M + 8}$$

$$E_M + 8 = 1.24 E_M$$

$$8 = 0.24 E_M$$

$$E_M = \frac{\cancel{8}}{\cancel{0.24}\ 3} \times 100 = 33.3$$

76. (b) Equivalent weight of metal

$$= \frac{\text{At. wt. of metal}}{\text{Valency (n factor)}}$$

$$37.2 = \frac{\text{At. wt. of M}}{2}$$

74.4 = At.wt of M

Mol. Formula of metal chloride

$$= M^{2+}\ Cl^{-} = MCl_2$$

Mol. Weight of this chloride = At. wt. of M + 2 × At. wt. of chlorine

At.wt. of Cl $= 74.4 + 2 \times 35.5$

$$= 74.4 + 71 = 145.4\text{g}$$

77. (b) Vapour density of metal chloride = 59.25

Molar mass of metal chloride $= 2 \times 59.25$

$= 118.50$

Let valency of element (M) be x

So, formula becomes $M^{x+}\ Cl^{-} = MCl_x$

Molar mass of MCl_x = 118.50

At. wt. of metal + x × At. wt. of Chlorine

$= 118.50$ ____(1)

Now, equivalent weight of metal = 4

$$\text{Equivalent weight of metal} = \frac{\text{At. wt. of metal}}{\text{Valency}}$$

$$4 = \frac{\text{At. wt. of metal}}{x}$$

$4x$ = At wt. of metal

Putting this value in equation 1–

$$4x + x \times 35.5 = 118.50$$

$$39.5x = 118.50$$

$$x = \frac{118.50}{39.5} - 3$$

So, valency of element = 3

78. (c) No. of gm equivalents of acid = No. of gm equivalents of base

$$\frac{W_{acid}}{\text{Eq. wt}_{acid}} = N \times V_{(L)}$$

$$\frac{0.16}{\text{Eq. wt}_{acid}} = \frac{1}{10} \times \frac{\cancel{25}}{\cancel{1000}\ 40}$$

$$0.16 \times 400 = \text{Eq. wt}_{acid}$$

$$64 = \text{Eq. wt}_{acid}$$

$$\text{Eq. wt of acid} = \frac{\text{Mol. wt. of acid}}{2\ (\text{n factor})}$$

$$64 = \frac{\text{Mol. wt. of acid}}{2}$$

$$128\text{g} = \text{Mol. wt. of acid}$$

79. (a) M_2O_3

Let at. Wt of M = x

$$\text{wt. \% of metal} = \frac{2x}{2x+48} \times 100 = 40$$

$$200x = 80x + 48 \times 40$$

$$120x = 48 \times 40$$

$$x = \frac{48 \times 40}{120}$$

Now, for 2nd oxide

Let $M^{x+}O^{2-}$, the formula will be M_2O_x

wt. %age of metal = 50%

$$\frac{32}{32+16x} \times 100 = 50$$

$$3200 = 1600 + 50 \times 16x$$

$$1600 = 800x$$

$$x = 2$$

Now, formula is

$M^{2+}\ O^{2-} = MO$

80. (a) Metal + Chlorine $\longrightarrow$ Metal Chloride

Chlorine = 52.8 g

Metal = 47.2 g

No. of equivalent of metal react = No. of equivalent of metal chloride formed

$$\frac{47.2}{E_M} = \frac{100}{E_M + 35.5}$$

$$47.2E_M + 1675.6 = 100E_M$$

$$1675.6 = 52.8E_M$$

$$E_M = 31.73$$

81. (a) $N = M \times \text{n factor}$

$N = 1 \times 3 = 3N$

82. (a) Given wt. of water = 54g

Given wt. of ethanol = 100 – 54 = 46g

Mole Fraction of alcohol (ethanol)

$$= \frac{n_{\text{ethanol}}}{n_{\text{ethanol}} + n_{\text{water}}}$$

C_2H_5OH
(M.Mass=46g)

$$X_{\text{ethanol}}\ (\text{mole fraction}) = \frac{\frac{46}{46}}{\frac{46}{46} + \frac{54}{18}}$$

$$X_{\text{ethanol}} = \frac{1}{1+3} = \frac{1}{4} = 0.25$$

83. (a) Total wt. of sugar solution = 214.2g

Wt. of sugar = 34.2g

Wt. of solvent = 214.2 g – 34.2 g = 180 g

$$\text{No. of moles of sucrose} = \frac{34.2}{342} = 0.1$$

$$\text{Molality of solution} = \frac{\text{No. of moles of solute}}{\text{wt. of solvent (in kgs)}}$$

$$= \frac{0.1}{\frac{180}{1000}}$$

$$= \frac{0.1 \times 100\not{0}}{18\not{0}} = 0.55$$

84. (d) Apply $N_1V_1 = N_2V_2$

$$\frac{1}{2} \times 200 = \frac{1}{10} \times V_2$$

$$V_2 = 1000\,\text{ml}$$

85. (a) Apply $M_1V_1 + M_2V_2 = M_3V_3$

$$3 \times 25 + 4 \times 75 = M_3(100)$$

$$\frac{75 + 300}{100} = M_3$$

$$\frac{375}{100} = M_3 = 3.75\text{M}$$

86. (b) 20% $H_2SO_4\left(\frac{w}{w}\right)$

20gm H_2SO_4 is present in 100gm of solution

$$\text{Density of solution} = \frac{\text{wt. of solution}}{\text{vol. of solution}}$$

$$1.02\,g/cm^3 = \frac{100g}{\text{vol. of solution}}$$

$$\text{vol. of solution} = \frac{100\cancel{g}}{1.02\cancel{g}} \times cm^3$$

$$\text{vol. of solution} = \frac{100}{1.02}cm^3 = 98.03\,cm^3$$

$$M = \frac{\text{no. of moles of solute}}{\text{vol. of soln. (in litres)}}$$

$$M = \frac{\frac{20}{98}}{\frac{98.03}{1000}} = \frac{20}{98} \times \frac{1000}{98.03}$$

$$= \frac{20000}{9606.94} = 2.08M$$

87. (b) Mass percentage $= \frac{12}{44} \times 100$

88. 3 molalNaOH solution means

3 moles of NaOH are present per kg of solvent

3 moles of NaOH = 40 × 3 = 120 g solute (NaOH)

Wt. of solvent = 1000 g

$$\text{Density of Solution} = \frac{\text{wt. of solution}}{\text{vol. of solution}}$$

$$1.110\,g/ml = \frac{1120g}{\text{vol. of solution}}$$

(wt. of solution = wt. of solute + wt. of solvent)

$$\boxed{1120\,g = 120 + 1000}$$

$$\text{Vol. of soln} = \frac{1120}{1.110} = 1009\,ml$$

$$M = \frac{\text{no. of moles of solute}}{\text{Vol of solution (in litres)}}$$

$$= \frac{3}{\frac{1009}{1000}}$$

$$= \frac{3 \times 1000}{1009}$$

$$= \frac{3000}{1009} = 2.97M$$

89. (b)

90. $M = \frac{\text{no. of moles of solulte}}{\text{vol. of soln. (in litres)}}$

$$M = \frac{\frac{24.5}{98}}{\frac{250}{1000}}$$

$$= \frac{24.5}{98} \times \frac{\overset{4}{\cancel{100}\cancel{0}}}{\cancel{25}\cancel{0}} = 1M$$

$N = M \times n$ factor

$N = 1 \times 2 = 2N$

91. (a) M = 1.25 M

→ 1.25 Moles of H_3PO_4 are present in 1 ltr of solution.

→ Wt. of 1.25 moles of

$H_3PO_4 = 1.25 \times 98 = 122.5g$

Wt. % age of $H_3PO_4 = 98\%$

$$\therefore \quad \frac{122.5}{\text{Total wt. of solution}}\ 100 = 9\ 8$$

$$\text{Total wt. of solution} = \frac{122.5 \times 100}{9.8}$$

= 1250 g

Vol. of solution. = 1 ltr.

$$\text{Density of solution} = \frac{125\cancel{0}g}{100\cancel{0}\,ml} = 1.25\,g/ml$$

92. (b) Molar mass of $Na_2SO_4 x\ H_2O = 142 + 18x$

$$\text{Molarity} = \frac{\text{No. of moles of solute}}{\text{Vol. of soluion (in litres)}}$$

$$0.5 = \frac{\frac{644}{142+18x}}{\frac{4000}{1000}}$$

$$0.5 = \frac{644}{142+18x} \times \frac{1000}{4000}$$

$$142 + 18x = \frac{161}{0.5}$$

$$18x = 180$$

$$x = 10$$

93. On increasing dilution, mole fraction of solvent will increase.

94. (c) Moles of solute

$$(n_1) = \frac{w_1}{m_1}, \text{moles of solvent}_{(n_2)} = \frac{w_2}{m_2}$$

$$X_1(\text{solute}) = 0.1, X_2 = \underset{(\text{solvent})}{0.9}$$

$$\frac{x_1}{x_2} = \frac{n_1}{n_2} = \frac{w_1}{m_1} \times \frac{m_2}{w_2} = \frac{1}{9} \quad ....(1)$$

$$\text{Molarity} = \frac{\text{no. of moles of solute}}{\text{Vol. of soln.(in litres)}} \quad ....(2)$$

$$\text{\& volume} = \frac{\text{Total wt. of solution}}{\text{Density of soln.}}$$

$$V = \left(\frac{w_1 + w_2}{2}\right) \text{ml}$$

Putting this value in eqn. 2-

$$M = \frac{\frac{w_1}{m_1}}{\frac{\frac{w_1 + m_2}{2}}{1000}} \quad ....(3)$$

$$M = \frac{\text{no. of moles of solute}}{w_2} \times 1000$$

(Molarity)

$$M = \frac{w_1}{m_1} \times \frac{1000}{w_2} \quad ...(4)$$

Now, Molarity = molarity

$$\frac{w_1}{m_1} \times \frac{2000}{(w_1 + w_2)} = \frac{1000\, w_1}{m_1 w_2}$$

$$w_1 = w_2 = 1$$

$$\text{From eqn. (1)} \ \frac{m_1\ (\text{solut})}{m_2\ (\text{solvent})} = \frac{9}{1}$$

95. (c) $2\,As + 3\,Cl_2 \longrightarrow 2\,AsCl_3$

M mass of $AsCl_3 = 75 + 3 \times 35.5\ 181.5$ gm

$AsCl_3$ Cl_2

$2 \times 181.5\text{gm} \longrightarrow 3 \times 71\text{gm}$

So, to prepare 363 gm $AsCl_3$, chlorine reqd.

$= 3 \times 71 = 213\text{g}$

Chlorine HCl

$5 \times 71\text{g}$ Chlorine is obtained from HCl

$\longrightarrow 16 \times 36.5\text{g}$

213g Chlorine is obtained from HCl

$$\longrightarrow \frac{16 \times 36.5 \times 213\text{g}}{5 \times 71}$$

To obtain 20 g HCl, Its Solution required = 100 g

To obtain $\frac{16 \times 36.5 \times 213\text{g}}{5 \times 71}$,

its solution required

$$= \frac{100}{20} \times \frac{16 \times 36 \times 5 \times 213\text{g}}{5 \times 71} = 1752\text{g}$$

From density

1.2g HCl solution has vol= 1 ml

$$1752\text{g HCl solution has vol} = \frac{1}{1.2} \times 1752\text{ml}$$

= 1460 ml = 1.46 ltr.

96. (a) Urea present = 12g

Wt. of solution= 1200 g

Density of sol. = 1.2 g/ml

$$\text{Density of sol.} = \frac{\text{wt. of sol.}}{\text{vol. of sol.}}$$

1.2 g/ml = 1200 g

$$1\cdot2\,g/ml = \frac{1200g}{Vol. of sol.}$$

$$Vol. of sol. = \frac{1200}{12}\times10 = 1000\,ml$$

$$Molarity of sol. = \frac{\frac{12}{60}}{\frac{1000}{1000}} = 0\cdot2M$$

97. (a) Mol. Wt of HCl = 36.5 g

Mass of HCl in 1 L of $\frac{M}{2}$ HCl

$$= \frac{1}{2}\times36.5 = 18.25g$$

Wt. of HCl lost on heating = 2.675 g

Wt. of HCl left $= 18\cdot25 - 2\cdot2675 = 15\cdot575g$

No. of moles in $15\cdot575g$ HCl $= \frac{15.575}{36.5}$

= 0.4267

Vol. of Soln. left after heating = 750 ml

$$= \frac{750}{1000} = 0.75L$$

$\therefore$ Molarity of solution $= \frac{0\cdot4267}{0\cdot75} = 0\cdot569M$

No. of millmoles in 100 ml $= 100\times0.569 = 56.9$

98. (d) To prepare 250 ml of 2M HNO_3, no. of moles of HNO_3 reqd. are–

$$M = \frac{n}{V(L)}$$

$$\Rightarrow 2 = \frac{n}{\cancel{250}}\times\overset{4}{\cancel{1000}}$$

$$\Rightarrow \frac{2}{4} = n$$

$$\therefore n = \frac{1}{2} = 0.5$$

$$\frac{given\ wt.}{Molar\ mass} = 0.5$$

$$\frac{w}{63} = 0.5 \qquad (M.mass\ of\ HNO_3 = 63g)$$

$$w = 0.5\times63 = \frac{63}{2} = 31.5g$$

So, effectively HNO_3 reqd. = 31.5g

Now, HNO_3 is 70% $\left(\frac{w}{w}\right)$

$\underline{HNO_3}$

70gm is obtained if we pick soln. = 100g

31.5g is obtained if we pick soln.

$$= \frac{10\cancel{0}}{7\cancel{0}}\times31.5 = 45gm$$

99. (c) Moles of NO_3^- = 0.1 (from $AgNO_3$)

Total Vol. 2L

($\because$ equal volumes of 0.1M $AgNO_3$ & 0.1M NaCl are mixed)

$$M(of\ nitrate\ ions) = \frac{0.1}{2} = 0.05M$$

100. (b) Use the relation, $M = \frac{X_B}{(1-X_B)}\times\frac{1}{M_A}\times1000$

Where X_B = mole fraction of solute

M_A = molar mass of solvent.

$$0.11 = \frac{X_B}{(1-X_B)}\times\frac{1}{18}\times1000$$

$$1.99 = \frac{X_B}{(1-X_B)}\times1000$$

$$2 = \frac{X_B}{(1-X_B)}\times1000$$

$$\frac{1-X_B}{X_B} = 500$$

$$\frac{1}{X_B} - 1 = 500$$

$$\frac{1}{X_B} = 501$$

$$X_B = \frac{1}{501} = 0.002$$

101. (a) Formula of metal nitride will be M_3N_2.

Let atomic wt. of metal be y amu.

So, percentage by wt. of metal

$$= \frac{3y}{3y+28} \times 100 = 72$$

$$300y = 216y + 72 \times 28$$

$$84y = 72 \times 28$$

$$y = \frac{72 \times 28}{84} = 24 \text{ amu}$$

102. (c) Empirical formula is the simplest whole no. ratio in which atoms combine.

	C	H	O
$C_6H_{12}O_6 \longrightarrow$	1	2	1
$CH_3COOH \longrightarrow$ ($C_2H_4O_2$)	1	2	1

103. (b)

	Moles	
C = 40 gm	$\frac{40}{12} = 3.33$	$\frac{3.33}{3.33} = 1$
H = 6.67 gm	$\frac{6.67}{1} = 6.67$	$\frac{6.67}{3.33} = 2$
O = 53.33 gm	$\frac{53.33}{16} = 3.33$	$\frac{3.33}{3.33} = 1$

(100 – 46.67 = 53.33)

Ans. CH_2O

104. (a) Mass of 22.4L N_2 at STP = 28g

Mass of 1L N_2 at STP $= \frac{28}{22.4}$ g

Now, mass of 1L of organic gas at STP $= \frac{28}{22.4}$ g

Mass of 22.4L of organic gas at STP

$$= \frac{28}{22.4} \times 22.4$$

= 28g

∴ Molar mass of organic gas = 28g

Empirical formula = CH_2

Empirical formula wt. × n = Molar mass

14 × n = 28

n = 2

$(EF)_n$ = Molecular formula

$(CH_2)_2$ = Molecular formula

C_2H_4 = Molecular formula

105. (c) $W_{H_2O} = 0.72$g

18 gm H_2O has H = 2g

0.72 18 gm H_2O has $H = \frac{2}{18} \times 0.72 = 0.08$g

So, given wt. of H = 0.08 g

$W_{CO_2} = 3.08$g

44 gm CO_2 has C = 12 g

3.08 gm CO_2 has $C = \frac{12}{44} \times 3.08 = 0.84$g

So, given wt. of C = 0.84 g

	Moles	
C = 0.84 g	$\frac{0.84 \text{ g}}{12} = 0.07$	$\frac{0.07}{0.07} = 1$
H = 0.08 g	$\frac{0.08 \text{ g}}{1} = 0.08$	$\frac{0.08}{0.07} = 1.1428$

$(C_1H_{1.1428})_{\times 7}$

$\boxed{C_7H_8}$

106. 44 g CO_2 contains C = 12g

0.4431g CO_2 contains

$$C = \frac{12}{44} \times 0.4431 = 0.1208\text{g of C}$$

18 gm H_2O contains H = 2g

0.1462g H_2O contains

$$H = \frac{2}{18} \times 0.1462 = 0.0162\text{g}$$

of H. wt. of oxgen

= 0.2012 – (0.1208 + 0.0162)

= 0.0642 g

	C	H	O
wt. $\longrightarrow$	$0\cdot1208$	0.0162	0.0642
moles of $\longrightarrow$	$\frac{0\cdot1208}{12} = 0.01$	$\frac{0\cdot0162}{1}$	$\frac{0\cdot0642}{16}$
Simple ratio	$\frac{0\cdot01}{0\cdot004} = 2\cdot5$	$\frac{0\cdot0162}{0\cdot004} = 4\cdot05$	1
In whole no.	5	8	2

So, empirical formula is $C_5H_8O_2$. Empirical for .wt n = mol. Formula wt. $100 \times n = 100$

$= n = 1$

Now, $(EF)_n = MF$

$(C_5H_8O_2)_1 = MF$

$\therefore$ $MF = C_5H_8O_2$

107. (c) M.Mass of $H_4P_2O_7 = 178$

No. of moles of $H_4P_2O_7 = \frac{534}{178} = 3$ moles

No. of moles of $NaOH = \frac{30 \times 10^{23}}{6 \times 10^{23}} = 5$ moles

$$\underset{3\text{ moles}}{H_4P_2O_7} + \underset{5\text{ moles}}{2NaOH} \longrightarrow Na_2H_2P_2O_7 + 2H_2O$$

2 moles of NaOH react with moles of $H_4P_2O_7 = 1$

5 moles of NaOH react with moles of

$H_4P_2O_7 = \frac{1}{2} \times 5 = 2.5$ moles

So, 0.5 moles of $H_4P_2O_7$ are left.

$\therefore$ $H_4P_2O_7$ is excess reagent.

NaOH is limiting reagent.

$\therefore$ $Na_2H_2P_2O_7$ moles formed = 2.5

moles of H_2O formed = 5

Total moles formed (of products) = 5 + 2.5 = 7.5

Molecules of products formed $7\cdot5 \times N_A$

108. (c) Eq. wt. of metal $= \frac{\text{At. wt. of M}}{\text{Valency}}$

$20 = \frac{\text{At. wt}}{3}$

$\boxed{60 = \text{At. wt of M}}$ s

109. (a) Metal carbonate $\longrightarrow$ Metal oxide

No. of gm equivalent of metal carbonate reacted = No. of gm equivalents of metal oxide formed

$\frac{2}{\text{Eq. wt of metal carbonate}}$

$= \frac{1.5}{\text{Eq. wt of metal oxide}}$

$\frac{2}{E_m + E_{CO_2^{3-}}}$

$= \frac{1.5}{E_m + E_{O^{2-}}}$

$\frac{2}{E_m + E_{CO_2^{3-}}}$

$= \frac{1.5}{E_m + E_{O^{2-}}}$

Now, $E_{CO_3^{2-}} = \frac{60}{2} = 30$ &

$E_{O^{2-}} = \frac{16}{2} = 8$

$\frac{2}{E_m + 30} = \frac{1.5}{E_m + 8}$

$2E_m + 16 = 1.5\,E_m + 45$

$0.5\,E_m = 29$

$E_m = \frac{29}{0.5} = 58g$

110. (a) $2 \times V.D = M.wt$

$\boxed{2x = \text{M. mass of tetrabasic acid}}$

Now. Eq. mass of acid $= \frac{\text{M. mass of acid}}{4}$

Eq. mass $= \frac{2x}{4} = \frac{x}{2}$

111. (c) No. of gm eq. of solution reacted = No. of gm eq. of base react

$$\frac{40}{1000} \times N = \frac{28}{1000} \times 0 \cdot 1$$

$$N = \frac{2 \cdot 8}{40} = \frac{28}{400}$$

$$= \frac{7}{100} = 0 \cdot 07N$$

Changing the volume does not change the normality of soln.

$\therefore$ N of 1000 ml of a soln. = 0.07 N

So, gm eq. in 1 L of a soln. = $0 \cdot 07 \times 1$

$0 \cdot 07$ gm eq

Initial gm eq. of acid $= \frac{100}{1000} \times 1 = 0 \cdot 1$gm eq.

No. of eq. of acid reacted with metal $= 0 \cdot 1 - 0 \cdot 07$

= 0.03

No. of gm eq. of metal = 0.03

No. of gm eq. $= \dfrac{\text{given wt. of metal}}{\text{Eq. wt. of metal}}$

$$0 \cdot 03 = \frac{0 \cdot 36}{\text{Eq. wt. of metal}}$$

Eq. wt. of metal $= \dfrac{0 \cdot 36}{0 \cdot 03} = 12$

112. (d) Normality of $NaOH = X$, Normality of $Ba(OH)_2 = 2Y$

Meq.of Bases taken together = Meq. of acid reacted

$$V_1X + V_2 2Y = 100 \times 0 \cdot 1$$

$$\boxed{V_1 X + V_2 \cdot 2Y = 10} \quad .....(1)$$

Now, $\dfrac{V_1}{V_2} = \dfrac{1}{4}$ & $\dfrac{X}{Y} = 4$

$\therefore$ using eqn. (1)

$$V_1X + 2X\,4V_1 \cdot \frac{X}{4} = 10$$

$$3V_1X = 10$$

$$\boxed{V_1X = \frac{10}{3}}$$

$$V_1 = \frac{V_2}{4} \text{ \& } X = 4Y$$

$$\therefore \frac{V_2}{4} \times 4Y = \frac{10}{3}$$

$$\boxed{V_2Y = \frac{10}{3}}$$

Fraction of acid Neutralized by

$$Ba(OH)_2 = \frac{\text{gm eq. of } Ba(OH)_2 \text{ used}}{\text{Total gm eq. of both bases used collectively}}$$

$$= \frac{2YV_2}{2YV_2 + V_1X}$$

$$= \frac{2 \times \frac{10}{3}}{2 \times \frac{10}{3} + \frac{10}{3}}$$

$$= \frac{\frac{20}{3}}{\frac{30}{3}} = \frac{20}{3} \times \frac{3}{30} = 0 \cdot 67$$

113. (a) In order to prepare 1.5 litres of 0.2 M H_2SO_4, no of moles of H_2SO_4 required = 0.2×1.5

= 0.3 moles

So, H_2SO_4 effective wt. of reqd. = 98×0.3

= 29.4 g

Now, H_2SO_4 solution is 75% by mass which means

75 g H_2SO_4 is present in solution = 100 g

75g H_2SO_4 comes from solution = 100g

29.4 g H_2SO_4 comes from solution

$$= \frac{100}{75} \times 29.4 = 39.2g$$

So, wt. of soln. reqd. = 39.2 g

Density of sample = 1.8g/ml

1.8g solution is obtained from volume = 1 ml

39.2g solution is obtained from volume

$$= \frac{1}{1.8} \times 39.2$$

$$= \frac{39.2}{1.8} \text{ml}$$

= 21.7 ml

114. (a) $CaCO_3 \longrightarrow CaO + CO_{2(g)} \uparrow$

residueis that of CaO

56gm CaO is obtained from $CaCO_3$ = 100g

28g CaO is obtained from $CaCO_3 = \frac{100}{56} \times 28$

= 50 g

115. (a) 1000 ml solution has acid = 6 gm

12.5 ml solution has acid $= \frac{6}{1000} \times 12.5$

= 0.075 gm

So, wt. of acid that entered into reaction with base = 0.075 g

Since, it's a case of neutralization,

$\therefore$ gm eq. of acid reacted = No. of gm eq. of base reacted

(no. of moles × n factor) $\frac{0.075}{\text{M.Mass}} \times \text{n factor}$

$$= \frac{1}{10} \times \frac{10}{1000} (N \times V_L)$$

$$\frac{0.075}{\text{M.Mass}} \times 2 = \frac{1}{1000}$$

$\boxed{\text{M.Mass} = 150\text{g}}$

116. (b) 1 gm eq. of oxygen atom = 8 g of oxygen

Now, 16g O has atoms= N_A

8g O has atoms $= \frac{N_A}{\cancel{16}2} \times \cancel{8} = \frac{N_A}{2}$

117. (d) Moles of $N_2 = \frac{6.02 \times 10^{23}}{6.02 \times 10^{23}} = 1$

& 1 mole N_2 weights = 28 gm

Moles of $SO_2 = \frac{3.01 \times 10^{23}}{6.02 \times 10^{23}}$

$$= \frac{1}{2} = 0.5 \text{moles}$$

& 0.5 moles of SO_2 weight = 32 gm

Total wt. of the mixture = 28 + 32 = 60 g

118. (a) $\underset{(g)}{CO} + \underset{(g)}{\frac{1}{2}O_2} \longrightarrow \underset{(g)}{CO_2}$

11.207 litres of CO_2 will be formed form 11.207 litres of CO at STP

119. (d) *x*% by mass of H_2SO_4 means

x gm of H_2SO_4 is present in 100 gm solution density = y g/ml

y_g of solution has volume = 1 ml

100g of solution has volume $= \frac{1}{y} \times 100 \text{ml}$

$$= \frac{100}{y} \times \frac{1}{100}$$

$$= \frac{1}{10y} \text{L}$$

$$M = \frac{\text{no. of moles of solute}}{\text{Vol. of Soln. (in litres)}}$$

$$M = \frac{\frac{x}{98}}{\frac{1}{10y}}$$

$$= \frac{x}{98} \times 10y$$

$$= \frac{10xy}{98}$$

Normality = M × n factor

$$= \frac{10xy}{98} \times 2$$

$$= \frac{20xy}{98}$$

120. (a) $\frac{\text{I}^{\text{st}}\text{ Soln.}}{0.45\text{N}}$

Let 2L of this solution is chosen

So, moles of NaOH = 0.90 moles

$\frac{\text{2nd Solution}}{0.6\text{N}}$

Let 1L of this solution is chosen

So, moles of NaOH = 0.6 moles

After mixing, 3L vol. will contain moles of NaOH = 1.5

1L vol. will contain moles $= \frac{1.5}{3} = 0.5$ moles

wt. of 0.5 moles of NaOH = 20g

121. (d) 3 M $Na_2S_2O_3$ means 3 moles of $Na_2S_2O_3$ are present in 1 litre of solution.

Now, 1 ml of soln. weight = 1.25 g

1000 ml of soln. weight = $1 \cdot 25 \times 1000$

1250 g

M. Mass of $Na_2S_2O_3$ = 158g

So, % age by wt. of $Na_2S_2O_3 = \frac{3 \times 158}{1250} \times 100$ = 37.92

For mole fraction,

Total wt. of solution = 1250 g

Total wt. of solute = $3 \times 158\text{g} = 474\text{g}$

Wt. of solvent (water) = 1250 – 474 = 776 g

Moles of $H_2O = \frac{776}{18} = 43 \cdot 11$ moles

Mole fraction

$$Na_2S_2O_3 = \frac{\text{No. of moles of } Na_2S_2O_3}{\text{Total no. of moles}}$$

$$= \frac{3}{3 + 43 \cdot 11} = 0 \cdot 065$$

For molarity,

$$Na_2S_2O_3 \longrightarrow 2Na^+ + S_2O_3^{2-}$$

3M	0	0
0	6M	3M

122. (a) Let m be the atomic mass of metal

So, 16ygm of oxygen reacts with metal = m gm

8 gm of oxygen reacts with metal $= \frac{mx}{16y} \times 8$

$$\frac{8\ mx}{16y} = E$$

$$\boxed{\frac{2Ey}{x} = m}$$

123. (a) 0.1 mole of carbohydrate contains H = 1g

1 mole of carbohydrate contains $= \frac{1}{0 \cdot 1} = 10\text{g}$

10g H = 10 mole of H atoms

If empirical formula is = CH_2O

Then molecular formula will be = $C_5H_{10}O_5$

Only then 1 : 2 : 1 ratio of C : H : O will be achieved.

Smart Notes ....

Smart Notes

2

CHAPTER

Structure of Atom

Sub Atomic Particles

1. Which of the following is the correct statement regarding Cathode Rays?
   (a) Cathode rays carry negative charge
   (b) They cause ionization of the gas through which they pass
   (c) They possess penetration effect
   (d) All are correct

2. The $\frac{e}{m}$ ratio $\left(\frac{\text{charge}}{\text{mass}}\right)$ for Cathode rays
   (a) 2×10^8 coulombs/g
   (b) 6×10^{11} coulombs/g
   (c) 1.76×10^8 coulombs/g
   (d) None of these

3. An electron has mass nearly equal to
   (a) $\frac{1}{600}$th of that of Hydrogen atom
   (b) $\frac{1}{187}$th of that of H atom.
   (c) $\frac{1}{1837}$th of that of H atom
   (d) $\frac{1}{2653}$th of that of H atom.

4. The charge on electron was found by
   (a) R.A. Milliken (b) Strong
   (c) Mosely (d) Neil Bohr.

5. Cathode rays originate
   (a) Only from the material of cathode
   (b) Only from the gas inside the discharge tube
   (c) Only from air
   (d) First from the material of cathode & then from the gas inside the discharge tube.

6. Anode rays (or Canal rays) are known to possess.
   (a) Positive charge
   (b) partial negative charge
   (c) complete negative charge
   (d) no charge

7. Discharge tube experiment for a discovery of anode rays was performed by
   (a) Goldstein (b) J.J. Thomson
   (c) millikan (d) Sommerfiled

8. Which one of the following is INCORRECT regarding anode rays?
   (a) They travel is straight lines
   (b) $\frac{e}{m}$ ratio is constant for anode rays irrespective of the gas takes in discharge tube
   (c) They are positively charged
   (d) These are deflected in electric & magnetic fields

Models of Atom

9. Among the first line of Lyman, Balmer, Paschan & Brackett series in Hydrogen atomic spectra which one has a higher energy?

(a) Lyman (b) Balmer

(c) Paschen (d) Brackett.

10. If the speed of electron in the Bohr's first orbit of H is x, then the speed in 3rd orbit is

(a) $\frac{x}{9}$ (b) $\frac{x}{3}$

(c) $3x$ (d) $12x$

11. What transition in the H spectrum would have the same λ as the Balmer transition, n = 4 to n = 2 of He^+ spectrum?

(a) $n_1 = 2$ to $n_2 = 2$ (b) $n_1 = 1$ to $n_2 = 2$

(c) $n_1 = 4$ to $n_2 = 5$ (d) None of these

12. Let υ_1 be the frequency of series limit of Lyman series, υ_2 be the frequency of first line of Lyman series & υ_3 be the frequency of series limit of Balmer series, then

(a) $\upsilon_1 - \upsilon_2 = \upsilon_3$ (b) $\upsilon_1 = \upsilon_3 + \upsilon_2$

(c) $\upsilon_2 \times \upsilon_3 = \upsilon_1$ (d) $\upsilon_1 + \upsilon_2 = \upsilon_3$

13. An electron in H atom transits from 3rd excited state of ground state. The maximum number of different photons emitted are

(a) 3 (b) 6

(c) 4 (d) 5

14. The energy difference between the states of $n = 2$ & $n = 3$ is 'E' eV in H atom. The ionization potential of H atom is

(a) 3.6 E (b) 36 E

(c) 12.24 E (d) 7.2 E

15. In an atom, two electrons move around the nucleus in circular orbits of ratio R & 4R. The ratio of time taken by them to complete one revolution is

(a) 1 : 4 (b) 4 : 1

(c) 1 : 8 (d) 8 : 1

16. The potential energy of an electron in H atom is – 6.8 eV. Then, the electron is in 'Y' excited state. Y is

(a) 2nd (b) 3rd

(c) 1st (d) 5th

17. The distance between 4th & 3rd Bohr orbits of He^+ is

(a) 1.851×10^{-10} m (b) 1.32×10^{-10} m

(c) 1.32×10^{-10} m (d) 9.97×10^{-21} m

18. Which of the following transitions will require energy equivalent to the Ionization energy of H?

(a) $n = 3$ to $n = 4$ (b) $n = 1$ to $n = 2$

(c) $n = 2$ to $n = 5$ (d) $n = 1$ to $n = \infty$

19. Raisin pudding model or water melon model was introduced by

(a) J. J Thomson (b) Rutherford

(c) Bohr's (d) Sommerfields.

20. An α-particle is a particle which carries $\underline{x}$ units of positive charge & has mass $\underline{y}$ times that of a proton. x & y are respectively

(a) 4, 2 (b) 2, 4

(c) 3, 5 (d) 3, 6

21. Which one of the following is 'NOT' an observation made by Rutherford in his α-particles scattering experiment?

(a) Most of the α-particle (about 99.9%) passed through the foil without undergoing any deflection.

(b) Very few were deflected back.

(c) Few α-particles underwent deflection through small angles.

(d) None of these

22. Which of the following correctly relates the radius of nucleus (r) with mass number, A?

(where $R_0 = 1.4 \times 10^{-15}$ m)

(a) $r = R_0 A^{\frac{1}{5}}$ (b) $r = R_0 A^{\frac{1}{3}}$

(c) $r = R_0 A^{\frac{1}{2}}$ (d) $r = R_0 A^{1/10}$

23. Which observation led Rutherford to conclude that nucleus has a very small volume

(a) The number of α-particles deflected back was very large.

(b) The number of α-particles deflected were zero.

(c) The number of α-particles deflected were very small.

(d) None of these

24. Which of these statements is INCORRECT?

(a) The specific charge for positive rays is much less than specific charge for cathode rays because positive rays comprise of ionized atoms, whose mass is much higher.

(b) Bohr's model can explain spectrum of any atom or ion having one electron only

(c) $E_4 - E_3 < E_3 - E_2 < E_2 - E_1$ for H atom.

(d) None of the above is incorrect.

25. Select the correct match

Column A	Column B
(a) Stability of atom	Explained by Rutherford
(b) Rydberg's equation	Derived with the help of Bohr's theory.
(c) Line spectrum	Explained by J.J. Thom

(d) Both (b) & (c) are correct

26. With respect to BOHR'S THEORY which of the following statement is correct?

(a) Energy & angular momentum of an electron is quantized.

(b) Velocity of an electron in nth orbit is given by the relation, $V_n = 2.188 \times 10^8 \times \frac{Z}{n}$ cm/s

(c) Energy is emitted or absorbed only when the electrons jump from one orbit to the other.

(d) All are correct.

27. Choose the limitations of Bohr's model of an atom.

(a) It could not explain Zeeman effect & stark effect

(b) It was not is accordance with de Broglie's dual nature of matter & Heisenberg's uncertainty principle

(c) It could not explain the line spectra of multi electron atoms.

(d) All of these

28. The ionization energy of He^+ is 8.72×10^{-18} J $atom^{-1}$ Calculate the energy of first stationary state of Li^{2+}.

(a) -44.1×10^{-19} J $atom^{-1}$

(b) -19.62×10^{-18} J $atom^{-1}$

(c) -66.25×10^{-27} J $atom^{-1}$

(d) -109.89×10^{-27} J atom

29. If ΔE is the energy emitted in eV when an electronic transition occurs from higher energy to a lower energy level in H-atom, the λ of the line produced is approximately equal to

(a) $\frac{12375}{\Delta E}$ Å (b) $\frac{19800}{\Delta E}$ Å

(c) $\frac{26600}{\Delta E}$ Å (d) $\frac{22500}{\Delta E}$ Å

30. Which transition gives rise to least energy photon in H atom?

(a) $n = 6$ to $n = 5$ (b) $n = 6$ to $n = 1$

(c) $n = 5$ to $n = 4$ (d) $n = 2$ to $n = 1$

31. An electron in an atom jumps in such a way that its K.E. changes from x to $\frac{x}{4}$. Then, what is the change in P.E?

(a) $+\frac{3}{2}x$ (b) $-\frac{3}{2}x$

(c) $-\frac{3}{4}x$ (d) $-\frac{6}{7}x$

32. According to Bohr's theory, which one of the following values of angular momentum of H atom is not permitted

(a) $\frac{1.25h}{\pi}$ (b) $\frac{h}{\pi}$

(c) $\frac{1.5h}{\pi}$ (d) $\frac{0.5h}{\pi}$

33. If the energy of H-atom in the ground state is - E, the velocity of photo-electron emitted when a photon having energy Ep strikes a stationary Li^{2+} in ground state, is given by

(a) $V = \sqrt{\frac{2(E_p - E)}{m}}$ (b) $V = \sqrt{\frac{2(E_p + 9E)}{m}}$

(c) $V = \sqrt{\frac{2(E_p - 3E)}{m}}$ (d) $V = \sqrt{\frac{2(E_p - 9E)}{m}}$

34. The ratio of velocity of electrons in 2nd to that in 4th excited state of Be^{3+} will be -

(a) 5 : 3 (b) 3 : 5

(c) 1 : 2 (d) 2 : 1

Electromagnetic Wave and Hydrogen Spectrum

35. When a metal is irradiated with light of frequency $4 \times 10^{-16}\ s^{-1}$, the photoelectron emitted has six times the K.E. as the K.E. of photoelectron emitted when the metal was irradiated with light of frequency $2 \times 10^{16}\ s^{-1}$. Calculate the threshold frequency of the metal

(a) $3 \times 10^{16}\ s^{-1}$ (b) $1.6 \times 10^{16}\ s^{-1}$

(c) $2 \times 10^{16}\ s^{-1}$ (d) $4.2 \times 10^{16}\ s^{-1}$

36. In photoelectric effect on a metallic surface, if the work function is 4.4×10^{-20} J & K.E. of ejected electrons is 1.2×10^{-20} J, then what is the incident energy of photon?

(a) 8×10^{-20} J (b) 6×10^{-1} J

(c) 5.6×10^{-20} J (d) None of these

37. Photoelectric emission is observed from a surface for frequencies v_1 & v_2 & the K.E. in two cases are in the ratio 1 : K, then v_0 (threshold frequency)

(a) $\frac{v_2 - v_1}{K - 1}$ (b) $\frac{Kv_1 - v_2}{K - 1}$

(c) $\frac{Kv_2 - v_2}{K - 1}$ (d) None of these

38. When the frequency of light incident on a metallic plate is doubled, then the K.E. of the emitted photoelectron is

(a) Tripled

(b) half

(c) not charged

(d) Increased but more than doubled of the previous K.E.

39. According to EM wave theory.

(a) The radiations possess wave character & travel with a velocity less than velocity of light

(b) The radiations possess particles character & travel with a velocity equal to the velocity of light

(c) The radiations possess wave character & travel with a velocity equal to the velocity of light.

(d) The radiation consist of electric & magnetic fields oscillating parallel to each other & both perpendicular to the direction of propagation of radiation.

40. Which one of the following is not an electromagnetic wave?

(a) γ-rays (b) *x*-rays

(c) Infra-red (d) Cathode rays

41. Which of the following is/are limitations of EM wave theory?

(a) photoelectric effect

(b) Black body radiations

(c) line spectra of atoms

(d) All of these are limitation

42. The ratio of velocity of *x*-rays to γ-rays is

(a) = 1 (b) > 1

(c) < 1 (d) ≥ 1

43. Planck's Quantum theory suggests '*x*' nature of electromagnetic radiations '*x*' is

(a) particle

(b) wave

(c) both particle & wave

(d) neither particle nor wave

44. The energy possessed by 1 mole of photons is called

(a) 1 Coulomb of energy

(b) 1 watt of energy

(c) 1 Einstein of energy

(d) None of these

45. According to Planck's theory, the energy of each quantum is directly proportional to

(a) frequency of radiation

(b) wavelength of radiation

(c) Time period of radiation

(d) Both (b) & (c)

46. The correct value of Planck's constant (h) is

(a) $6 \cdot 626 \times 10^{-34}$ Js

(b) $6 \cdot 626 \times 10^{-27}$ erg sec

(c) $8 \cdot 66 \times 10^{-35}$ Js

(d) Both (a) & (b)

47. Planck's quantum theory explains

(a) Blacks body radiation

(b) Photoelectric effect

(c) Both (a) & (b)

(d) Spectra of multi electron species.

48. Which one of the following is/are correct?

(a) 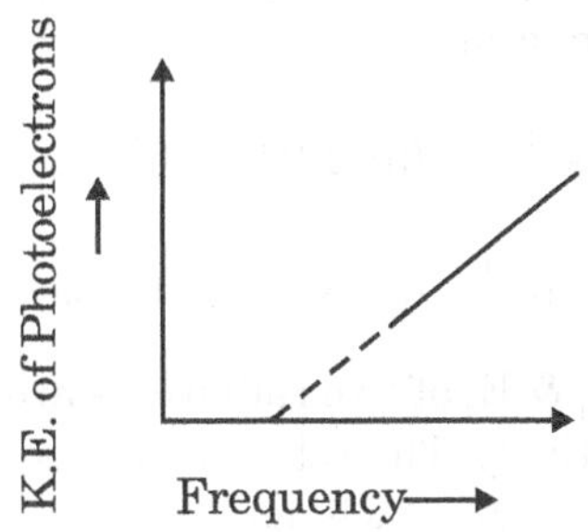

(b) 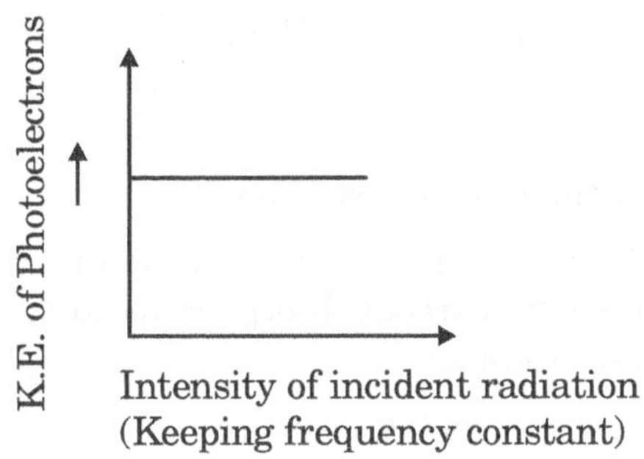

(c) Both (a) & (b)

(d) None of these

49. A 600 W lamp emits monochromatic radiation of $\lambda = 313$ nm. How many approximate photons are emitted from the lamp per second?

(a) 10^{21} (b) 10^{16}

(c) 10^{14} (d) 10^{57}

50. How long would it take for a radio wave of $v = 6 \times 10^3\ s^{-1}$ to travel from mass to Earth, a distance of 8×10^7 km ?

(a) 4 min 26 sec (b) 7 min 43 sec

(c) 8 min (d) 8 min 20 sec

51. The λ of a photon (in Å) having energy equal to one electron volt is

(a) 12.40×10^3 Å (b) 66.10×10^5 Å

(c) 5.7×10^2 Å (d) None of these

52. The number of photons of light having wave number $\frac{1}{y}$ in 1J of energy source is

(h = Planck's Constant, C = velocity of light)

(a) $\frac{y}{hc}$ (b) $\frac{hc}{y}$

(c) $\frac{1}{hcy}$ (d) hcy

53. An electromagnetic wavelength of 600 nm corresponds to

(a) v (frequency) = $0.5 \times 10^{15} s^{-1}$

(b) $\bar{v}$ (wave number) = $0.166 \times 10^{-1} m^{-1}$

(c) v (frequency) = $8 \times 10^{18} s^{-1}$

(d) Both (a) & (b)

54. A gas absorbs photon of λ 355 nm and then emits two wavelengths. If one is having $\lambda = 680$ nm, Calculate the other wavelength.

(a) 743 nm (b) 325 nm

(c) 518 nm (d) 1035 nm

55. The longest wavelength of light capable of breaking a single (Cl – Cl) bond in Cl_2 is, if

$Cl_{2(g)} \rightarrow 2Cl_{(g)}; \ \Delta H = 242\, KJ\ mol^{-1}$

(a) 494 nm (b) 494 pm

(c) 494 Å (d) 494 m

56. If R_H represents Rydberg constant, then the energy of electron in the ground state of H atom is

(a) $\frac{-hc}{R_H}$ (b) $-R_H ch$

(c) $\frac{-R_{HC}}{h}$ (d) Both (a) & (c)

57. The ratio of lowest wavelength of Lyman to highest wavelength of Lyman series is

(a) $\frac{3}{4}$ (b) $\frac{7}{6}$

(c) $\frac{8}{5}$ (d) $\frac{2}{3}$

58. If the Ionisation energy of H atom is x KJ. Then, what is the energy required for an e^- in kJ to jump from $n = 1$ to $n = 2$?

(a) $\frac{3x}{4}$ (b) $\frac{36x}{5}$

(c) $\frac{18x}{7}$ (d) $\frac{2x}{3}$

59. In a sample of H atoms electron make transition from 5th excited state to ground state, producing all possible photons, calculate the number of different lines in infra red region.

(a) 6 (b) 3

(c) 5 (d) 7

60. Select the correct statement

(a) Different number of visible line when sample of electron jumps from 5th shell to ground state in H spectrum is 3.

(b) The difference between the wave number of 1st line of Balmer series & last line of Paschen series for Li^{2+} ion is $\frac{R}{4}$

(c) The wavelength of a spectral line for an electronic transition is inversely related to the difference in the energy levels involved in transition

(d) All are correct

61. Select the incorrect match

Column A	Column B
(a) $n = 7$ to $n = 1$	Lyman series
(b) $n = \infty$ to $n = 1$	Limiting line of Lyman series
(c) $n = 3$ to $n = 2$	First line of Balmer series
(d) $n = \infty$ to $n = 5$	1st line of Brackett series

62. Which one of the following falls in visible region?

(a) Lyman series

(b) Balmer series

(c) Pfund series

(d) Paschan series

63. The spectrum of He^{+} is expected to be similar to that of

(a) N (b) H

(c) Li^{+} (d) Be^{2+}

64. Calculate the number of photons of light will wavelength 200 nm that carries 5J of energy.

(a) 5×10^{19} (b) 6×10^{25}

(c) 9×10^{27} (d) 80×10^{22}

65. A 100 watt bulb emits monochromatic light of $\lambda = 400$ nm. Calculate the number of photons emitted per second

(a) 3.676×10^{-15} (b) 2.012×10^{20}

(c) 18×10^{-27} (d) 5×10^{25}

66. The ionization energy of Na atom is 495.5 kJ mol^{-1}. The lowest possible frequency of light that ionizes a sodium atom is -

(a) $1.24 \times 10^{15}\ s^{-1}$ (b) 1.24×10^{15} s

(c) $1.24 \times 10^{15}\ ms^{-1}$ (d) None of these

67. The energies E_1 & E_2 of two radiations are 25 eV & 50 eV respectively. The relation between their wavelength i.e., λ_1 & λ_2 will be

(a) $\lambda_1 = 2\lambda_2$ (b) $\lambda_1 = \lambda_2$

(c) $2\lambda_1 = \lambda_2$ (d) $4\lambda_1 = \lambda_2$

68. Select the incorrect statement.

(a) Emission spectra may be continuous or discontinuous but absorption spectra is always discontinuous

(b) Line spectra (atomic spectrum) is also called as the finger print of an element.

(c) Absorption spectrum is obtained when white light is first passed through the substance & the transmitted light is analysed in spectroscope

(d) none of these

69. Pick the wrong relation.

(a) $\frac{1}{\lambda} = R\left(\frac{1}{n_1^2} - \frac{1}{n_2^2}\right)$

(b) $\bar{v} = R\left(\frac{1}{n_1^2} - \frac{1}{n_2^2}\right)$

(c) $\frac{1}{\lambda} = Rz^2\left(\frac{1}{n_1^2} - \frac{1}{n_2^2}\right)$ (for Li &Be)

(d) Value of R is 109678 cm^{-1}.

70. How many photons are emitted when an electron makes transition from $n = 1$ to $n = 3$ in a H atom?

(a) 1 (b) 2

(c) 3 (d) None of these

71. The ratio of slopes of KE_{max}, v/s υ & v_0 v/s υ curve in the photoelectric effect gives

(a) Planck's constt.

(b) Charge of electron

(c) Work function

(d) None of these

72. In a collection of H-atoms, all the electrons Jump from n= 5 to ground state, directly or indirectly. No line is observed in Balmer series. The total no. of possible different radiations are

(a) 6 (b) 7

(c) 8 (d) 10

73. Photo electron emission is observed for 3 different metal x, y & z. The kinetic energy of photoelectrons v/s 'v' (frequency) is plotted for each metal. Then, which of the following graph is correct?

(a) 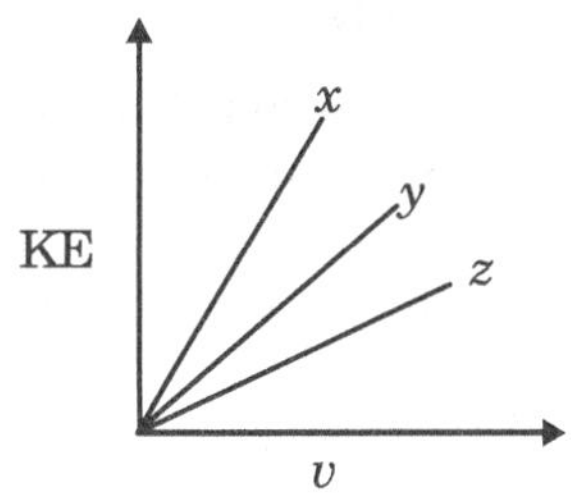

(b) 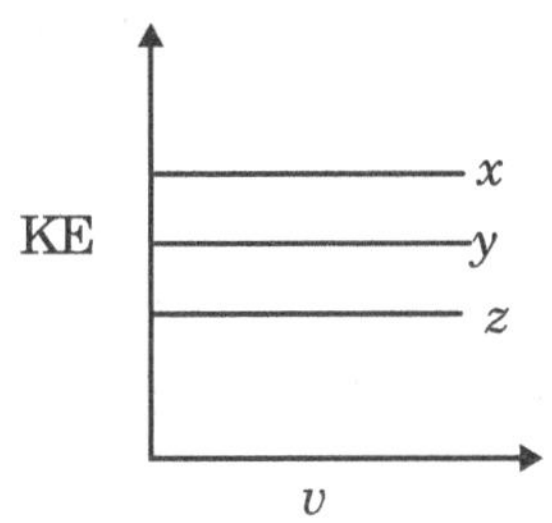

(c) 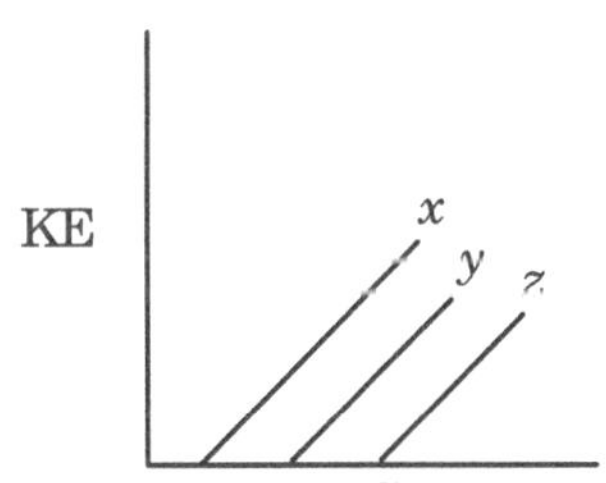

(d) Both (a) & (b)

74. If the wavelength of series limit of Lyman series for He⁺ ion is $y\,A^{0}$, what will be the wavelength of series limit of Balmer series for Li^{2+} ion?

(a) $\frac{16y}{9}$Å (b) $\frac{3}{7}$yÅ

(c) $\frac{5}{2}$yÅ (d) $\frac{4}{3}$yÅ

75. If the wave number of 1st line of Balmer series of H atom is 'Y', then

(a) Wave number of 1st line of Lyman series of He^{+} will be $\frac{108}{5}y$.

(b) The wavelength of 2nd line of Lyman series of H atom is $\frac{5}{32}y$.

(c) The wavelength of 2nd line of Balmer series of H atom is $\frac{5}{32}y$

(d) Both (a) & (b)

76. The photon emitted due to electronic transition from 5th excited state of 2nd excited state in Li^{2+} is used to excite He^{+} already in first excited state. He^{+} after absorbing the photon reaches in an orbit having total energy equal to

(a) −13.6 eV (b) −3.4 eV

(c) −27.4 eV (d) None of these

77. A proton accelerated from rest through a potential difference of 'V' volts has a wavelength λ associated with it. An alpha particle, in order to have same wavelength must be accelerated from rest through a potential difference of

(a) $\frac{V}{8}$ volt (b) 4 V volt

(c) 6 V volt (d) $\frac{3V}{7}$ volt

78. Consider the following transitions

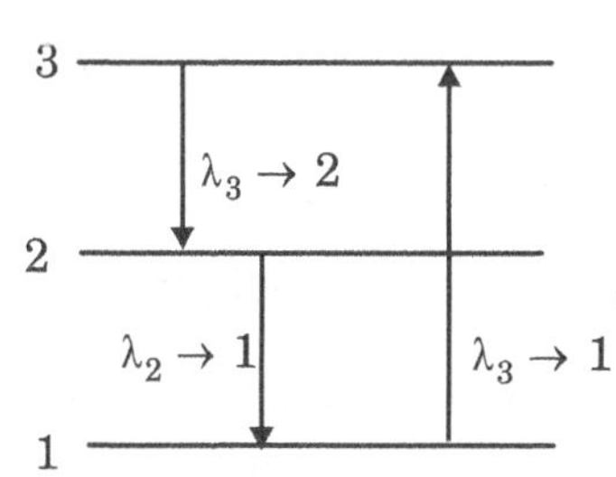

$\lambda_{3\to1}$: represents the wavelength of photon emitted when electronic transition occurred from $3 \to 1$. $\nu_{3\to1}$: represents the frequency of photon when electronic transition occurred from $3 \to 1$

Then, which is incorrect

(a) $\Delta E_{3\to1} = \Delta E_{3\to2} + \Delta E_{2\to1}$

(b) $\frac{1}{\lambda_{3\to1}} = \frac{1}{\lambda_{3\to2}} + \frac{1}{\lambda_{2\to1}}$

(c) $\lambda_{3\to1} = \lambda_{3\to2} + \lambda_{2\to1}$

(d) $\nu_{3\to1} = \nu_{3\to2} + \nu_{2\to1}$

79. A single electron is moving in an orbit of H atom from where there can be maximum of 6 spectral lines. A single electron is moving in an orbit of H from where there can be maximum of 3 Spectral lines. The ratio of the velocity of electron in these orbits will be

(a) 1 : 2 (b) 3 : 4
(c) 5 : 4 (d) 4 : 7

80. Photons emitted when electrons in a H - atom make transition from a higher energy state to lower energy state, whose difference in angular momentum is $\frac{h}{\pi}$, are made to incident on Na metal (work function = 2.3eV). The maximum possible K. E. of emitted photo electrons is -

(a) 7.9 eV (b) 0.25 eV
(c) 10.45 eV (d) 9.79 eV

81. The energy absorbed by each molecule A_2 of a substance is 4.4×10^{-19} J and bond energy per molecule is 4×10^{-19} J. The K. E. of the molecule per atom is-

(a) 2.2×10^{-19} J (b) 2×10^{-19} J

(c) 4×10^{-20} J (d) 2×10^{-20} J

De Broglie Dual Behaviour of Matter & Heisenberg's uncertainty principle

82. If the uncertainty in position & momentum are equal, than uncertainty in velocity is

(a) $\frac{1}{2m}\sqrt{\frac{h}{\pi}}$ (b) $\frac{1}{4m}\sqrt{\frac{h}{\pi}}$

(c) $2m\sqrt{\frac{h}{\pi}}$ (d) $3m\sqrt{\frac{h}{\pi}}$

83. If E_e, E_α & E_p represent the kinetic energies of an electron, alpha particle & a proton respectively, each moving with same de-Broglie wavelength then

(a) $E_e = E_\alpha = E_p$ (b) $E_e > E_p > E_\alpha$

(c) $E_e = E_p > E_\alpha$ (d) $E_e = E_\alpha > E_p$

84. The de-Broglie wavelength of Bohr's first orbit is λ. The circumference of Bohr's 4th orbits is

(a) 4λ (b) 16λ

(c) 12 λ (d) 2λ

85. If the uncertainty in the position of an electron is 5Å then calculate the uncertainty is speed.

(a) 2.2×10^{16} ms^{-1} (b) 1.14×10^{15} ms^{-1}

(c) 4.42×10^{18} ms^{-1} (d) 6.65×10^{7} ms^{-1}

86. If the wavelength of electron is numerically equal to the distance travelled by it in 1 sec, then

(a) $\lambda = \frac{h}{p}$ (b) $\lambda = \sqrt{\frac{h}{m}}$

(c) $\lambda = \sqrt{\frac{h}{p}}$ (d) None of these

87. Uncertainty in the position of an electron if $\Delta v = 3 \times 10^{-2}$ ms^{-1} is

(a) 1.92×10^{-2} m

(b) 19.2×10^{-3} m

(c) 6×10^{4} m

(d) 1.1×10^{11} m

88. The K.E. of an electron is E, when the wavelength of incident radiation is λ. To increase the K.E. of electron of 2E, the incident wavelength must be

(a) 2

(b) $\frac{\lambda}{2}$

(c) $\frac{hc\lambda}{E\lambda + hc}$

(d) $\frac{hc\lambda}{E\lambda - hc}$

89. Be^{3+} & a proton are accelerated by the same potential, their de - Broglie wavelengths will have the ratio -

(a) $1:3\sqrt{3}$ (b) 2 : 16

(c) 1 : 5 (d) $1:2\sqrt{2}$

90. $\sqrt{V}$ on two particles A & B are plotted against de - Broglie wavelengths (were V is the potential on the particles). Which of the following relation is correct about the mass of particles?

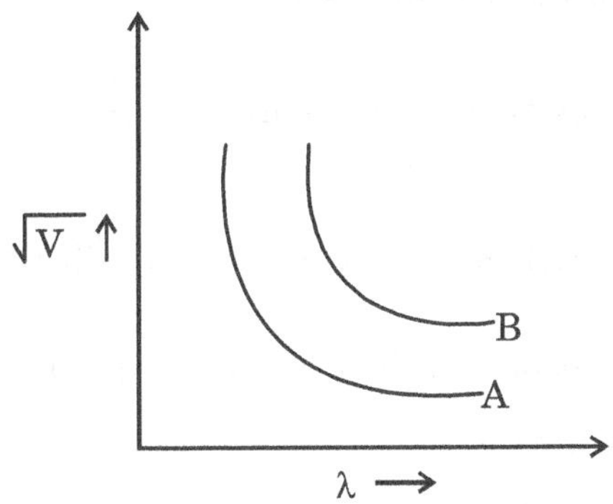

(a) $M_A = M_B$ (b) $M_A > M_B$

(c) $M_B > M_A$ (d) $M_A \le M_B$

Quantum Mechanical model of atom

91. In ψ_{321}, the sum of angular orbital momentum, spherical nodes & angular node is

(a) $\frac{\sqrt{6}h+4\pi}{2\pi}$ (b) $\frac{\sqrt{6}h+8\pi}{2\pi}$

(c) $\frac{\sqrt{6}h+2\pi}{2\pi}$ (d) None of these

92. Which set of quantum numbers is valid for a d-electrons?

(a) 4, 3, 2, $+\frac{1}{2}$ (b) 4, 2, 1, 0

(c) 4, 2, 1, $+\frac{1}{2}$ (d) 4, 3, –2, $\frac{1}{2}$

93. For principal quantum number 4, the number of subshells & orbitals are respectively

(a) 4 & 16 (b) 2 & 3

(c) 4 & 14 (d) 4 & 12

94. If an electron has $n = 4$ & $m = -2$, the n

(a) the electron is in 4th orbit

(b) the electron may be in s-orbital

(c) the electron may be in p-orbital

(d) Both 1 & 3

95. According to $(n + l)$ rule, after completing '*np*' level, the enters

(a) $(n+1)$s (b) $(n-1)$s

(c) $(n+1)$p (d) All of these

96. Which of the following sets of quantum number represent incorrect arrangement?

	n	l	m	S
(a)	3	2	-2	$\frac{1}{2}$
(b)	4	0	0	$\frac{1}{2}$
(c)	3	2	-3	$\frac{1}{2}$
(d)	5	3	0	$\frac{1}{2}$

97. The correct set of four quantum numbers for the outermost electron of Na (Z = 11) is

	n	l	m	S
(a)	3	0	0	$\frac{1}{2}$
(b)	3	1	1	$\frac{1}{2}$
(c)	3	0	1	$\frac{1}{2}$
(d)	4	0	0	$\frac{1}{2}$

98. The maximum numbers of electrons that have principal Q.no, $n = 3$ & spin quantum number = $\frac{+1}{2}$ are

(a) 9 (b) 8

(c) 7 (d) 29

99. If azimuthal quantum number could have value n also in addition to normal value, then electronic configuration of Cr (z = 24) is

(a) $1s^2\,1p^6\,2s^2\,2p^6\,3s^2\,2d^6$

(b) $1s^2\,1p^6\,2s^2\,2p^6\,3s^2\,3p^6$

(c) $1s^2\,2s^2\,2p^2\,3s^2\,3p^6\,4s^2 3d^4$

(d) $1s^2\,2s^2\,2p^5\,3s^2\,3p^6\,4s^1 3d^5$

100. Which one of the following has zero nodal plane?

(a) d_{xy}

(b) d_{yz}

(c) d_{z^2}

(d) P_x

101. Which graph is/are valid for 1s?

(a) 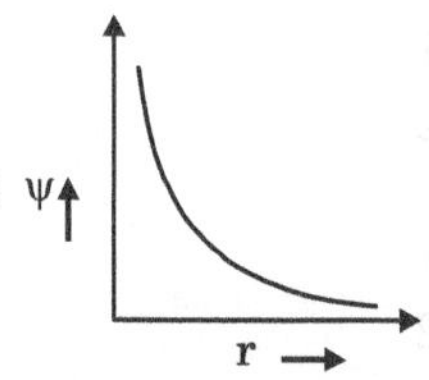

(b) 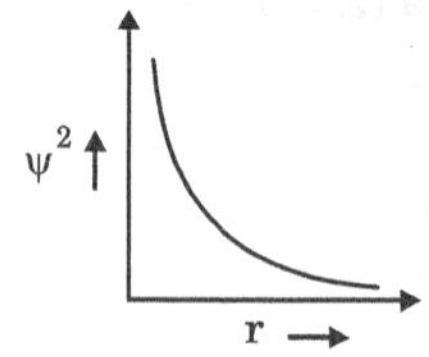

(c) 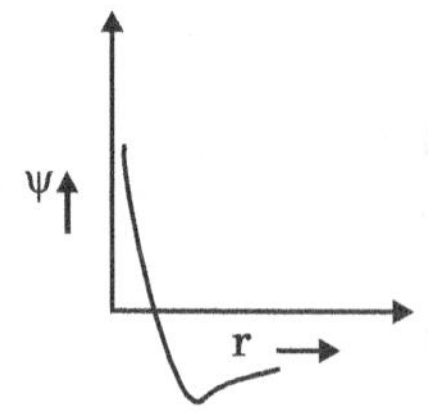

(d) Both (a) & (b)

102. Choose the incorrect relation

(a) No. of radial nodes = $n - l - 1$

(b) No. of angular nodes = l

(c) Total no. of nodes = $n - 1$

(d) p-orbitals are non-directional

103. Identify the incorrect relation

(a) Energies of orbitals in the same subshell decreases with increasing atomic number. i.e., $E_{2s(H)} > E_{2s(Li)} > E_{2s(Na)} > E_{2s(K)}$

(b) For multi electron species, the lower the value of (n + l), lesser is the energy of orbital

(c) The energy of electron in a unielectron specie is governed only by principal quantum number & energy of orbitals follows the order:

(d) None of these

104. The correct formula is/are

(a) Orbit Angular Momentum = $\frac{nh}{2\pi}$

(b) Orbital angular momentum = $\sqrt{l(l+1)}\frac{h}{2\pi}$

(c) No. of orbitals in a subshell = $2l + 1$

(d) All are correct

105. How many electrons in an atom having n = 4 will have m = –1?

(a) 6

(b) 2

(c) 4

(d) 8

106. How many elements would be in the third period (n = 3) of periodic table if spin quantum number could have the value $-\frac{1}{2}, 0, +\frac{1}{2}$?

(a) 12

(b) 6

(c) 15

(d) 101

107. The nodal plane for $d_{x^2-y^2}$ will have angle with x - axis of -

(a) 30°

(b) 45°

(c) 60°

(d) 70°

108. Which value of orbit angular momentum is not possible for H - atom?

(a) $\frac{h}{\pi}$ (b) $1.5\frac{h}{\pi}$

(c) $0.75\frac{h}{\pi}$ (d) $2.5\frac{h}{\pi}$

109. A compound of Vanadium has a magnetic moment (M) of 1.73BM. If the vanadium ion in the compound is present as V^{x+}, then calculate the value of x?

(a) 1 (b) 2

(c) 3 (d) 4

110. Not considering the electron spin, the degeneracy of the second excited state is 9, while the degeneracy of first excited state of H^- is -

(a) 1 (b) 3

(c) 3 (d) 4

111. The H like specie Li^{2+} is in a spherically symmetric state S_1 with one radial mode. Upon absorbing light, the ion undergoes transition to a state S_2. The state S_2 has one radial node & its energy is equal to the ground state energy of H atom. Then,

The state S_1 is -

(a) 1s (b) 2s

(c) 2p (d) 3s

112. Energy of the state S_1 in units of H atom ground state energy is -

(a) 0.75 (b) 0.95

(c) 2.25 (d) 3.35

113. The orbital angular momentum quantum number of the state S_2 is -

(a) 0 (b) 1

(c) 2 (d) 3

114. What is the maximum no. of electrons which can be accommodated in an atom in which the highest principal quantum no. is 4?

(a) 10 (b) 18

(c) 36 (d) 54

115. What is the maximum no. of electrons which can be accommodated in principal quantum no. is 4?

(a) 16 (b) 18

(c) 22 (d) 32

116. Not considering the electron spin, the degeneracy of the first excited state ($n = 2$) of H atom is -

(a) 2 (b) 3

(c) 4 (d) 8

117. No. of electrons in a shell having principal quantum 6 are -

(a) 6 (b) 12

(c) 36 (d) 72

Answers

1. (d)	**2.** (c)	**3.** (c)	**4.** (a)	**5.** (d)	**6.** (a)	**7.** (a)	**8.** (b)	**9.** (a)	**10.** (b)
11. (b)	**12.** (a)	**13.** (a)	**14.** (d)	**15.** (c)	**16.** (c)	**17.** (a)	**18.** (d)	**19.** (a)	**20.** (b)
21. (d)	**22.** (b)	**23.** (c)	**24.** (d)	**25.** (b)	**26.** (d)	**27.** (d)	**28.** (b)	**29.** (a)	**30.** (a)
31. (a)	**32.** (a)	**33.** (d)	**34.** (a)	**35.** (b)	**36.** (c)	**37.** (b)	**38.** (d)	**39.** (c)	**40.** (d)
41. (d)	**42.** (a)	**43.** (a)	**44.** (c)	**45.** (a)	**46.** (d)	**47.** (c)	**48.** (c)	**49.** (a)	**50.** (a)
51. (a)	**52.** (a)	**53.** (d)	**54.** (a)	**55.** (a)	**56.** (b)	**57.** (a)	**58.** (a)	**59.** (a)	**60.** (d)
61. (d)	**62.** (b)	**63.** (b)	**64.** (a)	**65.** (b)	**66.** (a)	**67.** (a)	**68.** (d)	**69.** (c)	**70.** (d)
71. (b)	**72.** (a)	**73.** (c)	**74.** (a)	**75.** (d)	**76.** (b)	**77.** (a)	**78.** (c)	**79.** (d)	**80.** (d)
81. (d)	**82.** (a)	**83.** (b)	**84.** (b)	**85.** (b)	**86.** (b)	**87.** (b)	**88.** (c)	**89.** (a)	**90.** (b)
91. (a)	**92.** (c)	**93.** (a)	**94.** (d)	**95.** (a)	**96.** (c)	**97.** (a)	**98.** (a)	**99.** (a)	**100.** (c)
101. (d)	**102.** (d)	**103.** (d)	**104.** (d)	**105.** (a)	**106.** (a)	**107.** (b)	**108.** (c)	**109.** (d)	**110.** (*)
111. (b)	**112.** (c)	**113.** (b)	**114.** (c)	**115.** (d)	**116.** (c)	**117.** (d)			

Explanations

11. (b) Balmer transition for He^+ spectrum -

$$\frac{1}{\lambda} = RZ^2\left(\frac{1}{n_1^2} - \frac{1}{n_2^2}\right)$$

$$\frac{1}{\lambda} = R \times 2^2\left(\frac{1}{2^2} - \frac{1}{4^2}\right)$$

$$\frac{1}{\lambda} = 4R\left(\frac{1}{4} - \frac{1}{16}\right)$$

$$\frac{1}{\lambda} = R\left(\frac{4}{4} - \frac{4}{16}\right) \text{\_\_\_\_(1)}$$

Now, for H, λ is same –

$$\frac{1}{\lambda} = R\left(\frac{1}{n_1^2} - \frac{1}{n_2^2}\right) \text{\_\_\_\_(2)}$$

comparing both equations

$$\frac{4}{4} = \frac{1}{n_1^2} \qquad -\frac{4}{16} = -\frac{1}{n_2^2}$$

$$n_1 = 1 \qquad \frac{1}{4} = \frac{1}{n_2^2}$$

$$n_2^2 = 4$$

$$n_2 - 2$$

12. (a) Series limit of Lyman series -

$n_1 = 1 \,\&\, n_2 = \infty$

$$\Delta E = 2.18 \times 10^{-18} \times \left(\frac{1}{1^2} - \frac{1}{\infty^2}\right)$$

$$\Delta E = 2.18 \times 10^{-18} J$$

Now, $\Delta E = hV$

$\Delta E = hV_1$ (given)

$$\frac{2.18 \times 10^{-18}}{h} = V_1 \text{\_\_\_\_(1)}$$

Similarly, first line of Lyman series,

$n_2 = 2$ to $n_1 = 1$

$$\Delta E = 2.18 \times 10^{-18} \times \left(\frac{1}{1^2} - \frac{1}{4}\right)$$

$$\Delta E = 2.18 \times 10^{-18} \times \frac{3}{4}$$

Now, $\Delta E = hV$

$\Delta E = hV_2$

$$\frac{2.18 \times 10^{-18} \times 0.75}{h} = V_2 \text{\_\_\_\_(2)}$$

For series limit of Balmer series,

$n_2 = \infty$ to $n_1 = 2$

$$\Delta E = 2.18 \times 10^{-18} \times \left(\frac{1}{2^2} - \frac{1}{\infty^2}\right)$$

$$\Delta E = \frac{2.18 \times 10^{-18}}{4}$$

Also, $\Delta E = hV$

$\Delta E = h \times V_3$

$$\frac{2.18 \times 10^{-18}}{4h} = V_3$$

$$\frac{2.18 \times 10^{-18}}{4h} \times 0.25 = V_3$$

Now, clearly, $V_1 - V_2 = V_3$

13. (a) An electron in H atom is transiting from 4th shell to first shell.

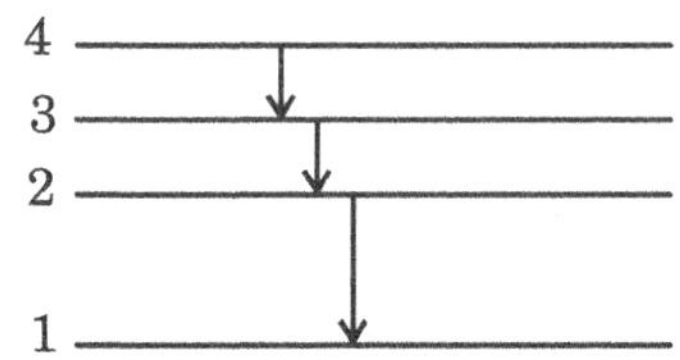

$\therefore$ Max. no. different photons emitted = 3

$(4 \to 3,\ 3 \to 2 \,\&\, 2 \to 1)$

14. (d) $E_3 - E_2 = E$

$$-13.6 \times \frac{1}{9} - \left(-13.6 \times \frac{1}{4}\right) = E$$

$$-\frac{13.6}{9} + \frac{13.6}{4} = E$$

$$13.6\left(\frac{1}{4} - \frac{1}{9}\right) = E$$

$$13.6 \times \frac{5}{36} = E$$

$$13.6 = \frac{36}{5} E$$

$$\text{Now, } IE_H = +13.6 \text{ eV} = +\frac{36}{5} E = 7.2E$$

15. (c) $R = 0.529 \times \frac{n_1^2}{z}$ _____(1)

$$4R = 0.529 \times \frac{n_2^2}{z} \text{ \_\_\_\_\_(2)}$$

$$\therefore \frac{n_1}{n_2} = \frac{1}{2}$$

$$\text{Now, } T \propto \frac{n^3}{z^2}$$

$$T_1 \propto \frac{n_1^3}{z^2}$$

$$T_2 \propto \frac{n_2^3}{z^2}$$

$$\frac{T_1}{T_2} = \left(\frac{n_1}{n_2}\right)^3 = \left(\frac{1}{2}\right)^3 = \frac{1}{8}$$

16. (c) $\text{Total energy} = \frac{PE}{2}$

$$T.E. = -\frac{6.8}{2}$$

$$T.E. = -3.4 \text{ ev}$$

$$T.E_n = -13.6 \times \frac{z^2}{n^2} \text{ ev/atom}$$

$$-3.4 = -13.6 \times \frac{1^2}{n^2}$$

$$n^2 = 4$$

$$n = 2$$

So, e^- is in first excited state.

17. (a) Distance b/w forth and third orbits = $r_4 - r_3$

$$= 0.529 \times \frac{4^2}{2} - \left(0.529 \times \frac{3^2}{2}\right)$$

$$= 0.529 \times \frac{16}{2} - 0.529 \times \frac{9}{2}$$

$$= \frac{0.529}{2}(16-9) = 0.529 \times \frac{7}{2} A^o$$

$$= 1.85 A^o = 1.85 \times 10^{-10} M$$

28. (b) I.E. of $He^+ = 8.72 \times 10^{-18}$ J atom^{-1}

$$I.E. = +x \times \frac{z^2}{n^2} \text{ J/atom}$$

$$8.72 \times 10^{-18} = x \times \frac{4}{1}$$

$$x = \frac{8.72 \times 10^{-18}}{4} = 2.18 \times 10^{-18}$$

For Li^{2+}

$$T.E. = -2.18 \times 10^{-18} \times \frac{z^2}{n^2} \text{ J/atom}$$

$$T.E. = -2.18 \times 10^{-18} \times \frac{3^2}{1}$$

$$= -2.18 \times 10^{-18} \times 9 \text{ J/atom}$$

$$= -19.62 \times 10^{-18} \text{ J/atom}$$

29. (a) $\Delta E (\text{in ev}) = \frac{hc}{\lambda}$

$$\Delta E \times 1.6 \times 10^{-19} J = \frac{19.87 \times 10^{-26} \text{ Js ms}^{-1}}{\lambda}$$

$$\Delta E \times 1.6 \times 10^{-19} J = \frac{19.87 \times 10^{-26} \text{ m}}{\lambda}$$

$$\lambda = \frac{19.87 \times 10^{-26} m}{1.6 \times 10^{-19} \times \Delta E} = \frac{12.41 \times 10^{-7}}{\Delta E} m$$

$$\lambda = \frac{12.41 \times 10^{-7}}{\Delta E} m$$

$$\lambda = \frac{12.41 \times 10^{-7}}{\Delta E} \times 10^{10} A^o$$

$$\lambda = \frac{12.41 \times 1000}{\Delta E} = \frac{12410}{\Delta E} A^o \sim \frac{12375}{\Delta E} A^o$$

31. (a)

	State I	State II
KE	x	$\frac{x}{4}$
P.E.	$-2x$	$\frac{-2x}{4}$

$$\Delta P.E. = \frac{-2x}{4} - (2x) = -\frac{x}{2} + 2x = \frac{-x + 4x}{2}$$

$$\Delta PE = \frac{3x}{2}$$

33. (d) Energy of H atom in ground state

$= -13.6$ eV $= -E$

I.E. of stationary Li^{2+} in ground state

$$= +13.6 \times \frac{Z^2}{n^2} eV/atom$$

$$= E \times \frac{3^2}{1^2} = 9E$$

Energy of photon = E_p

Now, $E_p - E = \frac{1}{2} mv^2$

$$\sqrt{\frac{2(E_p - E)}{m}} = v$$

34. (a) $V_n \propto \frac{n}{z}$

$$\frac{V_1}{V_2} = \frac{n_1}{n_2} = \frac{5}{3}$$

35. (b) $\nu_1 = 4 \times 10^{16} s^{-1}$

$\nu_2 = 2 \times 10^{16} s^{-1}$

$h\nu_1 - h\nu_0 = 6KE$ ____(1)

$h\nu_2 - h\nu_0 = KE$ ____(2)

Divide $\frac{1}{2}$

$$\frac{\nu_1 - \nu_0}{\nu_2 - \nu_0} = 6$$

$$\nu_1 - \nu = 6\nu_2 - 6\nu_0$$

$$5\nu_0 = 6\nu_2 - \nu_1$$

$$5\nu_0 = 6(2 \times 10^{16}) - 4 \times 10^{16}$$

$$5\nu_0 = 12 \times 10^{16} - 4 \times 10^{16}$$

$$\nu_0 = \frac{8 \times 10^{16}}{5} = 1.6 \times 10^{16} s^{-1}$$

36. (c) Incident energy - threshold energy = K. E. (work function)

Incident energy $- 4.4 \times 10^{-20} J = 1.2 \times 10^{-20} J$

Incident energy = $1.2 \times 10^{-20} + 4.4 \times 10^{-20}$

$= 10^{-20}(5.6) = 5.6 \times 10^{-20} J$

37. (b) $hv_1 - hv_0 = KE_1$ ____(1)

$hv_2 - hv_0 = KE_2$ ____(2)

Divide $\frac{1}{2}$

$$\frac{v_1 - v_0}{v_2 - v_0} = \frac{1}{K}$$

$$Kv_1 - Kv_0 = v_2 - v_0$$

$$Kv_1 - v_2 = -v_0 + Kv_0$$

$$Kv_1 - v_2 = Kv_0 - v_0$$

$$Kv_1 - v_2 = v_0(K - 1)$$

$$\frac{Kv_1 - v_2}{K - 1} = v_0$$

38. (d) Suppose incident freq. is $5s^{-1}$.

So, incident energy = 5h

Let, threshold energy = 2h

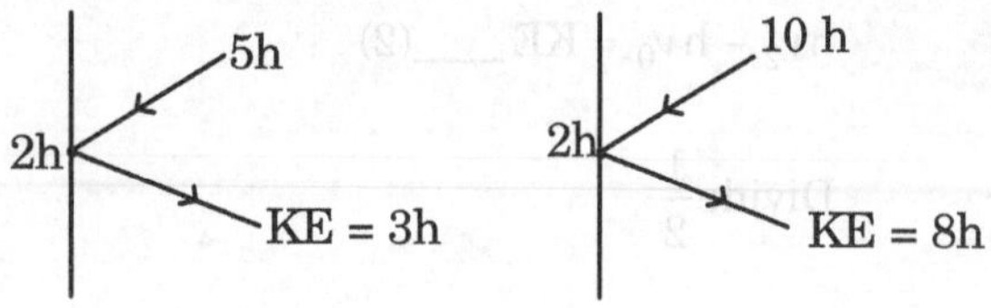

KE is more than doubled.

49. (a) Power of lamp = 600W

$\therefore$ Energy emitted by lamp = $600 Js^{-1}$

Apply, $E = \frac{nhc}{\lambda}$

$$600 = \frac{n \times 6.626 \times 10^{-34} Js \times 3 \times 10^{8} Ms^{-1}}{313.3 \times 10^{-9} M}$$

$$\frac{600 \times 313.3 \times 10^{-9}}{6.626 \times 3 \times 10^{-26}} = n$$

$$\frac{187980 \times 10^{-9}}{19.878 \times 10^{-26}} = n$$

$$9456 \times 10^{17} = n$$

$$\approx 10000 \times 10^{17} = n$$

$$10^{21} = n$$

50. (a) Speed $= \frac{\text{distance}}{\text{time}}$

$$\text{time} = \frac{\text{distance}}{\text{speed}} = \frac{8 \times 10^{7} \times 10^{3} M}{3 \times 10^{8} Ms^{-1}}$$

$$\text{time} = \frac{8}{3} \times 10^{2} s = \frac{800}{3} \text{seconds} = 266.66 \text{ sec}$$

or 4 min. 26 sec

51. (a) $E = 1eV = 1.6 \times 10^{-19} J$

$$E = \frac{hc}{\lambda}$$

$$1.6 \times 10^{-19} J = \frac{6.626 \times 10^{-34} Js \times 3 \times 10^{8} ms^{-1}}{\lambda}$$

53. (d) Apply, $= v = \frac{c}{\lambda}$ & then $\bar{v} = \frac{1}{\lambda}$

54. (a) $E = E_1 + E_2$

$$\frac{hc}{355} = \frac{hc}{680} + \frac{hc}{\lambda}$$

$$\frac{1}{355} - \frac{1}{680} = \frac{1}{\lambda}$$

$$\frac{680 - 355}{355 \times 680} = \frac{1}{\lambda}$$

$$\frac{1}{\lambda} = \frac{325}{241400}$$

$$\lambda = 743 \text{ nm}$$

55. (a) To break 6×10^{23} molecules, energy required

= 242KJ

To break 1 molecule, energy required

$$= \frac{242}{6 \times 10^{23}} KJ$$

Apply, $E = \frac{hc}{\lambda}$

$$\frac{242}{6 \times 10^{23}} \times 1000 J$$

$$= \frac{6.626 \times 10^{-34} Js \times 3 \times 10^{8} ms^{-1}}{\lambda}$$

$$\lambda = \frac{19.87 \times 10^{-26} \times 6 \times 10^{23}}{242 \times 1000} m$$

$$\lambda = \frac{0.492 \times 10^{-3}}{1000} = 0.492 \times 10^{-6} m$$

Now, $1m = 10^{9} nm$

$$\lambda = 0.492 \times 10^{-6} \times 10^{9} \text{ nm}$$

$$\lambda \approx 492 \text{ nm}$$

56. (b) Energy of e^- in ground state of H = –Rhc

57. (a) Lowest λ in Lyman series would correspond to transition, $\infty \to 1$.

$$\frac{1}{\lambda} = R\left(\frac{1}{n_1^2} - \frac{1}{n_2^2}\right)$$

$$\frac{1}{\lambda} = R\left(\frac{1}{1^2} - \frac{1}{\infty^2}\right)$$

$$\frac{1}{\lambda} = R \quad \lambda_1 = \frac{1}{R} \quad ....(a)$$

Highest λ in Lym an series would correspond to transition, $2 \rightarrow 1$.

$$\frac{1}{\lambda} = R\left(\frac{1}{n_1^2} - \frac{1}{n_2^2}\right)$$

$$\frac{1}{\lambda} = R\left(\frac{1}{1^2} - \frac{1}{2^2}\right)$$

$$\frac{1}{\lambda} = R\left(1 - \frac{1}{4}\right)$$

$$\frac{1}{\lambda} = R\left(\frac{4-1}{4}\right)$$

$$\frac{1}{\lambda} = \frac{3R}{4}$$

$$\lambda_2 = \frac{4}{3R} \quad ....(b)$$

$$\frac{\lambda_1}{\lambda_2} = \frac{\frac{1}{R}}{\frac{4}{3R}}$$

$$\frac{\lambda_1}{\lambda_2} = \frac{1}{R} \times \frac{3R}{4}$$

$$\frac{\lambda_1}{\lambda_2} = 3:4$$

58. (a) I.E. of H $= x$ KJ

$$\text{T.E.} = -x \times \frac{z^2}{n^2} \text{ KJ/atom}$$

$$\text{Energy reqd.} = E_2 - E_1 = -x \times \frac{1^2}{4} - \left(-x \times \frac{1^2}{1^2}\right)$$

$$= -\frac{x}{4} + x = x - \frac{x}{4} = \frac{3x}{4}$$

59. (a) 5th excited state refers to 6th shell.

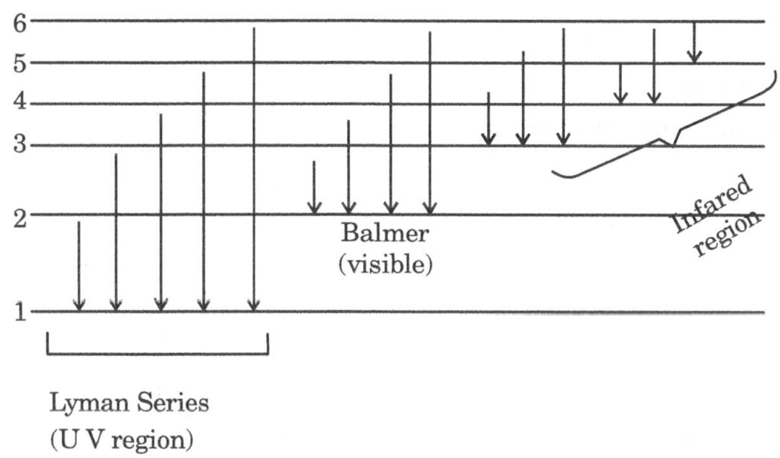

6 different lines in infrared

60. (d) (a)

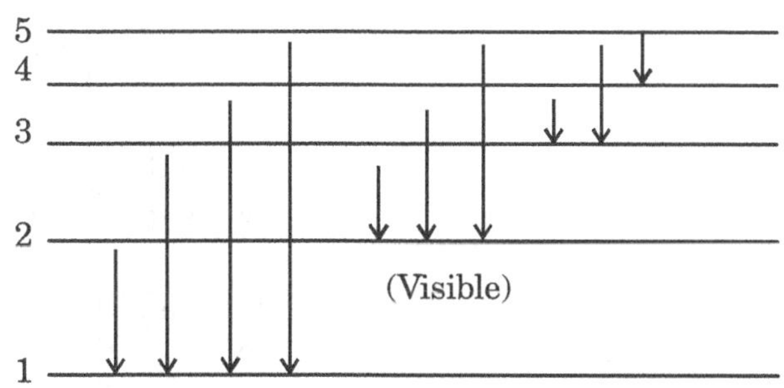

(b) 1st line of Balmer series, $3 \to 2$ transition.

Last line of Pachen series, $\infty \to 3$

wave No. $(\bar{v})$ of Ist line of Balmer series for Li^{2+}

$$\bar{v} = \frac{1}{\lambda} = Rz^2\left(\frac{1}{2^2} - \frac{1}{3^2}\right)$$

$$\bar{v} = Rz^2\left(\frac{1}{4} - \frac{1}{9}\right)$$

$$\bar{v} = 9R\left(\frac{5}{36}\right)$$

$$\bar{v} = \left(\frac{5R}{4}\right) \quad .....(1)$$

Last line of Paschen series, $\infty \to 3$

$$\bar{v} = \frac{1}{\lambda} = Rz^2\left(\frac{1}{3^2} - \frac{1}{\infty^2}\right)$$

$$\bar{v} = \frac{1}{\lambda} = R \times 9\left(\frac{1}{9}\right)$$

$$\bar{v} = R \quad .....(2)$$

Difference b/w the wave No.'s $= \frac{5R}{4} - R = \frac{R}{4}$

66. (a) I.E. of Na = 495.5 KJ/mol

To ionise 1 mole of Na atoms, energy required = 495.5 KJ

To ionise 6×10^{23} Na atoms, energy required = 495.5 KJ

To ionise 1 Na atoms, enrgy required

$$= \frac{495.5}{6 \times 10^{23}} \text{KJ}$$

Now, E = hv

$$\frac{495.5}{6 \times 10^{23}} \times 1000\,\text{J} = 6.626 \times 10^{-34}\ \text{Js} \times \text{v}$$

$$\frac{495.5}{6 \times 10^{23}} \times \frac{1000}{6.626 \times 10^{-34}} = \text{v}$$

$$\frac{495.5}{39.756} \times \frac{1000}{10^{-11}} = \text{v}$$

$$12.46 \times 10^{14}\ \text{s}^{-1} = \text{v}$$

$$1.24 \times 10^{15}\text{s}^{-1} = \text{v}$$

67. (a) $E = \frac{hc}{\lambda}$

$$25 = \frac{hc}{\lambda_1} \_\_\_\_(1)$$

$$50 = \frac{hc}{\lambda_2} \_\_\_\_(2)$$

$$\frac{\cancel{25}^{\,1}}{\cancel{50}_{\,2}} = \frac{hc}{\lambda_1} \times \frac{\lambda_2}{hc}$$

$$\frac{1}{2} = \frac{\lambda_2}{\lambda_1}$$

$$\lambda_1 = 2\lambda_2$$

74. (a) Series limit of Lyman series, $n = \infty$ to $n = 1$

Apply, $\frac{1}{\lambda} = RZ^2\left(\frac{1}{1^2} - \frac{1}{\infty^2}\right)$

$$\frac{1}{\lambda} = R \times 2^2(1 - 0)$$

$$\frac{1}{\lambda} = 4R$$

$$\frac{1}{y} = 4R \quad ....(a)$$

For Li^{2+} series limit of Balmer series, $n = \infty$ to $n = 2$

$$\frac{1}{\lambda} = RZ^2\left(\frac{1}{2^2} - \frac{1}{\infty^2}\right)$$

$$\frac{1}{\lambda} = R \times 9\left(\frac{1}{4}\right)$$

$$\frac{1}{\lambda} = \frac{9R}{4} \quad ...\ (b)$$

Divide $\frac{1}{2}\ \frac{1}{y} \times \lambda = \frac{4R}{9R/4}$

$$\frac{\lambda}{y} = \frac{4\cancel{R}}{9\cancel{R}} \times 4 = \frac{16}{9}$$

$$\boxed{\lambda = \frac{16y}{9}\ \overset{o}{A}}$$

75. (d) 1st line of Balmer series, $n_2 = 3$ to $n_1 = 2$

$$\bar{v} = \frac{1}{\lambda} = R\left(\frac{1}{4} - \frac{1}{9}\right)$$

$$y = R\left(\frac{9-4}{36}\right)$$

$$y = \frac{5R}{36}$$

$$R = \frac{36y}{5}$$

Wave no. of 1st line of Lyman series for He^+

$$\bar{v} = RZ^2\left(\frac{1}{1^2} - \frac{1}{2^2}\right)$$

$$\bar{v} = 4R\left(1 - \frac{1}{4}\right)$$

$$\bar{v} = \frac{4 \times 3}{4}R$$

$$\bar{v} = \frac{3}{\cancel{4}} \times \frac{\overset{9}{\cancel{36}}y}{5} \times 4 = \frac{27}{5}y \times 4 = \frac{108}{5}y$$

Wavelength of 2nd line of Lyman series for H -

$$\frac{1}{\lambda} = R\left(\frac{1}{1^2} - \frac{1}{3^2}\right)$$

$$\frac{1}{\lambda} = R\left(1 - \frac{1}{9}\right)$$

$$\frac{1}{\lambda} = \frac{8}{9}R$$

$$\frac{1}{\lambda} = \frac{8}{9} \times \frac{36y}{5} = \frac{32}{5}y$$

$$\boxed{\lambda = \frac{5}{32}y}$$

80. (d) Difference in angular momentum $= \frac{h}{\pi}$

$$\therefore (n_2 - n_1)\frac{h}{2\pi} = \frac{h}{\pi}$$

$$\therefore (n_2 - n_1) = 2 \text{ (Difference in shell no.)}$$

For photo electric effect to be observed

Energy of photon > work function (2.3 eV)

∴ 2 photons are possible in H - atom where difference in shell no. is 2 & energy > 2.3 eV.

∴ Energy photon = 12.09 eV (from 3 → 1 transition)

& 2.55 eV (from 4 → 2 transition)

Max. KE of photo electron will correspond to max. energy of incident photon.

$$\therefore KE_{max.} = 12.09 - 2.3 = 9.79\,eV$$

81. (d) Energy absorbed - Bond Energy = K. E. of the molecule

$$4.4 \times 10^{-19} J - 4 \times 10^{-19} J = \text{KE of molecule}$$

$$10^{-19} \times 0.4 J = \text{KE of molecule}$$

$$4 \times 10^{-20} J = \text{KE of molecule}$$

$$\text{K.E. per atom} = \frac{4 \times 10^{-20}}{2} = 2 \times 10^{-20} J$$

De Broglie Dual Behaviour of Matter & Heisenberg's uncertainty principle

82. (a) $\Delta x = \Delta p$ (given)

Apply

$$\Delta x . \Delta p = \frac{h}{4\lambda} \quad \text{(for numerical purpose)}$$

$$\Delta p^2 = \frac{h}{4\lambda}$$

$$(m\Delta V)^2 = \frac{h}{4\lambda}$$

$$m^2 . \Delta V^2 = \frac{h}{4\lambda} \Rightarrow \Delta V^2 = \frac{h}{4\lambda m^2}$$

$$\Rightarrow \Delta V = \sqrt{\frac{h}{4\lambda m^2}}$$

$$\Delta V = \frac{1}{2m}\sqrt{\frac{h}{\lambda}}$$

84. (b) De - Broglie's wavelength of 1st orbit = λ

$$n\lambda = 2\lambda r$$

$$1 \times \lambda = 2\lambda r_1 \text{ \_\_\_\_(1)}$$

For 4th orbit $4 \times \lambda' = 2\pi r_4$ ____(2)

Divide $\frac{1}{2}$

$\lambda = 2\pi r_1$ ______(3)

$4\lambda' = 2\pi r_4$ _____(4)

$$\frac{1}{4} \times \frac{\lambda}{\lambda'} = \frac{r_1}{r_4}$$

$$\frac{\lambda}{4\lambda'} = \frac{0.529 \times \frac{1^2}{z}}{0.529 \times \frac{4^2}{z}}$$

$$\frac{\lambda}{4\lambda'} = \frac{1}{16}$$

$$\frac{16\lambda}{4} = \lambda' \quad \Rightarrow \lambda' = 4\lambda$$

Now, circumference of forth orbit = $4\lambda'$

(from eqn. 2)

$= 4(4\lambda) = 16\lambda$

87. (b) $\Delta x . \Delta V = \frac{h}{4\pi m}$

$$\Delta V = \frac{0.005}{100} \times 600 = 3 \times 10^{-2} MJ^{-1}$$

$$\Delta x = \frac{h}{4xm\Delta V} = \frac{6.626 \times 10^{-34}}{4 \times 3.14 \times 9.1 \times 10^{-31} \times 3 \times 10^{-2}}$$

$$\Delta x = 1.92 \times 10^{-3} M$$

88. (c) $hv - hv_0 = K.E.$

$$h = \frac{c}{\lambda} - hv_0 = E$$

$$\frac{hc}{\lambda} - E = hv_0 \qquad \text{(a)}$$

$$hv - 2v_0 = K.E.$$

$$h\frac{c}{\lambda'} - hv_0 = 2E$$

$$\frac{hc}{\lambda'} - 2E = hv_0 \quad .... \text{(b)}$$

From 1 & 2

$$\frac{hc}{\lambda} - E = \frac{hc}{\lambda}, -2E$$

$$E = \frac{hc}{\lambda} - \frac{hc}{\lambda}$$

$$E = hc\left(\frac{1}{\lambda'} - \frac{1}{\lambda}\right)$$

$$\frac{E}{hc} = \frac{1}{\lambda'} - \frac{1}{\lambda}$$

$$\frac{E}{hc} + \frac{1}{\lambda} = \frac{1}{\lambda'}$$

$$\frac{E\lambda + hc}{hc\lambda} = \frac{1}{\lambda'}$$

$$\boxed{\lambda' = \frac{hc\lambda}{E\lambda + hc}}$$

89. (a) For Be^{3+}

$$\lambda_1 = \frac{h}{\sqrt{2 \times 9 \times 3 \times V}} \text{ ______(1)}$$

$$\frac{\lambda_1}{\lambda_2} = \frac{h}{\sqrt{2 \times 9 \times 3 \times V}} \times \frac{\sqrt{2 \times 1 \times 1 \times V}}{h}$$

$$\frac{\lambda_1}{\lambda_2} = \frac{1}{\sqrt{27}} \quad \Rightarrow \quad \frac{\lambda_1}{\lambda_2} = \frac{1}{3\sqrt{3}}$$

$$\Rightarrow \quad 3\sqrt{3}\,\lambda_1 = \lambda_2$$

For proton

$$\lambda_2 = \frac{h}{\sqrt{2 \times 1 \times 1 \times V}} \text{ ______(2)}$$

90. (b)

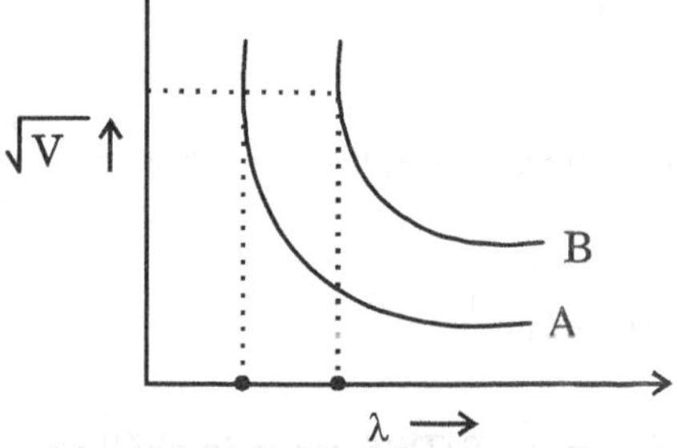

$$\lambda = \frac{h}{\sqrt{2mev}}$$

$\lambda b > \lambda A$

$\therefore M_A > M_B$

91. (a) In Ψ_{321}

$n = 3, l = 2, m = 1$

$$\text{Orbital angular momentum} = \sqrt{l(l+1)}\frac{h}{2\pi}$$

$$= \sqrt{2(2+1)}\frac{h}{2\pi}$$

$$= \sqrt{6}\frac{h}{2\pi} \quad ...(a)$$

$$\text{spherical nodes} = n - l - 1 = 3 - 2 - 1 = 3 - 3 = 0 \quad ....(b)$$

$$\text{Angular nodes} = l = 2 \quad ....(c)$$

(Nodal plane)

$$\text{Sum of all the three} = \sqrt{6}\frac{h}{2\pi} + 0 + 2$$

$$= \sqrt{6}\frac{h}{2\pi} + 2 = \frac{\sqrt{6}h + 4\pi}{2\pi}$$

93. (a) For n = 4

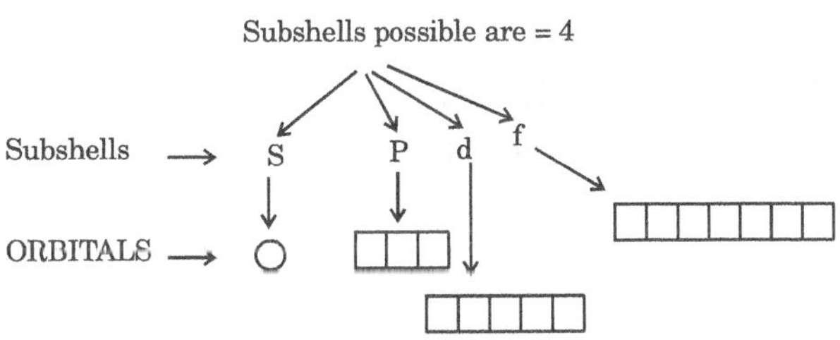

16 orbitals are possible

94. (d) If n = 4 & m = – 2, electron may be in d or f orbital.

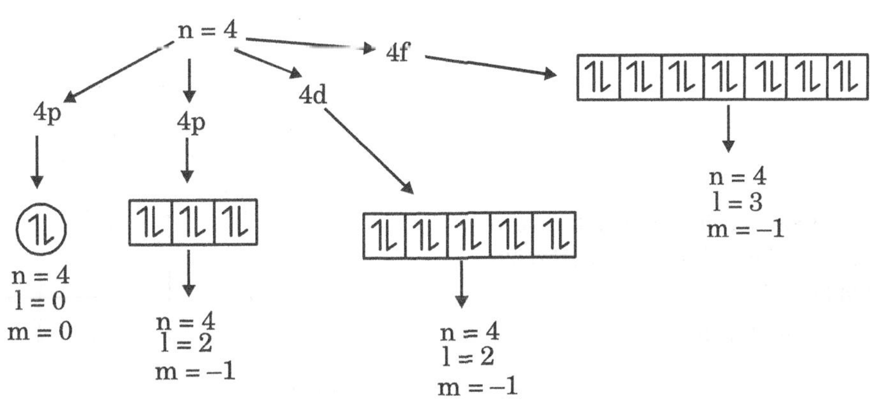

96. (c) If l = 2, m can have possible value = – 2, –1, 0, + 1, + 2 but not – 3

97. (a) Na(11): – $1s^2\ 2s^2\ 2p^6\ 3s^1$

For outermost electron in 3s -

n = 3

l = 0

m = 0

$s = \pm\frac{1}{2}$

99. (a) Let possible value of l = 0 – n

Then, for n = 1

value of l = 0 – 1

1s & 1p are possible

For n = 2

Value of l = 0 – 2 = 0, 1, 2

2s, 2p, 2d are possible

For n = 3

Value of l = 0 – 3 = 0, 1, 2, 3

3s 3p, 3d & 3f are possible

For n = 4

Value of l = 0 – 4 = 0, 1, 2, 3, 4

4s, 4p. 4d, 4f, 4g are possible

Now, checking their (n + l) values –

1s	1p	2s	2p	2d	3s	3p	3d	3f	4s	4p	4d	4f	4g
↓	↓	↓	↓	↓	↓	↓	↓	↓	↓	↓	↓	↓	↓
①	②	②	③	④	③	④	⑤	⑥	④	⑤	⑥	⑦	⑧

(n + l)

Following Aufbau's rule –

1s 1p 2s 2p 3s 2d 3p 4s 3d 4p

Cr(z = 24) – $1s^2\ 1p^6\ 2s^2\ 2p^6\ 3s^2\ 2d^6$

106. (a)

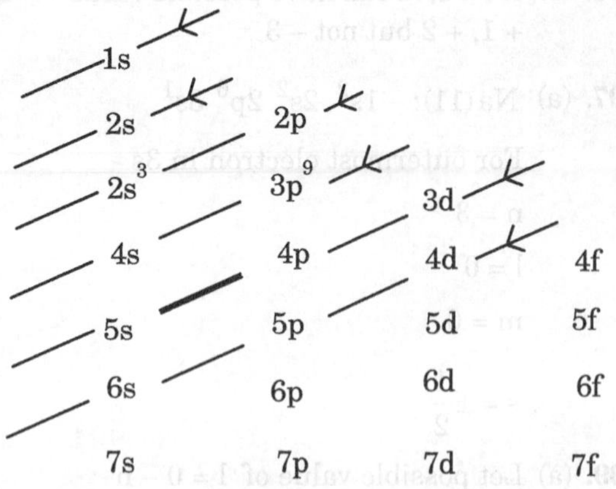

3rd period begins from 3s & ends at 3p. It spin quantum no. has 3 values, it means an orbital can accommodate . Now, Max elements in 3rd period will be 12.

107. (b)

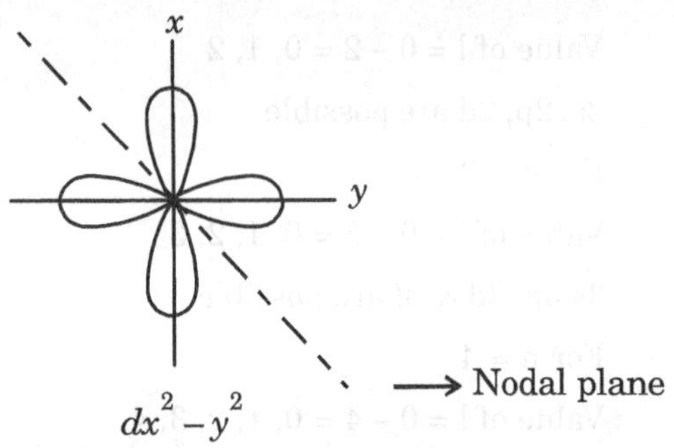

108. (c) Follow, $mvr = \frac{nh}{2\pi}$

109. (d) V^{x+}

$M = \sqrt{n(n+2)}$

$M = 1.73 \text{ B.M.}$

$\therefore n = 1$ (no. of unpaired $e^- = 1$)

$\underline{V(23):-}$ $(Ar)4s^2 3d^3$

$V^{4+}:$ $(Ar)3d^1$

So, $x = 4$

110. (*) $H^-(2):-1s^2$

It's a multi-electron specie first excited state will be 2s & 2nd excited state will be 2p. degeneracy of 2s is 1.

Degeneracy of 2p will be 3.

113. (b) No. of radial node = n – l – 1

1 = n – 0 – 1

n = 2

So, S_1 is 2S

S_2 has one radial node & its energy is equal to G. state energy of H.

So, $-13.6\text{ev} = -13.6\times\frac{3^2}{n^2}\text{ev}$

$n_2 = 9$

n = 3

For S_2

No. of radial nodes = n – l – 1

1 = 3 – l – 1

1 = 2 – l

l = 1

$\therefore S_2$ is 3p

ORBITAL ANGULAR MOMENTUM Q. NO. = 1

$\frac{\text{Energy of state } S_1}{\text{Energy of G.S. of H}} = \frac{-13.6\times\frac{3^2}{2^2}}{-13.6\times\frac{1^2}{1^2}} = \frac{9}{4} = 2.5$

114. (c) $1s^2 2s^2 2p^6 3s^2 3p^6 4s^2 3d^{10} 4p^6$

115. (d) 4 →

4s → 2

4p → 6

4d → 10

4f → 14

32

116. (c)

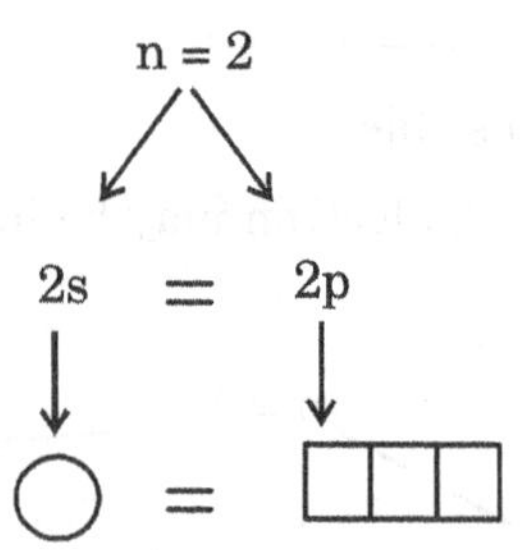

117. (d)

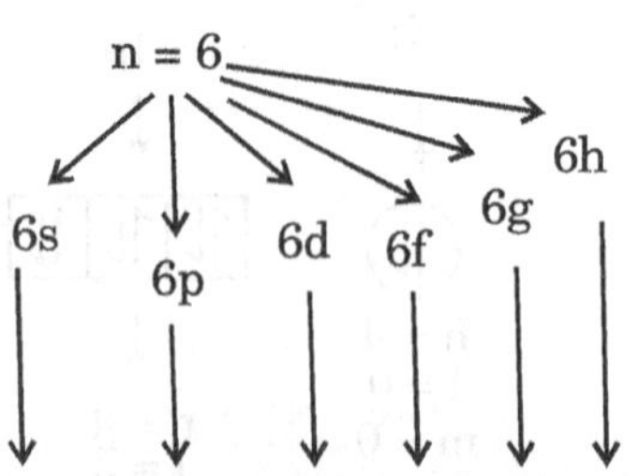

electrons – 2 + 6 + 10 + 14 + 18 + 22 = 72

Smart Notes

Smart Notes

3 CHAPTER

Classification of Elements and Periodicity in Properties

History of periodic table, IUPAC nomenclature & Identification of Groups, periods

1. If aufbau's is rule is not followed Ca will be placed in which of the following blocks?

   (a) S (b) P

   (c) d (d) f

2. Choose the incorrect match.

	Atomic No.	IUPAC Name
(a)	101	Unnilunium
(b)	105	Unnilpentium
(c)	104	Unilquadium
(d)	107	Unnilheptium

3. If an orbital could contain 3 electrons, total groups in the periodic table would have been

   (a) 20

   (b) 27

   (c) 36

   (d) 42

4. In the 4th period of the periodic table, how many elements have one or more 4d electrons?

   (a) 2

   (b) 16

   (c) 0

   (d) 12

5. An element X with Z = 112 has been recently discovered. To which group and period will it belong, respectively

   (a) 12, 7 (b) 10, 6

   (c) 8, 4 (d) 6, 6

6. If each orbital can take maximum of 3 electrons, what are number of elements in 2nd period & 4th period respectively

   (a) 11, 20 (b) 12, 27

   (c) 15, 25 (d) 16, 30

7. Element X belongs to 4th period. It contains 18 & 1 electron in the penultimate & ultimate orbit. Then, X should be

   (a) a normal element

   (b) a transition element

   (c) an inert gas

   (d) an inner-transition element

8. Choose the incorrect match

	Elements		**Common Names**
(a)	Gp. 16	→	Halogens
(b)	Elements placed after uranium	→	Trans uranium elements
(c)	f-block elements	→	Inner transition elements
(d)	None of these		

9. Lanthanoids are

(a) 14 elements in the 6th period (at. No. 90 to 103) that are filling 4f

(b) 14 elements in the 7th period (at. No. 90 to 103) that are filling 5f

(c) 14 elements in the 6th period (at. No. 58 to 71) that are filling 4f

(d) 14 elements in the 7th period (at. No. 58 to 71) that are filling 4f

10. Which of the element whose atomic numbers are given below, cannot be accommodated in the present set up of the long form of the periodic table?

(a) 107 (b) 118
(c) 126 (d) 102

11. The element having electronic configuration $1s^2\,2s^2\,2p^6\,3s^2\,3p^6\,3d^5\,4s^1$ will be placed in

(a) 2nd group, 4th period

(b) 6th group, 4th period

(c) 1st group, 4th period

(d) None of these

12. The correct order of filling up of subshells in seventh period is

(a) 5f, 4d, 6s, 7s (b) 7s, 5f, 6d, 7p
(c) 5f, 6d, 7s, 9p (d) 7s, 7p, 5f, 6d

13. The last electron of element X has quantum n = 2, l = 1, m =1, s = $-\frac{1}{2}$. The position of X in the periodic table is

(a) s-block, period II (b) p-block, period II
(c) d-block, period III (d) f-block, period III

14. The element with electronic configuration, $[Kr]\,4d^{10}\,4f^{14}\,5s^2\,5p^6\,5d^2\,6s^2$ belongs to

(a) d-block, 4th group, 6th period

(b) s-block, 2nd group, 5th period

(c) f-block, 3rd group, 6th period

(d) d-block, 10th group, 6th period

15. The IUPAC name of the element with atomic number 106 is

(a) Unnilhexium (b) Unnilennium
(c) Unnilbium (d) Unniloctium

16. The electronic configuration of element X is $[Ar]3d^{10}\,4s^2\,4p^3$. The chemistry of X is therefore likely to be similar to that of

(a) bromine (b) nitrogen
(c) sulphur (d) boron

17. Which block elements are inner transition elements?

(a) s-block elements (b) p-block elements
(c) d-block elements (d) f-block elements

18. The pair of elements containing same number of electrons in the valence shell is

(a) N, O (b) Na, Cl
(c) Ca, Cl (d) Cl, Br

19. Among the elements Br, Al, Cs and Xe, the element which was unknown at the time of Mendeleev was

(a) Cs (b) Al
(c) Br (d) Xe

20. The last element of lanthanoid series is

(a) Ce (b) Lu
(c) Lr (d) Th

21. The attempt for classifying elements by grouping some elements on the basis of their atomic masses in sets of three elements was made by

(a) Doebereiner (b) Newland
(c) Lother Meyer (d) Mendeleev

22. Which of the following property is not the characteristic of transition elements?

(a) Variable valency

(b) Their compounds are generally coloured

(c) All are non-metals

(d) Catalytic properties

23. The total number of elements in f-block are

(a) 14 (b) 40
(c) 30 (d) 28

24. The element that belongs to representative group of elements in the periodic table is

(a) lanthanum (b) neon
(c) iron (d) aluminium

25. Which set includes atomic numbers of elements having same valency?

(a) 31, 17 (b) 3, 11
(c) 4, 24 (d) 2, 4

26. In the long form of the periodic table, the block which contains metals, non-metals and metalloids is

(a) s-block (b) p-block

(c) d-block (d) f-block

27. Which of the following is the atomic number of a transuranic element?

(a) 96 (b) 90

(c) 76 (d) 57

28. The element whose position is anomalous in the periodic table is

(a) Chlorine (b) Argon

(c) Sodium (d) Hydrogen

Questions based on Electronegativity, Screening effect, Ionization energy & electron gain enthalpy

29. Which of the following isoelectronic ion has the lowest IE?

(a) Na^+ (b) F^-

(c) Mg^{2+} (d) O^{2-}

30. In which of the following pairs, the I.E. of the first species is less than that of the second?

(a) $O^{\ominus}, O^{2-}$ (b) S, P

(c) N, P (d) Be^+, Be

31. If atomic number of an inert gas is Z then an element with which of the following atomic number will have the highest I.E?

(a) Z – 1 (b) Z – 4

(c) Z + 1 (d) Z + 2

32. Out of N, O, Ne, Na, Na^+, select the species which have maximum & minimum ionization energy respectively.

(a) Na^+, Na (b) Ne, Na

(c) O, Ne (d) Ne, O

33. The successive ionization energies of an element 'A' are IE_1 = 6eV, IE_2 = 18.8eV, IE_3 = 28.4eV and IE_4 = 120eV. The number of valence electron in element 'A' are

(a) 3 (b) 4

(c) 5 (d) 6

34. The first ionization enthalpy values of the 3rd period elements, Na, Mg & Si are respectively 496, 737 & 786 KJ mol^{-1}. The first I.E. value for Al

(a) will be in between 737 & 786 kJ/mol

(b) will be more close to 575 kJ/mol

(c) will be more close to 786 kJ/mol

(d) Can't predict

35. The incorrect statement among the following is

(a) The IE_1 of Al is less than IE_1 of Mg

(b) IE_2 of Mg > IE_2 of Na

(c) IE_1 of Na < IE_1 of Mg

(d) IE_3 of Mg > IE_3 of Al

36. Which of the following statements is incorrect?

(a) The 2nd I.E. of S is greater than that of chlorine

(b) IE_3 of P > IE_3 of Cl

(c) IE_2 of Ne < IE_1 of N

(d) IE_2 of B < IE_2 of C

37. Which one of the following is a true statement?

(a) I.E. of O(g) is less than I.E. of $O^-_{(g)}$

(b) I.E. of $He_{(g)}$ is greater than I.E. of $He^+_{(g)}$

(c) I.E. of $N_{(g)}$ is greater than I.E. of $N^+_{(g)}$

(d) None of these

38. Select the Incorrect Match?

Outer Electronic Configuration

(a) $ns^1 \rightarrow$ 20, 195, 215, 230

(b) $ns^2 \rightarrow$ 23, 40, 200, 250

(c) $ns^2\,np^1 \rightarrow$ 25, 37, 45, 260, 270

(d) None of these

39. Ionization Energies of an element 'X' are given

$IE_1 \rightarrow$ 150 kJ/mol

$IE_2 \rightarrow$ 250 kJ mol

$IE_3 \rightarrow$ 1000 kJ/mol

If 'X' reacts with different elements, the which of the following compounds are not possible?

(a) XF_2 (b) X_3N_2

(c) X_2O (d) XO

40. What is correct about IE?

(a) IE_1 of N > IE_1 of O

(b) IE_2 of N > IE_2 of P

(c) IE_2 of Li > IE_2 of Ne

(d) Both (a) & (c)

41. Consider the following steps

$$Y_{(g)} \longrightarrow Y^{+}_{(g)} + e^{-};\ \Delta H = 100\,J\,mol^{-1}$$

$$Y_{(g)} \longrightarrow Y^{2+}_{(g)} + 2e^{-};\ \Delta H = 250\,J\,mol^{-1}$$

Then, the correct statements is/are

(a) IE_1 of $Y_{(g)}$ is 100 J mol^{-1}

(b) IE_1 of $Y^{+}_{(g)}$ is 150 J mol^{-1}

(c) IE_2 of $Y_{(g)}$ is 150 Jmol^{-1}

(d) All of correct

42. Minimum IE, is shown by which configuration

(a) $1s^2\,2s^2\,2p^5$

(b) $1s^2\,2s^2\,2p^6\,3s^2\,3p^2$

(c) $1s^2\,2s^2\,2p^6\,3s^1$

(d) $1s^2\,2s^2\,2p^6$

43. Which of the following statement is/are correct?

(a) The IE_1 of Al is less than IE_1 of Mg

(b) The IE_1 of Na is less than IE_1 of Mg

(c) The IE_3 of Mg is greater than IE_3 of Al

(d) All are correct

44. Screening effect is not observed in:

(a) He^+ (b) Li^{2+}

(c) Be^{3+} (d) In all the three

45. Select the endothermic reaction/s

(a) $S^{-}_{(g)} \rightarrow S^{2-}_{(g)}$

(b) $N_{(g)} \rightarrow N^{-}_{(g)}$

(c) $Al^{2+}_{(g)} \rightarrow Al^{3+}_{(g)}$

(d) All of these are endothermic

46. Which process requires absorption of energy?

(a) $F \rightarrow F^-$ (b) $Cl \rightarrow Cl$

(c) $O \rightarrow O^{2-}$ (d) All of these

47. Consider the following process

$$Na^{+}_{(g)} \xrightarrow{I} Na_{(g)} \xrightarrow{II} Na_{(s)}$$

(a) In step I, energy is released whereas in step II, energy is absorbed

(b) In both the steps, energy is evolved

(c) In both the steps, energy is absorbed

(d) Both (a) & (c)

48. The electron affinity of halogens are F = 322, Cl = 349, Br = 324 & I = 295 kJ mol^{-1}, the higher value for Cl as compared to that of F is due to

(a) Weaker electron - electron repulsion in Cl

(b) Higher atomic radius of F

(c) Smaller electronegativity of F

(d) More number of p-orbitals in Cl

49. Which of the following process is exothermic?

(a) $N^- \rightarrow N$ (b) $N \rightarrow N^+$

(c) $N \rightarrow N^-$ (d) $N^+ \rightarrow N^{2+}$

50. Which of the following reactions would not proceed in the forward direction spontaneously?

(a) $Xe + He^+ \rightarrow Xe^+ + He$

(b) $Si + Cl^+ \rightarrow Si^+ + Cl$

(c) $F^- + I \rightarrow F + I^-$

(d) $Be^- + B \rightarrow B^- + Be$

51. The first ionization energy of $Al_{(g)}$ is 577.5 kJ mol^{-1}. If 2nd & 3rd I.E. of Al are in the ratio 2 : 3 & If ΔH for $Al_{(g)} \longrightarrow Al^{3+}_{(g)} + 3e^{-}$ is 5140 kJ mol^{-1}, then IE_2 & IE_3 of Al respectively are

(a) 1825 kJ/mol, 2737.5 kJ mol^{-1}

(b) 2000 kJ/mol, 2523 kJ/mol

(c) 1600 kJ/mol^{-1}, 1665 kJ/mol^{-1}

(d) 5000 kJ/mol^{-1}, 5223 kJ/mol^{-1}

52. Following data is given

$$O_{(g)} + e^{-} \longrightarrow O^{\ominus}_{(g)}\ ;\ \Delta_{eg}H = -140.9\,kJ\,mol^{-1}$$

$$O^{\ominus}_{(g)} + e^{-} \longrightarrow O^{2-}_{(g)}\ ;\ \Delta_{eg}H = 744\,kJ\,mol^{-1}$$

Then, the first & second electron affinities of O are (in kJ/mol)

(a) 140.9, 744 (b) 140.9, –744

(c) –140.9, 744 (d) –140.9, –744

53. Out of N, O, Ne, Na, Na^+, select the species which have maximum & minimum ionization energy respectively

(a) Na^+, Na (b) Ne, Na

(c) O, Ne (d) Ne, O

54. Second electron gain enthalpy

(a) is always negative

(b) is always positive

(c) can be positive or negative

(d) is always zero

55. Select the correct statement

(a) First electron gain enthalpy of $O_{(g)}$ is negative

(b) First electron gain enthalpy of $Ne_{(g)}$ is positive

(c) Noble gases, Nitrogen & gp. II elements have positive electron gain enthalpies

(d) All are correct

56. $\Delta_{eg}H$ values of elements A, B, C & D are respectively –135, –60, –200 & –348 kJ mol^{-1}. The outer configuration of element B is

(a) $3s^2 3p^5$ (b) $3s^2 3p^4$

(c) $3s^2 3p^3$ (d) $3s^2 3p^2$

57. Select the correct order.

(a) S > Se > Te > O (Negative electron gain enthalpy)

(b) O > Te > Se > S (Electron gain enthalpy)

(c) Ne > Xe > He (Electron gain enthalpy)

(d) All are correct

58. Second electron gain enthalpy (in KJ/mol) of oxygen & sulphur respectively are

(a) –780, + 590 (b) –590, + 780

(c) +780, + 590 (d) +590, + 780

59. Which of the following arrangement represents the correct order of electron gain enthalpy (with negative sign) of the given atomic species?

(a) S > O > Cl > F (b) Cl > F > S > O

(c) Cl > F > S > O (d) S > O > Cl > F

60. Electron affinity of the elements or ions follows correct order in which of the following options

I. $S > O^{\ominus}$ II. $O > S^{\ominus}$

III. $O^{\ominus} > S^{\ominus}$ IV. $N^- > S$

(a) Only I is correct

(b) Only I & III are correct

(c) Only I, II are correct

(d) All the order are correct

61. The correct order of electron affinity follows the order :

(a) C > N < O < F

(b) N > O < C < F

(c) F > O < N < C

(d) None of these

62. Select the endothermic reaction/s

(a) $S^-_{(g)} \longrightarrow S^{2-}_{(g)}$

(b) $N_{(g)} \longrightarrow N^-_{(g)}$

(c) $Al^{2+}_{(g)} \longrightarrow Al^{3+}_{(g)}$

(d) All of these are endothermic

63. Choose the correct statement/s among the following

(a) 1st electron gain enthalpy is always positive

(b) Ionisation energy is always positive

(c) Second electron gain enthalpy is always positive

(d) Successive ionization energies are always higher

(a) Only a, b & c are correct

(b) Only c & d are correct

(c) all are correct

(d) None of these is correct

64. Select the incorrect match of property?

(a) F > Cl > Br > I (Electron affinity)

(b) Mg > Si > Al > Na (Ionisation Energy)

(c) $Al^{2+} > Al^{3+} > Al$ (Effective Nuclear charge)

(d) All are incorrect

65. The electronic configuration of four elements are

A. $1s^2 2s^2 2p^6 3s^2 3p^4$

B. $1s^2 2s^2 2p^6 3s^2 3p^6$

C. $1s^2 2s^2 2p^6 3s^2 3p^1$

D. $1s^2 2s^2 2p^6 3s^2 3p^2$

The element having maximum IE_2 is

(a) A (b) B

(c) C (d) D

66. The dipositive ion can be formed by the atom with valence configuration

(a) $ns^2 np^6$ (b) ns^1

(c) $ns^2 np^4$ (d) ns^2

67. The ionization enthalpies of transition elements are

(a) much smaller as compared to p-block elements

(b) intermediate between those of s and p-block elements

(c) much greater as those of s-block elements

(d) same as those of s and p-block elements

68. The second ionization energy (in KJ/mol) of an element X is the energy required to

(a) remove one mole of electron from one mole of monovalent gaseous cation of the element

(b) remove two moles of electrons from one mole of gaseous atoms

(c) remove one mole of electron from one mole of gaseous anion

(d) remove one mole of electron from one mole of gaseous cation of the element

69. The variation of first Ionisation enthalpy with atomic number (from 1 to 60) is shown below:

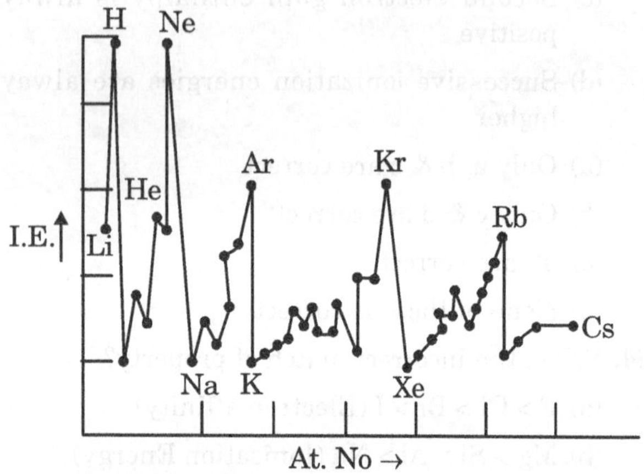

The position of which pair of elements is wrongly shown?

(a) Li and H (b) Cs and Rb

(c) Cs and He (d) all of these

70. The first ionization energies of alkaline earth metals are higher than those of boron family elements. This is because of

(a) higher nuclear charge in alkaline earth metals

(b) more penetration effect of s-orbitals than p-orbitals

(c) smaller size of boron family

(d) both 2 & 3

71. Match the electronic configuration (column-I) with the ionization energy (column-II)

Column I	Column II
a. ns^2	p. 2100 kj/mol
b. $ns^2 np^1$	q. 1400 kj/mol
c. $ns^2 np^3$	r. 800 kj/mol
d. $ns^2 np^6$	s. 900 kj/mol

(a) a - s, b - r, c - q, d - p

(b) a - q, b - p, c - r, d - s

(c) a - p, b - q, c - s, d - r

(d) a - r, b - s, c - q, d - p

72. The element with maximum ionization potential is

(a) Mg (b) Na

(c) Al (d) K

73. Successive ionization energies of an element X are given below (in K cal).

IP_1	IP_2	IP_3
165	195	556

Electronic configuration of element X is

(a) $1s^2 2s^2 2p^6 3s^2 3p^2$

(b) $1s^2 2s^1$

(c) $1s^2 2s^2 2p^2$

(d) $1s^2 2s^2 2p^6 3s^2$

74. The correct sequence of decreasing order of ionization energy is

(a) Li > H > He (b) H > He > Li

(c) He > Li > H (d) He > H > Li

75. The successive ionization energies of an element 'A' are IE_1 = 6.0 eV, $I.E_2$ = 18.8 eV, $I.E_3$ = 30.2 eV and $I.E_4$ = 120 ev. Formula of oxide of element A is

(a) AO_3 (b) A_2O_3

(c) AO_2 (d) AO

76. Among the following which has the highest first ionization energy?

(a) K (b) Na

(c) B (d) He

77. Arrange S, P As in order of increasing ionisation energy

(a) S < O < As (b) P < S < As

(c) As < S < P (d) As < P < S

78. From which of the following sub-shell, removal of electron will require highest energy?

(a) 4s

(b) 4p

(c) 4d

(d) 4f

79. I.E., of Ar is less than that of Ne. An explanation of this fact is that

(a) The effective nuclear charge experienced by a valence electron in Ar is much larger than in Ne

(b) The effective nuclear charge experienced by a valence electron in Ar is much smaller than in Ne

(c) The atomic radius of Ar is smaller than Ne

(d) The number of neutrons in Ar nucleus is larger than in Ne nucleus

80. The most electropositive element among the following elements of 3rd period is

(a) Al (b) Mg

(c) Cl (d) Si

81. Which of the following configuration is of the atom with the lowest second ionization energy?

(a) $1s^2 2s^2 2p^4$ (b) $1s^2 2s^2 2p^6$

(c) $1s^2 2s^2 2p^6 3s^1$ (d) $1s^2 2s^2 2p^6 3s^2$

82. The element which have no tendency to accept the incoming electron is/are

(a) N (b) Be

(c) P (d) both 1 and 2

83. Of the following pairs, the one containing example of electronegative elements in the periodic table is

(a) Sodium and potassium

(b) Fluorine and chlorine

(c) Calcium and magnesium

(d) Boron and silicon

84. The energy evolved is highest for which of the following reactions?

(a) $F_{(g)} + e^- \rightarrow F^-_{(g)}$

(b) $Cl_{(g)} + e^- \rightarrow Cl^-_{(g)}$

(c) $S_{(g)} + 2e^- \rightarrow S^{2-}_{(g)}$

(d) $O_{(g)} + 2e^- \rightarrow O^{2-}_{(g)}$

85. Increasing order of electron affinity is

(a) N < O < CI < Ne (b) O < N < Ne < CI

(c) Ne < N < O < CI (d) CI < N < O < Ne

86. In which pair, the second atom has larger electron gain enthalpy than first?

(a) Br, CI (b) Na, Mg

(c) Sr, Ca (d) None of these

87. Which of the following element has maximum electron gain enthalpy?

(a) Oxygen (b) Carbon

(c) Nitrogen (d) Phosphorous

88. Choose the species with highest electron affinity

(a) F (b) O

(c) O^- (d) Na

89. Along the period in the periodic table, the electronegativity of the elements

(a) decreases

(b) increases

(c) first increases then decreases

(d) first decreases then increases

90. Which of the following property is associated with an atom of high electronegativity?

(a) Large atomic radius

(b) High ionization energy

(c) Low electron affinity

(d) Low ionization energy

91. The element with least negative electron gain enthalpy is

(a) S (b) F

(c) Cl (d) P

Questions Based on Atomic Radii & Hydration Energy

92. The radii of N, N^{3-}, O & O^{2-} are in the order of

(a) $O^{2-} > N^{3-} > O > N$

(b) $N^{3-} > O^{2-} > N > O$

(c) $O^{2-} = O > N^{3-} > N$

(d) None of these

93. Atomic radii of F & Ne in Angstrom are respectively given by

(a) 0.72, 1.60

(b) 1.60, 1.60

(c) 0.72, 0.72

(d) 1.60, 0.72

94. Atomic radii of alkali metals follows the order Rb > K > Na > Li but ionic radii in aqueous solution follows the reverse order $Li^+ > Na^+ > K^+ > Rb^+$. The reason for this is

(a) Ionisation energy

(b) Hydration energy

(c) Lattice energy

(d) Metallic character

95. The size of following species F^-, Ne, Na^+ is affected by

(a) effective nuclear charge

(b) principal shell

(c) electron-electron interaction in outer orbitals

(d) none of the factors because their size is same

96. Four elements P, Q, R & S have ground state electronic configuration as

$P \rightarrow 1s^2\, 2s^2\, 2p^6\, 3s^2\, 3p^3$

$Q \rightarrow 1s^2\, 2s^2\, 2p^6\, 3s^2\, 3p^1$

$R \rightarrow 1s^2\, 2s^2\, 2p^6\, 3s^2\, 3p^6\, 3d^{10}\, 4s^2\, 4p^3$

$S \rightarrow 1s^2\, 2s^2\, 2p^6\, 3s^2\, 3p^6\, 3d^{10}\, 4s^2\, 4p^1$

Which of the following option/s represent the correct order?

(a) Size of P < size of Q

(b) Size of R < size of S

(c) Size of P < size of R (appreciable difference)

(d) All of these

97. Among the elements Al, Si, P, Mg the smallest and largest atoms respectively are

(a) Al and Mg (b) Al and Si

(c) P and Si (d) P and Mg

98. Choose the correct statement.

(a) For Cesium, the atomic radius < ionic radius; but for bromine, the atomic radius > ionic radius

(b) For Cesium and bromine both, the atomic radii > ionic radii

(c) For Cesium and bromine both, the atomic radii < ionic radii

(d) For Cesium, the atomic radius > ionic radius but for bromine, the atomic radius < ionic radius

99. Which of the following ions has the smallest radius?

(a) Be^{2+} (b) Li^+

(c) O^{2-} (d) F^-

Miscellaneous

100. An element 'M' has atomic number 19. Which of the following statement is correct.

(a) it is a soft metal

(b) it forms unipositive ion

(c) it has larger size than element with atomic number 20

(d) all of these are correct

101. The incorrect statement for the element having electronic configuration $1s^2\, 2s^2\, 2p^6\, 3s^2$

(a) It is divalent and electropositive

(b) It forms basic oxide

(c) The element is sodium

(d) It has low electron affinity

102. The property of the elements which does not show periodic behaviour is

(a) atomic radius

(b) elecronegativity

(c) ionization potential

(d) neutron proton ratio

103. Which of the following have no unit?

i. Atomic radius

ii. Electron gain enthalpy

iii. Ionisation enthalpy

iv. Metallic character

(a) i only

(b) iv only

(c) i and iv

(d) ii and iii

104. On moving down the group, reactivity of alkali metals .... And reactivity of halogens..........

(a) increase, decrease

(b) increase, increase

(c) decrease, decrease

(d) decrease, increase

105. Incorrect order of properties indicated is

i. $ns^2 np^1 > ns^2 np^3$ ionisation energy

ii. $p^6 \approx p3$ – ionisation energy

iii. $p^5 \approx p^4$ – electron affinity

iv. $[Ar]3d^{10}4s^2 > [Ar]3d^{10}4s^2 4p^2$ –size

(a) ii, iv only (b) i, ii only

(c) iii, iv only (d) i and iv only

106. An atom of an element has electronic configuration 2, 8, 7. Which of the following statement is incorrect?

(a) The element's valency is 7

(b) The element exists as a diatomic molecule

(c) The element is of non-metallic nature

(d) The element forms a basic oxide

107. A. $1s^2 2s^2 2p^6 3s^1$

B. $1s^2 2s^2 2p^6$

C. $1s^2 2s^2 2p^5$

D. $1s^2 2s^2 2p^6 3s^2 3p^1$

E. $1s^2 2s^2 2p^6 3s^2 3p^5$

Which statement are correct?

i. A & D are metals.

ii. Element with smallest size is B.

iii. Element E has largest size.

iv. Element A shows oxidation state of +1.

v. Element C has highest electron affinity.

(a) i, ii, iii, iv (b) i, iv only

(c) ii, iii, iv only (d) i, iv, v only

108. Which of the following orders is incorrect with respect to the property indicated?

(a) $NH_3 < PH_3 < AsH_3$ -- acidic behaviour

(b) Li < Be < B < C -- first ionisation energy

(c) $Al_2O_3 < MgO < Na_2O < K_2O$ -- basic character

(d) $Li^+ < Na^+ < K^+ < Cs^+$ -- ionic radius

109. Which one of the following is not considered as transition element?

(a) Zn

(b) Cd

(c) Hg

(d) All of these

110. Consider two changes

$A \rightarrow A^+ + e^-; E_1$ and $A+ \rightarrow A^{2+} + e^-; E_2$

Where, E_1 and E_2 are the two energies required to pull out two electrons. The correct order of E_1 and E_2 would be

(a) $E_1 < E_2$ (b) $E_1 = E_2$

(c) $E_1 > E_2$ (d) None of these

111. The first ionisation energy of elements increases across the period. The first ionisation energy of nitrogen and oxygen will be respectively.

(a) 13.6 eV, 14.6 eV (b) 14.6eV, 13.6eV

(c) 13.6eV, 13.6 eV (d) 14.6eV, 14.6eV

112. The correct order of covalent, Van der Waals's and crystal radii is

(a) $r_{covalent} < r_{crystal} < r_{Van\ der\ Wall}$

(b) $r_{covalent} < r_{Van\ der\ Wall} < r_{crystal}$

(c) $r_{crystal} < r_{covalent} < r_{Van\ der\ Wall}$

(d) $r_{crystal} < r_{Van\ der\ Wall} < r_{covalent}$

113. Alkali metals in each period have

(a) smallest size

(b) highest electropositive character

(c) lowest ionization potential

(d) both (b) & (c)

114. S^{2-} has higher ionic radius than Ca^{2+} because

(a) Ca atom has less number of neutrons than S

(b) Ca^{2+} has a higher nuclear charge than S^{2-}

(c) Their electronegativities are different

(d) S^{2-} has higher effective nuclear charge than Se

115. In the periodic table, which of the following properties remains unchanged on descending a group?

(a) number of electrons in penultimate shell

(b) Density

(c) Valence electrons

(d) both 1 & 3

116. Which of the following property increases down a group in periodic table?

(a) Electronegativity (b) Electron affinity

(c) Ionization potential (d) Size of the atom

117. On moving from group-13 to group-17, the property which increases is

(a) metallic character

(b) electronegativity

(c) gram atomic volume

(d) size of the atoms

118. Which of the following pairs of periodic properties show reverse trends on moving along a period from left to right and from top to bottom in a group?

(a) Nuclear charge and electron affinity

(b) Atomic radius and electron affinity

(c) Ionisation energy and electron affinity

(d) electronegativity and ionization energy

119. Out of first 100 elements, number of elements having electrons in 3d-orbitals are

(a) 80 (b) 10

(c) 100 (d) 60

120. An element X has configuration [He] $2s^2\ 2p^4$. Which of the following element will have property similar to X?

(a) [Ne] $3s^2\ 3p^2$ (b) [Ne] $3s^2\ 3p^4$

(c) [Ne] $3s^2\ 3p^5$ (d) [Ne] $3s^2\ 3p^3$

121. Choose the correct statement for elements of 3rd period in periodic table.

a. The highest principal quantum number of filled shell of all elements is 3.

b. It contains no element with filled 3d subshell.

c. It contains the elements of first transition series

d. The total number of elements in this period are 8.

(a) a, b & d (b) b, c & d

(c) a, c & d (d) a, b & c

122. An element X belonging to first short period has an outer electronic configuration $ns^2\ np^1$. The formula of its nitride is

(a) XN_3 (b) XN

(c) X_2N (d) XN_2

123. If the differentiating electron in an atom goes to the 4p orbital, then atomic number of the element is in the range of

(a) 13 – 18 (b) 81 – 86

(c) 31 – 36 (d) 49 – 54

124. Which of the following is correct decreasing order of the ionisation energy?

(a) Na < Cl > P > S > Mg > Al

(b) Cl > P > Ne > S > Mg > Al

(c) Ne > S > Mg > Cl > P > Al

(d) Mg > Al > Ne > Cl > P > S

125. Which of the following sequences is correct for decreasing order of the ionisation energy?

(a) Ge, Se, As, Br

(b) Br, Se, As, Ge

(c) Br, As, Se, Ge

(d) Ge, As, Se, Br

126. The species with highest ionisation potential is

(a) Li^+ (b) Mg^+

(c) Al^+ (d) Ne

127. The atomic number of elements A, B, C and D are Z – 1, Z, Z +1 and Z + 2 respectively. If B is a noble gas, choose the correct answer form the following statements

a. A is an element with very high electron affinity

b. C is an element which forms tripositive ion

c. A^{-1}, B, C^{+1}, D^{+2} are isoelectronic species

(a) a & b (b) only c

(c) a & c (d) a, b & c

128. Identify the most stable ion among following

(a) Li^+ (b) Be^-

(c) B^- (4) C^-

129. Choose the correct statement regarding element Z with atomic number 21.

a. The atomic number of elements lying below Z in periodic table is 39.

b. The sudden jump in I.E. values of Z is between IE_2 and IE_3.

c. Z is paramagenetic in nature

d. Preceeding element of Z is an alkaline earth metal

(a) c & d (b) a, c & d

(c) a, b & c (d) a, b, c & d

130. The electronic configuration of four elements are

i. [Xe] $6s^2$ ii. [Xe] $4f^{14}\ 5d^1\ 6s^2$

iii. [Ne] $3s^2\ 3p^5$ iv. [Ar] $3d^7\ 4s^2$

Which of the following statement about these elements is incorrect?

(a) (i) forms dipositive ion

(b) (ii) is an f-block element

(c) (iii) has highest negative electron gain enthalpy

(d) (iv) is a metal with low melting point

131. Choose the correct statement for element $^{19}_{9}M$

a. Formation of M^- is endothermic process

b. Its anion is isoelectronic with neon

c. It has smallest size in its period

d. It has highest electron gain enthalpy in its group

(a) both a & (b) both c & d

(c) a, b & d (d) both b & c

132. Select the incorrect statement for the atoms with valence shell configuration of $ns^2\ np^5$

(a) They belong to group-17

(b) The element of this group in 8th period is Uus

(c) The atoms have highest ionization energy in their respective periods

(d) The size of the atom of this group is smaller than succeeding noble gas atom

133. The statement which is not correct for periodic classification of elements is

(a) the properties of elements are a periodic function of their atomic numbers

(b) non-metallic elements are less in number than metallic elements

(c) the first ionisation energies of elements along a period do not vary in a regular manner with increase in atomic number

(d) generally electronegativity and electron affinity increase across the period

134. Match the coloumn

Scientist	Periodic Table
(I) Duma	(A) Octave rule
(II) Newland	(B) Atomic volume curve
(III) Lother Meyer	(C) Homologous series
(IV) Doberneir	(D) Triad rule

The correct option is

(a) I-A, II-B, III-C, IV-D (b) I-C, II-A, III-B, IV-D

(c) I-C, II-A, III-D, IV-B (d) I-D, II-B, III-A, IV-C

135. It has been found that the atoms combines to form bonds in order to keep 8 valence electrons and attain stability. Hypothetically, if stability is attained with 7 electrons instead of 8 electrons, the formula of the stable nitride ion would be

(a) N^+ (b) N^{2-}

(c) N^{2+} (d) N^{3-}

136. Which of the following statements is not true about the long form of modern periodic table?

(a) it reflects the sequence of filling of electrons in order to sub-energy levels s, p, d and f

(b) it helps to predict the stable valency states of the elements

(c) it reflects trends in physical and chemical properties of the elements

(d) both (a) & (c)

137. The correct order of electron affinity of B, C, N and O is

(a) $O > C > N > B$ (b) $B > N > C > O$

(c) $O > C > B > N$ (d) $O < B > C > N$

138. The sizes of given species are such that

(a) $I > I^+ > I^-$ (b) $I^+ > I^- > I$

(c) $I^- > I > I^+$ (d) $I > I^- > I^+$

139. Which correspond to most electropositive characters?

(a) $[He]\,2s^2$ (b) $[He]\,2s^1$

(c) $[Xe]\,6s^2$ (d) $[Xe]\,6s^1$

140. Consider atom with the following configurations

I. $1s^2\,2s^2\,2p^6$ II. $1s^2\,2s^2\,2p^6\,3s^1$

III. $1s^2\,2s^2\,2p^6\,3s^2$

The atom which has the largest first ionisation potential and lowest first ionisation potential is respectively

(a) I and II (b) II and I

(c) I and III (d) III and I

141. First three ionisation energy values of an element A are 6.0, 18.6, 28.3 eV and 208.3 eV. The unstable oxidation states are

(a) A^{2+} and A^{3+} (b) A and $A^{\oplus}$

(c) A^+ and A^{2+} (d) $A^{\oplus}$ and A^{3+}

142. Which of the following involves maximum amount of energy

(a) $Mg^-(g) \rightarrow Mg(g)$

(b) $Mg^{2+}(g) \rightarrow Mg^{3+}(g)$

(c) $Mg^-(g) \rightarrow Mg^+(g)$

(d) $Mg^+(g) \rightarrow Mg^{2+}(g)$

143. Element having highest I.E. is -

(a) Ne (b) He

(c) Be (d) N

Answers

1. (c)	2. (d)	3. (b)	4. (c)	5. (a)	6. (b)	7. (b)	8. (a)	9. (c)	10. (c)
11. (b)	12. (b)	13. (b)	14. (a)	15. (a)	16. (b)	17. (d)	18. (d)	19. (d)	20. (b)
21. (a)	22. (c)	23. (d)	24. (d)	25. (b)	26. (b)	27. (a)	28. (d)	29. (d)	30. (b)
31. (a)	32. (a)	33. (a)	34. (b)	35. (b)	36. (c)	37. (d)	38. (d)	39. (c)	40. (d)
41. (d)	42. (c)	43. (d)	44. (d)	45. (d)	46. (c)	47. (b)	48. (a)	49. (a)	50. (a)
51. (a)	52. (b)	53. (a)	54. (b)	55. (d)	56. (c)	57. (d)	58. (d)	59. (d)	60. (c)
61. (a)	62. (d)	63. (b)	64. (d)	65. (b)	66. (d)	67. (b)	68. (a)	69. (d)	70. (d)
71. (a)	72. (a)	73. (d)	74. (d)	75. (b)	76. (d)	77. (c)	78. (a)	79. (b)	80. (b)
81. (d)	82. (d)	83. (b)	84. (b)	85. (c)	86. (b)	87. (c)	88. (a)	89. (b)	90. (b)
91. (d)	92. (b)	93. (a)	94. (b)	95. (a)	96. (d)	97. (d)	98. (d)	99. (a)	100. (d)
101. (c)	102. (d)	103. (b)	104. (a)	105. (b)	106. (d)	107. (b)	108. (b)	109. (d)	110. (a)
111. (b)	112. (a)	113. (d)	114. (b)	115. (c)	116. (d)	117. (b)	118. (b)	119. (a)	120. (b)
121. (a)	122. (b)	123. (c)	124. (a)	125. (c)	126. (a)	127. (c)	128. (a)	129. (b)	130. (d)
131. (d)	132. (c)	133. (c)	134. (b)	135. (b)	136. (b)	137. (c)	138. (c)	139. (d)	140. (a)
141. (c)	142. (b)	143. (b)							

Explanations

1. (c) Ca (20) : $1s^2, 2s^2, 2p^6, 3s^2, 3p^6, 3d^2$

2. (d) At No. 107 is unilseptium.

3. (b) $s^1s^2s^3$ $p^1p^2p^3p^4p^5p^6p^7p^8p^9$

$d^1d^2d^3d^4d^5d^6d^7d^8d^9d^{10}d^{11}d^{12}d^{13}d^{14}d^{15}$

4. (c)

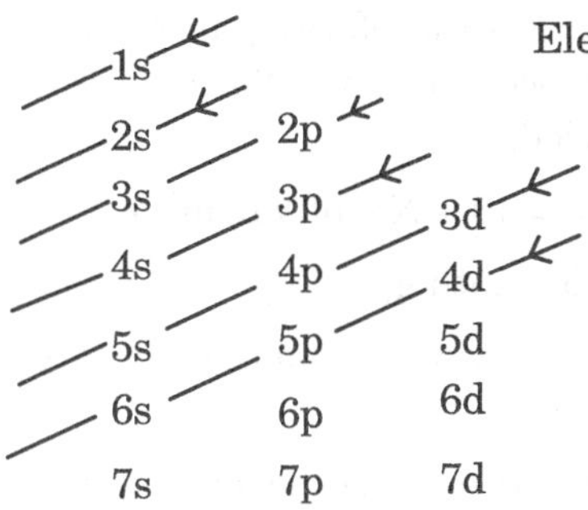

4th period begins from 4s & ends at 4p

Elements having 3d electron will also be placed in 4th period

But, element having 4d electrons will be placed in 5th period & NOT 4th period

5. (a)

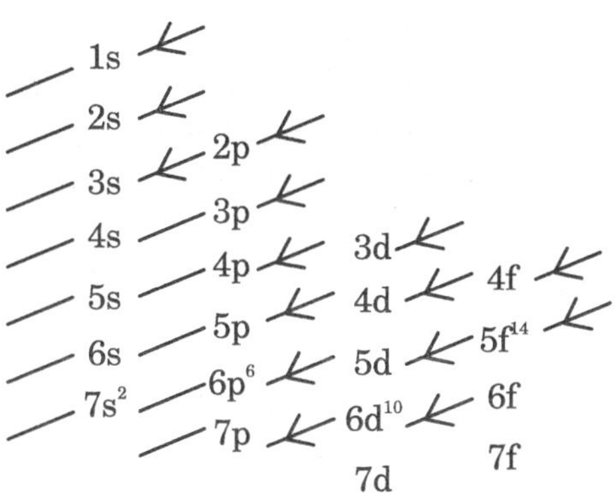

$Rn(86)5f^{14}\ 6d^{10}\ 7s^2$

At. No. 112 will have Configuration

$Rn(86)$-..........$6p^6$

$(Rn)^{86}7s^2\ 5f^{14}\ 6d^{10}$

It's a d-block element belongs to 7th period & 12th group

6. (b)

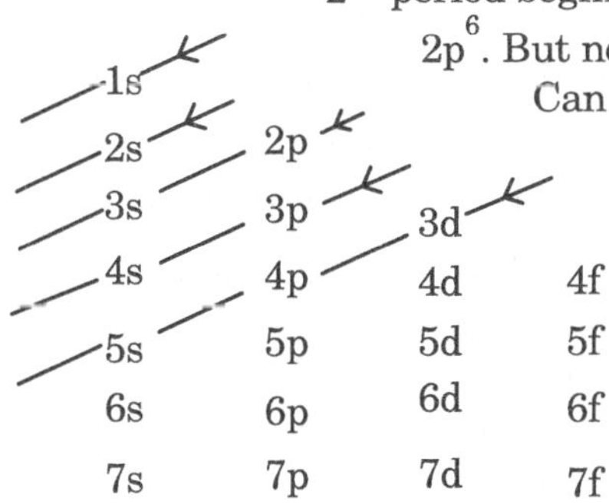

2nd period begins at 2s & ends at $2p^6$. But now since each orbital Can take maximum of 3 electrons, it begins at 2s & ends at $2p^9$. So, 12 elements are there is 2nd period.

For 4th period.

$4s^3\ 3d^{15}\ 4p^9$

$\Rightarrow$ 27 elements

7. (b) Element X is Copper

Cu(29): $1s^2 2s^2 2p^6 3s^2 3p^6 4s^1 3d^{10}$

Here, all conditions are satisfied. Cu has 1 electron in ultimate orbit 4S. It has 18 electrons in penultimate orbit ($3s^2\ 3p^6\ 3d^{10}$). It also belongs to 4th period.

10. (c) The present set of periodic table has 7 periods

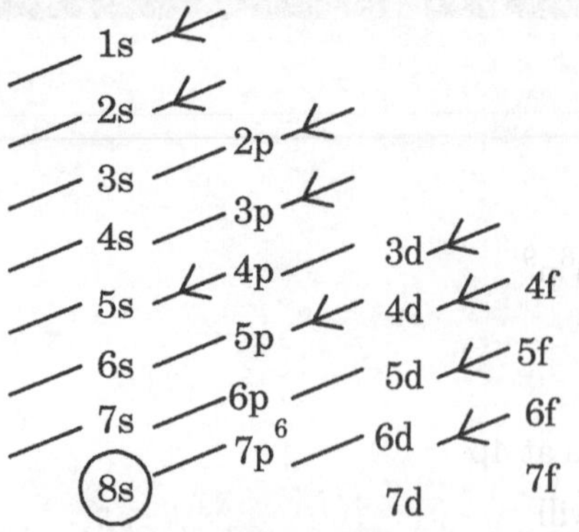

11. (b) Actual configuration

$1s^2\ 2s^2\ 2p^6\ 3s^2\ 3p^6\ 4s^1\ 3d^5$

It's a d-block element

It belongs to 4th period (highest shell occupied), Gp. No. = 6 (ns electrons + (n – 1) d electrons)

12. (b)

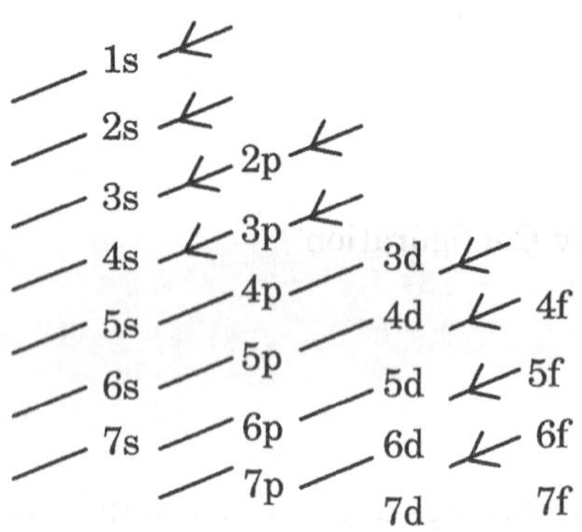

7th period begins from 7s & ends at 7p

13. (b) Since last electron of element X has quantum No's

$n = 2, l = 1, m = 1, s = \frac{1}{2}$, e^- is in 2p subshell.

Configuration will be $1s^2\ 2s^2\ 2p^y$ (where y = No. of e^- in 2p subshell)

The element 'X' is a p-block element belongs to 2nd period (highest shell occupied)

14. (a) (Kr) $4d^{10}$ $4f^{14}$ $5s^2$ $5p^6$ $5d^2$ $6s^2$

↑ ↓ ↓ ↓ ↓

n = 4 n = 4 n = 5 n = 5 n = 5 n = 6

l = 2 l = 3 l = 0 l = 1 l = 2 l = 0

Correct way of writing configuration is (Kr) $5s^2\ 4d^{10}\ 5p^6\ 6s^2\ 4f^{14}\ 5d^2$

Now, since last electron enters the d-orbitals, it is a d-block element.

Highest shell occupied is 6 i.e., it belongs to 6th period.

Gp. No. for d-block element = $ns^{e-} + (n-1)d^{e-}$

$\Rightarrow 6s^2 + 5d^2$

$\Rightarrow 2 + 2 = 4^{th}$ gp.

15. (a) Unnilhexium

16. (b) Given configuration of X is (Ar) $3d^{10}\ 4s^2\ 4p^3$ Actual way of writing configuration is (Ar) $4s^2\ 3d^{10}\ 4p^3$. X is therefore a p-block element with highest occupied shell No. 4. So, it belongs to 4th period.

Gp. No. = 10 + No. of Valence e^-

= 10 + 5 = 15th gp.

Its chemistry will be similar to N (also belongs to 15th gp)

29. (d) All the four ions are isoelectronic

Na^+ → $10e^-$, 11p; F^- → $10e^-$, 9p; Mg^{2+} → $10e^-$, 12p; O^{2-} → $10e^-$, 8p

O^{2-} has the least number of proton to hold electrons. So removal of electron is easier from O^{2-}.

30. (b) Li Be B C N O F Ne

→

Na Mg Al Si P S Cl Ar

IE_1 of P > IE_1 of S

(It is because P has exactly half filled configuration)

P(15) : $1s^2\ 2s^2\ 2p^6\ 3s^2\ 3p^3$

S(16) : $1s^2\ 2s^2\ 2p^6\ 3s^2\ 3p^4$

31. (a) ⟶ I. E increases

Li Be B C N O F Ne

32. (a) Both Ne & Na^+ are isoelectronic with 10 electrons each. But Na^+ has 11 protons, so effective number charge is more in comparison to Ne which has 10 protons. So, Na^+ has maximum IE.

Na metal, being alkali metal is highly electropositive So, it has the minimum ionization energy.

33. (a) Since biggest jump occurs between , therefore the number of valence electrons in this atom are 3.

34. (b) H He

Li Be B C N O F Ne

$\xrightarrow{\text{IE}\uparrow \text{SES}}$

Na Mg Al Si P S Cl Ar

Expected Order of IE_1: Si > Al > Mg > Na

35. (b) Once on electron is removed from Na, It has noble gas configuration. So, IE_2 of Na > IE_2 of Mg.

36. (c) IE_1 of Ne > $1E_1$ of N

$\therefore IE_2$ of Ne > IE_1 of N

(Because, IE_2 of Ne > IE_1 of N)

37. (d) I.E. of $O^-_{(g)}$ is less than $O_{(g)}$

Successive ionization energies are always higher i.e.

$IE_3 > IE_2 > IE_1$

39. (c) There is maximum jump between the values of IE_2 & IE_3. So, element 'X' has 2 valence electrons. Hence, compounds like XF_2, X_3N_2 & XO are possible

41. (d) $Y_{(g)} \longrightarrow Y^+_{(g)} + e^-$; $\Delta H = 100\,J\,mol^{-1}$ ....(1)

$Y^+_{(g)} \longrightarrow Y^{2+} + e^-$; $\Delta H = 150\ J\,mol^{-1}$ ....(2)

Adding (1) & (2), We get

$Y_{(g)} \longrightarrow Y^{2+}_{(g)} + 2e^-$; $\Delta H = 250\ J\,mol^{-1}$

42. (c) Option 1 is F, option 2 is Si, Option 3rd is Na & 4th is Ne.

H He

Li Be B C N O F Ne

Na Mg Al Si P S Cl Ar

Of all these elements listed above, Na possesses lowest IE_1 Owing to its metallic nature

43. (d) Al(13) : $1s^2\,2s^2\,2p^6\,3s^2\,3p^1$

Mg(12) : $1s^2\,2s^2\,2p^6\,3s^2$

IE_1 of Mg > IE_1 of Al (Because of penetration effect)

Na(11) : $1s^2\,2s^2\,2p^6\,3s^1$

IE_1 of Mg > IE_1 of Na

IE_3 of Mg > IE_3 of Al

($\because$ after loss of $2e^-$, Mg^{2+} will possess noble gas configuration)

44. (d) All the three are unielectron species, So no chance of shielding effect (interelectronic repulsion)

45. (d) $S^{\ominus}_{(g)} + e^- \longrightarrow S^{2-}_{(g)}$; $\underset{eg}{\Delta H} = +ve$

$N_{(g)} + e^- \longrightarrow N^{\ominus}_{(g)}$; $\underset{(eg)}{\Delta H} = +ve$

$Al^{2+}_{(g)} \longrightarrow Al^{3+}_{(g)} + e^-$; $\underset{IE}{\Delta H} = +ve$

46. (c) (a) $O_{(g)} + e^- \longrightarrow O^{\ominus}_{(g)}$; $\underset{eg_1}{\Delta H} = -\ ve$

(b) $O^{\ominus}_{(g)} + e^- \longrightarrow O^{2-}_{(g)}$; $\underset{eg_2}{\Delta H} = +ve$

2nd electron gain enthalpy of O is +ve because once $O^{\ominus}$ is formed (the no. of electrons exceed the No. of protons), so addition of 2nd electrons is tougher.

47. (b) $Na_{(g)} \longrightarrow Na^+_{(g)} + e^-$; $\Delta H = +\,ve$

$Na^+_{(g)} + e^- \longrightarrow \underset{(g)}{Na}$; $\Delta H = -ve$

$Na_{(s)} \longrightarrow Na_{(g)}$; $\Delta H = +ve$ sublimation

$Na_{(g)} \longrightarrow Na_{(s)}$; $\Delta H = -\,ve$

49. (a) $\underset{(g)}{N} + e \longrightarrow \underset{(g)}{N^{\ominus}}$; $\Delta H = +ve$

$\underset{(g)}{N^-} \longrightarrow \underset{(g)}{N} + e^-$; $\Delta H = -ve$

50. (a) 'I⁻' will have lesser I.E. than F^-, rather electron can flow from I^- to 'F'.

51. (a) IE

$Al_{(g)} \longrightarrow Al^+_{(g)} + e^-$; $\Delta_1 H = 577.5\,kJ\,mol^{-1}$

$Al^+_{(g)} \longrightarrow Al^{2+}_{(g)} + e^-$; $\Delta_2 H = 2x$

$Al^{2+}_{(g)} \longrightarrow Al^{3+}_{(g)} + e^-$; $\Delta_3 H = 3x$

$Al_{(g)} \longrightarrow Al^{3+}_{(g)} + 3e^-$; $\Delta H = 5140\,kJ\,mol^{-1}$

577.5 + 2x + 3x = 5140

577.5 + 5x = 5140

5x = 5140 – 577.5

$x = \frac{4562.5}{5} = 912.5$

IE_2 of Al = 2x = 2 × 912.5 = 1825 kJ mol^{-1}

IE_3 of Al = 3x = 3 × 912.5 = 2737.5 kJ mol^{-1}

52. (b) Electron gain enthalpy & electron affinity have opposite signs.

53. (a) Both Ne & Na^+ are isoelectronic with 10 electrons each. But Na^+ has 11 protons, So effective nuclear charge is more in comparison to Ne which has 10 protons.

So, Na^+ has maximum IE.

Na metal, being alkali metal is highly electropositive. So, it has the minimum ionization energy.

56. (c) $\Delta_{eg}H$ value of $-348\ kJ\ mol^{-1} \rightarrow 3s^2\ 3p^5$ (Chlorine)

$-200\ kJ\ mol^{-1} \rightarrow 3s^2\ 3p^4$ (Sulphur)

$-135\ kJ\ mol^{-1} \rightarrow 3s^2\ 3p^2$ (Silicon)

$-60\ kJ\ mol^{-1} \rightarrow 3s^2\ 3p^3$ (Phosphorous)

58. (d) Addition of 1st electron in O is tougher in comparison to S. So, addition of 2nd electron will be still more tougher. So, more energy will be required to add an additional electron in Oxygen.

59. (d) Cl has maximum negative electron gain enthalpy i.e. $-348\ kJ\ mol^{-1}$ followed by F. Members of Oxygen family will be placed after halogens.

60. (c) $S_{(g)} + e^- \longrightarrow S_{(g)}^{\ominus};\ \Delta_{eg}H = -ve$

$O_{(g)} + e^- \longrightarrow O_{(g)}^{\ominus};\ \Delta_{eg}H = -ve$

$O_{(g)}^{\ominus} + e^- \longrightarrow O_{(g)}^{2-};\ \Delta_{eg}H = +ve$

$S_{(g)}^{-} + e^- \longrightarrow S_{(g)}^{2-};\ \Delta_{eg}H = +ve$

61. (a) Li Be B C N O F

'N' has least electron affinity because of already stable half filled configuration.

62. (d) $S_{(g)}^{\ominus} + e^- \longrightarrow S_{(g)}^{2-};\ \Delta_{eg}H = +ve$

$N_{(g)} + e^- \longrightarrow N_{(g)}^{\ominus};\ \Delta_{eg}H = +ve$

$Al_{(g)}^{2+} \longrightarrow Al_{(g)}^{3+} + e^-;\ \Delta_{IE}H = +ve$

63. (b) (a) 1st electron gain enthalpy is not always positive. For eg. It is negative for halogens & positive for noble gases.

$Cl_{(g)} + e^- \longrightarrow Cl_{(g)}^{-},\ \Delta_{eg}H = -349\ kJ\ mol^{-1}$

$Ne_{(g)} + e^- \longrightarrow Ne_{(g)}^{-},\ \Delta_{eg}H = +ve$

Also, wrt Ionisation. Energy

(b) $M_{(g)} \longrightarrow M_{(g)}^{+} + e^-;\ \Delta H_i = +ve$

But for ions like,

$Ne_{(g)}^{-} \longrightarrow Ne_{(g)} + e^-;\ \Delta H_i = -ve$

(Here, I.E. of ions like Ne^- is negative)

64. (d) Cl > F > Br > I (Correct order for electron affinity)

Si > Mg > Al > Na (Correct order for ionization energy)

$Al^{3+} > Al^{2+} > Al$ (Correct order for decreasing effective nuclear charge)

actual order:

Si > Mg > Al > Na

Mg(12) : $1s^2\ 2s^2\ 2p^6\ 3s^2$

Al(13) : $1s^2\ 2s^2\ 2p^6\ 3s^2\ 3p^1$

($\because$ of penetration effect)

68. (a) $X_{(g)} \longrightarrow X_{(g)}^{+} + e^-;\ \Delta H = IE_1$

$X_{(g)}^{+} \longrightarrow X_{(g)}^{2+} + e^-;\ \Delta H = IE_2$

71. (a) Gp. 2 elements have valence configuration ns^2

Gp. 13 elements have valence configuration $ns^2\ np^1$

Gp. 15 elements have valence configuration $ns^2\ np^3$

Gp. 18 elements have valence configuration $ns^2 np^6$

Order of IE.

$ns^2\ np^6 > ns^2 np^3 > ns^2 > ns^2 np^1$

(↓ pentration effect)

72. (a) Mg has max. I.E. among these.

73. (d) Maximum jump is between IP_2 & IP_3, $\therefore$ it has 2 valence electrons. ANSWER will be 4th potion.

75. (b) Max. jump is between IE_3 & IE_4. It has 3 valence e^-. So, Valency of A is 3. Formula of its oxides is

$A^{3+}\ \ O^{2-}$ (cross) → A_2O_3

76. (d) He is a noble gas & smallest among all noble gases.

77. (c) Li Be B C N O F Ne
(P)(S) Cl Ar
(As)

78. (a) S-Subshell has maximum penetration effect.

81. (d) $1s^2\,2s^2\,2p^4 \rightarrow O$
$1s^2\,2s^2\,2p^6 \rightarrow Ne$
$1s^2\,2s^2\,2p^6\,3s^1 \rightarrow Ne$
$1s^2\,2s^2\,2p^6\,3s^2 \rightarrow Mg$
IE_2 follow the order. Na > Ne > O > Mg

82. (d) (a) N has exactly half filled configuration & small size So, it has no tendency to accept an electron.
(b) Be has configuration $1s^2, 2s^2$, the incoming electron has to enter higher energy level that is 2p. That's why Be is reluctant to gain an e^-.

84. (b) E. Affinity of Cl is maximum. So, energy released is maximum.

86. (b) Mg has +ve electron gain enthalpy which Na has –ve E.G.E.

87. (c) N has positive electron gain enthalpy

91. (d) Order of negative E.G.E (Electron gain enthalpy) is Cl > F > S > P

92. (b) Anions have large size in comparison to corresponding atoms.
$N^{3-} > N$ & $O^{2-} > O$

Also,
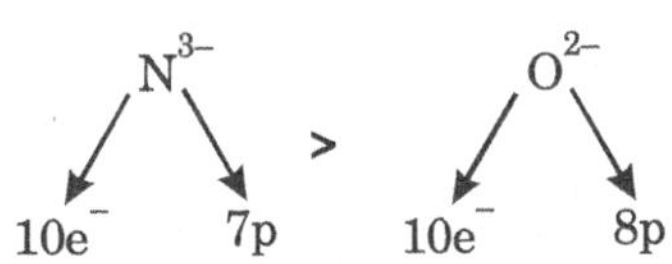

93. (a) Atomic size of Ne > Atomic size of F.

95. (a)

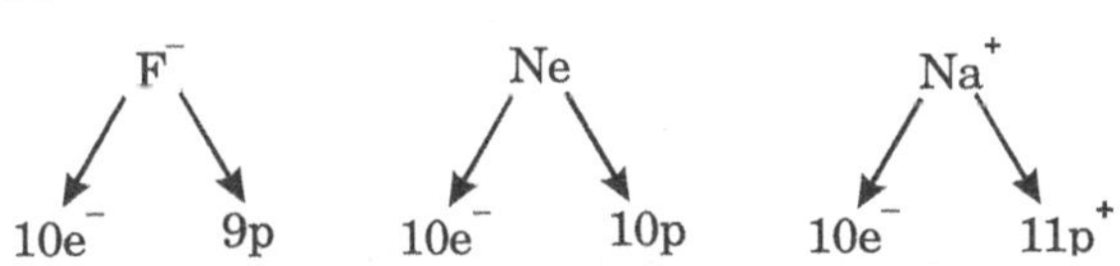

Effective nuclear charge is maximum in Na^+ & min in F^-

96. (d) P is phosphorous, Q is aluminium, R is arsenic, S is Gallium.

Li Be B C N O F Ne
Ne Mg Al Si (P)(S) Cl Ar
(Ga) (As)

Size of Q (Al) > size of P (phosphorous)
Size of S (Ga) > size of R (Arsenic)
Size of R (As) > size of P (Phosphorous)

97. (d) Li Be B C N O F Ne
⟶ Al. size ↓ ses
(Mg) (Al)(Si)(P)

Largest is Mg & smallest is P.

98. (d) Cesium ion is a cation i.e., Cs^+
So, size of Cs > Cs^+
(atomic radius > radius of cation)
Bromide ion is an anion ∴ e. Br^-
So, size of Br^- > Br.

100. (d) Element 'M' is Potassium. It is unipositive & has size greater than Ca (at. No. 20)

101. (c) Element is Mg. It forms basic oxide.

107. (b) A is 'Na', B is 'Ne', 'C', is 'F', D is 'Al', E is 'Cl'.

109. (c) Zn, Cd, Hg do not have partly filled d-orbitals either in their normal state as in their common oxdn. state (+2)

114. (b)
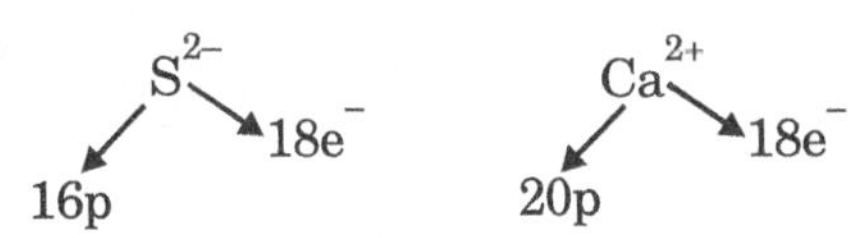

Ca^{2+} has higher effective nuclear charge. So, size of Ca^{2+} is smaller than S^{2-}.

119. (a) $1s^2\,2s^2\,2p^6\,3s^2\,3p^6\,4s^2 \rightarrow$ this configuration is for atomic No. 20.

120. (b) For [He] $2s^2\,2p^4$ & [Ne] $3s^2\,3p^4$, valence e^- are same.

121. (a) Elements will 3d electrons will belong to 4th period.

122. (b) Element X has valence configuration $ns^2\,np^1$ & hence valence e^- = 3
So, It valency = 3

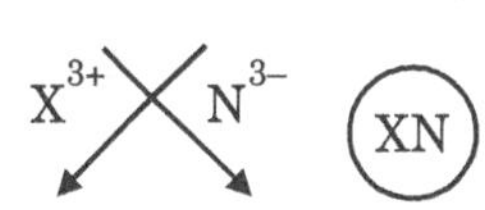

123. (c) If the differentiating electron goes to 4p, then $1s^2\,2s^2\,3p^6\,3s^2\,3p^6\,4s^2\,3d^{10}\,4p^1$ to $1s^2\,2s^2\,2p^6\,3s^2\,3p^6\,4s^2\,3d^{10}\,4p^6$ (At. No. 31-36)

126. (a) Li^+ has noble gas configuration & is also smaller in size.

127. (c) B - noble gas with atomic no. Z
A - halogen with atomic No. Z - 1
C - alkali metal with atomic No. Z + 1
D - alkaline earth metal with atomic No. Z + 1

143. (b) He is an inert gas & has a small size.

Smart Notes

4

CHAPTER

Chemical Bonding and Molecular Structure

Concept Questions (Based on ionic bond)

1. The correct statement explaining ionic bond is-
   (a) ionic bond involves mutual sharing of electrons
   (b) ionic bond involves one sided sharing of electrons
   (c) ionic bond involves the complete transfer of electrons
   (d) Both 1 & 2
2. Which of tho following has ionic bond?
   (a) $Na^+ Cl^-$
   (b) $MgBr_2$
   (c) $NH_4{}^+Cl^-$
   (d) All of these
3. Non-favorable condition for the formation of an ionic bond is-
   (a) Low ionisation energy of the metal
   (b) High electron affinity of non-metal
   (c) low lattice energy
   (d) high lattice energy
4. Lattice energy depends upon the charge on ions and the size of ions. Then, match the following correctly.

Column A (ionic compounds)	Column B (Lattic Energies)
(1) NaCl	(p) – 2254 kJ / mol
(2) $CaCl_2$	(q) – 3932 kJ / mol
(3) MgO	(r) – 758 kJ / mol

   (a) 1– r, 2 – p, 3 – q (b) 3 – r, 2 – q, 3 – p
   (c) 1, 2 – r, 3 – p (d) None of these
5. Which of the following do not characterize ionic compounds?
   (a) These compounds usually exist in solid state
   (b) These posses high melting & boiling points
   (c) These are insoluble in solvents with high dielectric constant
   (d) These are good conductors of electricity
6. Choose the incorrect statement amongst the following.
   (a) Lattice enthalpy is determined by Born-Haber cycle
   (b) The number of electrons lost or gained during the formation of an electrovalent linkage is termed as electrovalency of an element
   (c) Lattice energy depends upon coulombic force of attraction
   (d) None of these

7. CaO and NaCl have the same crystal structure & approximately the same ionic radii. If 'U' is the lattice energy of NaCl, the approximate lattice energy CaO is –

(a) $\frac{U}{3}$ (b) U

(c) $\frac{U}{2}$ (d) 4U

8. Which of the following has maximum lattice energy?

(a) NaF (b) KF

(c) RbF (d) CsF

9. The compound with maximum negative lattice energy is-

(a) NaF (b) NaCl

(c) AlF_3 (d) MgF_2

10. Which of these will show variable electrovalency

(a) Cu (b) Co

(c) Fe (d) All of these

Questions Based on Octet Rule & Formal Charge

11. Which one of the following follows the octet rule?

(a) $BeCl_2$ (b) $AlCl_3$

(c) BCl_3 (d) None of these

12. The formal charge on *S* in HSO_4^- is

(a) +1 (b) +2

(c) +3 (d) 0

13. The formal charges on different atoms in the Lewis structure of N_3^- cannot be-

(a) –1, +1, 0 (b) –2, +1, 0

(c) 0, +1, –2 (d) –1, +1, –1

14. Which of the following is not an odd electron molecule?

(a) NO (b) NO_2

(c) O_2^- (d) None of these

Valence Bond Theory (VBT)

15. Chose the correct statement amongst the following.

(a) The formal charge on Cl in $HClO_4$ is +3

(b) $BeCl_2$ violates octet rule

(c) Xe in XeF_2 shows expanded octet

(d) All are correct

16. According to VBT, a covalent bond is formed by -

(a) partial overlap of two half filled atomic orbitals containing electrons with same spins

(b) complete overlap of two half filled atomic orbitals containing electrons with same spins

(c) partial overlap of two half filled atomic orbitals containing electrons with opposite spins

(d) partial overlap of two fully filled atomic orbitals containing electrons with opposite spins.

17. Select the correct statement regarding sigma bonds.

(a) It is the first bond to be formed

(b) It primarily decides the direction of covalent bond

(c) It is a weaker bond is comparison to a π bond

(d) Both 1 & 2

18. Which of the following overlaps can result in σ bond formation?

(if z-axis is opted as internuclear axis)

(a) S + S (b) $P_z + P_z$

(c) $S - P_z$ (d) All of the above

19. Choose the incorrect statement about p-orbitals

(a) These can form both σ & π bonds

(b) These are directional in nature

(c) F_2 molecule is formed by s – p overlap

(d) All of these

20. The following graph is meant for He_2 molecule

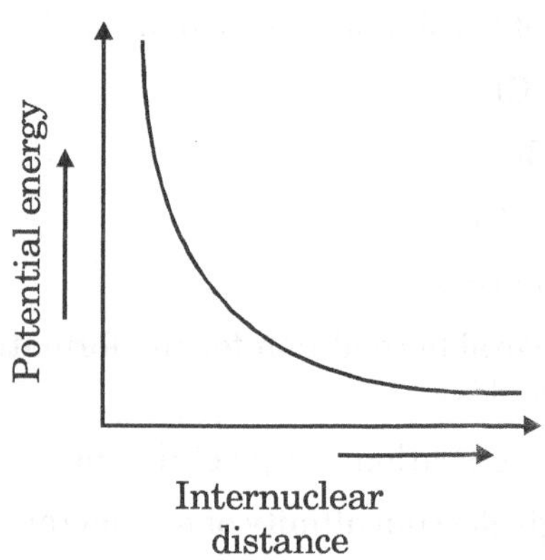

According to the graph given above,

(a) the potential energy increases as two He atoms approach each other leading to stability in the molecule

(b) the potential energy increases as two He atoms approach and formation of helium molecule is not possible, energetically

(c) the potential energy decreases as two He atoms approach & formation of molecule (He_2) occurs

(d) None of these

21. If z-axis is chosen as internuclear axis, then zero overlap will occur in the following-

(a) s + px (b) s + py

(c) px + py (d) All of these

22. Select the correct statement about (pie) bond ?

(a) It is formed by sideways or lateral overlap

(b) S-orbitals cannot form a pie bond

(c) In π bond formation, half of the πe^- cloud is present above the internuclear axis & half is present below the internuclear axis

(d) All are correct

Vsepr & Hybridistion

23. Select the correct match

(a) 1 – c, 2 – b. 3 – a

(b) 3 – c, 2 – a, 1 – b

(c) 1 – a, 2 – b. 3 – c

(d) Either b or c

24. Hybridization is not possible in -

(a) CH_4 (b) C_2H_6

(c) NH_3 (d) Na^+Cl^-

25. If CH_4 (methane) is formed according to VBT, $\angle H-C-H$ (angles) will be at –

(a) 60° (b) 120°

(c) 90° (d) 180°

26. The number of bond pairs around C in CO_2 according to VSEPR are –

(a) 2 (b) 3

(c) 4 (d) 5

27. In CH_4, the number of $sp^3 - s$ overlaps are –

(a) 1 (b) 2

(c) 2 (d) 4

28. Select the incorrect statement

(a) In dsp^3 hybridization $d_{x^2-y^2}$ participates

(b) In sp^3d^2 hybridization $d_{x^2-y^2}$ & d_{z^2} participate

(c) In sp^3d hybridization d_{z^2} participates

(d) None of these

29. Select the correct statement about ClO_3^- & SO_3^{2-}

(a) Both are isostructural but not isoelectronic

(b) Both are isoelectronic but not isostructural

(c) Both are neither isoelectronic nor isostrucural

(d) Both are isoelectronic & isostrucutral

30. Identify the mismatch

	Hybrid orbitals	**% age S –character**
(a)	sp	→ 50%
(b)	sp^2	→ 33.3%
(c)	sp^3	→ 25%
(d)	sp^3d	→ 15%

31. Arrange the following orbitals in correct order of their sizes-

sp, sp^2, sp^3

(a) $sp > sp^2 > sp^3$

(b) $sp^3 > sp^2 > sp$

(c) $sp = sp^2 > sp^3$

(d) $sp^3 = sp^2 > sp$

32. Which order is correct?

(a) $sp > sp^2 > sp^3$ (Electro negativity)

(b) $sp^3 - s > sp^2 - s > sp - s$ (Bond length)

(c) $sp^3 > sp^2 > sp$ (% age p – character)

(d) All orders are correct

33. Which of the following statements is incorrect?

(a) XeO_2F_2 & SF_4 have see-saw shape

(b) Among IF_7 & PCl_5, PCl_5 has lesser number of atoms in XY plane

(c) SF_4 is planar molecule

(d) O atom in H_2O is sp^3 hybridized

34. Which of the following molecule has maximum 90° Angles?

(a) PCl_5 (b) SF_6

(c) ClF_3 (d) CH_4

35. Non-Planar molecule is –

(a) XeF_4 (b) ClF_3

(c) XeF_5^+ (d) H_2O

36. Identify the number of planar species among the following-

I_3^-, XeF_2, XeF_3^+, BF_3, PCl_5, ICl_2^+

(a) 6 (b) 5

(c) 4 (d) 3

37. SO_3 and BF_3 have –

(a) Same shapes but different hybridizations

(b) different shapes but same hybridisations

(c) different shapes & different hybridisations

(d) same shapes & same hybridisations

38. Calculate the fractional charge on each atom in HBr molecule. Given that dipole moment of HBr = 0.70 D. Bond length of HBr = 1.41 A°

(a) 0.25 (b) 0.47

(c) 0.11 (d) None of these

Questions on bond angle, bond strength, Bond length & Molecular Orbital Theory

39. Which of the following correctly represents the direction of dipole in Pyrrole & furan?

(a)

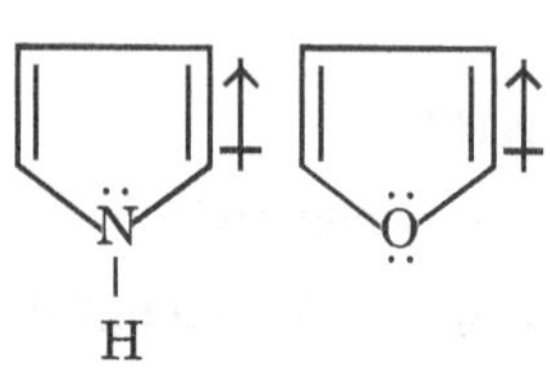

(b)

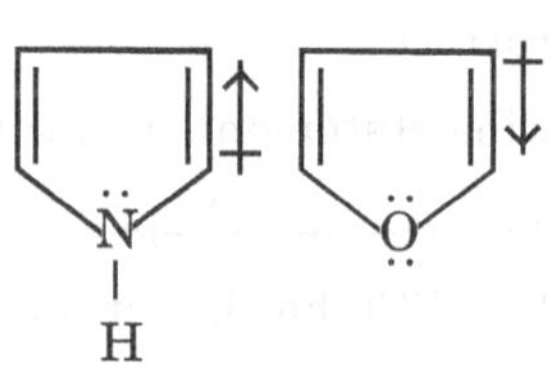

(c)

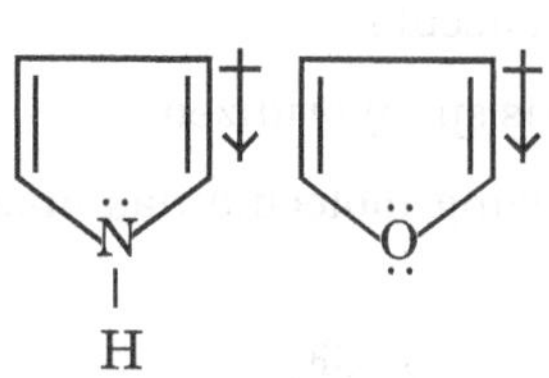

(d) None of these

40. Bond angle is maximum in -

(a) NO_2^+ (b) NO_2

(c) NO_2^- (d) NH_3

41. O–O bond length is larger in H_2O_2 than O_2F_2 because-

(a) according to bents rule, there is more p-character in O-F bond than O-H bond in, H_2O_2 So s-character in O-O bond is greater between O_2F_2.

(b) of lone pair - lone pair repulsion between O-O in H_2O_2. In O_2F_2 there is greater l.p – l.p. repulsion between O-F rather than in O-O

(c) Both of these

(d) None of these

42. Assuming that 2s - 2p mixing is not operative, the paramagnetic species among the following is

(a) Be_2 (b) B_2

(c) C_2 (d) N_2

43. Which of the following is a wrong order w.r.t. the property mentioned against each?

(a) $O_2^{2-} > O_2 > O_2^+$ (magnetic moment)

(b) $NO^{\ominus} > NO > NO^+$ (Bond length)

(c) $H_2 > H_2^+ > He_2^+$ (bond energy)

(d) $NO_2^+ > NO_2 > NO_2^-$ (bond length)

44. Choose the correct option regarding angles in formaldehyde.

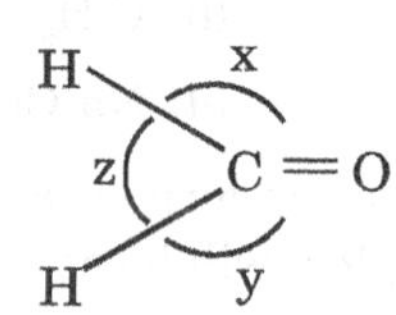

(a) x = y > z (b) x = z > y

(c) y > x > z (d) x > y > z

45. Arrange the following in decreasing order of bond angles.

COF_2, $COCl_2$, COBr, COI_2

(a) $COF_2 > COCl_2 > COBr_2 > COI_2$

(b) $COI_2 > COBr_2 > COCl_2 > COF_2$

(c) $COF_2 = COCl_2 > COBr_2 > COI_2$

(d) None of these

46. Which of the following group 14 compounds is most ionic?

(a) CCl_4 (b) $SiCl_4$

(c) $PbCl_2$ (d) $PbCl_4$

47. CuCl is more covalent than NaCl because

(a) Cu^+ has noble gas configuration but Na^+ has pseudonoble gas Configuration.

(b) Both Cu^+ & Na^+ have noble gas configuration.

(c) Both Cu^+ & Na^+ have pseudonoble gas configuration.

(d) Na^+ has noble gas configuration but has pseudonoble gas configuration.

48. The magnetic moment of KO_2 at room temperature is 'x' BM. The value of 'x' is

(a) 1.41 (b) 1.73

(c) 2.23 (d) 2.64

49. Which of the following order is correct regarding bond angles?

(a) $NH_3 > NF_3$ (b) $NH_3 > CO_2$

(c) $PF_3 > PH_3$ (d) Both 1 & 2

50. Assuming that Hunds rule is violated, the Bond order & the magnetic nature of diatomic molecules B_2 is

(a) 1 & diamagnetic (b) 0 & diamagnetic

(c) 1 & paramagnetic (d) 0 & paramagneti

51. Arrange the following in decreasing order of their stabilities.

$$Li_2, Li_2^-, Li_2^+$$

(a) $Li_2^+ > Li_2 > Li_2^-$

(b) $Li_2^- > Li_2^+ > Li_2$

(c) $Li_2 > Li_2^+ > Li_2^-$

(d) $Li_2^+ > Li_2^- > Li_2$

52. Select the correct order regarding bond angles.

(a) $BF_3 > BBr_3 > BCl_3 > BI_3$

(b) $BF_3 > BCl_3 > BBr_3 > BI_3$

(c) $BCl_3 > BBr_3 > BI_3 > BF_3$

(d) $BF_3 = BCl_3 = BBr_3 = BI_3$

53. Arrange the following bonds in the decreasing order of their bond lengths

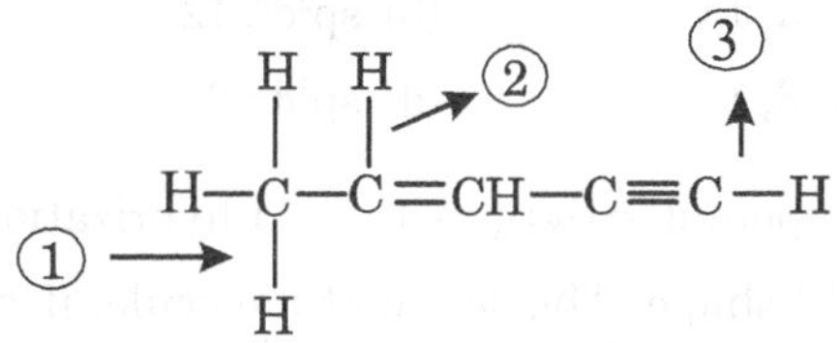

(a) 1 > 2 > 3 (b) 3 > 2 > 1

(c) 2 = 3 > 1 (d) 1 > 3 > 2

54. The bond angle in Cl_2O is 110°8' whereas in $(SiH_3)_2O$ is 114°, 3dπ - 2pπ type of back bonding occurs in both. It is because

(a) back bonding is more severe in Cl_2O.

(b) back bonding is more severe in $(SiH_3)_2O$.

(c) No back bonding is observed in $(SiH_3)_2O$.

(d) Back bonding is not observed in both

55. Which one of the following pairs of species have same bond order?

(a) NO^+ & CN^+ (b) CN^+ & NO^+

(c) CN^+ & CN^+ (d) O_2^- & $\overline{C}N$

56. In $XeOF_4$, let x be the total number of bond pairs around Xe, y be the total number of lone pairs and z be the total number of 90° OXeF angles, then $x + y + z$ is

(a) 21 (b) 25

(c) 28 (d) 26

57. Which of the following statement is true for

$IO_2F_2^-$

(a) The electron pairs are located at the corners of trigonal bipyramidal but one of the equatorial pairs is unshared

(b) It has sp^3d-hybridisation and is T-shaped

(c) Its shape is analogous to SF_4

(d) Both (a) and (c)

58. (i) $CH_3 - CH = CH_2$

(ii) $CH_3 - CH = CH - CH_3$

If number of atoms lying in one plane in (i) is x and in (ii) is y then ratio x : y is

(a) 3 : 2 (b) 3 : 1

(c) 7 : 8 (d) 1 : 2

59. Hybridization of iodine and total number of lone pairs in the structure of IF_4^+ are

(a) sp^3d, 13 (b) sp^3d^2, 12

(c) sp^3d, 1 (d) sp^3d^3, 3

60. A compound AB_xL_y with sp^3d hybrization of A has 'T' shape. The shape of molecule, if A form AB_yL_x type is (L = lone pair)

(a) 'T' shape

(b) linear shape

(c) see-saw

(d) triangular bipyramidal

61. Consider the following three species and identify the incorrect order

(i) ClO_4^-

(ii) SO_4^{2-}

(iii) PO_4^{3-}

(a) i> ii > iii (number of maximum possible dative bonds)

(b) i = ii = iii (% age of p-character in orbitals of central atom)

(c) i > ii > iii % age p-character

(d) Both 1 & 2

Brain Storming Questions

62. Match column I and Column II

Column – I (bond formed)	Column – II (Type of overlapping involved)
a. $NH_3 + H^+ \rightarrow NH_4^+$	i. $p\pi - p\pi$
b. $H + F \rightarrow H - F$	ii. $sp^3 - s$
c. $N + N \rightarrow N \equiv N$	iii. $sp^3 - p$
d. $C + 4Cl \rightarrow CCl_4$	iv. $s - p$

(a) a - ii , b - iv, c - iii, d - i

(b) a - iii, b - iv, c - i, d - ii

(c) a - ii, b - iv, c - iv, d - iii

(d) a - ii, b - iv, c - i, d - iii

63. Octet rule is violated in

(a) SF_6 (b) IF_7

(c) ClF_3 (d) All of these

64. The molecules/ion which are polar & planar

(a) $SnCl_2$ (b) $\overset{\ominus}{NH_2}$

(c) IF_3 (d) All of these

65. If ethylene molecule lies in XY plane, then nodal plane of Pie bond will be lie in

(a) XZ plane (b) YZ plane

(c) XY plane (d) All of these

66. In which of the following molecules/ion all the bonds are not equal?

(a) XeF_4 (b) BF_4^-

(c) SF_4 (d) SiF_4

67. In which of the following molecules, all atoms possess expanded octet

(a) PCl_5 (b) SF_6

(c) IF_7 (d) None of these

68. In which of the following processes, the Bond order has increased & the magnetic behavior has changed

(a) $NO \rightarrow NO^+$ (b) $O_2 \rightarrow O_2^+$

(c) $N_2 \rightarrow N_2^+$ (d) $C_2 \rightarrow C_2^+$

69. Which of the following statements are true about $I(CN)_2^-$?

(a) It is having pπ-pπ bond & dipole moment = 0

(b) It is planar

(c) The central atom is sp^3d hybridized

(d) All are correct

70. Which statement(s) are valid for XeO_2F_2 & SOF_4?

(a) Both have pπ-dπ bond.

(b) Both have dipole moment, ≠ O

(c) Both have central atom sp^3d hybridized

(d) All are correct

71. Of the species O_2^+, O_2^-, O_2 & O_2^{2-} which would have the maximum strongest bond?

(a) O_2^- (b) O_2

(c) O_2^2 (d) O_2^{2-}

72. Select the incorrect statement

(a) In $CF_2 = C = CF_2$, all the atoms are is one plane

(b) Super oxides are coloured & paramagnetic

(c) F_3^- is not possible

(d) Dipole moment of acetylene is zero

73. $ZnCO_3$ is thermally less stable than $MgCO_3$ because

(a) Mg^{2+} has greater polar sing power than Zn^{2+}

(b) Zn^{2+} has greater polarizing power than Mg^{2+}

(c) Mg^{2+} & Zn^{2+} have same polarizing power

(d) Zn^{2+} & Mg^{2+} have same polarisibility

74. How may of the following molecules have dipole moment $\neq 0$?

$BF_3, CO_2, SO_3, CF_4, SF_6, XeF_2$, trans-but-2-ene, trans-pent-2-ene.

(a) 1 (b) 11

(c) 8 (d) 6

75. Which of the following compound are planar is well as non-polar?

(a) XeF_4 (b) XeF_2

(c) XeF_5^- (d) All of these

76. The shape of (ClF_4^-) & $(ClF_2)^-$ ions is respectively

(a) Linear & linear

(b) See-saw &bent

(c) square planar & linear

(d) Tetrahedral

77. Which of the following pairs is iso- structural?

(a) NO_2^+ & CO_2 (b) HCN & $SnCl_2$

(c) SO_3 & CH_4 (d) None of these

78. PCl_xF_{5-x} molecule will be polar & non - polar for what values of x respectively?

(a) 2, 3, 5 & 0, 4, 1

(b) 0, 2, 5 & 3, 1, 4

(c) 4, 3, 0 & 2, 0, 4

(d) 2, 1, 4 & 3, 0, 5

79. Which one of the following is a planar molecule?

(a) Al_2Cl_6 (b) Al_2Br_6

(c) I_2Cl_6 (d) Both (a) & (b)

Dipole Moment (μ)

80. Which of the following molecules have dipole moment, $\mu = 0$?

(a) BF_3 (b) XeF_4

(c) XeF_2 (d) All of these

81. Which of the following posses both polar & non-polar bonds?

(a) H_2O_2 (b) CH_4

(c) H_2 (d) Cl_2

82. The dipole moment of fluorobenzene is less than that of Chlorobenzene because

(a) F is more electronegative than Cl

(b) Cl is more electronegative than F.

(c) More effective mesomeric effective in case of F (2pπ–2pπ) as compared to that of Cl (3pπ – 2pπ)

(d) None of these

83. Arrange ortho, meta&paradichlorobenzene in decreasing order of their dipole moments.

(a) O>M>P (b) P>M>O

(c) M > O > P (d) M = O > P

84. In which of the following molecule Dipole moment (Observed) is found to be greater than Dipole moment (theoretical)?

(a) 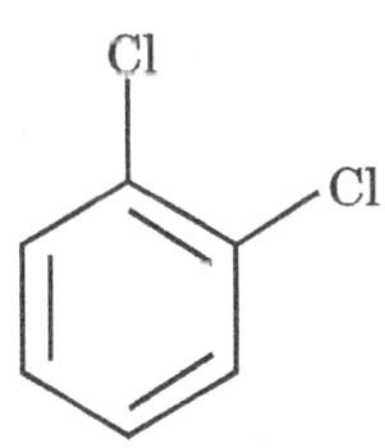

(b) 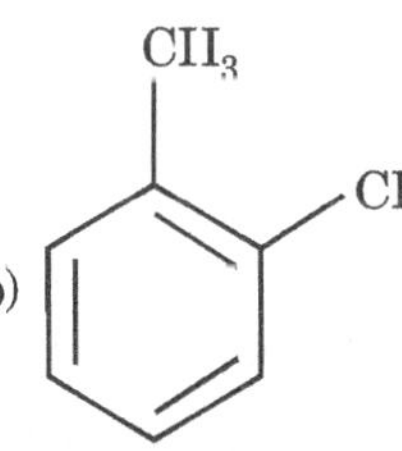

(c) 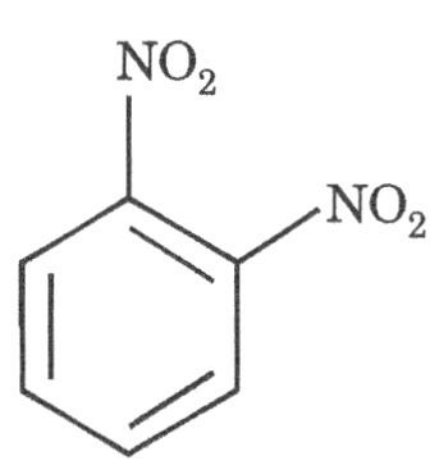

(d) 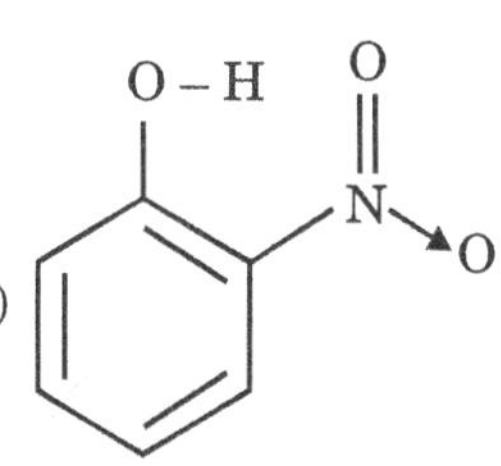

Fajan's Rule, Bond Angle & Dipole Moment

85. Which one of the following has greatest polarisibility?

(a) F^- (b) Cl^-
(c) Br^- (d) I^-

86. Which is most ionic?

(a) LiF (b) Li_2O
(c) Li_3N (d) All are equally ionic

87. Out of the following molecules given, which all have the same bond angles?

$CH_4, CCl_4, PF_3, PH_3, NH_3, \overset{+}{N}H_4$

(a) PF_3 & PH_3
(b) PF_3, NH_3 & PH_3
(c) CH_4 & PH_3
(d) CH_4, CCl_4 & $\overset{+}{N}H_4$

88. Calculate the fractional charge on each atom in HBr molecule.

Given that dipole moment of HBr = 0.78 D. Bond length of HBr = 1.141Å

(a) 0.25 (b) 0.47
(c) 0.11 (d) None of these

89. For PCl_3F_2 & PF_3Cl_2, the dipole moment ( ) values are, respectively

(a) $\mu = 0$ & $\mu \neq 0$ (b) $\mu = 0$ & $\mu = 0$
(c) $\neq 0$ & $\neq 0$ (d) Both 2 & 3

90. In NH_3,

(a) The bond moments & orbital moments point in same direction
(b) The bond moment & orbital moment point in different direction
(c) The bond moment is zero
(d) The orbital moment is zero

91. Select the correct order of dipole moment.

(a) 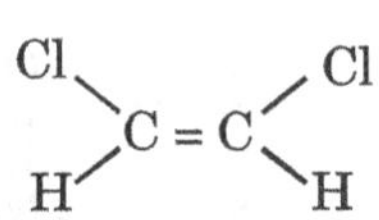 > 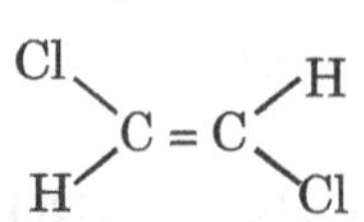

(b) 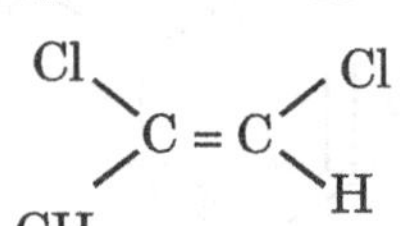 > 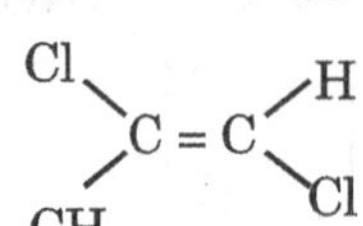

(c) $CH_3Cl > CH_2Cl_2 > CHCl_3 > CCl_4$

(d) All are correct

92. In molecules of the type AX_2L_n where L represents the Lone pair & 'n' its number, there exists a bonds between A & X. Then $\angle XAX$ –

(a) will be maximum for n = 3
(b) will be less than 180° for n = 0
(c) will always decreases if 'n' increases
(d) will always increase if 'n' increases

93. Which one has maximum dipole moment

(a) 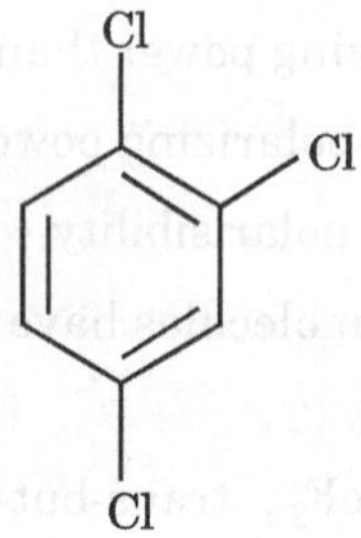 (b) 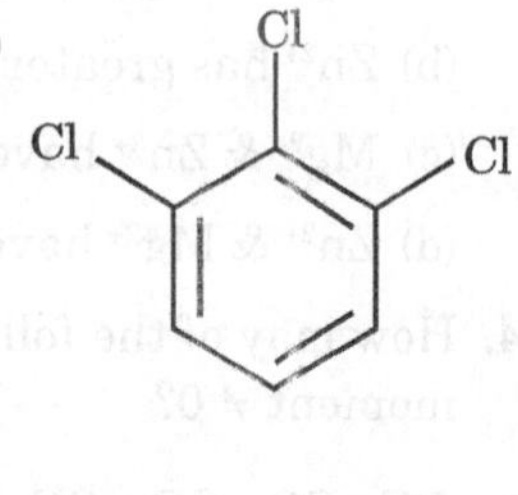

(c) 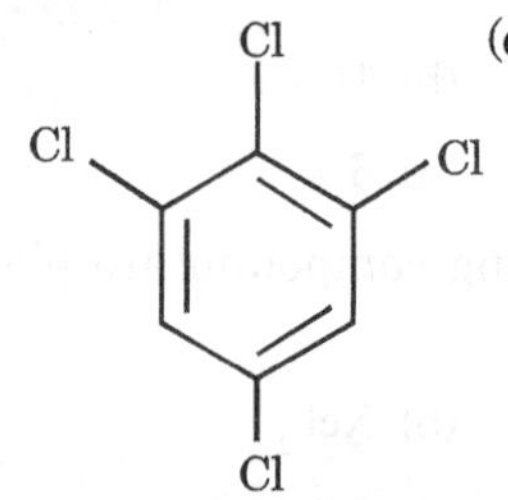 (d) 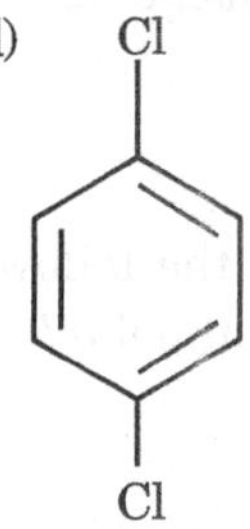

94. Calculate the % ionic character of HCl. Observed dipole moment is 1.03 D & Bond length is 1.275 Å

(a) 16.83 (b) 19%
(c) 83.37% (d) 82%

95. Select the correct of order solubility in polar solvent.

(a) AgI > AgBr > AgCl > AgF
(b) AgF > AgCl > AgBr > AgI
(c) AgCl > AgBr > AgF > AgI
(d) AgF > AgCl = AgBr > AgI

96. Which of the following combinations will have the highest polarization?

(a) Na^+, Br^- (b) Al^{3+}, Br^-
(c) Al^{3+}, I^- (d) Na^+, Cl^-

97. Chose the correct statement among the following.

(a) The Lone pair has no contribution to the dipole moment if the Lone pair is present in pure s or pure p orbital

(b) $H-F > H-Cl > H-Br > H-I$
(Dipole moment order)

(c) $CH_3-Cl > CH_3-F > CH_3-Br > CH_3-I$
(Dipole moment order)

(d) All are correct

98. How many statements are correct?

(1) There are total 6 HCH bond angles in CH_4

(2) In PCl_3F_2, F atoms prefer to occupy axial position.

(3) The structures,

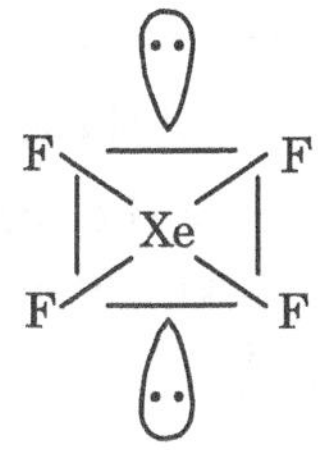

and

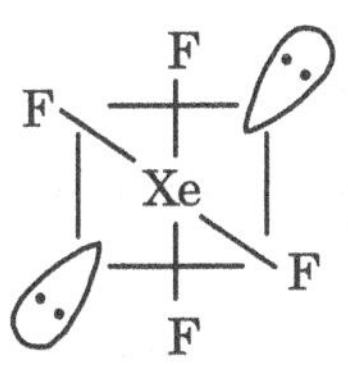

have same shapes.

(4) There are 12, 90° angles in SF_6

(a) 1 (b) 2
(c) 3 (d) 4

99. In which of the following pair , all the bonds are not of equal length?

(a) $BeCl_2, H_2O$ (b) CH_4, CCl_4
(c) IF_5, PCl_5 (d) PCl_3, NH_3

100. What is correct about MY_2 , if x is the number of lone pair (s) on M :

(a) It is always polar for any value of x.

(b) As x increases, dipole moment increases.

(c) Dipole moment of MY_2 follows the order $(x = 2) > (x = 1) > (x = 3)$

(d) Dipole moment of MY_2 follows the order $(x = 1) > (x = 2) > (x = 3)$

101. Which of the following statement is true for $IO_2F_2^-$?

(a) The electron pairs are located at the corners of a triagonal bipyramidal structure but one of the equatorial pair is unshared.

(b) Central atom has sp^3d & is T – shaped

(c) Its shape is analogous to SF_4

(d) Both 1 & 3 are correct.

102. Correct order of bond angle is -

(a) $PF_3 > PCl_3 > PBr_3 > PI_3$

(b) $CI_4 > CBr_4 > CCl_4 > CF_4$

(c) $Br_2O > Cl_2O > H_2O$

(d) $BF_3 > BeF_2 > XeF_2$

103. Match the column I with column II

	Column I	Column II
(a)	Molecule with all atoms in same plane.	p. BF_2Cl
(b)	Non – polar molecule	q. SF_4
(c)	Planar molecule	r. XeO_4
(d)	Px, Py & Pz orbitals are involved in hybridization of central atom.	

(a) a - r ; b - p, r ; c - p ; d - q , r

(b) a - p ; b - r ; c - q ; d - r

(c) a - p ; b - r ; c - p ; d - q , r

(d) a - q ; b - p, r ; c - p , q ; d - p

104. The percentage of S – character in orbitals of central atom of a molecule 'A' is greater than that in XeO_2F_2 & less than that in CS_2. Then what is true about A?

(a) 'A' can have bond angle = 109°28'

(b) 'A' can have bond angle < 109°28'

(c) 'A' can have bond angle = 120°

(d) 'A' can have bond angle greater than 90° & less than 120°.

(a) a & c only (b) c & d only
(c) a , b & c only (d) a , b , c & d

105. The hybrid orbital having the bulkiest lobe is -

(a) Sp^2 (b) Sp^3d
(c) Sp (d) Sp^3

106. The number of F - Br - F angles at 90° in BrF_5 is-

(a) 5 (b) 4
(c) 10 (d) 0

107. Out of $ZnCl_2$ & $MgCl_2$, which is more soluble in organic solvents ?

(a) $ZnCl_2$

(b) $MgCl_2$

(c) $ZnCl_2$ & $MgCl_2$ have same solubility

(d) $ZnCl_2$ & $MgCl_2$ are insoluble in organic solvents

108. What should be the correct statement wrt XeF_5^- ?

(a) Xe has sp^3d hybridization.

(b) It is square planar.

(c) There are two non-bonding electron pairs, one above the plane & other below the plane.

(d) It is an odd electron specie

109. Consider the following statements

(1) The hybridization found in cation of solid PCl_5 is sp^3.

(2) In AB_2L_2 type, the BAB bond angle is always greater than the normal tetrahedral angle.

(3) In ClO_3^-, NH_3 & XeO_3, the hybridization & the no. of lone pairs on central atom are same.

(4) In P_4 molecule, there are 6 P – P bonds & 4 lone pairs of electrons of all these statements,

(a) I, II & III are correct only

(b) I, III & IV are correct only

(c) III & IV are correct only

(d) All are correct

110. Which of the following compound contain both ionic & covalent bonds ?

(a) NH_4Cl (b) KCN

(c) NaOH (d) All of these

111. Which statement is/are correct about hybridization ?

(a) Concept of hybridization is valid for covalent bonding.

(b) In hybridization, half filled, fully filled or even vacant orbitals can also participate

(c) Hybrid orbitals are directional

(d) All are correct

112. Identify the correct option (s)

(a) $NH_4^+ > NH_3 > \overset{\ominus}{N}H_2$ – order of bond angle

(b) In NH_4Br, N is sp^2 hybridised

(c) NO_2 is paramagnetic

(d) Both 1 & 3 are correct

113. Which of the following are polar?

(a) XeF_4 (b) XeF_6

(c) $XeOF_4$ (d) Both 2 & 3

114. Match the species given in column – I with the type of hybridization given in Column – II.

	Column I		Column II
(a)	$IO_2F_2^-$	(p)	sp^3d
(b)	F_2SeO	(q)	sp^3
(c)	$ClOF_3$	(r)	sp^2
(d)	XeF_5^+	(s)	sp^3d^2

(a) a – p; b – q; c – p; d – s

(b) a – q; b – p; c – s; d – s

(c) a – q; b – s; c – p; d – p

(d) None of these

115. Geometry (i.e. arrangement of electron pairs around central atom) of $ClOF_3$ is similar to -

(a) XeF_6 (b) $SOCl_2$

(c) I_3^- (d) ClO_4^-

116. Consider the following statements and arrange in the order of true/false.

Statement 1 :- In $SnCl_2$, the bonding takes place in ground state & the bond angle Cl – Sn – Cl is slightly less than 120°.

Statement 2 :- Molecular geometry of XeF_7^+ is pentagonal bipyramidal.

Statement 3 :- In SF_4, the bond angles, instead of being 90° & 180° are 89° & 177° respectively due to the presence of a lone pair.

Statement 4 :- BF_3 has B in excited state.

(a) T T T T (b) T F T T
(c) F F F T (d) F F F F

117. Which of the following statements is incorrect ?

(a) In ClF_3, the axial Cl – F bond length is larger than equatorial Cl – F bond length.

(b) In SF_4, F – S – F equatorial bond angle is not 120°.

(c) In $(ICl_4)^-$, the Cl – I – Cl bond angle is 90°.

(d) In OBr_2, the bond angle is less than OCl_2

118. In which of the following molecules, vacant orbital participates in hybridization?

(a) B_2H_6 (b) BF_4^-

(c) CH_4 (d) Both 1 & 2

119. No. of π & σ bonds in naphthalene respectively are–

(a) 6, 19 (b) 4, 20

(c) 5, 19 (d) 5, 20

120. In which of the following, hybridization of N is different form others ?

(a) 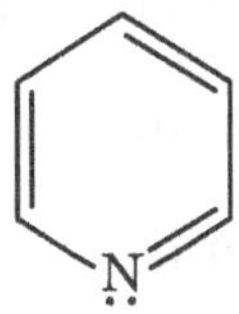 (b) $(CN)_2$

(c) $CaCN_2$ (d) 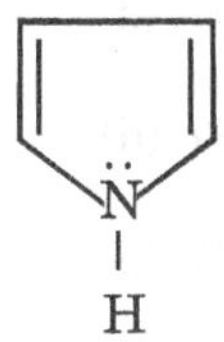

121. Identify the mismatch

		Column – I Molecular orbitals	Column – II Nodal planes
a.	1.	$\sigma 2s$	0
b.	2.	$\sigma^* 2pz$	1
c.	3.	$\pi^*(2px)$	2
d.	4.	$\pi(2px)$	0

122. Which one of the following does not have inter-molecular H - bonding?

(a) H_2O (b) O – nitrophenol

(c) HF (d) CH_3COOH

123. According to molecular orbital theory,

(a) O_2^+ is paramagnetic & has bond order greater than that of O_2

(b) O_2^+ is paramagnetic & has bond order less than that of O_2

(c) O_2^+ is diamagnetic & has bond order greater than that of O_2

(d) O_2^+ is diamagnetic & has bond order more than that of O_2

124. Which specie has the maximum number of lone pairs of electrons on the central atom?

(a) ClO_3^- (b) XeF_4

(c) SF_4 (d) I_3^-

125. The geometry around the central atom in ClF_4^+ is -

(a) Square planar (b) Square pyramidal

(3) Octahedral (4) Triagonalbipyramidal

126. The xenon compounds that are isostructural with IBr_2^- & BrO_3^-, respectively are –

(a) Linear XeF_2 & pyramidal XeO_3

(b) Bent XeF_2 & pyramidal XeO_3

(c) Bent XeF_2 & planar XeO_3

(d) Linear XeF_2 & tetrahedral XeO_3

127. Which of the following is paramagnetic?

(a) NO (b) NO_2

(3) O_2 (4) All of these

128. A diatomic molecule xy has a dipole moment of 1.2D. If its bond distance is 100 pm, calculate the percentage ionic character in xy –

(a) 50% (b) 25%

(c) 100% (4) 10%

129. Which of the following hydrocarbons has the lowest dipole moment>?

(a) 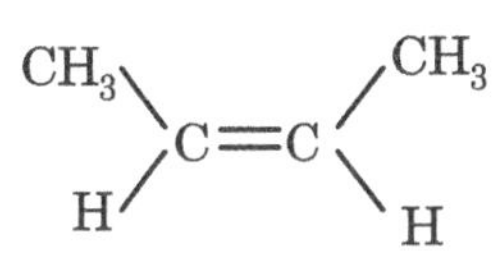

(b) $CH_3 - C \equiv C - CH_3$

(c) $CH_3CH_2C \equiv CH$

(d) $CH_2 = CH - C \equiv CH$

130. The nodal plane of π bond in ethene is located in–

(a) The molecular plane

(b) A plane parallel to molecular plane

(c) A plane perpendicular to the molecular plane which bisects the C – C σ bond at right angle

(d) None of these

131. Which of the following molecular specie has unpaired electron (s)?

(a) N_2 (b) F_2

(c) O_2^- (d) O_2^{2-}

132. Which has highest dipole moment?

(a) CH_3Cl (b) CH_2Cl_2

(c) $CHCl_3$ (d) CCl_4

Answers

1. (c)	2. (d)	3. (c)	4. (a)	5. (c)	6. (d)	7. (d)	8. (a)	9. (c)	10. (d)
11. (d)	12. (b)	13. (a)	14. (d)	15. (d)	16. (c)	17. (d)	18. (d)	19. (c)	20. (b)
21. (d)	22. (d)	23. (a)	24. (d)	25. (c)	26. (a)	27. (d)	28. (d)	29. (d)	30. (d)
31. (b)	32. (d)	33. (c)	34. (b)	35. (c)	36. (b)	37. (a)	38. (c)	39. (b)	40. (a)
41. (c)	42. (c)	43. (a)	44. (a)	45. (b)	46. (c)	47. (d)	48. (b)	49. (d)	50. (a)
51. (c)	52. (d)	53. (a)	54. (b)	55. (b)	56. (a)	57. (d)	58. (c)	59. (a)	60. (b)
61. (c)	62. (d)	63. (d)	64. (d)	65. (c)	66. (c)	67. (d)	68. (b)	69. (d)	70. (d)
71. (a)	72. (a)	73. (b)	74. (a)	75. (d)	76. (c)	77. (a)	78. (d)	79. (c)	80. (d)
81. (a)	82. (c)	83. (a)	84. (d)	85. (d)	86. (a)	87. (d)	88. (c)	89. (a)	90. (a)
91. (d)	92. (a)	93. (b)	94. (a)	95. (b)	96. (c)	97. (d)	98. (d)	99. (c)	100. (c)
101. (d)	102. (c)	103. (c)	104. (d)	105. (c)	106. (d)	107. (a)	108. (c)	109. (b)	110. (d)
111. (d)	112. (d)	113. (d)	114. (a)	115. (c)	116. (a)	117. (d)	118. (d)	119. (c)	120. (b)
121. (d)	122. (b)	123. (a)	124. (d)	125. (d)	126. (a)	127. (d)	128. (b)	129. (b)	130. (a)
131. (c)	132. (a)								

Explanations

7. (d) $F_{NaCl} = \dfrac{K \times 1 \times 1}{r^2}$ ...(1)

$F_{CaO} = \dfrac{K \times 2 \times 2}{r^2}$ ...(2)

$F_{CaO} = \dfrac{4K}{r^2}$

So, $U_{CaO} = 4U_{NaCl}$

37. (a)

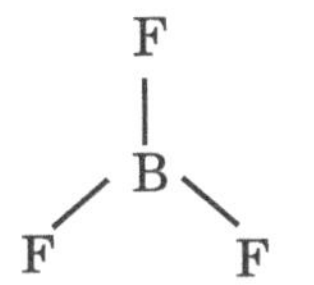

Boron – sp^2

Shape – triagonal plane

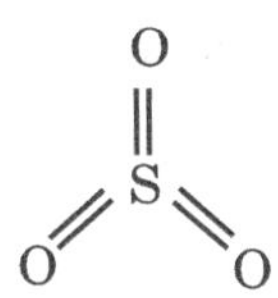

Sulphur – sp^2

Shape – triagonal planas

38. (c) $m = q \times d$

$0.78D = q \times 1.41$ A

0.78×10^{-18} esu cm $= q \times 1.41 \times 10^{-8}$ cm

$q = \dfrac{0\cdot 78 \times 10^{-18}}{1\cdot 41 \times 10^{-8}}$ esu $= 0\cdot 55 \times 10^{-10}$ esu

Fractional charge $= \dfrac{0\cdot 55 \times 10^{-10}}{4\cdot 8 \times 10^{-10}} = 0.\cdot 11$

40. (a) 180°

$O = \overset{+}{N} = O$

41. (c)

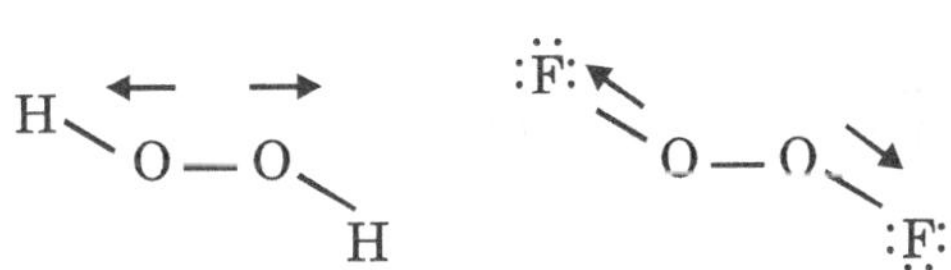

42. (c) C_2 (12 eletrons)

$\sigma 1s^2 \sigma^* 1s^2 \sigma 2s^2 \sigma^* 2s^2 \sigma^* 2p^2{}_z\ \pi 2p^1{}_x$

$= \pi 2p^1{}_y\ \pi^* 2p_x = \pi^* 2p_y\ \sigma^* p_z$

(without 2s – 2p mixing)

So, it has 2 unpaired electrons. Hence, paramagnetic.

43. (a) $O_2^{2-}(18): \sigma 1s^2\ \sigma^* 1s^2\ \sigma 2s^2\ \sigma^* 2s^2\ \sigma 2p^2 z\ \pi 2p^2{}_x$

$= \pi 2p_y^2\ \ \pi^* 2p_x^2\ = \pi^{*2} 2p_y\ \ \sigma^* 2p_z$

$O_2(16): \sigma 1s^2\ \sigma^* 1s^2\ \sigma 2s^2\ \sigma^* 2s^2\ \sigma 2p_z^2\ \pi 2p_x^2$

$= \pi 2p_y^2\ \pi^* 2p_x^1\ = \pi^* 2p_y^1$

$O_2^+(15): \sigma 1s^2 \sigma^* 1s^2 \sigma 2s^2\ \sigma^* 2s^2\ \sigma^* 2p_z^2\ \pi 2p_x^2$

$= \pi 2p_y^2\ \ \pi^* 2p_x^1\ = \pi^* 2p_y$

$M = \sqrt{n(n+2)}$ B.M. where n = number of unpaired electrons

Magnetic moment order : $O_2 > O_2^+ > O_2^{2-}$

44. (a) The C atom in formaldehyde is sp^2 hybridized. Bond angles, expected were120°. But the repulsion between double bond and single bonds dominate over single bond single bond repulsions. So,

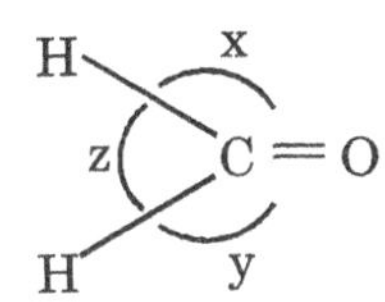

x = y > 120°

Z < 120°

45. (b)

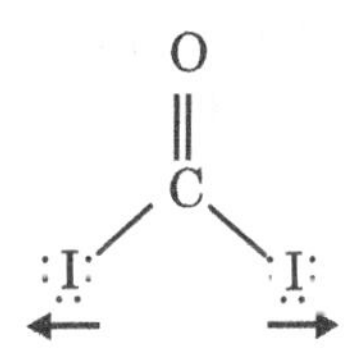

lp – lp repulsions dominate and bond angle is increased (Steric Crowding) and larger size of I.

46. (c) Greater the size of cation, greater is ionic character. Out of Pb^{2+} & pb^{2+} is greater. So, $PbCl_2$ is maximum ionic.

48. (b) KO_2 has O_2^- ion

$O_2^-(17): \sigma 1s^2 \sigma^* 1s^2 \sigma 2s^2 \sigma^* 2s^2\ \sigma 2p_z^2\ \pi 2p_x^2$

$= \pi 2p_y^2\ \pi^* 2p_x^2\ = \pi^* 2p_y^1\ \sigma^* 2p_z$

It has one unpaired electron.

So, $M = \sqrt{n(n+2)} = \sqrt{1(1+2)}$

$= \sqrt{3}$ B.M $= 1.73$ B.M.

49. (d) 

⟶ Bond pairs are converged.

⟶ greater bp-bp repulsions, so bond angle is increased

(N is more electro -ve than H)

⟶ Bond pairs are diverged.

⟶ lesser bp-bp repulsions.

(F is more electro -ve than N)

50. (a) $B_2(10): \sigma 1s^2, \sigma^* 1s^2, \sigma 2s^2, \sigma^* 2s^2, \pi 2p_x^2 = \pi 2p_y, \sigma 2p_z, \pi^* 2p_x = \pi^* 2p_y$

(VIOLATED HUND'S RULE)

$$\text{Bond Order} = \frac{N_b - N_a}{2} = \frac{6-4}{2} = \frac{2}{2} = 1$$

All e^- are paired up, so diamagnetic.

52. (d) All trihalides of Boron have regular trigonal planar geometries. So, bond angles are same i.e. 120°.

53. (a) Bond 1 is formed by $sp^3 -$ S overlap

Bond 2 is formed by $sp^2 -$ S overlap

Bond 3 is formed by sp – S overlap

B. Length

$sp^3 - S > sp^2 > S > sp - S$

54. (b) 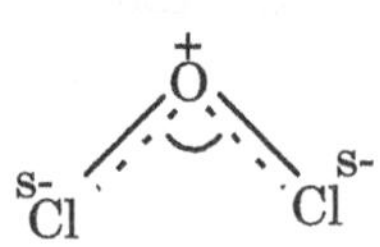

110'8°

⟶ Because of 3 lone pairs on Cl, the back bonding from O is not sufficient as the lone pairs of Cl repell the electron cloud coming from O

55. (b) Both $\bar{C}N$ & NO^+ have same bond order because they are isoelectronic species.

56. (a) There will be NO 90° angle because of distortion created by lone pair resting on Xenon.

80. (d) 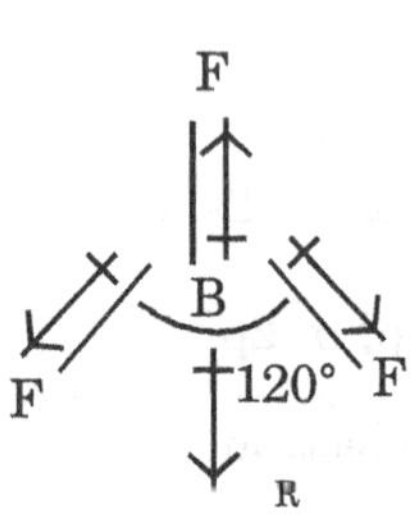

$\boxed{\text{Net} = 0}$

$$\mu_R^2 = \mu_1^2 + \mu_2^2 + 2\mu_1\mu_2 \cos\theta$$

$$\mu_R^2 = \mu^2 + \mu^2 + 2\mu\mu\cos 120^\circ$$

$$\mu_R^2 = \mu^2 + \mu^2 + 2\mu\mu \times \frac{-1}{2}$$

$$\mu_R^2 = \mu^2 + \mu^2 - \mu^2$$

$$\mu_R^2 = \mu^2$$

$$\boxed{\mu_R = \mu}$$

XeF_4

Shape of XeF_4 is Square Planar

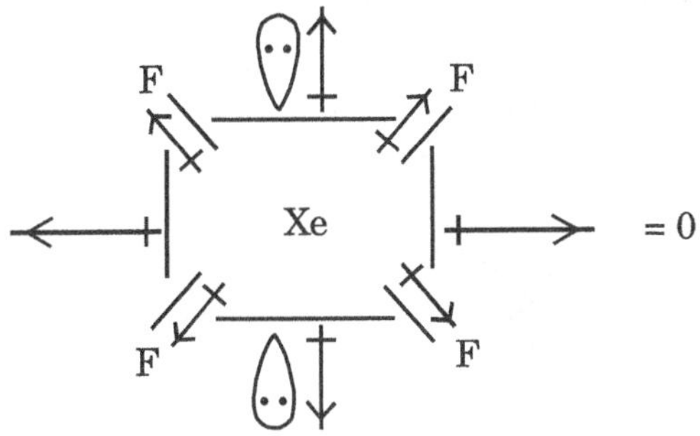

XeF_2 is a linear molecule

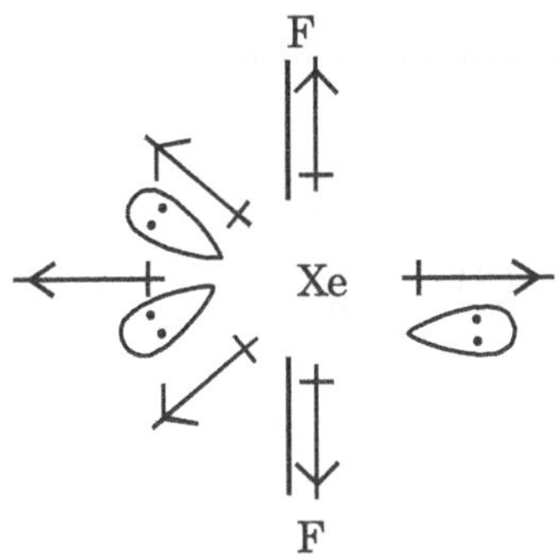

81. (a)

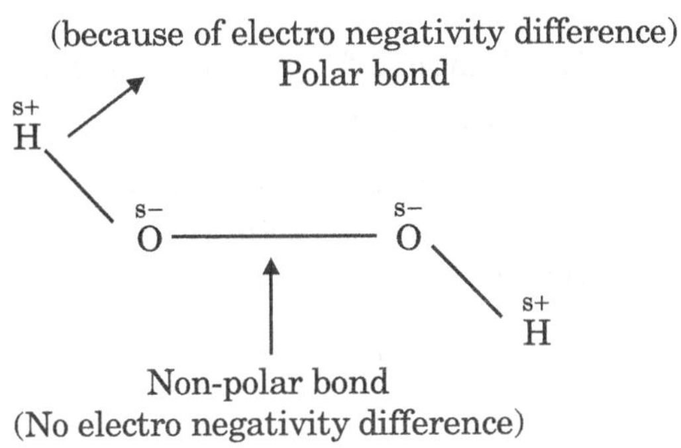

82. (c)

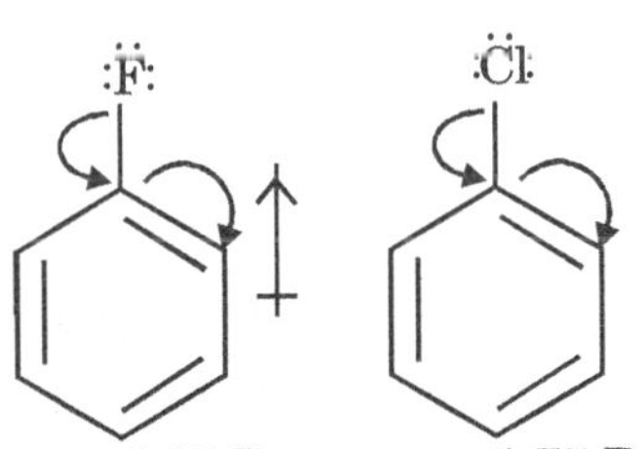

= 1.63 D = 1.75 D

83. (a)

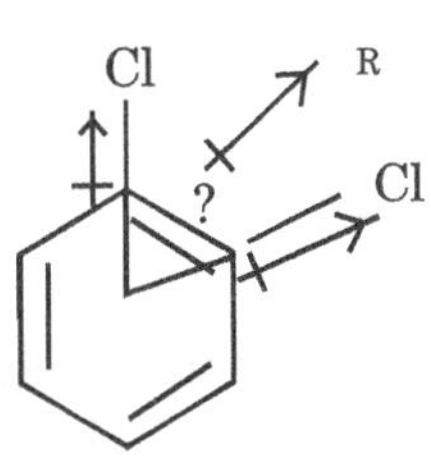

? = 60°

$$\mu_R^2 = \mu_1^2 + \mu_2^2 + 2\mu_1\mu_2 \cos 60^\circ$$

$$\mu_R^2 = \mu^2 + \mu^2 + 2\mu_1 \times \frac{1}{2}$$

$$\mu_R^2 = 3\mu^2$$

$$\boxed{\mu_R = \sqrt{3}\mu}$$

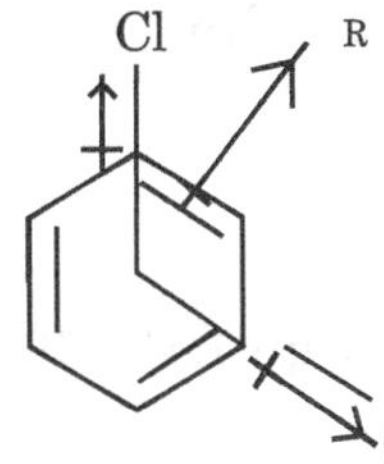

? = 120°

$$\mu_R^2 = \mu_1^2 + \mu_2^2 + 2\mu_1\mu_2 \cos 120^\circ$$

$$\mu_R^2 = \mu^2 + \mu^2 - \mu^2$$

$$\mu_R^2 = \mu^2$$

$$\boxed{\mu_R = \mu}$$

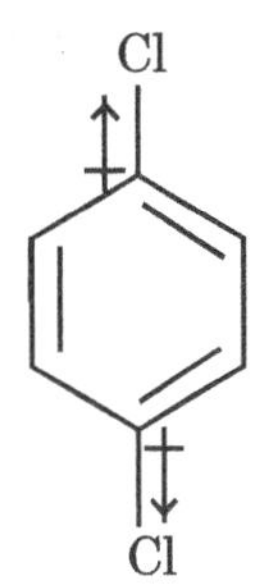

$\boxed{= 0}$

84. (d) In ortho nitrophenol, intramolecular H-bonding occurs. The bond angle gets squared & hence, dipole moment is increased

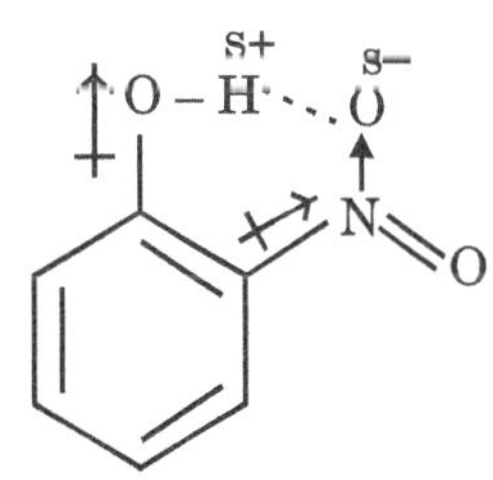

85. (d) I^- is biggest in size among halide ions & has highest polarisibility.

86. (a) Cation is same in all the three i.e., Li^+.

Comparing F^-, O^2 & N^{3-}

F^- → $10e^-$, $9p$ O^{2-} → $10e^-$, $8p$ N^{3-} → $10e^-$, $7p$

Effective nuclear charge is maximum in F^-. So, smaller is the ANION, greater is ionic character.

87. (d) CH_4, CCl_4 & $\overset{+}{N}H_4$ all have regular tetrahedral geometry

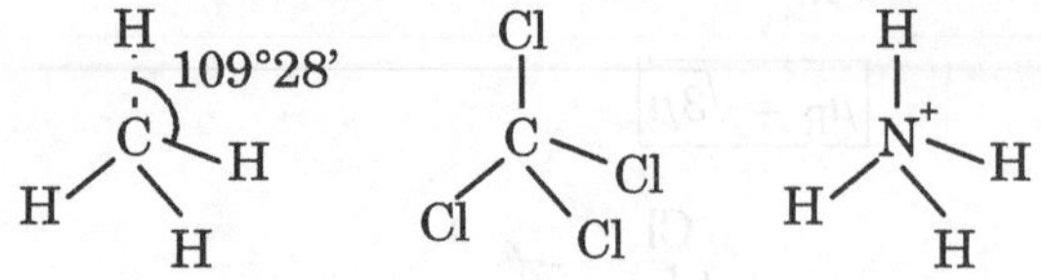

88. (c) m = q × d

0.78D = q × 1.41 A

0.78×10^{-18} esu cm = q × 1.41 × 10^{-8} cm

$$q = \frac{0\cdot78\times10^{-18}}{1\cdot41\times10^{-8}} \text{esu} = 0\cdot55\times10^{-10}\text{esu}$$

$$\text{Fractional charge} = \frac{0\cdot55\times10^{-10}}{4\cdot8\times10^{-10}} = 0\cdot11$$

89. (a)

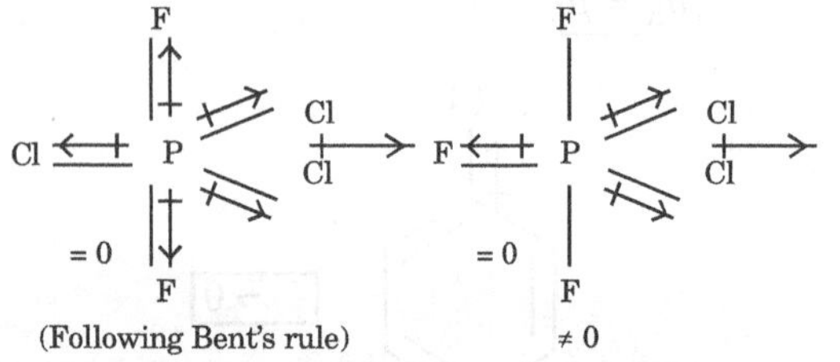

90. (a)

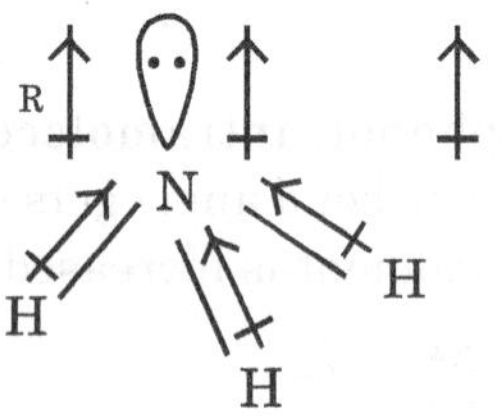

92. (a) Take an example of XeF_2

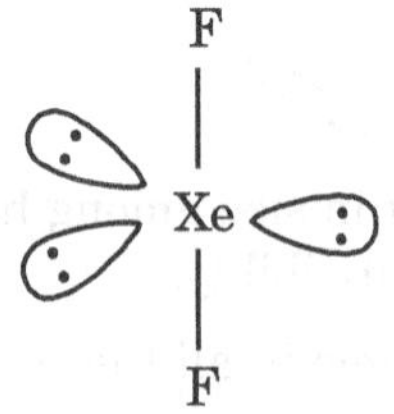

$XeF_2L_2\,(AX_2L_n)$

Bond angle = 180°

(maximum for n = 3)

93. (b)

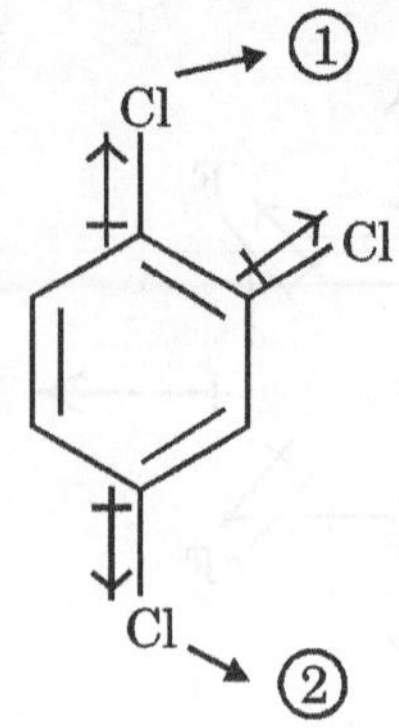

Vector 1 & 2 will cancel out.

Net Dipole Moment =

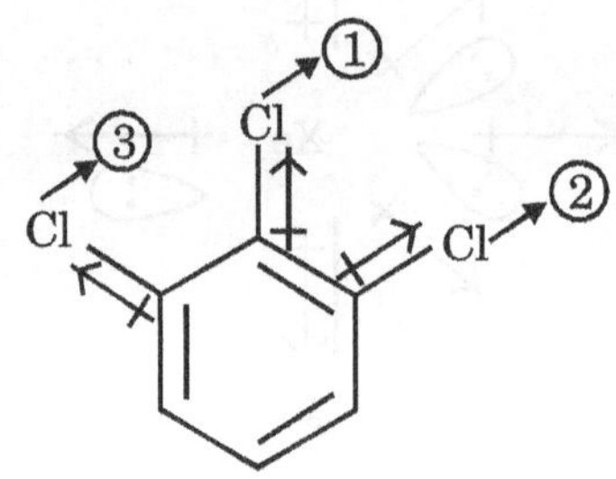

Resultant of 2 & 3 will add to 1, Net Dipole Moment = 2

94. (a) Had HCl been totally ionic, situation would have been H^+Cl^- (both would have one unit charge)

$$\mu_{\text{ionic (total)}} = q \times d = 4\cdot8 \times 10^{-10}\,\text{esu} \times 1\cdot275\times 10^{-8}\,\text{cm}$$

$$= 4\cdot8\times1\cdot275\times10^{-18}\,\text{esucm}$$

$$\mu_{\text{observed}} = 1\cdot03\text{ D}$$

$$\%\text{ age ionic character} = \frac{\mu_{\text{obs.}}}{\mu_{\text{ionic}}}\times100$$

$$= \frac{1\cdot03\text{D}}{4\cdot8\times1\cdot275\text{D}}\times100$$

$$= \frac{1\cdot03\text{D}}{6\cdot12\text{D}}\times100 = 16.83\%$$

95. (b) AgF is most ionic and hence more soluble in polar solvent.

96. (c) Smaller . Size of cation & greater the size of anion, greater is the polarisation

99. (c)

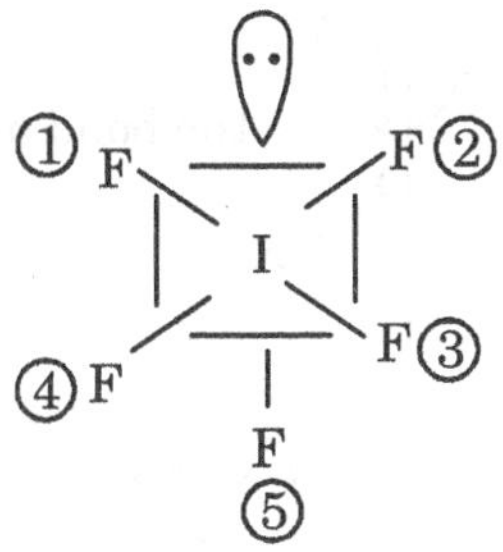

Bonds, labeled 1, 2, 3 & 4 suffer more repulsions from lone pair than from bond 5. So, bond lengths of these bonds is ↑ sed.

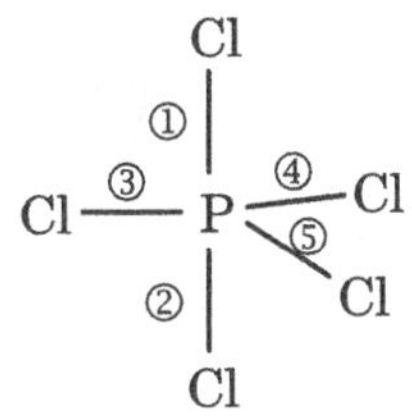

→ Bonds labeled 1 &2 , suffer more significant b.p – b.p. repulsions (3 , at 90° Angles) while bonds labeled as 3, 4 & 5, suffer less significant (2, at 90°) b.p – b.p. repulsions. So, bond length of 1 & 2 (Axial bonds) is increased.

100. (c) MY_2 (when x = 2) → H_2O

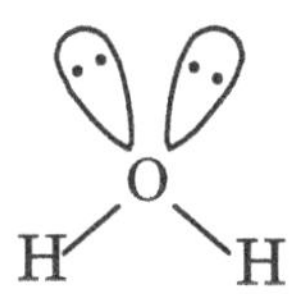

MY_2 (when x = 1) could be $SnCl_2$

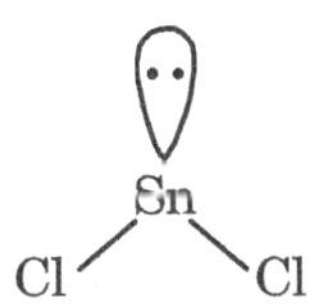

MY_2 (when x = 3) could be XeF_2

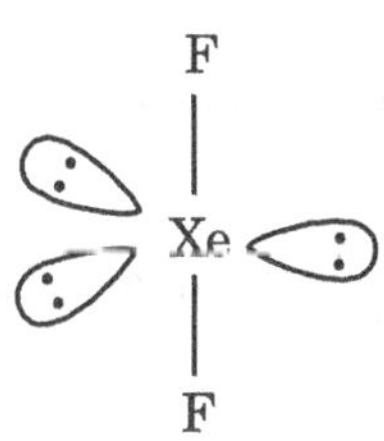

101. (d)

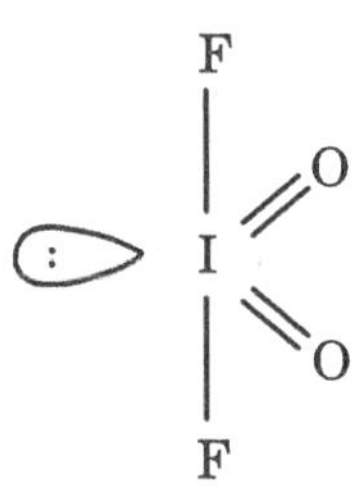

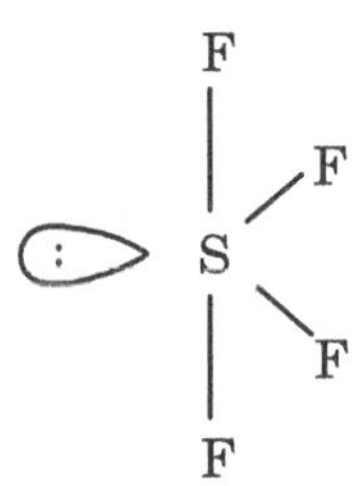

102. (c)

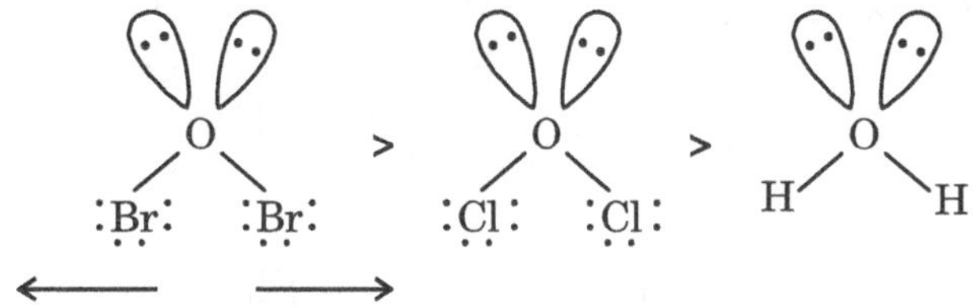

(prominent lone pair – lone pair repulsion wrt Br will ↑ se the bond angle, maximum)

103. (c) # In BF_2Cl, all the four atoms are in same plane

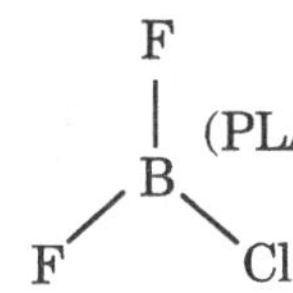

(PLANAR MOLECULE)

Non polar molecule is XeO_4 (regular tetrahedral geometry)

#

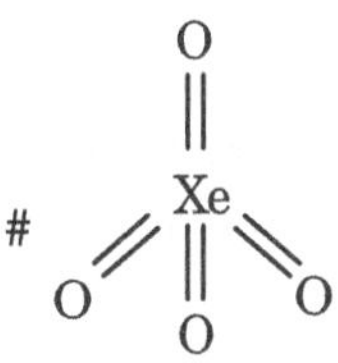

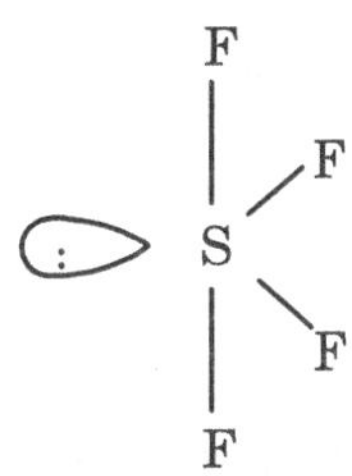

'S' is sp^3d hybridised

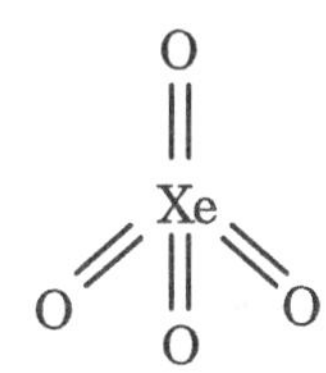

'Xe' is sp^3d hybridised

104. (d)

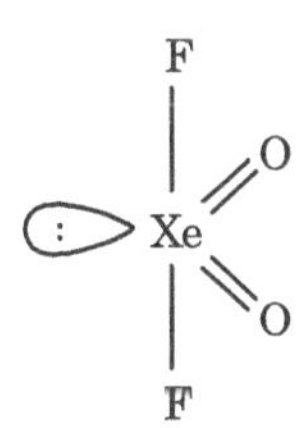

S = C = S

Sp^3d hybridisation of Xe	Sp hybridisation of C
S percentage $= \frac{1}{5} \times 100$	S percentage $= \frac{1}{2} \times 100$
= 20%	= 50%

Now, according to question , 'A' has S percentage > XeO_2F_2 < CS_2

It could be Sp^2 (s = 33.3%) or sp^3 (s = 25%) 'A' could be -

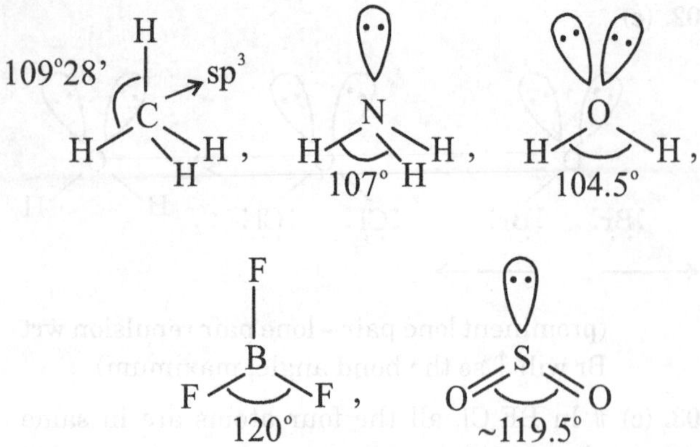

106. (d) 'Br' in BrF5 is sp³d² hybridized.

Expected Structure ⟶ 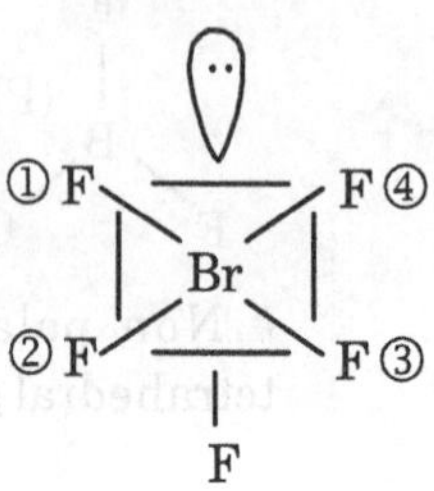

The bonds numbered 1 , 2, 3 & 4 will suffer more repulsions from lone pair of Br rather than the bond pair lying below the plane.

Actual Structure ⟶ 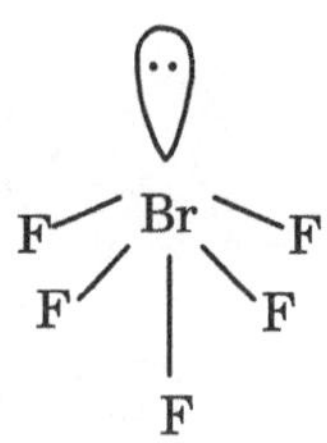

None of the bond angles is 90°.

107. (a) $ZnCl_2$ is more covalent because Zn^{2+} has pseudonoble gas configuration.

108. (c)

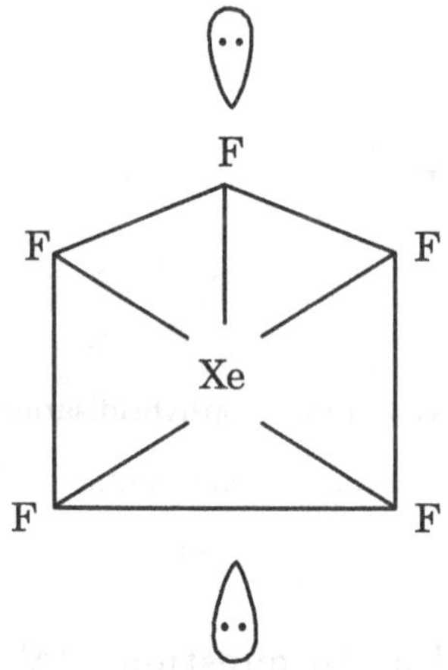

109. (b) Solid PCl_5 exists as $[PCl_4]^+[PCl_6]^-$. In $[PCl_4]^+$, P is sp³ hybridised.

In ethers like 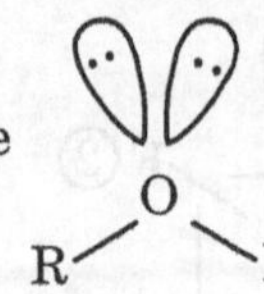 the bond angle is > 190°28'

But in molecule like 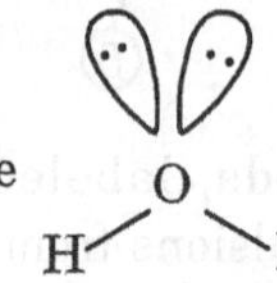 the bond angle is < 109°28'.

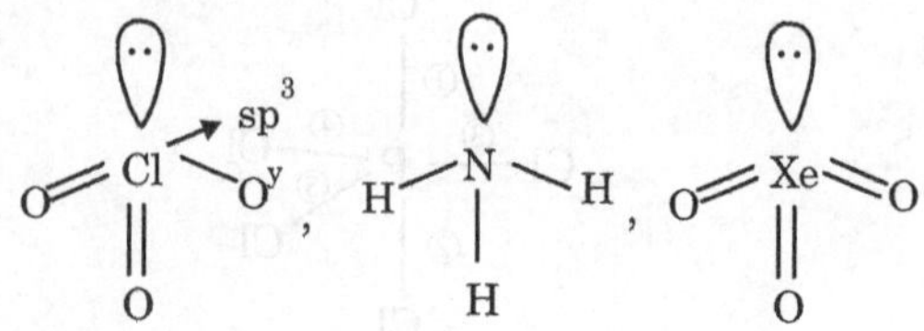

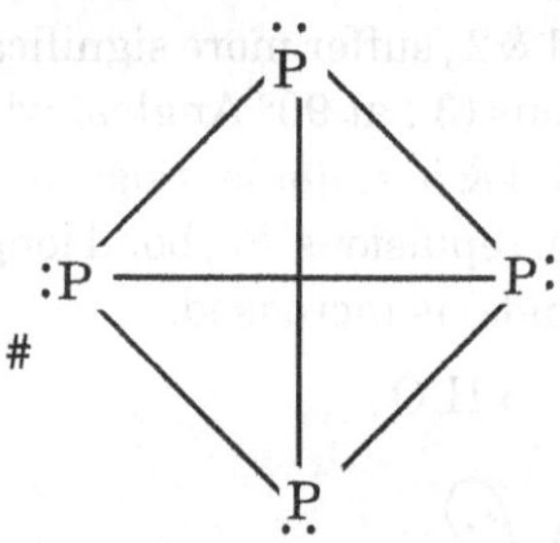

110. (d) 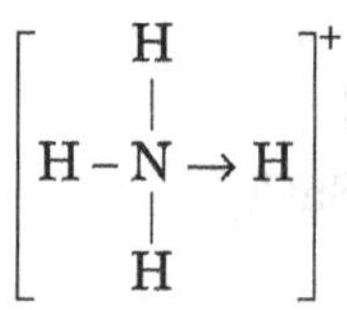 Cl^- , $K^+\ \overline{C} \equiv \ddot{N}$, $Na^+\ \overline{O}-H$

112. (d) # 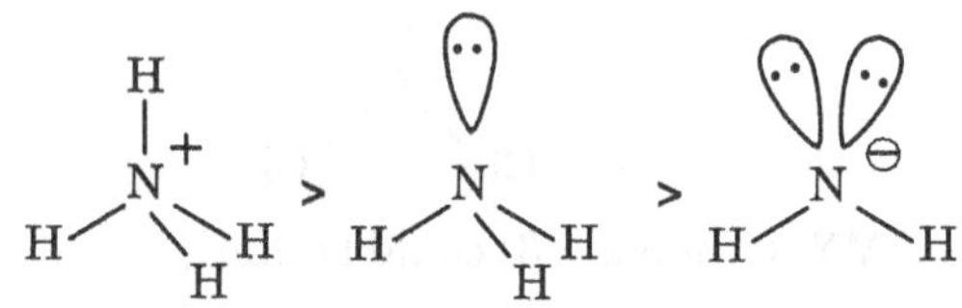

In $NH_4^+Br^-$, N is sp² hybridised

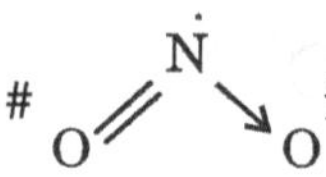 is paramagnetic (it has unpaire⁻)

113. (d)

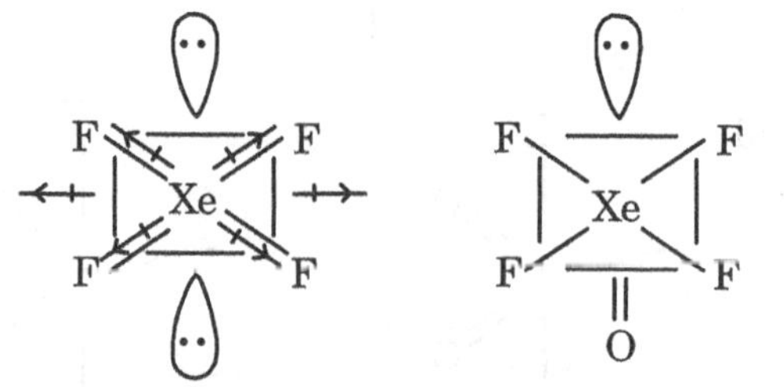

= 0 ≠ 0

114. (a) $IO_2F_2^-$

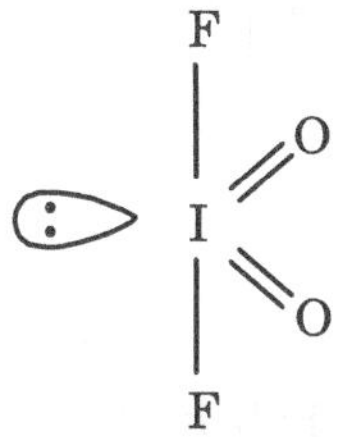

sp^3d

F_2SeO

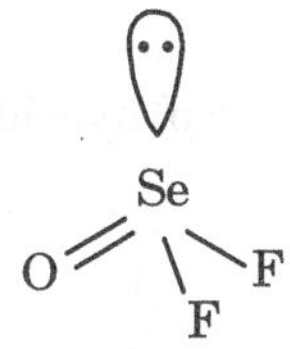

sp^3

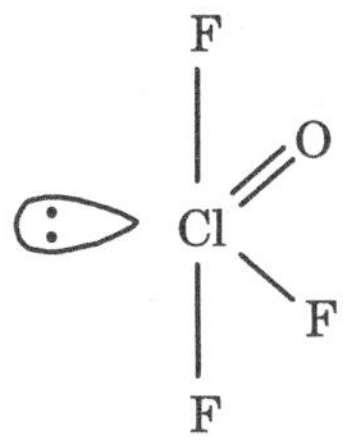

sp^3d

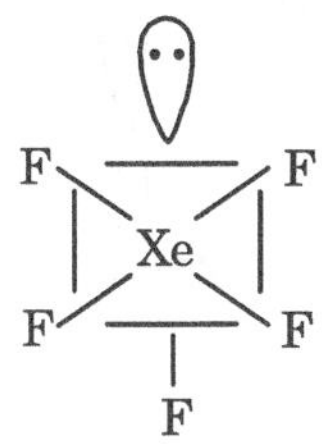

sp^3d^2

115. (c)

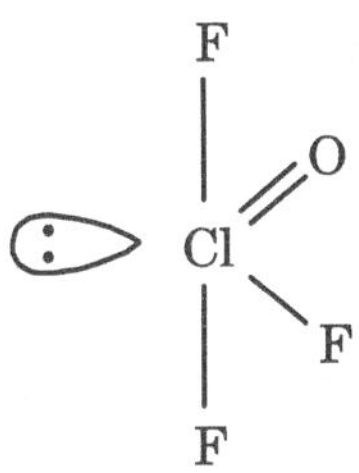

($ClOF_3$)

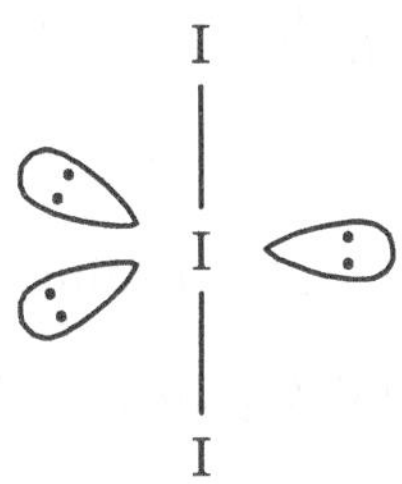

(I_3^-)

116. (a) # Sn has valence configuration $5s^2$ $5p^2$.

$5s^2$ $5p^2$

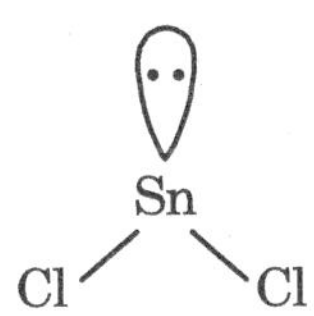

XeF_7^+

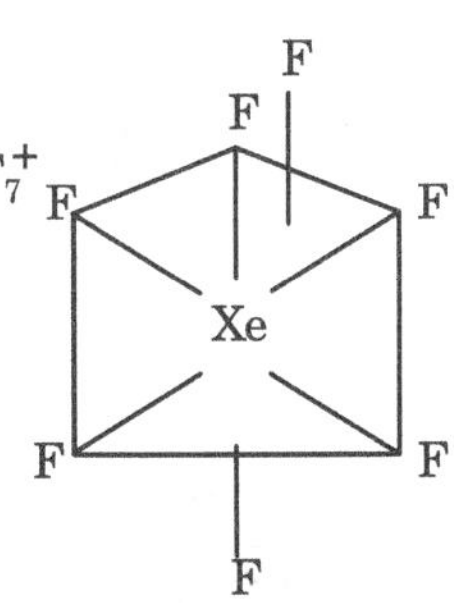

SF_4

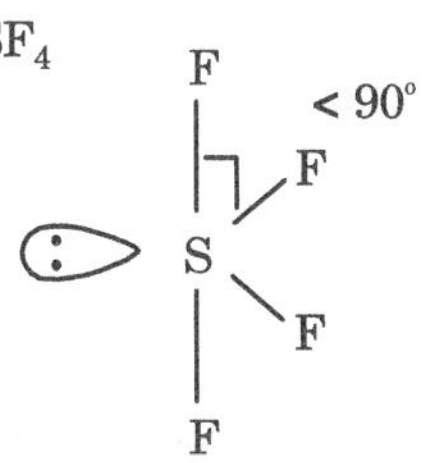

#

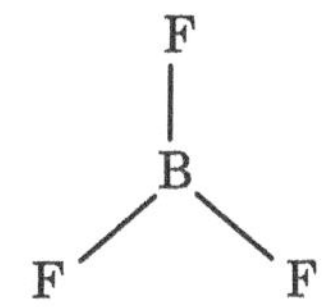

B (5) :- $1s^2$ $2s^2$ $2p^1$

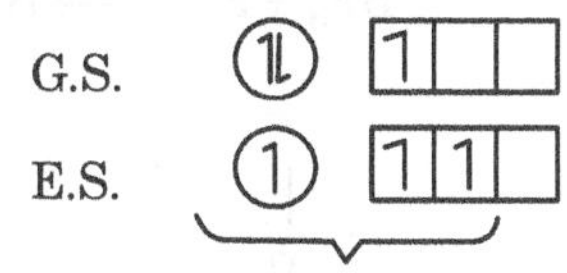

B can form 3 covalent bonds

117. (d) #

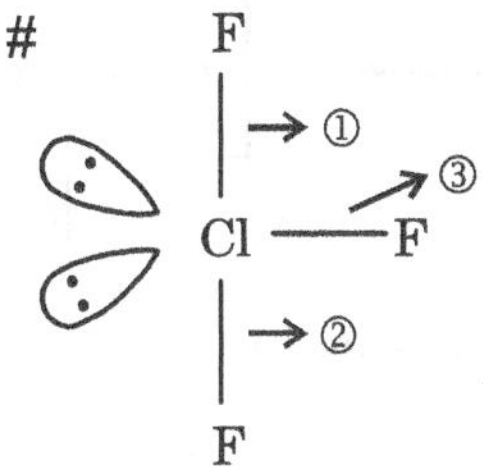

Bond lengths
1 & 2 > 3

#

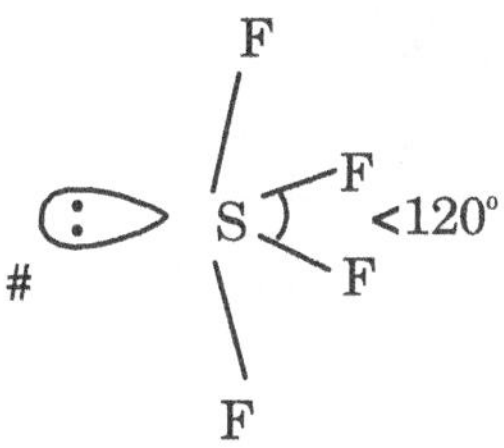

#

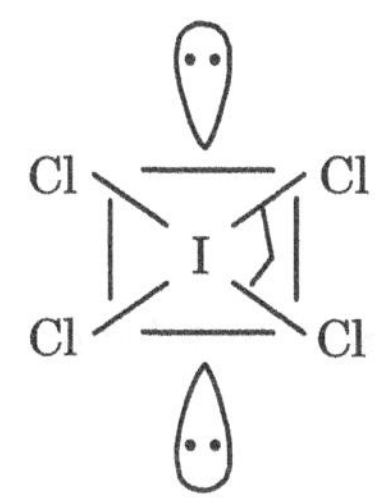

#

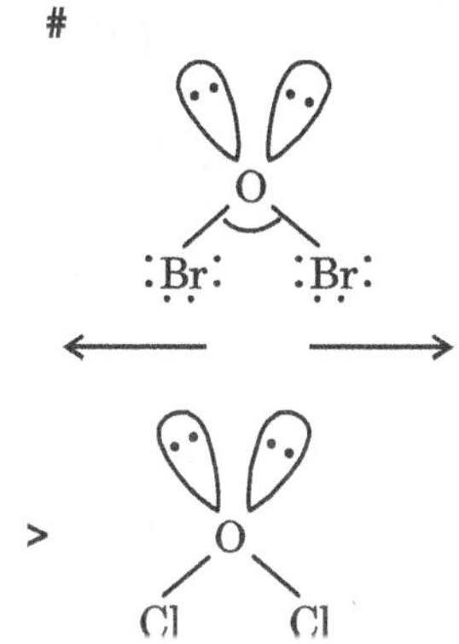

Bond angle (due to bigger size of Br, l.p – l.p. repulsions will ↑ se the bond angle).

118. (d)

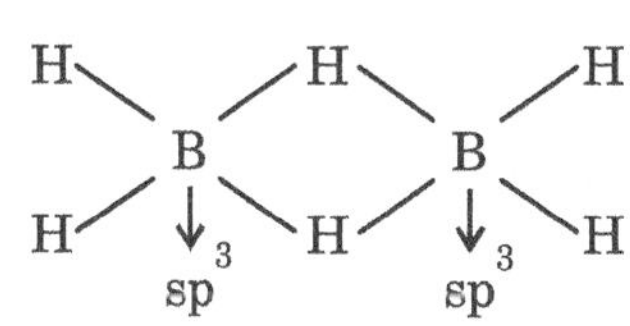

B (5) :- $1s^2$ $2s^2$ $2p^1$

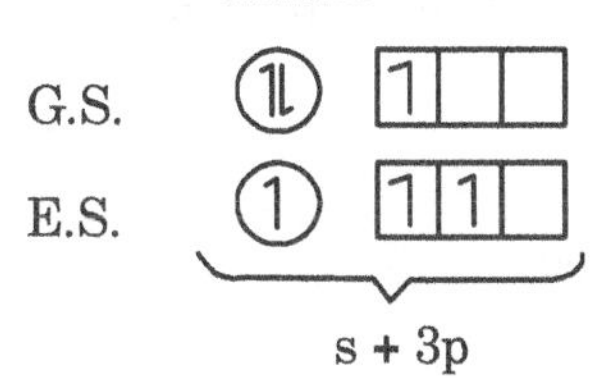

BF_4^-

$BF_3 + F^- \longrightarrow BF_4^-$

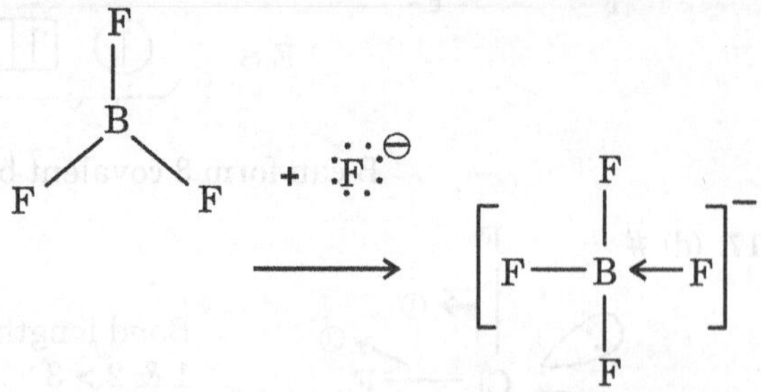

In BF_4^-, B is sp³ hybridised

B (5) :- $1s^2\ 2s^2\ 2p^1$

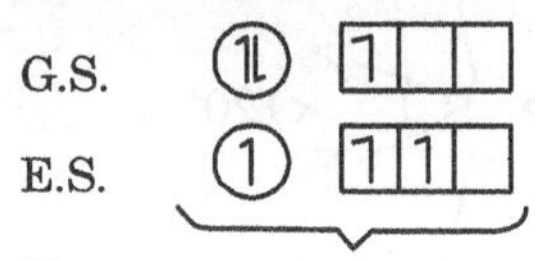

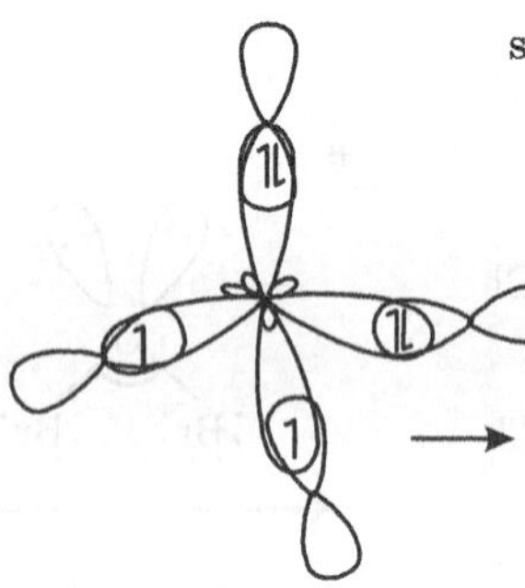

fully filled p-orbital of F

119. (c)

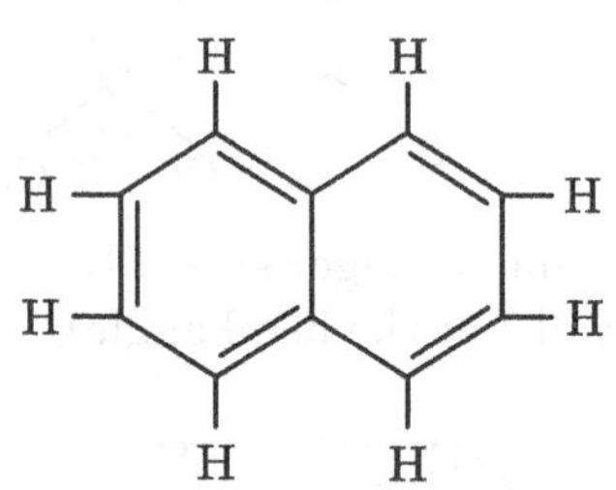

π bonds = 5
σ bonds = 19

120. (b) In pyridine, N is Sp² hybridized

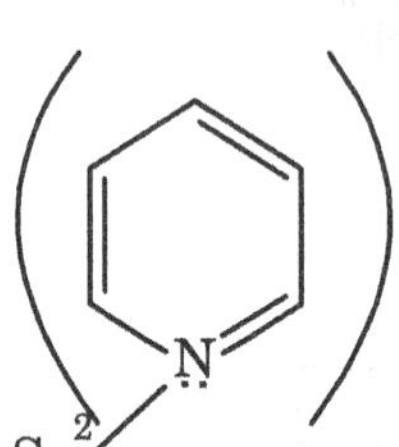

hybridized

in cyanogen

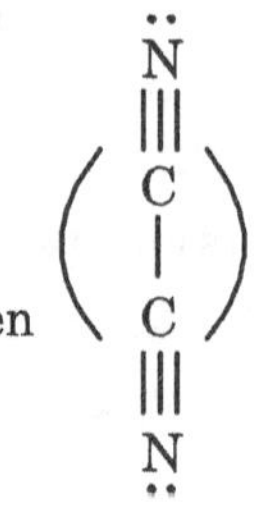

N is Sp

In $CaCN_2$, $Ca^{2+}\,CN_2^{2-}\left[\overset{\ominus}{:}N=C=\overset{\ominus}{N}:\right]$, N is Sp² hybridized

In

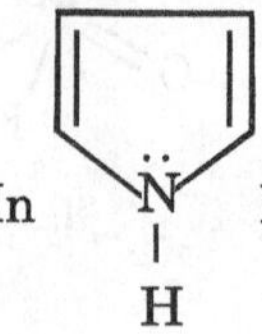

N is Sp² hybrized.

121. (*)

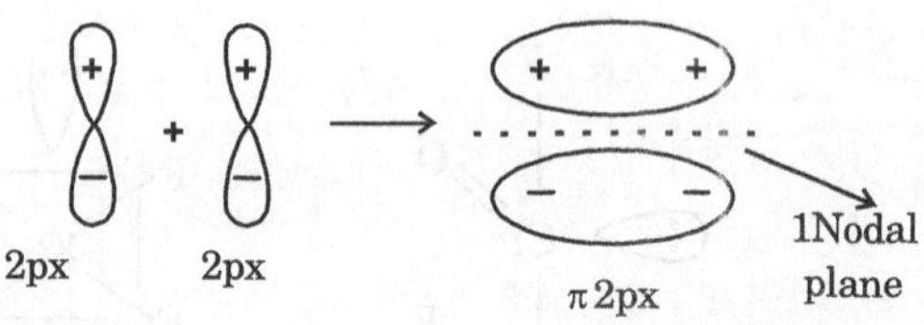

122. (b)

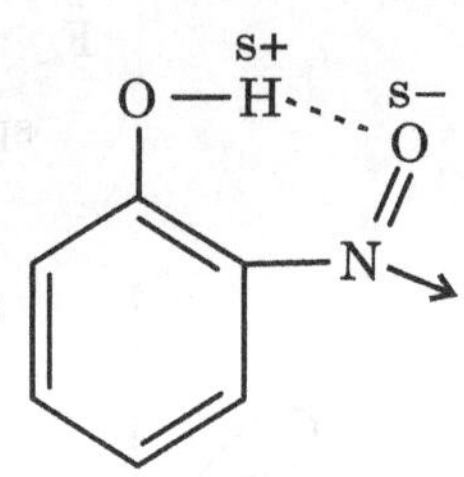

, it has intra-molecular H – bonding.

123. (a) $O_2(16): -\ \sigma 1s^2\, \sigma^* 1s^2\, \sigma 2s^2\, \sigma^* 2s^2\ \pi 2p^2x$

$= \pi 2p^2y\, \pi 2p^2z\ \pi^* 2p^1x = \pi^* 2p^1y$

B.o. $\frac{N_b - N_a}{2} = \frac{10-6}{2} = \frac{4}{2} 2$

$O_2^+(15): -\ \sigma 1s^2\, \sigma^* 1s^2\, \sigma 2s^2\, \sigma^* 2s^2\ \pi 2p^2x$

$= \pi 2p^2y\, \pi 2p^2z\, \pi^* 2p^1x = \pi^* 2py$

B.o. $\frac{N_b - N_a}{2} = \frac{10-5}{2} = \frac{5}{2} = 2.5$

124. (d)

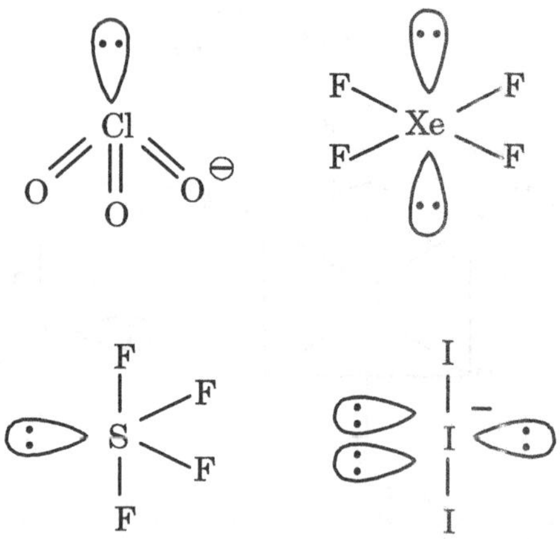

125. (d) ClF_4^+

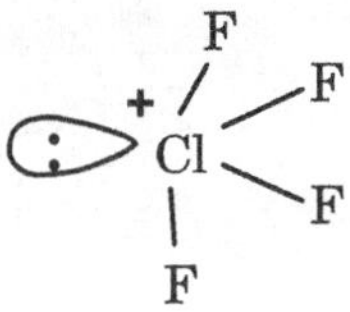

126. (a)

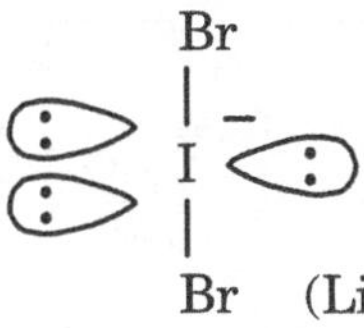

(Linear)

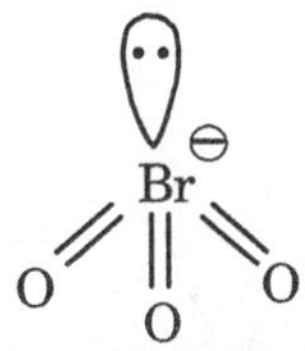

(Pyramidal)

127. (d)

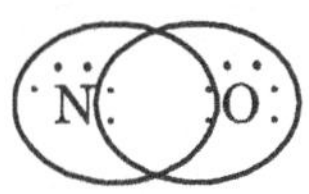

$\dot{N} = \ddot{O}:$,

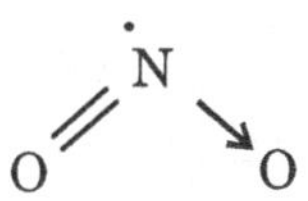

O_2 is paramagnetic based on MOT.

128. (b) Percentage ionic character $= \dfrac{M_{observed}}{M_{ionic}} \times 100$

Percentage ionic character

$$= \frac{1.2D}{4.8 \times 10^{-10}\,\text{esu} \times 100 \times 10^{-10}\,\text{cm}} \times 100$$

$$= \frac{1.2D}{4.8 \times 100 \times 10^{-2} \times 10^{-18}\,\text{esu cm}} \times 100$$

$(\because 1D = 10^{-18}\,\text{esu cm}) \qquad = \dfrac{1.2D}{4.8D} \times 100$

$$= \frac{12}{48} \times 100 = 25\%$$

(48 reduced to 4)

129. (b) $CH_3 - C \equiv C - CH_3$

130. (a)

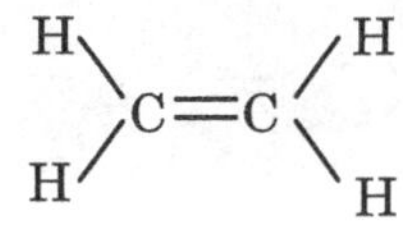

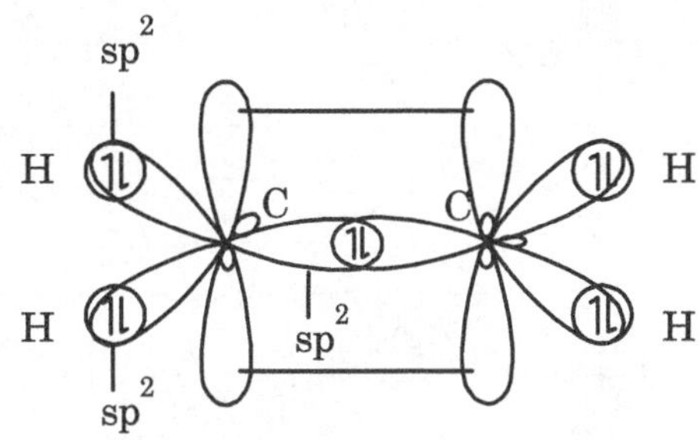

Molecular plane contains 'No π electron cloud'

131. (c) O_2^- has unpaired e^-.

O_2^- (17) :– $\sigma 1s^2 \sigma^* 1s^2 \sigma 2s^2 \sigma^* 2s^2\ \sigma 2p^2 z \pi 2p^2 x$

$= \pi 2p^2 y \pi^* 2p^2 x = \pi^* 2p^1 y$

132. (a) Dipole moment follows the order –

$CH_3Cl > CH_2Cl_2 > CHCl_3 > CCl_4$

Smart Notes

5

CHAPTER

States of Matter

Questions based on Gas Laws & Ideal Gas equation

1. For the reaction, $N_2O_{5(g)} \rightleftharpoons 2NO_{2(g)} + \frac{1}{2}O_{2(g)}$, calculate the mole fraction of $N_2O_{5(g)}$ at a constant volume & temperature; if the initial P is 600 mm Hg & the P at any time is 960 mm Hg. (assume ideal gas behavior)

(a) 0.5 (b) 0.975

(c) 0.375 (d) 0.175

2. An evacuated glass vessel weighs 50g when empty, 148g when filled with a liquid of density 0.98 g/mol& 50.5g when filled with an ideal gas at 760 mm Hg at 300 K. The molar mass of gas is

(a) 123 g (b) 246 g

(c) 75 g (d) 97 g

3. A vessel has N_2 gas & H_2O vapour at a total P of 1 atm. The partial Pressure of H_2O vapour is 0.3 atm. When the contents of this vessel are transferred to other vessel having volume equal to $\frac{1}{3}$rd Volume of original container, the total P of the system in the new vessel is (Keep T Constant)

(a) 2.4 atm (b) 1 atm

(c) 3.33atm (d) 0.093 atm

4. Two glass bulbs A and B are connected by a very small tube having a stop cork. Bulb 'A' has a volume of 100 ml & contained the gas while Bulb 'B' was empty & has a volume of 150 ml. On opening the stop cork, the P of the gas in Bulb 'A' will fall down to

(a) 80% (b) 60%

(c) 40% (d) 20%

5. A gas in an open container is heated form 27°C to 127°C. The fraction of original amount of gas remaining in the container will be

(a) $\frac{3}{4}$ (b) $\frac{1}{2}$

(c) $\frac{1}{4}$ (d) $\frac{1}{8}$

6. Figure shows two vessels A & B with rigid walls containing gases. The pressure, temperature & the volume are P_A, T_A & V in the vessel A & P_B, T_B & V in vessel B. The vessels are now connected through a small tube. $\frac{P}{T}$ value is equal to –

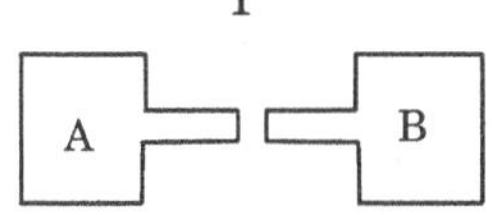

(a) $\frac{P}{T} = \frac{1}{2}\left(\frac{P_A}{T_A} + \frac{P_B}{T_B}\right)$ (b) $\frac{P}{T} = \frac{1}{2}\left(\frac{P_A}{T_B} + \frac{P_B}{T_A}\right)$

(c) $\frac{P}{T} = \left(\frac{P_A}{T_B} + \frac{P_B}{T_A}\right)$ (d) $\frac{P}{T} = \left(\frac{P_A}{T_A} + \frac{P_B}{T_B}\right)$

7. A volume V of a gas at a temperature T_1 & at pressure P is enclosed in a sphere. It is connected to another sphere of volume $\frac{V}{2}$ by a tube & stop cork. The second sphere is initially evacuated & the stop cork is closed. If the stop cork is opened the temperature of gas in second sphere becomes T_2. The first sphere is maintained at 'T_1', what is the final P within the apparatus

(a) $\frac{2PT_2}{2T_2+T_1}$ (b) $\frac{2PT_2}{T_2+2T_1}$

(c) $\frac{2PT_2}{2T_2+2T_1}$ (d) $\frac{2PT_2}{T_1+T_2}$

8. A mixture of propane & methane exerts a pressure of 320 mm Hg at a temperature T kelvin in a V litre flask. On complete combustion, gaseous mixture contains CO_2 only & exerts a pressure of 448 mm Hg under identical conditions. Hence, mole fraction of propane in the mixture is -

(a) 0.2 (b) 0.8

(c) 0.25 (d) 0.75

9. A spherical balloon of volume 3 litres is filled with H_2 at 1 atm, 300 K. A cylinder contains H_2 gas at 10 atm & 400 K. The volume of cylinder is 4 litres. The number of balloons required to fill the gas is/are-

(a) 11 (b) 12

(c) 13 (d) 9

10. If density of a gas A is twice that of other gas B at the same temperature then their pressures P_A & P_B & molecular weights M_A & M_B should have been related as–

(a) $M_A = 2M_B$ when $P_A = P_B$

(b) $M_A = 2M_B$ when $P_A = 2P_B$

(c) $2M_A = 2M_B$ when $P_A = P_B$

(d) None of these

11. A closed vessel contains He & O_3 at a pressure of P atm. The ratio of He & O atoms is 1:1. If He is removed from the vessel, the pressure of the system will reduce to -

(a) 0.5 P (b) 0.75 P

(c) 0.25 P (d) 0.33P

12. Consider the reaction $2A_{(g)} + 3B_{(g)} \longrightarrow C_{(g)}$ where gases A & B are insoluble & inert to water & C forms a basic solution. In an experiment, 3 moles of each A & B are allowed to react in 15 L flask at 500 K. When the reaction is complete, 5L of H_2O is added to the flask & temperature is reduced to 300 K. The pressure in the flask is (neglect aqueous tension)

(a) 1.64 atm (b) 2.46 atm

(c) 4.92 atm (d) 3.28 atm

13. Gas A (1 mole) dissociates in a closed rigid container of volume 0.16 litres as per following reaction–

$$2A_{(g)} \longrightarrow 3B_{(g)} + 2C_{(g)}$$

If degree of dissociation of A is 0.4 (40%) & remains constant in entire range of temperature, then the correct Pv/sT graph is– (R = 0.082 litreatm/mol/k)

(a)

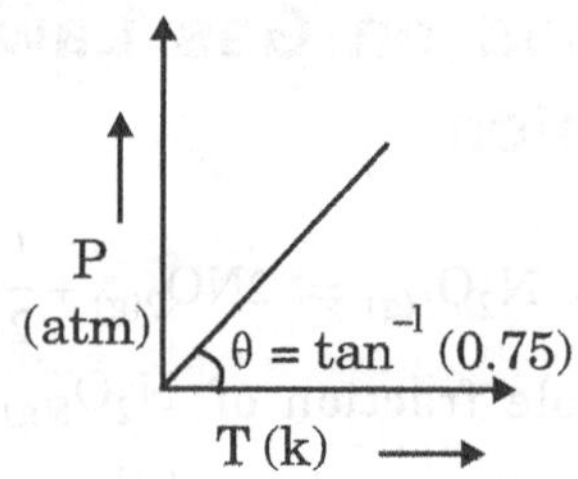

(b)

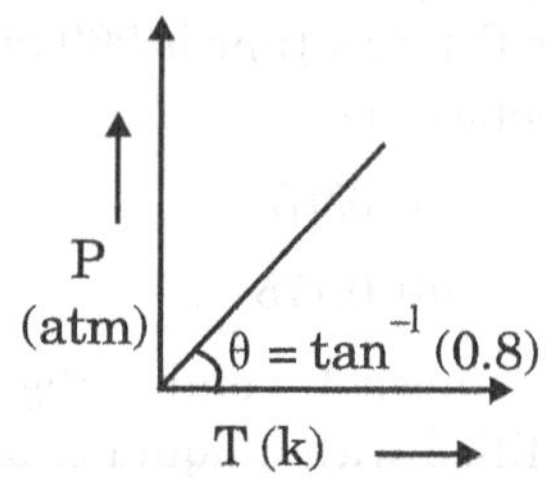

(c)

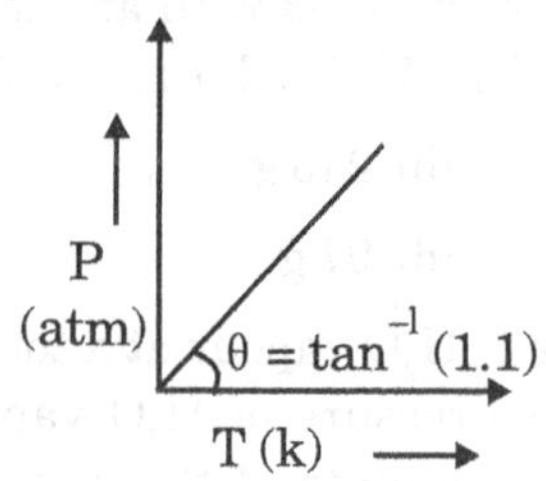

(d)

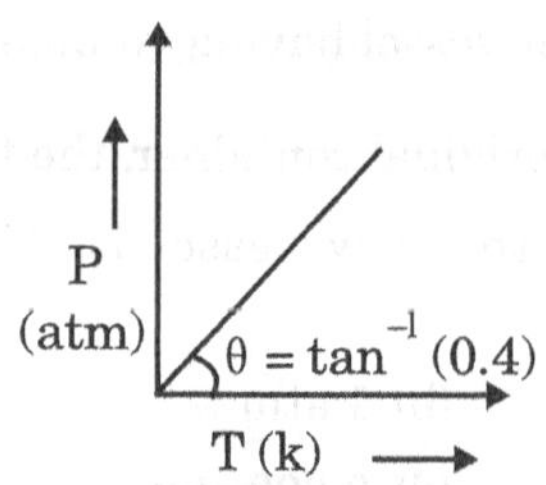

14. Select the correct observation for a 8.21 litre container, filled with 2 moles of the He at 300 K.

(a) It has a pressure 6 atm

(b) If it is closed non-rigid (like skin of a balloon); its volume increases to 16.42 litres on heating to 600K

(c) Both 1 & 2 are correct

(d) None of these are correct

15. At constant temperatures of 273 K, $\frac{1}{V}$ v/s P are plotted for 2 ideal gases A & B as shown below. Find out the no. of moles of A & B.

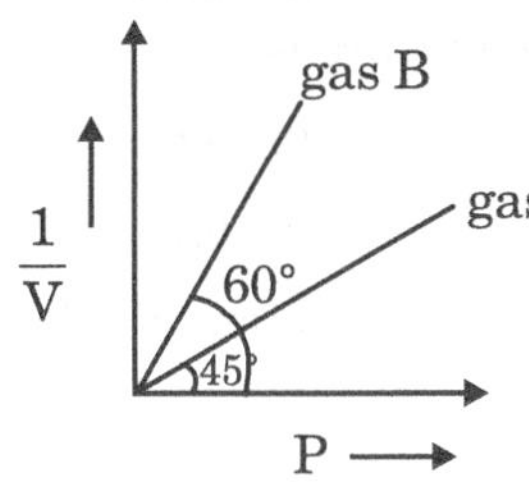

(a) $n_A = \frac{1}{22 \cdot 4}, n_B = \frac{1}{22 \cdot 4\sqrt{3}}$

(b) $n_A = \frac{1}{11 \cdot 2}, n_B = \frac{1}{5 \cdot 6}$

(c) $n_A = \frac{1}{22 \cdot 4}, n_B = \frac{1}{22 \cdot 4}$

(d) $n_B = \frac{1}{11 \cdot 2}, n_A = \frac{1}{5 \cdot 6}$

16. Consider the gaseous reaction,

$N_2 + 3H_2 \longrightarrow 2NH_3$ 1 mol N_2 and 4 mole H_2 taken in a 15L flask at 27°C. After the completion of the reaction, 5L of H_2O is added to the reaction vessel, P set up in the flask would be

(a) $\frac{3 \times 0.0821 \times 300}{15}$ atm

(b) $\frac{2 \times 0.0821 \times 300}{10}$ atm

(c) $\frac{1 \times 0.0821 \times 300}{15}$ atm

(d) $\frac{1 \times 0.0821 \times 300}{10}$ atm

17. Under what conditions will a pure sample of an ideal gas not only exhibit a pressure of 1 atm but also a concentration of 1 mole litre^{-1}. (R = 0.082 ltratm mol^{-1} degree^{-1})

(a) At STP

(b) When V = 22.4 litre

(c) When P = 12.18 K

(d) Impossible under any condition

18. What % of a sample of nitrogen must be allowed to escape so that its temperature, pressure and volume are to be changed as given –

273°C, 3atm, 1.65L	→	0°C, 0.75atm, 0.55L

(a) 16.67% (b) 75%

(c) 25% (d) 83.3%

19. A vessel of 5L capacity contains 1.4 g nitrogen at 1800 K. Assuming that at this temperature, 40% of molecules are dissociated into atoms & the gas is ideal, what is the gas pressure?

(a) 2.07 atm (b) 1.467 atm

(c) 1.05 atm (d) 2.67 atm

20. A quantity of 4g of oxygen occupies 10 L at a particular pressure & temperature. If the pressure of gas is doubled & absolute temperature is halved, in order to maintain constant volume

(a) 3g gas should be removed

(b) 3g gas should be added

(c) 16g gas should be added

(d) 12g gas should be added

21. A spherical air bubble is rising from the depth of a lake when pressure is P atm& temperature is TK. The % age increase in its radius when it comes to the surface of a lake will be (Assume temperature & pressure at surface to be 2T kelvin & $\frac{P}{4}$ respectively)

(a) 100% (b) 50%

(c) 40% (d) 20%

22. A chemist has synthesized a greenish yellow gaseous compound of chlorine & oxygen & finds that its density is $7.71\frac{g}{L}$ at 36°C & 2.88 atm. The molecular formula of the compound is

(a) ClO_3 (b) ClO_2

(c) ClO (d) Cl_2O_2

23. A 2.24L cylinder of oxygen at NTP is found to develop a leakage. When the leakage was plugged, the pressure dropped to 570 mm to Hg. The no. of moles of gas that escaped will be

(a) 0.025 (b) 0.050

(c) 0.075 (d) 0.09

24. A closed container of Vol. = 30 L contains a mixture of N_2 & O_2 gases at a temperature of 27°C & pressure of 4 atm. The total mass of the mixture is 148 g. The moles of individual gases in the container are (Take R = 0.08 it atm mole)

(a) $n_{N_2} = 2$ mole, $n_{O_2} = 3$ mole

(b) $n_{N_2} = 3$ mole, $n_{O_2} = 2$ mole

(c) $n_{N_2} = 4$ mole, $n_{O_2} = 1$ mole

(d) $n_{N_2} = 2 \cdot 5$ mole, $n_{O_2} = 2 \cdot 5$ moles

25. The pressure of an ideal gas contained in a closed container increases by 0.4% when it is heated by 1°C. Then, the final temperature of gas is

(a) 250°C (b) –22°C

(c) 250 k (d) 22°C

26. The slope of the graph between 'PV' & 'P' at constant temperature for an ideal gas is

(a) zero (b) 1

(c) $\frac{1}{2}$ (d) 0.77

27. At 0°C, the density of gaseous oxide ZO at 2 bar is same as that of dinitrogen at 5 bar. The molecular mass of ZO is

(a) 60 (b) 11.2

(c) 70 (d) 156

28. A mixture of 2 mole of $O_{2(g)}$ & 1 mole of $N_{2(g)}$ is maintained in a rigid container at constant room temperature. When additional $N_{2(g)}$ is put into the container at constant temperature, the pressure doubles. How many moles of N_2 are in the final mixture?

(a) 2 (b) 3

(c) 4 (d) 5

29. A gas mixture contains twice as many moles of O_2 as N_2. Addition of 0.2 moles of argon to this mixture increases the pressure from 0.8 to 1.1atm. The moles of O_2 in the mixture are -

(a) 0.679 moles (b) 0.35 mole

(c) 0.89 moles (d) 3.2 moles

30. The pressure in a container of some volume at some temperature is found to be P atm when it contained 1 mole of NH_3 gas. What would be the final pressure if 2 moles of H_2, 2 moles of HCl gas & 1 mole of O_2 is introduced into the container under same condition. (Assume reaction b/w NH_3 &HCl only)

(a) 6 P (b) 3P

(c) 4P (d) P

31. A mixture of C_3H_8 & CH_4 exerts a pressure of 320 mm Hg at temperature T kelvin in a V litre flask. On complete combustion , gaseous mixture contains CO_2 only & exerts a pressure of 448 mm Hg under identical conditions. So, mole fraction of C_3H_8 in the mixture is -

(a) 0.2 (b) 0.8

(c) 0.25 (d) 0.75

32. Select the correct graph among the following –

(a)

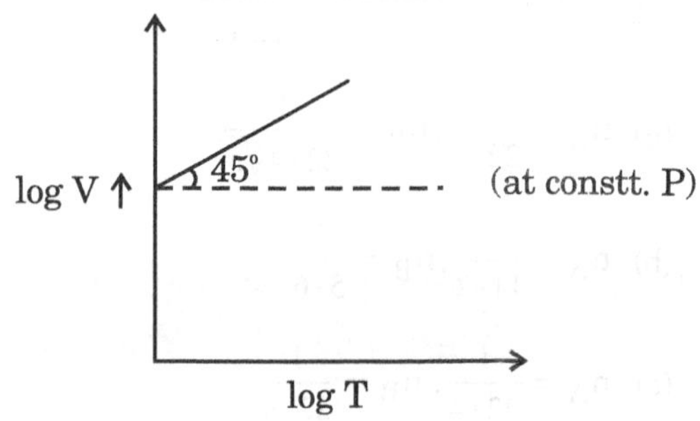

(b)

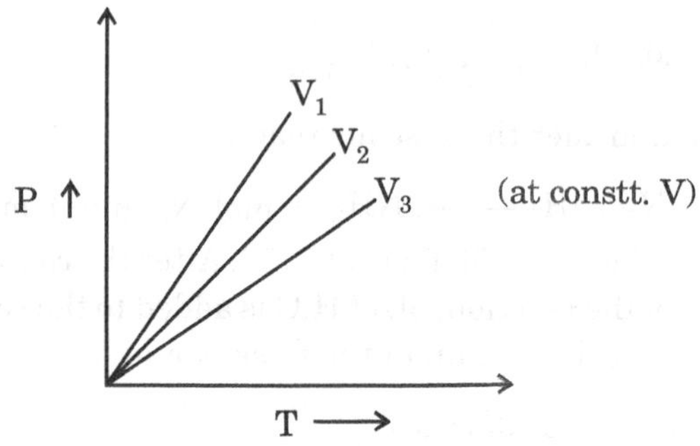

(c)

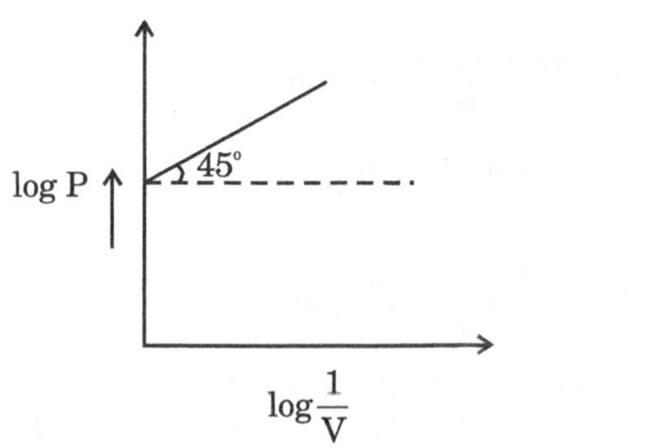

(d) All are correct

33. Three different gases X, Y & Z of molecular masses 4, 18, & 72 were enclosed in a vessel at constant temperature till equilibrium is reached. Which of the following statement is true?

(a) Gas Z will be at top of vessel

(b) Gas X will be at top of vessel

(c) Gas Y will be at top of vessel

(d) Gases will form homogenous mixture

34. Which of the following curves is/are correct according to Charle's law?

(a)

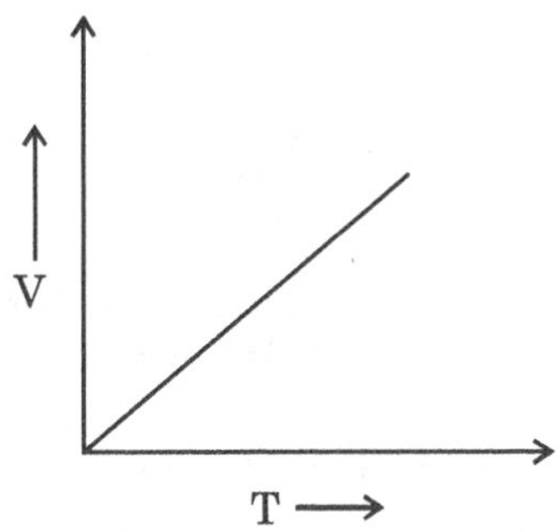

(b)

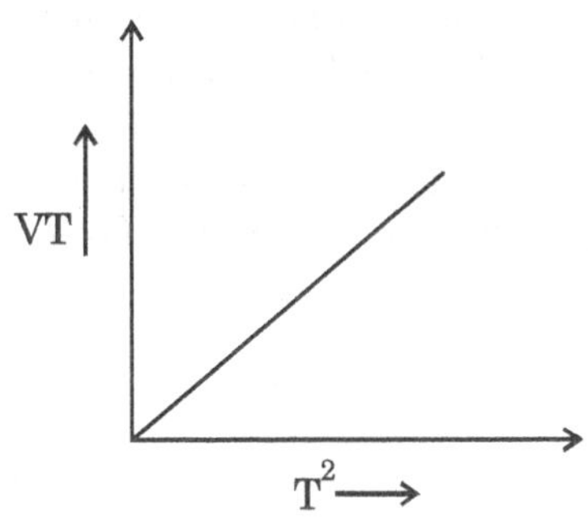

(c)

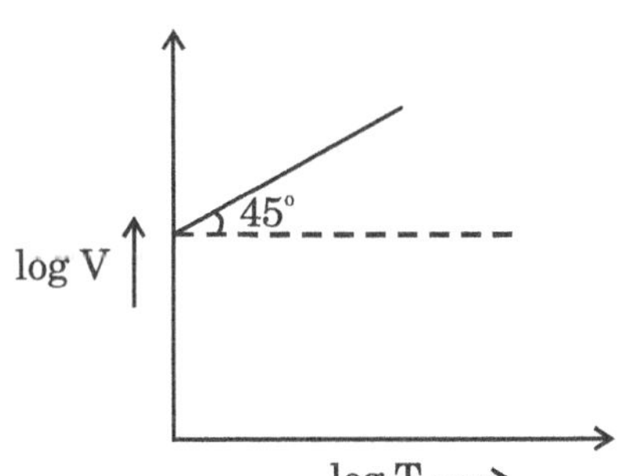

(d) All are correct

35. Which of these graphs is/are correct.

(a)

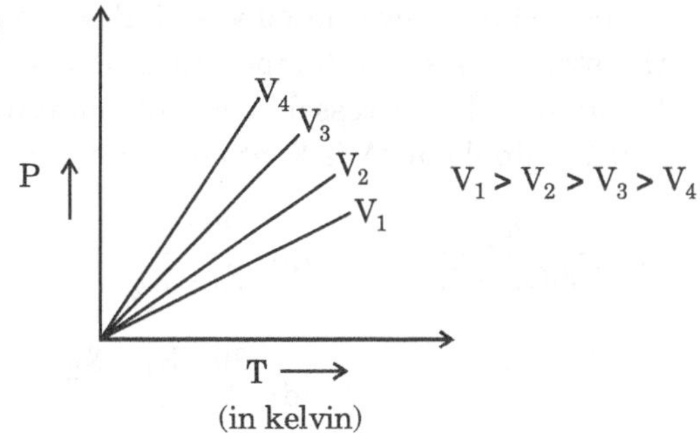

(b)

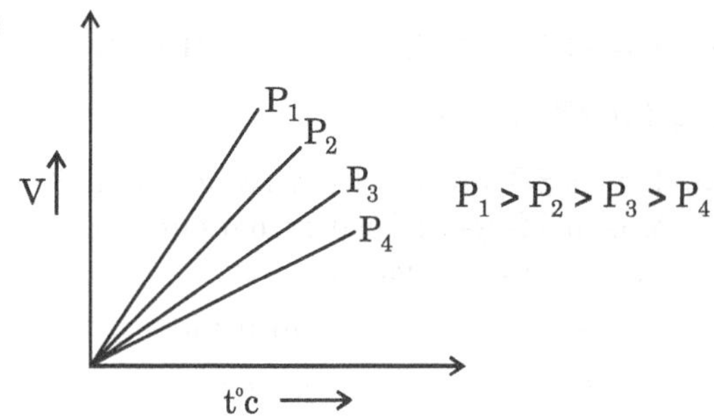

(c)

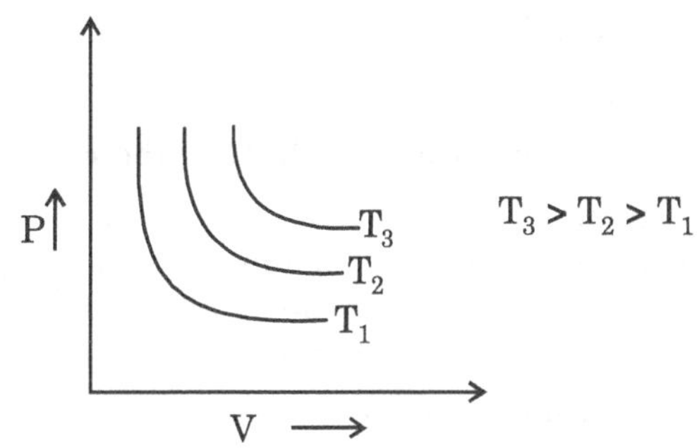

(d) Both (a) & (c) correct

36. A gas cylinder contains 370g oxygen at 30 atm pressure & 25°C. What mass of oxygen will escape if the cylinder is first heated to 75°C & then the valve is held open until pressure become 1 atm. Temperature remain constant at 75°C?

(a) 359.4g (b) 321.26g

(c) 317.37g (d) 281g

37.

2L Argon (gas) 1 atm	2L HCl (gas) 1 atm	2L NH_3 (gas) 1 atm

In the above experiment, when the partitions are removed (keeping T constt.), Then

(a) Final P is 1 atm

(b) No. of moles of gas inside are 6/RT finally.

(c) Partial pressure of NH_3 is $\frac{1}{3}$ atm.

(d) Partial pressure of argon is $\frac{1}{3}$ atm

38. A vessel 'A' containing gas at 300K & X_1 atm is connected to an evacuated vessel 'B'. On opening the stop – cork, the temperature in vessel B is found to be T_2 but vessel 'A' is still maintained at 300K. If final P in 'A' is X_2 atmosphere, then T_2 is–

(a) $\frac{X_2}{300(X_1 - X_2)}$ (b) $\frac{300X_2}{X_1 - X_2}$

(c) $\frac{X_1 - X_2}{300X_2}$ (d) $\frac{300(X_1 - X_2)}{X_2}$

Questions based on Dalton's Laws & Graham's Law

39. 4 gm of SO_2 diffuses from a container in 8 minutes. Mass of He gas diffusing from same container over same time interval is

(a) 1g (b) 0.4 g

(c) 16 gm (d) 64 gm

40. Rate of diffusion of a gas depends upon

(a) Its molar mass (b) Pressure

(c) Temperature (d) All of these

41. A bottle of dry ammonia & a bottle of dry HCl connected through a long tube are opened simultaneously at both ends. The white ammonium chloride formed will be first appearing

(a) at centre of tube

(b) near the HCl bottle

(c) Near the NH_3 bottle

(d) throughout the length of tube.

42. The reaction between gaseous NH_3& HBr produces a white solid NH_4Br. A small quantity of gaseous NH_3 & gaseous HBr at some pressure & temperature are introduced simultaneously through holes in the opposite ends of an open tube which is one metre long. Calculate the distance of white solid formed from the end which was used to introduce NH_3.

Given : $\sqrt{\frac{18}{17}} = 2.18$

(a) 155.55 cm (b) 32.26 cm

(c) 47 cm (d) 68.55 cm

43. At same temperature & pressure the rate of diffusion of H_2 gas is $3\sqrt{3}$ times that of an alkyne. What is molecular formula of the alkyne?

(a) C_4H_{10} (b) C_4H_6

(c) C_3H_2 (d) C_3H_4

44. 80 ml of O_2 takes 2 minutes to pass through the hole. What volume of SO_2 will pass through the hole in 3 minutes?

(a) $\frac{120}{\sqrt{2}}$ (b) $\frac{140}{\sqrt{2}}$

(c) 65 (d) None of these

45. A 3 : 2 molar mixture of N_2 & CO is present in a vessel at 500 bar Pressure. Due to a hole in the vessel, the gas mixture leaks out. The composition of mixture effusing out initially is

(a) $\frac{n_{NO_2}}{n_{CO}} = \frac{1}{2}$ (b) $\frac{n_{N_2}}{n_{CO}} = \frac{6}{1}$

(c) $\frac{n_{CO}}{n_{N_2}} = \frac{1}{2}$ (d) $\frac{n_{CO}}{n_{N_2}} = \frac{2}{3}$

46. Two identical porous containess are filled, one with H_2 gas & other with He gas at the same temperature & pressure. After one day 2L of He gas had leaked from the container. The volume of H_2 gas that had leaked from its container is

(a) 2L (b) 2.82 L

(c) 4 L (d) 0.712

47. One mole of N_2 gas at 0.8 atm takes 38s to diffuse through a pin hole, whereas one mole of an unknown compound of xenon with fluorine at 1.6 atm takes 57 seconds to diffuse through the same hole. Then, the molecular formula of compound is

(a) XeF_2 (b) XeF_4

(c) XeF_6 (d) XeF_5^+

48. Equal weights of CH_4 & O_2 are mixed in an empty container at 25°C. The fraction of total pressure exerted by O_2 is

(a) $\frac{1}{3}$ (b) $\frac{1}{2}$

(c) $\frac{2}{3}$ (d) $\frac{1}{3} \times \frac{273}{298}$

49. Equal moles of CO, B_2H_6, H_2 & CH_4 are placed in a container. If a hole was made in container, after 5 minutes the partial P of gases in container would be (at. Wt. of C, O, B and H are 12, 16, 11, 1 respectively)

(a) $P_{CO} = P_{B_2H_2} > P_{CH_2} > P_{H_2}$

(b) $P_{CO} = P_{B_2H_6} > P_{CH_4} > P_{H_2}$

(c) $P_{CO} = P_{B_2H_6} = P_{CH_4} > P_{H_2}$

(d) None of these

50. At STP, a container has 1 mole He, 2 mole CH_4, 3 moleO_2 & 4 moleN_2. Without changing total pressure if 2 mole O_2 is removed, the partial pressure of O_2 will decrease by

(a) 26% (b) 40%

(c) 58.33 % (d) 66.66%

51. A vessel is filled with a mixture of O_2 & N_2. At what ratio of partial pressures will the mass of gases be identical?

(a) $P(O_2) = 0.785\ P(N_2)$

(b) $P(O_2) = 8.75\ P(N_2)$

(c) $P(O_2) = 11.4\ P(N_2)$

(d) $P(O_2) = 0.875\ P(N_2)$

52. A certain quantity of a gas occupies a volume of 0.1L when collected over water at 288K and 0.92 bar pressure. The same gas occupied a volume of 0.085L at STP in dry condition. Calculate the aqueous tension at 288K.

(a) 30 bar (b) 0.6 bar

(c) 0.24 bar (d) 0.0031 bar

53. Dalton's law (at normal conditions) of partial pressure is applicable to

(a) H_2 & O_2 Mixture (b) H_2 & Cl_2 Mixture

(c) H_2 & N_2 Mixture (d) H_2 & CO_2 Mixture

54. A mixture of He & CH_4 at 1.4 bar pressure contains 20% by mole of He. The partial pressure of He will be

(a) 0.7 bar (b) 0.28 bar

(c) 0.56 bar (d) 0.8 bar

55. 2L of moist H_2 gas were collected over water at 26°C at total P of one atm and the quantity of H_2 collected was 0.0788 mole. What is the mole fraction of H_2 in the moist gas?

(a) 0.989 (b) 0.897

(c) 0.967 (d) None of these

56. A gaseous mixture of 3 gases A, B & C has a pressure of 10 atm. The total number of moles of all the gases is 10. If the partial pressures of A & B are 3 & 1 atm respectively & If C has moleculas wt. of 2, what is the weight of C present in the mixture?

(a) 6 (b) 8

(c) 12 (d) 3

57. Carbon monoxide (CO) & oxygen (O_2) react according to the equation

$$2CO_{(g)} + O_{2(g)} \longrightarrow 2CO_{2(g)}$$

Assuming that the reaction goes to completion, identify the limiting reagent & the partial pressures of gases left after the valve is opened. Assume temp. to be constant throughout = 300 K.

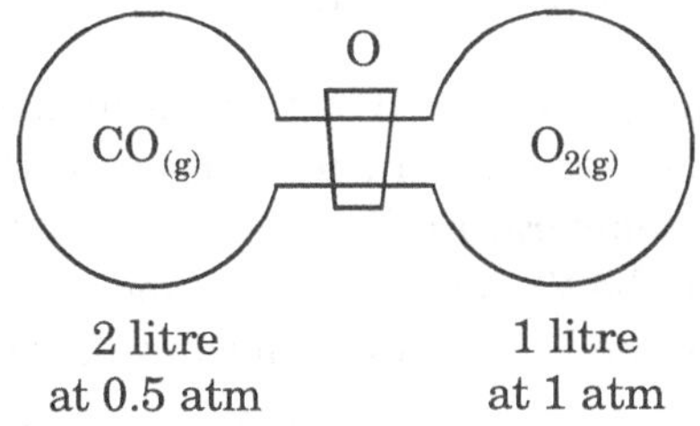

(a) O_2 is the limiting reagent & $P_{CO} = 0\cdot8$ atm

(b) CO is the limiting reagent & $P_{CO} = P_{O_2} = 0\cdot4$ atm

(c) CO is the limiting reagent & $P_{CO} = 0\cdot33$ atm $P_{O_2} = 0\cdot17$ atm

(d) O_2 is the limiting reagent & $P_{CO} = P_{O_2} = 0\cdot8$ atm

58. Calculate the molecular weight of a gas X which diffuses four times as fast as another gas Y, which in turn diffuses twice as fast as another gas Z. (Given : - molecular wt. of gas Z is 128)

(a) 6 (b) 12

(c) 2 (d) 4

59. A mixture of 50 ml of H_2 & 50 ml of O_2 is allowed to effuse through a small hole till the residual gas occupies 90 ml. Volume of the residual H_2 gas is -

(a) 46 ml (b) 44 ml

(c) 48 ml (d) 42 ml

60. A mixture of & CO_2 diffuses $\frac{1}{4}$ times as rapidly as H_2. The molar ratio of to CO_2 in the mixture is

(a) 1 : 1 (b) 2 : 1

(c) 3 : 1 (d) 3 : 2

Questions based on Kinetic theory of Gases & Maxwell distribution of velocities

61. The RMS velocity of H_2 is $\sqrt{7}$ times the RMS velocity of N_2. If T is the temperature of the gas, then

(a) $T_{H_2} = T_{N_2}$ (b) $T_{H_2} > T_{N_2}$

(c) $T_{N_2} > T_{H_2}$ (d) $T_{H_2} = \sqrt{7}T_{N_2}$

62. For two gases A & B with molecular weights M_A & M_B respectively, it is observed that at a certain temperature T, the mean velocity of A is equal to the V_{rms} of B. Thus, the mean velocity of A can be made equal to the mean velocity of B, if-

(a) A is a temperature T & B is at T' such that T' > T

(b) Heat energy is supplied to B

(c) Both A & B are raised to a higher temperature

(d) Both a & b are correct

63. If R.M.S. speed of gaseous molecules is x cm/sec at a Pressure of P atm then RMS at a pressure of 2P atm and constant temperature will be

(a) x

(b) 2n

(c) 4n

(d) $\frac{n}{4}$

64. At what temperature will the total K.E. of 0.30 moles of He be same as total K.E. of 0.40 moles of Ar at 400 K?

(a) 400 K

(b) 300 K

(c) 533 K

(d) 573 K

65. If temperature changes from 27°C to 127°C, the relative % age change in RMS velocity is

(a) 15.6

(b) 3.86

(c) 17.72

(d) 82.6

66. Choose the mismatch w.r.t. the postulate (of KTG) and the property of gases.

Sr. No.	Postulate	Property
(a)	There is no force of attrac-tion b/w the particles of a gas at ordinary temperature and pressure.	Gases expand and occupy all the space avai-lable to them.
(b)	Particles of a gas are always in constant (rapid) and random motion	Gases do not have fixed shape.
(c)	Collisions of gas molecules are perfectly elastic which means that total energy before and after collisions remains the same.	Gases do not settle down under the influence of gravity.
(d)	None of these.	

67. Following graph, also called the Maxwell - Boltzmann distribution curve is given

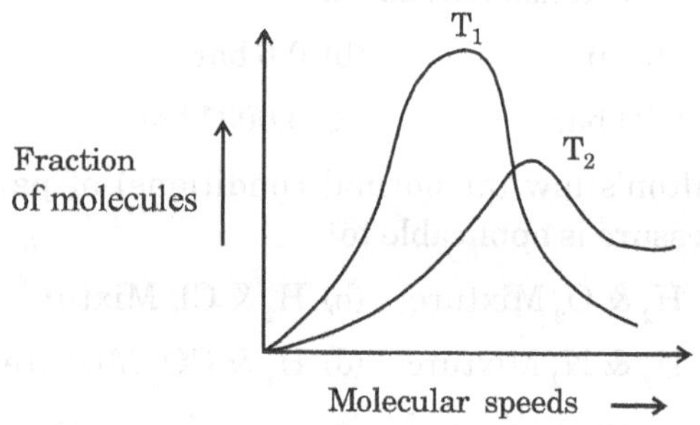

Then, which statement is/are correct wrt above graph

(a) $T_2 > T_1$

(b) The fraction of molecules with very low or very high speeds is small.

(c) At higher temperature, the most probable speed increases but the fraction of molecule having most probable speed decreases.

(d) All are correct

68. The correct formula for average kinetic energy is

(a) $KE_{Av.} = \frac{3}{2}nRT$

(b) $KE_{Av.} = \frac{3}{2}RT$

(c) $K.E._{Average} = \frac{3}{2}K_bT$ (where k_b is boltzman constant)

(d) None of these

69. Which of the following is correct ?

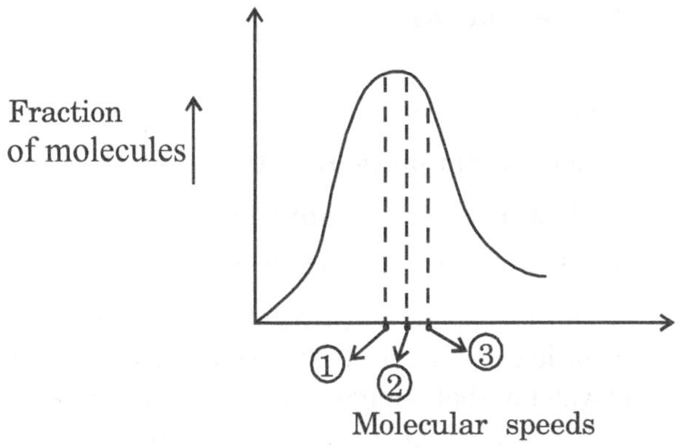

(a) 1 = Most probable velocity, 2 = average velocity, 3 = RMS velocity

(b) 1 = RMS velocity, 2 = Most probable velocity, 3 = average velocity

(c) 1 = Most probable velocity, 2 = average velocity, 3 = Mean velocity

(d) None of these

70. The RMS speed of N_2 molecules in a gas is u. If the temperature is doubled and the nitrogen molecules dissociates into nitrogen atom, RMS speed becomes

(a) $\frac{u}{2}$

(b) 2u

(c) 4u

(d) 14u

71. If the V_{RMS} is $30R^{\frac{1}{2}}$ at 27°C, then calculate the molar mass of gas in Kgs.

(a) 1

(b) 2

(c) 4

(d) 0.001

72. If only one gas molecule is taken in a container, the correct relation between average speed (v), root means square (u) & most probable speed (α) is

(a) $u > v > \alpha$

(b) $\alpha = v = u$

(c) $u > \alpha > v$

(d) None of these

73. In the following graph of Maxwell Boltzmann's distribution of molecular velocities

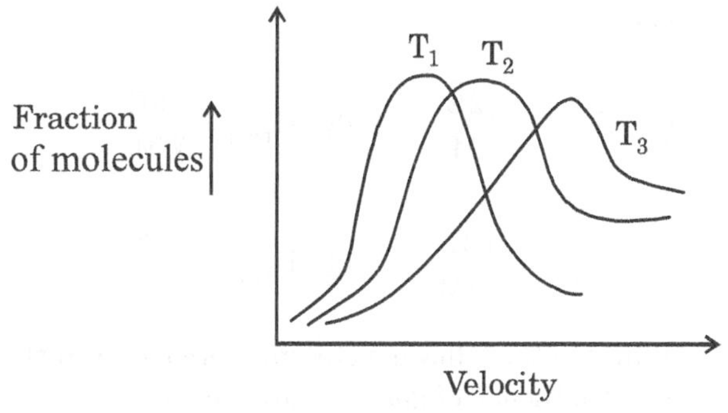

(a) $T_3 > T_2 > T_1$

(b) $T_3 > T_1 > T_2$

(c) $T_3 = T_2 = T_1$

(d) None of these

74. Indicate the correct statement for equal volume of N_2 (g) & CO_2 (g) at 298 k & 1 atm.

(a) The average K.E. per molecule is same for N_2 & CO_2

(b) The density of N_2 is less than that of CO_2

(c) The total KE of both N_2 & CO_2 is same

(d) All are correct.

75. $\left(\frac{1}{N}\cdot\frac{dN}{dV}\right)$

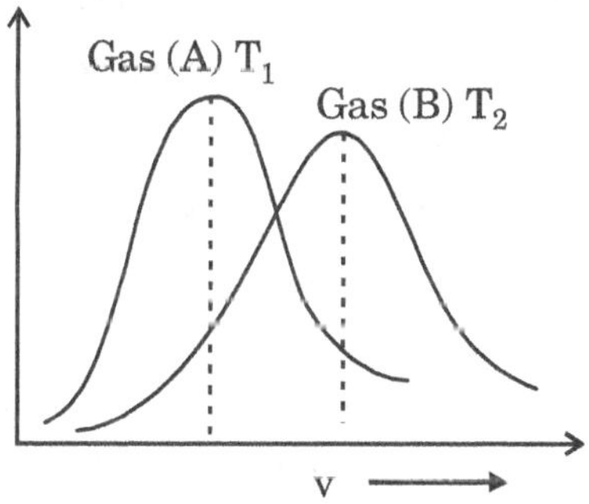

For the above graph, drawn for two different samples of gases at two different temperature, T_1 & T_2. Which of the following statements is necessarily true?

(a) If $T_2 > T_1$, M_B is necessarily greater than M_A

(b) If $T_1 > T_2$, M_A is necessarily greater than M_B

(c) $\frac{T_2}{M_B} > \frac{T_1}{M_A}$

(d) Both (b) & (c) are correct.

76. For a monoatomic gas, kinetic energy is E. The relation of root mean square velocity in terms of kinetic energy is (here M is the molar mass of Gas)

(a) $U_{rms} = \sqrt{\frac{2E}{M}}$ (b) $U_{rms} = \sqrt{\frac{3E}{M}}$

(c) $U_{rms} = \sqrt{\frac{E}{2M}}$ (d) $U_{rms} = \sqrt{\frac{E}{3M}}$

77. Which of the following statements concerning the kinetic theory of gases is /are correct?

I. Molecules make elastic collisions with each other & with the walls of their container.

II. The average kinetic energy of a large number of molecules of mass M, is proportional to $\sqrt{M}$ at a given temperature.

III. The molecules of a gas are in constant random motion.

IV. All the molecules of a gas have the same kinetic energy at a given temperature.

(a) I, II, III, IV (b) I, II, III

(c) I, III, IV (d) I, III

78. A balloon filled with a gas shrinks when cooled. When it is done shrinking, which of the following is the same as it was originally?

(a) RMS velocity of gas particles

(b) Density of gas

(c) Pressure in the balloon

(d) K. E. of gas particles

79. 2 identical flasks are at the same temperature. One is filled with 2g of H_2, the other will 28g of N_2. Which property would be different for two samples?

(a) Pressure

(b) Average kinetic energy

(c) Density

(d) The number of molecules in each container

80. If T_1, T_2 & T_3 are temperatures at which the $U_{rms}, U_{average}, U_{mp}$ of a gas are all equal to 1200 m/s then the correct statement is–

(a) $T_1 > T_2 > T_3$ (b) $T_3 > T_2 > T_1$

(c) $T_1 = T_2 > T_3$ (d) $T_3 > T_2 = T_1$

81. 1 mole of A_2 at 300 K is kept in a closed vessel at 1 atm pressure. It is heated 600 K, when 20% of A_2 dissociates as

$$A_{2(g)} \longrightarrow 2A_{(g)}$$

(The resultant pressure in the vessel is

(a) 12 atm (b) 24 atm

(c) 16 atm (d) 2 atm

82. A sample of air saturated with H_2O vapour recorded a pressure of 640 torr. If vapour pressure of water at that temperature is x toss & the sample contains N_2 & O_2 in the molar ratio of 3:1, the partial pressure of N_2 is found to be 450 torr.. Then x is –

(a) 54 torr (b) 45 torr

(c) 49 torr (d) 40 torr

83. Which of the following gas has a density of 18 g/L at 1 atm pressure & 27°C?

(a) O_2

(b) CO_2

(c) NH_3

(d) SO_2

84. What is true regarding the following graphs plotted for a gas 'A'?

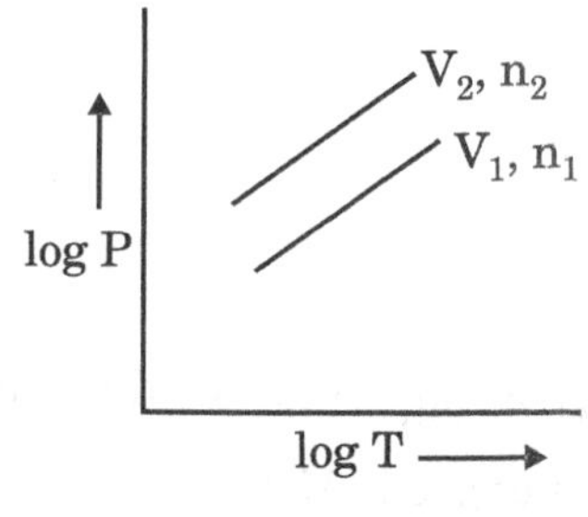

(a) If $n_1 = n_2$ then $V_2 > V_1$

(b) The slope of each graph depends on the given mass of gas A

(c) The slope of graph is constant i.e. +1

(d) Both 1 & 2 are correct

85. Average kinetic energy per molecule of an ideal gas at 300K is $6 \cdot 21 \times 10^{-y}$ J. Then Y is

(a) 10

(b) 20

(c) 21

(d) 30

86. If constants in Boyles law, Gay-Lussac's law & Avogadro's law are *x, y* & *z* respectively then units of (*x.y.z*) are –

(a) L atm $mol^{-1} k^{-1}$

(b) atm^2 mol k^{-1}

(c) L atm

(d) L^2 atm^2 $mol^{-1} k^{-1}$

Questions based on Reals Gases & Compressibility factor Z

87. A gas obeys P (V – b) = RT. Which of the following are correct about this gas?

(I) Isochoric curves have slope = $\frac{R}{V-b}$

(II) Isobaric curves have slope = $\frac{R}{P}$ & intercept b.

(III) For the gas, Z (compressibility factor) = $1 + \frac{Pb}{RT}$

(IV) Repulsive forces > Attractive forces

(a) Only (I) is correct

(b) Only (II)&(III) are correct

(c) Only (III)&(IV) are correct

(d) All are correct

88. Select the correct reaction-

(a) $P_{ideal} = P_{Real} + \frac{an^2}{V^2}$

(b) $V_{ideal} = V_{real} - nb$

(c) Z > 1 (for H_2 & He)

(d) All are correct

89. Which of the following statements correctly differentiate ideal & real gases?

(a) Ideal gases are hypothetical but real gases exist practically

(b) Ideal gases cannot be liquefied but real gases can be liquefied at low temperature & high pressure

(c) Ideal gases do not possess intermolecular forces of attraction

(d) All the above can differentiate

90. For one mole of vanderwaals gas when b = 0 and T = 300 K, the PV v/s $\frac{1}{V}$ plot is shown below.

The value of van der waals constant 'a' is

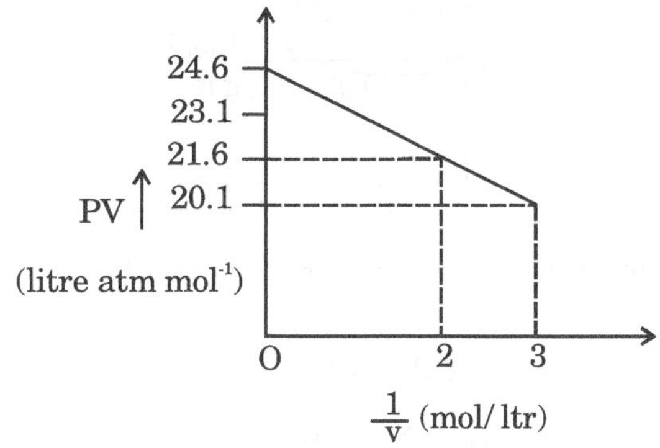

(a) 1 (b) 4.5

(c) 1.5 (d) 3.0

91. Select the correct relation among the following.

(a) $Z = \frac{V_{real}}{V_{ideal}}$

(b) For positive deviation, V_M (molar volume) > 22.4 litres

(c) For negative deviation, V_M < 22.4 litres

(d) All are correct

92. The compressibility factor for N_2 at 223 K and 800 atm is 1.95 and at 373 K and 200 atm is 1.10. A certain mass of N_2 occupies a volume of 1 dm^3 at 223 K and 800 atm. Calculate the volume occupied by the same quantity of N_2 at 373 K and 200 atm.

(a) 10 L (b) 6 L

(c) 4 L (d) 3.77 Ltr.

93. At low pressure vanderwaals equation for 3 moles of a real gas will have which of the following forms

(a) $\frac{PV}{RT + Pb} = 3$ (b) $\frac{PV}{RT - \frac{3a}{V}} = 3$

(c) $\frac{PV}{RT + 3Pb} = 1$ (d) $\frac{PV}{RT - \frac{a}{V}} = 3$

94. If 'b' is excluded volume then available for 'N' molecules in 'V' litre container is –

(a) (V – b) (b) $\left(V - \frac{N_0}{N} b\right)$

(c) $\left(V - \frac{N}{N_0} b\right)$ (d) None of these

95. Consider the graph between Z (compressibility factor) & pressure P.

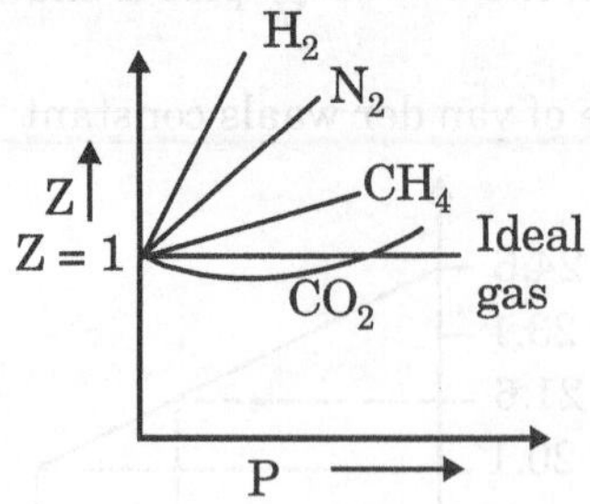

The correct order of ease of liquefaction of the gases is

(a) $H_2 < N_2 < CH_4 < CO_2$

(b) $CO_2 < CH_4 < N_2 < H_2$

(c) $H_2 < CH_4 < N_2 < CO_2$

(d) $CH_4 < H_2 < N_2 < CO_2$

96. Choose the conclusion which are appropriate for the observation stated.

(a) All conclusions are true

(b) conclusions I, II & IV are true

(c) conclusion III && IV are true

(d) None of these

97. A gas described by vanderwaals equation

(a) Behave similar to an ideal gas in the limit of large molar volumes

(b) Is characterized by vanderwaals coefficients that are dependent on the identity of gas but are independent of temperature

(c) Has the pressure that is lower than the pressure exerted by the same gas behaving ideally

(d) All these are correct

98. At moderate pressure, the compressibility factor for a particular gas is given by -

$$Z = 1 + 0.34P - \frac{160}{T}P$$

(Where p is in bar & T is in kelvin)

What is the Boyle's temperature of this gas?

(a) 298 K (b) 340 K

(c) 470 K (d) 920 K

99. A graph between Z & P is demonstrated as follows-

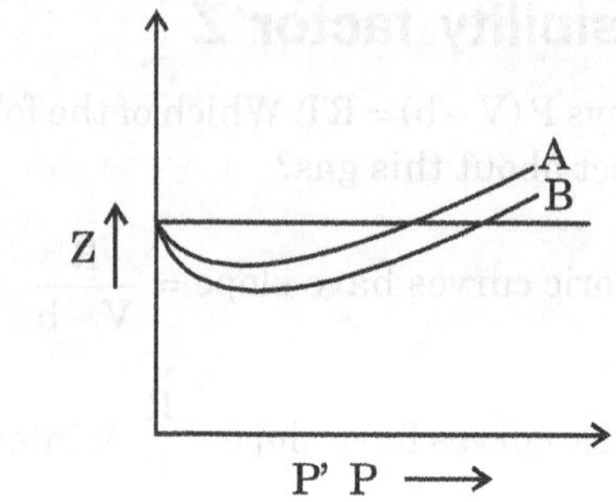

For two gases A & B, at pressure P' marked in the graph -

(a) Attractive forces are greater in gas A than in B.

(b) Attractive forces are greater in gas B than in A.

(c) Gas A can be easily liquefied.

(d) Gas A has greater value of Vander Waals constant 'a'.

100. The figure shows the effect of pressure on compressibility factor Z of a gas –

The correct statement is /are –

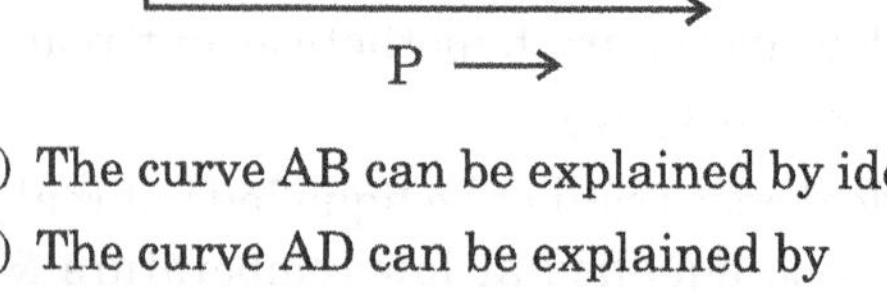

(a) The curve AB can be explained by ideal gas

(b) The curve AD can be explained by

$$PV = RT - \frac{a}{V}$$

(c) All the curves show ideal behavior

(d) Both 1 & 2 are correct

101. Z for critical state is -

(a) 1 (b) Negative

(c) $\frac{3}{8}$ (d) $\frac{6}{8}$

102. Select the correct statement (s).

(a) When 2 real gases have the same value of 'b' but different value of 'a', gas having larger value of 'a' will occupy lesser volume.

(b) When 2 real gases have the same value of 'a' but different value of 'b', gas having greater value of 'b' will occupy greater volume.

(c) b (excluded volume) = 4 × vol. occupied by 1 mole of a gas.

(d) All are correct.

103. For 1 mole of a Vander waals gas when b = 0 & T = 300K, the PV v/s 1/V plot is shown below. The value of Vander waals constant a (atm L mol- 1) is –

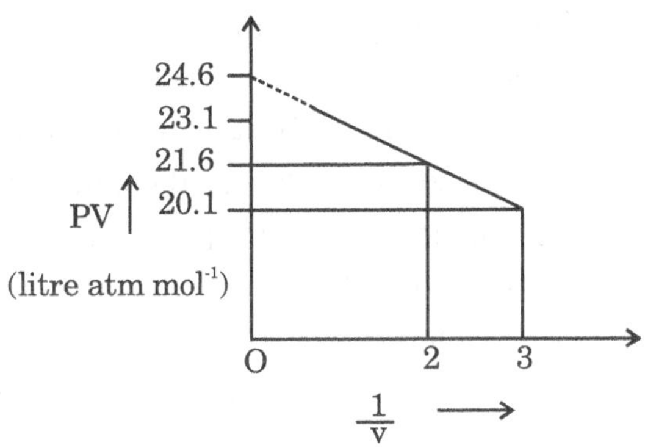

(a) 1 (b) 4.5

(c) 1.5 (d) 3

104. Select the correct graph –

(a)

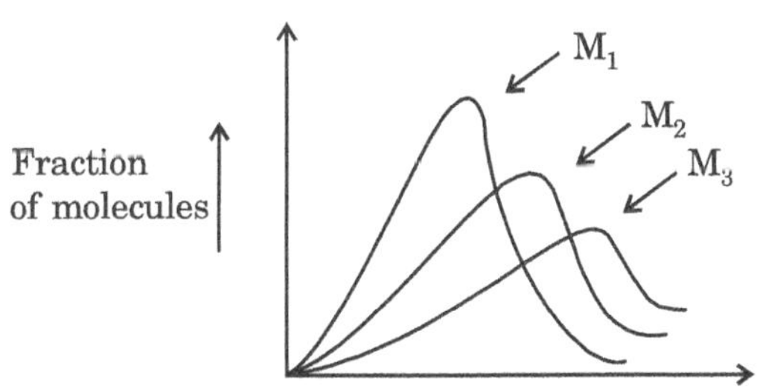

(M_1, M_2, M_3 are different molar masses at same temperature)

(b)

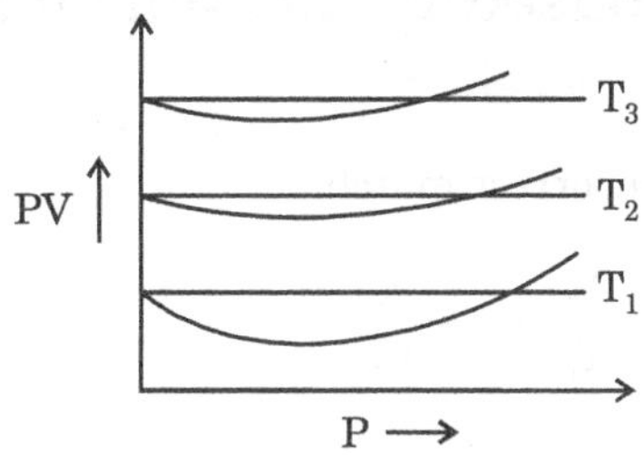

($T_3 > T_2 > T_1$)

(c)

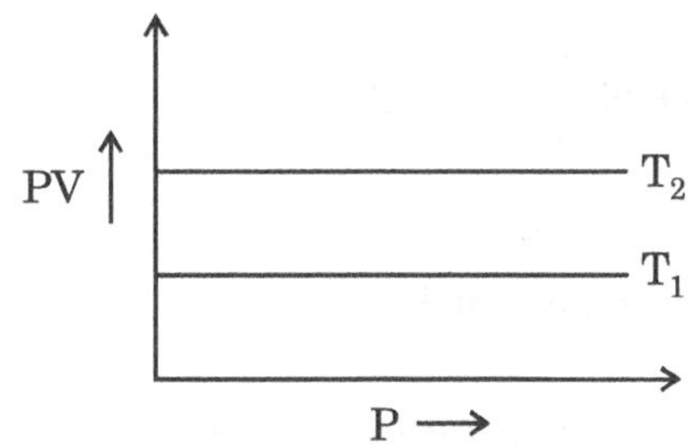

($T_2 > T_1$)

(d) All are correct

105. Compressibility factor (Z) for a certain mass of O_2 is given below -

Case	T	P	Z	V
I	330 K	800 atm	1.9	1 L
II	570 K	200 atm	1.1	V_2

The value of V_2 is –

(a) 1 L

(b) 2 L

(c) 3 L

(d) 4 L

105. The Vander waal's equation for 0.5 moles of real gas is -

(a) $\left(P+\frac{a}{4V^2}\right)\left(\frac{V-b}{2}\right)=\frac{2RT}{2}$

(b) $\left(P+\frac{a}{4V^2}\right)(2V-b)=RT$

(c) $\left(P+\frac{a}{4V^2}\right)=\frac{RT}{2(V-2b)}$

(d) $\left(P+\frac{a}{4V^2}\right)=\frac{2RT}{2(V-2b)}$

Questions based on Liquefaction of gases

106. Select the incorrect match.

(a) Value of $\frac{T_c}{T_i} \longrightarrow \frac{4}{27}$

(b) Value of $\frac{P_c V_c}{R T_c} \longrightarrow \frac{7}{8}$

(c) Value of $\frac{T_i}{T_b} \longrightarrow 2$

(d) None of these

107. Choose the mismatch

(a) Critical temperature $(T_c) \longrightarrow \frac{8a}{27bR}$

(b) Critical temperature $(P_c) \longrightarrow \frac{a}{27b^2}$

(c) Critical temperature $(v_c) \longrightarrow 3b$

(d) None of these

108. If the critical temperature of a gas is 300K, then its Boyle's temperature will be -

(a) 1012.5K (b) 1250K

(c) 1350K (d) 1400.5K

109. Consider the following statements: If the vanderwaal's parameter for 2 gases are given

	a. (atm lit^2 mol^{-2})	b. (lit mol^{-1})
Gas X :	6.5	0.056
Gas Y:	8	0.011

Then,

(a) $V_c(x) < V_c(y)$

(b) $P_c(x) < P_c(y)$

(c) $T_c(x) < T_c(y)$

Select the correct option

(a) a alone

(b) a & b

(c) a, b & c

(d) b & c

111. For a real gas, the P – V curve was experimentally plotted & it had the following appearance. For this Andrew's graph, choose the correct statement-

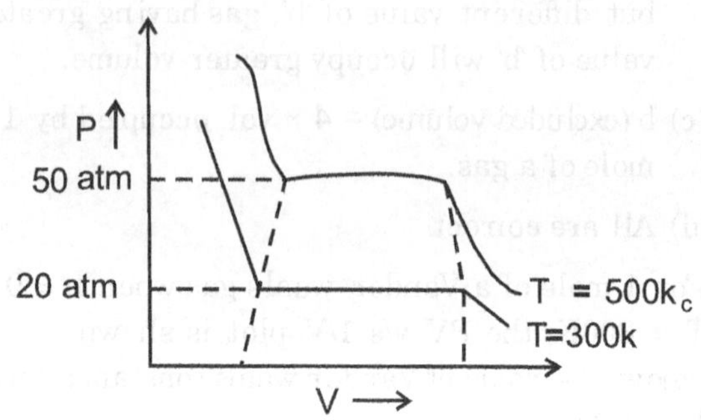

(a) At T = 500 K, P = 40 atm, the state will be liquified

(b) At T = 300 K, P = 50 atm, the state will be gas

(c) At T = 300 K, P = 20 atm, the state will be gas

(d) At 300 K < T < 500K, P > 50 atm, the state will be liquefied

112. Select the incorrect statement (s).

(a) We can condense vapour by applying pressure.

(b) To liquefy a gas one must lower the temperature below Tc & also apply pressure.

(c) At Tc, there is no distinction b/w liquid & vapour state, so density of liquid is nearly equal to density of vapour.

(d) None of these is incorrect

Answers

1. (c)	**2.** (a)	**3.** (a)	**4.** (c)	**5.** (a)	**6.** (a)	**7.** (a)	**8.** (a)	**9.** (d)	**10.** (a)
11. (c)	**12.** (b)	**13.** (b)	**14.** (c)	**15.** (a)	**16.** (d)	**17.** (c)	**18.** (d)	**19.** (a)	**20.** (d)
21. (a)	**22.** (b)	**23.** (a)	**24.** (a)	**25.** (b)	**26.** (a)	**27.** (c)	**28.** (c)	**29.** (b)	**30.** (c)
31. (a)	**32.** (d)	**33.** (d)	**34.** (d)	**35.** (d)	**36.** (a)	**37.** (d)	**38.** (b)	**39.** (a)	**40.** (d)
41. (b)	**42.** (d)	**43.** (b)	**44.** (a)	**45.** (d)	**46.** (b)	**47.** (c)	**48.** (a)	**49.** (a)	**50.** (c)
51. (d)	**52.** (c)	**53.** (b)	**54.** (b)	**55.** (c)	**56.** (c)	**57.** (c)	**58.** (c)	**59.** (d)	**60.** (c)
61. (c)	**62.** (d)	**63.** (a)	**64.** (c)	**65.** (a)	**66.** (d)	**67.** (d)	**68.** (c)	**69.** (a)	**70.** (b)
71. (d)	**72.** (b)	**73.** (a)	**74.** (d)	**75.** (d)	**76.** (a)	**77.** (d)	**78.** (c)	**79.** (c)	**80.** (b)
81. (b)	**82.** (d)	**83.** (b)	**84.** (c)	**85.** (c)	**86.** (d)	**87.** (d)	**88.** (d)	**89.** (d)	**90.** (c)
91. (d)	**92.** (d)	**93.** (b)	**94.** (c)	**95.** (a)	**96.** (c)	**97.** (d)	**98.** (c)	**99.** (b)	**100.** (d)
101. (c)	**102.** (d)	**103.** (c)	**104.** (d)	**105.** (d)	**105.** (d)	**106.** (d)	**107.** (d)	**108.** (a)	**109.** (d)
110. (b)	**111.** (d)	**112.** (d)							

Explanations

1. (c)

$N_2O_5 \rightleftharpoons 2NO_2 + \frac{1}{2}O_2$	$P_i V = n_i RT$
Att = 0 a 0 0	$\frac{600}{760} \times V = a \times RT$ ....(a)
Att = t $a - x$ $2x$ $\frac{x}{2}$	$P_f V = n_f RT$
Total gaseous moles = $a - x + 2x + 0.5x$	$\frac{960}{760} \times V = \left(a + \frac{3}{2}x\right)RT$ ....(b)

Now, $\frac{2}{5}a = x$

$\therefore$ Mole fraction of N_2O_5 at time

$$t = \frac{a - x}{a + 1.5x} = \frac{a - 0.4a}{a + 1.5 \times 0.4a} = \frac{0.6a}{a + 0.6a}$$

Mole fraction of $N_2O_5 = \frac{0.6\cancel{a}}{1.6\cancel{a}} = \frac{6}{16} = 0.375$

2. (a) Weight of evacuated vessel = 50g

Weight of vessel (with liquid) = 148 g

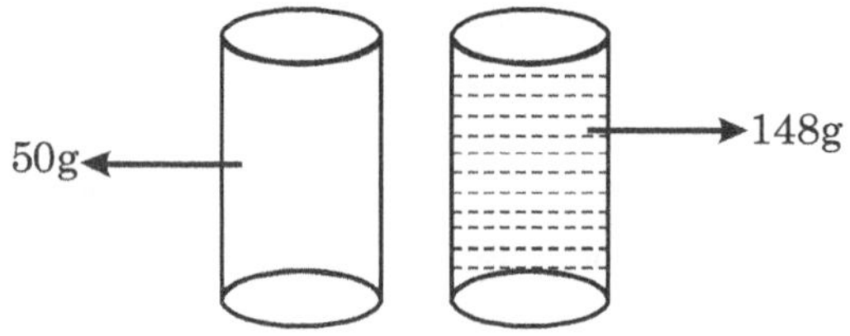

wt. of liquid = 148-50 = 98 gm

of liquid = 0.98 /ml

0.98 g liquid occupies volume = 1 ml

98 liquid occupies volume

$$= \frac{1}{0.98} \times 98 = 100\,\text{ml}$$

So, volume of container = Vol. of liquid

= 100 ml

= 0.1 L

Wt. of container along with gas = 50.5 g

Wt. of evacuated container = 50g

Therefore, wt. of gas = 50. 5 – 50 = 0.5 g

Apply, PV = nRT

(760 mm Hg = 10 atm)

$$1 \times 0.1 = \frac{0.5}{M} \times 0.0821 \times 300$$

$$M = 1500 \times 0.0821 = 123.15\text{g}$$

3. (a) $P_{Total} = P_{N_2} + P_{H_2O}$ vapour

$1 = P_{N_2} + 0.3$

$P_{N_2} = 0.7\,\text{atm}$

Apply. PV = nRT

$0.7 \times V = n \times R \times T$ ...(a)

After transferring the content

PV = nRT

$P \times \frac{V}{3} = nRT$ ....(b)

From (a) & (b)

$$0.7 \times \cancel{V} = P \times \frac{\cancel{V}}{3}$$

$\boxed{P = 2.1\text{ atm}}$

This is Just the P of dry N_2 gas

Total P of the system = $P_{N_2} + P_{H_2O}$ vapour

$= 2.1 + 0.3 = 2.4\,\text{atm}$

4. (*)

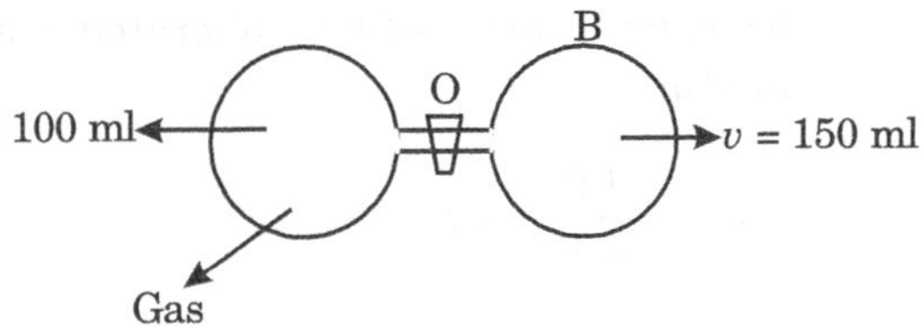

For chamber A, $PV = nRT$

$P \times 0.1 = n \times R \times T$ ...(1)

After opening the stop cork

$PV = nRT$

$P' \times (0.1 + 0.15) = nRT$

$P' \times 0.25 = nRT$ ...(2)

From 1 & 2

$P \times 0.1 = P' \times 0.25$

$$\overset{2}{1\cancel{00}} \times \frac{P \times 011}{1\cancel{0} \times 0.\underset{5}{\cancel{25}}} = P'$$

$$\boxed{P' = 0.4\ P}$$

$$\boxed{\text{i.e. } P' = \frac{40}{100} \times P}$$

5. (a) Apply $n_1 T_1 = n_2 T_2$

$n_1 \times 300 = n_2 \times 400$

$\boxed{\frac{3}{4} n_1 = n_2} \longrightarrow$ no. of mole of gas left in container

Fraction remaining inside the container

$$= \frac{\frac{3}{4} n_1}{n_1} = \frac{3}{4}$$

6. (a)

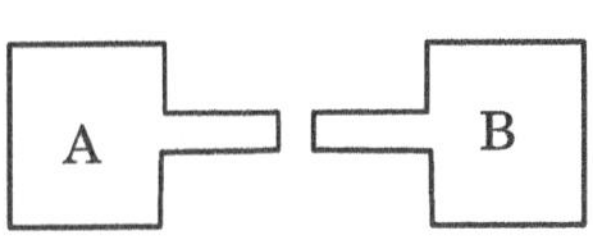

For Chamber A

(before opening & joining the tubes)

$P_A \times V = n_A \times R \times T_A$ ...(a)

For chamber B

$P_B \times V = n_B \times R \times T_B$ ...(b)

After joining the tubes

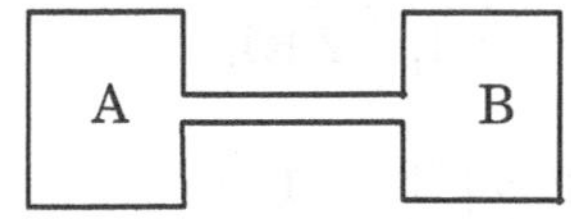

$\Rightarrow P \times 2V = (n_A + n_B) \times R \times T$

$$\frac{P \times 2V}{RT} = n_A + n_B$$

Placing the value of n_A & n_B from equations 1& 2

$$\frac{P \times 2V}{RT} = \frac{P_A \times V}{RT_A} + \frac{P_B V}{RT_B}$$

$$\frac{P \times 2V}{RT} = \frac{V}{R}\left(\frac{P_A}{T_A} + \frac{P_B}{T_B}\right)$$

$$\frac{2P}{T} = \frac{P_A}{T_A} + \frac{P_B}{T_B}$$

$$\frac{P}{T} = \frac{1}{2}\left(\frac{P_A}{T_A} + \frac{P_B}{T_B}\right)$$

7. (a)

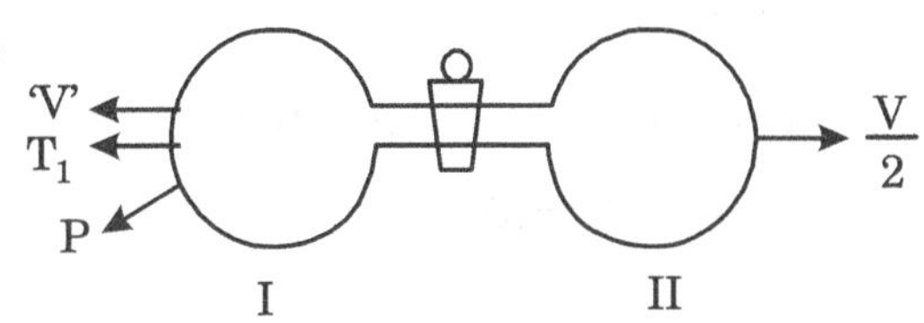

Initially, II sphere is evacuated. Gas is only in sphere I, initially.

$\therefore$ For sphere-I

$PV = nRT$

$PV - n \times R \times T_1$

$$\frac{PV}{RT_1} = n \quad \text{...(a)}$$

After opening the stop cork. Let 'x' moles move to sphere II –

So, no. of moles in sphere $I = n - x$

No. of moles in sphere $II = x$

For sphere I –

(now) $P' \times V = (n - x)\ RT_1$ ...(b)

For sphere II –

(now) $P' \times \frac{V}{2} = x\ RT_2$ ....(c)

$n = (n - x) + (x)$

(Total moles are conserved)

So, $\frac{PV}{RT_1} = \frac{p'V}{RT_1} + \frac{p'V}{2RT_2}$

$$\frac{PV}{RT_1} = \frac{P'V}{R}\left(\frac{1}{T_1} + \frac{1}{2T_2}\right)$$

$$\frac{P}{T_1} = P'\left(\frac{1}{T_1} + \frac{1}{2T_2}\right)$$

$$\frac{P}{T_1} = P'\left(\frac{2T_2 + T_1}{2T_1T_2}\right)$$

$$\Rightarrow P' = \frac{P}{T_1} \times \frac{2T_1T_2}{2T_2 + T_1}$$

$$P' = \frac{2T_2P}{2T_2 + T_1}$$

8. (*) $\underset{(g)}{C_3H_8} + \underset{(g)}{5O_2} \longrightarrow \underset{(g)}{3CO_2} + 4H_2O$ ...(a)

$\underset{(g)}{CH_4} + \underset{(g)}{2O_2} \longrightarrow \underset{(g)}{CO_2} + 2H_2O$ ...(b)

For a mixture of C_3H_8 & CH_4,

Apply PV = nRT

$$\frac{32\phi}{76\phi} \times V = nRT$$

$\frac{32V}{76RT} = n$ Total moles of mixture

Let No. of moles of propane = x (in mixture)

Let No. of moles of methane $= \frac{32V}{76RT} - x$

Now, from equation 1,

1 mole propane on combustion liberates CO_2 = 3 moles

X mole propane on combustion liberates CO_2 = $3x$ moles

Now, from equation 2,

1 mole methane on combustion gives CO_2 = 1 mole

$\left(\frac{32v}{76\,RT} - x\right)$ mole methane on combustion gives $CO_2 = \left(\frac{32v}{76\,RT} - x\right)$

Now, total moles of CO_2 liberated are (from mixture)

$$= 3x + \frac{32V}{76\,RT} - x$$

$$= \frac{32V}{76\,RT} + 2x$$

Apply PV = nRT (for CO_2)

$$\frac{448}{760} \times V = \left(\frac{32V}{76RT} + 2x\right)RT$$

$$\frac{448}{760} \times V = \frac{32V}{76} + 2x\,RT$$

$$\left(\frac{448V}{760} - \frac{32V}{760}\right) \times \frac{1}{2\,RT} = x$$

$$\left(\frac{448V - 320V}{760}\right) \times \frac{1}{2\,RT} = x$$

$$\frac{128V}{760} \times \frac{1}{2RT} = x$$

$$x = \frac{64\,V}{760\,RT}$$

Mole fraction of $\underset{\text{(Propane)}}{C_3H_3} = \frac{x}{\frac{32V}{76\,RT}}$

$$\frac{\frac{64V}{760RT}}{\frac{32V}{76\,RT}} = \frac{64V}{760RT} \times \frac{76\,RT}{32V} = 0.2$$

9. (d) No. of gaseous moles in cylinder PV = nRT

$$10 \times 4 = n \times 0\cdot0821 \times 400$$

$$\frac{40}{400 \times 0\cdot0821} = n$$

$\frac{1}{0\cdot821} = n = 1\cdot218$ gaseous moles

But all these gaseous moles can't be transferred in balloons. (without using external force)

Some gaseous moles will be transferred from cylinder to balloons (till the P becomes equal in both)

Again, PV = nRT

$$1 \times 4 = n \times 0 \cdot 0821 \times 400$$

$$\frac{4}{400 \times 0 \cdot 0821} = n$$

$$= \frac{1}{8 \cdot 21} = 0 \cdot 121$$

So, 0.121 mole will be left inside the cylinder, moles that will be transferred in balloons = 1.21 – 0.121 = 1.097

Now, no. of moles accommodated in 1 gaseous balloon, PV = nRT

$$\Rightarrow 1 \times 3 = n \times 0 \cdot 821 \times 300$$

$$\frac{3}{300 \times 0 \cdot 821} = n$$

$$= \frac{1}{8 \cdot 21} = 0 \cdot 121$$

So, 0.121 moles are accommodated in balloons = 1

1.097 moles are accommodated in balloons

$$= \frac{1}{0 \cdot 121} \times 1 \cdot 097 = 9$$

10. (*) PV = nRT

$$\Rightarrow PV = \frac{w}{m} RT$$

$$\Rightarrow PM = \frac{w}{v} RT \Rightarrow PM = dRT$$

$$d_A = d_B \times 2$$

$$P_A M_A = d_A RT_A \quad ...(a)$$

$$P_B M_B = d_B RT_B \quad ...(b)$$

$$\therefore \frac{P_A M_A}{d_A T_A} = \frac{P_B M_B}{d_B T_B}$$

$$\Rightarrow \frac{P_A M_A}{2d_B \times T_A} = \frac{P_B M_B}{d_B T_B} \text{ (since T are same)}$$

$$\frac{P_A M_A}{2} = P_B M_B \Rightarrow \text{Now, when}$$

$$P_A = P_B, M_A = 2M_B$$

11. (c) Since atomic ratio of He & O = 1 : 1

So, molar ratio of He & O_3 = $1 : \frac{1}{3}$

Moles of He = x

Moles of $O_3 = \frac{x}{3}$

PV = nRT

$$P \times V = \left(x + \frac{x}{3}\right) RT \quad ....(1)$$

$$\text{Also, } P' \times V = \frac{x}{3} \times R \times T \quad ...(2)$$

From 1 & 2

$$\frac{4x}{3} \frac{RT}{p} = \frac{xRT}{3P}$$

$$P' = \frac{P}{4} = 0 \cdot 25P$$

12. (b) $\overset{3}{\underset{(g)}{2A}} + \overset{3}{\underset{(g)}{3B}} \longrightarrow \underset{(g)}{C}$

When 3 moles of B react, 2 moles of A are reacted & 1 mole of C is formed.

So, 1 mole of A will be left & 1 mole C will be formed. Now, C is soluble in H_2O since it forms a basic solution.

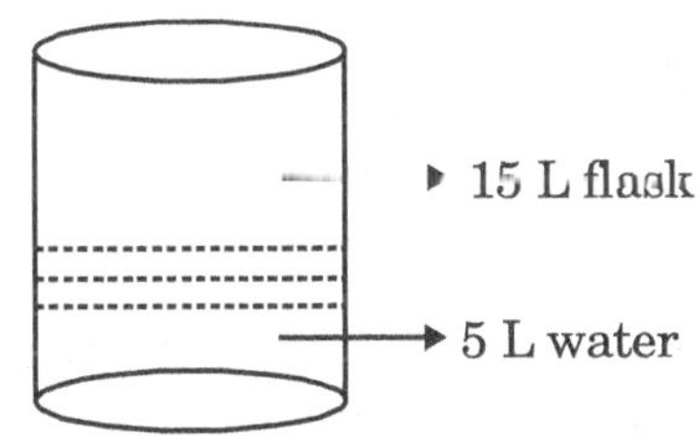

Gas molecules will roam is 10 L volume. Apply, PV = nRT

$$P \times 10 = 1 \times 0 \cdot 0821 \times 300$$

$$P = \frac{0 \cdot 821 \times 300}{10} = 2 \cdot 46 \text{ atm}$$

14. (c) Apply PV = nRT

$$P \times 8 \cdot 21 = 2 \times 0 \cdot 0821 \times 300$$

$$P = \frac{600 \times 0 \cdot 0821}{8 \cdot 21} = 6 \text{atm}$$

Also, $\frac{V_1}{T_1} = \frac{V_2}{T_2}$

$\Rightarrow \frac{8\cdot 21}{300} = \frac{V}{600}$

$\Rightarrow V = 16\cdot 42$ litres

15. (a) $PV = nRT$

$$\frac{P}{nRT} = \frac{1}{V}$$

$$\frac{1}{V} = P \times \frac{1}{nRT} + 0$$

$$\downarrow \quad \downarrow \quad \downarrow$$

$$\text{y axis} \quad mx \quad +C$$

Slope $= \frac{1}{nRT}$

$\Rightarrow \frac{1}{nRT} = \tan 45°$

$\Rightarrow n_A = \frac{1}{RT} = \frac{1}{22\cdot 4}$

Similarly, $\frac{1}{n_B RT} = \tan 60°$

$\Rightarrow \frac{1}{RT\sqrt{3}} = n_B$

$\Rightarrow n_B = \frac{1}{22.4\sqrt{3}}$

16. (d) $\underset{(g)}{\overset{(1)}{N_2}} + \underset{(g)}{\overset{(4)}{3H_2}} \longrightarrow \underset{(g)}{2NH_3}$

Here, N_2 is the limiting reagent.

When 1 mole of N_2 reacts, 3 moles of H_2 react and 2 mole of NH_3 are formed. Also, 1 mole H_2 is left.

$n_{NH_3} = 2$

$n_{H_2} = 1$

Now, volume of vessel = 15L

Volume of H_2O added = 5L

Volume left = 10L (for movement of gas)

n_{gas} = 1 mole of H_2 left

NH_3 will be soluble in H_2O

$\therefore$ Apply $PV = nRT$

$P \times 10 = 1 \times 0.0821 \times 300$

$P = \frac{1 \times 0.0821 \times 300}{10}$ atm

17. (c) Apply, $PV = nRT$

$1 = \frac{n}{V} \times R \times T$

$1 = 1 \times 0.0821 \times T$

$T = 12.18K$

18. (d) $n_1 = \frac{P_1V_1}{RT_1} = \frac{3 \times 1.65}{0.082 \times 546} = 0.11$

$n_2 = \frac{P_2V_2}{RT_2} = \frac{0.75 \times 0.55}{0.082 \times 273} = 0.0184$

% gas escaped $= \frac{n_1 - n_2}{n_1} \times 100 = 83.3\%$

19. (a)

	$N_{2(g)}$	$\longrightarrow$	$2N_{(g)}$
At t = 0	$\frac{1\cdot 4}{28} = 0\cdot 05$ moles		0
At t = t	$0\cdot 05 - \frac{40}{100} \times 0\cdot 05$		$2 \times \frac{40}{100} \times 0\cdot 05$

Apply, $PV = nRT$

moles of undissoicated N_2 left

$= 0\cdot 05 - \frac{40}{100} \times 0\cdot 05$

$= 0.05 - 0.02$

$= 0..03$

moles of atoms of N formed $= 2 \times \frac{40}{100} \times 0\cdot 05$

$= 0.04$

Apply, $PV = nRT$

$P \times 5 = (0\cdot 03 + 0\cdot 04) \times 0\cdot 0821 \times 1800$

$P \times 5 = 0\cdot 07 \times 0\cdot 0821 \times 1800$

P = 2.07 atm

20. (d) Apply, $PV = nRT$

$P \times 10 = \frac{4}{32} \times R \times T$ ....(a)

$2P \times 10 = n \times Rx = 0\cdot 05T$ ...(b)

Divide $\frac{1}{2}$

$$\frac{1}{2}=\frac{\frac{1}{8}}{n}\times\frac{1}{0.5}$$

$$\frac{1}{2}=\frac{1}{8}\times\frac{1}{n}\times\frac{1}{0.5}$$

$$\boxed{n=\frac{4}{8}-\frac{1}{2}}$$

Now, the no. of moles of O_2 should be = 0.5

i.e. O_2 should be added (since 4gm O_2 is already there)

21. (a) Apply PV = nRT

$$P\times V_1 = n\times R\times T \quad ...(a)$$

$$\frac{P}{4}\times V_2 = n\times R\times 2T \quad ...(b)$$

$$\frac{V_1}{V_2}=\frac{1}{8}$$

$$\boxed{V_2 = 8V_1}$$

$$\frac{4}{3}\pi r_2^3 = 8\times\frac{4}{3}\pi r_1^3$$

$$\left(\frac{r_2}{r_2}\right)^3 = 8$$

$$\frac{r_2}{r_1}=2$$

$$\Rightarrow \boxed{r_2 = 2r_1}$$

$$\%\text{ age increase in radius } = \frac{2r_1 - r_1}{r_1}\times 100$$

$$\frac{r_1}{r_1}\times 100 = 100\%$$

22. (b) Apply, PM = dRT

$$2.88\times M = 7.71\times 0.0821\times 309$$

$$M = 67.91 \text{ or } 68g \text{ (approx)}$$

23. (a) 22.4 L $\longrightarrow$ 1 mole of ideal gas at NTP

$$22.4\text{ L}\longrightarrow\frac{1}{22.4}\times 2.24 = 0.1 \text{ moles}$$

Vol. is Cornett. For moles left inside the container

So, P × n

$$\frac{P_1}{n_1}=\frac{P_2}{n_2}$$

$$\frac{760}{0.1}=\frac{570}{n_2}$$

$$n_2 = \frac{570\times 0.1}{760} = 0.075 \text{ moles}$$

↓

left inside the container

moles of gas that escaped = 0.1 – 0.075

= 0.025

24. (a) Let mass of N_2 = $x$gm

Mass of O_2 = (148 – x) gm

Apply, PV = nRT

$$4\times 30 = \left(\frac{x}{28}+\frac{148-x}{32}\right)\times 0.0821\times 300$$

$$\frac{120}{24.63}=\frac{x}{28}+\frac{148-x}{32}$$

$$4.87 = 0.0357x + 0.03125\,(148-x)$$

$$4.87 = 0.0357x + 4.625 - 0.03125\,x$$

$$0.245 = 0.00445\,x$$

$$\boxed{x = 55.05 \text{ g}}$$

$$\text{So, no. of moles of } N_2 = \frac{55.05}{28} = 1.96 \simeq 2 \text{ moles}$$

No. of moles of O_2 = 3 moles

25. (b) Let Initial pressure = P

Final Pressure = P + 0.4% of P

$$= P + \frac{0.4}{100}\times P = P + \frac{4}{1000}P$$

$$= P + \frac{4}{1000}P = \frac{1004P}{1000} = 1.004P$$

Initial temp. = Tk

Final temp. = (T + 1)k

$$\text{Apply, } \frac{P_1}{T_1}=\frac{P_1}{T_2}$$

$$\frac{P}{T}=\frac{1.004P}{(T+1)}$$

$T+1=1.004T$

$1=0.004\,T$

So, Final temp. $(T+1)k = 251k = -22°C$

26. (a)

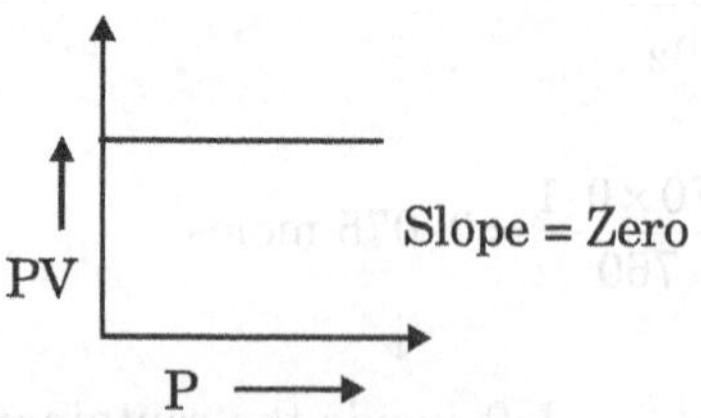

27. (c) For ZO,

Apply, PM = dRT

$2\times M = d\times R\times 273$ ...(a)

For N_2,

PM = dRT

$5\times 28 = dR\times 273$ ...(b)

$2\times M = 5\times 28$

$\boxed{M = 5\times 14 = 70g}$

28. (c) Apply PV = nRT

$P\times V = (2+1)RT$ ______(1)

Let additional moles of N_2 added = x

$2P\times V = (2+1+x)RT$ ______(2)

Divide $\frac{1}{2}$

$$\frac{P\times V}{2P\times V}=\frac{3RT}{(3+x)RT}$$

$$\frac{1}{2}=\frac{3}{3+x}$$

$3+x=6$

$x=3$

So, final moles of N_2 in mixture = 3 + 1 = 4

29. (b) Let no. of moles of $N_2 = x$

Let no. of moles of $O_2 = 2x$

Apply, PV = nRT

$0.8\times V = 3x\times R\times T$ ______(1)

After addition of argon,

$1.1\times V = (3x+0.2)\times R\times T$ ______(2)

$$\frac{0.8\times V}{1.1\times V}=\frac{3x}{3x+0.2}$$

$$\frac{8}{11}=\frac{3x}{3x+0.2}$$

$24x+1.6=33x$

$1.6=9x$

$x=\frac{1.6}{9}=0.177$

No. of moles of $O_2 = 2x = 2\times 0.177 = 0.355$

30. (c) Apply PV = nRT

$P\times V = 1\times R\times T$ ______(1)

When HCl, H_2 and O_2 are introduced, NH_3 will react with HCl.

$$\underset{(g)}{NH_3} + \underset{(g)}{HCl} \longrightarrow NH_4Cl_{(s)}$$

At t = 0	1	2	0
After completion	×	2 – 1	1 (Moles)

NH_3 is the limiting reagent. 1 mole of $HCl_{(g)}$ will be formed & 1 mole of $NH_4Cl_{(s)}$ will be formed. However H_2 & O_2 won't react till a spark is given.

So, now no. of moles of gas = 1 mole (HCl) + 2 moles (H_2) + 1 mole (O_2) = 4 mole mixtur

$P'\times V = 4RT$ ______(2)

$$\frac{P}{P'}=\frac{1}{4}$$

$4P = P'$

31. (a) Let no. of moles of $C_3H_8 = x$

Let no. of moles of $CH_4 = y$

PV = nRT

$320\times V = (x+y)R\times T$ ______(1)

Now,

$$C_3H_{8(g)} + SO_{2(g)} \longrightarrow 3CO_{2(g)} + 4H_2O_{(l)}$$

1 mole C_3H_8 gives CO_2 = 3 moles

x mole C_3H_8 gives CO_2 = 3x moles

$$\underset{(g)}{CH_4} + \underset{(g)}{2O_2} \longrightarrow \underset{(g)}{CO_2} + \underset{(l)}{2H_2O}$$

1mole CH_4 gives CO_2 = 1mole

y mole CH_4 gives CO_2 = y mole

Pressure due to CO_2 = 448 mm Hg

PV = nRT

$448 \times V = (3x + y)RT$ ________(2)

Divide $\frac{1}{2}$

$$\frac{320 \times \cancel{V}}{448 \times \cancel{V}} = \frac{(x+y)\cancel{RT}}{(3x+y)\cancel{RT}}$$

$$\frac{320}{448} = \frac{x+y}{3x+y}$$

x = 64

y = 256

$$X_{C_3H_8} = \frac{64}{256+64} = \frac{64}{320}$$

32. (d) For graph

$V \propto T$ (At constt. P & constt. no. of moles)

$V = KT$ (K is a constt.)

$\log V = \log(K.T)$

$$\underset{y}{\log V} = \underset{c}{\log K} + \underset{mx}{\log T}$$

Intercept = log K

Slope = +1

For graph 2

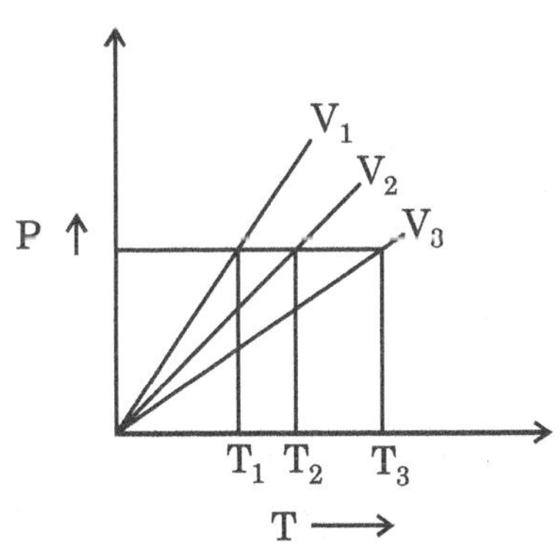

At same P, T_3 is maximum. So, V_3 is also maximum.

Hence $V_3 > V_2 > V_1$.

For graph 3 –

$P \propto \frac{1}{V}$ (Boyle's law)

$$P = \frac{K}{V}$$

$$P = K \times \frac{1}{V}$$

$$\underset{y}{\log P} = \underset{c}{\log K} + \underset{mx}{\log \frac{1}{V}}$$

34. (d) $V \propto T$ (At constt. P & no. of moles)

$V = KT$

$$\frac{V}{T} = K$$

For (2)

Since, V = KT

Multiply 'T' on both sides

$$\underset{y}{VT} = \underset{mx}{KT^2} + \underset{+c}{o}$$

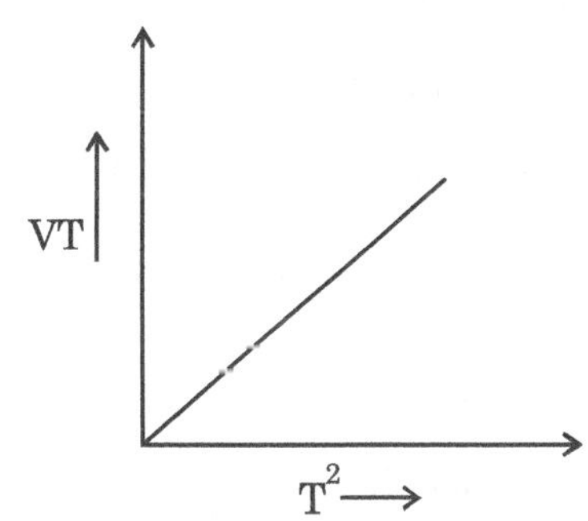

For (3)

V = KT

$$\underset{y}{\log V} - \underset{c}{\log K} + \underset{mx}{\log T}$$

36. (a) Initial moles, $n = \frac{370}{32} = 11.6$ moles

PV = nRT

$30 \times V = 11.6 \times 0.0821 \times 298$

V = 9.432 L

Final moles leftinside the container =

$$n = \frac{PV}{RT} = \frac{1\times 9.43}{0.082\times 348} = 0.330\,\text{moles}$$

wt. of O_2 left inside the container

$= 0.330 \times 32 = 10.6g$

wt. of O_2 that escaped $= 370 - 10.6 = 359.4g$

37. (d) For argon, apply (before removing partition)

$PV = nRT$

$1\times 2 = n\times R\times T$

$n = \frac{2}{RT} \rightarrow$ No. of moles of argon

For HCl, apply

$PV = nRT$

$1\times 2 = n\times R\times T$

$n = \frac{2}{RT} \rightarrow$ No. of moles of HCl

For NH_3, (before removing partition)

$PV = nRT$

$1\times 2 = n\times R\times T$

$n = \frac{2}{RT} \rightarrow$ No. of moles of NH_3

After removing the partition,

NH_3 & HCl react to form NH_4Cl (Solid)

$$NH_3 + HCl \longrightarrow NH_4Cl_{(s)}$$

	NH_3	HCl	$NH_4Cl_{(s)}$
At t = 0 No. of moles	$\frac{2}{RT}$	$\frac{2}{RT}$	0
After reaction	X	X	$\frac{2}{RT}$ (Solid NH_4Cl moles)

Apply, $PV = nRT$

$P\times 6 = \frac{2}{RT}\times RT$ (Only argon gas is left)

$P = \frac{2}{6} = \frac{1}{3}$ atm

This pressure is pressure exerted by argon.

38. (b) For gas in vessel A (before opening the stop cork)

Apply $PV = nRT$

$x_1 \times 1 = n\times E\times 300$

$n = \frac{x_1}{300R}$ ____(1)

Moles left in container $A = \frac{x_1}{300R} - y$

Apply. $PV = nRT$

$$X_2 \times 1 = \left(\frac{x_1}{300R} - y\right)R\times 300$$

$\frac{X_2}{300} = \frac{x_1}{300} - yR$ ____(3)

Putting the value of 'Y' from eqn. 2 in eqn. 3 -

$$\frac{X_2}{300} = \frac{x_1}{300} = yR$$

$$\frac{1}{300}(X_2 - X_1) = \frac{-X_2}{RT_2}\times R$$

$$\frac{X_2 - X_1}{300} = \frac{-X_2}{T_2}$$

$$\frac{-(X_2 - X_1)}{300} = \frac{X_2}{T_2}$$

$$\frac{X_1 - X_2}{300} = \frac{X_2}{T_2}$$

$$T_2 = \frac{X_2\,300}{X_1 - X_2}$$

39. (a) Apply $\frac{n_1}{t_1}\times\frac{t_2}{n_2} = \sqrt{\frac{M_2}{M_1}}$

Here, n_1 & n_2 are the moles of gases diffused out

$$\frac{4}{64}\times\frac{1}{n}\times 4\sqrt{\frac{4}{64}}$$

$\boxed{n = 1gm}$

41. (b)

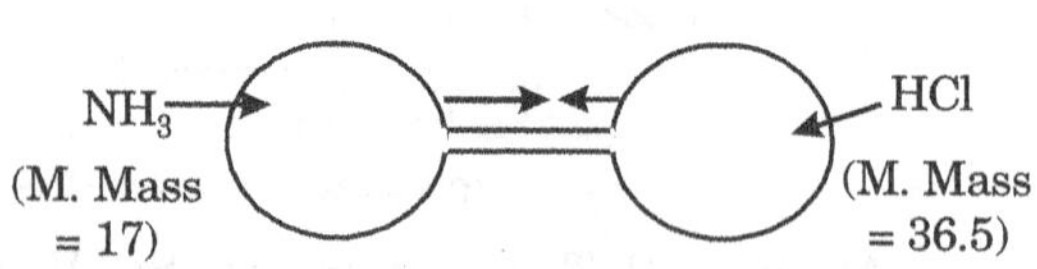

Since, M.Mass of NH_3 is less, it will diffuse readily.

42. (d)

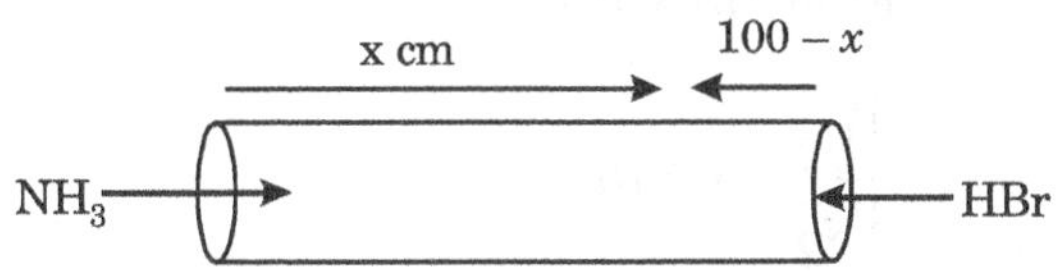

$$\frac{r_{NH_3}}{r_{HBr}} = \frac{x}{100-x} = \sqrt{\frac{M_{HBr}}{M_{NH_3}}}$$

$$\frac{x}{100-x} = \sqrt{\frac{81}{17}}$$

$$\frac{x}{100-x} = 2.18$$

$$\boxed{x = 68.55\,\text{cm}}$$

43. (b) $\frac{r_{H_2}}{r_{alkyne}} = \sqrt{\frac{M_{alkyne}}{M_{H_2}}}$

$$\frac{3\sqrt{3}\, r_{\cancel{alkyne}}}{r_{\cancel{alkyne}}} = \sqrt{\frac{M_{alkyne}}{2}}$$

Squaring both sides,

$$27 = \frac{M_{alkyne}}{2}$$

$$\Rightarrow \boxed{M_{alkyne} = 54}$$

44. (a) Apply $\frac{V_1}{t_1} \times \frac{t_2}{V_2} = \sqrt{\frac{M_2}{M_1}}$

$$\frac{80}{2} \times \frac{3}{V_2} = \sqrt{\frac{64}{32}}$$

$$\frac{120}{V_2} = \sqrt{2}$$

$$\boxed{V_2 = \frac{120}{\sqrt{2}}\,\text{ml}}$$

45. (d) Apply $\frac{r_{N_2}}{r_{CO}} = \frac{r_{N_2}}{n_{CO}} \sqrt{\frac{M_{CO}}{M_{N_2}}}$

$$\frac{r_{N_2}}{r_{CO}} = \frac{3}{2}\sqrt{\frac{\cancel{28}}{\cancel{28}}}\ 1$$

$$\boxed{\frac{r_{N_2}}{r_{CO}} = \frac{3}{2}} \text{ or } \boxed{\frac{r_{CO}}{r_{N_2}} = \frac{2}{3}}$$

46. (b) Applying, $\frac{v_1}{t_1} \times \frac{t_2}{v_2} = \sqrt{\frac{M_2}{M_1}}$

Since time taken is same

$$\frac{v_1}{v_2} = \sqrt{\frac{M_2}{M_1}}$$

$$\frac{2}{v_2} = \sqrt{\frac{2}{4}}$$

$$\Rightarrow 2\sqrt{2} = v_2$$

$$\Rightarrow 2 \times 1.414 = v_2$$

$$\boxed{v_2 = 2.82\ \text{L}}$$

47. (c) $\frac{r_1}{r_2} = \frac{P_1}{P_2}\sqrt{\frac{M_2}{M_1}}$

$$\frac{\frac{n_1}{t_1}}{\frac{n_2}{t_2}} = \frac{P_1}{P_2}\sqrt{\frac{M_2}{M_1}}$$

$$\frac{n_1}{t_1} \times \frac{t_2}{n_2} = \frac{P_1}{P_2}\sqrt{\frac{M_2}{M_1}}$$

$$\frac{1}{38} \times \frac{57}{1} = \frac{0.8}{1.6}\sqrt{\frac{M}{28}}$$

$\therefore$ Molarmass = 252

Now, $Xe + (F)_x = 252$

$131 + 20x = 252$

So, compounds XeF_6

48. (a) $P.P.O_2 = P_{Total} \times O_2$

$$P.P.O_2 = P.\ \frac{\frac{W_{O_2}}{32}}{\frac{W_{O_2}}{32} + \frac{W_{CH_4}}{16}}$$

$$\Rightarrow P.P_{O_2} = P.\ \frac{\frac{W}{32}}{\frac{W}{32} + \frac{W}{16}}$$

$$\Rightarrow P.P_{O_2} = P.\ \frac{\frac{W}{32}}{\frac{W + 2W}{32}}$$

$$P.P.O_2 = P.\frac{\frac{W}{32}}{\frac{3W}{32}}$$

$$\Rightarrow P.P.O_2 = P.\frac{W}{32}\times\frac{32}{W} = \frac{P}{3}$$

$$\boxed{\therefore\ P.P.O_2 = \frac{1}{3}\times P}$$

49. (a) M. mas of CO & B_2H_6 = 28

M. mass of CH_4 = 16

M. mass of H_2 = 2

So, CO & B_2H_6 will diffuse least & will have the maximum pressure inside the container.

50. (*) $P.PO_2 = P.XO_2$

$$P.P.O_2 = \frac{P.3}{10} = \frac{3P}{10} \quad ....(a)$$

After removal of 2 mole O_2

$$P.P.O_2 = \frac{P.1}{8} = \frac{1P}{8} \quad ....(b)$$

$$\frac{P_{O_2}}{P_{N_2}} = \frac{0.3125}{0.0357}$$

$$\boxed{P_{O_2} = 0.875\ P_{N_2}}$$

$$\%\text{ age decrease in partial} = \frac{\frac{3P}{10}-\frac{1P}{8}}{\frac{3}{10}P}\times 100$$

= 58.33%

51. (d) $$P_{O_2} = P.\frac{\frac{W_{O_2}}{32}}{\frac{W_{O_2}}{32}+\frac{W_{N_2}}{28}}$$

$$\boxed{\text{Now, } W_{O_2} = W_{N_2}}$$

$$\therefore\ P_{O_2} = P.\frac{\frac{W}{32}}{\frac{W}{32}+\frac{W}{28}} \quad ...(a)$$

$$P_{N_2} = P.\frac{\frac{W}{28}}{\frac{W}{28}+\frac{W}{32}} \quad ...(b)$$

From (a) & (b),

$$\frac{P_{O_2}}{P_{N_2}} = 0.875$$

52. (c) Let aq. tension at 288K be p bar

P dry gas at 288K = (0.92 – p) bar

Now, P_1 = (0.92 – p) bar P_1 = 1 bar

V_1 = 0.1L $\quad V_2$ = 0.085L

T_1 = 288K $\quad T_2$ = 273.15K

$$\frac{P_1V_1}{T_1} = \frac{P_2V_2}{T_2}$$

$$P_1 = \frac{P_2V_2T_1}{V_1T_2}$$

$$0.92 - P = \frac{1\times 0.085\times 288}{0.1\times 273.15} = 0.896$$

$$P = 0.024\ \text{bar}$$

54. (b) Let no. of moles of He = 20

So, no. of moles of CH_4 = 80

$$P\cdot P_{He} = P_{Total}\times He = 1\cdot 4\times\frac{20}{20+80}$$

$$= 1\cdot 4\times\frac{20}{100} = 0.28\ \text{bar}$$

55. (c) Apply, PV = nRT (for H_2)

$$P_{H_2}\times 2 = 0\cdot 0788\times 0\cdot 0821\times 299$$

$$\boxed{P_{H_2} = 0\cdot 967\ \text{atm}}$$

Applying Daltons law

$$P_{H_2} = P_{total}X_{H_2}$$

$$0\cdot 967 = 1X_{H_2}$$

$$\boxed{X_{H_2} = 0\cdot 967}$$

56. (c) Applying Dalton's law

P_A = 3 atm

$$P_A = P_{Total}X_A$$

$$3 = 10\ X_A$$

$$\boxed{X_A = 0\cdot 3}$$

$P_B = P_{Total} X_B$

$1 = 10\ X_B$

$\boxed{X_B = 0.1}$

Now, $X_A + X_B + X_C = 1$

$0.3 + 0.1 + X_C = 1$

$\boxed{X_C = 0.6}$

$$X_C = \frac{n_c}{\text{Total moles}}$$

$$0.6 = \frac{n_c}{10}$$

$\boxed{6 = n_c}$

$$n_c = \frac{\text{given wtc}}{\text{mol. wtc}}$$

$$6 = \frac{W_c}{2}$$

$\boxed{12g = W_c}$

57. (*) For CO, (Before opening the value)

PV = nRT

$0.5 \times 2 = n \times 0.0821 \times 300$

$$\frac{1}{0.0821 \times 300} = n = \boxed{0.040}$$

↑ initial moles of CO

For O_2 (before opening the value)

PV = nRT

$1 \times 1 = n \times 0.0821 \times 300$

$$\frac{1}{0.0821 \times 300} = n = \boxed{0.040}$$

↓ initial moles of O_2

$$2\,CO_{(g)} + O_{2(g)} \longrightarrow 2\,CO_{2(g)}$$

0.040 0.040 0

→ CO is the limiting reagent

So, no. of moles of CO_2 formed = 0.040

No. of moles of O_2 left = 0.040 – 0.020

$\boxed{n_{O_2} = 0.020}$

$\boxed{P_{Total} = 0.5 \text{ atm}}$

$PP_{CO} = 0.5\ X_{CO}$

$$PP_{CO} = 0.5 \frac{0.04}{0.04 + 0.02} = 0.33 \text{ atm}$$

$P.PO_2 = 0.5 - 0.33 = 0.17$ atm

58. (c) $r_y = 2r_z$

$$\frac{r_y}{r_z} = \frac{2r_z}{r_z} = \sqrt{\frac{M_z}{M_y}}$$

$$2 = \sqrt{\frac{128}{M_y}}$$

Sq. both sides

$$4 = \frac{128}{M_y}$$

$M_y = 32$

Now, $r_x = 4r_y$

$$\frac{r_x}{r_y} = \sqrt{\frac{M_y}{M_x}}$$

$$\frac{4r_y}{r_y} = \sqrt{\frac{32}{M_x}}$$

$$16 = \frac{32}{M_x}$$

$M_x = 2g$

59. (d) $\frac{r_{H_2}}{r_{O_2}} = \sqrt{\frac{32}{2}}$

$$\frac{r_{H_2}}{r_{O_2}} = \sqrt{16} \Rightarrow r_{H_2} = 4r_{O_2}$$

Let volume of O_2 effused = V ml

In same time, volume of H_2 effused = 4V ml

4V + V = 10

(Since total vol. was 100 ml & vol. left =90 ml)

5V + 10

V = 2 ml

Hence, vol. of H_2 effused = 4 V ml

= 4 × 2 ml = 8 ml

Initial vol. of H_2 = 50 ml

So, vol. of H_2 left = 50 – 8 = 42 ml

60. (c) Rate of diffusion (of mixture of gases) = ¼ rate of diffusion of H_2

$$\frac{r_{Mixture}}{r_{H_2}} = \sqrt{\frac{M_{H_2}}{M_{Mixture}}}$$

$$\frac{\frac{1}{4}r_{H_2}}{r_{H_2}} = \sqrt{\frac{2}{M_{Mixture}}}$$

$$\frac{1}{4} = \sqrt{\frac{2}{M_{Mixture}}}$$

Sq. both sides –

$$\frac{1}{16} = \frac{2}{M_{Mixture}}$$

$$M_{Mixture} = 32 gm$$

Let have be x moles of C_2H_4 moles of CO_2 in mixture.

So, $(x + y)$ mole of mixture weigh

$$= (x \times 28 + y \times 44) \text{ gms}$$

$$\text{1mole of mixture weigh} = \frac{28x + 44y}{x + y}$$

$$\therefore \frac{28x + 44y}{x + y} = 32$$

$28x + 446 = 32x + 32y$

$12y = 4x$

$3y = x$

Molar ratio of C_2H_4 to CO_2

$$= \frac{\text{no. of moles of } C_2H_4}{\text{no. of moles of } CO_2}$$

$$= \frac{x}{y} = \frac{3y}{y} = 3 : 1$$

61. (c) $RMS_{H_2} = \sqrt{7}\ RMS_{N_2}$

$$\sqrt{\frac{3RT_{H_2}}{2}} = \sqrt{7}\sqrt{\frac{3RT_{N_2}}{28}}$$

$$\frac{\cancel{3}\cancel{R}T_{H_2}}{2} = \left(\sqrt{7}\right)^2 \frac{\cancel{3}\cancel{R}T_{N_2}}{28}$$

$$\frac{T_{H_2}}{T_{N_2}} = \frac{\sqrt{7} \times \sqrt{7} \times 2}{28} = \frac{\cancel{7} \times \cancel{2}}{\cancel{28}\ \cancel{4}\ 2}$$

$$\boxed{2T_{H_2} = T_{N_2}}$$

62. (d) Mean velocity of A = RMS of B

$$\sqrt{\frac{8RT}{\pi M_A}} = \sqrt{\frac{3RT}{M_B}}$$

$$\text{So, } \frac{8}{\pi M_A} = \frac{3\pi}{8}$$

$$\frac{M_B}{M_A} = \frac{3\pi}{8}$$

$$\therefore\ M_B > M_A$$

Now, mean velocity of A = mean velocity of B

$$\text{Which means, } \sqrt{\frac{8RT_A}{\pi M_A}} = \sqrt{\frac{8RT_B}{\pi M_B}}$$

Now, $M_B > M_A$, So, Temp. of B has to be raised to make both values equal.

63. (a) $U_{RMS} = \sqrt{\frac{3RT}{M}} = \sqrt{\frac{3PV}{M}}$

As temperature is constant, the product PV is also a constant. So, RMS will remain constant.

64. (c) $K.E._{Total} = \frac{3}{2}RT$

$$\frac{3}{2} \times 0.3 \times R \times T = \frac{3}{2} \times 0.4 \times R \times 400$$

$$0.3 \times T = 0.4 \times 400$$

$$T = \frac{4}{3} \times 400 = 533 K$$

65. (a) $U_{RMS_1} = \sqrt{\frac{3RT}{M}} = \sqrt{\frac{3 \times R \times 300}{M}}$ ...(a)

$U_{RMS_2} = \sqrt{\frac{3 \times R \times 400}{M}}$ ...(b)

$\%\text{ age increase} = \frac{U_{RMS_2} - U_{RMS_1}}{U_{RMS_1}} \times 100 = 15.6\%$

68. (c) $K.E. = \frac{3}{2}nRT$ (for n moles of a gas)

$KE = \frac{3}{2}RT$ (for 1 mole of a gas)

$\therefore$ For N_A molecules, $KE = \frac{3}{2}RT$

For 1 molecules, $KE = \frac{3}{2}\frac{RT}{N_A} = \frac{3}{2}k_bT$

$\left(k_b = \frac{R}{N_A}\right)$
↑
Boltzmann's constant

69. (a) $V_{MP} : V_{av} : V_{RMS} = 1 : 1.128 : 1.224$

$\because V_{RMS} > V_{av} > V_{MP}$

70. (b) $RMS = \sqrt{\frac{3RT}{M}}$

$\frac{u}{u'} = \sqrt{\frac{3RT}{28}} \times \sqrt{\frac{14}{3RT \times 2}}$

$u = \sqrt{\frac{3RT}{28}}$ ______(1)

$\frac{u}{u'} = \frac{1}{\sqrt{4}}$

after dissociation –

$u' = \sqrt{\frac{3RT}{14} \times 2}$ ___(2)

$2u = u'$

71. (d) $U_{rMS} = \sqrt{\frac{3RT}{M}}$

$30R^{1/2} = \sqrt{\frac{3RT}{M}}$

squaring both sides –

$900R = \frac{3RT}{M}$

$M = \frac{300}{300} = 1g = 0.001Kg$

72. (*) Since there is just a single molecule, all the velocities will be equal.

74. (d) Av. $kE = \frac{3}{2}kT$

where k = Boltzmann's constt.

So, av. $k \cdot E \propto T$ (absolute temp.)

Also, PM = dRT

Molar mass of gas $\propto$ density

75. (d) As is evident from the graph on increasing temperature, most probable speed increases.

$\propto$ (most probable speed) $= \sqrt{\frac{2RT}{M}}$

Now, $\propto$ for gas B (at T_2) $>$ $\propto$ for gas A (at T_1)

i.e., $\sqrt{\frac{2RT_2}{M_B}} > \sqrt{\frac{2RT_1}{M_A}}$

$$\boxed{\frac{T_2}{M_B} > \frac{T_1}{M_A}}$$

76. (a) $U_{rms} = \sqrt{\frac{3PV}{M}}$ or $\sqrt{\frac{3RT}{M}}$ ______(1)

$KE = \frac{3}{2}RT$

Placing the value of $RT = \frac{2}{3}KE$ in 1 –

$U_{rms} = \sqrt{\frac{3}{M} \times \frac{2}{3}KE}$

$U_{rms} = \sqrt{\frac{2KE}{M}}$ or $\sqrt{\frac{2E}{M}}$

78. (c) Cooling will decrease the temperature which affects the volume.

79. (c) Moles of $H_2 = \frac{2}{2} = 1$

Moles of $N_2 = \frac{28}{28} = 1$

Density will be different.

$$\text{Density} = \frac{\text{given wt.}}{\text{Volume}}$$

80. (b) $U_{MP} = \sqrt{\frac{2RT_3}{M}}$ ____(1)
$(\propto)$

$U_{Average} = \sqrt{\frac{8RT_2}{M}}$ ____(2)
(V)

$U_{rms} = \sqrt{\frac{3RT_1}{M}}$ ____(3)
(u)

$\propto : v : u = 1 : 1 : 128 : 1.225$

Magnitude: – $U_{rms} > V_{av} > \propto$

To make them equal, T of $\propto$ should be raised

Max. $\therefore T_3 > T_2 > T_1$

81. (b) Apply, PV = nRT

$1 \times V = 1 \times R \times 300$ ...(a)

$$A_{2(g)} \longrightarrow 2A_{(g)}$$

At t = 0	1	0
At t = t	1 – 0.2	0.4

Total no. of gaseous moles = 0.8 + 0.4 = 1.2

Apply, PV = nRT

$P \times V = 1{\cdot}2 \times R \times 600$ ...(b)

Divide $\frac{1}{2}$ –

$$\frac{1 \times V}{P \times V} = \frac{1 \times R \times 300}{1{\cdot}2 \times R \times 600}$$

2.4 atm = P

82. (d) $P_{\text{wet gas}} = P_{\text{dry gas}} + \text{Aq.Tension}$

$640 = P_{\text{dry gas}} + x$

$(640 - x)$ atm $P_{\text{dry gas}}$ (dry air)

Let no. of moles of O_2 = y

No. of moles of O_2 = 3y

$P_{N_2} = P_{Total} \times N_2$

$$450 = (640 - x) \cdot \frac{3y}{3y + y}$$

$$450 = (640 - x) \cdot \frac{3y}{4y}$$

$$450 = (640 - x) \cdot \frac{3}{4}$$

$$\frac{4 \times 450}{3} = 640 - x$$

$150 \times 4 = 640 - x$

$600 = 640 - x$

x = 40 torr

$$\frac{PV}{K_b T} = \text{no. of particles of gas}$$

83. (b) Apply, PM = dRT

$1 \times M = 1{\cdot}8 \times 0{\cdot}0820 \times 300$

M = 44 g

So, gas is CO_2.

84. (c)

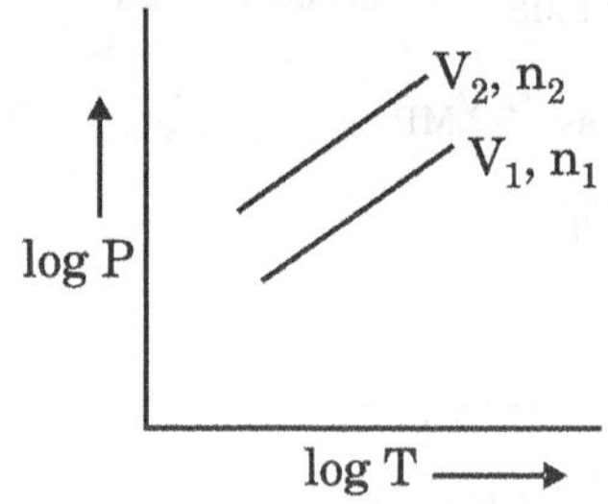

PV = nRT

$$\frac{P}{T} = \frac{nR}{V}$$

$$\frac{nR}{V} = k \text{ (constt.)}$$

$$\frac{P}{T} = k$$

Take log on both sides –

$$\log \frac{P}{T} = \log k$$

$$\log P - \log T = \log k$$

$$\log P = \log k + \log T$$

OR

$$\underset{y}{\log P} = \underset{c}{\log K} + \underset{m}{1} \times \underset{x}{\log T}$$

Slope of the graph $= \log \frac{nR}{V}$

$\therefore$ Clearly from the graph, intercept of V_2, n_2 is greater than that for V_1, n_1.

So, $\frac{n_2R}{V_2} > \frac{n_1R}{V_1}\ \frac{n_2}{V_2} > \frac{n_1}{V_1}$

If $n_1 = n_2$, $V_1 > V_2$.

85. (c) Av. K.E. $= \frac{3}{2} k_b\ T$

$$K_b \text{ (Boltzmans constant)} = \frac{R}{N_A} = \frac{8.314}{6.02 \times 10^{23}}$$

$$K_b = 1 \cdot 38 \times 10^{-23}$$

$$\frac{3}{2} \times 1 \cdot 38 \times 10^{-23} \times 300 = 6 \cdot 21 \times 10^{-y}$$

$$621 \times 10^{-23} == 6 \cdot 21 \times 10^{-y}$$

$$\therefore\ 10^{-21} = 10^{-y}$$

$y = 21$

86. (d) For Boyles law

$PV = nRT$

$$P = nRT \times \frac{1}{V} \qquad ...(a)$$

For Gay-Lucssais law

$PV = nRT$

$$P = T \times \frac{nR}{V}$$

For avogadros law –

$Pv = nRT$

$$V = n\frac{RT}{P}$$

Constt. For avogadros law

Multiplying all the three constants, we get

$$\Rightarrow\ nRT \times \frac{nR}{V} \times \frac{RT}{P}$$

$$\Rightarrow\ \frac{n^2R^3T^2}{PV}$$

$$= \frac{mol^2 \times \frac{atm^3L^3}{mol^3k^3} \times k^2}{atm \times L}$$

$$= mol^2 \times \frac{atm^3L^3}{mol^3k^3} \times k^3 \times \frac{1}{atm\ L}$$

$$\Rightarrow\ atm^2L^2\ mol^{-1}k^{-1}$$

87. (d) Option 3 $\rightarrow P(V - b) = RT$

$PV - Pb = RT$

$PV = Pb + RT$

Divide both sides by RT

$$\frac{PV}{RT} = \frac{Pb}{RT} + \frac{RT}{RT}$$

$\boxed{Z = 1 + \frac{Pb}{RT}}$ Z > 1 (Rep. forces> Att. Forces)

For option 1, Isochoric curves means Vconstt, so

$$\frac{P}{T} = \frac{R}{V - b}$$

(For P v/s T graph, Slope = $\frac{R}{V - b}$)

Similarly for option 2.

90. (c) For one mole of a van der waals gas –

$$\left(P + \frac{a}{V^2}\right)(V - b) = RT$$

$b = 0$ To calculate 'a', slope has to be determined

$$\left(P + \frac{a}{V^2}\right)(V) = RT$$

$$PV + \frac{a}{V} = RT$$

$$\text{So}, \tan\theta = \frac{P}{B} = \frac{24.6 - 21.6}{2} = \frac{3}{2} = 1.5$$

$$\underset{\downarrow}{PV} = RT - \frac{a}{V}$$

$$\underset{Y}{PV} = \underset{C}{RT} - \underset{MX}{a \times \frac{1}{V}}$$

91. (d) $Z=\frac{V_{real}}{V_{ideal}} \Rightarrow$ For + ve deviation,

$$Z>1 \Rightarrow \frac{V_{real}}{V_{ideal}}>1$$

Now, V_{ideal} (molar volume) = 22.4L

$$\therefore \frac{V_{real}}{22.4}>1$$

$V_{real} > 22.4$ litres

Similarly, for negative deviation, $V_M < 22.4$ litres.

92. (d) $Z=\frac{PV}{nRT}$

$$1.95=\frac{800\times 1}{n\times 0.0821\times 223}$$

n = 22.4 mole

So, volume of N_2 at 373 K and at 200 atm would be –

$$V=\frac{nZRT}{P}$$

$$=\frac{22.4\times 1.10\times 0.0821\times 373}{200}$$

V = 3.77 Ltr.

93. (b) $\left(P+\frac{an^2}{V^2}\right)(V-nb)=nRT$

$$\left(P+\frac{9a}{V^2}\right)(V-3b)=3RT \qquad (\because n=3)$$

At low P, 'b' is ignored

$$\therefore \left(P+\frac{9a}{V^2}\right)(V)=3RT$$

$$PV+\frac{9a}{V}=3RT$$

$$PV=3RT-\frac{9a}{V} \Rightarrow PV=3\left(RT-\frac{3a}{V}\right)$$

$$\frac{PV}{RT-\frac{3a}{V}}=3$$

94. (c) N_o – Avogadro's No.

Available volume = $(V-nb)=\left(V-\frac{N}{N_o}b\right)$

95. (a) Z < 1 means gas is more easily compressible & can be liquefied with ease.

At moderate P

$$\boxed{Z=1-\frac{a}{RTV}}$$

Lesser the volume of z, more is value of 'a' & more is ease of liquefaction.

98. (c) At Boyle's temperature, Z = 1

$$0.34=\frac{160}{T}P$$

$$T=\frac{160}{0.34}=470K$$

99. (b) P' is moderate P & we know that at moderate pressure, 'a' can't be ignored but 'b' is ignored.

$$\left(P+\frac{a}{V^2}\right)(V-b)=RT$$

$$\left(P+\frac{a}{V^2}\right)(V)=RT$$

$$PV+\frac{a}{V}=RT$$

$$PV=RT-\frac{a}{V}$$

Divide both sides by 'RT' –

$$\frac{PV}{RT}=\frac{RT}{RT}-\frac{a}{RTV}$$

$$Z=1-\frac{a}{RTV}$$

More the volume of 'a', greater will be dip in 'Z'.

101. (c) $Z=\frac{P_cV_c}{RT_c}$

103. (c) $\left(P+\frac{a}{V^2}\right)(V-b)=RT$

b = 0

$$\text{So,}\left(P+\frac{a}{V^2}\right)(V) = RT$$

$$PV+\frac{a}{V} = RT$$

$$PV = RT-\frac{a}{V}$$

$$PV = RT - a\times\frac{1}{V}$$

$$\begin{array}{ccc}\downarrow & & \downarrow \\ y & c & mx\end{array}$$

Slope = – a

$$\text{Slope} = a = \frac{P}{B} = \frac{24.6-20.1}{3}$$

$$= \frac{4.5}{3} = 1.5$$

105. (d) No. of moles of gas are constt. 'Z' is recorded at 2 different conditions here for the same gas.

$$Z = \frac{PV}{nRT}$$

$$1.9 = \frac{800\times1}{n\times.0821\times330}$$

$$n = \frac{800}{1.9\times0.0821\times330}$$

$$n = \frac{800}{51.47} = 15.54$$

For Case II

$$Z = \frac{PV}{nRT}$$

$$1.1 = \frac{200\times V_2}{15.54\times0.0821\times570}$$

$$\frac{1.1\times15.54\times0.0821\times570}{200} = V_2$$

$$4\text{ L} = V_2$$

105. (d) $\left(P+\frac{an^2}{V^2}\right)(V-nb) = nRT$

$$\left(P+\frac{a(0.5)^2}{V^2}\right)(V-0.5b) = 0.5RT$$

$$\left(P+\frac{a}{V^2}\times0.25\right)(V-\frac{b}{2}) = \frac{RT}{2}$$

$$\left(P+\frac{a}{4V^2}\right)\left(\frac{2V-b}{2}\right) = \frac{RT}{2}$$

$$\left(P+\frac{a}{4V^2}\right) = \frac{RT}{(2V-b)}$$

106. (d)

$$\underset{\text{(Boyles temp.)}}{T_B} = \frac{a}{Rb},\ T_i = \underset{\text{(Inversion temp.)}}{\frac{2a}{Rb}},\ T_c = \frac{8a}{27b^2}$$

108. (a) $T_c = \frac{8}{27}T_b$

$$\frac{300\times327}{8} - T_b$$

$$T_b = 1012.5\text{K}$$

109. (d) $V_c = 36,\ P_c = \frac{a}{27b^2},\ T_c = \frac{8a}{27Rb}$

Smart Notes

6

CHAPTER

Chemical Thermodynamics

1. When 1 mole of a gas is heated at constant volume, temperature is raised from 298 K to 308 K. Heat supplied to the gas is 500 J. Then which statement is correct?

(a) $q = \Delta U = 500\,J,\ W = 0$

(b) $q = W = -500\,J,\ \Delta U = 0$

(c) $q = W = 500\,J,\ \Delta U = 0$

(d) None of these

2. No heat is absorbed by the system from the surroundings, but work is done on the system. What type of wall does the system have?

(a) $\Delta U = w$ adiabatic, wall is insulated

(b) $\Delta U = -q$, thermally conducting walls

(c) $\Delta U = q - w$, closed system

(d) $\Delta U = q - w$, open system

3. The Internal energy when a system goes from state A to state B is 40 KJ/mol. If the system goes from A to B by a reversible path and returns to state A by an irreversible path. What would be the net change in internal energy?

(a) 40 KJ (b) > 40 KJ

(c) < 40 KJ (d) Zero

4. Select the incorrect match

(a) Molar entropy ⟶ Intensive property

(b) Heat capacity ⟶ Extensive property

(c) Boiling point ⟶ Intensive property

(d) Internal energy ⟶ intensive property

5. 2 moles of zinc are dissolved in HCl at 25°C. The work done in open vessel is

(a) –2.47 KJ (b) –4.95 KJ

(c) 0.0489 KJ (d) –5.66 KJ

6. The correct expression for work done in reversible isothermal ideal gas expansion is/are

(a) $w = -2.303\,nRT\log\frac{v_2}{v_1}$

(b) $w = -2.303\,nRT\log\frac{P_1}{P_2}$

(c) $w = nRT\ln\left(\frac{P_2}{P_1}\right)$

(d) All are correct

7. What is correct for adiabatic expansion of non-ideal gas into vacuum ?

(a) $w = 0, q = 0, \Delta U = 0$

(b) $w<0, q>0, \Delta U = 0$

(c) $w = q \neq \Delta U$

(d) None of these

8. 2 mole of an ideal gas at 27°C is compressed reversibly and isothermally from a pressure of $1.01 \times 10^5 \, Nm^{-2}$ to $5.05 \times 10^6 Nm^{-2}$, then which statement is correct?

(a) Work done on the gas is $1.95 \times 10^4 J$

(b) Work done on the gas is $1.19 \times 10^7 J$

(c) Work done by the gas is $1.95 \times 10^4 J$

(d) Work done by the gas is $1.19 \times 10^7 J$

9. Work done by a system is 8 joule, when 40 Joule heat is supplied to it. What is the increase in internal energy of system?

(a) 25 J (b) 30 J

(c) 32 J (d) 26 J

10. Which one of the following equations does not correctly represent the first law of thermodynamics for given process?

(a) Cyclic process : $q = -w$

(b) Isochoric process : $\Delta E = q$

(c) Adiabatic process : $\Delta E = -q$

(d) None of these

11. A system has internal energy equal to U_1, 450 J of heat is taken out of it and 600 J of work is done on it. The final energy of the system will be

(a) $U_1 + 150$ (b) $U_1 + 1500$

(c) $U_1 - 150$ (d) None of these

12. 2.8 g N_2 gas at 300 K and at 20 atm was allowed to expand isothermally against a constant external pressure of 1 atm. What is correct for this process? (assume gas to be ideal)

(a) $w = -236.95$ J

(b) $q = 236.95$ J

(c) $\Delta U = 0$

(d) All are correct

13. Choose the correct statement among the following?

(a) Absolute value of internal energy can be measured.

(b) Neither q nor w is a state of function but q + w is a state function

(c) Hot water placed in perfectly insulated closed container is an example of isolated system

(d) Both 2 and 3 are correct.

14. Which one of the following statement differentiates reversible and irreversible process correctly?

(a) Both the processes take finite time for completion

(b) A reversible process is not in equilibrium whereas an irreversible process is at equilibrium at all stages

(c) A reversible process takes infinite time for completion and is at equilibrium at all the stages whereas an irreversible process takes finite time and is not in equilibrium at all stages

(d) None of these

15.

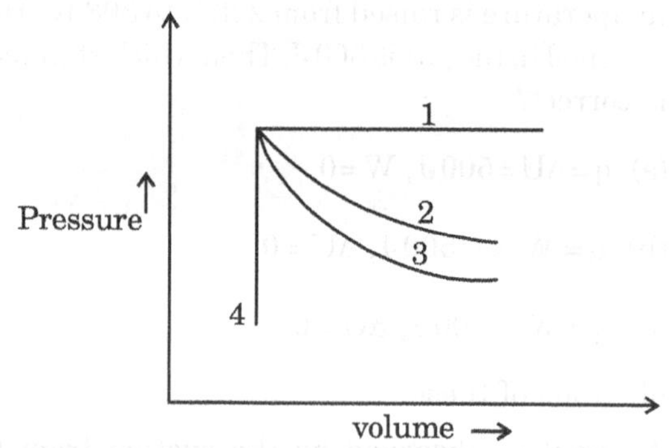

Select the correct match wrt graph above

(a) 1- isothermal
2-adiabatic
3- isochoric
4- isobaric

(b) 1- isobaric
2 -isothermal
3 - adiabatic
4- isochoric

(c) 1- isothermal
2-adiabatic
3- isobaric
4- isochoric

(d) None of these

16. Assuming that water vapour is an ideal gas, the internal energy change (ΔU) when 1 mol of water is vapourised at 1 bar pressure and 100°C is-

(Given :- Molar enthalpy vapourisation of water at 1 bar and 373K = 41KJ mol^{-1} and R = 8.314 J mol^{-1} K^{-1}) will be-

(a) 37.904 KJ mol^{-1} (b) 41 KJ mol^{-1}

(c) 4.100 KJ mol^{-1} (d) 3.7 KJ mol^{-1}

17. Ethyl chloride (C_2H_5Cl) is prepared by the reaction of ethylene with hydrogen chloride.

$$C_2H_{4(g)} + HCl_{(g)} \longrightarrow C_2H_5Cl_{(g)} \quad ; \Delta H = -72.3 KJ/mol$$

What is the volume of ΔE (in KJ), if 98g of ethylene and 109.5g of HCl are allowed to react at 300K?

(a) –64.81 (b) –190.71

(c) –209.41 (d) –224.38

18. The reversible expansion of ideal gas under adiabatic and isothermal conditions is shown in the figure. Which of the following statements is / are correct?

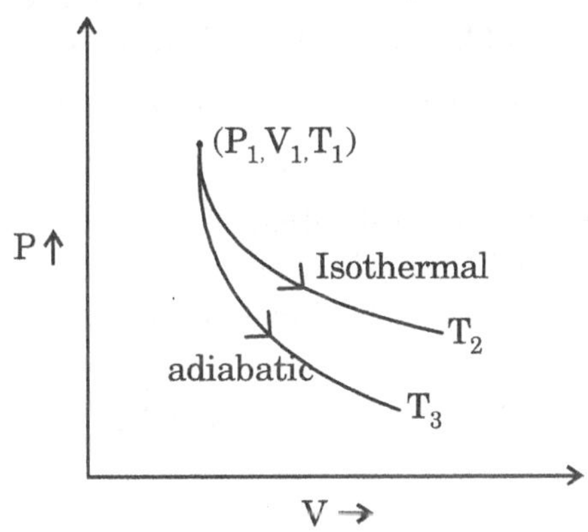

(a) $T_1 = T_2$

(b) $\Delta U_{isothermal} > \Delta U_{adiabatic}$

(c) $T_3 > T_1$

(d) Both 1 and 2 are correct

19. A container with piston is filled with 0.04 mol of an ideal gas and expands reversibly from 50ml to 375ml at a constant temperature of 34°C. As it does so, it absorbs 208 J of heat. The values of q and w for the process will be – (R = 8.314 J/mol/k) (ln 7.5 = 2.01)

(a) q = – 208 J, w = – 208 J

(b) q = –208 J, w = + 208 J

(c) q = + 208 J, w = + 208 J

(d) q = + 208 J, w = – 208 J

20. For the reaction, $A_{(g)} + B_{(g)} \rightarrow C_{(s)} + D_{(l)}$ which is true ?

(a) $\Delta H = \Delta E$ (b) $\Delta H < \Delta E$

(c) $\Delta H > \Delta E$ (d) None of these

21. 1 mole of an ideal gas undergoes a change of state (3atm, 3L) to (3atm, 7L) with a change in internal energy (ΔE)= 40 L atm. The change in enthalpy (ΔH) of the process in L atm is -

(a) 52 L atm (b) 65 L atm

(c) 77 L atm (d) 82 L atm

22. Consider the reaction at 300 K -

$$C_6H_{6(l)} + \frac{15}{2}O_{2(g)} \longrightarrow 6CO_{2(g)} + 3H_2O_{(l)} \quad ; \Delta H = -3271 KJ$$

What is ΔE (change in internal energy) for the combustion of 3.5 moles of C_6H_6 at 300k?

(a) – 11435.41 KJ

(b) – 3267.26 KJ

(c) – 3838.56 KJ

(d) – 4000 KJ

23. When 1 mole of a gas is heated at constant volume, temperature is raised from 298 K to 308 K. Heat supplied to the gas is 500 J. Then which statement is correct?

(a) $q = \Delta U = 500 J, w = 0$

(b) $q = w = -500 J, \Delta U = 0$

(c) $q = w = 500 J, \Delta U = 0$

(d) None of these

24. 1 mole of a monoatomic gas is expanded through path ABC as shown in figure below-

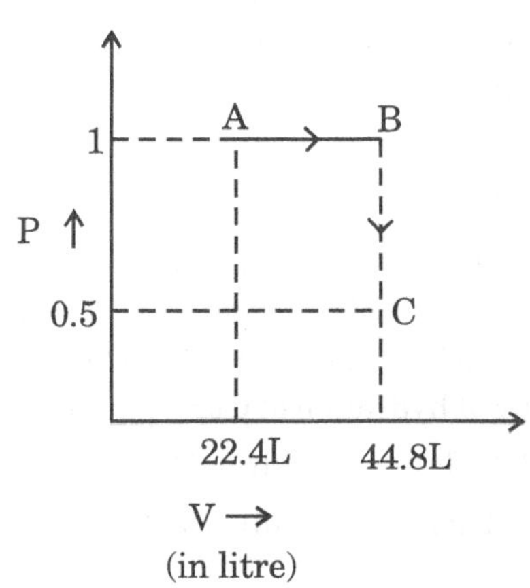

Select the correct statements.

(1) If C_p & C_v are 0.125 & 0.075 cal/g, respectively, the mol. mass of the gas is 40.

(2) Temperature at points A, B, C are 273, 546 & 273 K, respectively.

(3) Path A to B depicts isobaric process.

(4) ΔV for overall cycle, $A \to B \to C \to A$ is 3.4 KJ.

(a) 3, 4, 5 (b) 1, 2, 4

(c) 1, 3, 4 (d) 1, 2, 3

25. When the gas is ideal & the process is isothermal, then

(a) $P_1V_1 = P_2V_2$ (b) $\Delta U = 0$

(c) $\Delta H = 0$ (d) All are correct

26. The maximum work done when pressure of n moles of H_2 was reduced from 20 atm to 1 atm at constt. temperature of 273K is found to be 8180 calories. Then, the value of n is -

(a) 5 (b) 7

(c) 2 (d) 2.5

27. Select the correct match -

Column - I	Column - II
(A) Isothermal Process	(p) Temperature changes
(B) Isobaric Process	(q) Pressure changes
(C) Isochoric Process	(r) Volume changes
(D) Cyclic Process	(s) Initial & Final states are same

(a) A - q, r ; B - p, r ; C - p, q ; D - s

(b) A - q, s ; B - p, q ; C - s ; D - q, r

(c) A - q, r ; B - s, r ; C - p, q ; D - s

(d) None of these

Enthalpy of Different Types of Reactions & Calorimeter

28. 2 moles of an ideal gas undergo reversible isothermal expansion from 1L to 10L at 300K. The enthalpy change of the gas is -

(a) 4.98 KJ (b) 11.47 KJ

(c) – 11.47 KJ (d) O KJ

29. The enthalpies of solution of anhydrous $CuSO_4$, hydrated $CuSO_4.5H_2O$ are -66.5 and 11.7 KJ mol^{-1} respectively. Calculate the enthalpy of hydration of $CuSO_4$ to $CuSO_4.5H_2O$

(a) – 33.2 kJ (b) – 78.2 kJ

(c) – 786.2 kJ (d) – 99 kJ

30. The lattice enthalpy of solid NaCl is 772 kJ mol^{-1} and the enthalpy of solution is 2 kJ mol^{-1}. If the hydration enthalpy of Na^+ and Cl^- are in ratio of 3:2.5, what is the enthalpy of hydration of chloride ion

(a) –140 kJ mol-1 (b) –350 kJ/mol

(c) –820 kJ/mol (d) None of these

31. Use the given standard enthalpies of formation (in kJ/mol) to determine the enthalpy change of reaction given below

$$NH_{3(g)} + 3F_{2(g)} \longrightarrow NH_{3(g)} + 3HF_{(g)}$$

$$\Delta_f^o H(NH_3) = -46.2 \;;\; \Delta_f^o H(NF_3) = -113\;;$$

$$\Delta_f^o H(HF) = -269$$

(a) – 335. 8 kJ/mol (b) – 873.8 kJ/mol

(c) – 697.2 kJ/mol (d) – 890. 4 kJ/mol

32. The enthalpy of combustion at 25°C of H_2, cyclohexane (C_6H_{12}) and cyclohexene (C_6H_{10}) are –241, –3920 and -3800 kJ/mol respectively. The heat of hydrogenation of cyclohexene is

(a) –121 kJ mol^{-1} (b) + 121 kJ mol^{-1}

(c) – 242 kJ mol^{-1} (d) + 242 kJ mol^{-1}

33. The heat of neutralization of HCN by NaOH is - 55.9 kJ mol^{-1}. If the heat of neutralization of HCN by NaOH is - 12.1 kJ/mol. Then the energy of dissociation of HCN is

(a) – 43.8 kJ (b) + 43.8 kJ

(c) 68 kJ (d) – 68 kJ

34. For the given heat of reaction,

(a) $C_{(s)} + O_{2(g)} \longrightarrow CO_{2(g)} + 97\,kcal$

(b) $CO_{2(g)} + C_{(s)} \longrightarrow 2CO_{(g)} - 39\,kcal$

The heat of combustion of is

(a) 68 kcal (b) – 68 kcal

(c) + 48 kcal (d) – 48 kcal

35. If the bond dissociation energies of XY, X_2 and Y_2 (all diatonic molecules) are in the ratio 1 : 1 : 0.5 and $\Delta_f H$ for the formation of XY is –200 kJ mol^{-1}, the bond dissociation energy of X2 will be

(a) 100 kJ mol^{-1} (b) 200 kJmol^{-1}

(c) 400 kJ mol^{-1} (d) 800 kJ mol^{-1}

36. The enthalpy of neutralization of oxalic acid by a strong base is - 25.4 kcal mol^{-1}. The enthalpy of neutralization of strong acid and strong base is – 13.7 kcal $equivalent^{-1}$. The enthalpy of dissociation of acid is

(a) 1 kcal mol^{-1} (b) 2 kcal mol^{-1}

(c) 11.3 kcal mol^{-1} (d) 22.6 kcal mol^{-1}

37. 50 ml of 0.1M $Ba(OH)_2$ is added to 50 ml of 0.1M H_2SO_4. The rise in temperature is ΔT_1. If the experiment is repeated by taking 100 ml of each Sol., the rise in temperature is ΔT_2. Then

(a) $\Delta T_1 = \Delta T_2$ (b) $2\Delta T_2 = \Delta T_1$

(c) $3\Delta T_2 = \Delta T_1$ (d) None of these

38. Enthalpy of neutralization of H_3PO_3 acid is - 106.68 kJ / mol using NaOH. If enthalpy of neutralization of HCl by NaOH is - 55.84 kJ / mol, calculate enthalpy of dissociation of H_3PO_3 into its ions

(a) 50.84 kJ / mol (b) 5 kJ / mol

(c) 2.5 kJ / mol (d) 7.5 kJ / mol

39. In the reaction $CS_{2(l)} + 3O_{2(g)} \longrightarrow CO_{2(g)} + 2SO_{2(g)}$: $\Delta H = -265\,kcal$ The enthalpy of formation of CO_2 and SO_2 are both negative and are in the ratio 4:3.

The enthalpy of formation of CO_2 is + 26 kcal / mol. What is the enthalpy of formation of SO_2?

(a) – 71.7 kcal mol^{-1} (b) – 200 kcal mol^{-1}

(c) – 227 kcal mol^{-1} (d) + 227 kcal mol^{-1}

40. Which of the following will produce the highest rise in temperature?

(a) 67 ml of 1M NaOH + 33 ml of 0.5M H_2SO_4

(b) 33 ml of 1M NaOH + 67 ml of 0.5M H_2SO_4

(c) 40 ml of 1M NaOH + 60 ml of 0.5M H_2SO_4

(d) 50 ml of 1M NaOH + 50 ml of 0.5M H_2SO_4

41. The enthalpy of formation of $H_2O_{(l)}$ is - 285 kJ mol^{-1} and enthalpy of neutralization of a strong acid and a strong base is – 55 kJ mol^{-1}. What is the enthalpy of formation of $\bar{O}H$?

(a) $-620\,kJ/mol$ (b) $-840\,kJ/mol$

(c) $-230\,kJ\,mol^{-1}$ (d) None of these

42. Boron can undergo the following reactions with the given enthalpy changes

$$2B_{(s)} + \frac{3}{2}O_{2(g)} \longrightarrow B_2O_{3(s)} \; ; \Delta H = -1260\,kJ$$

$$2B_{(s)} + 3H_{2(g)} \longrightarrow B_2H_{6(g)} \; ; \Delta H = 30\,kJ$$

If in a container (operating at constant pressure) which is isolated from the surrounding, mixture of H_2 and O_2 is passed over excess of B(s), calculate the molar ratio of ($O_2 : H_2$) so that temperature of the container does not change

(a) 15 : 3 (b) 42 : 1

(c) 1 : 84 (d) None of these

43. The average Xe – F bond energy is 34 kcal / mol, IE_1 is 279 kcal / mol, electron affinity of F is 85 kcal / mol and bond dissociation energy of F_2 is 38 kcal / mol. Then, the enthalpy change for the reaction is

(a) 357 kcal / mol (b) 420 kcal / mol

(c) 292 kcal / mol (d) 392 kcal / mol

44. Ethanol undergoes decomposition to form two sets of products

$$C_2H_5OH_{(g)} \nearrow C_2H_4 + H_2O \; ; \Delta H = 45.54\,kJ$$

$$C_2H_5OH_{(g)} \searrow CH_3CHO + H_2 \; ; \Delta H = 68.91\,kJ$$

If the molar ratio of C_2H_4 to CH_3CHO is 8 : 1, then energy involved in the decomposition of 1 mole of C_2H_5OH is

(a) 48.13 kJ

(b) 66.25 kJ

(c) 71.83 kJ

(d) 96 kJ

45. When a certain amount of ethylene was combusted, 6226 kJ heat was evolved. If heat of combustion of ethylene is 1411 kJ, the volume of O_2 (at NTP) that entered into reaction is

(a) 296.5 ml

(b) 296.5 litres

(c) 6226 × 44.8 litres

(d) 11.2 litres

46. Find out the heat evolved in combustion of 112 litres of water gas at NTP (water gas is a mixture of equal volume $H_{2(g)}$ of $CO_{(g)}$ and)

$$\underset{(g)}{H_2} + \frac{1}{2}\underset{(g)}{O_2} \longrightarrow H_2O_{(g)} \ ; \Delta H = -241.8\,kJ$$

$$CO_{(g)} + \frac{1}{2}\underset{(g)}{O_2} \longrightarrow CO_{2(g)} \ ; \Delta H = -283\,kJ$$

(a) 1312 kJ (b) 1500 kJ

(c) 47.66 kJ (d) 98 kJ

47. The reaction

$$\underset{(g)}{CH_4} + \underset{(g)}{Cl_2} \longrightarrow \underset{(g)}{CH_3\,Cl} + HCl_{(g)}$$

has $\Delta H = -25\,kcal$

Given :	Types of bonds	Bond energy (in kcal)
	C – Cl	84
	H – Cl	109
	C – H	x
	Cl – Cl	y
	Also, $\frac{x}{y} = \frac{9}{5}$	

What is the bond enthalpy of Cl – Cl bond ?

(a) 70 kcal (b) 80 kcal

(c) 67.75 kcal (d) 60 kcal

48. The heats of atomization of $PH_{3(g)}$ and $P_2H_{4(g)}$ are 953 kJ / mol and 1485 kJ / mol respectively. The P – P bond energy in kJ / mol is

(a) 213 (b) 426

(c) 318 (d) 1272

49. The standard enthalpy of formation of NH_3 is – 46 kJ mol^{-1}. If the enthalpy of formation of H_2 from its atoms is – 436 kJ / mol and that of N_2 is – 712 kJ mol^{-1}, the average bond enthalpy of N - H bond in NH_3 is

(a) – 964 kJ / mol (b) + 352 kJ / mol

(c) + 1056 kJ / mol (d) – 1102 kJ / mol

50. Calculate the enthalpy change for the process $CCl_{4(g)} \longrightarrow C_{(g)} + 4Cl_{(g)}$ and calculate the bond enthalpy of C – Cl in $CCl_{4(g)}$

Given:- $\underset{vap.}{\Delta H}(CCl_4) = 30.5\,kJ\,mol^{-1}$

$\Delta_f H(CCl_4) = -135.5\,kJ\,mol^{-1}$

$\Delta_a H(C) = 715\ kJ\,mol^{-1}$

$\Delta_a H(Cl_2) = 242\,kJ\,mol^{-1}$

Here $\Delta_a H$ is the enthalpy of atomization

(a) $1304\,kJ\,mol^{-1}, 326\,kJ\,mol^{-1}$

(b) $1000\,kJ\,mol^{-1}, 250\,kJ\,mol^{-1}$

(c) $1200\,kJ\,mol^{-1}, 300\,kJ\,mol^{-1}$

(d) None of these

51. The enthalpy of combustion of propane (C_3H_8) gas in terms of given data is-

Bond energy (in KJ/mol)

$E_{C-H} = +X_1 \quad E_{O=O} = +X_2 \quad E_{C=O} = +X_3$

$E_{O-H} = +X_4 \quad E_{C-C} = +X_5$

Resonance energy of

CO_2 is – z kJ / mol &

$\underset{vaporization}{\Delta H}\ [H_2O_{(l)}]$ is y kJ /mol

(a) $8x_1 + 2x_5 + 5x_2 - 6x_3 - 8x_4 - 4y - 3z$

(b) $6x_1 + x_5 + 5x_2 - 3x_3 - 4x_4 - 4y - 3z$

(c) $8x_1 + 2x_5 + 5x_2 - 6x_3 - 8x_4 - y - z$

(d) $8x_1 + x_5 + 5x_2 - 6x_3 - 8x_4 - 4y + 3z$

52. The heat liberated when 1.89g of benzoic acid is burnt in a bomb calorimeter at 25°C increases the temperature of 18.94 kg of water by 0.0632°C. If the specific heat of water at 25°C is 0.998 Cal/g/degree, the value of heat of combustion of C_6H_5COOH is-

(a) 881.1 kcal (b) 771.4 kcal

(c) 981.1 kcal (d) 871.2 kcal

53. Heat of combustion of CH_4, C_2H_4, C_2H_6 are – 890, – 1411 & – 1560 KJ/mol respectively. Which of these has the highest calorific value in kJ/gm?

(a) CH_4

(b) C_2H_4

(c) C_2H_6

(d) All of these has equal calorific values

54. 36gm of pure water takes 100 seconds to evaporate from a vessel & a heater is connected to an electric source which delivers 806 watt.

The ΔH vaporization of H_2O is -

(a) 0.403KJ/mol (b) 32.66KJ/mol

(c) 40.3KJ/mol (d) 3.40 KJ/mol

55. The enthalpy change for a reaction depends upon-

(a) The physical states for reactants & products.

(b) The use of different reactants for the same products.

(c) The differences in initial or final temperatures of involved substances.

(d) All of the above

56. Consider the following reactions -

$H_2O_{(l)} \longrightarrow H_2O_{(g)}$; $\Delta H_1 = 44KJ$

$2CH_3OH_{(l)} + 3O_{2(g)} \longrightarrow 4H_2O_{(l)} + 2CO_{2(g)}$;

$\Delta H_2 = -1453KJ$

What is the value of ΔH for the second reaction if water vapour instead of liquid water is formed as product ?

(a) –1409KJ (b) –1277KJ

(c) –2839.51KJ (d) –2861- 66KJ

57. The standard enthalpy of formation of $NH_{3(g)}$ is – 91.8KJ/mol. The amount of heat required to decompose 34g of $NH_{3(g)}$ into its elements is -

(a) 183.6KJ (b) 91.8 KJ

(c) 46 KJ (d) 59.9 KJ

58. The enthalpy of fusion for ice is x KJ/mol. What is the energy released when y gm of water (liquid) is converted into ice?

(a) $\frac{xy}{18}$KJ (b) $\frac{y}{18x}$KJ

(c) $\frac{18y}{x}$KJ (d) $\frac{x}{18y}$KJ

59. 1 mole of a non-ideal gas undergoes a change of state from (1 atm, 3 L, 200k) to (5 atm, 6 L, 250k) with a change in internal energy (ΔU) = 40 L atm. The change in enthalpy (ΔH) of the process in L atm is -

(a) 27 (b) 47

(c) 57 (d) 67

60. The difference between heats of combustion of C_6H_5COOH at constant pressure & at constant volume at 27°C is-

(a) – 1447 KJ (b) – 1247 KJ

(c) 1247 KJ (d) 1447 KJ

61. Consider the following reactions-

(a) $C_{(s)} + \frac{1}{2}O_{2(g)} \longrightarrow CO_{(g)}$;

$\Delta H = -26\,kcal/mol$

(b) $CO_{(g)} + \frac{1}{2}O_{2(g)} \longrightarrow CO_{(g)}$;

$\Delta H = -68\,kcal/mol$

Which of the following statement is/are incorrect?

(a) Heat of formation of $C_{(s)}$ is – 26 kcal mol^{-1}

(b) Heat of formation of $CO_{2(g)}$ is – 194 kcal mol^{-1}

(c) Heat of formation of $CO_{(g)}$ is – 268 kcal mol^{-1}

(d) All are incorrect

62. Select the correct relation (s) among the following-

(a) $\Delta°H$ formation of $CO_{2(g)}$ = $\Delta°H$ combustion of graphite

(b) $\Delta°H$ formation of $B_2O_{3(s)}$ = $2 \times \Delta°H$ combustion of boron (s)

(c) $\Delta°H$ sublimation of sodium$_{(s)}$ = $\Delta°H$ atomisation of sodium

(d) All are correct

63. The standard molar enthalpies of formation of cyclohexane (l) & C_6H_6 (l) at 25°C are - 156 & + 49 KJ mol $^{-1}$ respectively. The standard enthalpy of hydrogenation of cyclohexane (l) at 25°C is - 119 KJ. What is the resonance energy of benzene?

(a) – 357 KJ mol^{-1} (b) – 205 KJ mol^{-1}

(c) – 152 KJ mol^{-1} (d) – 606 KJ mol^{-1}

64. Which of the following has a standard enthalpy of formation equal to zero?

(a) $HF_{(aq.)}$ (b) $F_{2(g)}$

(c) H_2SO_4 (d) Both 2 & 3

65. A person inhales 640g of oxygen per day. If all the oxygen is used up for converting sugar into CO_2 & H_2O, the heat evolved in one day will be (given ΔH combustion of sucrose = – 5645 KJ/Mol).

(a) 4704 KJ/mol (b) 9408 KJ/mol

(c) 470.4 KJ/mol (d) 784 KJ/mol

66. Which of the following reaction defines $\Delta_f^o H$?

(a) $C_{(diamond)} + O_2 \longrightarrow CO_2$

(b) $\frac{1}{2}H_{2(g)} + \frac{1}{2}F_{2(g)} \longrightarrow HF_{(g)}$

(c) $N_{2(g)} + 3H_{2(g)} \longrightarrow 2NH_{3(g)}$

(d) $CO_{(g)} + \frac{1}{2}O_{2(g)} \longrightarrow CO_{2(g)}$

67. Which of the following reaction represents enthalpy of formation?

(a) $2C + O_2 \longrightarrow 2CO$

(b) $CO + \frac{1}{2}O_2 \longrightarrow CO_2$

(c) $\frac{1}{2}N_2 + \frac{3}{2}H_2 \longrightarrow NH_3$

(d) All of these

68. Work done in expansion of an ideal gas from 4 litre to 6 litre against a constant external pressure of 2.1 atm was used to heat up 1 mole of H_2O at 293K. If specific heat of H_2O is $4.2\ J g^{-1} K^{-1}$, what is the final temperature of H_2O?

(a) 273 K (b) 393 K

(c) 332K (d) 299K

69. For the reaction,

$CH_{4(g)} \longrightarrow C_{(g)} + 4H_{(g)}$; ΔH = X KJ/mol

(a) ΔH is enthalpy of formation of methane

(b) ΔH is enthalpy of atomisation of methane

(c) ΔH is enthalpy of hydrogenation of methane

(d) ΔH is enthalpy of ionization of methane

70. In which of the following solid species, enthalpy of sublimation of 2 different species is required for calculating lattice enthalpy using Born Haber's cycle?

(a) $MgBr_2$

(b) KI

(c) NaCl

(d) CaO

71. Given,

$Mg_{(s)} \longrightarrow Mg_{(g)}$; ΔH = S

(enthalpy of sublimation)

$Mg_{(g)} \longrightarrow Mg^{2+}_{(g)} + 2e^-$; ΔH = I

(sum of 1st and 2nd ionisation enthalpy)

$Br_{2\,(g)} \longrightarrow 2Br_{(g)}$; ΔH = D

(Bond dissociation enthalpy)

$Br_{(g)} + e^- \longrightarrow Br^-_{(g)}$; $\Delta_{eg}H = -E$

$Mg^{2+}_{(g)} + 2Br^-_{(g)} \longrightarrow MgBr_{2(s)}$;

ΔH = – U (lattice enthalpy)

Then, which of the following is correct for $\Delta_f H_{MgBr_2}$?

(a) $\Delta_f H = S + I + D + (-E) + (-U)$

(b) $\Delta_f H = S + 2I + D + (-E) + (-U)$

(c) $\Delta_f H = S + I + D + (-2E) + (-U)$

(d) $\Delta_f H = S + I + D + (-E) + U$

72. The lattice energy of $Na_2CO_{3(s)}$ is 205 KJ/mol & the hydration energies of Na^+ & CO_3^{2-} ions are - 80 & - 40 KJ/mol, respectively. Which of the following statements is correct about solubility of Na_2CO_3 in H_2O?

(Assume ΔH independent of temperature)

(a) Solubility of Na_2CO_3 increases with increase in temperature.

(b) Solubility decreases with increase in temperature.

(c) Solubility first decreases gradually & then sharply with increase in temperature.

(d) Solubility is not affected by temperature at all.

73. In which of the following, the bond enthalpy & bond dissociation enthalpy are identical?

(a) H –H bond enthalpy in $H_{2(g)}$

(b) O –H bond enthalpy in $H_2O_{(g)}$

(c) C –H bond enthalpy in $CH_{4(g)}$

(d) N –H bond enthalpy in $NH_{3(g)}$

74. The enthalpy of combustion of CH_4 is – 890 KJ. The volume of CH_4 at 0°C & 1 atm to be burnt to produce 2670 KJ heat is -

(a) 67.2 L

(b) 22.4 L

(c) 44.8 L

(d) 11.2 L

75. For the allotropic change represented by the equation : -

$$C_{(graphite)} \longrightarrow C_{(diamond)} \ ; \ \Delta H = +1.9\,KJ\,mol^{-1}$$

If 6g of graphite & 6g of diamond is burnt completely in separate experiments, then the heat liberated will be higher by -

(a) 1.9 KJ in case of graphite

(b) 1.9 KJ in case of diamond

(c) 0.95 KJ in case of diamond

(d) 0.95 KJ in case of graphite

76. Ethyl chloride is prepared by reaction of ethylene (ethene) with hydrogen chloride as :-

$$C_2H_{4(g)} + HCl_{(g)} \longrightarrow C_2H_5Cl_{(g)} \ ;$$

$$\Delta H = -72.3\,KJ\,mol^{-1}$$

What is the value of energy change (in KJ) if 70g of ethylene & 73g of HCl are allowed to react ?

(a) – 142.12 KJ (b) – 155 KJ

(c) – 180 KJ (d) – 786 KJ

77. The combustion of 1g of sucrose ($C_{12}H_{22}O_{11}$) in a bomb calorimeter causes the temperature to rise from 25°C to 28.5°C. The heat of combustion of sucrose is -

(Sp. Heat capacity of calorimeter is $4.90 KJ\,g^{-1}\,K^{-1}$)

(a) $-5.86 \times 10^3\,KJ\,mol^{-1}$

(b) $-1.26 \times 10^{13}\,KJ\,mol^{-1}$

(c) $6.66 \times 10^{12}\,KJ\,mol^{-1}$

(d) None of these

78. In the reaction,

$$2H_{2(g)} + O_{2(g)} \longrightarrow 2H_2O_{(g)} \ ; \Delta H = -x\,kJ$$

(a) x kJ is the heat of formation of H_2O

(b) x kJ is the heat of reaction

(c) $\frac{x}{2}$ kJ is the heat of formation of H_2O

(d) Both two and three are correct

Spontaneity, Gibbs Energy Change and Equilibrium - Black

79. The value of $\log_{10}K$ for a reaction, $A \rightleftharpoons B$ is-

(Given: $\Delta_r H^o_{298k} = -54.07\,KJ\,mol^{-1}, \Delta_r S^o_{298k} = 10 JK^{-1}\,mol^{-1}$ & $R = 8.314 JK^{-1}\,mol^{-1}$; $2.303 \times 8.304 \times 298 = 5705$)

(a) 5 (b) 10

(c) 95 (d) 100

80. For an ideal gas, consider only P-V work in going from an initial state X to the final state Z. The final state Z can be reached by either of the two paths shown in the figure. Which of the following choices is /are correct?

(Take ΔS as change in entropy and W as work done)

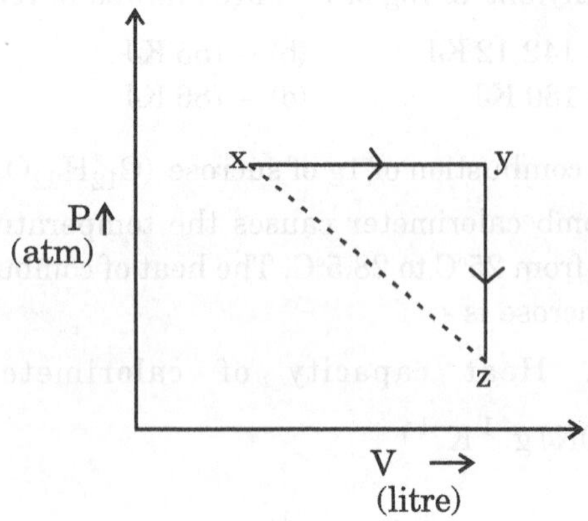

(1) $\Delta S_{x\to z} = \Delta S_{x\to y} + \Delta S_{y\to z}$

(2) $W_{x\to z} = W_{x\to y} + W_{y\to z}$

(3) $W_{x\to y\to z} = W_{x\to y}$

(4) $\Delta S_{x\to y\to z} = \Delta S_{x\to y}$

(a) Only (1) is correct

(b) Both (1) & (3) are correct

(c) All are correct

(d) None of these

81. The direct conversion of A to B is difficult. So, it is carried out by the following path-

$\Delta S_{(A\to C)} = 50$ e.u.

$\Delta S_{(C\to D)} = 30$ e.u.

$\Delta S_{(B\to D)} = 20$ e.u.

Where e.u. is entropy unit, then $\Delta S_{(A\to B)}$ is -

(a) + 100 e.u.

(b) + 60 e.u.

(c) – 100 e.u.

(d) – 60 e.u.

82. $C_2H_{6(g)} + \frac{7}{2}O_{2(g)} \longrightarrow 2CO_{2(g)} + 3H_2O_{(g)}$

$\Delta S_{vap.}(H_2O) = x_1$ cal/k (Boiling point $= T_1$)

$\Delta_f H\,[H_2O, l] = x_2$, $\Delta_f H(CO_2) = x_3$,

$\Delta_f H\ C_2H_6 = x_4$

What is the heat of reaction given above ?

(a) $2x_3 + 3x_2 - x_4$

(b) $2x_3 + 3x_2 - x_4 + 3x_1T_1$

(c) $2x_3 + 3x_2 - x_4 + 4x_1T_1$

(d) None of these

83. For a reaction, $A_{(g)} \rightleftharpoons B_{(g)}$ at equilibrium, the partial pressure of B is found to be $\frac{1}{4}^{th}$ of partial pressure of A. The value of ΔG^o of the reaction $A \to B$ is -

(a) $-RT \ln 4$

(b) $RT \ln 4$

(c) $-RT \log_{10} 6$

(d) $-RT \log_{10} 3$

84. The enthalpy change for a given reaction at 298K is $-x$ J mol^{-1} (x being positive). If the reaction occurs spontaneously at 298K, the entropy change at that temperature.

(a) Can be negative but numerically larger than $\frac{x}{298}$

(b) Can be negative but numerically smaller than $\frac{x}{298}$

(c) Can't be negative

(d) Can't be positive

85. 90g of water at 0°C is frozen at constant pressure. If the enthalpy of fusion of ice is 6 KJ mol^{-1}, the entropy change in the process is -

(a) 109.9 JK^{-1}

(b) – 109.9 JK^{-1}

(c) – 211.91 JK^{-1}

(d) + 211.9 JK^{-1}

86. Entropy changes with -

(a) Change in temperature

(b) Change in pressure

(c) Change in volume

(d) All of these

87. Select the correct order of entropy.

(a) 1 mol of gas $x >$ 1 mol of liquid $x >$ 1 mole of solid x

(b) 1 mol of gas $x >$ 1 mol of liquid $x =$ 1 mole of solid x

(c) 1 mol of liquid $x >$ 1 mole of solid $x >$ 1 mol of gas x

(d) None of these

88. The correct signs of entropy change for the following process, respectively are -

(a) A liquid crystallises into a solid

(b) Temperature of a crystal is raised

(c) $CaCO_{3(s)} \longrightarrow CaO_{(s)} + CO_{2(g)}$

(d) $\underset{(1\,atm)}{N_{2(g)}} \longrightarrow N_{2(g)}$; (0.2 atm)

(a) – ve , + ve, + ve, – ve

(b) – ve , + ve, + ve, + ve

(c) – ve , – ve, – ve, + ve

(d) None of these

89. Which of the following statement (s) is /are true for Gibbs energy?

(1) Gibbs energy is a state function.

(2) It is an extensive property.

(3) It decreases during a spontaneous change.

(4) Change in Gibbs energy represents the maximum useful work that can be obtained from the process.

(a) Only 1 & 2 are correct

(b) Only 2, 3 & 4 are correct

(c) Only 2 & 4 are correct

(d) All are correct

90. Select the correct statement.

(a) $\Delta^o G$ is 0 when system equilibrium

(b) ΔG measures the rate of reaction.

(c) $\Delta G = \Delta^o G$ when the system is at standard conditions

(d) Both 1 and 2

91. Which of the following combination of ΔH & ΔS 'Can' lead to spontaneous reaction?

(a) $\Delta H = -$ ve, $\Delta S = +$ve

(b) $\Delta H = +$ve, $\Delta S = +$ve

(c) $\Delta H = -$ ve, $\Delta S = -$ve

(d) All of these

92. The temperature dependence of equilibrium constant of a reaction is given by -

$$\ln k_{eq.} = 4.8 - \frac{2059}{T}$$

Then the value of $\Delta°S$, $\Delta°H$ & $\Delta°G$ respectively are -

(a) 5.31 KJ, 17.12 KJ, 39.9 JK^{-1}

(b) 6.62 KJ, 20.53 KJ, 44 JK^{-1}

(c) 17.12 KJ, 39.9 JK^{-1}, 5.31 KJ

(d) 20.53 KJ, 6.62 KJ, 44 JK^{-1}

93. Calculate $\Delta G°$ for the reaction -

$$CO_{(g)} + \frac{1}{2}O_{2(g)} \longrightarrow CO_{2(g)}$$

Given : - $\Delta_r^0 H = -282.8\,KJ, \Delta°S$

for CO_2, CO, O_2 are 213.6,197.6,

205 $JK^{-1}\,mol^{-1}$

(a) – 257.023 KJ mol^{-1} (b) – 237.12 KJ mol^{-1}

(c) – 299 KJ mol^{-1} (d) – 476 KJ mol^{-1}

94. For a particular reversible reaction at temperature T, ΔH & ΔS were found to be both +ve. If T_e is the temperature at equilibrium, the reaction would be spontaneous when -

(a) T_e is 5 time T (b) $T = T_e$

(c) $T > T$ (d) $T > T_e$

95. The correct expression is /are -

(a) $K_{eqm} = e^{\frac{-\Delta G^0}{RT}}$

(b) $\Delta G_{system} = -T\Delta S_{Total}$

(c) In isothermal process,

$$W_{reversible} = -nRT \ln\frac{v_f}{v_i}$$

(d) All are correct

96. For the reaction

$2O_3 \rightleftharpoons 3O_2 \ \Delta H_f = -16$ KJ/mole

We can say that

(a) ozone is more stable than oxygen

(b) ozone is less stable than oxygen and ozone decomposes forming oxygen readily

(c) oxygen is less stable than ozone and oxygen readily forms ozone

(d) none of the above

97. The value log10K for a reaction $A \rightleftharpoons B$ is -

(Given $\Delta_r H$ at 298K = −54.07KJ mol^{-1} & $\Delta_r S$ = 10JK^{-1}mol^{-1} R = 8.314 JK^{-1}mol^{-1}, $2.303 \times 8.314 \times 298 = 5705$)

(a) 5 (b) 10

(c) 95 (d) 100

98. Which one of the following is essentially zero at equilibrium?

(a) ΔS_{System} (b) ΔS_{total}

(c) $\Delta°G$ (d) ΔH

99. Identify the incorrect statement.

(a) Gibbs energy is an extensive property & a state function.

(b) ΔG has units of energy because both ΔH & $T\Delta S$ are energies

(c) ΔS_{System} is the energy which is available to do useful work.

(d) Magnitude of free energy change of a chemical reaction helps in prediction of useful work.

100. For the reaction,

$CaO_{(s)} + SO_{3(g)} \longrightarrow CaSO_{4(s)}$; $\Delta G^o = 345$ KJ

Calculate ΔG for this reaction at 298K when pressure of SO_3 is 0.5 atm.

(a) − 343.28 KJ]

(b) − 477 KJ

(c) + 471 KJ

(d) + 477 KJ

101. A quantity of 0.5 mole of an ideal gas at 300K expands isothermally against a constant pressure of 2 atm from 1 L to 5 L. So, $\Delta S_{universe}$ is -

(a) 4 JK^{-1} mol^{-1}

(b) 7 JK^{-1} mol^{-1}

(c) 9.5 JK^{-1} mol^{-1}

(d) None of these

102. For the process, (1 bar, 373K) $H_2O_{(l)} \longrightarrow H_2O_{(g)}$ (1 bar, 373K), the correct set of thermodynamic parameters is

(a) $\Delta G = 0, \Delta S = +ve$

(b) $\Delta G = 0, \Delta S = -ve$

(c) $\Delta G = +ve, \Delta S = 0$

(d) $\Delta G = -ve, \Delta S = +ve$

Answers

1. (a)	**2.** (a)	**3.** (d)	**4.** (d)	**5.** (b)	**6.** (d)	**7.** (a)	**8.** (a)	**9.** (c)	**10.** (c)
11. (a)	**12.** (d)	**13.** (d)	**14.** (c)	**15.** (b)	**16.** (a)	**17.** (c)	**18.** (d)	**19.** (d)	**20.** (b)
21. (a)	**22.** (a)	**23.** (a)	**24.** (d)	**25.** (d)	**26.** (a)	**27.** (a)	**28.** (d)	**29.** (b)	**30.** (b)
31. (b)	**32.** (a)	**33.** (b)	**34.** (b)	**35.** (d)	**36.** (b)	**37.** (a)	**38.** (b)	**39.** (a)	**40.** (d)
41. (c)	**42.** (c)	**43.** (c)	**44.** (a)	**45.** (b)	**46.** (a)	**47.** (d)	**48.** (a)	**49.** (b)	**50.** (a)
51. (a)	**52.** (b)	**53.** (a)	**54.** (c)	**55.** (d)	**56.** (b)	**57.** (a)	**58.** (a)	**59.** (d)	**60.** (b)
61. (d)	**62.** (d)	**63.** (c)	**64.** (b)	**65.** (b)	**66.** (b)	**67.** (c)	**68.** (d)	**69.** (b)	**70.** (b)
71. (c)	**72.** (a)	**73.** (a)	**74.** (a)	**75.** (c)	**76.** (a)	**77.** (a)	**78.** (d)	**79.** (b)	**80.** (b)
81. (b)	**82.** (b)	**83.** (b)	**84.** (b)	**85.** (b)	**86.** (d)	**87.** (a)	**88.** (b)	**89.** (d)	**90.** (d)
91. (*)	**92.** (a)	**93.** (a)	**94.** (d)	**95.** (d)	**96.** (b)	**97.** (b)	**98.** (b)	**99.** (c)	**100.** (a)
101. (a)	**102.** (a)								

Explanations

1. (a) At Constt. Volume, w = 0

$\Delta E = q + w$ (from 1st law)

ΔU or $\Delta E = q + 0$

$\Delta U = q$

2. (a) ΔE or $\Delta U = Q + w$

Now, q = 0

$\Delta E = w$

3. (d) It's a cyclic process and $\Delta E = 0$ for cyclic process.

4. (d) Internal energy is an extensive property and depends upon matter.

5. (b) $Zn + 2HCl \longrightarrow ZnCl_2 + H_2 \uparrow$

$W = -P\Delta V = -n_{H_2}RT$

$= \dfrac{-2 \times 8.34 \times 298}{1000} = -4.955 KJ$

7. (a) For an adiabatic process, q = 0

w = 0 (expansion in vacuum)

$\Delta E = q + w$

$\Delta E = 0 + 0 = 0$

8. (a) $w_{rev} = -2.303 nRT \log \dfrac{p_1}{p_2}$

$= -2.303 \times 2 \times 8.314 \times 300 \log \dfrac{1.01 \times 10^5}{5.05 \times 10^6}$

$w = +1.95 \times 10^4 J$

$w_{compression}$ = w. done on the system

9. (c) $\Delta E = q + w$

$\Delta E = 40 + (-8)$

$\Delta E = 32J$

10. (c)

1) ΔE is a state function	2) Isochoric process has constant volume	(c) For adiabatic Process
For cyclic process, $\Delta E = 0$	$\Delta V = 0$, So w=0	q = 0
$\Delta E = q + w$	$\Delta E = q + w$	$\Delta E = q + w$
q = w	$\Delta E = q$ $\Delta E = q$	$\Delta E = w$

11. (a) Initial Internal energy = U_1

w done = 600 J

q = – 450 J

$\Delta U = q + w = -450 + 600 = 150 J$

Total energy = $U_1 + 150$

12. (d) Initially for N_2 : Apply PV = nRT

$20 \times V_1 = \dfrac{2.8}{28} \times 0.0821 \times 300$

$V_1 = 0.123 L$

Finally for N_2

$1 \times V_2 = \dfrac{2.8}{28} \times 0.0821 \times 300$

$V_2 = 2.463 L$

$w = -P_{ext} \cdot \Delta V = -1(2.463 - 0.123)$

$= -2.340 L atm$

$w = \dfrac{-2.340 \times 2}{0.0821}$

$= \dfrac{-2.340 \times 2 \times 4.18 J}{0.0821} = -236.95 J$

$\Delta E = q + w$

($\Delta E = 0$ for isothermal process in case of ideal gas)

$0 = q + w$

$q = -w$

$\therefore q = +236.95 J$

16. (a) $H_2O_{(l)} \longrightarrow H_2O_{(g)}$

$\Delta H = \Delta U + \Delta ngRT$

$41 = \Delta U + 1 \times \frac{8.314}{1000} \times 373$

$41 = \Delta U + 3.101$

$\Delta U = 37.9\,KJ\,mol^{-1}$

17. (c) $C_2H_{4(g)} + HCl_{(g)} \longrightarrow C_2H_5Cl_{(g)}$

Apply, $\Delta H = \Delta E + \Delta ng\,RT$

$-72.3 = \Delta E + (-1) \times \frac{8.314}{1000} \times 300$

$-72.3 + 2.5 = \Delta E$

$-69.8\ KJ\,mol^{-1} = \Delta E$

Now, ${}^{n}C_2H_{4(g)} = \frac{98}{28} = 3.5$ moles

(No, of moles of ethylene)

${}^{n}HCl = \frac{109.5}{36.5} = 3$ moles

Limiting reagent is HCl,

So, 3 moles of ethylene will eact wit 3 moles of HCl to form 3 moles of ethyl chloride

So, ΔE wrt 3 moles $= -69.8 \times 3 = -209.4\,KJ$

18. (d) $T_1 = T_2$ (isothermal process)

$T_3 > T_1$ is in correct ($\because$ cooling occurs in adiabatic expansion)

$\Delta U_{isothermal} = 0$ & is $> \Delta U_{adiabatic} = -ve$

19. (d) T he process is isothermal reversible expansion,

So, $\Delta E = 0$

$\Delta E = q + w$

$q = -w$

If $q = +208J$

$w = -208\,J$

20. (b) Apply, $\Delta H = \Delta E + \Delta ng\,RT$

$\Delta H = \Delta E + (-2)RT$

$\Delta H = \Delta E - 2RT$

$\Delta H + 2RT = \Delta E$

$\Delta E > \Delta H$

21. (a) $\Delta H = \Delta E + P\Delta V$

$\Delta H = 40 + 3(7-3)$

$\Delta H = 40 + 3 \times 4$

$\Delta H = 52\,L\,atm$

22. (a) $C_6H_{6(l)} + \frac{15}{2}O_{2(g)} \longrightarrow 6CO_{2(g)} + 3H_2O_{(l)}$;

$\Delta H = -3271\,KJ/mol$

Apply, $\Delta H = \Delta E + \Delta ngRT$

$-3271 = \Delta E + (-1.5) \times \frac{8.314}{1000} \times 300$

$-3271 = \Delta E - 1.5 \times \frac{8.314}{10\not{0}\not{0}} \times 3\not{0}\not{0}$

$-3271 = \Delta E - 3.74$

$\Delta E = -3271 + 3.74 = -3267.26\,KJ/mol$

So, $\Delta E = -3267.26\,KJ/mol$

For 1 mol, $\Delta E = -3267.26$

For 3.5 mol, ΔE

$= -3267.26 \times 3.5 = -11435.41\,KJ$

23. (a) At Constt. Volume, w = 0

$\Delta E = q + w$ (from 1st law)

ΔU or $= \Delta E = q + 0$

$\Delta U - q$

24. (d) (1) $C_p - C_v = \frac{R}{M}$

$0.125 - 0.075 = \frac{2}{M}$

$\frac{2}{0.05} = M$

M = 40g

(2) At A,

P = 1 atm

V = 22.4 L

$\therefore$ T = 273K

At B,

P = 1 atm

V = 44.8L

$\therefore$ T = 546K

At C,

P = 0.5 atm

V = 44.8 L

$\therefore$ T = 273 K

25. (d) For ideal gas, Boyle's law is valid at constt. temp.

$\therefore P_1V_1 = P_2V_2$

U is a function of temp. if temp. is constt, $\Delta U = 0$

Also, $\Delta H = \Delta E + \Delta(PV)$

$\Delta H = 0 + 0$

$\Delta H = 0$

26. (a) $W = -2.303nRT\log\frac{P_1}{P_2}$

$-8180 = -2.303n \times 2 \times 273 \log\frac{20}{1}$

$n = 5$

28. (d) $H = E + PV$

$\Delta H = \Delta E + \Delta(PV)$

Isothermal condition means T is constt.

$PV = K \rightarrow$ Constt. (Boyle's law valid)

$\Delta E = 0$ (for ideal gas constt. T)

$\therefore \Delta H = 0 + 0$

$\Delta H = 0$

29. (b) Using Hess's law

$\Delta H_1 = \Delta H_2 + \Delta H_3$

$-66.5 = \Delta H_2 + 11.7$

$-66.5 - 11.7 = \Delta H_2$

$-78.2\,kJ = \Delta H$

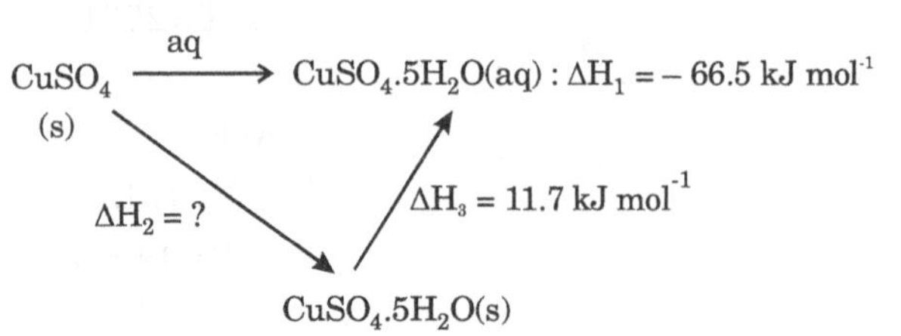

30. (b) $\Delta H_{Sol.} = \Delta H_{hyd.} + \Delta H_{LE}$

$2 = \Delta H_{hyd.} + 772$

$\Delta H_{hyd.} = -770\,kJ/mole$

Now, $\Delta H_{hyd.} = H.E_{Na^+} + H.E_{Cl^-}$

$-770 = n + \frac{2.5n}{3}$

$-770 = \frac{3n + 2.5n}{3}$

$-2310 = 5.5n$

$\boxed{n = -420\,kJ/mole}$

$H.E.\ of\ Cl^- = \frac{2.5}{3} \times n = \frac{2.5}{3} \times -420$

$= -350\,kJ/mol$

31. (b) $NH_3 + 3F_2 \longrightarrow NF_3 + 3HF$

$\Delta_f H = \sum \Delta_f^o H_{(P)} - \sum \Delta_f H_{(R)}$

$= (-113 + 3 \times -269) - (-46.2 + 0)$

$= (-113 + (-807)) + 46.2$

$= (-113 - 807) + 46.2$

$= -920 + 46.2 = -873.8\,kJ/mol$

32. (a) $H_{2(g)} + \frac{1}{2}O_{2(g)} \longrightarrow H_2O_{(l)}$ .... (a)

$;\Delta H = -241\,kJ\,mol^{-1}$

$C_6H_{10} + \frac{17}{2}O_{2(g)} \longrightarrow 6CO_2 + 5H_2O_{(l)}$ ...(b)

$;\Delta H = -3800\,kJ\,mol^{-1}$

$C_6H_{12} + 9O_{2(g)} \longrightarrow 6CO_{2(g)} + 6H_2O_{(l)}$ . . . .

(c) Now, adding equation a and b

$$H_{2(g)} + \frac{1}{2}O_{2(g)} \longrightarrow H_2O_{(l)} \quad ;\Delta H = -241\,kJ\,mol^{-1} \quad .....(a)$$

$$(+)\; C_6H_{10} + \frac{17}{2}O_{2(g)} \longrightarrow 6CO_{2(g)} + 5H_2O_{(l)} \;;\; \Delta H = -3800\,kJ\,mol^{-1} \;......(b)$$

$$C_6H_{10} + H_{2(g)} + 9O_{2(g)} \longrightarrow 6CO_{2(g)} + 6H_2O_{(l)} \;;\; \Delta H = -4041\,kJ\,mol^{-1}$$

$$C_6H_{10} + H_{2(g)} + 9O_{2(g)} \longrightarrow 6CO_{2(g)} + 6H_2O_{(l)};\; \Delta H = -4041\,kJ\,mol^{-1}$$

$$\underset{(-)}{C_6H_{12}} + \underset{(-)}{9O_{2(g)}} \longrightarrow \underset{(-)}{6CO_{2(g)}} + \underset{(-)}{6H_2O_{(l)}} \;;\; \Delta H = -3920\,kJ\,mol^{-1}$$

$$\Rightarrow \quad C_6H_{10} + H_2 - C_6H_{12} \longrightarrow 0$$

$$C_6H_{10} + H_2 \longrightarrow C_6H_{12}\,, \quad \Delta H = -121\,kJ$$

33. (b) Enthalpy of dissociation = (–12.1) – (– 55.9)

= – 12.1 + 55.9 = 43.8 kJ

Or

Enthalpy of dissociation = 55.9 – 12.1 = + 43.8 kJ

34. (b) (a) $C_{(s)} + O_{2(q)} \longrightarrow CO_{2(q)};\Delta H = -97\,kcal$

(b) $\underset{(-)}{CO_{2(g)}} + \underset{(-)}{C_{(s)}} \longrightarrow \underset{(-)}{2CO_{(g)}};\Delta H = +39\,kcal$

$$C_{(s)} + O_{2(g)} - CO_{2(g)} - C_{(s)} \rightarrow CO_{2(g)} - 2CO_{(g)}$$

$$O_{2(g)} - CO_{2(g)} \longrightarrow CO_{2(g)} - 2CO_{(g)}$$

$$2CO_{(g)} + O_{2(g)} \longrightarrow 2CO_{2(g)};\; \Delta H = -136\,kcal$$

($\because$ Heat of combustion is defined for 1 mole)

$$CO_{(g)} + \frac{1}{2}O_{2(g)} \longrightarrow CO_{2(g)};\Delta H = -68\,kcal$$

35. (d) $\frac{1}{2}X_{2(g)} + \frac{1}{2}Y_{2(g)} \longrightarrow XY_{(g)}$;

$\Delta H = -200\,kJ\,mol^{-1}$

Let B.E. of XY = n kJ / mol

B.E. of X_2 = n kJ / mol

B.E. of $Y_2 = \frac{n}{2}$ kJ / mol

$$\Delta_f H = \sum BE_R - \sum BE_P$$

$$-200 = \left(\frac{n}{2} + \frac{0.5n}{2}\right) - n$$

$$-200 = 0.75n - n$$

$$-200 = -0.25n$$

$$\frac{200}{0.25} = n$$

$$n = 800\,kJ/mol$$

36. (b) Enthalpy of neutralization of oxalic acid by a strong base = – 25.4 kcal mol^{-1} = –12.7 kcal equivalent^{-1}

So, enthalpy of dissociation = 13.7 – 12.7 = 1 kcal equivalent^{-1} = 2 kcal mol^{-1}

37. (a) When 100 ml of each, $Ba(OH)_2$ and H_2SO_4 will be picked, the heat evolved will be doubled but rise in temperature will be same. If 'q' is doubled, 'M' is also doubled.

($q = MC\Delta T$)

38. (b) Structure of H_3PO_3

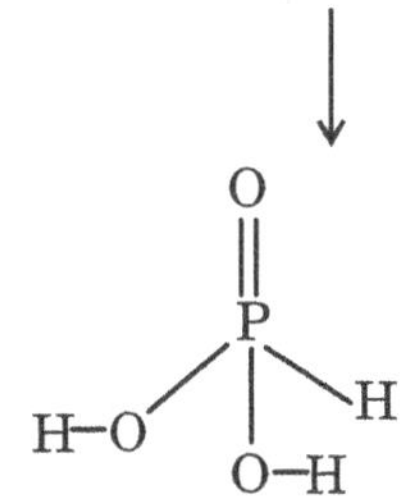

(It is a dibasic acid)

$$H_3PO_3 + 2NaOH \longrightarrow Na_2HPO_3 + 2H_2O$$

Since, It is a dibasic acid, so, energy released should have been $= 55.84 \times 2 = 111.68$

But actual energy released = 106.68kJ

So, enthalpy of ionization of $H_3PO_3 = 111.68 - 106.68 = 5$ kJ / mol

39. (a) $CS_2 + 3O_2 \longrightarrow CO_2 + 2SO_2$;

$\Delta_f H = -265$ kcal ....(1)

$C + O_2 \longrightarrow CO_2$; $\Delta H = -4x$ ....(2)

$S + O_2 \longrightarrow SO_2$; $\Delta H = -3x$ ....(3)

wrt reaction 1,

$$\Delta_f H = \sum \Delta_f H_{(P)} - \sum \Delta_f H_{(R)}$$

$$-265 = (-4x + (-6x)) - (26))$$

$$-265 = -4x - 6x - 26$$

$$-265 + 26 = -10x$$

$$-239 = -10x$$

$$x = 23.9 \text{ kcal/mol}$$

Enthalpy of formation of

$SO_2 = -3x = -3 \times 23.9 = -71.7$ kcal/mol

40. (d) [No. of gm eq. = N × $V_{(l)}$]

(a) 67 ml of 1M NaOH + 33 ml of 0.5M H2SO4

NaOH: 67 ml, 1N (=1M) → 0.067 gm eq.
+
H_2SO_4: 33 ml, 1N (0.5M) → 0.033 gm eq.

Neutralisation of 0.033 gm eq. occurs. Energy released is 57.1 . 0.033 kJ

(b)

33 ml of 1M NaOH: 33 ml, 1N → 0.033 gm eq.
+
67 ml of 0.5 M H_2SO_4: 67 ml, 1N → 0.067 gm eq.

Neutralisation of 0.033 gm eq. occurs. Energy released is 0.033 × 57.1 kJ

(c) 40 ml of 1M NaOH + 60 ml of 0.5M H_2SO_4

40 ml, IN → 0.04 g eq.
60ml, IN → 0.06 g eq.

Neutralisation of 0.04 gm eq. occurs. Energy released is 57.1× 0.04 kJ

(d) 50 ml of 1M NaOH + 50 ml of 0.5M H_2SO_4

50 ml, IN → 0.05 gm eq.
50ml, IN → 0.05 g eq.

Neutralisation of 0.05 gm eq. occurs. Energy released is 57.1× 0.05 kJ

Since, energy released is maximum in last option, So temperature rise is also maximum

41. (c) $H^+_{(aq)} + \overline{O}H_{(aq)} \longrightarrow H_2O_{(l)}$

$$\Delta H_{f(H^+)} = 0$$

$$\Delta_f H_{(H_2O)} = -285 \text{ kJ mol}^{-1}$$

$$\Delta H_{(neutralization)} = \Delta_f H_{(H_2O, l)} - \Delta_f H_{(\overline{O}H, aq.)}$$

So, $\Delta_f H_{(\overline{O}H, aq.)} = (-285 - (-55))$

$= -230$ kJ mol^{-1}

42. (c) To ensure that temperature does not change, energy released should be equal to energy absorbed. $= \frac{3}{2}$

1260 kJ energy is liberated by moles of $O_2 = \frac{3}{2}$

30 kJ energy is absorbed when moles of H2 reacted = 3

1260 kJ energy is absorbed when moles of H2 reacted $= \frac{3}{30} \times 1260 = 126$ moles

molar ratio of

$$\frac{O_2}{H_2} = \frac{\frac{3}{2}}{\frac{126}{1}} = \frac{3}{2} \times \frac{1}{126} = \frac{1}{84} = 1:84$$

43. (c) Energy absorbed in breaking 4 mole of Xe – F bonds = 4 × 34 = 136 kcal

Energy absorbed as ionization energy = 279 kcal

Energy released as electron gain enthalpy = 85 kcal

Energy released during formation of F2 = 38 kcal

Total energy absorbed = 136 + 279 = 415 kcal

Total energy released = 85 + 38 = 123 kcal

Net energy absorbed = 415 – 123 = 292 kcal

44. (a)

$$C_2H_5OH \nearrow \underset{8x}{C_2H_4} + H_2O \quad ;\underset{\text{—step 1}}{\Delta H} = 45.54\,kJ$$

$$C_2H_5OH \searrow \underset{x}{CH_3CHO} + H_2 \quad ;\underset{\text{—step 2}}{\Delta H} = 68.91\,kJ$$

Energy absorbed in step 1 = 45.54 × 8*x* kJ

Energy absorbed in step 2 = 68.92 × *x* kJ

The energy absorbed when 9*x* moles of C_2H_5OH react = 45.54 × 8 *x* + 68.91*x*

The energy absorbed when 1 moles of C2H5OH react

$$= \frac{45.54 \times 8x + 68.91x}{9x}$$

$$= \frac{364.32x + 68.91x}{9x}$$

$$= \frac{433.23x}{9x} = 48.13\,kJ$$

15. (b) $C_2II_4 + 3O_2 \longrightarrow 2CO_2 + 2II_2O$,

$\Delta H = -1411\,kJ/mol$

1411 kJ energy is evolved when amount of ethylene burnt = 1 mole

6226 kJ energy is evolved when amount of ethylene burnt

$$= \frac{1}{1411} \times 6226 \text{ mole} = 4.412 \text{ moles}$$

∴ the number of moles of ethylene reacted = 4.412

No. of moles of O_2 reacted = 13.23 (4.412 ×3)

Volume of O_2 reacted at NTP = 13.23 × 22.4 = 296.91 litres

46. (a) $H_{2(g)} + \frac{1}{2}\underset{(g)}{O_2} \longrightarrow \underset{(g)}{H_2O}\ ; \Delta H = -241.8\,kJ$

$$\underset{(+)}{CO_{(g)}} + \frac{1}{2}\underset{(+)}{O_{2(g)}} \longrightarrow \underset{(+)}{CO_{2(g)}}\ ; \Delta H = -283\,kJ$$

$$CO_{(g)} + H_{2(g)} + O_{2(g)} \longrightarrow CO_{2(g)} + H_2O_{(g)}\ ;$$

$\Delta H = -524.8\,kJ$

44.8 litres of water gas liberate energy = 524.8 kJ

112 litres of water gas liberate energy

$$= \frac{524.8}{44.8} \times 112\,kJ = 1312\,kJ$$

47. (d) $\Delta_f H = \sum B.E_R - \sum B.E_P$

$-25 = (4x + y) - (3x + 84 + 109)$

$-25 = 4x + y - 3x - 84 - 109$

$-25 = x + y - 84 - 109$

$-25 + 84 + 109 = x + y$

$168 = x + y$ ___(1)

Also, $\frac{x}{y} = \frac{9}{5} = \frac{9}{5}y$

Now, from equation 1

$$168 = \frac{9}{5}y + y$$

$$168 = \frac{14y}{5}$$

$y = 60$

$$x = \frac{9}{5}y$$

$$x = \frac{9}{5} \times 60$$

$x = 108$

48. (a) For, PH_3, P – H bond energy = heat of atomization of $\frac{PH_3}{3}$

For P_2H_4,

Heat of atomization = P – P bond energy + 4 × P – H B.E.

∴ P – P B.E = 1485 – (4 × 317.67) = 213 kJ /mol

49. (b) $2H_{(g)} \longrightarrow H_{2(g)}\ ;\Delta^oH = -436\,kJ/mol$

$2N_{(g)} \longrightarrow N_{2(g)}\ ;\Delta^oH = -712\,kJ/mol$

$\frac{1}{2}\underset{(g)}{N_2}+\frac{3}{2}\underset{(g)}{H_2} \longrightarrow NH_{3(g)}\ ;\Delta^o_f H = -46\,kJ/mol$

$(\Delta^0_f H = \sum B.E_R - \sum B.E_P)$

$-46 = \left(\frac{1}{2}\times 712+\frac{3}{2}\times 436\right)-(3x)$

$-46 = (356+654)-3x$

$x(B.E_{N\text{-}H}) = 352kJ/mol$

50. (a) (1) $CCl_{4(l)} \longrightarrow CCl_{4(g)}$;

$\Delta H = 30.5\,kJ\,mol^{-1}$ ___(1)

(2) $C_{(s)} + 2Cl_{2(g)} \longrightarrow CCl_4$;

$\Delta H = -135.5\,kJ\,mol^{-1}$ ___(2)

(3) $C_{(s)} \longrightarrow C_{(g)}$; $\Delta H = 715\,kJ\,mol^{-1}$ ___(3)

(4) $Cl_{2(g)} \longrightarrow 2Cl_{(g)}$; $\Delta H = 242\,kJ\,mol^{-1}$ ___(4)

The resultant equation is obtained as follows 3rd equation + 2 × 4th equation – 1st equation – 2rd equation so, $\Delta H = 1304\,kJ/mol$

Bond enthalpy of (C – Cl) bond in

$CCl_{4(g)} = \frac{1304}{4} = 326\,kJ/mol$

51. (a) $C_3H_8 + 5O_2 \rightarrow 3CO_2 + 4H_2O$

Energy absorbed $= (2x_5 + 8x_1 + 5x_2)$

Energy released $= (6x_3 + 8x_4 + 3z + 4y)$

$\Delta H = E_{absorbed}$ – Energy released

$= (2x_5 + 8x_1 + 5x_2) - (6x_3 + 8x_4 + 3z + 4y)$

$= 2x_5 + 8x_1 + 5x_2 - 6x_3 - 8x_4 - 3z - 4y$

52. (b) Heat absorbed,

$q = mc\Delta T = 18940 \times 0.998 \times 0.632$
$= 11946.14\,Cal$

∴ on burning 1.89g of C_6H_5COOH,

heat evolved = 11946.14 Cal

on burning 122g of C_6H_5COOH,

heat evolved $= \frac{11946.14}{1.89} \times 122 = 771.12kcal$

53. (a) For CH_4

Energy released on burning 16gm = 890 KJ

Energy released on burning 1gm

$= \frac{890}{16} KJ = 55.625KJ$

For C_2H_4

Energy released on burning 28gm C_2H_4 = 1411 KJ

Energy released on burning 1gm C_2H_4

$= \frac{1411}{28} KJ = 50.39KJ$

For C_2H_6

Energy released on burning 30gm C_2H_6 = 1560 KJ

Energy released on burning 1gm C_2H_6

$= \frac{1560}{30} KJ = 52KJ$

So, CH_4 has the highest calorific value

54. (c) In 1 second, energy derived = 806J

In 100 second, energy derived

$= 806 \times 100 = 80600J$

So, 36gm water requires energy = 80600J

18gm water requires energy

$= \frac{80600}{\cancel{36}\ 2} \times \cancel{18} = 40.3KJ/mol$

56. (b) $H_2O_{(l)} \longrightarrow H_2O_{(g)}$; $\Delta H_1 = 44KJ$

Multiply equation by '4'

(1) $4H_2O_{(l)} \longrightarrow 4H_2O_{(g)}$; $\Delta H_1 = 44 \times 4KJ$

(2) $2CH_3OH_{(l)} + 3O_{2(g)} \longrightarrow 4H_2O_{(l)} + 2CO_{2(g)}$;

$\Delta H_2 = -1453KJ$

Add eqn. 1 & 2: – $2CH_3OH_{(l)} + 3O_{2(g)} \longrightarrow 4H_2O_{(g)}$

$+2CO_{2(g)}$; $\Delta H = 176 + (-1453KJ) = -1277KJ$

57. (a) $\frac{1}{2}N_{2(g)} + \frac{3}{2}H_{2(g)} \rightleftharpoons NH_{3(g)}$; $\Delta H = -91.8KJ/mol$

The amount of heat released when 17gm NH_3 is formed = 91.8KJ

The amount of heat required when 17gm NH_3 is decomposed = 91.8KJ

The amount of heat required when 34gm NH_3

is decomposed = $\frac{91.8}{\cancel{17}} \times \overset{2}{\cancel{34}}KJ$

= 183.6KJ

58. (a) $Ice_{(solid)} \rightleftharpoons Water_{(liquid)}$; $\Delta H = x KJ/mol$

So, x KJ energy is observed in converting 18gm ice to liquid water.

Then, Energy released on converting 18gm water to ice = x KJ

Energy released on converting y gm water to

ice = $\frac{x}{18} \times y$

59. (d) H = E + PV

$H_1 = E_1 + 1 \times 3$ ____(1)

$H_2 = E_2 + 5 \times 6$ ____(2)

2 – 1

$H_2 - H_1 = (E_2 + 30) - (E_1 + 3)$

$\Delta H - E_2 + 30 - E_1 - 3$

$\Delta H = \Delta E + 27$

$\Delta H = 40 + 27 = 67 L atm$

60. (b) $C_6H_5COOH_{(l)} + \frac{15}{2}O_{2(g)} \longrightarrow 7CO_{2(g)} + 3H_2O_{(l)}$

$\Delta H = \Delta E + \Delta ngRT$ [$\Delta H = qp$ & $\Delta E = qv$]

$\Delta H - \Delta E = \Delta ngRT$

$\Delta H - \Delta E = -\frac{1}{\cancel{2}} \times 8.314 \times \overset{150}{\cancel{300}}$

$\Delta H - \Delta E = -1247.1 J$

61. (d) There is 'No' reaction depicting the heat of formation of C. However, the first reaction depicts the heat of formation of

CO & $\Delta_f H$ for $CO_{(g)} = -26 kcal/mol$.

Adding equations 1 & 2, we get

$C_{(s)} + O_{2(g)} \longrightarrow CO_{2(g)}$; ΔH

$= -26 + (-68) = -26 - 68 = -94 kcal/mol$

So, heat of formation of $CO_{2(g)}$ is –94 kcal/mol

62. (d)

(a) $C_{(graphite)} + O_{2(q)} \longrightarrow CO_{2(q)}$; $\Delta°H = x J/mol$ ↑

This enthalpy change represents standard enthalpy of combustion of C as well as standard emthalpy of formation of CO_2.

(b) Also, $2B_{(s)} + \frac{3}{2}O_{2(q)} \longrightarrow B_2O_{3(s)}$; $\Delta°H = y J/mol$

Here, enthalpy of formation of $B_2O_3 = y$ J/mol

& enthalpy of combustion of B = $\frac{y}{2}$ J/mol

(c) $Na_{(s)} \longrightarrow Na_{(g)}$; $\Delta H = z J/mol$ ↓

Enthalpy of sublimation of sodium

& enthalpy of atomization of sodium.

63. (c) If C_6H_6 (benzene) is considered cyclohexatriene, the observed enthalpy is -

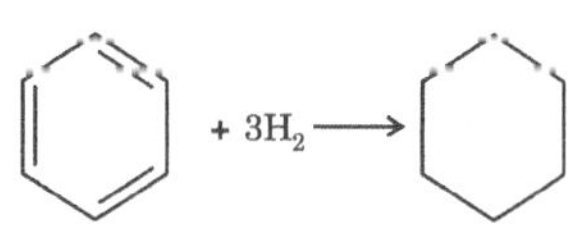

; NH = 3 × 119 = – 357 KJ/mol

Calculated enthalpy of above reaction

$\Delta_r H = \Delta_f^0 H_{(C_6H_{14})} - [\Delta_f^0 H_{(C_6H_6)} + 3\Delta_f^0 H(H_2)]$

$= -156 - (49 + 0) = -205 KJ mol^{1}$

Resonance energy = Observed

$\Delta_r H^0$ – Calculated $\Delta_r^0 H$

$= (-357) - (205) = -152 KJ/mol$

65. (b) $C_{12}H_{22}O_{11} + 12O_2 \longrightarrow 12CO_2 + 11H_2O$

$640 g\ O_2 = \frac{640}{32} = 20 moles (given)$

Now, 12 mole O_2 react with sucrose = 1 mole

20 moles O_2 will react with sucrose

$= \frac{1}{12} \times 20 = 1.66$ moles

Heat liberated on burning 1 mole sucrose

= 5645KJ

Heat liberated on burning 1.66 mole sucrose

$= 5645 \times 1.66 = 9408$ KJ/mol

67. (c) Enthalpy of formation represents the enthalpy change when 1 mole of a substance is formed from its constituent elements.

68. (d) $W = -P_{ext}.\Delta V$

$W = -2.1(6-4) = -4.2$ atm litre

$W = -4.2 \times 101.325$ J

So, these many Joules are used to heat up water.

Specific heat of $H_2O = 4.2\ J g^{-1} k^{-1}$

For 1 mole, energy required to raise the temp. by 1°

$= 4.2 \times 18 = 75.6\ J mol^{-1} k^{-1}$

$4.2 \times 101.325 = 75.6(T - 293)$

T = 299 K

70. (b) Both K & I_2 are in solid state.

Both require sublimation energy.

71. (c) (a) $Mg_{(s)} \longrightarrow Mg_{(g)}$; $\Delta H = S$

(b) $Mg_{(g)} \longrightarrow Mg^{2+}_{(g)} + 2e^-$; $\Delta H = I$

(c) $Br_{2\,(g)} \longrightarrow 2Br_{(g)}$; $\Delta H = D$

(d) $Br_{2\,(g)} + 2e^- \longrightarrow 2Br^-_{(g)}$; $\Delta H = -2E$

(5) $Mg^{2+}_{(g)} + 2Br^-_{(g)} \longrightarrow MgBr_{2(s)}$; $\Delta H = -U$

So, $\Delta_f H = S + I + D + (-2E) + (-U)$

72. (a) $Na_2CO_3 \longrightarrow 2Na^+ + CO_3^{2-}$

H.E. for Na^+ = – 80 KJ/mol

H.E. for 2 mole $Na^+ = -80 \times 2 = -160$ KJ

H.E. for 1 mole $CO_3^{2-} = -40$ KJ

Total H.E. = – 160 + (– 40) = – 200 KJ/mol

$\Delta H_{L.E.} = 205$ kJ / mol

$\Delta H_{dissolution} = \Delta H_{L.E.} + \Delta H_{H.E.}$

$\Delta H_{dissolution} = 205 + (-200)$

$\Delta H_{dissolution} = 5$ KJ/mol

So, solubility of Na_2CO_3 is endothermic.

So, solubility increases with increase in temperature.

74. (a) $CH_{4(g)} + 2O_{2(g)} \longrightarrow CO_{2(g)} + 2H_2O_{(l)}$

890 KJ heat is liberated by burning CH_4 = 1 mole

2670 KJ heat is liberated by burning

$CH_4 = \frac{1}{890} \times 2670 = 3$ moles

For volume of CH_4

Apply PV = nRT

$1 \times V = 3 \times 0.0821 \times 273$

$V = 3 \times 22.4$

V = 67.2L

75. (c) $C_{(graphite)} \longrightarrow C_{(diamond)}$;

$\Delta H = +1.9\ KJ mol^{-1}$

To convert 12g graphite to diamond, energy required = 1.9 KJ

To convert 6g graphite to diamond, energy required $= \frac{1.9}{12} \times 6 = 0.95$ KJ

When 6g diamond will be converted to 6g graphite, energy released = 0.95 KJ

76. (a) $C_2H_{4(g)} + HCl_{(g)} \longrightarrow C_2H_5Cl_{(g)}$;

$\Delta H = -72.3\ KJ mol^{-1}$

No. of moles of $C_2H_4 = \frac{70}{28} = 2.5$

No. of moles of $HCl = \frac{73}{36.5} = 2$

Clearly, limiting reagent is HCl.

So, finally 2 moles of HCl will react with 2 moles of C_2H_4 to form 2 moles of C_2H_5Cl.

(0.5 moles of C_2H_4 are unreacted)

So, energy liberated wrt 2 moles = 72.3×2

= 144.6 KJ

So, $\Delta H_{(\text{wrt 2 moles})} = -144.6$ KJ

77. (a) Apply,

$q = mc\Delta T$

$q = 1 \times 4.90 \times (28.5 - 25)$

$q = 4.90 \times 3.5$

$q = 17.15$

So, energy released per g of sucrose = 17.15 KJ

Energy released wrt 342 g of sucrose =58.65.3 KJ
(1 mole sucrose)

So, $\Delta H_{\text{Combusiton}} = -5.865 \times 10^3 \text{ KJ mol}^{-1}$

78. (d) Heat of formation is defined for 1 mole.

79. (b) $\Delta^o G = \Delta^o H - T\Delta^o S$

$\Delta^o G = -54.07 - 298 \times \frac{10}{1000}$

(Standard state, $\therefore T = 25^o C$)

$\Delta^o G = -54.07 - 2.98 = -57.05 \text{KJ mol}^{-1}$

Now, $\Delta_r^o G = -2.303 \times R \times T \log_{10} k_{eq.}$

-57.05×1000

$= -2.303 \times 8.314 \times 298 \log_{10} k_{eq.}$

$-57050 = -5705 \log_{10} k_{eq.}$

$\frac{-57050}{-5705} = \log_{10} k_{eq.}$

$\therefore \log_{10} k_{eq.} = 10$

80. (b)

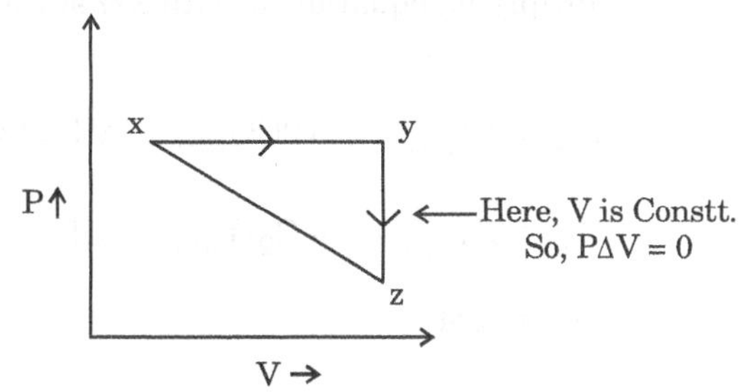

$\Delta S_{x \to z} = \Delta S_{x \to y} + \Delta S_{y \to z}$

(it is correct because it is a state function)

$W_{x \to z} = W_{x \to y} + W_{y \to z}$

(incorrect because it is a path function)

$W_{x \to y \to z} = W_{x \to y}$ (correct $\because W_{x \to z} = 0$)

$\Delta S_{x \to y \to z} = \Delta S_{x \to y}$

(incorrect $\because$ it is a state function)

81. (b) $\Delta S_{A \to C} = S_C - S_A = 50 \text{ eu}$ __(1)

$\Delta S_{C \to D} = S_D - S_C = 30 \text{ eu}$ __(2)

$\Delta S_{B \to D} = S_D - S_B = 20 \text{ eu}$ __(3)

$\Delta S_{A \to B} = S_B - S_A$

$= \text{Eqn (i)} + \text{Eqn (ii)} - \text{Eqn (iii)}$

$= 50 + 30 - 20 = 60$ eu

82. (b) $\Delta S_{vap.} = \frac{\Delta H_{vap.}}{T_b}$

$X_1 = \frac{\Delta H_{vap.}}{T_1} \Rightarrow \Delta H_{vap.} = T_1 x_1$

$\therefore H_2O_{(l)} \longrightarrow H_2O_{(g)}$; $\Delta H = T_1 x_1$ ______(1)

$H_{2(g)} + \frac{1}{2}O_{2(g)} \longrightarrow H_2O_{(l)}$; $\Delta H = x_2$ ______(2)

$C_{(s)} + O_{2(g)} \longrightarrow CO_{2(g)}$; $\Delta H = x_3$ ______(3)

$2C_{(s)} + 3H_{2(g)} \longrightarrow C_2H_{6(g)}$; $\Delta H = x_4$ ____(4)

Add eqn. (1) & (2)

$H_2O_{(l)} \longrightarrow H_2O_{(g)}$; $\Delta H = T$

$H_{2(g)} + \frac{1}{2}O_{2(g)} \longrightarrow H_2O_{(l)}$; $\Delta H = x_2$

$H_{2(g)} + \frac{1}{2}O_{2(g)} \longrightarrow H_2O_{(g)}$; $\Delta H = T_1 x_1 + x_2$

------(5)

Multiplying equation '3' with 2 & subtracting eqn. 4 from 3

$$2C_{(s)} + 2O_{2(g)} \longrightarrow 2CO_{2(g)} \;;\; \Delta H = 2x_3$$

$$2C_{(s)} + 3H_{2(g)} \longrightarrow C_2H_{6(g)} \;;\; \Delta H = x_4$$

Sub., we get

$$2C_{(s)} + 2O_{2(g)} \longrightarrow 2CO_{2(g)} \;;\; \Delta H = 2x_3$$

$$2C_{(s)} + 3H_{2(g)} \longrightarrow C_2H_{6(g)} \;;\; \Delta H = x_4$$

$$\underset{(-)}{2C_{(s)}} + \underset{(-)}{3H_{2(g)}} \longrightarrow \underset{(-)}{C_2H_{6(g)}} \;;\; \Delta H = x_4$$

$$2O_2 - 3H_2 \longrightarrow 2CO_2 - C_2H_6$$

$$C_2H_6 + 2O_2 \longrightarrow 2CO_2 + 3H_2; \Delta H = 2x_3 - x_4$$

Finally, we have

Eqn. 5 $\longrightarrow H_{2(g)} + \frac{1}{2}O_{2(g)} \longrightarrow H_2O_{(g)}; \Delta H = T_1x_1 + x_2$

Eqn. 6 $\longrightarrow C_2H_{6(g)} + O_{2(g)} \longrightarrow 2CO_{2(g)} + 3H_{2(g)}; \Delta H = 2x_3 - x_4$

Multiplying eqn. (5) with 3 and then add 5 and 6.

$$3H_2 + \frac{3}{2}O_{2(g)} \longrightarrow 3H_2O_{(g)}; \Delta H = 3T_1x_1 + 3x_2$$

$$C_2H_6 + 2O_{2(g)} \longrightarrow 2CO_{2(g)} + 3H_{2(g)}; \Delta H = 2x_3 - x_4$$

On addition,

$$C_2H_6 + \frac{7}{2}O_2 \longrightarrow 2CO_2 + 3H_2O; \Delta H = 3T_1x_1 + 3x_2 + 2x_3 - x_4$$

83. (b) $A_{(g)} \rightleftharpoons B_{(g)}$

At eqm :- $P \quad \frac{P}{4}$

$$Kp = \frac{(P^B)'}{(P^A)'} = \frac{\frac{P}{4}}{\frac{P}{1}} = \frac{P}{4} \times \frac{1}{P} = \frac{1}{4}$$

Now, $\Delta G^o = -RT \ln K$

$$= -RT \ln\frac{1}{4} = -RT(\ln 1 - \ln 4) = RT \ln 4$$

84. (b) $\Delta G = (-x) - \left(298\frac{-2x}{298}\right)$

$$\Delta G = -x + 2x$$

$$\Delta G = +x$$

$$\left(\text{If } \Delta S < \frac{x}{298}\right)$$

$$\Delta G = \Delta H - T\Delta S$$

$$\Delta G = (-x) - \left(298\frac{-x}{2 \times 298}\right)$$

$$\Delta G = -x + \frac{x}{2} = \frac{-x}{2}$$

85. (b) $\underset{(s)}{\text{Ice}} \rightleftharpoons H_2O_{(l)} \;;\; \Delta_{fusion}H = 6$ KJ/mol

$$n = \frac{90}{18} = 5$$

For melting 1 mole Ice, $\Delta H = 6$ KJ

For melting 5 mole Ice, $\Delta H = 6 \times 5 = 30$ KJ

When 5 mole (90gm) water is frozen,

energy released = 30 KJ

i.e. $\Delta H = -30$ KJ

$$\Delta S = \frac{\Delta H}{T_f} = \frac{-30000}{273} = -109.89\,JK^{-1}$$

90. (d) Reaction is spontaneous when -

$\Delta H = -$ ve, $\Delta S = +$ ve (at all temperature)

$\Delta H = +$ ve, $\Delta S = +$ ve (at high temperature)

ΔH – ve, $\Delta S = -$ve (at low temperature)

92. (a) $\Delta G^o = -2.303RT \log k_{eq.}$

$$\Delta G^o = -RT \ln k_{eq.}$$

$$\Delta G^o = \Delta^o H - T\Delta S^o$$

$$-RT \ln k_{eq.} = \Delta^o H - T\Delta S^o$$

$$RT \ln k_{eq.} = -\left(\Delta^o H - T\Delta S^o\right)$$

$$RT \ln k_{eq.} = -\Delta^o H + T\Delta S^o$$

$$\ln k_{eq.} = \frac{T\Delta S^o}{RT} - \frac{\Delta^o H}{RT}$$

$$\ln k_{eq.} = \frac{\Delta S^o}{R} - \frac{\Delta^o H}{RT} \text{ \_\_(1)}$$

$$\text{Also, } \ln k_{eq.} = 4.8 - \frac{2059}{T} \text{ (Given)\_\_(2)}$$

Relating the two equation,

$$\frac{\Delta^o S}{R} = 4.8 \Rightarrow \Delta^o S = 4.8 \times 8.314 = 39.9\,JK^{-1}$$

$$\frac{-2059}{T} = \frac{-\Delta^o H}{RT}$$

$$2059 = \frac{\Delta^o H}{R}$$

$$2059 \times 8.314 = \Delta^o H$$

$$17.12\,KJ = \Delta^o H$$

Apply, $\Delta^o G = \Delta^o H - T\Delta S^o$

$$= 17.12 - 298 \times 39.9 \times 10^{-3} = 5.31\,KJ$$

93. (a) $\Delta_r^o S = \Sigma\ \Delta^o S_P - \Sigma\ \Delta^o S_R$

$$= (219.6 - 197.6 - \frac{1}{2} \times 205) = -86.5\,JK^{-1}$$

$$\Delta_r^o H = -282.8\,KJ = -282800\,JK^{-1}$$

$T = 25 + 276 = 298\,k$ (standard state)

Apply, $\Delta^o G = \Delta^o H - T\Delta^o S$

$$= -282800 - 298 \times (-86.5)$$

$$= -257023\,J\,mol^{-1}$$

$$= -257.023\ KJ^{-1}\,mol^{-1}$$

94. (d) For eqm, $\Delta G = 0$

$$\Delta G = \Delta H - T\Delta S$$

$$\Delta H = T\Delta S$$

$$\Delta H = T_e \Delta S$$

ΔH & ΔS are +ve

For spontaneous reaction,

$\Delta G = -$ ve, So $T\Delta S > \Delta H \therefore T > T_e$

97. (b) $\Delta^\circ G = \Delta^\circ H - T\Delta^\circ S = -54.07 - 298 \times 10$

$$= -54070 - 2980$$

$$= -57050\ Jmol^{-1}$$

$$\Delta G^\circ = -RT\ln k = -2.303\,RT \log_{10} k$$

$$-57050 = -2303 \times 8.314 \times 298\ \log_{10} k$$

$$\frac{-57050}{-5705} = \log_{10} k$$

$$10 = \log_{10} k$$

99. (c) ΔG_{System} is the energy available to do useful work.

100. (a) $CaO_{(s)} + SO_{3(g)} \longrightarrow CaSO_{4(s)}$;

Apply, $\Delta G = \Delta G^o + 2.30.3\,RT \log Q_c$

$$Q_c = \frac{1}{(pSO_3)} = \frac{1}{0.5} = 2$$

$$\Delta G = -345\,KJ + 2.303 \times 8.314 \times 10^{-3} \log Q_c$$

$$= (-345 + 1.72) = -343.28\,KJ$$

101. (a) $\Delta S_{universe} = \Delta S_{sorr.} + \Delta S_{sys.}$

$$\Delta S_{sys.} = 2.303 nR \log \frac{V_2}{V_1}$$

$$= 2.303 \times 0.5 \times 8.314 \log \frac{5}{1} = 6.7\,JK^{-}$$

$$\Delta S_{sorrounding.} = \frac{q_{sorrounding}}{T}$$

For isothermal process (for ideal gas) $\Delta U = 0$

Apply, $\Delta U = q + w$

$$0 = q + w$$

$$q = -w = +p\Delta V = +2(5-1)$$

$$q = +8\,L\,atm = +8 \times 101.3\,J$$

$$q = +810.4\,J$$

$$q_{system} = 810.4\,J$$

$$q_{sorrounding} = -810.4J$$

$$\Delta S_{sorr.} = \frac{-810.4}{300} = -2.7\,JK^{-1}$$

$$\Delta S_{universe} = \Delta S_{sys.} + \Delta S_{sorr.}$$

$$= 6.7 - 2.7 = 4\,JK^{-1}\,mol^{-1}$$

102. (a) At B.pt, equilibrium is established,

So, $\Delta G = 0$

$\Delta S_{system} = +$ ve

Smart Notes

CHAPTER 7

Chemical Equilibrium

1. The following equilibrium constants are given

$N_2 + 3H_2 \rightleftharpoons 2NH_3; K_1$

$N_2 + O_2 \rightleftharpoons 2NO; K_2$

$H_2 + \frac{1}{2}O_2 \rightleftharpoons H_2O; K_3$

The equilibrium constant for the oxidation of by oxygen to give NO is:

(a) K_1K_2/K_3 (b) $K_2K_3^3/K_1$

(c) $K_2K_2^2/K_1$ (d) $K_2^2K_3/K_1$

2. Consider the following reactions in which all the reactants and the products are in gaseous state

$2PQ \rightleftharpoons P_2 + Q_2; K_1 = 2.5 \times 10^5$

$PQ + \frac{1}{2}R_2 \rightleftharpoons PQR; K_2 = 5 \times 10^{-2}$

The value of K_3 for the equilibrium

$\frac{1}{2}P_2 + \frac{1}{2}Q_2 + \frac{1}{2}R_2 \rightleftharpoons PQR$

(a) 2.5×10^{-2} (b) 2.5×10^3

(c) 1×10^{-4} (d) 5×10^3

3. In a reaction, $A + 2B \rightleftharpoons 2C$, 2.0 mole of 'A' 3.0 mole of 'B' and 2.0 mole of 'C' are placed in a 2.0 L flask and the equilibrium concentration of 'C' is 0.5 mole/L. The equilibrium constant (K) for the reaction is

(a) 0.073 (b) 0.147

(c) 0.05 (d) 0.026

4. 9.2 grams of N_2O_4 (g) is taken in a closed one litre vessel and heated till the following equilibrium is reached

$N_2O_4(g) \rightleftharpoons 2NO_2(g)$

At equilibrium, 50% N_2O_4 (g) is dissociated. What is the equilibrium constant (in mol^{-1})

(Molecular weight of N_2O_4 = 92)

(a) 0.1 (b) 0.2

(c) 0.4 (d) 2

5. The equilibrium constant at 298 K for the reaction $A + B \rightleftharpoons C + D$ is 100. If the initial concentration of all the four species were 1 M each, then equilibrium concentration of D (in mol L^{-1}) will be

(a) 0.182 (b) 0.818

(c) 1.818 (d) 1.182

6. $NH_4COONH_2(s) \rightleftharpoons 2NH_3(g) + CO_2(g)$.

If equilibrium pressure is 3 atm for the above reaction, K_p for the reaction is

(a) 4 (b) 27

(c) $\frac{4}{27}$ (d) $\frac{1}{27}$

7. A mixture of N_2 and H_2 in the molar ratio 1 : 3 attains equilibrium when 50% of the mixture has reacted. If P is the total pressure of the mixture, then partial pressure of NH_3 formed is

(a) $\frac{P}{2}$ (b) $\frac{P}{3}$

(c) $\frac{P}{4}$ (d) $\frac{P}{6}$

8. The equilibrium constant (K_p) for the decomposition of gaseous H_2O

$$H_2O(g) \rightleftharpoons H_2(g) + \frac{1}{2}O_2(g)$$

is related to degree of dissociation (α) at a total pressure *p*, given by

(a) $K_p = \dfrac{\alpha^3 p^{\frac{1}{2}}}{(1+\alpha)(2+\alpha)^{\frac{1}{2}}}$

(b) $K_p = \dfrac{\alpha^3 p^{\frac{3}{2}}}{(1-\alpha)(2+\alpha)^{\frac{1}{2}}}$

(c) $K_p = \dfrac{\alpha^{\frac{3}{2}} p^2}{(1-\alpha)(2+\alpha)^{\frac{1}{2}}}$

(d) $K_p = \dfrac{\alpha^{\frac{3}{2}} p^{\frac{1}{2}}}{(1-\alpha)(2+\alpha)^{\frac{1}{2}}}$

(e) $K_p = \dfrac{\alpha^2 p}{\sqrt{2}}$

9. 'a' moles of PCl_5 are heated in a closed container to equilibrate $PCl_5(g) \rightleftharpoons PCl_3(g) + Cl_2(g)$ at a pressure of *p* atm. If x moles of PCl_5 dissociate at equilibrium, then

(a) $\frac{x}{a} = \left(\frac{K_p}{p}\right)^{\frac{1}{2}}$ (b) $\frac{x}{a} = \frac{K_p}{K_p + p}$

(c) $\frac{x}{a} = \left(\frac{K_p}{K_p + P}\right)^{\frac{1}{2}}$ (d) $\frac{x}{a} = \left(\frac{K_p + p}{K_p}\right)^{\frac{1}{2}}$

10. The dissociation equilibrium of a gas AB_2 can be represented as

$$2AB_2(g) \rightleftharpoons 2AB(g) + B_2(g)$$

The degree of dissociation is '*x*' and is small compared to1. The expression relating the degree of dissociation (*x*) with equilibrium constant K_p and total pressure P is

(a) $\left(2\frac{K_p}{P}\right)$ (b) $\left(2\frac{K_p}{P}\right)^{\frac{1}{3}}$

(c) $\left(2\frac{K_p}{P}\right)^{\frac{1}{2}}$ (d) $\left(\frac{K_p}{P}\right)$

11. 5 moles of SO_2 and 5 moles of O_2 are allowed to react. At equilibrium, it was found that 60% SO_2 is used up. If the pressure of the mixture is one atmosphere, the partial pressure of O_2 is

(a) 0.52 atm (b) 0.21 atm

(c) 0.41 atm (d) 0.82 atm

12. For the reaction $C(s) + CO_2(g) \longrightarrow 2CO(g)$. K_p = 63 atm at 100 K. If at equilibrium, $P_{CO} = 10P_{CO2}$ the total pressure of the gases at equilibrium is

(a) 6.3 atm (b) 6.93 atm

(c) 0.63 atm (d) 0.693 atm

13. The value of K_{P_1} and K_{P_2} for the reactions

$X \rightleftharpoons Y + Z$ ...(1)

and $A \rightleftharpoons 2B$ ...(2)

are in the ratio of 9 : 1. If the degree of dissociation of X and A be equal, then total pressure at equilibrium (1) and (2) are in the ratio

(a) 3 : 1 (b) 1 : 9

(c) 36 : 1 (d) 1 : 1

14. At temperature T, a compound $AB_2(g)$ dissociates according to the reaction, $2AB_2(g)$ with degree of dissociation α, which is small compared with unity. The expression for K_p, in terms of α and the total pressure, P_T is

(a) $\frac{P_T\alpha^3}{2}$ (b) $\frac{P_T\alpha^2}{3}$

(c) $\frac{P_T\alpha^3}{3}$ (d) $\frac{P_T\alpha^2}{2}$

15. In a system, $A(s) \rightleftharpoons 2B(g) + 3C(g)$. If the concentration of C at equilibrium is increased by a factor 2, it will cause the equilibrium concentration of B to change to

(a) two times of its original value.

(b) one-half of its original value.

(c) $2\sqrt{2}$ times of its original value.

(d) $\frac{1}{2\sqrt{2}}$ times of its original value.

16. For the equilibrium:

$LiCl.\,3NH_3(g) \rightleftharpoons LiCl.NH_3(g) + 2NH_3(g)$, At 40°C, a 5 litre vessel contains 0.1 mole of LiCl. NH_3. How many mole of NH_3 should be added to the flask at this temperature to drive the backward reaction for completion?

(a) 0.7837 mole (b) 1.7837 mole

(c) 3.2 mole (d) 2.00 mole

17. One of the reaction that takes place in producing steel from iron ore is the reduction of iron(II) oxide by carbon monoxide to give iron metal

$CO_2.FeO(s) + CO(g) \rightleftharpoons Fe(s) + CO_2(g)$; K_p = 0.265 and atm at 1050 K. What are the equilibrium partial pressures of CO and CO_2 respectively at 1050 K if the initial partial pressures are:

$$p_{co} = 1.4 \text{ atm and } P_{co_2} = 0.80 \text{ atm ?}$$

(a) $[P_{co}] = 1.739$ atm and $P_{co_2} = 0.80$ atm

(b) $[P_{co}] = 1.739$ atm and $P_{co_2} = 0.461$ atm

(c) $[P_{co}] = 4.79$ atm and $P_{co_2} = 0.46$ atm

(d) $[P_{co}] = 2.739$ atm and $P_{co_2} = 1.461$ atm

18. For the reaction,

$$2HI(g) \rightleftharpoons H_2(g) + I_2(g)$$

The degree of dissociation (α) of HI(g) is related to equilibrium constant, K_p by the expression

(a) $\frac{1+2\sqrt{K_p}}{2}$ (b) $\sqrt{\frac{1+2K_p}{2}}$

(c) $\sqrt{\frac{2K_p}{1+2K_p}}$ (d) $\frac{2\sqrt{2K_p}}{1+2\sqrt{K_p}}$

19. 23 g of $N_2O_4(g)$ was placed in a 1 L container at 400 K and allowed to attain equilibrium $N_2O_4(g) \rightleftharpoons 2NO_2(g)$. The total pressure at equilibrium was found to be 11.6 bar, then what will be the partial pressure of NO_2 at equilibrium?

(a) 9.15 bar (b) 498 bar

(c) 6.6 bar (d) 8.3 bar

20. What is the expression of K_p for the following equilibrium $2NH_3(g) \rightleftharpoons N_2(g) + 3H_2(g)$; if initially '*a*' mole of NH_3 is present and at equilibrium '*x*' mole of it decomposed and total equilibrium pressure is *p*?

(a) $\frac{27p^2x^4}{16(a-x)^2(a+x)^2}$ (b) $\frac{9px^2}{8(a-x)(a+x)}$

(c) $\frac{3px^2}{4(a-x)^2}$ (d) $\frac{3p^2x^2}{8(a^2-x^2)^2}$

21. $A + 3B \rightleftharpoons 2C + D$

For this hypothetical reaction, initial moles of A is twice that of B. If at equilibrium moles of B and C are equal then percent of B reacted is:

(a) 10% (b) 20%

(c) 40% (d) 60%

22. In which of the following case reaction goes farthest to completion?

(a) $K = 10^3$ (b) $K = 10^{-2}$

(c) $K = 10$ (d) $K = 10^0$

23. When 0.087 mol of NO and 0.0437 mol of Br_2 are mixed in a closed container at constant temperature, 0.0518 mole of NOBr is obtained at equilibrium. The equilibrium amount of NO and Br_2 are:

(a) NO = 0.352 mol, Br_2 = 0.0178 mol

(b) NO = 0.0352 mol, Br_2 = 0.0178 mol

(c) NO = 0.352 mol, Br_2 = 0.178 mol

(d) NO = 0.0352 mol, Br_2 = 0.0178 mol

24. One mole of H_2O and one mole of CO are taken in 10 L vessel and heated to 725 K. At equilibrium 40% of water (by mass) reacts with CO according to the equation.

$$H_2O + CO \longrightarrow CO_2 + H_2 \uparrow$$

What is the value of equilibrium constant (KC) for the reaction?

(a) 44 (b) 4.4

(c) 0.44 (d) 2.22

25. $X_2 + X^- \rightleftharpoons X_3^- (x = \text{iodine})$

This reaction is set up in aqueous medium. We start with 1 mol of X_2 and 0.5 mol of X in 1 L flask. After equilibrium is reached, excess of $AgNO_3$ gave 0.25 mol of yellow ppt. Calculate this equilibrium constant

(a) 1.33 (b) 2.66

(c) 2.00 (d) 3.00

26. For the reaction $N_2O_4 \rightleftharpoons 2NO_2$ degree of dissociation α. The number of moles at equilibrium will be

(a) $(1 - \alpha)^2$ (b) $(1 + \alpha)$

(c) 3 (d) 1

27. At a certain temperature, $2HI \rightleftharpoons H_2 + I_2$ only 50% HI is dissociated at equilibrium. The equilibrium constant is

(a) 1.0 (b) 3.0

(c) 0.5 (d) 0.25

28. For the reaction, $C(s) + CO_2(g) \rightleftharpoons 2CO(g)$, the partial pressure of CO_2 and CO are 2.0 and 4.0 atm respectively at equilibrium. The K_p for the reaction is

(a) 0.5 (b) 4.0

(c) 8.0 (d) 32.0

29. 6 gram hydrogen reacts with 9.023×10^{23} chlorine molecules to form HCI. If total pressure of the reaction is 800 mm then what will be the partial pressure of HCI?

(a) 400 mm (b) 533 mm

(c) 600 mm (d) 153.3 mm

30. In the reaction, $C(s) + CO_2(g) \rightleftharpoons 2CO(g)$ the equilibrium pressure is 12 atm. If 50% of CO_2 reacts, calculate K_p for reaction

(a) 12 atm (b) 8 atm

(c) 16 atm (d) 4 atm

31. At a certain temperature, the equilibrium constant K_c is 16 for the reaction, $SO_2(g) + NO_2(g) \rightleftharpoons SO_3(g) + NO(g)$. If 1.0 mol each of all the four gases is taken in a 1 L container the concentration of NO_2 at equilibrium will be

(a) 16 mol L^{-1} (b) 0.8 mol L^{-1}

(c) 0.4 mol L^{-1} (d) 0.6 mol L^{-1}

32. XY_2 dissociates as $XY_2(g) \rightleftharpoons XY(g) + Y(g)$. When the initial pressure is 600 mm of Hg, the total pressure at equilibrium developed is 800 mm of Hg. Therefore pressure of Y at equilibrium is

(a) 200 (b) 50

(c) 100 (d) 150

33. A mixture of 1.57 mol of N_2, 1.92 mol of H_2, and 8.13 mol of NH_3 is introduced into a 20 L reaction vessel at 500 K. At this temperature, the equilibrium constant, K_C for the reaction $N_2(g) + 3H_2(g) \rightleftharpoons 2NH_3(g)$ is 1.7×10^2. Is the reaction mixture at equilibrium?

(a) Not at equilibrium, forward shift

(b) Not at equilibrium, backward shift

(c) Cannot be predicted

(d) In equilibrium

34. Equilibrium constant, K_C for the reaction

$N_2(g) + 3H_2(g) \rightleftharpoons 2NH_3(g)$ at 500 K is 0.061

At a particular time, the analysis shows that composition of the reaction mixture is 3.0 mol $L^{-1} N_2$, 2.0 mol L^{-1} H_2 and 0.5 mol L^{-1} NH_3. Is the reaction at equilibrium and if not in which direction does the reaction tend to proceed to reach equilibrium?

(a) At equilibrium

(b) Not at equilibrium, backward shift

(c) Not at equilibrium, forward shift

(d) Cannot be predicted

35. A reaction mixture containing H_2, N_2 and NH_3 has partial pressures 2 atm, 1 atm and 3 atm, respectively at 725 K. If the value of K_p for the reaction $N_{2(g)} + 3H_{2(g)} \rightleftharpoons 2NH_{3(g)}$ is 4.28×10^{-5} atm^{-2} at 725 K, In which direction the net reaction will go

(a) Forward

(b) Backward

(c) No net reaction

(d) Direction of reaction cannot be predicted

36. The equilibrium:

$$P_4(g) + 6Cl_2(g) \rightleftharpoons 4PCl_3(g)$$

is attained by mixing equal moles of P_4 and Cl_2 in an evacuated vessel. Then equilibrium:

(a) $[Cl]_2 > [PCl_3]$ (b) $[Cl]_2 > [P_4]$

(c) $[P_4] > [Cl_2]$ (d) $[PCl_3] > [P_4]$

37. For the chemical equilibrium,

$$CaCO_3(s) \rightleftharpoons CaO(s) + CO_2(g),$$

$\Delta H°$ Can be determined from which one of the following graphs?

(a) 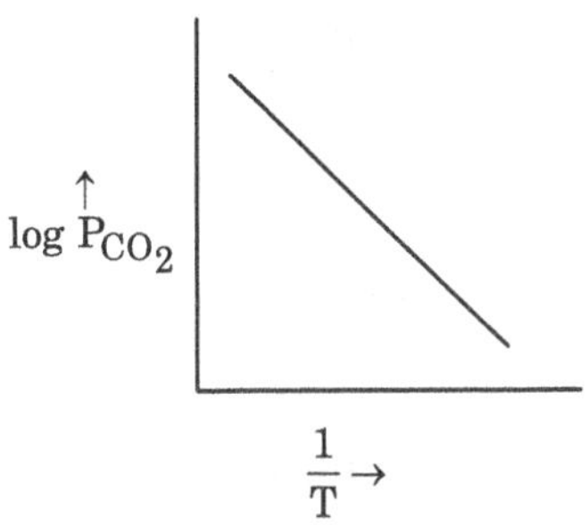

(b) 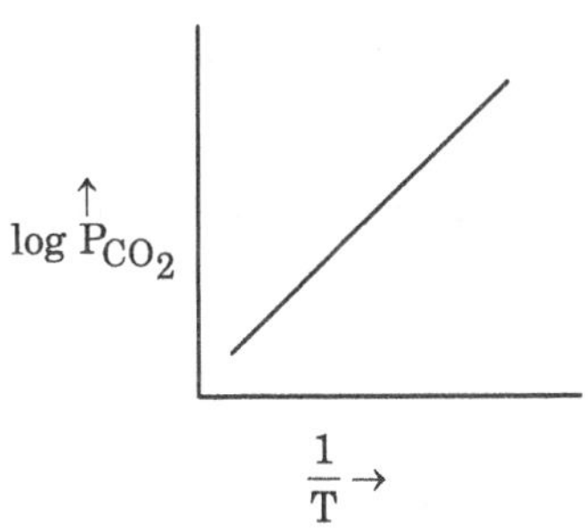

(c) 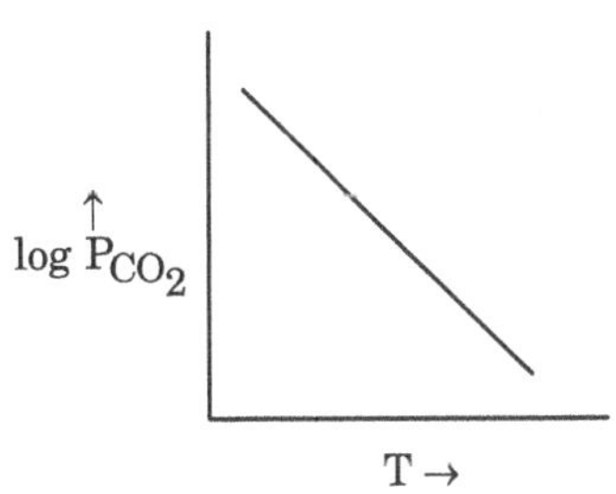

(d) 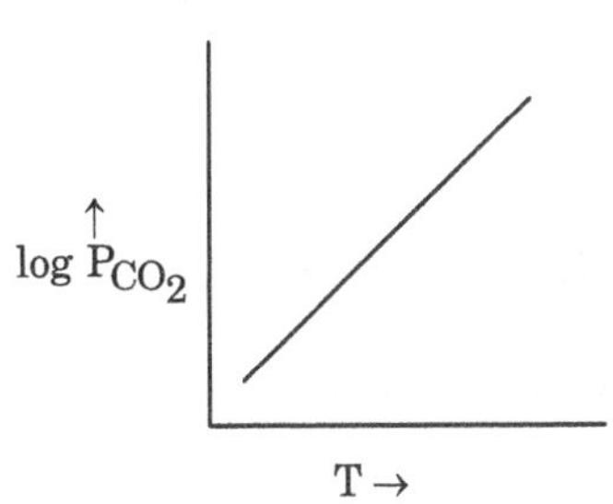

38. Consider the following reactions

$$\frac{1}{2}N_2(g) + O_2(g) \rightleftharpoons NO_2(g) \quad K_1$$

$$2NO_2(g) \rightleftharpoons N_2O_4(g) \quad K_2$$

Using above equations, write down expression for K of the following reaction:

$$N_2O_4(g) \rightleftharpoons N_2(g) + \frac{1}{2}O_2(g)$$

(a) K_1K_2 (b) $\frac{K_2^2}{K_1}$

(c) $\frac{1}{K_2K_1^2}$ (d) $\frac{1}{K_1^2K_2}$

39. Cu^{2+} ions reacts with Fe^{2+} ions according to the following reaction:

$$Cu^{2+} + Fe^{2+} \rightleftharpoons Cu + 2Fe^{3+}$$

At equilibrium, the concentration of Cu^{2+} ions is not changed by the addition of:

(a) Cu^{2+} (b) Fe^{2+}

(c) Cu (d) Fe^{3+}

40. $N_2(g) + 3H_2(g) \rightleftharpoons 2NH_3(g)$

Initially the molar ratio of N_2 and H_2 was 1:3. At equilibrium, 50% of each has reacted. If the equilibrium pressure is P, the partial pressure of NH_3 at equilibrium is:

(a) $\frac{P}{3}$ (b) $\frac{P}{4}$

(c) $\frac{P}{6}$ (d) $\frac{P}{8}$

41. 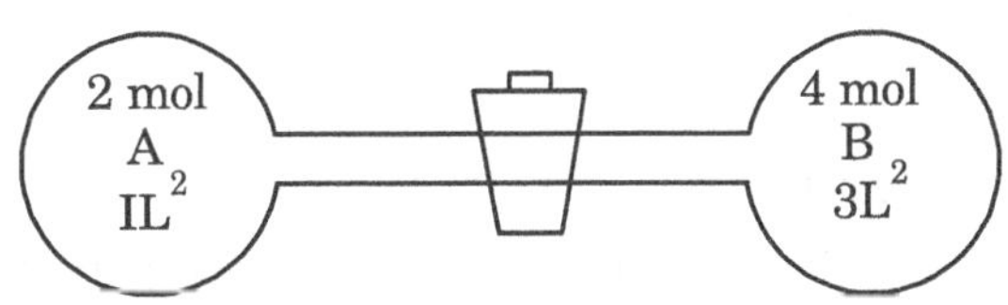

When A_2 and B_2 are allowed to react, the equilibrium constant of the reaction at 27°C is found ($K_c = 4$)

$$A_2(g) + B_2(g) \rightleftharpoons 2AB(g)$$

What will be the equilibrium concentration of AB?

(a) 1.33 M (b) 2.66 M

(c) 0.66 M (d) 0.33 M

42. In the reaction, $ACO_3(s) \rightleftharpoons AO(s) + CO_2(g)$, the equilibrium pressure of CO_2 at 127°C and 147°C are 0.14 and 0.34 bar respectively. The enthalpy of the reaction is:

(a) 64 kJ/mol (b) 35.8 kJ/mol
(c) 31.1 kJ/mol (d) 79.8 kJ/mol

43. Equilibrium constant K_p increases with increase in temperature. The ΔH for the reaction would be

(a) Positive (b) negative
(c) zero (d) cannot be predicted

44. 20% of N_2O_4 molecules are dissociated in a sample of gas at 27°C and 760 torr. Mixture has the density atequilibrium equal to:

(a) 1.48 g/L (b) 1.84 g/L
(c) 2.25 g/L (d) 3.12 g/L

45. A reaction at equilibrium involving 2 moles each of PCl_5, PCl_3 and Cl_2 is maintained at 250°C and a total pressure of 3 atm. The value of K_p is: (Consider dissociation of PCl_5)

(a) 2 (b) 3
(c) 4 (d) 1

46. For the reaction,

$CuSO_4 \cdot 5H_2O(s) \rightleftharpoons CuSO_4 \cdot 3H_2O(s) + 2H_2O(g)$

Which one is the correct representation?

(a) $K_p = [PH_2O]^2$ (b) $K_c = [H_2O]^2$
(c) $K_p = K_c (RT)^2$ (d) All of these

47. The equilibrium of the reaction

$N_2(g) + 3H_2(g) \rightleftharpoons 2NH_3(g)$

will be shifted right when:

(a) $K_p > 1$ (b) $Q > K_p$
(c) $Q = K_p$ (d) $Q = 2K_p$

48. K_1 and K_2 are respective equilibrium constant for reactions :

(i) $XeF_6(g) + H_2O(g) \rightleftharpoons XeOF_4(g) + 2HF(g);\ K_1$

(ii) $XeO_4(g) + XeF_6(g) \rightleftharpoons XeOF_4(g) + XeO_3F_2;\ K_2$

The equilibrium constant for the reaction

$XeO_4(g) + 2HF(g) \rightleftharpoons XeO_3F_2(g) + H_2O(g)$

will be

(a) $\frac{K_1}{K_2^2}$ (b) $K_1 - K_2$
(c) $\frac{K_1}{K_2}$ (d) $\frac{K_2}{K_1}$

49. In the reaction:

$NH_2COONH_4(s) \rightleftharpoons 2NH_3(g) + CO_2(g)$

the equilibrium pressure was 3 atm at 1000 K. The K_p of the reaction:

(a) 27 (b) 4
(c) $\frac{4}{27}$ (d) $\frac{27}{4}$

50. If the concentration of $\overline{O}H$ ions in the reaction,

$Fe(OH)_3(s) \rightleftharpoons Fe^{3+}(aq.) + 3OH^-(aq.)$

is decreased to $\frac{1}{4}$ times, then equilibrium concentration of Fe^{3+} will increase

(a) 64 times (b) 4 times
(c) 8 times (d) 16 times

51. In the preparation of CaO from $CaCO_3$ using the equilibrium,

$CaCO_3(s) \rightleftharpoons CaO(s) + CO_2(g)$,

K_p is expressed as $\log K_P = 7.282 - \frac{8500}{T}$.

For complete decomposition of $CaCO_3$ the temperature in Celsius to be used is:

(a) 1167
(b) 894
(c) 8500
(d) 850

52. In which of the following reactions, K_c and K_p are not equal?

(a) $2NO(g) \rightleftharpoons N_2(g) + O_2(g)$
(b) $SO_2(g) + NO_2(g) \rightleftharpoons SO_3(g) + NO(g)$
(c) $H_2(g) + I_2(g) \rightleftharpoons 2HI(g)$
(d) $2C(s) + O_2(g) \rightleftharpoons 2CO(g)$

53. Given that the equilibrium constant for the reaction $2SO_2(g) + O_2(g) \rightleftharpoons 2SO_3$ has a value of 300 at a particular temperature. What is the value of the equilibrium constant for the following reaction at same temperature?

$$SO_3(g) \rightleftharpoons SO_2(g) + \frac{1}{2}O_2(g)$$

(a) 1.8×10^{-3} (b) 3.6×10^{-3}

(c) 5.7×10^{-2} (d) 1.3×10^{-5}

54. For the gas phase reaction,

$$C_2H_4(g) + H_2(g) \rightleftharpoons C_2H_6(g);$$

$\Delta H = -136\ 8$ kJ mol^{-1}

Carried out in a vessel, the equilibrium concentration of C_2H_4 can be increased by:

(a) increase in temperature

(b) decrease in temperature

(c) removing some H_2

(d) All fo these

55. For the synthesis of ammonia by the reaction $N_2 + 3H_2 \rightleftharpoons 2NH_3$ in the Haber's process, the attainment of equilibrium is correctly predicted by the curve

(a)

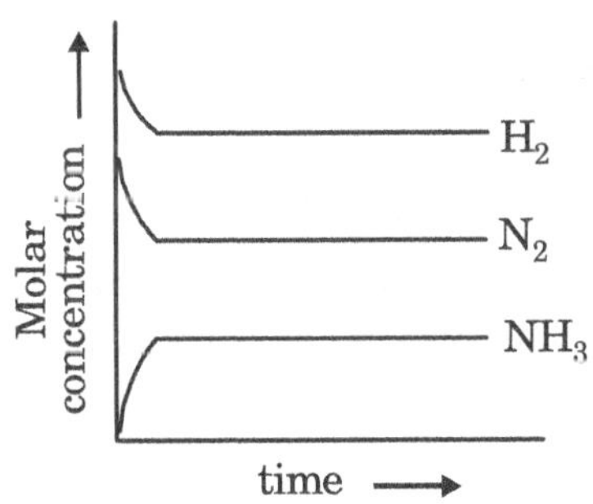

(b)

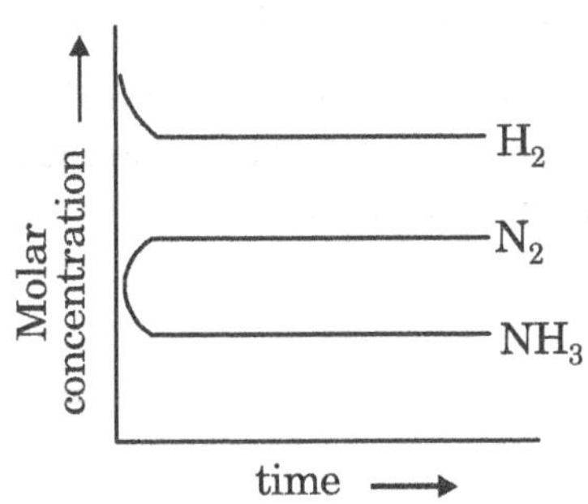

(c)

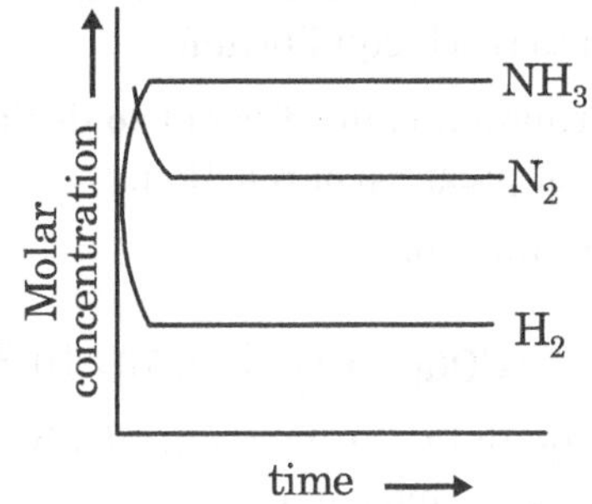

(d)

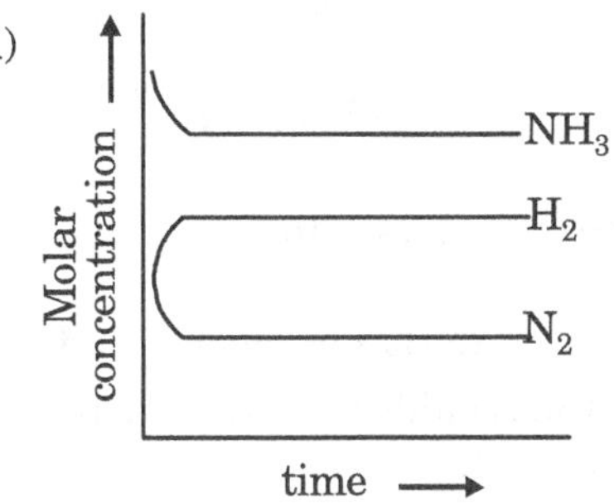

56. The equilibrium constant for the reaction $N_2(g) + O_2(g) \rightleftharpoons 2NO(g)$ is 4×10^{-4} at 200 K. In presence of a catalyst, equilibrium is attained ten times faster. Therefore, the equilibrium constant in presence of the catalyst at 200 K is :

(a) 40×10^{-4}

(b) 4×10^{-4}

(c) 40×10^{-3}

(d) difficult to compute without more data

57. Two moles of NH_3 when put into a previously evacuated vessel (one litre) partially dissociated into N_2 and H_2. If at equilibrium one mole of NH_3 is present, the equilibrium constant is :

(a) $\frac{3}{4}$ $mol^2 litre^{-2}$ (b) $\frac{27}{64}$ $mol^2 litre^{-2}$

(c) $\frac{27}{32}$ $mol^2 litre^{-2}$ (d) $\frac{27}{16}$ $mol^2 litre^{-2}$

58. At 1400 K, $K_c = 2.5 \times 10^{-3}$ for the reaction

$$CH_4(g) + 2H_2S(g) \rightleftharpoons CS_2(g) + 4H_2(g).$$

A 10.0L reaction vessel at 1400 K contains 2.00 mole CH_4, 3.0 mole of CS_2, 3.0 mole of H_2 and 4.0 mole of H_2 S. Then

(a) This reaction is at equilibrium with above concentrations.

(b) The reaction will proceed in forward direction to reach equilibrium.

(c) The reaction will proceed in backward direction to reach equilibrium

(d) The information is insufficient to decide the direction of progress of reaction.

59. K_p for the equilibrium

$COCl_2(g) \rightleftharpoons CO(g) + Cl_2$ is 6.70×10^{-9} atm at 100°C Calculate the fraction of phosgene dissociated when 1 mole of phosgene is taken in a 100 litre vessel and allowed to reach equilibrium.

(a) 6.7×10^{-9} (b) 6.7×10^{-7}

(c) 2.48×10^{-4} (d) 1.48×10^{-4}

60. For $N_2 + 3H_2 \rightleftharpoons 2NH_3$ one mole N_2 and 3 moles H_2 are at 4 atm. Equilibrium pressure is found to be 3 atm. Hence K_p is

(a) $\frac{1}{(0.5)(0.15)^2}$ (b) $\frac{1}{(0.5)(1.5)^3}$

(c) $\frac{3 \times 3}{(0.5)(1.5)^3}$ (d) None

61. The following equilibrium is set up at 400°C is 1litre flask, $2NO_{(g)} + O_{2(g)} \rightleftharpoons 2NO_{2(g)}$ K = 10. If moles of NO and NO_2 are equal at equilibrium determine no. of mole of O_2 under equilibrium conditions

(a) 0.1 (b) 1

(c) 100 (d) None

62. For the reaction $H_{2(g)} + I_{2(g)} \rightleftharpoons 2HI_{(g)}$ the value is equilibrium constant is 64. If the volume of the container is reduced to one-fourth of its original value of equilibrium constant will be

(a) 64 (b) 32

(c) 276 (d) 16

63. For the reaction $AB_2(g) \rightleftharpoons AB(g) + B(g)$, the degree of dissociation (β) of $AB_2(g)$ is related with P or V as under

(a) $\beta \propto \frac{1}{P}$ (b) $\beta \propto \frac{1}{V}$

(c) $\beta \propto \frac{1}{\sqrt{P}}$ (d) $\beta \propto \sqrt{V}$

64. 1 mole ethyl alcohol was treated with 1 mol of acetic acid at 25°C. At equilibrium, two-thirds of the acid converted into ester. The equilibrium constant for the reaction is

(a) 4 (b) 3

(c) 2 (d) 1

65. Assume that the decomposition of HNO_3 can be represented by the following equation

$4HNO_3(g) \rightleftharpoons 4NO_2(g + 2H_2O(g) + O_2(g)$ and at the given temperature 400 K and pressure 30 atm the reaction approaches equilibrium. At equilibrium partial pressure of HNO_3 is 2 atm. Find K_c in $(mol/lit)^3$.

(a) 30 (b) 12

(c) 18 (d) 16

66. Equilibrium constant for the reaction,

$2NO + O_2 \rightleftharpoons 2NO_2$ is K_{C1}

$NO_2 + SO_2 \rightleftharpoons SO_3 + NO$ is K_{C2} and

$2SO_3 \rightleftharpoons 2SO_2 + O_2$ is K_{C3}

then correction reaction is :

(a) $K_{C3} = K_{C1} \times K_{C2}$

(b) $K_{C3} \times K_{C1} \times K^2_{C2}$

(c) $K_{C3} \times K_{C1} \times K^2_{C2} = 1$

(d) $K_{C3} \times K^2_{C1} \times K_{C2} = 1$

67. In a reaction $AB_{(g)} \rightleftharpoons A_{(g)} + B_{(g)}$, the value of K_p is numerically equal to one third of the total equilibrium pressure P. The percent dissociation of AB is

(a) 25 (b) 30.5

(c) 40 (d) 50

68. Consider the reaction : $AB_{2(g)} \rightleftharpoons AB_{(g)} + B_{(g)}$. If the initial pressure of AB_2 is 500 torr and equilibrium pressure is 600 torr, the equilibrium constant K_p in terms of torr is :

(a) 20

(b) 50

(c) 25

(d) 100

69. The ratio of equilibrium constant in terms of partial pressures and that in terms of molar concentrations for the reaction: $NH_2COONH_{4(g)} \rightleftharpoons 2NH_{3(g)} + CO_{2(g)}$ at 27°C is:

(a) $(8.314 \times 27)^3$ (b) $(8.314 \times 300)^3$

(c) $(0.082 \times 300)^2$ (d) $(0.082 \times 300)^3$

70. Sulphide ion reacts with solid sulphur to form $S_2^{2-}{}_{(aq)}$ and $S_3^{2-}{}_{(aq)}$ with equilibrium constants 12 and 132 respectively. The equilibrium constant for the formation of $S_3^{2-}{}_{(aq)}$ from $S_2^{2-}{}_{(aq)}$ and sulphur is

(a) 132×12 (b) $\frac{1}{11}$

(c) 11 (d) None of these

71. 1.0 mole of $AB_{5(g)}$ is placed in a closed container under one atmosphere and at 300K. It is heated to 600K when 20% by mass of it dissociates as : $AB_{5(g)} \rightarrow AB_{(g)} + 2B_{2(g)}$.

The resultant pressure is

(a) 1.2 atm (b) 2.4 atm
(c) 2.8 atm (d) 1.4 atm

72. The equilibrium constant for the reaction ;

$N_2(g) + O_2(g)\ f\!\rightleftharpoons 2NO(g)$ at temperature

T is 4×10^{-4} The value of K_c the reaction.

$NO(g) \rightleftharpoons \frac{1}{2}N_2(g) + \frac{1}{2}O_2(g)$

at the same temperature is :

(a) 0.02 (b) 50

(c) 4×10^{-4} (d) 2.5×10^{-2}

73. 56 g of nitrogen and 8 g hydrogen gas are heated in a closed vessel. At equilibrium 34 g of ammonia are present. The equilibrium number of moles of nitrogen, hydrogen and ammonia are respectively

(a) 1, 2, 2

(b) 2, 2, 1

(c) 1, 1, 2

(d) 2, 1, 2

74. One mole of a compound AB reacts with one mole of a compound CD according to the equation, $AB + CD \rightleftharpoons AD + CB$. When equilibrium had been established it was found that $\frac{3}{4}$ mole each of reactant AB and CD had been converted to AD and CB. There is no change in volume. The equilibrium constant for the reaction is

(a) $\frac{9}{16}$ (b) $\frac{1}{9}$

(c) $\frac{16}{9}$ (d) 9

Topic Question Based on Le-Chatelier's Principle

75. Which can be explained as applications of Le-Chatelier's principle?

(a) Transport of oxygen by haemoglobin in blood

(b) Removal of CO_2 from tissues by blood

(c) Tooth decay due to use of sweet substances

(d) All of the above

76. The exothermic formation of ClF_3 is represented by the equation:

$$Cl_{2(g)} + 3F_{2(g)} \rightleftharpoons 2ClF_{3(g)}; D_rH = -329\,kJ$$

Which of the following will increase the quantity of in an equilibrium mixture of Cl_2, F_2 and ClF_3?

(a) Removing Cl_2

(b) Increasing the temperature

(c) Adding F_2

(d) Increasing the volume of the container

77. A cylinder fitted with a movable piston contains liquid water in equilibrium with water vapour at 25°C. Which operation result in a decrease in the equilibrium vapour pressure?

(a) Moving the piston downward a short distance

(b) Removing a small amount of vapour

(c) Removing a small amount of the liquid water

(d) Dissolving salt in the water

78. When a bottle of cold drink is opened, the gas comes out with a fizzle due to:

(a) decrease in pressure suddenly which results in a decrease in solubility of CO_2 gas in water

(b) decrease in temperature

(c) increase in pressure

(d) none of the above

79. What is the effect of a tenfold increase in pressure on K_p in the reaction:

$N_2(g) + 3H_2(g) \rightleftharpoons 2NH_3(g)$

at equilibrium?

(a) A tenfold increase (b) A tenfold decrease

(c) No change (d) Equal to K_c

80. In a flask of colourless N_2O_4 is in equilibrium with brown coloured NO_2. At equilibrium, when the flask is heated to 100°C the brown colour deepens on cooling, the brown colour became less coloured. The change in enthalpy for the system is:

(a) negative (b) positive

(c) zero (d) not defined

81. Le Chatelier's principle is not applicable to:

(a) $Fe(s) + S(s) \rightleftharpoons FeS(s)$

(b) $H_2(g) + I_2(g) \rightleftharpoons 2HI(g)$

(c) $N_2(g) + 3H_2(g) \rightleftharpoons 2NH_3(g)$

(d) $N_2(g) + O_2(g) \rightleftharpoons 2NO(g)$

82. Consider the reactions,

(i) $PCl_5(g) \rightleftharpoons PCl_3(g) + Cl_2(g)$

(ii) $N_2O_4(g) \rightleftharpoons 2NO_2(g)$

The addition of an inert gas at constant volume:

(a) will increase the dissociation of PCl_5 as well as N_2O_4

(b) will reduce the dissociation of PCl_5 as well as N_2O_4

(c) will increase the dissociation of PCl_5 and step up the formation of N_2O_4

(d) will not disturb the equilibrium of the reactions

83. For the reaction,

$N_2(g) + 3H_2(g) \rightleftharpoons 2NH_3(g)$; $\Delta H^o = -ve$

the number of moles of H_2 at equilibrium will increase when:

(a) volume of vessels is increased

(b) volume of vessels is decreased

(c) Ne gas is added at constant volume

(d) NH_3 is removed

84. The following reactions are known to occur in the body:

$CO_2 + H_2O \rightleftharpoons H_2CO_3 \rightleftharpoons H^+ + HCO_3^{-1}$

If CO_2 escapes from the system

(a) pH will decrease

(b) Hydrogen ion concentration will diminish

(c) H_2CO_3 concentration will remain unchanged

(d) The forward reaction will be favoured

85. In a vessel containing N_2, H_2 and NH_3 at equilibrium, some helium gas is introduced so that total pressure increase while temperature and volume remain constant. According to Le Chatelier's principle, the dissociation of NH_3 :

(a) Increases

(b) decreases

(c) remains unaltered

(d) changes unpredictably

86. The equilibrium,

$SO_2Cl_2(g) \rightleftharpoons SO_2(g) + Cl_2(g)$ is attained at 25°C in a closed container and an inert gas, helium is introduced. Which of the following statement are correct?

(a) Concentrations of SO_2, Cl_2 and SO_2Cl_2 are changed.

(b) More chlorine is formed

(c) Concentration of SO_2 is reduced

(d) No change occurs

87. When pressure is applied to the equilibrium system 'ice $\rightleftharpoons$ water', which of the following phenomenon will happen?

(a) More ice will be formed.

(b) Water will evaporate

(c) More water will be formed.

(d) Ice sublimes.

88. The unit of equilibrium constant K_c of a reaction is $mol^{-2}l^2$. For this reaction, the product concentration increases by

(a) Increasing the pressure

(b) Lowering the temperature

(c) Lowering the pressure

(d) Both B and C

89. For the reaction $2NO(g) \rightleftharpoons N_2(g) + O_2(g)$

$\Delta H = -180 KJ\ mol^{-1}$

Which of the following fact does not hold good?

(a) The pressure change at constant temperature do not effect the equilibrium constant.

(b) The volume change at constant temperature do not affect the equilibrium constant.

(c) The dissociation of NO is favoured more at high temperature

(d) The dissociation of NO is favoured more at high temperature.

90. The concentration of a pure solid or liquid phase is excluded in the expression of equilibrium constant

(a) Solid and liquid concentration are independent their quantities

(b) Solids and liquids react slowly

(c) Solids and liquids at equilibrium do not with gaseous phase

(d) Both b & c

91. In which of the following equilibrium reactions, the equilibrium would shift to right if total pressure is increased?

(a) $N_2(g) + 3H_2(g) \rightleftharpoons 2NH_3(g)$

(b) $H_2(g) + I_2(g) \rightleftharpoons 2HI(g)$

(c) $H_2(g) + Cl_2(g) \rightleftharpoons 2HCl(g)$

(d) $N_2O_4(g) \rightleftharpoons 2NO_2(g)$

92. For a reaction $2SO_2(g) + O_2(g) \rightleftharpoons 2SO_3(g)$; $\Delta H -188.3 kJ$. The number of moles of SO_3 found is increased if:

(a) temperature is increased at constant volume

(b) inert gas is added to the mixture

(c) O_2 is removed from the mixture

(d) volume of the reaction flask is decreased

93. At the equilibrium of the reactions, $2X_{(g)} + Y_{(g)} \longrightarrow X_2Y_{(g)}$, the number of moles of X_2Y at equilibrium is affected by the:

(a) temperature and pressure

(b) temperature only

(c) pressure only

(d) temperature, pressure and catalyst used

94. On doubling P and V with constant temperature, the equilibrium constant will:

(a) remain constant (b) become double

(c) become one-fourth (d) none of these

95. What will be the effect of addition of catalyst at constant temperature?

(a) The equilibrium constant will remain constant

(b) ΔH of the reaction will remain constant

(c) k_f and k_b will increase upto same extent

(d) All of these are ocrrect

96. For the reaction : $Cl_2(g) + 3F_2(g) \rightleftharpoons 2ClF_3(g)$; $\Delta H = -329 kJ$, dissociation of $ClF_3(g)$ will be favoured by :

(a) increasing the temperature

(b) increasing the volume of the container

(c) adding of F_2 gas

(d) All are correct

97. Increase in the pressure for the following equilibrium results in the:

$H_2O(l) \rightleftharpoons H_2O(g)$

(a) formation of more $H_2O(l)$

(b) formation of more $H_2O(g)$

(c) increasing in b.p. of $H_2O(l)$

(d) decrease in b.p. of $H_2O(l)$

98. $N_2(g)$ and $H_2(g)$ are allowed to react in a closed vessel at given temp. and pressure formation of $NH_3(g)[N_2(g) + 3H_2(g) \rightleftharpoons 2NH_3(g) + 22.4 kcal]$. If He(g) is added equilibrium at constant pressure then which is/are correct ?

(a) Concentration of $N_2(g)$, $II_2(g)$ and $NII_3(g)$ decrease.

(b) Moles of $NH_3(g)$ decreases.

(c) The extent of cooling depends on amount of He(g) added.

(d) All are correct

99. When NaN_3 is heated in a closed vessel, O_2 is liberated and $NaNO_2$ is left behind. At equilibrium

(a) addition of $NaNO_3$ favours forward reaction

(b) addition of $NaNO_2$ favours reverse reaction

(c) increasing pressure favours reverse reaction

(d) all (a), (b), (c) are correct

100. Consider a saturated solution, (i.e., a solution in equilibrium with solid solute) of sodium chloride. If the temperature is raised, then for

$$NaCl(g) + aq \rightleftharpoons NaCl(aq);\ DH = +ve;$$

(a) more solid will dissolve

(b) some solid will precipitate out from the solution

(c) the solution will become super saturated

(d) solution concentration will remain unchanged

101. If little heat is added to ice $\rightleftharpoons$ liquid equilibrium in a sealed container, then:

(a) pressure will rise (b) temperature will rise

(c) temperature will fall (d) no change in P and T

102. The equilibrium constant $Br_2 \rightleftharpoons 2Br$ at 500 K and 700 K are 10^{-10} and 10^{-5} respectively. The reaction is:

(a) endothermic (b) exothermic

(c) fast (d) slow

103. If pressure is applied to the following equilibrium, Liquid $\rightleftharpoons$ Vapour. The boiling temperature of the liquid:

(a) will decrease (b) will increase

(c) may not change (d) will not change

Question Based on Vapor Density

104. For the dissociation of PCl_5, the variation of $(1 + x)$ against $\left(\frac{D}{d}\right)$ is represented as:

(where x = degree of dissociation)

(a) $(1+x)\uparrow$

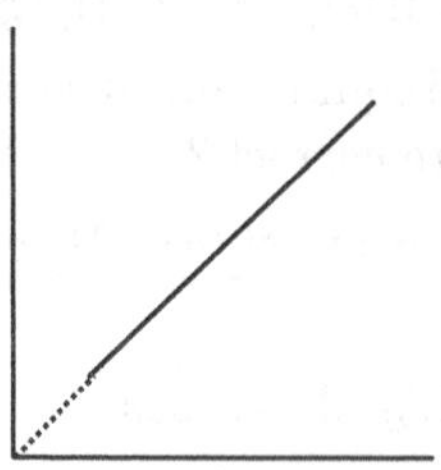

$\frac{D}{d}\rightarrow$

(b) $(1+x)\uparrow$

$\frac{D}{d}\rightarrow$

(c) $(1+x)\uparrow$

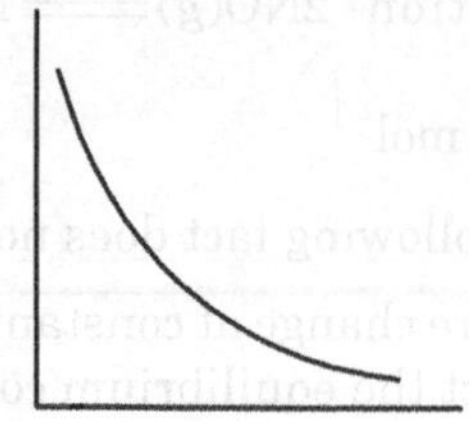

$\frac{D}{d}\rightarrow$

(d) $(1+x)\uparrow$

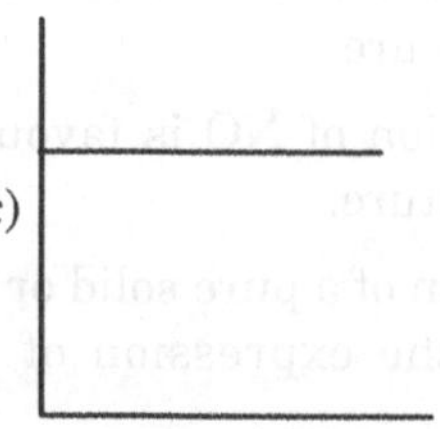

$\frac{D}{d}\rightarrow$

105. Vapour density of the equilibrium mixture of the reaction $SO_2Cl_2 \rightleftharpoons SO_{2(g)} + Cl_{2(g)}$ is 50.0 percent dissociation of SO_2Cl_2 is:

(a) 33.33 (b) 35.0

(c) 30.0 (d) 66.67

106. The degree of dissociation $= \frac{D_0 - D_{th}}{(n-1)D_{th}}$ is correct matched for the equilibrium (where is the observed V.D. & theoretical V.D.)

(a) $X \rightleftharpoons \frac{n}{3}Y + \frac{2n}{3}Z$ (b) $X \rightleftharpoons \frac{n}{2}Y + \frac{n}{2}Z$

(c) $X \rightleftharpoons \frac{4n}{5}Y + \frac{n}{5}Z$ (d) All are correct

107. At the equilibrium of the reaction $N_2O_{4(g)} \rightleftharpoons 2NO_{2(g)}$, the observed molecular mass is at 350 K. The percentage of dissociation of $N_2O_{4(g)}$ at 350 K is:

(a) 10% (b) 15%

(c) 20% (d) 18%

(e) 13 %

108. Decomposition of $N_2O_4(g)$ takes places as follows:

$$N_2O_4(g) \rightleftharpoons 2NO_2(g)$$

'D' is the vapour density at initial stage and '*d*' is the vapour density at equilibrium. We get following graph when the degree dissociation 'α' is plotted against $\left(\frac{D}{d}\right)$

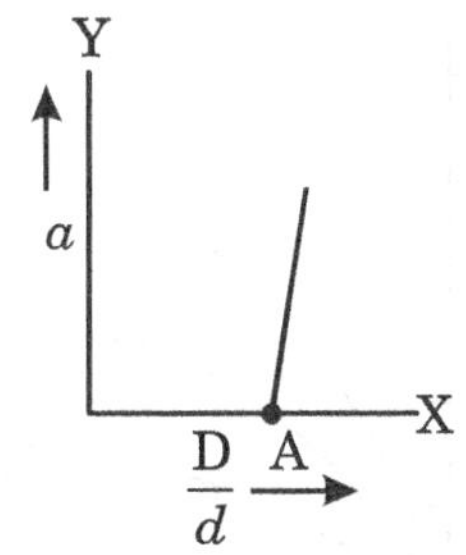

What is the value of $\left(\frac{D}{d}\right)$ at A?

(a) 0 (b) 0.5

(c) 1 (d) 1.5

109. For the reaction, $N_2O_4(g) \rightleftharpoons 2NO_2(g)$; if percentage dissociation of N_2O_4 are 25%, 50%, 75% and 100%, then the sequence of observed vapour densities will be:

(a) $d_1 > d_2 > d_3 > d_4$ (b) $d_4 > d_3 > d_2 > d_1$

(c) $d_1 = d_2 = d_3 = d_4$ (d) $(d_1 = d_2) > (d_3 = d_4)$

110. Acetic acid undergoes dimerization, when dissolved in benzene

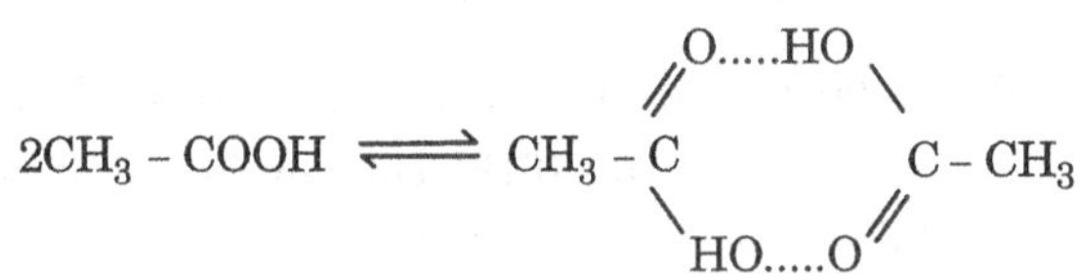

Molecular mass of acetic acid is found 120. Which among the following relation is correct?

(a) $\alpha = 2\left(\frac{D-d}{d}\right)$ (b) $\alpha = 2\left(\frac{D-d}{D}\right)$

(c) $\alpha = 2\left(\frac{d-D}{d}\right)$ (d) $\alpha = \frac{2d}{D-d}$

where, d = Observed vapour density

D = Theoretical vapour density

111. Dissociation of PCl_5 takes place on heating as follows:

$$PCl_5(g) \rightleftharpoons PCl_3(g) + Cl_2(g)$$

'D' is the vapour density of PCl_5 at initial state 'd' is the vapour density of the equilibrium mixture. Which of the following graph represents correct variation of the degree of dissociation 'α' against $\left(\frac{D}{d}\right)$?

(a)

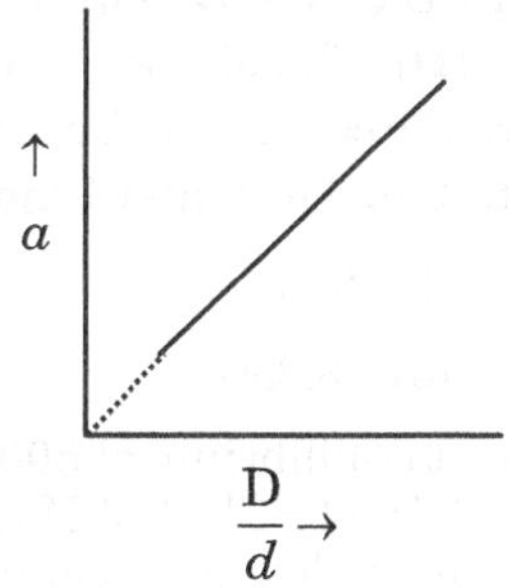

(b)

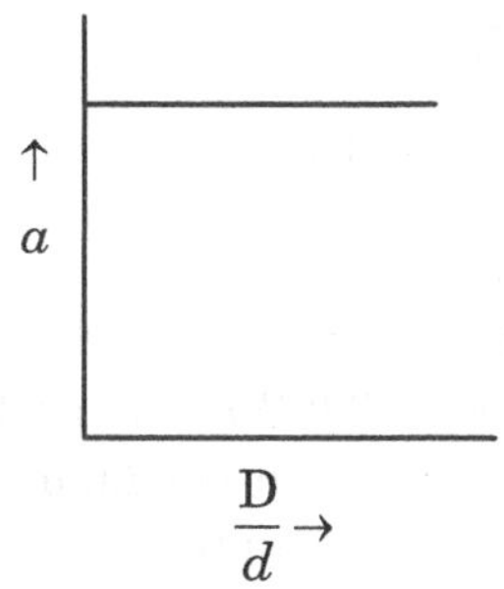

(c)

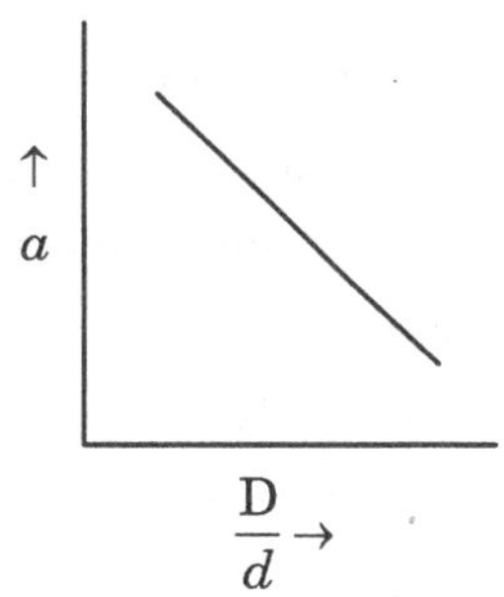

(d)

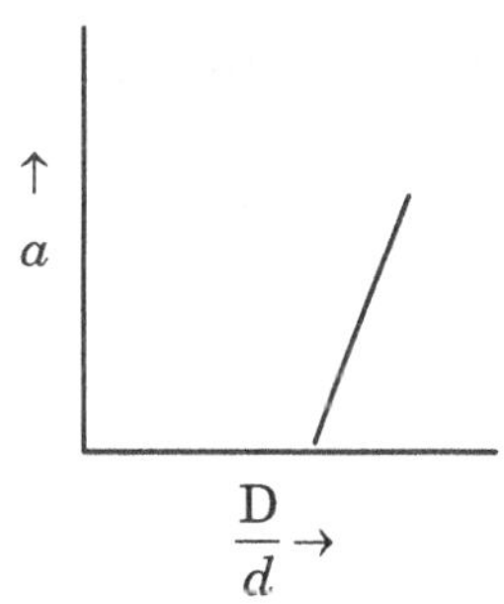

Topic Miscellaneous Questions

112. PCl_5 is 40% dissociated when pressure is 2 atm. It will be 80% dissociated when pressure is approximately

(a) 0.2 atm (b) 0.5 atm

(c) 0.3 atm (d) 0.6 atm

113. 2 mol PCl_5, 1 mol Cl_2 and 1 mol PCl_3 are taken in 1 L flask: When equilibrium is setup, PCl_5 is found to 50% dissociated into the products. Hence K_c is

(a) 4 mol L^{-1} (b) 1.5 mol L^{-1}

(c) 1 mol L^{-1} (d) 0.17 mol L^{-1}

114. For the reaction $A+B \rightleftharpoons C+D$, equilibrium concentration of [C] = [D] = 0.5 M if we start with 1 mol each of A and B. Percentage of A converted into C if we start with 2 mol of A and 1 mol B, is

(a) 25% (b) 40%

(c) 66.66% (d) 33.33%

115. PCl_5, PCl_3 and Cl_2 are at equilibrium at 500K and having concentration 1.59 M PCl_3, 1.59M Cl_2 and 1.41 M PCl_5. Then, k_c for following reaction is:

$$PCl_5 \rightleftharpoons PCl_3 + Cl_2$$

(a) 2.79 (b) 1.79

(c) 9 (d) 7.7

116. For the reaction $3A(g)+B(g) \rightleftharpoons 2C(g)$ at a given temperature, K_c = 9.0 What must be the volume of the flask, if a mixture of 2.0 mol each of A, B and C exist in equilibrium?

(a) 6 L (b) 9 L

(c) 36 L (d) None of these

117. For the following gases equilibrium, $N_2O_4(g) \rightleftharpoons 2NO_2(g)$ is found to be equal to K_c. This is attained when

(a) K_c (b) 273 K

(c) 1 K (d) 12.19 K

118. For the reaction: $R \rightleftharpoons P$

Variation of concentration is plotted against time:

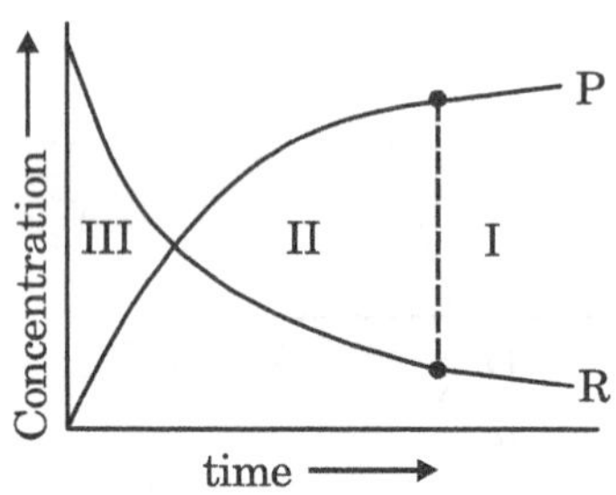

Which of the following regions show(s) equilibrium?

(a) III (b) II

(c) I (d) II and III both

119. The preparation of $SO_3(g)$ by reaction $SO_2(g)+\frac{1}{2}O_2(g) \rightleftharpoons SO_3(g)$ is an exothermic reaction. If the preparation follows the following temperature-pressure relationship for its % yield, then for temperatures T_1, T_2 and T_3. The correct option is:

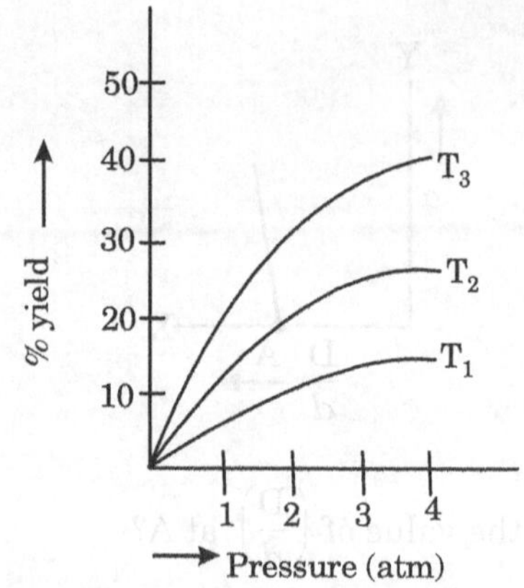

(a) $T_3 > T_2 > T_1$

(b) $T_1 > T_2 > T_3$

(c) $T_1 = T_2 = T_3$

(d) Nothing could be predicted about temperature through given information.

120. What is ΔG° for the following reaction?

$$\frac{1}{2}N_2(g)+\frac{3}{2}H_2(g) \rightleftharpoons NH_3(g);$$

$$K_p = 4.42\times10^4 \text{ at } 25°C$$

(a) –26.5 KJ/mol^{-1} (b) –11.5 KJ/mol^{-1}

(c) – 2.2 KJ/mol^{-1} (d) –0.9 KJ/mol^{-1}

121. Consider the reaction $2A \rightleftharpoons B$. The equilibrium constant is 1×10^2. If the concentration of 'A' is 0.12375 mol L^{-1}, the concentration of 'B' in mol litre^{-1} at equilibrium is:

(a) 0.10 (b) 0.09

(c) 0.02 (d) 0.05

122. Which is/are correct?

(a) $2.303 \log K = -\frac{\Delta H^o}{RT} + \frac{\Delta S^o}{R}$

(b) $\Delta G^o = -2.303 RT \log K$

(c) $-2.303 \log K = -\frac{\Delta H^o}{RT^2} + \frac{\Delta S^o}{R}$

(d) $2.303 \log K = \frac{1}{RT}(\Delta H^o + \Delta S^o)$

123. The equilibrium constant K_c for $A_{(g)} \rightleftharpoons B_{(g)}$ is 1.1 means

(a) [A]eq > [B]eq (b) [A]eq < [B]eq

(c) [Aeq] = 1.1M (d) [Beq] = 1.1M

124. The value of equilibrium constant of the reaction,

$HI(g) \rightleftharpoons \frac{1}{2}H_2(g) + \frac{1}{2}I_2(g)$ is 8.0.

The equilibrium constant of the reaction

$H_2(g) + I_2(g) \rightleftharpoons 2HI(g)$ will be:

(a) $\frac{1}{64}$ (b) 16

(c) $\frac{1}{8}$ (d) $\frac{1}{16}$

125. $\Delta G°$ for the dissociation of the dimer ($A_2 \rightleftharpoons 2A$) in benzene solution at 27°C is 6.909 kcal/mol. If 8 moles of A is dissolved in 10 litre of benzene at 27°C, then what is the equilibrium concentration of dimer (A_2)?

(Given: R = 2 Cal/mol K)?

(a) 2×10^{-3} (b) 0.4
(c) 0.04 (d) 0.02

126. If we start with one mole of N_2, 3 moles of H_2 and two moles of NH_3 in a one litre container at 500°C (where) at equilibrium

(a) The number of moles of N_2, H_2 and NH_3 will be in the ratio 1 : 3 : 2

(b) The number of moles of N_2, H_2 will be in the ratio 1 : 3

(c) The number of moles of N_2, will be one

(d) The total number of moles will be six

127. In an aqueous solution of volume 500 mL, when the reaction of $2Ag^+ + Cu \rightleftharpoons Cu^{+2} + 2Ag$ reached equilibrium. The $[Cu^{+2}]$ was XM. When 500 mL of the water is further added, at the equilibrium $[Cu^{+2}]$ will be

(a) 2XM

(b) XM

(c) Between XM and $\frac{X}{2M}$

(d) Less than $\frac{X}{2M}$

128. Solid ammonium carbamate dissociates to give product according to equation

$NH_2COONH_4(s) \rightleftharpoons 2NH_3(g)$.

At equilibrium, ammonia is added such that partial pressure of NH_3, now equal to the original total pressure. The ratio of total pressure now, to the original total pressure is

(a) $\frac{27}{31}$ (b) $\frac{31}{27}$

(c) $\frac{4}{9}$ (d) $\frac{9}{4}$

129. 0.2 mole of H_2 and 0.5 mole I_2 of are mixed in a reaction vessel and heated until equilibrium is reached $H_2(g) + I_2(g) \rightleftharpoons 2HI$ which of the following statement is true?

(a) If x mole HI is present at equilibrium the quantity of HI present is 0.5 - x mole

(b) If x mole HI is present at equilibrium the quantity H_2 of present is x – 0.2 mole.

(c) If y mole of I_2 is used up in the reaction the quantity of HI present at equilibrium is 2y mole.

(d) If y mole of I_2 is used up in the reaction the quantity of H_2 present at equilibrium $0.2 - \left(\frac{y}{2}\right)$ mole.

130. N_2 and H_2 are taken in 1:3 molar ratio in a closed vessel to attain the following equilibrium

$N_2(g) + 3H_2(g) \rightleftharpoons 2NH_3(g)$ Find K_p for reaction at total pressure of 2P if P_{N_2} equilibrium is $\frac{P}{3}$

(a) $\frac{1}{3P^2}$ (b) $\frac{4}{3P^2}$

(c) $\frac{4P^2}{3}$ (d) None

131. The reaction $A(g) + B(g) \rightleftharpoons C(g) + D(g)$ is studied in a one litre vessel at 250°C. The initial concentration of A was 3n and that of B was n. When equilibrium was attained, equilibrium concentration of C was found to the equal to the equilibrium concentration of B. What is the concentration of D at equilibrium?

(a) $\frac{n}{2}$ (b) $\left(3n - \frac{1}{2}\right)$

(c) $\left(n - \frac{n}{3}\right)$ (d) n

132. The equilibrium constant for,

$2H_2S(g) \rightleftharpoons 2H_2(g) + S_2(g)$ is 0.0118 at 1300 k while the heat of dissociation is 597.4kJ. The standard equilibrium constant of the reaction at 1200 K is:

(a) 1.180×10^{-4}

(b) 11.80

(c) 118.0

(d) cannot be calculated from given data

133. Starting with 100 mm (Hg) pressure of SO_2Cl_2 in a closed container, the following equilibrium is set up at some temperature: $SO_2Cl_{2(g)} \rightleftharpoons SO_{2(g)} + Cl_{2(g)}$. If mole fraction of SO_2Cl_2 at equilibrium is $\frac{1}{3}$, K_p for the reaction is:

(a) 25 mm (b) 50 mm
(c) 100 mm (d) 150 mm

134. 1 mol N_2 and mol H_2 are placed in a closed container at a pressure of 4 atm. The pressure falls to 3 atm at the same temperature when the following equilibrium is attained. $N_{2(g)} + 3H_{2(g)} \rightleftharpoons 2NH_{3(g)}$. The equilibrium constant K_p for dissociation of NH_3 is:

(a) $\frac{1}{0.5} \times (1.5)^3 \, atm^{-2}$

(b) $0.5 \times (1.5)^3 \, atm^2$

(c) $\frac{0.5 \times (1.5)^3}{3 \times 3} atm^2$

(d) $\frac{3 \times 3}{0.5 \times (1.5)^3} atm^{-2}$

135. For the reaction, $C_{(s)} + CO_{2(g)} \longrightarrow 2CO_{(g)}$, $K_p = 63$ atm at 1000 K. If at equilibrium; $P_{CO} = 10P_{CO_2}$, then total pressure of the gases at equilibrium is:

(a) 6.3 atm (b) 6.93 atm
(c) 0.63 atm (d) 0.693 atm
(e) 69.3 atm

136. A hypothetical reaction :

$A_{(g)} + B_{(g)} \rightleftharpoons C_{(g)} + D_{(g)}$ occurs in a single step, the specific rate constant of forward reaction at TK is 2.0×10^{-3} mol L^{-1}L s^{-1}. When start is made with equimolar amounts of A and B, it is found that concentration of A is twice that of C at equilibrium. The specific rate constant of the backward reaction is

(a) 5.0×10^{-4} mol^{-1} Ls^{-1}

(b) 8.0×10^{-3} mol^{-1} Ls^{-1}

(c) 1.5×10^{2} mol^{-1} Ls^{-1}

(d) none of these

137. In the following hypothetical reaction $A + 3B \rightleftharpoons 2C + D$, initial moles of A is twice that of B. If at equilibrium moles of B and C are equal, % of B reacted is

(a) 10% (b) 20%
(c) 40% (d) 60%

138. Consider the following equilibria at

$25^\circ C$, $2NO_{(g)} \rightleftharpoons N_{2(g)} + O_{2(g)}; K_1 = 4 \times 10^{30}$

and

$NO_{(g)} + \frac{1}{2}Br_{2(g)} \rightleftharpoons NOBr_{(g)}; K_2 = 1.4\,mol^{-1/2}L^{1/2}$.

The value of K_c for the reaction (at the same temperature)

$\frac{1}{2}N_{2(g)} + \frac{1}{2}O_{2(g)} + \frac{1}{2}Br_{2(g)} \rightleftharpoons NOBr_{(g)}$ is:

(a) 3.5×10^{-31} (b) 2.8×10^{-15}
(c) 7.0×10^{-16} (d) 5.6×10^{30}

139. a mole of PCl_5 is heated in a closed container to equilibriate $PCl_{5(g)} \rightleftharpoons PCl_{3(g)} + Cl_{2(g)}$ at a pressure of P atm. If x moles of PCl_5 dissociate at equilibrium, then

(a) $\frac{x}{a} = \frac{K_p}{K_p + P}$ (b) $\frac{x}{a} = \left(\frac{K_p + P}{K_p}\right)^{\frac{1}{2}}$

(c) $\frac{x}{a} = \left(\frac{K_p}{P}\right)^{\frac{1}{2}}$ (d) $\frac{x}{a} = \left(\frac{K_p}{K_p + P}\right)^{\frac{1}{2}}$

140. Reactions that have standard free energy changes less than zero always have equilibrium constant equal to:

(a) unity (b) greater than unity

(c) less than unity (d) zero

141. Calculate ΔG° for conversion of oxygen to ozone $\frac{3}{2}O_{2(g)} \longrightarrow O_{3(g)}$ at 298 K, if K_p for this conversion is 2.47×10^{-29}:

(a) 163 kJ mol^{-1} (b) 2.4 $\times$ 10^2 kJ mol^{-1}

(c) 1.6 kJ mol^{-1} (d) 2.38 $\times$ 10^6 kJ mol^{-1}

142. For the following equilibrium $N_2O_4 \rightleftharpoons 2NO_2$ in gaseous phase, NO_2 is 50% of the total volume when equilibrium is set up. Hence percent of dissociation of N_2O_4 is

(a) 50% (b) 25%

(c) 66.66% (d) 33.33%

Level-II

143. An arbitrary compound P_2Q decomposes according to reaction:

$$2P_2Q(g) \rightleftharpoons 2P_2(g) + Q_2(g)$$

If one starts the decomposition reaction with 4 moles of P_2Q and value of equilibrium constant K_p is numerically equal to total pressure at equilibrium. Then which option is/are correct at equilibrium:

(a) moles of $n_{P_2Q} = n_{Q_2}$

(b) moles of $n_{P_2} = (\frac{8}{3})$

(c) degree of dissociation is $\alpha = \left(\frac{2}{3}\right)$

(d) all are correct

144. For the following eqm,

$H_{2(q)} \rightleftharpoons 2H_{(q)}$, there is 0.1% dissociation at eqm. Pressure of 1atm. If P is 4 atm, percentage dissociation will be -

(a) 0.4%

(b) 0.05%

(c) 0.025%

(d) 0.2%

145. When 20 g of $CaCO_3$ were put into 10 litre flask and heated to 800°C, 35% $CaCO_3$ remained unreacted at equilibrium. K_p for decomposition of $CaCO_3$ is:

(a) 1.145 atm (b) 0.145 atm

(c) 2.145 atm (d) 3.145 atm

146. N_2O_4 is dissociated to 33% and 40% at total pressure P_1 and P_2 atm respectively. Then the ratio $\frac{P_1}{P_2}$ is:

(a) $\frac{7}{4}$ (b) $\frac{7}{3}$

(c) $\frac{8}{3}$ (d) $\frac{8}{4}$

147. The equilibrium constant for the reaction $CO(g) + H_2O(g) \rightleftharpoons CO_2(g)$ is 5. How many moles of CO_2 must be added to 1 litre container already containing 3 moles each of CO and H_2O to make 2M equilibrium concentration of CO?

(a) 15 (b) 19

(c) 5 (d) 20

148. $\log \frac{K_P}{K_c} + \log RT = 0$ is a relationship for the reaction:

(a) $PCl_3 \rightleftharpoons PCl_3 + Cl_2$

(b) $2SO_2 + O_2 \rightleftharpoons 2SO_3$

(c) $H_2 + I_2 \rightleftharpoons 2HI$

(d) $N_2 + 3H_2 \rightleftharpoons 2NH_3$

149. In the following reaction started only with A_8,

$$2A_8(g) \rightleftharpoons 2A_3(g) + 3A_2(g) + A_4(g)$$

mole fraction of A_2 is found to 0.36 at a total pressure of 100 atm at equilibrium. The mole fraction of $A_8(g)$ equilibrium is :

(a) 0.28 (b) 0.72

(c) 0.18 (d) None of these

150. In the equilibrium $2SO_2(g) + O_2(g) \rightleftharpoons 2SO_3(g)$, the partial pressure of SO_2, O_2 and SO_3 are 0.662, 0.10 and 0.331 atm respectively. What should be the partial pressure of oxygen so that the equilibrium concentrations of SO_2 and SO_3 are equal?

(a) 0.4 atm (b) 1.0 atom

(c) 0.8 atm (d) 0.25 atm

Answers

1. (b)	**2.** (c)	**3.** (c)	**4.** (b)	**5.** (c)	**6.** (a)	**7.** (b)	**8.** (d)	**9.** (c)	**10.** (b)
11. (c)	**12.** (b)	**13.** (c)	**14.** (a)	**15.** (d)	**16.** (a)	**17.** (b)	**18.** (d)	**19.** (c)	**20.** (a)
21. (d)	**22.** (a)	**23.** (d)	**24.** (c)	**25.** (a)	**26.** (b)	**27.** (d)	**28.** (c)	**29.** (b)	**30.** (c)
31. (c)	**32.** (a)	**33.** (b)	**34.** (c)	**35.** (b)	**36.** (c)	**37.** (a)	**38.** (d)	**39.** (c)	**40.** (a)
41. (c)	**42.** (a)	**43.** (a)	**44.** (d)	**45.** (d)	**46.** (d)	**47.** (b)	**48.** (d)	**49.** (b)	**50.** (a)
51. (b)	**52.** (d)	**53.** (c)	**54.** (d)	**55.** (a)	**56.** (b)	**57.** (d)	**58.** (c)	**59.** (d)	**60.** (b)
61. (a)	**62.** (a)	**63.** (c, d)	**64.** (a)	**65.** (a)	**66.** (b)	**67.** (d)	**68.** (c)	**69.** (d)	**70.** (c)
71. (c)	**72.** (b)	**73.** (c)	**74.** (d)	**75.** (d)	**76.** (c)	**77.** (d)	**78.** (a)	**79.** (c)	**80.** (b)
81. (a)	**82.** (d)	**83.** (a)	**84.** (b)	**85.** (c)	**86.** (d)	**87.** (c)	**88.** (a)	**89.** (c)	**90.** (a)
91. (a)	**92.** (d)	**93.** (a)	**94.** (a)	**95.** (d)	**96.** (d)	**97.** (a, c)	**98.** (d)	**99.** (c)	**100.** (a)
101. (d)	**102.** (a)	**103.** (b)	**104.** (a)	**105.** (b)	**106.** (d)	**107.** (b)	**108.** (c)	**109.** (a)	**110.** (c)
111. (d)	**112.** (a)	**113.** (a)	**114.** (d)	**115.** (b)	**116.** (a)	**117.** (d)	**118.** (c)	**119.** (b)	**120.** (a)
121. (d)	**122.** (b)	**123.** (b)	**124.** (a)	**125.** (b)	**126.** (b)	**127.** (d)	**128.** (b)	**129.** (c)	**130.** (b)
131. (a)	**132.** (a)	**133.** (b)	**134.** (b)	**135.** (b)	**136.** (b)	**137.** (d)	**138.** (c)	**139.** (d)	**140.** (b)
141. (a)	**142.** (d)	**143.** (d)	**144.** (b)	**145.** (a)	**146.** (d)	**147.** (b)	**148.** (b)	**149.** (a)	**150.** (a)

Explanations

1. (b) $N_2 + 3H_2 \rightleftharpoons 2NH_3$

$$k_1 = \frac{(NH_3)^2}{(N_2)(H_2)^3} \quad \text{——— (1)}$$

$N_2 + O_2 \rightleftharpoons 2NO$

$$k_2 = \frac{(NO)^2}{(N_2)(O_2)} \quad \text{——— (2)}$$

$H_2 + \frac{1}{2}O_2 \rightleftharpoons H_2O$;

$$k_3 = \frac{(H_2O)^1}{(H_2)^1 (O_2)^{\frac{1}{2}}} \quad \text{——— (3)}$$

AIM $\longrightarrow$ $2NH_3 + \frac{5}{2}O_2 \longrightarrow 2NO + 3H_2O$

$$k' = \frac{(NO_2)^2 (H_2O)^3}{(NH_3)^2 (O_2)^{\frac{5}{2}}}$$

$$\therefore \frac{k_2 k_3^3}{k_1} = k'$$

2. (c) $2PQ \rightleftharpoons P_2 + Q_2$; $k_1 = 2.5 \times 10^5$

Divide this eqn. by 2, the eqm. constt. Will become square root of the previous -

$PQ \rightleftharpoons \frac{1}{2}P_2 + \frac{1}{2}Q_2$; $K_1' = \sqrt{25 \times 10^4} = 5 \times 10^2$

$PQ + \frac{1}{2}R_2 \rightleftharpoons PQR$; $K_2 = 5 \times 10^{-2}$

Reaction

reqd. $\longrightarrow$ $\frac{1}{2}P_2 + \frac{1}{2}Q_2 + \frac{1}{2}R_2 \rightleftharpoons PQR$

$$K_3 = \frac{(PQR)}{(P_2)^{\frac{1}{2}} (Q_2)^{\frac{1}{2}} (R_2)^{\frac{1}{2}}}$$

$$K_3 = \frac{K_2}{K_1'} = \frac{5 \times 10^{-2}}{5 \times 10^2} = 10^{-4}$$

3. (c) $A + 2B \rightleftharpoons 2C$

At $t = 0$ 2   3   2

At eqm $2 - x$ $3 - 2x$ $2 + 2x$

$$\text{Eqm. Conc. of C} = \frac{2 + 2x}{2} = 0.5$$

$2 + 2x = 1$

$2x = -1$

$x = -\frac{1}{2}$ (moles can't be negative)

So, $2C \rightleftharpoons A + 2B$

At $t = 0$ 2   2   3

At eqm $2 - x$ $2 + \frac{x}{2}$ $3 + x$

Now, $\frac{2 - x}{2} = 0.5$

$2 - x = 1$

$x = 1$

$$\text{So, } K_C = \frac{(A)(B)^2}{(C)^2} = \frac{\left(\frac{\frac{5}{2}}{2}\right)\left(\frac{4}{2}\right)^2}{\left(\frac{1}{2}\right)^2}$$

$$K_C = \frac{5}{\cancel{2}} \times \frac{1}{\cancel{2}} \times \cancel{4}^{2} \times \frac{\cancel{4}^{2}}{1} = 20$$

But, K_C for $A + 2B \rightleftharpoons 2C = \frac{1}{20} = 0.05$

4. (b) $N_2O_{4(g)} \rightleftharpoons 2NO_{2(g)}$

0.1   0

$0.1 - 0.05$ 2×0.05

$$\left[\frac{9.2}{92} = 0.1\text{m mole } N_2O_4\right]$$

$$K_c = \frac{(0.1)^2}{0.05} = \frac{0.1 \times 0.1}{0.05} = 0.2$$

5. (c) $A + B \rightleftharpoons C + D$; $K = 100$

At t = 0   1   1   1   1

At t = eqm $1-x$ $1-x$ $1+x$ $1+x$

$$K = \frac{(1+x)(1+x)}{(1-x)(1-x)}$$

$$100 = \frac{(1+x)^2}{(1-x)^2}$$

$$10 = \frac{1+x}{1-x}$$

$$10 - 10x = 1 + x$$

$$9 = 11x$$

$$x = \frac{9}{11} = 0.818$$

Eqm. Conc. of D $= 1 + x = 1 + 0.818 = 1.818$

6. (a) $NH_4COONH_2 \rightleftharpoons 2NH_3 + CO_2$

(s)   (g)   (g)

$NH_4COONH_2 \rightleftharpoons 2NH_3 + CO_2$

$NH_4COONH_2 \rightleftharpoons 2NH_3 + CO_2$

(s)   (g)   (g)

At eqm   2P   P

$2P + P = 3$

$3P = 3$

$P = 1$

$Kp = (2P)^2 (P) = 4P^3 = 4 \times 1 = 4$

7. (b) $N_2 + 3H_2 \rightleftharpoons 2NH_3$

(g)   (g)   (g)

x $3x$ 0

$x - y$ $3x - 3y$ $2y$

Initial N_2 & H_2 mixture $= 4x$

Mixture reacted = 50% of $4x = 2x$

Now, N_2 reacted $= y$ moles

H_2 reacted $= 3y$ moles

So, $4y = 2x$

$$y = \frac{x}{2}$$

$$P.P._{NH_3} = P_{Total} X_{NH_3}$$

$$= P.\frac{2y}{4x - 2y} \quad \left[\text{Total moles at eqm} = x - y + 3x - 3y + 2y = 4x - 2y\right]$$

$$= P.\frac{2y}{4(2y) - 2y}$$

$$= P.\frac{2y}{6y} = P.\frac{1}{3} = \frac{P}{3}$$

8. (d) $H_2O_{(g)} \rightleftharpoons H_{2(g)} + \frac{1}{2}O_{2(g)}$

At t = 0   1   0   0

At t = eqm $1-\alpha$ α $\frac{\alpha}{2}$

$$Kp = \frac{(p^{H_2})^1 (p^{O_2})^{\frac{1}{2}}}{(pH_2O)^1}$$

$$\text{Total moles at eqm.} = 1 - \alpha + \alpha + \frac{\alpha}{2} = 1 + \frac{\alpha}{2}$$

$$Kp = \frac{\left(P.\frac{\alpha}{1+\frac{\alpha}{2}}\right)^1 \left(P.\frac{0.5\alpha}{1+\frac{\alpha}{2}}\right)^{\frac{1}{2}}}{\left(P.\frac{1-\alpha}{1+\frac{\alpha}{2}}\right)^1}$$

$$Kp = \frac{\alpha^{\frac{3}{1}} . P^{\frac{1}{2}}}{(1-\alpha)(2+\alpha)^{1/2}}$$

9. (c) $PCl_5 \rightleftharpoons PCl_3 + Cl_2$

(g)   (g)   (g)

a   0   0

$a - x$ x x

$$Kp = \frac{(P^{PCl_3})^1 (P^{Cl_2})^1}{(P^{PCl_5})^1}$$

$$Kp = \frac{\left(P.\frac{x}{a+x}\right)^1 \left(P.\frac{x}{a+x}\right)^1}{\left(P.\frac{a-x}{a+x}\right)^1}$$

$$Kp = (x.)\left(P.\frac{x}{a+x}\right) \times \frac{1}{a-x}$$

$$Kp = \frac{Px^2}{a^2 - x^2}$$

$$\frac{Kp}{P} = \frac{x^2}{a^2 - x^2}$$

Reciproacing both sides

$$\frac{P}{Kp} = \frac{a^2 - x^2}{x^2}$$

$$\frac{P}{Kp} = \frac{a^2}{x^2} - 1$$

$$\frac{P}{Kp} + 1 = \frac{a^2}{x^2}$$

$$\frac{P + Kp}{Kp} = \frac{a^2}{x^2}$$

$$\sqrt{\frac{P+Kp}{Kp}} = \frac{a}{x}$$

$$\text{Now, } \frac{x}{a} = \sqrt{\frac{Kp}{P+Kp}}$$

10. (b) $2AB_{2(g)} \rightleftharpoons 2AB_{(g)} + B_{2(g)}$

$$\begin{array}{llll} \text{At } t = 0 & 1 & 0 & 0 \\ \text{At } t = \text{eqm} & 1 - x & x & \frac{x}{2} \end{array}$$

$$\text{Total moles at eqm.} = 1 - x + x + \frac{x}{2} = 1 + 0.5x$$

$$Kp = \frac{(pAB)^2 (pB_2)^1}{(pAB_2)^2}$$

$$= \frac{\left(P.\frac{x}{1+0.5x}\right)^2 \left(P.\frac{0.5x}{1+0.5x}\right)^1}{\left(P.\frac{1-x}{1+0.5x}\right)^2}$$

$= \text{Since } x << 1$

$$Kp = \frac{\frac{P^2x^2}{1} \times \frac{P.\,0.5x}{1}}{\frac{P^2}{1}} = P^2x^2 \times \frac{Px}{2} \times \frac{1}{P^2}$$

$$Kp = \frac{Px^3}{2}$$

$$\left(\frac{2Kp}{P}\right)^{\frac{1}{3}} = x$$

11. (c) $2SO_{2(g)} + O_{2(g)} \rightleftharpoons 2SO_{3(g)}$

$$\begin{array}{llll} \text{At } t = 0 & 5 & 5 & 0 \\ \text{At = eqm} & 5 - x & 5 - \frac{x}{2} & x \end{array}$$

SO_2 used up = 60%

i.e. $\frac{60}{100} \times 5 = 3$

So, $x = 3$

$$P.PO_2 = P_{Total}XO_2 = \frac{5 - \frac{x}{2}}{5 - x + 5 - \frac{x}{2} + x} \times 1$$

$$P.PO_2 = \frac{5 - 1.5}{5 - 3 + 5 - 1.5 + 3}$$

$$P.PO_2 = \frac{3.5}{13 - 4.5} = \frac{3.5}{8.5} = \frac{35}{85} = 0.41$$

12. (b) $C_{(s)} + CO_{2(g)} \longrightarrow 2CO_{(g)}$; $Kp = 63$ atm

$$\begin{array}{lll} \text{At } t = 0 & P & 0 \\ \text{At } t = t & P - p' & 2P' \end{array}$$

Now, at eqm., $P_{CO} = 10P_{CO_2}$

$2P' = 10(P - p')$

$2P' = 10P - 10p'$

$12p' = 10P$

$$P' = \frac{10P}{12} = \frac{5P}{6}$$

$$Kp = \frac{(2p')^2}{P - p'} = \frac{4p'^2}{P - p'}$$

$$63 = \frac{4 \times \left(\frac{5P}{6}\right)^2}{P - \frac{5P}{6}}$$

$$63 = \frac{4 \times \frac{25P^2}{36}}{\frac{P}{6}}$$

$$63 = \frac{4 \times 25P^{\not{2}}}{\underset{6}{\not{36}}} \times \frac{\not{6}}{\not{P}}$$

$$\frac{63 \times 6}{100} = P$$

$$\frac{378}{100} = P = 3.78\,atm$$

So, $P = 3.78\,atm$

$$P' = \frac{5P}{6} = \frac{5}{6} \times 3.78 = 3.15\,atm$$

Now, Total P at eqm. $= P - p' + 2p'$

$= P + p'$

$= 3.78 + 3.15$

6.93 atm

13. (c)

	$X \rightleftharpoons Y + Z$			$A \rightleftharpoons 2B$	
At t = 0	1	0	0	1	0
At t = eqm	$1-x$	x	x	$1-x$	$2x$

$$Kp_1 = \frac{(PY)^1 (PZ)^1}{(PX)^1} \qquad Kp_2 = \frac{(PB)^2}{(PA)}$$

$$Kp_1 = \frac{\left(\frac{x}{1+x}.P_1\right)\left(\frac{x}{1+x}.P_1\right)}{\left(\frac{1-x}{1+x}.P_1\right)}$$

$$Kp_2 = \frac{\left(P_2.\frac{2x}{1+x}\right)^2}{\left(P_2.\frac{1-x}{1+x}\right)^1}$$

$$Kp_1 = \frac{P_1 x^2}{(1+x)(1-x)}$$

$$Kp_2 = P_2.\frac{4x^2}{(1+x)(1-x)}$$

$$\frac{Kp_1}{Kp_2} = \frac{\frac{P_1 x^2}{1-x^2}}{\frac{P_2.4x^2}{1-x^2}}$$

$$\frac{Kp_1}{Kp_2} = \frac{P_1 x^2}{1-x^2} \times \frac{1-x^2}{P_2 4x^2}$$

$$\frac{9}{1} = \frac{P_1}{P_2} \times \frac{1}{4}$$

$$36 : 1 = \frac{P_1}{P_2}$$

14. (a) Similar as question 10

15. (d) $A_{(s)} \rightleftharpoons 2B_{(g)} + 3C_{(g)}$

At eqm $\qquad 2x \qquad 3x$

$$K_C = (B)^2 (C)^3 = (2x)^2 (3x)^3 = 108x^5$$

Now, when concentration of C is doubled then,

$A \rightleftharpoons 2B + 3C$

At eqm $\qquad y \qquad 6x$

$$K_C = (B)^2 (C)^3 = y^2 (6x)^3$$

$$K_C = y^2 \times 216x^3$$

K_c will remain constant

So, $108x^5 = y^2 \times 216x^3$

$$x^2 = y^2 \times 2$$

$$\frac{x^2}{2} = y^2$$

$$\frac{x}{\sqrt{2}} = y$$

$$z \times 2x = \frac{x}{\sqrt{2}}$$

$$z = \frac{1}{2\sqrt{2}}$$

16. (a) $\underset{(g)}{LiCl.3NH_3} \rightleftharpoons \underset{(g)}{LiCl.NH_3} + \underset{(g)}{2NH_3}$

$Kp = 9\,atm^2$

$\underset{(g)}{LiCl.NH_3} + \underset{(g)}{2NH_3} \rightleftharpoons \underset{(g)}{LiCl.3NH_3}$;

$Kp = \frac{1}{9} atm^{-2}$

At t = 0	0.1	a	0
At eqm	0	(a − 0.2)	0.1

Let initial moles of NH_3 be 'a' to bring completion of reaction

At eqm,

$$Kp_1 = \frac{1}{(P.NH_3)^2}$$

$$\frac{1}{9} = \frac{1}{(P.NH_3)^2}$$

$P'NH_3 = 3$

Applying,

$PV = nRT$

$3 \times 5 = n \times 0.0821 \times 313$

$n = 0.5837$

Tota moles at eqm. = a − 0.2 = 0.5837

a = 0.5837 + 0.2

a = 0.7837

17. (b) $\underset{(s)}{FeO} + \underset{(g)}{CO} \rightleftharpoons \underset{(s)}{Fe} + CO_{2(g)}$; $Kp = 0.265\,atm$

Initially → 1.4 0.80

$$Qp = \frac{(P^{CO_2})^1}{(P^{CO})^1} = \frac{0.80}{1.4} = 0.5714$$

$Q_p > K_p$,

So reaction will going backward direction

$\underset{(s)}{Fe} + CO_{2(g)} \rightleftharpoons \underset{(s)}{FeO} + CO_{(g)}$; $Kp' = \frac{1}{0.265}$

Initially	0.80	1.4
At t = eqm	0.80 − p	1.4 + p

$$K'p = \frac{1.4 + P}{0.80 - P}$$

$$\frac{1}{0.265} = \frac{1.4 + P}{0.80 - P}$$

$0.80 - P = 0.265(1.4 + P)$

$0.80 - P = 0.371 + 0.265P$

$0.80 - 0.371 = 1.265P$

$\frac{0.419}{1.265} = P$

$0.34 = P$

Eqm. P of $CO_2 = 0.80 - 0.34 = 0.46\,atm$

Eqm. P of $CO = 1.4 + P = 1.4 + 0.34 = 1.74\,atm$

18. (d) Discussed earlier in several questions.

19. (c) $N_2O_{4(g)} \rightleftharpoons 2NO_{2(g)}$

At t = 0 $\frac{23}{92}$ moles = 0.25 moles

If we apply, PV = nRT

$P \times 1 = 0.25 \times 0.0821 \times 400$

P = 8.21 atm

$N_2O_{4(g)} \rightleftharpoons 2NO_{2(g)}$

At t = 0	8.21	0
At t = eqm	8.21 − P	2P

Total P at eqm = 11.6 bar

$\therefore\ 8.21 - P + 2P = 11.6$

$8.21 + P = 11.6$

$P = 11.6 - 8.21$

$P = 3.39\,atm$

$P.P.NO_2 - 2P$

20. (a) Discussed similar questions earlier

21. (d) $A + 3B \rightleftharpoons 2C + D$

	A	B	C	D
At t = 0	$2x$	x	0	0
At t = eqm.	$2x - y$	$x - 3y$	$2y$	y

At eqm. moles of B = no. of moles of C

$\therefore\ x - 3y = 2y$

$x = 5y$

$\frac{x}{5} = y$

Percentage of B reacted $= \frac{3y}{x} \times 100$

$= \frac{3 \times \frac{x}{5}}{x} \times 100$

$= \frac{3x}{5} \times \frac{1}{x} \times 100$

$= 60\%$

24. (c) $H_2O + CO \rightleftharpoons CO_2 + H_2$

At t = 0	1	1	0	0
At eqm	1 − 0.4	1 − 0.4	0.4	0.4

$$K_c = \frac{\frac{0.4}{10} \times \frac{0.4}{10}}{\frac{0.6}{10} \times \frac{0.6}{10}}$$

$$K_c = \frac{0.4 \times 0.4}{100} \times \frac{100}{0.6 \times 0.6}$$

$$K_c = \frac{16}{36} = 0.44$$

25. (a) $I_{2(aq.)} + I^-_{(aq.)} \rightleftharpoons I^-_{3(aq.)}$

At t = 0	1	0.5	0
At t = eqm	1 − x	0.5 − x	x

$AgNO_3$ reacts with I^- & form AgI ↓.

$$AgNO_3 + I^- \longrightarrow AgI_{(s)} + NO^-_{3(aq.)}$$

$0.5 - x = 0.25$

$x = 0.25$

$$K_c = \frac{x}{(1-x)(0.5-x)} = \frac{0.25}{0.75 \times 0.25}$$

$$K_c = \frac{100}{75} = 1.33$$

31. (c) $SO_{2(g)} + NO_{2(g)} \rightleftharpoons SO_{3(g)} + NO_{(g)}$;

$K_C = 16$

$$K_c = \frac{(1+x)^2}{(1-x)^2}$$

$$16 = \frac{(1+x)^2}{(1-x)^2}$$

$$4 = \frac{1+x}{1-x}$$

$4 - 4x = 1 + x$

$3 = 5x$

$\frac{3}{5} = x = 0.6$

$(NO_2) = 1 - x = 1 - 0.6 = 0.4$

33. (b) $N_{2(g)} + 3H_{3(g)} \rightleftharpoons 2NH_{3(g)}$; $K_C = 1.7 \times 10^2$

$$Q_C = \frac{(NH_3)^2}{(N_2)(H_2)^3} = \frac{\left(\frac{8.13}{20}\right)^2}{\left(\frac{1.57}{20}\right)\left(\frac{1.92}{20}\right)^3}$$

$$Q_C = \frac{8.13 \times 8.13}{20 \times 20} \times \frac{20 \times 20 \times 20 \times 20}{1.57 \times 1.92 \times 1.92 \times 1.92}$$

$$Q_C = \frac{400 \times 8.13 \times 8.13}{1.57 \times 1.92 \times 1.92 \times 1.92}$$

$Q_C > K_C$

Reaction is not in eqm, rushes backward.

37. (a) For the given reaction,

$K_P = P_{CO_2}$

Acc, to Van't Hoff equation-

$$\frac{d}{dT} \ln k_p = \frac{\Delta H}{RT^2}$$

$$\ln k_p = -\frac{\Delta H}{RT} + 1$$

$$2.303 \log_{10} kp = \frac{-\Delta H}{RT} + 1$$

$$2.303\log_{10} P_{Co_2} = \frac{-\Delta H}{RT} + 1$$

$$\log_{10} P_{Co_2} = \frac{-\Delta H}{2.303RT} + \frac{1}{2.303}$$

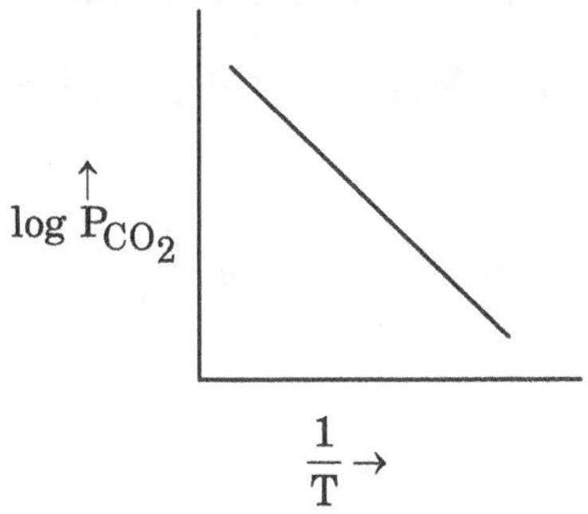

40. (a) $N_2 + 3H_2 \rightleftharpoons 2NH_3$

At t = 0   P'   3P'   0

At eqm $P' - \frac{P'}{2}$ $3P' - \frac{3P'}{2}$ P'

Total P at eqm. = P

$$\therefore P' - \frac{P'}{2} + 3P' - \frac{3P'}{2} + P' = P$$

$$5P' - \frac{3P'}{2} - \frac{P'}{2} = P$$

$$5P' - \frac{4P'}{2} = P$$

$$\frac{6P'}{2} = P$$

3P' = P'

P.P of NH_3 at eqm. = $P' = \frac{P}{3}$

41. (c)

$$A_{2(g)} + B_{2(g)} \rightleftharpoons 2AB_{(g)}$$

At t = 0   2   4   0

At t = eqm   2 − x   4 − x   2x

$$k_C = \frac{(AB)^2}{(A_2)'(B_2)'} = \frac{\left(\frac{2x}{4}\right)^2}{\left(\frac{2-x}{4}\right)',\left(\frac{4-x}{4}\right)',}$$

$$4 = \frac{4x^2}{16} \times \frac{16}{(2-x)(4-x)}$$

$$(2-x)(4-x) = x^2$$

$$8 - 2x - 4x + x^2 = x^2$$

$$8 - 6x = 0$$

$$8 = 6x$$

$$x = \frac{\cancel{8}\,4}{\cancel{6}\,3}$$

$$\text{Eqm. Conc. of AB} = \frac{2x}{4} = \frac{2}{4} \times \frac{4}{3} = \frac{2}{3} = 0.66\text{M}$$

42. (a) $ACO_{3(s)} \rightleftharpoons AO_{(s)} + CO_{2(g)}$

Eqm. P of CO_2 (at 127°C) = kp_1 = 0.14 bar

Eqm. P of CO_2 (at 147°C) = kp_2 = 0.34 bar

$$\log \frac{k_2}{k_1} = \frac{\Delta H}{2.303R} \frac{(T_2 - T_1)}{T_1T_2}$$

$$\log \frac{0.34}{0.14} = \frac{\Delta H}{2.303 \times 8.314 \times 10^{-3}} \frac{(420 - 400)}{400 \times 420}$$

$$\log 2.42 = \frac{\Delta H}{2.303 \times 8.314 \times 10^{-3}} \times \frac{20}{400 \times 420}$$

$$\frac{0.38 \times 2.303 \times 8.314 \times 10^{-3} \times 400 \times \cancel{420}\,21}{\cancel{20}} = \Delta H$$

$\Delta H \approx 64$ KJ/mol

49. (b) $NH_2COONH_{4(S)} \rightleftharpoons 2NH_{3(g)} + CO_{2(g)}$

At eqm.   2P   P

3P = 3 atm

P = 1 atm

$$Kp = (2P)^2(P) = 4P^3 = 4 \times 1 = 4 \text{ atm}^3$$

50. (a) $Fe(OH)_{3(s)} \rightleftharpoons Fe^{3+}_{(aq.)} + 3\overline{OH}_{(aq.)}$

At eqm. x $3x$

$$K_C = (Fe^{3+})(\overline{OH})^3 = (x)(3x)^3 = 27x^4$$

$Fe(OH)_{3(s)} \rightleftharpoons Fe^{3+}_{(aq.)} + 3\overline{OH}_{(aq.)}$

At eqm. $y \quad \frac{3x}{4}$

$K_C = (y)\left(\frac{3x}{4}\right)^3$

$27x^4 = y \times \frac{27x^3}{64}$

$x = \frac{y}{64}$

$y = 64x$

51. (b) For the given reaction, in case of complete decomposition of $CaCO_3$, Kp = 1

$\log Kp = 7.282 - \frac{8500}{T}$

$\log 1 = 7.282 - \frac{8500}{T}$

$\frac{8500}{T} = 7.282$

$T = 1167K = 894°C$

52. (d) Apply $Kp = K_C(RT)^{\Delta ng}$

For d, $\Delta ng = +1$

$Kp = K_C(RT)$

53. (c) For, $2SO_{2(g)} + O_{2(g)} \rightleftharpoons 2SO_{3(g)}$; $k_c = 300$

$2SO_{3(g)} \rightleftharpoons 2SO_{2(g)} + O_{2(g)}$; $k_c = \frac{1}{300}$

Now, eqn. is divided by 2, kc will be square root of the previous

$SO_{3(g)} \rightleftharpoons SO_{2(g)} + \frac{1}{2}O_2$;

$k_c = \sqrt{\frac{1}{300}} = 0.0573$

56. (b) Using a catalyst will only increase the rate of reaction.

58. (c) At 1400*k*,

$CH_{4(g)} + 2H_2S_{(g)} \rightleftharpoons CS_{2(g)} + 4H_{2(g)}$;

$K_C = 2.5 \times 10^{-3}$

$$Q_C = \frac{(CS_2)(H_2)^4}{(CH_4)(H_2S)^2} = \frac{\frac{3}{10} \times \left(\frac{3}{10}\right)^4}{\frac{2}{10} \times \left(\frac{4}{10}\right)^2} \begin{matrix} = 0.075 \\ = 7.5 \times 10^{-2} \end{matrix}$$

$Q_C > K_C$, reaction goes backwards

61. (a) $2NO_{(g)} + O_{2(g)} \rightleftharpoons 2NO_{2(g)}$; $K_C = 10$

At t = 0 $\quad a \quad a \quad 0$

At t = eqm. $\quad a - x \quad a - \frac{x}{2} \quad x$

$a - x = x$

$a = 2x$

$x = \frac{a}{2}$

$k_c = 10$

$K_C = \frac{x^2}{(a-x)^2\left(a - \frac{x}{2}\right)}$

$10 = \frac{x^2}{(2x-x)^2\left(2x - \frac{x}{2}\right)}$

$10 = \frac{x^2}{x^2 \times \frac{3x}{2}}$

$10 = \frac{1}{3x}x^2$

$3x = \frac{2}{10} \Rightarrow x = \frac{2}{30} = 0.066$

No. of moles of O_2 at eqm $= a - \frac{x}{2}$

$= 2x - \frac{x}{2} = \frac{3x}{2}$

$= \frac{3}{2} \times 0.066 = 0 \cdot 1$

65. (a) $4HNO_{3(g)} \rightleftharpoons 4NO_{2(g)} + 2H_2O_{(g)} + O_{2(g)}$

At t = 0 $\quad P \quad 0 \quad 0 \quad 0$

At t = eqm $\quad P - P' \quad P' \quad \frac{P'}{2} \quad \frac{P'}{4}$

Total P at eqm = 30 atm

$\therefore\ P - P' + P' + \frac{P'}{2} + \frac{P'}{4} = 30$

$P + 0.75P' = 30$ ______(1)

Also, $p\ HNO_3 = 2atm = P - P'$

$P - p' = 2atm$

$P = 2 + p'$ ______(2)

Putting value of P in eqn.1 –

$2 + p' + 0.75P' = 30$

$2 + 1.75P' = 30$

$1.75P' = 28$

$P' = \frac{28}{1.75} = 16$

$\therefore P = 18$ ____using eqn.2

$$Kp = \frac{(P')^4 \times \left(\frac{P'}{2}\right)^2 \times \frac{P'}{4}}{(P - p')^4}$$

$$= \frac{(16)^4 \times (8)^2 \times 4}{(2)^4}$$

$= (16)^3 \times 64 \times 4$

$Kp = (16)^3 \times 256 = 1048576$

Now, $Kp = k_c (RT)^{\Delta ng}$

$1048576 = k_c (0.0821 \times 400)^3$

$\frac{1048576}{(32.84)^3} = k_c$

$\frac{1048576}{35416.81} = k_c \approx 30$

67. (d) $AB_{(g)} \rightleftharpoons A_{(g)} + B_{(g)}$

At t = 0	P'	0	0
At t = eqm	P' – P"	P'	P"

Now,

$P' - P'' + P'' + P'' = P \rightarrow$ Total eqm. pressure

$P' + P'' = P$

$Kp = \frac{P'' \times P''}{P' - p''} = \frac{1}{3} \times P$ (given)

$\frac{P''^2}{P' - p''} = \frac{P' + P''}{3}$

$3p''^2 = (P' + p'')(P' - p'')$

$3p''^2 = P'^2 - p''^2$

$4P''^2 = P'^2$

$2p'' = P'$

$P'' = \frac{P'}{2}$

Now % age dissocation of AB $= \frac{P''}{P'} \times 100$

$= \frac{\frac{P'}{2}}{P'} \times 100 = \frac{P'}{2} \times \frac{1}{P'} \times 100$

$= 50\%$

70. (c) $S^{2-}_{(aq.)} + \underset{(s)}{S} \rightleftharpoons S^{2-}_{2\,(aq.)}$; $kc = 12$

$S^{2-}_{(aq.)} + 2\underset{(s)}{S} \rightleftharpoons S^{2-}_{3\,(aq.)}$; $Kc_2 = 132$

$Kc_1 = \frac{(S_2^{2-})}{(S^{2-})}$; $Kc_2 = \frac{(S_3^{2-})}{(S^{2-})}$

AIM.- $\underset{(aq.)}{S_2^{2-}} + \underset{(s)}{S} \rightleftharpoons S^{2-}_{3\,(aq.)}$

$K_c' = \frac{Kc_2}{Kc_1} = \frac{132}{12} = 11$

88. (a) $\underset{(g)}{A} + \underset{(g)}{B} \rightleftharpoons C_{(s)}$

$K_C = \frac{1}{(A)(B)}$

$= \frac{1}{(mol\,ltr^{-1})(mol\,ltr^{-1})} = mol^{-2}ltr^2$

Increasing the P will move reaction towards lesser 2. Of gaseous mole i.e. Conc. Of product is used.

99. (c) Addition of solid ($NaNO_3$ or $NaNO_2$) will not affect the equilibrium.

102. (a) High temp. favors endothermic reaction.

108. (c) $\propto = \frac{D-d}{d(n-1)}$

For, $N_2O_{4(g)} \rightleftharpoons 2NO_{2(g)}$

$n = 2$

$\propto = \frac{D-d}{d}$

$\propto = \frac{D}{d} - \frac{d}{d}$

$\Rightarrow \propto = \frac{D}{d} - \frac{d}{d} - 1$

↓ ↓ ↓

y = mx c

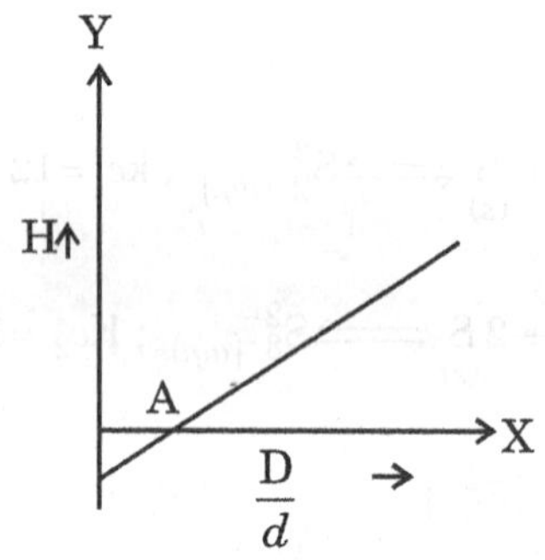

At A, $\propto = 0$

$\therefore \frac{D}{d} = 1$

110. (c) $n\,A_{(g)} \rightleftharpoons (A)_{n(g)}$

At t = 0   1   0

At t = t $1-\propto$ $\frac{\propto}{n}$

Let V be the volume of each gaseous mole.

Total No. of gaseous moles, after association

$= 1 - \propto + \frac{\propto}{n}$

Total volume of gases after association

$= \left(1 - \propto + \frac{\propto}{n}\right)V$

$d \propto \frac{1}{\left(1 - \propto + \frac{\propto}{n}\right)V}$ ____ (2)

$\frac{D}{d} = \frac{\frac{1}{V}}{\frac{1}{\left(1 - \propto + \frac{\propto}{n}\right)V}}$

$\frac{D}{d} = \frac{1}{V} \times V\left(1 - \propto + \frac{\propto}{n}\right)$

$\frac{D}{d} = 1 - \propto + \frac{\propto}{n}$

$\frac{D}{d} - 1 = \frac{\propto}{n} - \propto$

$\frac{D}{d} - 1 = \propto\left(\frac{1}{n} - \frac{1}{1}\right)$

$\frac{D-d}{d\left(\frac{1}{n} - 1\right)} = \propto$

For given question -

$\frac{D-d}{d \times \frac{-1}{2}} = \propto$

$\frac{D-d}{-\frac{d}{2}} = \propto$

$\frac{2(D-d)}{-d} = \propto$

$2D - 2d = \propto \times -d$

$\propto = \frac{2(d-D)}{d}$

144. (b) $H_{2(q)} \rightleftharpoons 2H_{(q)}$

$kp = \frac{4\propto^2}{1-\propto^2} \cdot P$

$\propto \approx \sqrt{\frac{kp}{2p}} \Rightarrow \frac{\propto_1}{\propto_2} = \sqrt{\frac{P_2}{P_1}}$

$\propto_2 = \frac{0.1\%}{2} = 0.05\%$

Smart Notes

Smart Notes

CHAPTER 8

Ionic Equilibrium

1. Which is not correct for Lewis acids?
   (a) They contain at least one vacant orbital
   (b) They have a tendency to accept electrons
   (c) The smaller ion has greater acidic strength
   (d) In case of ions, the strength of acid is inversely proportional to its charge

2. The conjugate base of $[Al(H_2O)_3(OH)_3]$
   (a) $[Al(H_2O)_3(OH)_2]$
   (b) $[Al(H_2O)_3(OH)_2O]^-$
   (c) $[Al(H_2O)_3(OH)_3]^-$
   (d) $[Al(H_2O)_2(OH)_4]^-$

3. Ostwald's dilution law is applicable in the case of the:
   (a) CH_3COOH (b) NaCl
   (c) NaOH (d) H_2SO_4

4. The conjugate acid $S_2O_8^{2-}$ is:
   (a) $H_2S_2O_8$ (b) H_2SO_4
   (c) HSO_4^- (d) $HS_2O_8^-$

5. Which equilibrium can be described as Lewis acid-base reaction but not Bronsted acid-base reaction?
   (a) $H_2O + CH_3COOH \rightleftharpoons H_3O^+ + CH_3COO^-$
   (b) $2NH_3 + H_2SO_4 \rightleftharpoons 2NH_4^+ + SO_4^{2-}$
   (c) $NH_3 + CH_3COOH \rightleftharpoons NH_4 + CH_3COO^-$
   (d) $[Cu(H_2O)_4]^{2+} + 4NH_3 \rightleftharpoons [Cu(NH_3)_4]^{2+} + 4H_2O$

6. The two Bronsted bases in the reaction are
   $HC_2O_4^- + PO_4^{3-} \rightarrow HPO_4^{2-} + C_2O_4^{2-}$
   (a) $HC_2O_4^-$ and PO_4^{3-}
   (b) HPO_4^{2-} and $C_2O_4^{3-}$
   (c) PO_4^{3-} and $C_2O_4^{2-}$
   (d) $HC_2O_4^-$ and HPO_4^{2-}

7. The conjugate base of $H_2PO_4^-$ is:

(a) HPO_4^{2-} (b) P_2O_5

(c) H_3PO_4 (d) PO_4^{3-}

8. Which of the following is a Bronsted acid as well as Bronsted base?

(a) Na_2CO_3 (b) H_2O

(c) Na_2CO_3 (d) BF_3

9. K_a for formic acid and acetic acid are 2.1×10^{-4} and 1.1×10^{-5} respectively. The relative strength of acids is:

(a) 2 : 1 (b) 2.3:1

(c) 1:2.1 (d) 4.36:1

10. At infinite dilution, the percentage ionisation for both strong and weak electrolytes is:

(a) 1% (b) 20%

(c) 50% (d) 100%

11. When rain is accompanied by a thunder storm, the collected rain water will have a pH value

(a) depending on the amount of dust in air

(b) slightly lower than that of rain water without thunderstorm

(c) slightly higher than that when the thunder storm is not there

(d) uninfluenced by occurrence of thunderstorm

12. For weak electrolyte, AB, the degree of ionization would be (V = volume of solution having 1 mole electrolyte and K is the ionization constant of the electrolyte)

(a) $\frac{K}{V^2}$ (b) K.V

(c) $\frac{K}{V}$ (d) $\sqrt{K.V}$

13. Which of the following is incorrect representation of aqueous ?

(a) H_3O^+

(b) $H_5O_2^+$

(c) $H_7O_3^+$

(d) $H_9O_5^+$

14. Choose the incorrect pair?

Species	Conjugate Acid	Conjugate Base
(a) H_2O	H_3O^+	$\bar{O}H$
(b) HCO_3^-	H_2CO_3	CO_3^{2-}
(c) NH_3	NH_4^+	NH_3^-

(d) None of these

15. Choose the correct statement(s) amongst the following-

(1) Electron deficient species like $AlCl_3$, Co^{3+}, Mg^{2+} etc, act as Lewis acids while species like $NH_3, \bar{O}H$ which can donate a pair of electrons can act like lewis bases

(2) BCl_3 & H^+ both are lewis acids

(3) The acid base pair that differs only by one proton is called a conjugate acid base pair

(4) The equilibrium of the reaction,

$$\underset{\text{Acid}}{HA_{(aq)}} + \underset{\text{Base}}{H_2O_{(\ell)}} \rightleftharpoons \underset{\text{Conjugate Acid}}{H_3O^+_{(aq)}} + \underset{\text{Conjugate base}}{A^-_{(aq)}}$$

moves in the direction of formation of weaker & weaker base.

(5) Strong acids also have strong conjugate bases.

(a) All are correct

(b) Only 1, 2, 3 are correct

(c) Only 2, 3 & 4 are correct

(d) Only 1, 2, 3 & 4 are correct

16. If pK_b for fluoride ion at 25°C is 10, the ionisation constant of hydrofluoric acid in water at this temperature is:

(a) 1.74×10^{-5} (b) 3.52×10^{-3}

(c) 10^{-4} (d) 5.38×10^{-2}

17. The degree of dissociation of 1.0 M weak acid, HA, is 0.5%. If 2 ml of 1.0 M HA solution is diluted to 32 ml, the degree of dissociation of the acid and H_3O^+ ion concentration in the resulting solution will be respectively:

(a) 0.02 and 3.125×10^{-4}

(b) 0.02 and 1.25×10^{-3}

(c) 1.25×10^{-3} and 0.02

(d) 0.02 and 8.0×10^{-12}

Questions Based on pH, pOH & Ionic product of water

18. A solution has a pH = 9, it is 1000 times more basic than the original solution. What was the pH of the original solution?

(a) 12 (b) 6

(c) 9 (d) 10

19. To a 10 mL of 10^{-3} N H_2SO_4 solution water has been added to make the total volume of one litre. Its pOH would be

(a) 3 (b) 12

(c) 9 (d) 5

20. pH of a strong diprotic acid (H_2A) at concentrations:

(a) 10^{-4}M (b) 10^{-4}N

Are respectively:

(a) 3.7 and 4.0 (b) 4 and 3.7

(c) 4 and 4 (d) 3.7 and 3.7

21. 10^{-5}M NaOH solution at 25°C is diluted 1000 times. The pH of the resultant solution will:

(a) be equal to 8 (b) lie between 7 and 8

(c) lie between 6 and 7 (d) remain unchanged

22. 4.0g of NaOH and 4.9 g of H_2SO_4 are dissolved in water and volume is made upto 250 mL. The pH of this solution is:

(a) 7.0 (b) 1.0

(c) 2.0 (d) 12.0

23. At 90°C, pure water has $[H^+] = 10^{-6}$ M, if 100 mL of 0.2 M HCl is added to 200 mL of 0.1 M KOH at 90°C then pH of the resulting solution will be:

(a) 5 (b) 6

(c) 7 (d) None of these

24. Given K_a values of 5.76×10^{-10} and 4.8×10^{-10} for NH_4^+ and HCN respectively. What is the equilibrium constant for the following reaction?

$$NH^+_4(aq.) + CN^-(aq.) \rightleftharpoons NH_3(aq.) + HCN(aq.)$$

(a) 0.83 (b) 1.2

(c) 8.0×10^{-11} (d) 27.6×10^{-10}

25. What is the hydronium ion concentration of a 0.25 M HA solution?

(a) 10^{-4} M (b) 10^{-5} M

(c) 10^{-7} M (d) 10^{-10} M

26. 0.01 M HA (aq.) is 2% ionized; $[OH^-]$ of solution is:

(a) 2×10^{-4} (b) 10^{-8}

(c) 5×10^{-11} (d) 5×10^{-12}

27. If degree of ionization is 0.01 of decimolar solution of weak acid HA then pK_a of acid is:

(a) 2 (b) 3

(c) 5 (d) 3

28. What is $[NH^+_4]$ in a solution that contain 0.02 M NH_3 ($K_b = 1.8 \times 10^{-5}$) and 0.01 M KOH?

(a) 9×10^{-6} (b) 1.8×10^{-5}

(c) 3.6×10^{-5} (d) None of these

29. 6.0 gm weak acid HA(mol. Wt. = 60g/mol) is dissolved in water and formed 10 m³ solution. If $K_{a(HA)} = 10^{-9}$, then pOH of solution is:[Given : log 4 = 0.6]

(a) 6.7

(b) greater than 6.7 and less than 7.0

(c) greater than 7.0 and less than 7.3

(d) greater than 7.3

30. The value of the ionic product constant for water, (k_w) at 60°C is $9.6 \times 10^{-14} M^2$. What is the $[H_3O^+]$ of a neutral aqueous solution at 60°C.

(a) 5×10^{-9} M

(b) 2×10^{-7} M

(c) 6×10^{-11} M

(d) 3.1×10^{-7} M

31. For pure water:

(a) pH increases while pOH decreases with rise in temperature

(b) pH decreases while pOH increases with rise in temperature

(c) both pH and pOH decreases with rise in temperature

(d) both pH and pOH increases with rise in temperature

32. 10 ml of $\frac{M}{200} H_2SO_4$ is mixed with 40 ml of $\frac{M}{200} H_2SO_4$. The pH of the resulting solution will be

(a) 1 (b) 2

(c) 2.3 (d) None of these

33. A solution with *p*H 2.0 is more acidic than the one with *p*H 6.0 by a factor of:

(a) 3 (b) 4

(c) 3000 (d) 10000

34. The first and second dissociation constants of an acid H_2A are 1.0×10^{-5} and 5.0×10^{-10} respectively. The overall dissociation constant of the acid will be

(a) 5.0×10^{-5} (b) 5.0×10^{15}

(c) 5.0×10^{-15} (d) 0.2×10^{5}

35. The pK_a for acid A is greater than pK_a for acid B. The strong acid is:

(a) acid A (b) acid B

(c) are equally strong (d) None of these

36. Given, $HF + H_2O \xrightarrow{K_a} H_3O^+ + F^-$. Which relation is correct?

(a) $K_b = K_a$ (b) $K_b = \frac{1}{K_a}$

(c) $K_a \times K_b = K_w$ (d) $\frac{K_a}{K_b} = K_w$

37. The unit of ionic product of water (K_w) is:

(a) $mol^{-1}\ litre^{-1}$ (b) $mol^{-1}\ litre^{-2}$

(c) $mol^{-2}\ litre^{-1}$ (d) $mol^{2}\ litre^{-2}$

38. If 50 ml of 0.2M KOH is added to 40 ml of 0.5M HCOOH. The pH of the resulting solution is

$(Ka = 1.8 \times 10^{-4}, \log 18 = 1.26)$

(a) 3.74 (b) 5.64

(c) 7.57 (d) 6.62

39. The resultant *p*H of a solution on mixing 200 mL of aqueous solution of HCl (*p*H= 2.0) with 300 mL of an aqueous solution of NaOH (*p*H = 12.0) is

(a) 11.0310 (b) 11.3010

(c) 10.000 (d) None of these

40. Number of H^+ ions present in 10 mL of a solution of pH= 3 is:

(a) 10^{13} (b) 6.02×10^{18}

(c) 6.02×10^{13} (d) 6.02×10^{10}

41. What molar concentration of NH_3 provides a $[OH^-]$ of 1.5×10^{-3} M? ($K_b = 1.8 \times 10^{-5}$)

(a) 0.125 M

(b) $(0.125 + 1.5 \times 10^{-3})$M

(c) $(0.125 - 1.5 \times 10^{-3})$M

(d) (1.5×10^{-3})M

42. The *p*H of a solution is 5.0. To this solution sufficient acid is added to decrease the *p*H to 2.0. The increase in hydrogen ion concentration is:

(a) 1000 times (b) $\frac{5}{2}$ times

(c) 100 times (d) 5 times

43. The values of dissociation constant of some acids (at 25°C) are given below. The strongest acid in water is:

(a) 1.4×10^{-2} (b) 1.6×10^{-4}

(c) 4.4×10^{-10} (d) 4.3×10^{-7}

44. Which solution will have pH closer to 1.0 ?

(a) 100 mL of $\left(\frac{M}{10}\right)$ HCl + 100 mL of $\left(\frac{M}{10}\right)$ NaOH

(b) 55 mL of $\left(\frac{M}{10}\right)$ HCl + 45 mL of $\left(\frac{M}{10}\right)$ NaOH

(c) 10 mL of $\left(\frac{M}{10}\right)$ HCl + 90 mL of $\left(\frac{M}{10}\right)$ NaOH

(d) 75 mL of $\left(\frac{M}{5}\right)$ HCl + 25 mL of $\left(\frac{M}{5}\right)$ NaOH

45. 0.04 g of pure NaOH is dissolved in 10 litre of distilled water. The *p*H of the solution is:

(a) 9 (b) 10

(c) 11 (d) 12

46. A solution of *p*H 9.0 is one thousand times as a solution of *p*H:

(a) 6 (b) 7

(c) 4 (d) 10

47. How many litres of water must be added to 1 litre of an aqueous solution of HCl with a pH of 1 to create an aqueous solution with pH of 2?

(a) 2.0 L (b) 9.0 L

(c) 0.1 L (d) 0.9 L

48. In which of the following cases pH is greater than 7?

(a) 50 ml of 0.1 M HCl + 50 ml of 0.1 M NaCl

(b) 50 ml of 0.1 M H_2SO_4 + 50 ml of 0.2 M NaOH

(c) 50ml of 0.1 M CH_3COOH + 50 ml of 0.1 M KOH

(d) 50 ml of 0.1 M HNO_3 + 50 ml of 0.1 M NH_3

49. pH of $Ba(OH)_2$ solution is 13. Millimoles of $Ba(OH)_2$ present in 10 ml of solution would be

(a) 1.00 (b) 0.50

(c) 10.00 (d) 15.00

50. Calculate the pH of a solution obtained by mixing 2 ml of HCl of pH = 2 and 3 ml of solution of KOH of pH = 12

(a) 10.30 (b) 3.70

(c) 11.30 (d) None of these

51. What is OH^- ion concentration is 0.01 M solution of aniline hydrochloride (Kb for aniline = 4×10^{-10})?

(a) 2×10^{11} M (b) 2×10^{-10} M

(c) 5×10^{-4} M (d) 5×10^{-5} M

52. The degree of dissociation of water at 25°C is $1.8 \times 10^{-7}\%$ and density is 1.0 g cm^{-3}. The ionic constant for water is

(a) 1.0×10^{-14} (b) 2.0×10^{-16}

(c) 1.0×10^{-16} (d) 1.0×10^{-8}

53. What is the pH of a neutral solution at 375°C, where K_w equals 2.5×10^{-14} ? (log 2 = 0.3)

(a) 7.0 (b) 13.6

(c) 6.8 (d) 6.6

54. The hydronium ion concentration in an aqueous solution of H_2SO_4 is 2.0×10^{-4} M at 25°C. The hydroxide ion concentration in the solution is

(a) 0 (b) 2.0×10^{-4} M

(c) 5×10^{3} M (d) 5×10^{-11} M

55. The number of hydronium ions in 1 ml an aqueous solution of pH 12.0 at 25°Cis

(a) 0.01 (b) 10^{-12}

(c) 6.02×10^{8} (d) 6.02×10^{11}

56. The pH of 4.0×10^{-4} M HNO_3 solution is (log 2 = 0.3)

(a) 4.6 (b) 3.4

(c) 3.6 (d) 4.0

57. How many grams of HCl should be dissolved in sufficient water to get 500 ml of an aqueous solution of pH = 2.0?

(a) 0.01 (b) 0.005

(c) 0.1825 (d) 0.365

58. What mass of NaOH should be dissolved in sufficient water to get 20 m^3 of an aqueous solution of pH = 7.3, at 25°?

(a) 0.16 g (b) 1.6×10^{-4} g

(c) 0.04 g (d) 0.12 g

59. What is the pH of solution made by mixing equal volumes of 0.1 N – H_2SO_4, 0.1 N – HNO_3, 0.1N – HCl?

(a) 1 (b) 2

(c) 3 (d) 4

60. The concentration of acetate ions in 1 M acetic acid ($K_a = 2 \times 10^{-5}$) solution containing 0.1 M-HCl is

(a) $2 \times 10\ ^{1}$ M (b) $2 \times 10\ ^{3}$ M

(c) 2×10^{-4} M (d) 4.4×10^{-3} M

61. The pKa of acetylsalicylic acid (aspirin) is 3.5. The pH of gastric juice in human stomach is about 2 to 3 and the pH in the small intestine is about 8. Aspirin will be

(a) unionized in the small intestine as well as in the stomach

(b) completely intestine the small intestine as well as in the stomach

(c) ionized in the stomach and almost unionized in the small intestine

(d) ionized in the small intestine and almost uniomized in the stomach

62. How much water must added to 300 ml of 0.2 M solution of CH_3COOH, so the degree of dissociation is doubled K_a for the acetic acid = 1.8×10^{-5}.

(a) 1200 ml (b) 300 ml
(c) 600 ml (d) 900 ml

63. The correct order of increasing in the following aqueous solution is

(a) 0.01 M - $NaHCO_3 < 0.01\ M - NaCN < 0.01M - KCl$

(b) 0.01 M- $KCl < 0.01\ M - NaCN < 0.01M - NaHCO_3$

(c) 0.01 M- $KCl < 0.01M - NaHCO_3 < 0.01M - NaCN$

(d) 0.01 M- $NaCN < 0.01M - KCl < 0.01M - NaHCO_3$

64. The acid ionization constant of Zn^{2+} is 2.0×10^{-10}. What is the pH of 0.001 M solution of $ZnCl_2$? (log 2 = 0.3)

(a) 9.7 (b) 4.85
(c) 6.35 (d) 3.35

65. The acid ionization constant of Zn^{2+} is 2.0×10^{-10}. What is the basic dissociation constant of $Zn(OH)^+$?

(a) 5×10^{-5} (b) 2.0×10^4
(c) 2×10^{-10} (d) 5×10^9

66. The dissociation constant of a weak monoprotic acid is numerically equal to the dissociation constant of its conjugate base. What is pH of 0.01 M solution of this acid?

(a) 7.0 (b) 6.0
(c) 8.0 (d) 4.0

67. The pH of 0.005M Codeine ($C_{18}H_{21}NO_3$) solution is 9.95. Its K_b

(a) 8×10^{-9} (b) 3×10^{-7}
(c) 1.6×10^{-6} (d) None of these

68. At 25°C the dissociation constants of acid HA and base BOH in aqueous solution is same. The pH of 0.01 M solution of HA is 5.0. The pH of 0.1 M solution of BOH is

(a) 5.0 (b) 9.0
(c) 9.5 (d) 8.5

69. An amount of 0.16 g of N_2H_4 is dissolved in water and the total volume made up to 500 ml. What is the percentage of N_2H_4 that has reacted with water in this solution? K_b for $N_2H_4 = 4.0 \times 10^{-6}$.

(a) 0.02% (b) 0.014%
(c) 2% (d) 2.82%

Salt Hydrolysis

70. What is the concentration of aqueous ammonia in the solution prepared by dissolving 1.8 moles of ammonium acetate in 10 L solution (K_a of $CH_3COOH = K_b$ of $NH_4OH = 1.8 \times 10^{-5}$)

(a) 1.0×10^{-3} M (b) 0.01 M
(c) 3.24×10^{-3} M (d) 3.09×10^{-4} M

71. The pK_a of a weak acid HA, is 4.80. The pK_b of a weak base, BOH, is 4.78. The pH of an aqueous solution of corresponding salt, BA, will be

(a) 9.22 (b) 9.58
(c) 4.79 (d) 7.01

72. Consider the following salts. Which one(s) when dissolved in water will produce an acidic solution?

1. NH_4Cl 2. NaCN
3. KNO_3

(a) 1 and 3 (b) only 1
(c) only 3 (d) 2 and 3

73. From separate solutions of four sodium salts NaW, NaX, NaY and NaZ had pH had pH 7.0, 9.0, 10.0 and 11.0 respectively, when each solution was 0.1 M, the weakest acid is:

(a) HW (b) HX
(c) HY (d) HZ

74. If salt of strong acid and weak base hydrolyses appreciably (α = 0.1), which of the following formula is to be used to calculate degree of hydrolysis 'α'?

(a) $\alpha = \sqrt{\frac{k_w}{k_b}.c}$ (b) $\alpha = \sqrt{\frac{k_w}{k_b.c}}$

(c) $\alpha = \sqrt{\frac{K_w}{K_a.c}}$ (d) None of these

75. The correct formula to calculate the hydroxyl ion concentration of an aqueous solution of $NH_4^+NO_3^-$ is:

(a) $\sqrt{\frac{C \times K_w}{K_b}}$ (b) $\sqrt{\frac{K_w . k_b}{C}}$

(c) $\sqrt{\frac{C \times K_w}{K_b}}$ (d) $\sqrt{\frac{K_a \times K_w}{C}}$

76. $[H^+] = \sqrt{\frac{K_w K_a}{C}}$ is suitable for

(a) NaCl, NH_4Cl

(b) CH_3COONa, NaCN

(c) CH_3COONa, $(NH_4)_2SO_4$

(d) CH_3COONH_4, $(NH_4)_2CO_3$

77. The degree of hydrolysis of 0.1 $RNH_3^+Cl^-$ solution is 1.0%. If the concentration of $RNH_3^+Cl^-$ is made 0.4 M, what is the new degree of hydrolysis (in percentage)?

(a) 0.01 (b) 0.001

(c) 0.2 (d) 0.5

78. What will be the pH of an aqueous solution of 1.0 M ammonium formate? Given $pK_a = 3.8$ and $pK_b = 4$.

(a) 7.5 (b) 3.4

(c) 6.9 (d) 10.2

79. The percentage degree of hydrolysis of a salt of weak acid (HA) and weak base (BOH) in its 0.1 M solution is found to be 10%. If the molarity of the solution is 0.05 M, the percentage hydrolysis of the salt should be:

(a) 5% (b) 10%

(c) 20% (d) None of these

80. 1 cc of 0.1 N HCl is added to 99 cc solution of NaCl. The pH of the resulting solution will be

(a) 7 (b) 3

(c) 4 (d) 1

81. The degree of hydrolysis of a salt of weak acid and weak base in it's 0.1 M solution is found to be 50%. If the molarity of the solution is 0.2 M, the percentage hydrolysis of the salt should be

(a) 100% (b) 50%

(c) 25% (d) 10%

82. The pK_a of a weak acid, HA, is 4.80. The pK_b of a weak base, BOH, is 4.78. The pH of an aqueous solution of the corresponding salt, BA, will be:

(a) 8.58 (b) 4.79

(c) 7.01 (d) 9.22

83. What will be the pH at the equivalence point during the titration of a 100 mL 0.2 M solution of CH_3COONa with 0.2 M solution of HCl?

$K_a = 2 \times 10^{-1}$

(a) $3 - \log\sqrt{2}$ (b) $3 + \log\sqrt{2}$

(c) $3 - \log 2$ (d) $3 + \log 2$

84. For which salt the pH of its solution does not change with dilution?

(a) NH_4Cl (b) CH_3COONH_4

(c) CH_3COONa (d) None of these

85. Which is not an acid salt?

(a) NaH_2PO_2 (b) NaH_2PO_3

(c) NaH_2PO_4 (d) NaH_SO_3

86. Hydrolysis of oxide ion in water produces:

(a) H^+ (b) OH^-

(c) O_2 (d) H_2O

87. The reverse process of neutralization is:

(a) hydrolysis (b) decomposition

(c) dehydration (d) synthesis

88. The hydrolysis of the salt of strong acid and weak base is

(a) anionic hydrolysis

(b) cationic hydrolysis

(c) amphoteric hydrolysis

(d) None of these

89. pH for the solution of salt undergoing anionic hydrolysis (say CH_3COONa) is given by:

(a) $pH = \frac{1}{2}[pK_w + pK_a + \log c]$

(b) $pH = \frac{1}{2}[pK_w + pK_a - \log c]$

(c) $pH = \frac{1}{2}[pK_w + pK_b - \log c]$

(d) None of the above

90. The use of methyl orange as an indicator in the volumetric determination of the equivalent mass of a weak acid would lead to:

(a) a low value for the equivalent mass

(b) a high value for the euqilvalent mass

(c) no error in the value

(d) improved accuracy

91. The pink colour of phenolphthalein in alkaline medium is due to:

(a) negative ion (b) positive ion

(c) OH^- ions (d) neutral form

92. The aqueous solution of a salt is alkaline. This shows that salt is made from:

(a) a strong acid and strong base

(b) a strong acid and weak base

(c) a weak acid and weak base

(d) a weak acid and strong base

93. The salt that does not hydrolyse, is:

(a) $SnCl_2$ (b) $FeCl_3$

(c) $SnCl_4$ (d) $CaCl_2$

94. What is the pH of a $1M CH_3COONa$ solution?

K_a of acetic acid $= 1.8 \times 10^{-5}$, $K_w = 10^{-14}$

(a) 2.4 (b) 3.6

(c) 4.8 (d) 9.4

95. A certain weak acid has a dissociation constant 1.0×10^{-4}. The equilibrium constant for its reaction with a strong base is:

(a) 1.0×10^{-4} (b) 1.0×10^{-10}

(c) 1×10^{10} (d) 1.0×10^{-14}

96. The initial rate of hydrolysis of methyl acetate (1 M) by a weak acid (HA. 1M) is $\frac{1}{100}^{th}$ of that of a strong acid (HA. 1M) at 25°C. The K_a of H.A is:

(a) 1×10^{-4} (b) 1×10^{-5}

(c) 1×10^{-6} (d) 1×10^{-3}

97. Ionization constants of a weak acid HA and weak base are 3.0×10^{-7} each at 298 K. The percent degree of hydrolysis of their salt

(a) 40 (b) 50

(c) 75 (d) 33 33

98. pH of 0.5 M $Ba(CN)_2$ solution (pK_b of CN^- =9.30)is:

(a) 8.35 (b) 3.35

(c) 9.35 (d) 9.50

99. The pH of solutions of both ammonium acetate and sodium chloride is 7 due to

(a) hydrolysis in both case

(b) the former hydrolyses and not the latter

(c) no hydrolysis in both

(d) hydrolysis of the latter but not the former

100. A salt of strong acid and a weak base is dissolved in water. Its hydrolysis is

(a) not affected by heating

(b) increased by adding the strong acid

(c) suppressed by adding strong acid

(d) suppressed by dilution

101. The pH value of 0.1 M solutions of CH_3COONa (I), CH_3COOH (II), CH_3COONH_4 (III), NaOH(IV) and HCl (V) is in the order

(a) I < II < III < IV < V

(b) V < IV < III < II < I

(c) V < II < III < I < IV

(d) V < II < I < III < IV

102. The pH at the equivalence point when solution of 0.01 M- CH_3COOH is titrated with a solution of 0.01 M-NaOH, is (pK_a of CH_3COOH = 4.7, log 5 = 0.7)

(a) 10.5 (b) 3.5

(c) 10.35 (d) 3.65

Buffer Solution & Common Ion effect

103. The base imidazole has K_b of 1.0×10^{-7}. What volumes of 0.02 M HCl and 0.02 M imidazole should be mixed to make 150ml of a buffer of pH7?

(a) 100ml and 50ml (b) 50ml and 100ml

(c) 75ml and 75ml (d) 60ml and 90ml

104. pH of $0.01M-(NH_4)_2SO_4$ and $0.02M-NH_4OH$ buffer (pK_a of NH_4^+ = 9.26) is

(a) 9.26 + log 2 (b) 9.26 - log 2

(c) 4.74 + log 2 (d) 9.26

105. What happens to pH of a solution when NH_4Cl crystal is added to a dilute solution of NH_4OH?

(a) Decreases

(b) Increases

(c) Remains unaffected

(d) All of these

106. If 1 M CH_3COONa is added to 1M CH_3COOH:

(a) pH of the solution increases

(b) pH decreases

(c) pH does not change

(d) None of the above

107. The addition of HCl does not suppresses the ionization

(a) acetic acid (b) benzoic acid

(c) H_2S (d) H_2SO_4

108. 50% neutralization of a solution of formic acid ($K_a = 2 \times 10^{-4}$) with NaOH would result in a solution having a hydrogen ion concentration of:

(a) 2×10^{-4} (b) 3.7

(c) 2.7 (d) 1.85

109. The pH of a weak mono acid base at 80% neutralization with a strong acid in a dilute solution is 7.40. The ionization constant of the base is:

(a) 1.0×10^{-6} (b) 1.6×10^{-7}

(c) 1.0×10^{-5} (d) None of these

110. The degree of dissociation of acetic acid in an aqueous solution of the acid is practically unaffected

(a) by adding a pinch of NaCl

(b) by adding a drop of HCl

(c) by adding water

(d) by raising the temperature

111. How many grams of NaOH should be added in 500 ml of 2 M acetic acid solution to get a buffer solution of maximum buffer capacity?

(a) 20.0 (b) 10.0

(c) 40.0 (d) 30.0

112. CH_3NH_2 (0.1 mole, $Kb = 5 \times 10^{-4}$) is added to 0.08 moles of HCl and the solution is diluted to 1 litre, resulting (H^+) of solution is

(a) 1.6×10^{-11} (b) 8×10^{-11}

(c) 5×10^{-5} (d) 8×10^{-2}

113. 0.1 M formic acid solution is titrated against 0.1 M NaOH solution. What would be the difference in pH between $\frac{1}{5}$ and $\frac{4}{5}$ stages of neutralization of acid?

(a) $2 \log \frac{3}{4}$ (b) $2 \log \frac{1}{5} 5$

(c) $\log \frac{1}{3}$ (d) $2 \log 4$

114. A buffer solution is made up of acetic acid [$pK_a = 5$] having concentration = 1.5 M and sodium acetate having conc. = 0.15 M. What is the number of $\overline{O}H$ ions present in 1 litre solution?

(a) $10^{-10} N_A$ (b) $10^{-4} N_A$

(c) $10^{-3} N_A$ (d) $10^{-6} N_A$

115. 1 MNaCl and 1 M HCl are present in an aqueous solution. The solution is

(a) Not a buffer solution and with pH < 7

(b) Not a buffer solution with pH > 7

(c) A buffer solution with pH < 7

(d) A buffer solution with pH > 7

116. A solution is 0.1 MCH_3COOH and 0.1 MCH_3COONa. Which of the following will change the pH significantly, of the solution?

(a) Addition of water

(b) Addition of small amount of CH_3COONa without change in volume

(c) Addition of small amount of CH_3COOH without change in volume

(d) None will change the pH significantly

117. What is the difference in pH for $\frac{1}{3}$ and $\frac{2}{3}$ stages of neutralization of 0.1M CH_3COOH with 0.1M NaOH.

(a) $-2 \log 3$ (b) $2 \log \left(\frac{1}{4}\right)$

(c) $2 \log \left(\frac{2}{3}\right)$ (d) $-2 \log 2$

118. To prepare a buffer of pH 8.26 amount of $(NH_4)_2SO_4$ to be added to 500 ml of 0.01M $NH_4^+OH^-$ solution $[pK_a(NH_4^+) = 9.26]$ is

(a) 0.05 mole (b) 0.025 mole

(c) 0.10 mole (d) 0.005 mole

119. The relation for calculating pH of a solution containing acid and its salt is

(a) $pH = pK_a + \log\frac{[\text{Conjugate base}]}{[\text{Acid}]}$

(b) $pH = pK_a - \log\frac{[\text{Conjugate base}]}{[\text{Acid}]}$

(c) $pH = pK_a + \log\frac{[\text{Conjugate base}]}{[\text{Salt}]}$

(d) $pOH = pK_a - \log\frac{[\text{Acid}]}{[\text{Conjugate base}]}$

120. The correct statement about buffer solution is

(a) It contains a weak acid and its conjugate base

(b) It contains a weak base and its conjugate acid

(c) It shows little change in pH on adding small amount of an acid or base

(d) All of the above

121. The Handerson's equation used to calculate the pOH of basic buffer is

(a) $pOH = pK_b + \log\frac{[\text{Conjugate base}]}{[\text{base}]}$

(b) $pOH = pK_b - \log\frac{[\text{Conjugate base}]}{[\text{base}]}$

(c) $pOH = pK_b + \log\frac{[\text{Conjugate base}]}{[\text{base}]}$

(d) $pOH = pK_b - \log\frac{[\text{base}]}{[\text{Conjugate base}]}$

122. A buffer solution of CH_3COOH and CH_3COONa is used

(a) In the preparation of potash alum

(b) In the removal of PO_4^{3-} ions

(c) In increasing the pH value of a solution

(d) In precipitation of $Cr(OH)_3$ from $CrCl_3$

123. Which may be added to 1 litre of water to act as a buffer?

(a) 1 mole of $HC_2H_3O_2$ and 1 mole of HCl

(b) 1 mole of NH_4OH and 1 mole of NaOH

(c) 1 mole of NH_4Cl and 1 mole of HCl

(d) 1mole of $HC_2H_3O_2$ and 0.5 mole of NaOH

124. To a mixture of acetic acid and sodium acetate a further amount of sodium acetate is added. The pH of the mixture

(a) Increases (b) Decreases

(c) Remain unchanged (d) Not predictable

125. For preparing a buffer solution of pH 6 by mixing sodium acetate and acetic acid, the ratio of concentration of salt and acid ($K_a = 10^{-5}$) should be

(a) 1 : 10 (b) 10 : 1

(c) 100 : 1 (d) 1 : 100

126. Solution prepared by dissolving equal number of mole of HOCl ($K_a = 3.2 \times 10^{-8}$) and NaOCl is a buffer of pH

(a) 8.0 (b) 3.2

(c) 7.5 (d) 4.8

127. In a mixture of CH_3COOH and CH_3COONa the ratio of salt to acid concentration is increased by 10 folds. The pH of the solution will increase by

(a) 0 (b) 1

(c) 2 (d) 3

128. pH of a solution containing 0.3 MHA and 0.1 MA^- (K_b for $A^- = 1.0 \times 10^{-5}$) is

(a) 5 + log 3 (b) 5 - log 3

(c) 9 + log 3 (d) 9 - log 3

129. A buffer solution contains 500 ml of 0.2M CH_3COONa and 500 ml of 0.1M CH_3COOH. 1 L of water is added to this buffer. pH before and after dilution : (pK_a of CH_3COOH = 4.74)

(a) 5.04, 3.74 (b) 5.04, 5.04

(c) 5.04, 4.89 (d) 9.56, 9.56

130. In which of the following , the buffer action ceases to act ?

(a) 100 ml of 0.1M CH_3COOH + 25ml of 0.1M NaOH

(b) 100 ml of 0.1M CH_3COOH + 50ml of 0.1M NaOH

(c) 100 ml of 0.1M CH_3COOH + 80ml of 0.1M NaOH

(d) 100 ml of 0.1M CH_3COOH + 50ml of 0.3M NaOH

131. On addition of 10 ml of 1.0M HCl , which of the following is no more a buffer solution?

(a) 100 ml containing 0.2M NH_3 and NH_4Cl each

(b) 100 ml containing 0.2M NH_3and0.1 M NH_4Cl

(c) 100 ml containing 0.05M $NH_3$and NH_4Cl each

(d) 100 ml containing 0.1 5M NH_3 and NH_4Cl each

132. Which of the following molar ratio of NH_3 and HCl in aqueous solution will constitute a buffer?

(a) 1 : 2 (b) 1 : 3

(c) 2 : 1 (d) 2 : 1

133. Calculate the *p*H of a solution containing 0.1mol of , 0.2 mol of CH_3COOH and CH_3COONa 0.05 mol of NaOH in 1 L (pK_a for CH_3COOH = 4.74)

(a) 4.74 (b) 5.04

(c) 5.44 (d) 5.20

Questions based on solubility product

134. K_{sp} of AgCl is 1×10^{-10}. Its solubility in 0.1M KNO_3 will be

(a) 10^{-5} moles /litre (b) > 10^{-5}moles /litre

(c) < 10^{-5}moles/litre (d) None of these

135. The solubility of a sparingly soluble salt AxBy in water is '*S*' moles per litre. The solubility product has the value

(a) S^2 (b) $x^y y^x . S^{x+y}$

(c) $x^x y^y . S^{x+y}$ (d) S^{x+y}

136. Which of the following is most soluble in water?

(a) $Ba_3(PO_4)_2 (K_{sp} = 6 \times 10^{-10})$

(b) $ZnS(K_{sp} = 7 \times 10^{-16})$

(c) $Fe(OH)_3 (K_{sp} = 6 \times 10^{-38})$

(d) $Ag_3(PO_4)(K_{sp} = 1.8 \times 10^{-18})$

137. What is the pH of a saturated solution of $Cu(OH)_2$ ($K_{sp} = 2.6 \times 10^{-19}$)

(a) 6.1 (b) 7.30

(c) 8.42 (d) 7.90

138. Which of the following would increase the solubility of $Pb(OH)_2$?

(a) Add hydrochloric acid

(b) Add a solution of $Pb(NO_3)_2$

(c) Add a solution of NaOH

(d) None of the above the solubility of a compound is constant at constant temperature

139. What is the concentration of Pb^{2+} when $PbSO_4$ ($K_{sp} = 1.8 \times 10^{-8}$) begins to precipitate from a solution that is 0.0045 M in SO_4^{2-}?

(a) 4×10^{-8} M (b) 1.0×10^{-6} M

(c) 2.0×10^{-8} M (d) 4.0×10^{-6} M

140. If K_{sp} for $HgSO_4$ is 6.4×10^{-5}, then solubility of this substance in mole per m^3 is

(a) 8×10^{-3} (b) 6.4×10^{-5}

(c) 8×10^{-6} (d) None of these

141. If the solubility of lithium sodium hexafluoride aluminate, $Li_3Na_3(AIF_6)_2$ is 'S' mole lts, its solubility product is equal to

(a) 729 S8 (b) 12 S8

(c) 18 S8 (d) 2916 S8

142. The solubility product $Mg(OH)_2$ in water at 25°C is $2.56 \times 10^{-13} (mol\,dm^{-3})^3$ while that of $Al(OH)_3$ is $4.32 \times 10^{-34} (mol\,dm^{-3})^4$ and S_2 are the solubilities of $Mg(OH)_2$ and $Al(OH)_3$ in water in mol dm^{-3} at 25°C, what is the order of ratio $\frac{S_1}{S_2}$:

(a) 10^5 (b) 10^4

(c) 10^6 (d) 10^3

143. The solubility of CaF_2 ($K_{sp} = 3.4 \times 10^{-11}$) in 0.1 m solution of NaF would be

(a) 34×10^{-12} M (b) 3.4×10^{-10} M

(c) 3.4×10^{-9} M (d) 3.4×10^{-13} M

144. The solubility of Ag_2CO_3 in water at 25°C is 1×10^{-4} mole/litre. What is its solubility in 0.01 M Na_2CO_3 solution? Assume no hydrolysis of CO_3^2 - ion.

(a) 6×10^{-6} mole/litre

(b) 4×10^{-5} mole/litre

(c) 10^{-5} mole/litre

(d) 2×10^{-5} mole/litre

145. The solubility product of AgCl is 1.8×10^{-10}. Precipitation of AgCl will occur only when equal volumes of solution of

(a) 10^{-4} M Ag^+ and 10^{-4} M Cl^- are mixed

(b) 10^{-7} M Ag^+ and 10^{-7} M Cl^- are mixed

(c) 10^{-5} M Ag^+ and 10^{-5} M Cl^- are mixed

(d) 2×10^{-5} M Ag^+ and 2×10^{-5} M Cl^- are mixed

146. What is the solubility of $Al(OH)_3$, $K_{sp} = 1 \times 10^{-33}$, in a solution having $pH = 4$.

(a) 10^{-3} M (b) 10–6 M
(c) 10^{-4} M (d) 10^{-10} M

147. The solubility of $Fe(OH)_3$ would be maximum in

(a) 0.1 M NaOH (b) 0.1 M HCl
(c) 0.1 M KOH (d) 0.1 M H_2SO_4

148. Which of the following statements is correct for a solution saturated with AgCl and AgBr if their solubilities in moles per litre in separate solutions are x and y respectively?

(a) $[Ag^+] = x + y$ (b) $[Ag^+] = [Br^-] + [Cl^-]$

(c) $[Br^-] = y$ (d) $[Cl^-] = x$

149. At 25°C, the solubility product values of AgCl and AgCNS are 1.8×10^{-10} and 1.6×10^{-11} respectively. When water is saturated with both solids, calculate the ratio $\frac{[Cl^-]}{[CNS^-]}$ and also $[Ag^+]$ in the solution.

(a) 1.125, 4×10^{-6} M (b) 11.25, 1.4×10^{-5} M

(c) 1.25, 4×10^{-5} M (d) 1.25, 4×10^{-6} M

150. Arrange in increasing order of solubility of AgBr in following-

(i) 0.1 M NH_3 (ii) 0.1 M $AgNO_3$
(iii) 0.2 M NaBr (iv) pure water

(a) (iii) < (ii) < (iv) < (i)

(b) (iii) < (ii) < (i) < (iv)

(c) (iii) < (ii) = (i) < (iv)

(d) (ii) < (iii) < (iv) < (i)

151. What is the minimum pH when $Fe(OH)_3$ starts precipitating form a solution containing 0.1 M $FeCl_3$? K_{sp} of $Fe(OH)_3 = 8 \times 10^{-13}$ M^3.

(a) 3.7 (b) 5.7
(c) 10.3 (d) 8.3

152. The solubility product of AgCl is 10^{-10}. The minimum volume (in L) of water required to dissolve 1.722 mg of AgCl is (molecular weight of AgCl = 143.5)

(a) 10 (b) 2
(c) 1 (d) 20

153. When HCl gas is passed through a saturated solution of common salt, pure NaCl is precipitated because

(a) The impurities dissolve in HCl

(b) HCl is highly soluble in H_2O

(c) The product of $[Na^+]$ and $[Cl^-]$ exceeds the solubility product of NaCl

(d) The solubility product of NaCl is lowered by the chloride ion from aqueous HCl

154. The correct representation for the solubility product constant of Ag_2CrO_4 is

(a) $[Ag^+]^2[CrO_4^{2-}]$ (b) $[Ag^+][CrO_4^{2-}]$

(c) $[2Ag^+][CrO_4^{2-}]$ (d) $[2Ag^+]^2[CrO_4^{2-}]$

155. The solubility product of Hg_2I_2 is equal to

(a) $[Hg_2^+][I^-]$ (b) $[Hg^{2+}][I^-]$

(c) $[Hg_2^{2+}][I^-]^2$ (d) $[Hg^{2+}]^2[I^-]^2$

156. For a sparingly soluble salt $A_p B_p$, the relationship of its solubility product (L_s) with its solubility (s) is

(a) $L_s = s^{p+q}.p^p.q^p$ (b) $L_s = s^{p+q}.p^q.q^p$

(c) $L_s = s^{pq}.p^p.q^q$ (d) $L_s = s^{pq}.(pq)^{p+q}$

157. Let the solubilities of AgCl in H_2O, 0.01 M $CaCl_2$, 0.01 M NaCl and 0.05 M $AgNO_3$ be S_1, S_2, S_3 and S_4 respectively. What is the correct relationship between these quantities?

(a) $S_1 > S_2 > S_3 > S_4$ (b) $S_1 > S_2 = S_3 > S_4$
(c) $S_1 > S_3 > S_2 > S_4$ (d) $S_4 > S_2 > S_3 > S_4$

158. K_{sp} of $Al(OH)_3$ is 4.32×10^{-14}. pH of its saturated solution is about -

(a) 5.2 (b) 4.2
(c) 10.77 (d) 11.77

159. The solubility of in water is $Pb(OH)_2$. The solubility of $Pb(OH)_2$ in a buffer solution of $pH = 8$ is

(a) 8.64M (b) 2.16×10^{-16} M

(c) 8.64×10^{-16} M (d) 8.64×10^{-4} M

Answers

1. (d)	**2.** (d)	**3.** (a)	**4.** (d)	**5.** (d)	**6.** (c)	**7.** (a)	**8.** (b)	**9.** (d)	**10.** (d)
11. (b)	**12.** (d)	**13.** (d)	**14.** (d)	**15.** (d)	**16.** (c)	**17.** (b)	**18.** (b)	**19.** (c)	**20.** (a)
21. (b)	**22.** (a)	**23.** (b)	**24.** (b)	**25.** (a)	**26.** (c)	**27.** (c)	**28.** (c)	**29.** (c)	**30.** (d)
31. (c)	**32.** (b)	**33.** (d)	**34.** (c)	**35.** (b)	**36.** (c)	**37.** (d)	**38.** (a)	**39.** (b)	**40.** (b)
41. (a)	**42.** (a)	**43.** (a)	**44.** (d)	**45.** (b)	**46.** (a)	**47.** (b)	**48.** (c)	**49.** (b)	**50.** (a)
51. (a)	**52.** (a)	**53.** (c)	**54.** (d)	**55.** (c)	**56.** (b)	**57.** (c)	**58.** (a)	**59.** (a)	**60.** (c)
61. (d)	**62.** (d)	**63.** (c)	**64.** (c)	**65.** (a)	**66.** (d)	**67.** (c)	**68.** (c)	**69.** (c)	**70.** (a)
71. (d)	**72.** (b)	**73.** (d)	**74.** (d)	**75.** (b)	**76.** (b)	**77.** (d)	**78.** (c)	**79.** (b)	**80.** (b)
81. (b)	**82.** (c)	**83.** (a)	**84.** (b)	**85.** (a)	**86.** (b)	**87.** (a)	**88.** (b)	**89.** (a)	**90.** (a)
91. (a)	**92.** (d)	**93.** (d)	**94.** (d)	**95.** (c)	**96.** (a)	**97.** (d)	**98.** (c)	**99.** (b)	**100.** (c)
101. (c)	**102.** (a)	**103.** (b)	**104.** (d)	**105.** (a)	**106.** (a)	**107.** (d)	**108.** (a)	**109.** (a)	**110.** (a)
111. (a)	**112.** (c)	**113.** (d)	**114.** (a)	**115.** (a)	**116.** (d)	**117.** (d)	**118.** (b)	**119.** (a)	**120.** (d)
121. (a)	**122.** (b)	**123.** (d)	**124.** (a)	**125.** (b)	**126.** (c)	**127.** (b)	**128.** (d)	**129.** (b)	**130.** (d)
131. (c)	**132.** (d)	**133.** (c)	**134.** (a)	**135.** (c)	**136.** (d)	**137.** (d)	**138.** (a)	**139.** (d)	**140.** (d)
141. (d)	**142.** (b)	**143.** (c)	**144.** (c)	**145.** (a)	**146.** (a)	**147.** (d)	**148.** (b)	**149.** (b)	**150.** (a)
151. (c)	**152.** (c)	**153.** (c)	**154.** (a)	**155.** (c)	**156.** (a)	**157.** (c)	**158.** (c)	**159.** (d)	

Explanations

2. (d) Acid $- H^+ \longrightarrow$ Conjugate base

3. (a) Ostwald's dilution law is valid for weak electrolytes only.

4. (d) Base $+ H^+ \longrightarrow$ Conjugate acid

$\therefore\ S_2O_8^{2-} + H^+ \longrightarrow HS_2O_8^-$

5. (d) Lewis acid is an electron acceptor.

Lewis base is an electron donor specie.

6. (c) H+ acceptors are bronsted bases.

7. (a) $H_2PO_4^- \longrightarrow HPO_4^{2-} + H^+$

8. (b) H_2O as bronsted acid

$H_2O \rightleftharpoons H^+ + \bar{O}H$

Water as bronsted base

$H_2O + H^+ \rightleftharpoons H_3O^+$

9. (d) Apply,

$$\frac{\propto_1}{\propto_2} = \sqrt{\frac{ka_1}{ka_2}}$$

10. (d) At infinite dilution, the percentage ionization for both strong and weak electrolytes is 100%.

12. (d) $\propto = \sqrt{K.V.}$

13. (d) Option d, should have been

16. (c) $pKa + pK_b = 4$

$pKa + 10 = 14$

$pKa + 4$

$Ka = 10^{-4}$

18. (b) $pH = 9$

$poH = 5$

$(\bar{O}H) = 10^{-5} M$

Since, this solution is 1000 times more basic than original solution.

Now, $(H^+)(\bar{O}H) = 10^{-14}$

$(H^+)10^{-8} = 10^{-14}$

$(H^+) = 10^{-6} M$

$pH = 6$

19. (c) Gram eq. (moles of H+) initially present

$= N \times V_{(L)} = 10^{-3} \times \frac{10}{1000} = 10^{-5}$

Now, they are present in vol. = 1ltr

$(H^+)\text{now} = \frac{10^{-5}}{1} = 10^{-5} M$

$(H^+)(\bar{O}H) = k_w$

$10^{-5} \times (\bar{O}H) = 10^{-14}$

$(\bar{O}H) = 10^{-9} M$

$pOH = -\log(\bar{O}H) = -\log 10^{-9} = 9$

20. (a) $H_2A \longrightarrow 2H^+ + A^{2-}$

At t = 0	10^{-4}	0	0
At t = t	0	2×10^{-4}	10^{-4}

$(H^+) = 2 \times 10^{-4} M$

$pH = -\log(H^+) = -\log(2 \times 10^{-4})$

$pH = -(\log 2 + \log 10^{-4})$

$pH = -(0.3010 + (-4)\log 10)$

$pH = -(0.3010 - 4) = 4 - 0.3010 = 3.7$

When $H_2A = 10^{-4} N$

$(H^+) = 10^{-4}$

$pH = 4$

21. (b) NaOH soln. initially = 10^{-5} M

So, moles of $\bar{O}H = 10^{-5}$ in 1 litre

Now, $(\bar{O}H) = \frac{10^{-5}}{1000} = 10^{-8} M$

Resultant $(\bar{O}H) = 10^{-8} + 10^{-7}$ (from H_2O)

$= 10^{-7}(10^{-1} + 1) = 1.1 \times 10^{-7} M$

$pOH = -\log(\bar{O}H) = -\log(1.1 \times 10^{-7})$

$= -(\log 1.1 + \log 10^{-7})$

$= -(0.04 - 7) = 7 - 0.04 = 6.96$

$pH = 14 - 6.96 = 7.04$

22. (a) Wt. of NaOH = 4g

Moles of NaOH $= \frac{4}{40} = 0.1$ mole

Moles of $\bar{O}H$ furnished = 0.1(gm eq. of BASE)

Wt. of H_2SO_4 = 4.9gm

Moles of $H_2SO_4 = \frac{4.9}{98} = \frac{1}{20} = 0.05$

Moles of H+ furnished $= 2 \times 0.05 = 0.1$

(gm eq. of acid)

Now, gm eq. of acid = gm eq. of base

$\therefore pH = 7$

23. (b) At $90°C$, $(H^+) = 10^{-6}M$

So, neutral $pH = 6$

HCl	+	KOH
0.2N 100 ml		0.1N 200 ml

gm eq. of acid reacted = gm eq. of base reacted

Resultant solution is neutral & $pH = 6$

25. (a) $HA \rightleftharpoons H^+ + A^-$

(aq.) (aq.) (aq.)

	HA	H^+	A^-
At t = 0	C	0	0
At t = eqm	$C - C\alpha$	$C\alpha$	$C\alpha$

$(H^+) = C\alpha = C\sqrt{\frac{ka}{C}} = \sqrt{ka.C}$

$(H^+) = \sqrt{4 \times 10^{-8} \times 0.25} = 2 \times 10^{-4} \times 0.5$

26. (c) Calculate (H+) concentration

Then apply $(H^+)(\bar{O}H) = 10^{-14}$ (kw)

27. (c) Conc. of acid $= \frac{1}{10}M = 0.1M$

$\alpha = 0.01$

$HA \rightleftharpoons H^+ + A^-$

(aq.) *(aq.)* *(aq.)*

Now, $\alpha = \sqrt{\frac{ka}{C}}$

i.e. $C\alpha^2 = ka$

$0.1 \times (0.01)^2 = ka$

$10^{-5} = ka$

$pka = -\log ka = -\log 10^{-5} = 5$

28. (c) Solution of NH_3 refers to $NH_4^+\bar{O}H$

$NH_4^+\bar{O}H \rightleftharpoons NH_4^+ + \bar{O}H$

At t = 0	C	0	0
At t = t	$C - C\alpha$	$C\alpha$	$C\alpha$

Also, $KOH \longrightarrow K^+ + \bar{O}H$

At t = 0	0.01	0	0
After diss. →	0	0.01	0.01

(for NH_4OH) $K_b = \frac{(NH_4^+)(\bar{O}H)}{(NH_4OH)}$

$1.8 \times 10^{-5} = \frac{(NH_4^+)(10^{-2})}{(C - C\alpha)}$

$1.8 \times 10^{-5} = \frac{(NH_4^+)(10^{-2})}{C(1-\alpha)}$

Since its a weak base $\alpha << 1, 1-\alpha = 1$

$\frac{1.8 \times 10^{-5} \times C}{10^{-2}} = (NH_4^+)$

$\frac{1.8 \times 10^{-5} \times 0.02}{10^{-2}} = (NH_4^+) = 3.6 \times 10^{-5}M$

30. (d) If $k_w = 9.6 \times 10^{-14} M^2$ (at $60°C$)

$(H^+)(\bar{O}H) = 9.6 \times 10^{-14}$

$(H^+)^2 = 9.6 \times 10^{-14}$

$(H^+) = \sqrt{9.6 \times 10^{-14}} = 3.1 \times 10^{-7}$

31. (c) Dissociation of water is an endothermic process. With rise in temperature, H_2O will dissociate more. So, concentration of both (H^+) & $(\bar{O}H)$ will increase & hence pH & pOH, both decrease with rise in temp.

32. (b) $M_1V_1 + M_2V_2 = M_3V$

$$\frac{1}{200}\times 10 + \frac{1}{200}\times 40 = M_3 \times 50$$

$$\frac{1}{20}+\frac{1}{5} = M_3 \times 50$$

$$\frac{1+4}{20} = M_3 \times 50$$

$$\frac{5}{20\times 50} = M_3$$

$$\frac{1}{200} = M_3 \text{ (Final conc. of } H_2SO_4)$$

So, final conc. of $(H^+) = 2\times\frac{1}{200} = \frac{1}{100} = 10^{-2}M$

$(H^+) = 10^{-2}M$

$pH = 2$

33. (d) when $pH = 2$

$(H^+) = 10^{-2}M$

when $pH = 6$

$(H^+) = 10^{-6}M$

$x \times 10^{-6} = 10^{-2}$

$x = 10^{-2}\times 10^{6} = 10^{4}$

34. (c) $ka = ka_1 \times ka_2$

37. (d) $k_w = (H^+)(\bar{O}H) = (Mol\,ltr^{-1})(Mol\,ltr^{-1})$

$= Mol^2\,ltr^{-2}$

39. (b)

$pH = 2$	$pH = 12$
$(H^+) = 10^{-2}M$	$pOH = 2$
Vol. = 200 ml	$(\bar{O}H) = 10^{-2}$ M
$\because M = \frac{n}{V_{(L)}}$	V = 300 ml
$0.01 = \frac{n}{0.2}$	$\because M = \frac{n}{V_{(L)}}$
no. of moles of $H^+ = 0.002$	$0.01 = \frac{n}{0.3}$
	$n = 0.003$
	no. of moles of $\bar{O}H = 0.003$

So, resultant solution is basic

Moles of $\bar{O}H$ left = 0.001

Vol. = 500 ml

$$(\bar{O}H)\text{ left} = \frac{10^{-3}}{0.5} = 0.2\times 10^{-2}M$$

$pOH = -\log(2\times 10^{-3})$

$= 3 - \log 2 = 3 - 0.3010 = 2.7$

$pH = 14 - 2.7 = 11.3$

40. (b) $pH = 3$

$(H^+) = 10^{-3}M$

V = 10 ml

No. of moles of $H^+ = 10^{-3}\times\frac{10}{1000} = 10^{-5}$

No. of $H^+ = 10^{-5}\times 6\times 10^{23} = 6\times 10^{18}$

47. (b) Apply, $M_1V_1 = M_2V_2$

$10^{-1}\times 1 = 10^{-2}\times V_2$

$V_2 = 10L$

Vol. of water to be added = 9L

48. (c) CH_3COOH is a weak acid & KOH is a stronger base.

50. (a)

HCl	KOH
$pH = 2$	$pH = 12$
V = 2 ml	$pOH = 2$
$(H^+) = 10^{-2}$ml	$(\bar{O}H) = 10^{-2}$ M
	V = 3 ml

MilliMoles of $H^+ = 2\times 10^{-2}$

MilliMoles of $\bar{O}H = 3\times 10^{-2}$

MilliMoles of $\bar{O}H$ left $= 3\times 10^{-2} - 2\times 10^{-2} = 10^{-2}$

$$\text{Resultant }(\bar{O}H) = \frac{10^{-2}\times 10^{-3}}{5\text{(ml)}}\times 1000 = \frac{10^{-2}}{5}$$

$(\bar{O}H) = 2\times 10^{-2} = 2\times 10^{-3}$

$pOH = -\log(\bar{O}H) = 3 - \log 2 = 2.7$

$pH = 14 - 2.7 = 11.3$

54. (d) $(H_3O^+)(\overline{O}H) = 10^{-14}$

$(2\times10^{-4})(\overline{O}H) = 10^{-14}$

$(\overline{O}H) = 0.5\times10^{-10} = 5\times10^{-11}M$

58. (a) $20\,m^3 = 20\times10^6\ ml = 20000\,L$

$pH = 7.3$

$poH = 14 - 7.3 = 6.7$

$poH = -\log\dfrac{1}{(\overline{O}H)}$

$6.7 = \log\dfrac{1}{\dfrac{x}{20000}}$

(Antilog $6.7 = 5\times10^6$)

$5\times10^6 = \dfrac{20000}{x}$

$x = \dfrac{20000}{5\times10^6} = 4\times10^{-3}$ Moles

$4\times10^{-3} = \dfrac{10}{40}$ (for NaOH)

$W = 0.16\,g$

59. (a) Apply,

$N_1V_1 + N_2V_2 + N_3V_3 = N_4V_4$

$0.1\times V + 0.1\times V + 0.1\times V = N_4\times 3V$

$\dfrac{0.3V}{3V} = N_4$

$N_4 = \dfrac{0.3}{3} = 0.1$

So, final $(H^+) = \dfrac{0.3}{3} = 0.1$

$(H^+) = 10^{-1}$, So, $pH = 1$

60. (c) $\underset{(aq.)}{CH_3COOH} \rightleftharpoons \underset{(aq.)}{CH_3COO^-} + \underset{(aq.)}{H^+}$

$$\begin{array}{lll} HCl & \longrightarrow H^+ & + Cl^- \\ 0.1 & 0 & 0 \\ 0 & 0.1 & 0.1 \end{array}$$

(H^+) from HCl > (H^+) from CH_3COOH

$ka = \dfrac{(CH_3COO^-)(H^+)}{(CH_3COOH)}$

$2\times10^{-5} = \dfrac{(CH_3COO^-)(0.1)}{(CH_3COOH)}$

$2\times10^{-5} = \dfrac{(CH_3COO^-)\times1}{C - C\propto}$

$2\times10^{-5} = \dfrac{(CH_3COO^-)\times0.1}{C-(1-\propto)}$

$1-\propto\ \approx 1$

$2\times10^{-5} = \dfrac{(CH_3COO^-)\times0.1}{1}$

$2\times10^{-4}M = (CH_3COO^-)$

62. (d) $\propto = \sqrt{\dfrac{ka}{C}}$ ______(1)

$2\propto = \sqrt{\dfrac{ka}{C'}}$ ______(2)

divide $\dfrac{1}{2} = \sqrt{\dfrac{C'}{C}}$

$\dfrac{1}{4} = \dfrac{C'}{C}$

$\dfrac{C}{4} = C'$

$C = 0.2$

$C' = \dfrac{0.2}{4} = \dfrac{1}{20} = 0.05M$

$M_1V_1 = M_2V_2$

$0.2\times300 = 0.05\times V_2$

$\dfrac{60}{0.05} - V_2 - 1200\,ml$

So, volume of water to be added = 1200 – 300 = 900 ml

64. (c) $Zn^{2+} + H_2O \longrightarrow [Zn(OH)]^+ + H^+$

(from $ZnCl_2$)

$(H^+) = C\propto = C\sqrt{\dfrac{ka}{C}} = \sqrt{ka.c}$

$$(H^+) = \sqrt{2\times10^{-10}\times10^{-3}}$$

$$(H^+) = \sqrt{2\times10^{-13}}$$

$$pH = -\log(H^+) = -\log(2\times10^{-13})^{\frac{1}{2}}$$

$$pH = -\left[\frac{1}{2}(\log 2 + \log 10^{-13})\right]$$

$$= -\left[\frac{1}{2}(0.30.1 - 13)\right]$$

$$= -\left[\frac{1}{2}\times(-12.7)\right]$$

$$= \frac{12.7}{2} = 6.3$$

65. (a) $ka \times k_b = k_w$

$$2\times10^{-10}\times k_b = 10^{-14}$$

$$k_b = \frac{10^{-14}}{2\times10^{-10}} = 0.5\times10^{-4} = 5\times10^{-5}$$

66. (d) $ka = k_b$ (given)

$ka \times k_b = kw$

$ka^2 = kw$

$$ka = \sqrt{kw} = \sqrt{10^{-14}} = 10^{-7}$$

$$\underset{(aq.)}{HA} \rightleftharpoons \underset{(aq.)}{H^+} + \underset{(aq.)}{A^-}$$

$$\underset{(aq.)}{HA} \rightleftharpoons \underset{(aq.)}{H^+} + \underset{(aq.)}{A^-}$$

C	0	0
C − C∝	C∝	C∝

$$(H^+) = C\propto = C\sqrt{\frac{ka}{C}} = \sqrt{ka.c} = \sqrt{10^{-7}\times10^{-1}}$$

$$(H^+) = \sqrt{10^{-8}} = 10^{-4}M$$

$$pH = -\log(H^+) = -\log 10^{-4} = 4$$

67. (c) Codeine + $H_2O \rightleftharpoons$ Codeinium ion + $\bar{O}H$

pH = 9.95, So $(H^+) = 1.12\times10^{-10}M$

$$\bar{O}H = \frac{k_w}{(H^+)} = \frac{10^{-14}}{1.2\times10^{-10}} = 8.93\times10^{-5}M$$

$$k_b = \frac{(B^+)(\bar{O}H)}{(BOH)}$$

$$k_b = \frac{(8.93\times10^{-5})^2}{0.005}$$

$$k_b = 1.6\times10^{-6}$$

68. (c) $\underset{(aq.)}{HA} \rightleftharpoons \underset{(aq.)}{H^+} + \underset{(aq.)}{A^-}$

pH = 5 (of 0.01 M)

$$(H^+) = 10^{-5}M$$

$$C\propto = 10^{-5}$$

$$10^{-2}\propto = 10^{-5}$$

$$\propto = 10^{-5}\times10^{2} = 10^{-3}$$

$$Ka = C\propto^2 = 0.01(10^{-3})^2$$

$$Ka = 10^{-8}$$

$$BOH \rightleftharpoons B^+ + \bar{O}H$$

$$K_b = 10^{-8}\ (\because Ka = K_b, \text{given})$$

$$\propto \text{ of base} = \sqrt{\frac{k_b}{C}} = \sqrt{\frac{10^{-8}}{0.1}} = \sqrt{10^{-7}}$$

$$(\bar{O}H) = C\propto = 0.1\times\sqrt{10^{-7}} = 0.1\times10^{-3}\sqrt{\frac{1}{10}}$$

$$\bar{O}H) = \frac{10^{-4}}{3.16} = 0.316\times10^{-4}M$$

$$poH = -\log(\bar{O}H)$$

$$= -\log(0.316\times10^{-4})$$

$$= -(\log 0.316 + \log 10^{-4})$$

$$= 4 - \log 0.316$$

$$= 4 - (-0.5)$$

$$= 4.5$$

$$pH = 9.5$$

70. (a)

$$CH_3COONH_4 + H_2O \rightleftharpoons CH_3COOH + NH_4OH$$

At t = 0	C	0	0
At t = t	C − Ch	Ch	Ch

Moles of $CH_3COONH_4 = 1.8$

Vol. = 10L

$$C = \frac{1.8}{10} = 0.18$$

Now, $h = \sqrt{K_h}$ $\qquad (K_h \times ka \times kb = kw)$

$$h = \sqrt{\frac{kw}{ka\ k_b}}$$

$$h = \sqrt{\frac{10^{-14}}{1.8 \times 10^{-5} \times 1.8 \times 10^{-5}}}$$

$$h = \frac{10^{-7}}{1.8 \times 10^{-5}} = 0.55 \times 10^{-2}$$

Conc. of aq. $\frac{NH_3}{NH_4} OH = Ch$

$= 0.18 \times 0.55 \times 10^{-2}$

$= 0.1 \times 10^{-2}$

$= 10^{-3} M$

72. (b) NH_4Cl when dissolved in water will give NH_4OH & HCl (weak base & strong acid)

73. (d) Salt having weakest acid will have max. pH (because base is strong & same in all the salts)

75. (b) NH_4NO_3 is a salt of weak base & strong acid,

so $(H^+) = \sqrt{\frac{kw.c}{Kb}}$

Now, $(H^+)(\bar{O}H) = kw$

$$\sqrt{\frac{kw.c}{K_b}}(\bar{O}H) = kw$$

$$(\bar{O}H) = kw.\sqrt{\frac{K_b}{kw.c}} = \sqrt{\frac{kw.K_b}{C}}$$

76. (b) Valid for salts of weak acid & strong base.

77. (d) $RNH_3^+Cl^-$ is a salt of weak base (RNH_2) & (HCl) strong acid

$$h = \sqrt{\frac{K_h}{C}}$$

$$K_h = Ch^2$$

Apply $C_1h_1^2 = C_2h_2^2$

$$0.1 \times \left(\frac{1}{100}\right)^2 = 0.4 \times h_2^2$$

$$\frac{10^{-5}}{0.4} = h_2^2$$

$$\frac{10^{-4}}{4} = h_2^2$$

$$\frac{10^{-2}}{2} = h_2 = 0.5 \times 10^{-2} = 0.5\%$$

78. (c) Ammonium formate is a salt of weak acid and weak base.

Apply, $pH = \frac{1}{2}(pkw + pka - pk_b)$

$$= \frac{1}{2}(14 + 3.8 - 4) = \frac{1}{2 \times 13.8} = 6.9$$

79. (b) For a salt of weak acid and weak base, $h = \sqrt{K_h}$ degree of hydrolysis is independent initial concentration of salt.

80. (b) 1 cc of 0.1 N HCl

So, gm eq. of acid $= \frac{0.1}{10} \times \frac{1}{1000} = 10^{-4}$ gm eq.

10^{-4} gm eq. of acid $= 10^{-4}$ moles of H^+

After addition of 99 ml NaCl,

Total vol. = 100 ml

$$(H^+) = \frac{10^{-4}}{100} \times 1000 = 10^{-3} M$$

$$pH = -\log(H^+) = -\log(10^{-3}) = 3$$

81. (b) $h = \sqrt{k_h}$ (independent of concentration)

82. (c) Apply $pH = \frac{1}{2}(pkw + pka - pk_b)$

83. (a)

$$\underset{0.2M \quad 100ml}{CH_3COONa} + \underset{M=0.2 \quad V=?}{HCl} \longrightarrow CH_3COOH + NaCl$$

$$M = \frac{n}{V_{(L)}} \qquad 0.2 = \frac{0.02}{V_{(L)}}$$

$$0.2 = \frac{n}{100} \times 1000 \qquad V_{(L)} = \frac{0.02}{0.2}$$

$$n = 0.02 \qquad V = 0.1L$$

$$V = 100\,ml$$

Moles of $CH_3COONa = 0.02$

At equivalence point, moles of $HCl = 0.02$

$\therefore$ Moles of $CH_3COOH = 0.02$

Vol. = 200 ml

$$C \text{ of } CH_3COOH = \frac{0.02}{0.2} = 0.1M$$

$$(H^+) = C\alpha = C\sqrt{\frac{ka}{c}} = \sqrt{ka.c}$$

$$(H^+) = \sqrt{2 \times 10^{-5} \times 10^{-1}} = \sqrt{2} \times 10^{-3}$$

$$pH = 3 - \log\sqrt{2}$$

84. (b) CH_3COONH_4 is a salt of weak & weak base.

85. (a) NaH_2PO_2 is not an acid salt

$NaOH + H_3PO_2$

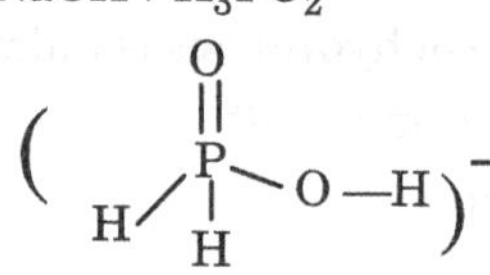

$\longrightarrow$ Monobasic acid

86. (b) $O^{2-} + H_2O \rightleftharpoons 2\bar{O}H$

95. (c) $HA + NaOH \rightleftharpoons NaA + H_2O$

(weak acid)

$$k_{eq.} = \frac{1}{K_h}$$

Now, $K_h \times ka = kw$

$$K_h \times 10^{-4} = 10^{-14}$$

$$K_h = 10^{-10}$$

$$k_{eq.} = 10^{10}$$

97. (d) $h = \sqrt{\frac{k_w}{ka.kb}}$

$$h = \sqrt{\frac{10^{-4}}{3 \times 10^{-7} \times 3 \times 10^{-7}}}$$

$$h = \frac{10^{-7}}{3 \times 10^{-7}} = 0.33$$

98. (c) $Ba(CN)_2$ is a salt of strong base & weak acid ($Ba(OH)_2$ & HCN).

It will undergo anionic hydrolysis.

$$Ba(CN)_2 \longrightarrow Ba^{2+} + 2\bar{C}N$$

At t = 0	0.5	0	0
At t = t	0	0.5	2 × 0.5

So, $(\bar{C}N) = 1M$

$[pka + pk_b = 14]$

$$pH = \frac{1}{2}(pkw + pka + \log 1)$$

$$pka + 9.30 = 14 = \frac{1}{2}(14 + 4.7 + \log 1)$$

$$pka = 4.7 = \frac{18.7}{2} = 9.35$$

99. (b)

100. (c) Salt of strong acid & weak base undergoes anionic hydrolysis.

$$A^- + H_2O \rightleftharpoons AOH + H^+$$

Now, adding strong acid will drive the reaction in backward direction.

101. (c) $CH_3COONa \longrightarrow$ salt of weak acid & strong base (resultant solution is basic)

$CH_3COOH \longrightarrow$ weakly acidic

$CH_3COONH_4 \longrightarrow$ may be acidic or basic depending upon ka & kb values

$NaOH \longrightarrow$ strongly basic

$HCl \longrightarrow$ strongly acidic

103. (b) pH of $0.01M-(NH_4)_2SO_4$ and

$0.02M-NH_4OH$ buffer (pK_a of $NH_4^+ = 9.26$) is

106. (a) $\underset{(aq.)}{CH_3COOH} \rightleftharpoons \underset{(aq.)}{CH_3COO^-} + \underset{(aq.)}{H^+}$ ______(1)

$CH_3COONA \longrightarrow CH_3COO^- + Na^+$ ______(2)

Presence of common ions $\left(CH_3COO^-\right)$ will drive the first Rxn. In backward direction.

So, $(H^+)\downarrow$ ses & $pH\uparrow$ ses.

107. (d) H_2SO_4 is itself a strong acid.

Concept of common ion effect is valid for weak electrolytes only.

108. (a) $HCOOH + NaOH \longrightarrow HCOONa + H_2O$

a

$a-\frac{a}{2}$ $\quad\quad \frac{a}{2}$

Acodoc buffer is formed, apply-

$$pH = pka + \log\frac{(Salt)}{(Acid)}$$

$$pH = pka + \log\left((Salt) = (Acid)\right)$$

$$pH = pka$$

$$-\log(H^+) = -\log ka$$

$$(H^+) = ka = 2\times10^{-4}$$

109. (a) $\underset{\text{(weak monoacidic base)}}{NH_4OH} + \underset{\text{(strong acid)}}{HCl} \longrightarrow NH_4Cl + H_2O$

At t = 0 $\quad$ a

At t = t $\quad$ a − 0.8a $\quad$ 0.8a

Its a basic buffer

$$\text{So, } pOH = pk_b + \log\frac{(Salt)}{(Base)}$$

$$pOH = 14 - 7.40$$

$$pOH = pK_b + \log\frac{0.8\not{a}}{0.2\not{a}}\,{}^4$$

$$6.60 = pk_b + \log 4$$

$$pK_b = 6$$

$$K_b = 10^{-6}$$

112. (c) $CH_3\ddot{N}H_2 + HCl \longrightarrow CH_3NH_3^+Cl^-$

At t = 0	0.1	0.08	0
At t = t	0.1 − 0.08	X	0.08

Basic buffer is formed

$$pOH = pK_b + \log\frac{(salt)}{(Base)}$$

$$-\log(\bar{O}H) = -\log K_b + \log\frac{0.08}{0.02}$$

$$\log(\bar{O}H) = \log K_b - \log 4$$

$$\log(\bar{O}H) = \log\frac{K_b}{4}$$

$$(\bar{O}H) = \frac{K_b}{4} = \frac{5\times10^{-4}}{4}$$

Now, $(H^+)(\bar{O}H) = 10^{-14}$

$$(H^+)\times\frac{5\times10^{-4}}{4} = 10^{-14}$$

$$(H^+) = 10^{-14}\times\frac{4}{5}\times\frac{1}{10^{-4}} = 0.8\times10^{-10}M$$

$$\boxed{(H^+) = 0.8\times10^{-10}M}$$

118. (b) $poH = pK_b + \log\frac{\left(NH_4^+\right)}{(NH_4OH)}$

$$5.74 = 4.74 + \log\frac{\left(NH_4^+\right)}{(0.01)}$$

$$\therefore \frac{\left(NH_4^+\right)}{(0.01)} = 10$$

$$\left(NH_4^+\right) = 0.1M$$

Moles of NH_4^+ in 500 ml solution = 0.05

Moles of $(NH_4)SO_4$ required = $\frac{0.05}{2} = 0.025$

125. (b) $pH = pka + \log\frac{(Salt)}{(Acid)}$

$$6 = 5 + \log\frac{(\text{Salt})}{(\text{Acid})}$$

$$1 = \log\frac{(\text{Salt})}{(\text{Acid})}$$

$$\frac{(\text{Salt})}{(\text{Acid})} = 10$$

126. (c) No. of moles of HOCl = No. of moles of NaOCl

$$pH = pka + \log\frac{(\text{Salt})}{(\text{Acid})}$$

$$pH = pka + \log 1 \qquad ((\text{Salt}) = (\text{Acid}))$$

$$pH = pka$$

$$pH = -\log ka = -\log 3.2 \times 10^{-8}$$

$$pH = -(\log 3.2 + \log 10^{-8}) \qquad (\log 3.2 \approx 0.5)$$

$$= -(0.5 - 8) = 8 - 0.5 = 7.5$$

127. (b) Acidic buffer is formed

$$pH = pka + \log\frac{(\text{Salt})}{(\text{Acid})}$$

$$\text{Let}\frac{(\text{Salt})}{(\text{Acid})} = 1$$

$$pH = pka + \log 1$$

$$pH = pka \_\_\_\_\_\_(1)$$

$$\text{Let}\frac{(\text{Salt})}{(\text{Acid})} = 10$$

$$pH = pka + 1 \_\_\_\_\_\_(2)$$

So, the pH↑ ses by 1 unit.

128. (d) (HA) = 0.3M

$(A^-) = 0.1M$

k_b for $A^- = 10^{-5}$

ka for HA = ?

$$ka \times k_b = k_w$$

$$ka \times 10^{-5} = 10^{-14}$$

$$ka = 10^{-9}$$

$$pH = pka + \log\frac{(\text{Salt})}{(\text{Acid})}$$

$$pH = 9 + \log\frac{0.1}{0.3}$$

$$pH = 9 + \log\frac{1}{3}$$

$$pH = 9 + \log 1 - \log 3$$

$$pH = 9 - \log 3$$

129. (b) $CH_3COONa + CH_3COOH$

0.2M 500 ml   0.1M 500 ml

$$pH = pka + \log\frac{(\text{Salt})}{(\text{Acid})}$$

$$pH = pka + \log\frac{0.2}{0.1}$$

$$pH = 4.74 + \log 2 = 4.74 + 0.30 = 5.04$$

Dilution has no effect on the *p*H of a buffer.

So, *p*H before & after dilution is same i.e. = 5.04

130. (d) In option (d)

$CH_3COOH + NaOH$

0.1M 100 ml   0.3M 50 ml

↓   ↓

10 millimoles   15 millimoles

CH_3COOH (weak acid) is a limiting reagent. So, acidic buffer can't be formed.

131. (c) In option C,

NH_3 = 0.05M, 100 ml = 5 millimole

NH_4Cl = 0.05M, 100 ml = 5 millimole

On, addition of 10 ml of 1M HCl = 10 millimoles

NH_3 will be exhausted, only HCl & NH_4Cl will be left. It will not be a buffer any more.

$$NH_3 + HCl \longrightarrow NH_4^+Cl^-$$

5 millimoles   10 millimoles

132. (d) $NH_3 + HCl \longrightarrow NH_4^+Cl^-$

In option d , $NH_3 = 2x$ moles

$HCl = x$ moles

HCl will be a limiting reagent.

x moles NH_4Cl will be formed.

x moles of NH_3 will react, so x moles of NH_3 will be left.

Finally, we have x moles of NH_3 & x moles of NH_4Cl . It will be basic buffer then.

133. (c) $CH_3COOH = 0.1$ mol

$CH_3COONa = 0.2$ mol

$NaOH = 0.05$ mol

$$CH_3COOH + NaOH \longrightarrow CH_3COONa + H_2O$$

	CH_3COOH	$NaOH$	CH_3COONa	H_2O
At t = 0	0.1	0.05	0	0
At t = t	0.1 − 0.05	×	0.05	0.05

Total moles of CH_3COONa now $= 0.2 + 0.05 = 0.25$

Total moles of CH_3COOH now $= 0.1 - 0.05 = 0.05$

$$pH = pka + \log\frac{(\text{Salt})}{(\text{Acid})}$$

$$pH = 4.74 + \log\left(\frac{0.25}{0.05}\right)$$

$$pH = 4.74 + \log 5$$

$$pH = 4.74 + 0.6990$$

$$pH = 5.44$$

135. (c) $A_xB_y \rightleftharpoons xA^{y+} + yB^{x-}$

At t = t $\quad xs \quad ys$

$$ksp = (A^{y+})^x (B^{x-})^y$$

$$= (xs)^x (ys)^y = x^x.s^x.y^y.s^y$$

$$ksp = x^x y^y.s^{x+y}$$

137. (d) $Cu(OH)_2 \longrightarrow Cu^{2+} + 2\bar{O}H$

$\qquad s \qquad 2s$

$$ksp = (Cu^{2+})(\bar{O}H)^2 = (s)(2s)^2 = 4s^3$$

$$2.6 \times 10^{-19} = 4s^3$$

$$\frac{2.6 \times 10^{-19}}{4} = s^3$$

$$\frac{26}{4} \times 10^{-20} = s^3$$

$$6.5 \times 10^{-20} = s^3$$

$$65 \times 10^{-21} = s^3$$

$$4.02 \times 10^{-7} = s$$

$$(\bar{O}H) = 2s = 2 \times 4.02 \times 10^{-7} = 8.04 \times 10^{-7}$$

$$poH = -\log(\bar{O}H) = -\log(8.04 \times 10^{-7})$$

$$poH = 7 - \log 8.04 = 7 - 0.90 = 6.1$$

$$pH = 14 - 6.1 = 7.9$$

138. (a) (1) $Pb(OH)_2 \rightleftharpoons Pb^{2+} + 2\bar{O}H$

(2) $HCl \longrightarrow H^+ + Cl^-$

Adding H^+ ions will lose the eqm. Conc. of $\bar{O}H$ ions & according to Le Chateliers principle, reaction 1 will rush in forward direction.

139. (d) For a compound to just start precipitating -

$IP = K_{sp}$

$$(Pb^{2+})(SO_4^{2-}) = 1.8 \times 10^{-8}$$

$$(Pb^{2+})(0.045) - 1.8 \times 10^{-8}$$

$$(Pb^{2+}) = \frac{1.8 \times 10^{-8}}{0.0045}$$

$$= 400 \times 10^{-8} = 4 \times 10^{-6} M$$

140. (d) $HgSO_4 \longrightarrow Hg^{2+} + SO_4^{2-}$

$$ksp = s^2$$

$$6.4 \times 10^{-5} = s^2$$

$$64 \times 10^{-6} = s^2$$

$$s = 8 \times 10^{-3} \text{ mol/ltr}$$

$$1 \text{ ltr} \longrightarrow 8 \times 10^{-3} \text{ mol}$$

$1\,ltr = 1000\,cm^3 = 10^{-3} M^3$

$10^{-3} M^3 \longrightarrow 8 \times 10^{-3} mol$

$1M^3 \longrightarrow \frac{8 \times 10^{-3}}{10^{-3}} = 8\,mol/M^3$

141. (d) $Li_3Na_3(AlF_6)_2 \longrightarrow 3Li^+ + 3Na^+ + 2AlF_6^{3-}$

$\quad 3s \quad 3s \quad 2s$

$ksp = (Li^+)^3(Na^+)^3(AlF_6^{3-})^2$

$= (3s)^3(3s)^3(2s)^2$

$= 27s^3 \times 27s^3 \times 4s^2$

$= 2916s^8$

142. (b) $Mg(OH)_2 \longrightarrow Mg^{2+} + 2\overline{O}H \qquad ksp = 4s_1^3$

$\quad s \quad 2s$

$2.56 \times 10^{-13} = 4s_1^3$

$\frac{2.56 \times 10^{-13}}{4} = s_1^3$

$0.64 \times 10^{-13} = s_1^3$

$64 \times 10^{-15} = s_1^3$

$s_1 = 8 \times 10^{-5}$

$Al(OH)_3 \longrightarrow Al^{3+} + 3\overline{O}H$

$ksp = 27s_2^4$

$4.32 \times 10^{-34} = 27s_2^4$

$\frac{4.32 \times 10^{-34}}{27} = s_2^4$

$0.16 \times 10^{-34} = s_2^4$

$16 \times 10^{-36} = s_2^4$

$2 \times 10^{-9} = s_2$

$\frac{s_1}{s_2} = \frac{8 \times 10^{-5}}{2 \times 10^{-9}} = 4 \times 10^4$

143. (c) $CaF_2 \longrightarrow Ca^{2+} + 2F^-$

$\quad s \quad 2s$

$NaF \longrightarrow Na^+ + F^-$

$0.1M \quad 0 \quad 0$

$0 \quad 0.1 \quad 0.1$

ksp of $CaF_2 = (Ca^{2+})(F^-)^2 = (s)(2s + 0.1)^2$

But $2s + 0.1 \approx 0.1$

ksp of $CaF_2 = (s)(0.1)^2$

$3.4 \times 10^{-11} = s \times (10^{-1})^2$

$3.4 \times 10^{-9} = s$

144. (c) $Ag_2CO_3 \longrightarrow 2Ag^+ + CO_3^{2-}$

$\quad 2s \quad s$

$ksp = (Ag^+)^2(CO_3^{2-}) = (2s)^2(s) = 4s^3$

$ksp = 4(10^{-4})^3 = 4 \times 10^{-12}$

'ksp will remain constt. for Ag_2CO_3 at contt. temp.'

$Na_2CO_3 \longrightarrow 2Na^+ + CO_3^{2-}$

$0.01 \quad 0 \quad 0$

$0 \quad 0.02 \quad 0.01$

$ksp = (Ag^+)^2 (CO_3^{2-})$

(For Ag_2CO_3)

$4 \times 10^{-12} = (s)^2(s + 0.01)$

$s + 0.01 \approx 0.01$

$4 \times 10^{-12} = s^2 (0.01)$

$\frac{4 \times 10^{-12}}{10^{-2}} = s^2$

$s = 2 \times 10^{-5}$

145. (a) Option (a)

For ppt. to occur, I.P. > ksp

I.P. for AgCl = $(Ag^+)(Cl^-)$

$$=\left(\frac{10^{-4}}{2}\right)\left(\frac{10^{-4}}{2}\right)$$

$$=\frac{1}{4}\times 10^{-8}$$

$$=0.25\times 10^{-8}$$

$$=25\times 10^{-10}$$

So, I.P. > ksp

∴ ppt. will be formed

157. (c)

I	II	III	IV
$AgCl \longrightarrow Ag^+_{s_1} + Cl^-_{s_1}$	$AgCl \longrightarrow Ag^+_{s_2} + Cl^-_{s_2}$	$AgCl \longrightarrow Ag^+_{s_3} + Cl^-_{s_3}$	$AgCl \longrightarrow Ag^+_{s_4} + Cl^-_{s_4}$
$H_2O \rightleftharpoons H^+ + \overline{O}H$	$CaCl_2 \longrightarrow Ca^{2+} + 2Cl^-$	$NaCl \longrightarrow Na^{3+} + Cl^-$	$AgNO_3 \longrightarrow Ag^+ + NO_3^-$
	0.01 0 0	0.01 0 0	0.05 0 0
	0 0.01 0.02	0 0.01 0.01	0 0.05 0.05

In case I	In case II	In case III	In case IV
ksp of AgCl = $(Ag^+)(Cl^-)$	ksp of AgCl = $(Ag^+)(Cl^-)$	ksp of AgCl = $(Ag^+)(Cl^-)$	ksp of AgCl = $(Ag^+)(Cl^-)$
$ksp = s_1 s_1$	$ksp = (s_2)(s_2 + 0.02)$	$ksp = (s_3)(s_3 + 0.01)$	$ksp = (s_4 + 0.05)(s_4)$
$ksp = s_1^2$	$ksp = s_2 \times 0.02$	$ksp = s_3 \times 0.01$	$ksp = 0.05 \times s_4$
$\sqrt{ksp} = s_1$	$50\,ksp = s_2$	$100\,ksp = s_3$	$20\,ksp = s_4$

'S_1' will be the biggest value becuase ksp is in 10^{-ve} power.

$S_1 > S_3 > S_2 > S_4$

158. (c) $Al(OH)_3 \longrightarrow Al^{3+} + 3\overline{O}H$

s   3s

$Ksp = 27s^4 = 4.32\times 10^{-14}$

$s = 2\times 10^{-4}$

$(\overline{O}H) = 3s = 3\times 2\times 10^{-4}$

$pOH = 4 - \log 6 = 3.22, pH = 10.77$

159. (d) Solubility of Pb $(OH)_2$ in winter $= 6\times 10^{-6}M$

$ksp = 4s^3 = 4\times(6\times 10^{-6})^3 = 4\times 216\times 10^{-18}$

In buffer

$pH = 8, POH = 6, (\overline{O}H) = 10^{-6}M$

$ksp = (Pb^{2+})(\overline{O}H)^2$

$4\times 216\times 10^{-18} = (5)(10^{-6})^2$

$4\times 216\times 10^{-18} = 5\times 10^{-12}$

$$\frac{4\times 216\times 10^{-18}}{10^{-12}} = 5$$

$864\times 10^{-6}M = 5$

or $8.64\times 10^{-4}M$

Smart Notes

CHAPTER 9

Redox Reactions

Calculations of oxidation states

1. Which of the following is correct with respect to Marshall's acid?
   (a) The actual oxidation state of S is + 6 in it.
   (b) Oxygen exhibits two different oxidation state.
   (c) It contains two peroxy linkages.
   (a) Both a & b
   (b) a, b & c
   (c) a only
   (d) Both b & c

2. An element exhibits two different oxidation states in the compound
   (a) NH_2OH
   (b) NH_4NO_3
   (c) N_2H_4
   (d) Both (a) & (c)

3. Select the correct statement
   (a) Oxidation state of oxygen in O_2^+ is $+\frac{1}{2}$
   (b) Oxidation state of oxygen in O_2^- is $-\frac{1}{2}$
   (c) Oxidation state of H in $LiAlH_4$ is – 1
   (d) All are correct

4. The oxidation numbers of the following underlined species are, respectively–

   $\underline{N}H_3, \underline{Pb}SO_4, Sb_2O_3, H\underline{N}O_3$

   (a) +3, +2, –3, –5
   (b) –3, +2, +3, +5
   (c) –3, +2, +5, +5
   (d) –1, –2, –3, –4

5. The correct oxidation states of Cr in CrO_5 and S in H_2SO_5 (Caro's acid) are respectively–
   (a) + 6 & + 6
   (b) + 6 & + 4
   (c) + 4 & + 4
   (d) + 6 & – 2

6. H_2SO_4 acts as an oxidizing agent, a dehydrating agent and an acid. Among each of the following reactions, select the incorrect behaviours–

Reaction		Behaviour of H_2SO_4
(a) $C_6H_{12}O_6 + \underset{\text{(conc.)}}{H_2SO_4} \rightarrow 6C + 6H_2O$	$\Rightarrow$	Dehydrating agent
(b) $5\underset{\text{(conc.)}}{H_2SO_4} + Zn \rightarrow H_2S + 4Zn^{2+} + 4SO_4^{2-} + 4H_2O$	$\Rightarrow$	Oxidising agent
(c) $\underset{\text{(dil)}}{H_2SO_4} + ZnCO_3 \rightarrow Zn^{2+} + CO_2 + SO_4^{2-} + H_2O$	$\Rightarrow$	Acid

(d) None of these

7. Which one of the following is not a redox reaction?

(a) $NH_4NO_2 \longrightarrow N_2 + 2H_2O$

(b) $MnO_4^- \longrightarrow MnO_4^{2-} + O_2$

(c) $2Pb(NO_3)_2 \longrightarrow 2Pb + 4NO_2 + O_2$

(d) $CaCO_3 \longrightarrow CaO + CO_2$

8. In which of the following, S shows an oxidation state of +6?

(a) $H_2S_2O_8$ (b) SO_4^{2-}

(c) SO_3 (d) All of these

9. The correct decreasing order of oxidation states is–

(a) $N_2O_5 > NO_2 > N_2O_3 > NO > NH_3$

(b) $NH_3 > NO > N_2O_3 > NO_2 > N_2O_5$

(c) $NO_2 > N_2O_3 > NO > NH_3 > N_2O_5$

(d) None of these

10. The oxidation states of S in $Na_2S_4O_6$ are respectively–

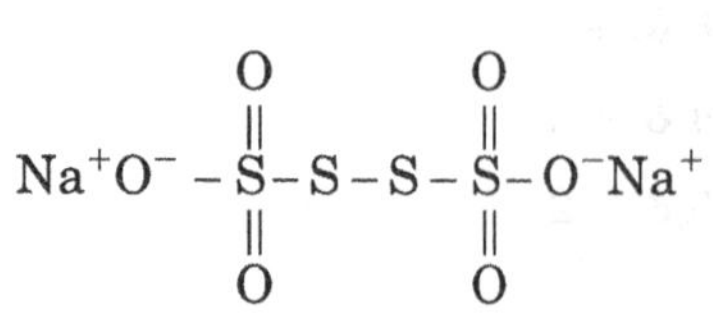

(a) +5, 0, 0, +5 (b) +6, 0, 0, +6

(c) +5, –2, –2, +5 (d) None of these

11. The average oxidation state of S in $Na_2S_2O_3$ is-

(a) +2 (b) +4

(c) +6 (d) –2

12. The average oxidation state of S in $Na_2S_4O_6$ is-

(a) +5 (b) +6

(c +2.65 (d) +3

13. Match list I (compounds) with list II (oxidation states of nitrogen) and select answer using the codes given below

List I	List II
(A) NaN_3	+5
(B) N_2H_2	+2
(C) NO	$-\frac{1}{3}$
(D) N_2O_5	–1

Code :–	(A)	(B)	(C)	(D)
(a)	3	4	2	1
(b)	4	3	2	1
(c)	3	4	1	2
(d)	4	3	1	2

14. The oxidation number of Fe in $Fe_4[Fe(CN)_6]_3$ is

(a) $\frac{6}{7}$ (b) +2 only

(c) + 1 & +3 (d) + 3 & +2

15. Which of the following have been arranged in correct order of oxidation numbers of nitrogen.

(a) $NH_3 < N_2O_5 < NO < N_2$

(b) $NO_2^+ < NO_3^- < NO_2^- > N_3^-$

(c) $NH_4^+ < N_2H_4 < NH_2OH < N_2O$

(d) $NO_2 < NaN_3 < NH_4^+ < N_2O$

16. Actual oxidation numbers of Pb in Pb_3O_4 are

(a) +2 & +4 (b) $\frac{-8}{3}$

(c) +4 & +8 (d) +2 & 3

17. The conversion of $K_4[Fe(CN)_6]$ to $K_3[Fe(CN)_6]$ by treating with H_2O_2 is

(a) Reduction, due to removal of electropositive element from potassium ferrocyanide

(b) Oxidation, due to removal to electropositive potassium from potassium ferrocyanide

(c) Reduction, due to addition of electronegative element potassium to potassium ferricyanide

(d) None of these

18. In which of the following compounds, chlorine exhibits positive oxidation state?

(a) HCl (b) ClF_3

(c) $HClO_4$ (d) Both (b) & (c)

19. In which of the following the oxidation state of metal is zero?

(a) Na (b) Na – Hg

(c) Zn – Cu (d) All of these

20. The correct order of oxidation number of Cl in the compounds is

(a) $ICl_3 > ClF_3 > Cl_2 > Cl_2O_7$

(b) $Cl_2O_7 > ClF_3 > Cl_2 > ICl_3$

(c) $Cl_2O_7 > Cl_2 > ClF_3 = ICl_3$

(d) None of these

Types of Redox reactions & balancing of equations

21. Hydrogen behaves as an oxidizing agent when it reacts with -

(a) N_2 to give NH_3 (b) Ca to give CaH_2

(c) Br_2 to give HBr (d) S to give H_2S

22. $\text{Sulphur} + 12\bar{O}H \longrightarrow 4S^{2-} + 2S_2O_3^{2-} + 6H_2O$

The molecular state of sulphur as reactant in the above reaction is –

(a) S_8^{2-} (b) $2S_4^{2-}$

(c) S_8 (d) S_8^-

23. $CrBr_3 \longrightarrow CrO_4^{2-} + BrO_4^-$

Total electron change in the above reaction is –

(a) 27 (b) $\frac{8}{3}$

(c) 11 (d) 8

24. Consider the following change -

$Ba(MnO_4)_2 \longrightarrow Mn^{2+}$

[M = mol. mass of $Ba(MnO_4)_2$]

Then the eqvt.of $Ba(MnO_4)_2$ is –

(a) $\frac{M}{5}$ (b) $\frac{M}{10}$

(c) $\frac{M}{7}$ (d) $\frac{M}{6}$

25. Which one of the following is an example of disproportionation reaction?

(a) $2H_2O_{2(aq.)} \longrightarrow 2H_2O_{(l)} + O_{2(g)}$

(b) $P_{4(s)} + 3\bar{O}H_{(aq.)} + 3H_2O_{(l)} \longrightarrow PH_{3(g)} + 3H_2PO^-_{2(aq.)}$

(c) $6Cl_{2(g)} + 12\bar{O}H_{(aq.)} \xrightarrow{hot} 10Cl^-_{(aq.)} + 2ClO^-_{3(aq.)} + 6H_2O$

(d) All are disproporotionation reactions.

26. Which one of the following reactions is an example conproportionation reactions?

(a) $2Mn_2O_7 \longrightarrow 4MnO_2 + 3O_2$

(b) $Pb_3O_4 + 8HCl \longrightarrow 3PbCl_2 + Cl_2 + 4H_2O$

(c) $5Br^- + BrO_3^- + 6H^+ \longrightarrow 3Br_2 + 3H_2O$

(d) $N_{2(g)} + O_{2(g)} \longrightarrow 2NO_{(g)}$

27. $\overset{-}{C}N$ is oxidised by NO_3^- in the presence of acid–

$$a\overset{-}{C}N + bNO_3^- + cH^+ \longrightarrow (a+b)NO + aCO_2 + \frac{c}{2}H_2O$$

What are the values of a, b, c, respectively-

(a) 3, 7, 7 (b) 3, 7, 10

(c) 3, 10, 7 (d) 3, 9, 3

28. A, B and C are three elements forming a part of compound in oxidation states of +2, +5 & –2 respectively. What would be the compound amongst the following?

(a) $A_2(BC)_2$ (b) $A_2(BC_4)_3$

(c) $A_3(BC_4)_2$ (d) ABC

29. Which of the following can act both as oxidising as well as reducing agents?

(a) HNO_2 (b) H_2O_2

(c) SO_2 (d) All

30. The equivalent mass of $MnSO_4$ is half of its molar mass when it is converted to-

(a) Mn_2O_3 (b) MnO_2

(c) MnO_4^- (d) MnO_4^{2-}

31. In the reaction, $N_{2(g)} + 3H_{2(g)} \longrightarrow 2NH_{3(g)}$; if E_1 & E_2 are the equivalent masses of NH_3 & N_2 respectively, then $E_1 - E_2$ is-

(a) 1 (b) 2

(c) 3 (d) 4

32. Where N_2 is converted into NH_3, the equivalent weight of N_2 is -

(a) 1.67 (b) 2.67

(c) 3.67 (d) 4.67

33. Which of the following represents reductions?

(a) $2Mg + O_2 \longrightarrow 2MgO$ (addition of oxygen)

(b) $2FeCl_{2(aq)} + Cl_{2(g)} \longrightarrow 3FeCl_3(aq.)$ (addition of electronegative element)

(c) $H_2S_{(g)} + Cl_{2(g)} \longrightarrow 2HCl_{(g)} + S_{(s)}$ (removal of hydrogen)

(d) $2HgCl_{2(aq.)} + SnCl_{2(aq.)} \longrightarrow Hg_2Cl_{2(s)} + SnCl_{4(aq.)}$ (addition of electropostive element)

34. Choose the incorrect match:-

Reaction	Reducing agents in the reaction
(a) $Fe_2O_3 + 2Al \longrightarrow 2Fe + Al_2O_3$	Al
(b) $5HNO_2 + 3H_2SO_4 + 2KMnO_4 \longrightarrow 5HNO_3 + K_2SO_4 + 2MnSO_4 + 3H_2O$	HNO_2
(c) $Fe_2O_3 + 3CO \longrightarrow 2Fe + 3CO_2$	Fe_2O_3
(d) $CuO + C \longrightarrow CO + Cu$	C

35. Which one of the following is not a redox reaction?

(a) $2KClO_{3(s)} \xrightarrow{\Delta} 2KCl_{(s)} + 3O_{2(g)}$

(b) $CH_{4(g)} + 2O_{2(g)} \longrightarrow CO_{2(g)} + 2H_2O_{(l)}$

(c) $2NaH \xrightarrow{\Delta} 2Na_{(s)} + H_{2(g)}$

(d) $KOH + HNO_3 \longrightarrow KNO_3 + H_2O$

36. For the following redox reaction –

$$2Cu_2O + Cu_2S \longrightarrow 6Cu + SO_2 \uparrow$$

Which of the following statements is incorrect?

(a) Copper is reduced from +1 to zero oxidation state.

(b) Sulphur is oxidized from – 2 to + 4

(c) Total electron change in the reaction is 12.

(d) Total electron change in the reaction is 6.

37. The total electron change that occurs in the following redox reaction is

$$3Fe_3O_4 + 8Al \longrightarrow 9Fe + 4Al_2O_3$$

(a) 24 (b) 12

(c) 16 (d) 10

38. The equivalent weight of complex ion $[Fe(CN)_6]^{4-}$ in the given reaction is

[M = molar mass of $[Fe(CN)_6]^{4-}$]

$[Fe(CN)_6]^{4-} \longrightarrow Fe^{3+} + CO_3^{2-} + NO_3^-$

(a) M (b) $\frac{M}{30}$

(c) $\frac{M}{61}$ (d) $\frac{M}{6}$

39. In the reaction,

$SO_2 + 2H_2S \longrightarrow 3S + 2H_2O$, which option has value twice that of equivalent mass of oxidizing agent (S = 32, O = 16)

(a) 64 (b) 32

(c) 48 (d) 16

40. $CrBr_3 \longrightarrow CrO_4^{2-} + BrO_4^-$

The total number of electron change in the above reaction is

(a) 27 (b) $\frac{8}{3}$

(c) 11 (d) 8

41. Match column I with Column II

Column – I	Column – II
(i) $(NH_4)_2Cr_2O_7 \rightarrow N_2 + Cr_2O_3 + 4H_2O$	(a) Intermolecular redox reaction
(ii) $PbO_2 + H_2O \rightarrow PbO + H_2O_2$	(b) Disproportionation
(iii) $Cr_2O_3 + 2Al \rightarrow Al_2O_3 + 2Cr$	(c) Intermolecular redox reaction
(iv) $Cl_2 + 2\bar{O}H \rightarrow Cl_2O^{\ominus} + Cl^- + H_2O$	(d) Metal displacement reaction

(a) (i) – c, (ii) – a, (iii) – a, d (iv) – b

(b) (i) – a, (ii) – c, (iii) – d, (iv) – b

(c) (i) – b, (ii), – a, (iii) – a, d (iv) – c

(d) Both (ii) & (iii)

42. Thiosulphate results differently with iodine & bromine in the reactions given below

$2S_2O_3^{2-} + I_2 \rightarrow S_4O_6^{2-} + 2I^-$

$S_2O_3^{2-} + 2Br_2 + 5H_2O \rightarrow 2SO_4^{2-} + Br^- + 10H^+$

Which of the following statements justifies the above dual behavior of thiosulphate?

(a) Bromine is a stronger oxidant than iodine

(b) Bromine is a weaker oxidant than iodine

(c) Thiosulphate undergoes oxidation by bromine & reduction by iodine in these reactions.

(d) Bromine undergoes oxidation & iodine undergoes reduction in these reactions.

43. The oxide Pb_3O_4 reacts differently with HCl & HNO_3 as shown

$Pb_3O_4 + 8HCl \rightarrow 3PbCl_2 + Cl_2 + 4H_2O$

$Pb_3O_4 + 4HNO_3 \rightarrow 2Pb(NO_3)_2 + PbO_2 + 2H_2O$

The correct reason(s) for this behavior is/are

(a) Pb_3O_4 is a mixed oxide : $2PbO.\ PbO_2$

(b) PbO is basic oxide & PbO_2 is an oxidizing agent

(c) HNO_3 is an oxidizing agent & therefore reaction between PbO_2 & HNO_3 is unlikely

(d) All are correct

44. Choose the correct statement when Pb_3O_4 is separately treated with HCl & HNO_3.

(a) Reaction of PbO with HCl is a redox reaction forming $PbCl_2$ & H_2O.

(b) Redox reaction occurs between PbO_2 & HCl

(c) PbO_2 & HNO_3 do not react with each other as both are oxidizing agents.

(d) Both (b) & (c) are correct

45. The correct stock notation for ferrous sulphate is

(a) $FeSO_4$ (b) $FeCl_3$

(c) Fe (II) SO_4 (d) Fe_2 (III) $(SO_4)_3$

46. Identify the correct statement wrt disproportionation reaction?

(a) The reacting species must contain an element which can exist in atleast three oxidation states.

(b) The same specie is simultaneously both oxidized & reduced

(c) The same element undergoes both increase & decrease in oxidation state.

(d) All of these

47. Which of the following reactions has/have spectator ions?

(a) $Zn + CuSO_4 \longrightarrow ZnSO_4 + Cu$

(b) $KIO_3 + KI + H_2SO_4 \longrightarrow KI_3 + K_2SO_4 + H_2O$

(c) $2KMnO_4 + 10KCl + 8H_2SO_4 \longrightarrow 5Cl_2 + 2MnSO_4 + 8H_2O + 6K_2SO_4$

(d) All have spectator ions

48. Which of the following reactions indicate the oxidizing behavior of H_2SO_4?

(a) $2PCl_5 + H_2SO_4 \rightarrow 2PCl_3 + HCl + SO_2Cl_2$

(b) $2NaOH + H_2SO_4 \rightarrow Na_2SO_4 + 2H_2O$

(c) $NaCl + H_2SO_4 \rightarrow NaHSO_4 + HCl$

(d) $2HI + H_2SO_4 \longrightarrow I_2 + SO_2 + H_2O$

49. $3Mg + N_2 \longrightarrow Mg_3N_2$

The total e^- change in the above redox reaction is

(a) 3 (b) 6

(c) 8 (d) 9

50. Match the reactions in Column I with their types in Column II

Column – I	Column – II
(i) $(NH_4)_2Cr_2O_7 \longrightarrow N_2 + Cr_2O_3 + 4H_2O$	(a) Intermolecular redox reaction.
(ii) $PbO_2 + H_2O \longrightarrow PbO + H_2O_2$	(b) Disproportionation
(iii) $Cr_2O_3 + 2Al \longrightarrow Al_2O_3 + 2Cr$	(c) Intermolecular redox
(iv) $Cl_2 + 2\bar{O}H \longrightarrow ClO_3^- + Cl^- + H_2O$	(d) Metal displacement

(a) (i) – c; (iii) – a; (iii) – a, d; (iv) – b

(b) (i) – a; (ii) – c; (iii) – d, (iv) – b

(c) (i) – b; (ii) – a; (iii) – a, d; (iv) – c

(d) (i) – a, b; (ii) – c; (iii) – a, d ; (iv) – b

51. $H_2S + H_2O_2 \longrightarrow S + 2H_2O$

The behavior of H_2O_2 depicted in the above reaction is

(a) acidic nature

(b) basic nature

(c) oxidizing nature

(d) reducing nature

52. Which of the following reactions is decomposition reaction but not a redox reaction?

(a) $2NO_2 \longrightarrow N_2O_4$

(b) $2NO_2 + H_2O \longrightarrow HNO_2 + HNO_3$

(c) $2KClO_3 \longrightarrow 2KCl + 3O_2$

(d) $CaCO_3 \longrightarrow CaO + CO_2$

53. $Fe + H_2O \longrightarrow Fe_3O_4 + H_2O$

In the above change, if the atomic weight of Fe is 56, then its equivalent weight will be

(a) 42 (b) 21

(c) 63 (d) 84

54. Which of the following is not an intra-molecular redox?

(a) $(NH_4)_2 Cr_2O_7 \longrightarrow N_2 + Cr_2O_3 + H_2O$

(b) $NH_4NO_3 \longrightarrow N_2O + 2H_2O$

(c) $CaCO_3 \longrightarrow CaO + CO_2 \uparrow$

(d) CHO
↓
CHO + $\overline{O}H \longrightarrow$ COO^-
↓
CH_2OH

55. Total electron change that occurs in the following redox reaction is –

$3Fe_3O_4 + 8Al \longrightarrow 9Fe + 4Al_2O_3$

(a) 24 (b) 12

(c) 16 (d) 10

Redox Titration

56. The equivalent wt. of Cl_2 is 47.33 grams in the following disproportionation reaction.

$Cl_2 + \overline{O}H \longrightarrow Cl^- + H_2O +$ oxidized product

(molar mass of Cl_2 = 71gm)

The oxidized specie is –

(a) ClO_2 (b) ClO_4^-

(c) $ClO^{\ominus}$ (d) ClO_3^-

57. If 1 mole of MnO_4^- oxidises 1.67 moles of M^{x+} to MO_3^-, then the value of x in the reactions is –

(a) 5

(b) 3

(c) 2

(d) 1

58. If the molecular weight of $Na_2S_2O_3$ & I_2 are M_1 & M_2 respectively, then what will be the equivalent weight of $Na_2S_2O_3$ & I_2 in the following reactions?

$2S_2O_3^{2-} + I_2 \longrightarrow S_4O_6^{2-} + 2I^-$

(a) M_1, M_2 (b) $M_1, \frac{M_2}{2}$

(c) $2\,M_1, M_2$ (d) $M_1, 2M_2$

59. How many moles of $KMnO_4$ are needed to oxidise a mixture of 1 mole of each $FeSO_4$ & FeC_2O_4 in acidic mediums

(a) $\frac{4}{5}$ (b) $\frac{5}{4}$

(c) $\frac{3}{4}$ (d) $\frac{5}{3}$

60. Hydroxyl amine reduces iron (III) according to following equation which statement is correct

$NH_2OH + Fe_2(SO_4)_3 \longrightarrow$

$N_2(g) + H_2O + FeSO_4 + H_2SO_4$

(a) n–factor for Hydroxyl amines is 1

(b) equivalent weight of $Fe_2(SO_4)_3$ is $\frac{M}{2}$

(c) 6 meq of $Fe_2(SO_4)_3$ is contained in 3 millimoles of ferric sulphate

(d) all of these

61. 0.5 g mixture of oxalic acid ($H_2C_2O_4$) and sodium oxalate ($Na_2C_2O_4$) with some impurities requires 40 ml of 0.1M NaOH for complete neutralization and 6 ml of 0.2 M $KMnO_4$ for complete oxidation. Calculate the % of $Na_2C_2O_4$ in the mixture

(a) 90% (b)

(c) 40% (d) 50%

62. A solution contains mixture of H_2SO_4, $H_2C_2O_4$. 20 ml of this Solution requires 40 ml of $\frac{M}{10}$ NaOH for neutralization and 20 ml of $\frac{N}{10} KMnO_4$ for oxidation. The molarity of $H_2C_2O_4$, H_2SO_4 are

(a) 0.1, 0.1 (b) 0.1, 0.05

(c) 0.05, 0.1 (d) 0.05, 0.05

63. 100 ml of 0.1M aqueous ferric alum Solution is completely reduced with Iron to give Ferrous Ions. The volume of 0.1M acidified $KMnO_4$ Sol. required for complete reaction with ferrous Ions in the Sol. will be

(a) 60 ml (b) 30 ml

(c) 40 ml (d) 100 ml

64. 10 mL of H_2O_2 Solution on treatment with KI and titration of liberated I_2, required 10 mL of 1 N hypo. Thus H_2O_2 is :

(a) 1N (b) 5.6 volume

(c) 17 gL^{-1} (d) all are correct

65. For decolourisation of 1 mol of $KMnO_4$, the number of moles of H_2O_2 used is

(a) $\frac{1}{2}$ (b) $\frac{3}{2}$

(c) $\frac{5}{2}$ (d) $\frac{7}{2}$

66. The equivalent weight of HNO_3 in the reaction

$$P + HNO_3 \longrightarrow HPO_3 + NO + H_2O$$

(a) $\frac{M}{2}$ (b) $\frac{M}{1}$

(c) $\frac{M}{3}$ (d) $\frac{M}{4}$

67. 1 mole mixture of equimolar ferrous oxalate and ferric oxalate required X moles of $KMnO_4$ in acidic medium. The value of 'X' is

(a) 0.9 (b) 4.5

(c) 3.8 (d) 5.6

68. If 1 mole of H_3PO_4 is reacted with 1 mole of $X(OH)_2$ as

$$H_3PO_4 + x(OH)_2 \longrightarrow XHPO_4 + 2H_2O, \text{ then}$$

(a) the equivalent weight of base is M.wt/3.

(b) the equivalent weight of H_3PO_4 is $\frac{98}{3}$.

(c) the resulting Sol. will require 1 mole NaOH for complete neutralization.

(d) 1 mole of $x(OH)_2$ more, is required for complete neutralization of $xHPO_4$

69. A 0.5 M Sol. of KI reacts with excess of H_2SO_4 & KIO_3 Sol. according to the equation

$$H^+ + I^- + IO_3^- \longrightarrow I_2 + H_2O$$

Then, which of the following statement is true?

(a) 200 ml of KI Sol. reacts with 0.1 mole KIO_3

(b) 100 ml of KI Sol. reacts with 0.060 moles of H_2SO_4

(c) 0.5 L of KI Sol. produces 0.15 mole of I_2

(d) None of the above

70. Volume of $Al_2(Cr_2O_7)_3$ of molarity 0.1 M required to react with 10 ml of 0.018 M $FeSO_4$ Sol. is

(a) 10 ml

(b) 18 ml

(c) 0.1 m

(d) 0.6 ml

71. Match the reactions is Column I with molar ratio of their respective Reaction

Reaction	Ratio of moles of reductant to oxidant
(a) $Bi_2O_3 + ClO^{\ominus} + \bar{O}H \longrightarrow BiO_3^- + Cl^-$	(p) 4
(b) $MnO_2 + PbO_2 + H^+ \longrightarrow MnO_4^- + Pb^{2+}$	(q) 2
(c) $TeO_3^{2-} + I^- + H^+ \longrightarrow Te + I_2$	(r) $\frac{1}{2}$
(d) $[Fe(CN)_6]^{3-} + Cr_2O_3 + \bar{O}H \longrightarrow [Fe(CN)_6]^{4-} + CrO_4^{2-}$	(s) $\frac{2}{5}$
(e) $COCl_2 + Na_2O_2 + \bar{O}H \longrightarrow CO(OH)_3$	(t) $\frac{1}{6}$

(a) a – q; b – s; c – P; d – e, e – r

(b) a – s; b – p; d – r; e – t, c – t

(c) a – s; b – s; c – r; d – t, e – q

(d) None of these

72. $KMnO_4$ reacts with oxalic acid according to the equation.

$$2MnO_4^- + 5C_2O_4^{2-} + 16H^+ \longrightarrow 2Mn^{2+} + 10CO_2 + 8H_2O$$

20 ml of 0.1 M $KMnO_4$ is equivalent to

(a) 120 ml of 0.25 M oxalic acid

(b) 150 ml of 0.1 M oxalic acid

(c) 50 ml of 0.1 M oxalic acid

(d) 50 ml of 0.2 M oxalic acid

73. How much volume of 0.1 N oxalic acid solute can be reduced by 250g of 8% (by mass) $KMnO_4$ Solution (M. wt. of $KMnO_4$ = 158)

(a) 6.3 L (b) 12.6 L

(c) 25.2 L (d) 0.63 L

74. Equivalents of MnO_4^- & $Cr_2O_7^{2-}$ per mole, when acting as oxidizing agent in acidic medium are in the ratio of

(a) 1 : 1 (b) 1 : 5

(c) 6 : 1 (d) 5 : 6

75. 0.1 mole of MnO_4^- in acidic medium can oxidize–

(a) 0.5 mole of Fe^{2+} (b) 0.166 mol of FeC_2O_4

(c) 0.25 moles of (d) All of these

76. When I^- is oxidised by MnO_4^- in alkaline medium, it is converted into –

(a) IO_3^- (b) I_2

(c) IO_4^- (d) IO–

77. 6 mol of a mixture of $Fe_2(SO_4)_3$ & Mohr's salt i.e. $FeSO_4.(NH_4)_2SO_4.6H_2O$ required 500 ml of 1M $K_2Cr_2O_7$ for complete oxidation in acidic medium. The molar ratio of Mohr's salt to $Fe_2(SO_4)_3$ salt in the mixture is –

(a) 1 : 4 (b) 1 : 1

(c) 3 : 1 (d) 5 : 1

78. 1 mole mixture of equimolar ferrous oxalate & ferric oxalate requires 'X' mole of $KMnO_4$ in acidic medium. Then, X is-

(a) 0.9 (b) 4.5

(c) 3.8 (d) 5.6

79. A Sol.X (25ml Fe^{2+} ions) required the Sol. Y $\left(20 \text{ ml of } \frac{MnO_4^-}{H^+}\right)$ for complete reaction. If Sol. X completely reacts with $\frac{MnO_4^-}{\overline{O}H}$, then the volume of $\frac{MnO_4^-}{\overline{O}H}$ Sol. required as–

(Keeping Molarity of MnO_4^- same in acidic & basic medium)

(a) 12 ml (b) 25 ml

(c) 33.3 ml (d) 35 ml

80. Which of the following is the strongest oxidizing agent?

(a) $\frac{BrO_3^-}{Br^-}$, $E^\circ = +1{\cdot}42V$

(b) $\frac{Fe^{3+}}{Fe^{2+}}$, $E^\circ = +0{\cdot}76V$

(c) $\frac{MnO_4^-}{Mn^{2+}}$, $E^\circ = +1{\cdot}52V$

(d) $\frac{Cr_2O_7^{2-}}{Cu^{3+}}$; $E^\circ = 1{\cdot}33V$

81. 6 moles of I_2 undergo disproportionation (involving 10 moles of electrons change), out of which 5 moles of I_2 are reduced. Oxidation no. of oxidized iodine atoms is -

(a) + 3 (b) + 5

(c) + 7 (d) + 1

82. A 10g sample of KI is oxidized with acidified $KMnO_4$. The liberated I_2 reacts with 100 ml of 0.2M hypo Sol. then percentage purity of sample is (M. wt. of KI = 166)

(a) 33.2% (b) 100%

(c) 66.4% (d) 16.6%

83. The Number of moles of $K_2Cr_2O_7$ in an acidic medium required to oxidise one mole of $Cu_3P$to CuO & H_3PO_4 is–

(a) $\frac{4}{3}$ (b) $\frac{11}{6}$

(c) $\frac{3}{4}$ (d) $\frac{6}{11}$

84. In which of the following reactions, the underlined reactants exhibited similar n – factor?

(a) $Cr_2O_7^{2-}$ as oxidizing agent in acidic medium.

(b) $Fe_2(C_2O_4)_3$ as reducing agent with $\frac{KMnO_4}{H^+}$

(c) $H_2C_2O_4$ as reducing agent with $\frac{Cr_2O_7^{2-}}{H^+}$.

(a) Only a (b) Both a & b

(c) Both b & c (d) None of these

85. Br_2 undergoes disproportionation toBr^- & BrO_3^- in the alkaline medium. Which of the following can be correct for the above reaction?

(a) The total electron change is 10

(b) There are 10 moles of $\overline{O}H$ in the balanced chemical equation on LHS.

(c) Equivalent weight of Br_2 as reductant is $\frac{M}{10}$ (M = molar mass of Br_2)

(d) n - factor of Br_2 is $\frac{5}{3}$.

(a) a , c & d (b) Both b & c

(c) a only (d) All are correct

Electro chemical Cell & potentials

86. The values of standard reduction potential of following reactions are given

$Cu^{2+} + 2e^- \longrightarrow Cu\ ;\ E^\circ = +0\cdot34\,V$

$Sn^{2+} + 2e^- \longrightarrow Sn\ ,\ E^\circ = -0\cdot14\,V$

$Zn^{2+} + 2e^- \longrightarrow Zn\ ,\ E^\circ = -0\cdot76\,V$

$Fe^{2+} + 2e^- \longrightarrow Fe\ ,\ E^\circ = -0\cdot40\,V$

Which one of the following is most easily reduced & oxidised, respectively?

(a) Cu^{2+}, Zn (b) Zn^{2+}, Sn

(c) Fe^{2+},Cu (d) Cu^{2+},Sn

87. The standard electrode potentials of two half reaction at 25°C are –

$Pb^{2+} + 2e^- \longrightarrow Pb\ ;\ \epsilon^\circ = -0\cdot13\,V$

$Ag^+ + e^- \longrightarrow Ag\ ;\ \epsilon^\circ + 0\cdot80\,V$

The possible cell reaction is

(a) $Pb^{2+} + 2Ag^+ \longrightarrow 2Ag + Pb$

(b) $Pb + 2Ag^+ \longrightarrow Pb^{2+} + 2Ag^+$

(c) $Pb + 2Ag^+ \longrightarrow Pb^{2+} + 2Ag$

(d) $Pb + 2Ag \longrightarrow Pb^{2+} + 2Ag^+$

88. The standard potential at 25°C for the following half reactions are given below

$Zn^{2+} + 2e^- \longrightarrow Zn\ ;\ \epsilon^\circ = -0\cdot762\ V$

$Mg^{2+} + 2e^- \longrightarrow Mg\ ;\ \epsilon^\circ = -2\cdot37\ V$

When zinc dust is added to the Sol. of $MgCl_2$:

(a) $ZnCl_2$ is formed

(b) Zinc dissolve in the Sol.

(c) No reaction takes place

(d) Mg is precipitated

89. Copper from $CuSO_4$ Sol. can be displaced by which of the following?

The standard reduction potentials of some electrodes are as follows

$\epsilon^\circ\ (Fe^{2+}, Fe) = -0\cdot44\,V$

$\epsilon^\circ\ (Zn^{2+}, Zn) = -0\cdot76\,V$

$\epsilon^\circ\ (Cu^{2+}, Cu) = +0\cdot34\,V$

$\epsilon^\circ\ (Cr^{3+}, Cr) = -0\cdot74\,V$

(a) Zn (b) Cr

(c) Fe (d) All of these

90. Given $E^\circ_{Al^{3+}/Al} = -1\cdot66\,v$, $E^\circ_{Cu^{2+}/Cu} = +0\cdot34\,V$

What will happen if we store aluminium nitrate in copper tumbler?

(a) the tumbler gets coated with aluminium.

(b) the Sol. becomes blue

(c) an alloy of copper &aluminium is formed

(d) there is no reaction

91. Given: $E^\circ_{Ag^+/Ag} = 0.80$ V, $E^\circ_{Mg^{2+}/Mg} = -2.37$ V,

$E^\circ_{Cu^{2+}/Cu} = +0.34$ V, $E^\circ_{Hg^{2+}/Hg} = +0.79$ V

Which of the following is correct?

(a) $AgNO_3$(aq.) can be stored in Cu vessel

(b) Cu $(NO_3)_2$(aq.) can be stored in Mg vessel

(c) $CuCl_2$ can be stored in Ag vessel

(d) Hg_2Cl_2 can be stored in Cu vessel

92. $E^\circ_{F_2/F} = 2.87$ V ; $E^\circ_{Li^+/Li} = -3.5$ V

$E^\circ_{Au^{3+}/Au} = 1.4$ V ; $E^\circ_{Br_2/Br^-} = 1.09$ V

On the basis of above data, which of the following statements is correct?

(a) F_2 is the best reducing agent

(b) Br^- can be oxidized by Au^{3+}

(c) Li^+ is the weakest reducing agent

(d) Au^{3+} is the strongest reducing agent

93. Standard reduction potentials of Mg, Zn, H_2, Cu & Ag are –2.37 V, – 0.76 V, – 0.25 V, 0V, +0.34V & 0.80 V, respectively. Which one of the following cells has the maximum voltage?

(a) $Mg^{2+} + 2Ag \longrightarrow Mg + 2Ag^+$

(b) $Cu^{2+} + H_2 \longrightarrow Cu + 2H^+$

(c) $Ni^{2+} + Zn \longrightarrow Ni + Zn^{2+}$

(d) $Mg + 2Ag^+ \longrightarrow Mg^{2+} + 2Ag$

94. (a) $Fe^{3+}_{(aq)} + e^- \longrightarrow Fe^{2+}_{(aq)}$; $E^\circ = +0.77$V

(b) $Al^{3+} + 3e^- \longrightarrow Al_{(s)}$; $E^\circ = -1.66$V

(c) $Br_2 + 2e^- \longrightarrow 2Br^-_{(aq)}$; $E^\circ = +1.08$V

Based on the date given above, the correct order of reducing power is

(a) $Br_2 < Fe^{3+} < Al$ (b) $Br^- < Fe^{2+} < Al$

(c) $Fe^{2+} < Al < Br^-$ (d) $Al < Br^- < Fe^{2+}$

95. $Ag \mid Ag^+_{(1M)} \parallel Cu^{2+}_{(1M)} \mid Cu$; $E^\circ = -0.46$V

$Zn \mid Zn^{2+}_{(1M)} \parallel Cu^{2+}_{(1M)} \mid Cu$; $E^\circ = +1.1$V

$Zn \mid Zn^{2+}_{(1M)} \parallel Ag^+_{(1M)} \mid Ag$; $E^\circ = x$V

The value of 'x' is

(a) 0.64 (b) 1.56

(c) –1.56 (d) – 0.64

96. Among the following the strongest oxidant & reductant respectively are

(A) $E^\circ_{BrO_3^-/Br_2} = +1.50$V

(B) $E^\circ_{Fe^{3+}/Fe^{2+}}$; $E^\circ = +0.76$ V

(C) $E^\circ_{MnO_4^-/Mn^{2+}}$; $E^\circ = +1.52$V

(D) $E^\circ_{Cr_2O_7^{2-}/Cr^{3+}}$; $E^\circ = +1.33$V

(a) BrO_3^-, Cr^{3+} (b) MnO_4^-; Fe^{3+}

(c) MnO_4^-, Fe^{2+} (d) $Cr_2O_7^{2-}$, Br_2

97. Most powerful oxidizing agent is -

(a) F_2 (b) Li

(c) H_2 (d) Mg

98. pH of HCl Sol. used in SHE at 298K is -

(a) 0.1 (b) 0

(c) 1 (d) 7

99. Red hot carbons remove oxygen from oxide XO & YO but not from ZO. Y can remove oxygen from XO. Based on this information, the order of reactivity of 3 metals X, Y & Z is -

(a) X >Y > Z

(b) Z >Y > X

(c) Y >X > Z

(d) Z >X > Y

100. A Solution contains mixture of H_2SO_4 & $H_2C_2O_4$. 20 ml of this Solution required 40 ml of $\frac{M}{10}$ NaOH for neutralization & 20 ml of $\frac{N}{10}$ $KMnO_4$ for oxidation. Molarity of H_2SO_4 & $H_2C_2O_4$ are, respectively–

(a) 0.1, 0.1 (b) 0.05, 0.05

(c) 0.05, 0.1 (d) 0.50, 0.05

101. The ratio of oxidation state of nitrogen in $C_6H_5NO_2$ to that in HNO_3 is –

(a) $\frac{5}{3}$ (b) $\frac{2}{5}$

(c) $\frac{4}{5}$ (d) $\frac{3}{5}$

102. For the given reaction,

$CuSO_4 + BaS_2O_3 \longrightarrow BaSO_4 + CuS_2O_3$

Which of the following is true?

(a) Oxidation no. of Ba changes

(b) Oxidation no. of Cu changes

(c) reaction is an example of redox reaction

(d) None of these

103. The n factor of MnO_4^- in acidic medium will be same as that in which of the following-

(a) N_2 to NO_2 (b) N_2H_4 to N_2O_3

(c) No to HNO_3 (d) NH_3 to NO

104. Total electron change in the following reactions are, respectively

(a) $Mg + HCl \longrightarrow MgCl_2 + H_2$

(b) $Zn + NO_3^- \longrightarrow N_2O + Zn^{2+}$

(a) 2, 8 (b) 2, 6

(c) 3, 4 (d) 5, 6

105. Which of the following is an incorrect statement?

(a) if an electrode acts as cathode with respect to S.H.E then standard reduction electrode potential is given a positive sign.

(b) If an electrode acts as anode w.r.t. S.H.E. then standard oxidation potential is given a negative sign.

(c) Metals with low reduction potential can displace metals with higher reduction potential from their salt solution.

(d) The concentration of HCl in S.H.E is 1M

106. Which of the following is correct about 'Electrochemical cell'?

(a) It converts chemical energy into electrical energy

(b) Anode is at – vepotential whereas cathode is at +ve potential

(c) Salt bridge is generally used

(d) All are correct

107. The following cell is set up $Cu_{(s)} | Cu^{2+}_{(aq)} || Ag^+_{(aq)} | Ag_{(s)}$. If the electrode potential are $E^\circ_{Cu^{2+}/Cu} = +0.34$ & $E^\circ_{Ag^+/Ag} = +0.8V$, What will be the emf of the cell?

(a) +1.14 V (b) +0.46 V

(c) – 0.46 V (d) –1.14 V

Answers

1. (a)	**2.** (b)	**3.** (d)	**4.** (b)	**5.** (a)	**6.** (d)	**7.** (d)	**8.** (d)	**9.** (a)	**10.** (a)
11. (a)	**12.** (c)	**13.** (a)	**14.** (d)	**15.** (c)	**16.** (a)	**17.** (b)	**18.** (b)	**19.** (d)	**20.** (b)
21. (b)	**22.** (c)	**23.** (a)	**24.** (b)	**25.** (d)	**26.** (c)	**27.** (b)	**28.** (c)	**29.** (d)	**30.** (b)
31. (a)	**32.** (d)	**33.** (d)	**34.** (c)	**35.** (d)	**36.** (c)	**37.** (a)	**38.** (c)	**39.** (b)	**40.** (a)
41. (a)	**42.** (a)	**43.** (d)	**44.** (d)	**45.** (c)	**46.** (d)	**47.** (d)	**48.** (d)	**49.** (b)	**50.** (b)
51. (a)	**52.** (d)	**53.** (b)	**54.** (c)	**55.** (a)	**56.** (a)	**57.** (c)	**58.** (b)	**59.** (a)	**60.** (d)
61. (b)	**62.** (d)	**63.** (c)	**64.** (d)	**65.** (c)	**66.** (c)	**67.** (a)	**68.** (c)	**69.** (c)	**70.** (c)
71. (a)	**72.** (c)	**73.** (a)	**74.** (d)	**75.** (d)	**76.** (a)	**77.** (b)	**78.** (a)	**79.** (c)	**80.** (c)
81. (b)	**82.** (a)	**83.** (b)	**84.** (b)	**85.** (a)	**86.** (a)	**87.** (c)	**88.** (c)	**89.** (d)	**90.** (d)
91. (c)	**92.** (b)	**93.** (d)	**94.** (b)	**95.** (b)	**96.** (c)	**97.** (a)	**98.** (b)	**99.** (b)	**100.** (b)
101. (d)	**102.** (d)	**103.** (d)	**104.** (a)	**105.** (b)	**106.** (d)	**107.** (b)			

Explanations

1. (a)

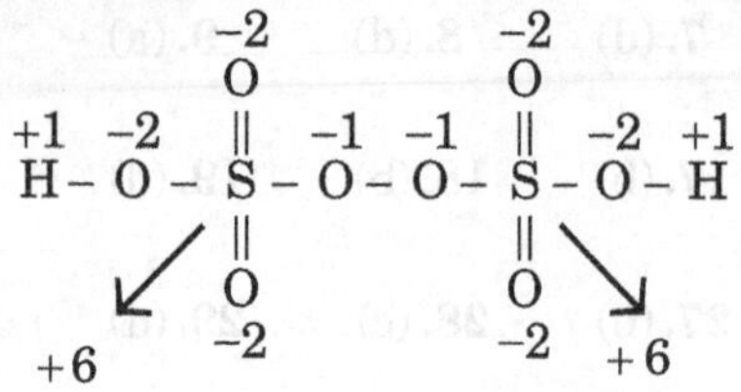

It contains only 1 peroxy linkage.

2. (b) NH_4^+ NO_3^-

$\downarrow$ $\downarrow$

-3 $+5$

4. (b) In NH_3

Let oxidation state of N = x

$x + (+3) = 0$

$x + 3 = 0$

$x = -3$

$PbSO_4$

$Pb^{2+}(SO_4)^{2-}$

Oxidation state of Pb = +2

Sb_2O_3

Let oxidation state of Sb = x

So, $2x + (-6) = 0$

($\because$ oxidation state of O = −2)

$2x - 6 = 0$

$x = +3$

HNO_3

Let oxidation on state of N = x

$+1 + x + (-6) = 0$ ($\because$ oxidation state of H = +1 & oxidation state of O = −2)

$x - 5 = 0$

$x = +5$

5. (a) CrO_5 has 2 peroxy linkages

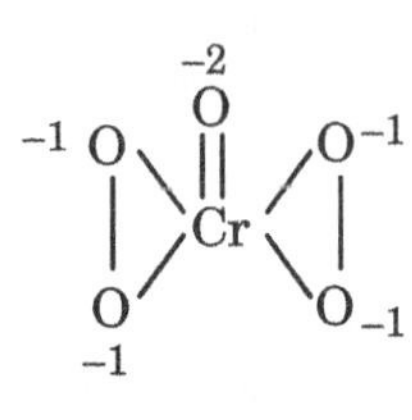

Let oxidation. State of Cr = x

$x + (-4) + (-2) = 0$

$x - 6 = 0$

$x = +6$

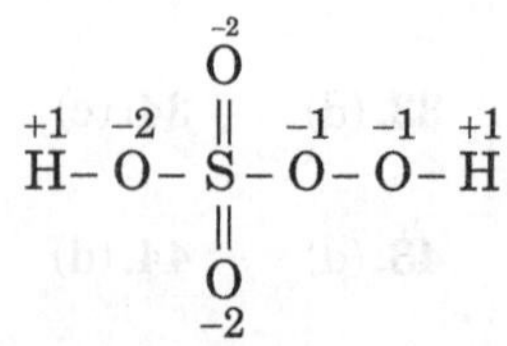

Let oxidation state of S = x

$x + (+1) + (+1) + (-1) + (-1) + (-2) + (-2) + (-2) = 0$

$x + 2 + (-2) + (-6) = 0$

$x + 2 - 8 = 0$

$x - 6 = 0$

$x = +6$

6. (d) All behaviours of H_2SO_4 are correct.

7. (d) $\overset{(+2)(+4)(-2)}{CaCO_3} \longrightarrow \overset{(+2)(-2)}{CaO} + \overset{(+4)(-2)}{CO_2}$

8. (d) $H_2S_2O_8$ is called Marshall's acid. It has one peroxy linkage.

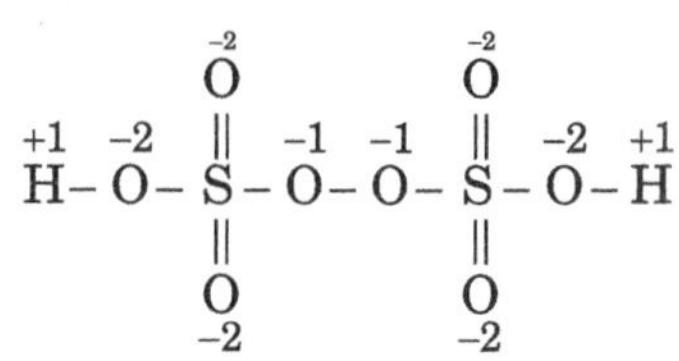

Let oxidation state of S = x

So, $2x + (-12) + 2 + (-2) = 0$

$2x - 12 = 0$

$2x = +12$

$x = +6$

In SO_4^{2-}

Let Oxdn. state of S = x

$x + (-8) = -2$

$x = -2 + 8 = +6$

In SO_3

Let Oxdn. State of S = x

$x + (-6) = 0$

$x - 6 = 0$

$x = +6$

9. (a) $\overset{+5}{N}_2O_5 > \overset{+4}{N}O_2 > \overset{+3}{N}_2O_3 > \overset{+2}{N}O > \overset{-3}{N}H_3$

11. (a) $Na_2S_2O_3$

Na is an alkali metal and has oxidation state = +1

So, $+2 + 2x(-6) = 0$

$2x - 4 = 0$

$\Rightarrow 2x = +4$

$\Rightarrow x = +2$

12. (c) Let oxidation state of S be x

$Na_2S_4O_6$

$(+1) + (a) + 4x + (-12) = 0$

$4x - 10 = 0$

$4x = +10$

$x = +2.5$

13. (a) NaN_3

Let oxdn. state of N = x

$+1 + 3x = 0$

$3x = -1$

$x = -\frac{1}{3}$

N_2H_2

Let oxdn. state of N = x

$2x + (2) = 0$

$2x = -2$

$x = -1$

NO

Let oxdn. state of N = x

$x + (-2) = 0$

$x - 2 = 0$

$x = +2$

N_2O_5

Let oxdn. state of N = x

$2x + (-10) = 0$

$2x - 10 = 0$

$2x = +10$

$x = +5$

14. (d) $Fe^{3+}[Fe(CN)_6]^{4-}$ Charge on CN = −1

$\downarrow \quad \downarrow$

$\underline{+3} \quad \underline{+2}$

15. (c) $\underset{\underline{N}H_4^+}{(-3)}$ $\underset{\underline{N}_2H_4}{(-2)}$ $\underset{\underline{N}H_2OH}{(-1)}$ $\underset{\underline{N}_2O}{(+1)}$

16. (a) Pb_3O_4 is a mixed oxide & it exists as $PbO_2.2PbO$

18. (b) $\underset{\underline{N}H_4^+}{(-3)}$ $\underset{\underline{N}O_3^-}{-5}$

19. (d) In isolated atoms & alloys, the oxidation state is taken to be zero.

20. (b) Cl_2O_7 — Cl = +7 oxidation state; ClF_3 — Cl = +3

Cl2 — Cl = 0; ICl_3 — Cl = −1

21. (b)

(a) $\overset{0}{N_2} + \overset{0}{3H_2} \longrightarrow \overset{(-3)(+1)}{2NH_3}$ → Oxdn. states

$\downarrow$ Reducing agent

(b) $\overset{0}{Ca} + \overset{0}{H_2} \longrightarrow \overset{(+2)(-1)}{CaH_2}$

$\downarrow$ Oxidising agent

(c) $\overset{0}{Br_2} + \overset{0}{H_2} \longrightarrow \overset{(+1)(-1)}{2HBr}$

$\downarrow$ Reducing agent

(d) $\overset{0}{8H_2} + \overset{0}{S_8} \longrightarrow \overset{(+1)(-2)}{8H_2S}$

$\downarrow$ Reducing agent

22. (c) Compare (i.e. balance) overall charge.

$$\underbrace{S_8 + 12\overline{O}H}_{-12\text{ charge on LHS}} \longrightarrow \underbrace{4S^{2-} + 2S_2O_3^{2-}}_{-12\text{ charge on RHS}} + 6H_2O$$

23. (a) In CrBr, Cr is in +3 odxn. state.

(1) $Cr^{3+} \longrightarrow CrO_4^{2-} + 3e^-$

(Cr = +6)

(2) $3Br^- \longrightarrow 3BrO_4^- + 24e^-$

(Br = +7)

Adding 1 & 2

$Cr^{3+} + 3Br^- \longrightarrow CrO_4^{2-} + 3BrO_4^- + 27e^-$

or $CrBr_3 \longrightarrow CrO_4^{2-} + 3BrO_4^- + 27e^-$

24. (b) In mole of $Ba(MnO_4)_2$ will give 2 moles of MnO_4^-.

$10e^- + 2MnO_4^- \longrightarrow 2Mn^{2+}$

(Mn = +7)

So, n factor for Ba $(MnO_4)_2$ = 10 &

Eq. wt. $= \frac{M}{10}$

(n factor for redox reactions is equal to electrons gained or lost per mole)

25. (d)

reduction ↓

$2H_2\overset{-1}{O_2} \longrightarrow 2H_2\overset{-2}{O} + \overset{0}{O_2}$

Oxidation ↑

reduction ↓

$\overset{0}{P_4} + 3\bar{O}H + 3H_2O \longrightarrow \overset{-3}{P}H_3 + 3H_2\overset{+1}{P}O_2^-$

Oxidation ↑

reduction ↓

$6\overset{0}{Cl_2} + 12\bar{O}H \longrightarrow 10\overset{-1}{Cl^-} + 2\overset{+5}{Cl}O_3^- + 6H_2O$

reduction ↑

26. (c) Conproportionation reaction is the reverse of disproportionation reaction.

27. (b) $a\,\bar{C}{\equiv}N + bNO_3^- + CH^+ \longrightarrow (a+b)NO + aCO_2 + \frac{C}{2}H_2O$

In $\bar{C} \equiv N$ oxdn. State of N = –3 and that of C = +2	In CO_2, oxdn. state of C = +4
So, $\overset{+2}{\bar{C}} \equiv \overset{-3}{N} \longrightarrow \overset{+2}{N}O + \overset{+4}{C}O_2$ ↓ Total of $7e^-$ are lost per mole	In NO, oxdn. state of N = +2

So, n factor of $\bar{C}N$ here = 7

$\overset{+5}{NO_3^-} \longrightarrow \overset{+2}{NO}$

'N' goes from + 5 to + 2 oxidation state.

Number of moles of electron gained per mol = 3

So, n factor of NO_3^- = 3

Now, criss – cross the n factors

$\bar{C}N + NO_3^- \longrightarrow NO + CO_2$

↓ ↓

n factor = 7 n factor = 3

So, $3\bar{C}N + 7NO_3^- \longrightarrow NO + CO_2$

Now, balance all atoms except O by merely looking at equation.

28. (c) For compd. $A_3(BC_4)_2$, the sum of oxidation no. is zero.

$A_3(BC_4)_2$

= +6 +2*x* + 5 + 8*x* – 2

= +6 + 10 – 16

= 16 – 16

= 0

29. (d) In HNO_2, in oxidation state of N is +3. N can Show oxidation state ranging from –3 to +5. Now since N is in +3, it can itself get either reduced or oxidized. So, can act both Oxidizing agent and Reducing agent.

30. (b)

$$\overset{+2}{\underline{Mn}}SO_4 \longrightarrow \overset{+4}{\underline{Mn}}O_2$$

n factor = 2

$$\text{Eq. wt of } MnSO_4 = \frac{\text{Mol.wt}}{\text{n factor}} = \frac{M}{2} = \frac{1}{2} \times M$$

31. (a)

$$\overset{0}{N_2} + 3H_2 \longrightarrow 2\overset{-3}{\underline{N}}H_3$$

n factor = 6

$(\text{wrt. } 1N, e^- \text{ gained} = 3$

$\text{wrt. } 2N, e^- \text{ gained} = 6)$

$$\underset{(E_2)}{\text{Eq. wt. of } N_2} = \frac{28}{6}$$

$$\overset{0}{N_2} + 3\overset{0}{H_2} \longrightarrow 2\overset{-3}{\underline{N}}\overset{+1}{H_3}$$

For NH_3, e^- lost wrt one N = 3

e^- lost wrt one N = 6

e^- gained wrt one H = 1

e^- gained wrt 6 H = 6

∴ No. of e^- gained or lost wrt 2 mole NH_3 = 6

No. of e^- gained or lost wrt 1 mole

$$NH_3 = \frac{6}{2} = 3$$

∴ n factor for NH_3 = 3

$$\underset{(E_1)}{\text{Eq. wt. of } NH_3} = \frac{17}{3}$$

$$\text{Now, } E_1 - E_2 = \frac{17}{3} - \frac{28}{6} = \frac{34-28}{6} = \frac{6}{6} = 1$$

32. (d)

$$\overset{0}{\underline{N}_2} \longrightarrow \overset{-3}{\underline{N}}H_3$$

n – factor for N_2 = 6

$$\text{Eq. wt } \frac{\text{Mol.wt}}{\text{n factor}} = \frac{28}{6} = 4.67$$

34. (c) In reaction (c), CO is the reducing agent.

35. (d) (a) $2\overset{+1}{K}\overset{+5}{Cl}\overset{-2}{O_3} \longrightarrow 2\overset{+1}{K}\overset{-1}{Cl} + 3\overset{0}{O_2}$

(b) $\overset{-4}{C}\overset{+1}{H_4} + 2\overset{0}{O_2} \longrightarrow \overset{+4}{C}\overset{-2}{O_2} + 2\overset{+1}{H_2}\overset{-2}{O}$

(c) $2\overset{+1}{Na}\overset{-1}{H} \longrightarrow 2\overset{0}{Na} + \overset{0}{H_2}$

(d) $\overset{+1}{K}\overset{-2}{O}\overset{+1}{H} + \overset{+1}{H}\overset{+5}{N}\overset{-2}{O_3} \longrightarrow \overset{+1}{K}\overset{+5}{N}\overset{-2}{O_3} + \overset{+1}{H_2}\overset{-2}{O}$

In option 4, there is no change in the oxidation state of any element. So, neither oxidation nor reduction occurs. Therefore, it is not a redox.

36. (c) $2Cu_2O + Cu_2S \longrightarrow 6Cu + SO_2 \uparrow$

In both Cu_2O & Cu_2S, Cu is in +1 oxidation state

In Cu_2S, S is in –2 oxidation state.

(1) $6e^- + 6Cu^+ \longrightarrow Cu^0$

(2) $S^{-2} \longrightarrow \overset{+4}{\underline{S}}O_2 + 6e^-$

So, total e^- change in the reaction is 6.

37. (a) Oxidation state of Fe in $Fe_3O_4 = +\frac{8}{3}$

$$\frac{8e^-}{3} + Fe^{+\frac{8}{3}} \longrightarrow Fe$$

Since 9 Fe are involved, So multiply the equation by 9.

$$24e^- + 9Fe^{+\frac{8}{3}} \longrightarrow 9Fe \qquad \text{....(a)}$$

Also, $Al \longrightarrow Al_2O_3$

$$2\overset{0}{Al} \longrightarrow 2Al^{3+} + 6e^-$$

Multiplying by 4

$$8M \longrightarrow 8Al^{3+} + 24e^- \qquad \text{....(b)}$$

∴ from(a) & (b), we can see that electron change is 24.

38. (c) In $[Fe(CN)_6]^{4-}$, the oxidation state of

$Fe = n + (-6) = -4$

$\boxed{n = +2}$

Oxidation No. wrt Fe in $(Fe(CN)_6)^4 \equiv +2$

In $\overset{..\ominus}{C \equiv N}$, oxidation number of N = –3

Now let oxidation No. of C = n

$[Fe(CN)_6]^{4-} \; +2 + 6n + 6(-3) = -4$

$2 + 6n - 18 = -14$

$6n - 16 = -4$

$6n = +12$

$\boxed{n = +2}$

So,

$$\overset{+2\ +2\ -3\ 4-}{[Fe(CN)_6]} \longrightarrow \overset{}{Fe^{3+}} + \overset{+4}{CO_3^{2-}} + \overset{+5}{NO_3^-}$$

e^- lostwrt Fe = 1

e^- lostwrt C = 2 × 6 = 12

e^- lostwrt N = 8 × 6 = 48

S, total e^- lost per mole

$[Fe(CN)_6]^{4-} = 1 + 12 + 48 = 61$

Now, Eq. wt. $= \dfrac{M}{61}$

39. (b)

$$\overset{+4}{SO_2} + \overset{-2}{H_2S} \longrightarrow \overset{0}{S} + H_2O$$

SO_2 ↓ Oxidising agent

Equivalent weight of $SO_2 = \dfrac{64}{4} = 16$

$\Rightarrow 2 \times 16 = 32$

40. (a)

$$\overset{+3}{Cr^{3+}} \longrightarrow \overset{+6}{CrO_4^{2-}} + 3e^- \quad \text{———} \; (1)$$

$$\overset{-1}{Br^-} \longrightarrow \overset{+6}{BrO_4^-} + 8e^- \times 3$$

$$\therefore \; 3Br^- \longrightarrow 3BrO_4^- + 24e^- \quad \text{———} \; 2$$

From eqn. (a) & (b),

Total electron change = 24 + 3 = $27e^-$

42. (a) Br_2 is a stronger oxidant than iodine & hence oxidises $S_2O_3^{2-}$ to SO_4^{2-} where S goes from +2 to +6 oxidation state.

47. (d) In all the three reactions, SO_4^{2-} is the spectator ion.

48. (d) In 4th option, I^- in the HI gets oxidized to I_2 & S in H_2SO_4 gets reduced to SO_2 (+ 6 to + 4 oxidation state).

49. (b) Mg goes from 0 to +2 oxidation state.

N goes from 0 to –3 oxidation state.

$3Mg \longrightarrow 3Mg^{2+} + 6e^-$

$N_2 + 6e^- \longrightarrow 2N^{3-}$

So, electron change is 6.

50. (b) (i)

Oxidation, Reduction

$$\overset{-3}{(NH_4)_2}\,\overset{+6}{Cr_2O_7} \longrightarrow \overset{0}{N_2} + \overset{+3}{Cr_2O_3} + 4H_2O$$

Here, oxidation & reduction occurs from within the same molecule. So, it is intermolecular redox.

(ii)

reduction, oxidation

$$\overset{+4}{PbO_2} + \overset{-2}{H_2O} \longrightarrow \overset{+2}{PbO} + \overset{-1}{H_2O_2}$$

It is an example of intermolecular redox.

(iii)

reduction, oxidation

$$\overset{+3}{Cr_2O_3} + \overset{0}{2Al} \longrightarrow \overset{+3}{Al_2O_3} + \overset{0}{2Cr}$$

It is an intermolecular redox & a metal displacement reaction also.

(iv) Oxidation: $\overset{0}{Cl_2} \rightarrow \overset{+5}{ClO_3^-}$; reduction: $Cl_2 \rightarrow Cl^-$

$$Cl_2 + 2OH \longrightarrow ClO_3^- + Cl^- + H_2O$$

It is a disproportionation reaction where same element undergoes reduction & oxidation simultaneously

51. (a) O in H_2O_2 goes from – 1 oxidation state to –2 in H_2O.

Also, S in H_2S gets oxidized from –2 to 0 in S.

52. (d) $$\overset{+2}{Ca}\,\overset{+4}{C}\,\overset{-2}{O_3} \longrightarrow \overset{+2}{Ca}\overset{-2}{O} + \overset{+4}{C}\overset{-2}{O_2}$$

None of the elements undergo reduction or oxidation.

53. (b) $$\overset{0}{Fe} + H_2O \longrightarrow \overset{+8/3}{Fe_3O_4} + H_2$$

$\downarrow$

$$\text{n factor} = \frac{8}{3}$$

$$\text{Eq.wt} = \frac{\text{At. wt}}{\text{n factor}} = \frac{56}{\frac{8}{3}} = \frac{56 \times 3}{8} = 21$$

54. (c) $CaCO_3 \longrightarrow CaO + CO_2$

It is not a redox reaction.

55. (a) Oxdn. state of Fe in $Fe_3O_4 = +\frac{8}{3}$

$$9\overset{+8/3}{Fe} \longrightarrow 9Fe$$

$$\text{Total charge carried} = +24\left(9 \times \frac{8}{3}\right)$$

$$9Fe + 24e^- \longrightarrow 9Fe$$

$$\overset{\circ}{Al} \longrightarrow Al^{+3} \ (\text{in } Al_2O_3)$$

For 8Al,

$$8Al \longrightarrow 8Al^{3+} + 24e^-$$

56. (a) $Cl_2 + \overline{O}H \longrightarrow Cl^- + H_2O + \text{oxidized product}$

eq. wt. in disproportionation

$$r \times n. = \frac{\text{M.wt}}{\text{n factor}_1} + \frac{\text{M.wt}}{\text{n factor}_2}$$

$$47.33 = \frac{71}{2} + \frac{71}{\text{n factor}}$$

$$47.33 = 35.5 + \frac{71}{\text{n factor}}$$

$$11.83 = \frac{71}{\text{n factor}}$$

$$\text{n factor} = \frac{71}{11.83} = 6$$

Which is possible only when Cl_2 changes to ClO_2^-

57. (c)

$$\overset{x+}{M^{x+}} + \overset{+7}{MnO_4^-} \longrightarrow \overset{+5}{MO_3^-} + Mn^{2+} + \frac{1}{2}O_2 \uparrow$$

n factor = 5 – x n factor = 5

No. of gm equivalent of M^{x+} reacted = No. of gm equivalent of MnO_4^- reacted

no. of moles x n factor = no. of moles x n factor

(of M^{x+}) (of M^+) (of MnO_4^-) (of MnO_4^-)

$$1.67 \times (5 - x) = 1 \times 5$$

$$1.67 \times 5 - 1.67x = 5$$

$$8.35 - 1.67x = 5$$

$$8.35 - 5 = 1.67x$$

$$\frac{3.35}{1.67} = x = 2$$

58. (b)

$$\overset{+2}{S_2O_3^{2-}} + \overset{0}{I_2} \longrightarrow \overset{+2.5}{S_4O_6^{2-}} + 2\overset{-1}{I^-}$$

n factor = 1 (wrt 2 mole S in 1 mole $S_2O_3^{2-}$) n factor = 2

$$\text{So, Eq. wt. of } Na_2S_2O_3 = \frac{M_1}{1}$$

$$\& \text{ Eq. wt. of } I_2 = \frac{M_2}{2}$$

59. (a) $2KMnO_4 + 10FeSO_4 + 8H_2SO_4 \rightarrow$

$K_2SO_4 + 2MnSO_4 + 5Fe_2(SO_4)_3 + 8H_2O$

10 moles $FeSO_4$ requires 2 moles of $KMnO_4$

1 mole $FeSO_4$ of required $= \frac{1}{5}$ moles of $KMnO_4$

$6KMnO_4 + 10FeC_2O_4 + 24H_2SO_4$

$\rightarrow 3K_2SO_4 + 6MnSO_4 + 5Fe_2(SO_4)_3 + 20CO_2 + 24H_2O$

10 mole of FeC_2O_4 require = 6 mole of $KMnO_4$

1 mole of FeC_2O_4 requires $= \frac{6}{10} = \frac{3}{5}$

Total moles of $KMnO_4$ required $= \frac{1}{5} + \frac{3}{5} = \frac{4}{5}$

62. (d) Let molarity of $H_2SO_4 = M_1$ and $H_2C_2O_4 = M_2$

No. of milli equivalents of

(H_2SO_4 + $H_2C_2O_4$) = No. of milli equivalent g NaOH

$20(2M_1 + 2M_2) = 40 \times \frac{1}{10}$

$40(M_1 + M_2) = 4$

$M_1 + M_2 = 0.1$ ...(a)

No.of milli equivalents of $H_2C_2O_4$ = No.of milli equivalents $KMnO_4$

$20(2M_2) = 20 \times \frac{1}{10}$

$M_2 = \frac{1}{20} = 0.05M$ ....(b)

$M_1 = 0.1 - 0.05 = 0.05M$

64. (d) No. of milli equivalent of H_2SO_4 = No. ofmilli equivalents of hypo

$10 \times N = 10 \times 1$

$N = 1\ N$

Volume strength = 5.6 XN = 5.6 × 1 = 56

$N = \frac{W}{Eq.wt} \times 1$

$W = N \times Eq.wt = 1 \times 17 = 17 g/lit$

65. (c) No.of equivalents $KMnO_4 = 1 \times 5 = 5$

No.of equivalents of $H_2O_2 = 5$

No.of moles of $H_2O_2 = \frac{5}{2}$

66. (c)

Reduction: +5 → +2

$P + H\underline{N}O_3 \longrightarrow HPO_3 + \underline{N}O + H_2O$

↓

n factor = 3

∴ Equivalent weight of $HNO_3 = \frac{M}{2}$

67. (a) $MnO_4^- + FeC_2O_4 \longrightarrow Mn^{2+} + Fe^{3+} + CO_2 \uparrow$
(ferrous oxalate)

$MnO_4^- + Fe_2(C_2O_4)_3 \longrightarrow Mn^{2+} + Fe^{3+} + CO_2 \uparrow$
(ferric oxalate)

No. of gm. Equivalent of MnO_4^- reacted	= No. of gm equivalent of Ferrous oxalate + No. of gm eq. of ferric oxalate + No. of gm eq. of ferric oxalate

$x \times 5 = 0.5 \times 3 + 0.5 \times 6$

$x \times 5 = 1.5 + 3$

$x \times 5 = 4.5$

$\boxed{x = 0\cdot 9}$

68. (c) The resulting Sol. has $XHPO_4$ which has an acidic H, so 1 mole NaOH will solve the purpose.

69. (c) $6H^+ + 5I^- + IO_3^- \longrightarrow 3I_2 + 3H_2O$

(a) Considering opting.

$KI \nearrow 0\cdot5M$, $\searrow 200\,ml$,

which means 0.1moles of KI are picked.

According to Stoichiometry,

5 mole of KI, react with KIO_3 = 1 mole

0.1 mole of KI, react with

$KIO_3 = \frac{1}{5} \times 0\cdot1 = 0\cdot02$ mole

$\therefore$ option 1 is wrong.

considering option 2

$KI \nearrow 0\cdot5M$, $\searrow 100ml$,

which means 0.05 moles of KI are picked.

According to balanced chemical eqn.

5 moles of KI react with moles of H^+ = 6

0.05 moles of KI react with moles of

$H^+ = \frac{6}{5} \times 0\cdot05$ = 0.06 moles of H^+

So, moles of H_2SO_4 reqd. = 0.03

$\therefore$ option 2 is wrong.

Considering option, 3,

$KI \nearrow 0\cdot5M$, $\searrow$ $0\cdot5L$, which means $0\cdot25$ moles of KI react.

According to balanced chemical equation,

5 moles of KI liberate $I_2 = \frac{3}{5} \times 0\cdot25 = 0.15$

Hence, option 3rd is correct.

70. (c) $3Cr_2O_7^{2-} + Fe^{2+} \longrightarrow Fe^{3+} + Cr^{3+}$

No. of gm eq. of $Al_2(Cr_2O_7)_3$ reacted = No. of gm eq. of $FeSO_4$ reacted

$$0\cdot1 \times 18 \times V_{(L)} = 0\cdot018 \times \frac{10}{1000}$$

Molarity × n factor × $V_{(L)}$)

$$V_{(L)} = 0\cdot018 \times \frac{10}{1000} \times \frac{1}{0\cdot1} \times \frac{1}{18}$$

$$V_{(L)} = 0\cdot0001 = 0\cdot1 ml$$

71. (a) (a) $2\bar{O}Cl + 2\bar{O}H + Bi_2O_3 \longrightarrow 2BiO_3^- + Cl^- + H_2O$

$2\bar{O}Cl$ ↓ oxidant; Bi_2O_3 ↓ Reductant

(b) $2MnO + 5PbO_2 + 8H^+ \rightarrow 2MnO_4^- + 5Pb^{2+} + 4H_2O$

MnO ↓ Reductant; PbO_2 ↓ oxidant

(c) $4I^- + 6H^+ + TeO_3^{2-} \longrightarrow 2I_2 + Te + 3H_2O$

I^- ↓ Reductant; TeO_3^{2-} ↓ oxidant

(d) $10\bar{O}H + 6[Fe(CN)_6]^{3-} + Cr_2O_3 \longrightarrow 6[Fe(CN)_6]^{4-} + 2CrO_4^{2-} + 5H_2O$

$[Fe(CN)_6]^{3-}$ ↓ Oxidant; Cr_2O_3 ↓ Reductant

(e) $2NaOH + 2COCl_2 + 2H_2O + Na_2O_2 \longrightarrow 2Co(OH)_3$

$COCl_2$ ↓ Reductant; Na_2O_2 ↓ oxidant

72. (c) No. of gm equivalents of $KMnO_4 = N \times V_{(L)}$

$= M \times n \text{ factor} \times V_{(L)}$

$= 0.1 \times 5 \times \frac{20}{1000}$

$= 0.01$

73. (a) $KMnO_4 + H_2C_2O_4 \longrightarrow$

Or

$MnO_4^- + C_2O_4^{2-} \longrightarrow Mn^{2+} + CO_2\uparrow$

Oxalic acid Sol. used = 0.1 N, V = ?

$KMnO_4$ Sol. used = 250 g

But it is 8% by mass which means

100 gmSol. has $KMnO_4$ = 8g

250gm Sol. has $KMnO_4 = \frac{8}{100} \times 250$

$= \frac{200}{10} = 20g$

So, actual wt. of $KMnO_4$ participated in the reaction = 20 g

No. of gm eq. $KMnO_4$ reacted = NO. of gm eq. of $H_2C_2O_4$ reacted

$\frac{20}{158} \times 5 = 0 \cdot 1 \times V_{(L)}$ (no. of moles of × n factor)

$\frac{100}{158} \times \frac{1}{0 \cdot 1} = V_{(L)}$

$\frac{1000}{158} = V_{(L)} = 6 \cdot 32L$

74. (d)

$MnO_4^- \longrightarrow Mn^{2+}$	$Cr_2O_7^{2-} \longrightarrow Cr^{3+}$
No. of gm eq. of MnO_4^- no. of moles × n factor = 1 × 5 = 5	No. of gm. Eq. of $Cr_2O_7^{2-}$ = No. of moles × n factor = 1 × 6 = 6

Ratio = 5 : 6

75. (d) $8H^+ + MnO_4^- + 5Fe^{2+} \longrightarrow Mn^{2+} + 5Fe^{3+} + 4H_2O$

1 mole of MnO_4^- oxidizes Fe^{2+} = 5 mole

0.1 mole of MnO_4^- oxidizes Fe^{2+} = 5 × 0.1 = 0.5 moles

$24H^+ + 3MnO_4^- + 5FeC_2O_4 \longrightarrow 5Fe^{3+} + 10CO_2 + 3Mn^{2+} + 12H_2O$

3 moles MnO_4^- oxidise FeC_2O_4 = 5 mole

0.1 moles MnO_4^- oxidise $FeC_2O_4 = \frac{5}{3} \times 0.1 = 0.166$ mol

$16H^+ + 2MnO_4^- + 5C_2O_4^{2-} \longrightarrow 2Mn^{2+} + 10CO_2 + 8H_2O$

2 moles MnO_4^- oxidise $C_2O_4^{2-}$ = 5 moles

0.1 moles MnO_4^- oxidise $C_2O_4^{2-} = \frac{5}{2} \times 0.1 = 0.25$ moles

77. (b) $Fe_2(SO_4)_3$ will not react with $K_2Cr_2O_7$

∵ neither Fe^{3+} ion nor SO_4^{2-} can be oxidized.

From Mohr's salt, only ferrous ions (Fe^{2+}) will get oxidized.

$$Cr_2O_7^{2-} + Fe^{2+} \longrightarrow Fe^{3+} + Cr^{3+}$$

(from Mohr'salt)

↓

Let x moles of Mohr's salt react

n factor = 6

No. of gm eq. of $K_2Cr_2O_7$ reacted = No. of gm eq. of Mohar's salt reacted

$$1 \times 6 \times \frac{500}{1000} = x \times 1$$

(N × V) (No. of moles × n factor)

$$6 \times \frac{500}{1000} = n$$

$x = 3$

In a mixture of 6 moles, moles of Mohr's salt = 3

Moles of $K_2Cr_2O_7$ = 3

Ratio of moles = 3 : 3 = 1 : 1

78. (a) Since, itsequimolar,

No. of moles of ferrous oxalate = 0.5

No. of moles of ferric oxalate = 0.5

$$\overset{2+}{Fe}C_2O_4^{2-} + MnO_4^- \longrightarrow Mn^{2+} + Fe^{3+} + C_2O_4^{2-}$$

(ferrous oxalate)

No. of gm equivalents of FeC_2O_4 reacted = No. of gm eq. of KMnO4 reacted

$0.5 \times 3 = x \times 5$

(no. of moles . n factor) (no. of moles . n factor)

$$x = \frac{0.5 \times 3}{5} = 0.3$$

So, $KMnO_4$ reqd. here = 0.3 Moles

$$Fe_2(C_2O_4)_3 + MnO_4^- \longrightarrow CO_2 + Mn^{2+} + Fe^{3+}$$

Only oxalate ions will get oxidized to CO_2.

No. of gm eq. of $Fe_2(C_2O_4)_3$ reacted = No. of gm eq. of MnO_4^- reacted

$0.5 \times 6 = x \times 5$ (No. of moles × n factor) So, $KMnO_4$ required here = 0.6 Moles Total moles of $KMnO_4$ required = 0.3 + 0.6 = 0.9

79. (c) In acidic medium

$$Fe^{2+} + MnO_4^- \longrightarrow Mn^{2+} + Fe^{3+}$$

(Soln. x) ↓

V=25ml V=20ml

No. of gm eq. of Fe^{2+} reacted = No. of gm eq. of MnO_4^- reacted

$$N_1V_1 = N_2V_2$$

$$M_1 \times 1 \times \frac{25}{1000} = M_2 \times 5 \times \frac{20}{1000} (N = M \times n \text{ factor})$$

$$M_1 \times 25 = M_2 \times 100$$

$$M_1 = 4M_2$$

In basic medium

$$Fe^{2+} + MnO_4^- \longrightarrow Fe^{3+} + MnO_2$$

(soln.x)

$$N_1V_1 = N_2V_2$$

$$M_1 \times 1 \times \frac{25}{1000} = M_2\ 3\ V_2$$

Now, $M_1 = 4M_2$ – from previous equation

$$4M_2 \times \frac{25}{1000} = M_2\ 3\ V_2$$

$$\frac{1}{10} = 3\ V_2$$

$$\frac{1}{30} L = V_2 = 33.3\,ml$$

80. (c) Strongest oxidizing agent will have the highest reduction potential.

81. (b) $I_2 \longrightarrow I^- + I^{n+}$

(1) $10e^- + 5I_2 \longrightarrow 10I^-$

(2) $I_2 \longrightarrow 2I^{n+} + 2x$

Since, change = 10 moles of electrons

In eqn. 1, 10 moles of e⁻ are being gained

From eqn. 2, 10 moles of e⁻ must be lost.

So, $2x = 10$

$x = +5$

82. (a) $MnO_4^- + I^- \longrightarrow I_2 + Mn^{2+}$ ______(1)

$I_2 + S_2O_3^{2-} \longrightarrow S_4O_6^{2-} + I^-$ ______(2)
(from hypo)

For I_2 For hypo

No. of moles × n factor = $N \times V_{(L)}$

$$x \times 2 = 0.2 \times 1 \times \frac{100}{1000}$$

$$x = \frac{0.1 \times 100}{1000} = 0.01$$

These 0.01 moles of I_2 came from reaction 1 –

For eqn. 1–

No. of eq. of KI = NO. of eq. of I_2

no. of moles × 1 = 0.01 × 2

No. of moles of pure KI = 0.02

$$\text{Wt. of pure KI} = 166 \times \frac{0.2}{100} = \frac{332}{100} = 3.32\text{g}$$

Percentage purity of

$$\text{Sample} = \frac{3.32}{10} \times 100 = 33.2\%$$

83. (b) $K_2Cr_2O_7 + Cu_3P \longrightarrow CuO + H_3PO_4 + Cr^{3+}$

No. of gm eq. of $K_2Cr_2O_7$ reacted = No. of gm eq. Cu_3P reacted

$x \times 6 = 1 \times 11$

(no. of moles × n factor)

$$x = \frac{11}{6}$$

84. (b) (a) $Cr_2O_7^{2-}$ as a oxidizing agent in acidic medium.

$$14H^+ + 6e^- + Cr_2O_7^{2-} \longrightarrow 2Cr^{3+} + 7H_2O$$

n factor for $Cr_2O_7^{2-} = 6$

(b) 1 mole $Fe_2(C_2O_4)_3$ will furnish 3 moles of $C_2O_4^{2-}$ which will be oxidized by $KMnO_4$. Fe^{3+} remains as such because it can't be oxidized.

$$3C_2O_4^{2-} + MnO_4^- \longrightarrow Mn^{2+} + CO_2$$

(+3)   (+4) → Oxdn. state of C

$$3C_2O_4^{2-} \longrightarrow 6CO_2 + 6e^-$$

(from $Fe_2(C_2O_4)_3$)

∴ n factor for $Fe_2(C_2O_4)_3$ will be 6

(c) $Cr_2O_7^{2-} + H_2C_2O_4 \longrightarrow CO_2 + Cr^{3+}$

Or

(+3)   (+4)

$$Cr_2O_7^{2-} + C_2O_4^{2-} \longrightarrow CO_2 + Cr^{3+}$$

↓

n factor = 2

n factor for $H_2C_2O_4 = 2$

85. (a) $Br_2 \longrightarrow Br^- + BrO_3^-$

(1) $Br_2 + 2e^- \longrightarrow 2Br^-$

(2) $12\overline{O}H + Br_2 \longrightarrow 2BrO_3^- + 10e^- + 6H_2O$
(Br = +5)

Multiplying equation 1 by '5' –

$5Br_2 + 10e^- \longrightarrow 10Br^-$ ______(3)

$12\overline{O}H + Br_2 \longrightarrow 2BrO_3^- + 10e^- + 6H_2O$ ______(4)

$6Br_2 + 12\overline{O}H \longrightarrow 10Br^- + 2BrO_3^- + 6H_2O$

\# Total electron change is 10, as is evident from 3 & 4th equation.

\# There are 12 moles of $\overline{O}H$ on LHS.

\# Br_2 is acting as reductant in eqn. 4 – its n – factor is 10. So, eqvt.wt = $\frac{M}{10}$

\# Total 6 moles of Br_2 resulted in exchange of e⁻ = 10

1 moles of Br_2 resulted in exchange of e⁻ = $\frac{10}{6} = \frac{5}{3}$

n – factor is calculated per mole so, n –factor of Br_2 is $\frac{5}{3}$

86. (a) SinceCu^{2+} has highest reduction potential, it is most easily reduced. Now, zinc has the highest oxidation potential (= + 0.76 v) it is most easily oxidized.

87. (c) Ag^+ has greater reduction potential, so Ag^+ is reduced &Pb is oxidised.

88. (c)

$$\overset{0}{Zn} + \overset{+2}{Mg}Cl_2 \longrightarrow \overset{+2}{Zn}Cl_2 + \overset{0}{Mg}$$

For above reaction to occur, zinc should get oxidized i.e., oxidation potential of zinc should be greater than that of Mg. But this is false. In fact, the oxidation potential of Mg > oxidation.Potential of Zinc.

89. (d)

$$\overset{0}{M} + \overset{+2}{Cu}SO_4 \longrightarrow \overset{+2}{M}SO_4 + \overset{0}{Cu}$$

In order to displace copper, the oxidation potential of M should be greater than that of Cu.

Oxidation potential of Fe = + 0.44 V

Oxidation potential of Zn = + 0.76 V

Oxidation potential of Cu = + 0.34 V

Oxidation potential of Cr = + 0.74 V

90. (d) $Cu + Al(NO_3)_3 \longrightarrow$

Displacement reaction will not occur because standard oxidation potential of Al (0.166 V) is more than that of Cu (–0.34 V).

91. (c) $CuCl_2 + Ag \longrightarrow AgCl + Cu$

Standard redn. Potential $\left(E^\circ_{\frac{Cu^{2+}}{Cu}}\right)$ for

$Cu^{2+} = +0 \cdot 34V$

Standard redn. Potential $\left(E^\circ_{\frac{Ag^+}{Ag}}\right)$ for

$Ag^+ = +0 \cdot 80V$

So, Ag^+ will have a greater tendency to get reduced & not Cu^{2+}. So, above reaction 'WILL NOT OCCUR' & hence $CuCl_2$ can be stored in Ag vessel.

92. (b) $Au^{3+} + Br^- \longrightarrow Br_2 + Au$

Au^{3+} has a greater standard, redn. Potential then Br_2 so Au^{3+} will get reduced.

93. (d) $E^\circ_{cell} = E^\circ_{\underset{(RP)}{C}} - E^\circ_{\underset{(RP)}{A}}$

To generate maximum $E^\circ_{cell}, E^\circ_{\underset{(RP)}{C}}$ should be higher & $E^\circ_{\underset{(RP)}{A}}$ should be lesser.

Max. Std. R.P. is for Ag^+, so it should be cathode & Min. Std. R.P. is for Mg^{2+}, so it should be anode.

Hence, Mg should get oxidized & Ag^+ should get reduced. So, 4th option.

94. (b) Best reducing agent will have the maximum oxidation potential which is for Al (+ 1.66 V) followed by

$Fe^{2+}(-0 \cdot 77V)$ & $Br^-(-1.08\ V)$

95. (b) $Ag \mid Ag^+_{(1M)} \parallel Cu^{2+}_{(1M)} \mid Cu\ ;\ E^\circ = -0 \cdot 46V$

$E^\circ_{cell} = E^\circ_{C\,(RP)} - E^\circ_{A\,(RP)}$

$-0 \cdot 46 = E^\circ_{\left(\frac{Cu^{2+}}{Cu}\right)} - E^\circ_{\left(\frac{Ag^+}{Ag}\right)}$ ...(a)

Similarly, $1 \cdot 1 = E^\circ_{\left(\frac{Cu^{2+}}{Cu}\right)} - E^\circ_{\left(\frac{Zn^{2+}}{Zn}\right)}$ ....(b)

Subtracting (b) from (a)

$-0 \cdot 46 - 1 \cdot 1 = E^\circ_{\left(\frac{Ag^+}{Ag}\right)} + E^\circ_{\left(\frac{Zn^{2+}}{Zn}\right)}$

$-1 \cdot 56 = E^\circ_{\left(\frac{Ag^+}{Ag}\right)} + E^\circ_{\left(\frac{Zn^{2+}}{Zn}\right)}$

$$1.56 = E^\circ_{\left(\frac{Ag^+}{Ag}\right)} + E^\circ_{\left(\frac{Zn^{2+}}{Zn}\right)} \quad ...(c)$$

Also,

$$X = E^\circ_{\left(\frac{Ag^+}{Ag}\right)} - E^\circ_{\left(\frac{Zn^{2+}}{Zn}\right)} \quad ....(d)$$

From 3 & 4

$$\boxed{X = 1.56\ V}$$

98. (b) HCl used = 1M

pH = log (H+) = 0

99. (b) (1) $XO + C \longrightarrow X + CO\uparrow$

(2) $YO + C \longrightarrow Y + CO\uparrow$

(3) $ZO + C \longrightarrow \text{No Reaction}$

(4) $XO + Y \longrightarrow YO + X$

In 1, 'X' goes form +2 oxidation state in XO to 'Zero' in 'X'. Same stands valid for 'Y' in second reaction.

X^{2+} & Y^{2+} are ready to get reduced. This means Z is a more reactive metal which will be most easily oxidized.

In fourth reaction, 'Y' is getting oxidized means Y is more reactive metal than X.

Max. reactive metal is Z followed by Y & then X.

100. (b) $H_2SO_4 + H_2C_2O_4 + NaOH \longrightarrow$

No. of gm eq. of acid = No. of gm eq. of base reacted

$$N_{H_2SO_4} \times x + N_{H_2SO_4} \times \frac{(20-x)}{1000} = \frac{40}{1000} \times \frac{1}{10}$$

$$N_{H_2SO_4} \times x \text{ vol. of } H_2SO_4 + N_{H_2C_2O_4} \times (20-x)$$

$= 4$ .....(a)

$KMnO_4$ will be able to oxdise $H_2C_2O_4$ only.

$$KMnO_4 + H_2C_2O_4 \longrightarrow Mn^{2+} + CO_2$$

$$N_{H_2C_2O_4} \times V_{H_2C_2O_4} = \frac{1}{10} \times \frac{20}{1000}$$

$$N_{H_2C_2O_4} \times \frac{(20-x)}{1000} = \frac{20}{10 \times 1000}$$

$$N_{H_2C_2O_4} = \frac{2}{20-x}$$

Putting this value in eq. (a)

$N_{H_2SO_4} \times x$ vol. of

$$H_2SO_4 + \frac{2}{20-x} \times 20 - x = 4$$

$$N_{H_2SO_4} \times X_{\text{vol. of } H_2SO_4 = 2}$$

So, no. of milli eq. of $H_2SO_4 = 2$

No. of milli eq. of $H_2C_2O_4 = 2$

gm eq. of $H_2SO_4 = 2 \times 10^{-3}$

gm eq. of $H_2C_2O_4 = 2 \times 10^{-3}$

Moles of H_2SO_4 = mole of $H_2C_2O_4 = 10^{-3}$

$= 0.001$

$$\text{Molarity of } H_2SO_4 = \frac{0.001}{20} \times 1000 = 0.05$$

$$\text{Molarity of } H_2C_2O_4 = \frac{0.001}{20} \times 1000 = 0.05$$

101. (d) $C_6H_5NO_2$ is –

'N' here has oxidation state = +3

(Based upon electronegativity difference)

Oxidation state of 'N' in HNO_3 = +5

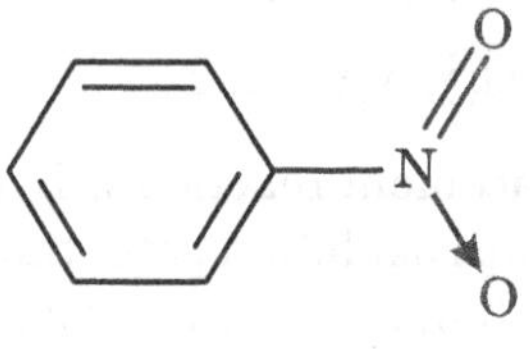

103. (d) In acidic acid $\overset{+1}{MnO_4^-} \longrightarrow Mn^{2+}$

n factor = 5

$$\overset{-3}{NH_3} \longrightarrow \overset{+2}{NO}$$

n factor = 5

104. (a) (1) $Mg \longrightarrow Mg^{2+} + 2e^-$

$2H^+ + 2e^- \longrightarrow H_2$

e^- change =2

$Zn \longrightarrow \overset{2+}{Zn} + 2e^-$ ....(i)

$8e^- + 2NO_3^- + 10H^+ \longrightarrow N_2O + 5H_2O$ ...(ii)

Multiply eqn. 1 by '4'

$4Zn \longrightarrow 4Zn^{2+} + 8e^-$

$8e^- + 2NO_3^- + 10H^+ \longrightarrow N_2O + 5H_2O$

Total e^- change = 8

107. (b) $\underset{(s)}{Cu} \longrightarrow \underset{(aq)}{Cu^{2+}} + 2e^-$

$\underset{(aq)}{Ag^+} + e^- \longrightarrow \underset{(s)}{Ag} \rightarrow$ Multiplying with 2

$\underset{(s)}{Cu} \longrightarrow \underset{(aq)}{Cu^{2+}} + 2e^-$

$2Ag^+_{(aq)} + 2e^- \longrightarrow 2Ag$

Adding

$2Ag^+_{(aq)} + Cu_{(s)} \longrightarrow Cu^{2+}_{(aq)} + 2Ag$

$\therefore E^\circ_{cell} = E^\circ_{C(RP)} - E^\circ_{A(RP)}$

$= (+0.80) - (+0.34)$

$= 0.80 - 0.34$

$= +0.46$ V

Smart Notes

10 CHAPTER

Hydrogen

1. When 1 mole of zinc reacts with 2 moles of NaOH, sodium zincate is formed, in which zinc exists in-
   (a) Anion part
   (b) Cationic part
   (c) Non-polar molecular part
   (d) None of these
2. Which of the following statement is correct about LiH?
   (a) Conducts electricity when melted.
   (b) Liberates H_2 at anode on electrolysis in molten condition
   (c) Produces H_2 when dissolved in water.
   (d) All are correct
3. Which one of the following metals evolves H_2 gas when it reacts with cold dilute HNO_3?
   (a) Mg (b) Al
   (c) Zn (d) Pt
4. The acid most suitable for preparing H_2 from Zn is -
   (a) Conc. H_2SO_4 (b) Conc. HCl
   (c) dil. H_2SO_4 (d) HNO_3
5. Select the correct statement regarding preparation of H_2.
   (a) Na reacts with cold water to evolve H_2.
   (b) Mg reacts with boiling water to evolve H_2.
   (c) Fe reacts with steam / hot water to liberate H_2.
   (d) All are correct.
6. How many of the following reactions can be used to prepare H_2?
   (1) $NaH + H_2O \longrightarrow$
   (2) $LiH + H_2O \longrightarrow$
   (3) $CaH_2 + H_2O \longrightarrow$
   (4) $Sn + NaOH \longrightarrow$
   (5) $Pb + NaOH \longrightarrow$
   (6) $Zn + NaOH \longrightarrow$
   (7) $Al + NaOH \longrightarrow$
   (8) $F_2 + H_2O \longrightarrow$
   (a) 5 (b) 3
   (c) 6 (d) 7

7. In the laboratory, for the preparation of dihydrogen gas from granular zinc, which one of the following can be used?

(a) Conc. H_2SO_4 (b) Conc. HCl

(c) Conc. HNO_3 (d) None of these

8. Lanes process involves the following reactions -

(1) $3Fe + 4H_2O \xrightarrow[1072K]{1023-} Fe_3O_4 + 4H_2 \uparrow$

(2) $Fe_3O_4 + 4H_2 \longrightarrow 3Fe + 4H_2O$

(3) $Fe_3O_4 + 4CO \longrightarrow 3Fe + 4CO_2 \uparrow$

Choose the correct statement -

(a) Reaction 1 involves oxidation of Fe.

(b) Reactions 2 & 3 involve reduction of Fe_3O_4.

(c) Reactions 2 & 3 are used to regenerate iron from Fe_3O_4.

(d) All are correct.

9. On passing steam through a mass of red hot coke or coal at 1273K, in the presence of Ni as catalyst, water gas ($CO + H_2$) is produced .This reaction is called -

(a) Bosch's process (b) Lanes process

(c) Serpeck's process (d) Mond's process

10. Select the correct statement (s).

(a) $CO + H_2$ (1 : 1mixture) is also called syn gas.

(b) To separate CO from water gas, it (water gas) is mixed with steam (at 673K) in the presence of $FeCrO_4$ as catalyst. It is called water gas shift reaction.

(c) CO is soluble in ammoniacal cuprous formate solution.

(d) All are correct.

11. Hydrogen can be prepared by the action of dil. H_2SO_4 on-

(a) Copper (b) Iron

(c) Lead (d) Mercury

12. Which of the following combinations will produce gas?

(a) Cu metal & conc. HNO_3

(b) Zn metal & NaOH (aq.)

(c) Au metal & NaCN(aq.)

(d) Fe metal with Conc. HNO_3

13. Pure hydrogen (99.9%) can be made by which of the following processes?

(a) Mixing natural hydrocarbons of high molecular mass

(b) electrolysis of water

(c) reaction of salt like hydrides with H_2O

(d) reaction of CH_4 with steam

14. An ionic hydride of an alkali metal has significant covalent character & is almost unreactive towards oxygen & chlorine. It is also used in the synthesis of other useful hydrides. The alkali metal is-

(a) Na (b) K

(c) Li (d) Be

15. Select the incorrect match.

Column-I	Column-II
(a) Synthesis gas	Used in synthesis of methanol
(b) Calgon	Has formula $Na_2[Na_4(PO_3)_6]$
(c) Heavy water	Obtained by prolonged electrolysis of water

(d) None of these

16. Select the correct reactions

(a) $CaH_2 + 2H_2O \longrightarrow Ca(OH)_2 + 2H_2$

(b) $2F_2 + 2H_2O \longrightarrow 4HF + O_2$

(c) $3F_2 + 3H_2O \longrightarrow 6HF + O_3$

(d) All are correct

17. Hydrogen is prepared in the laboratory, by treating granular zinc with dilute H_2SO_4. Which of the following statements are true?

(a) Dilute H_2SO_4 does not react with pure zinc.

(b) KOH solution removes the impurities of PH_3 & AsH_3 from evolved H_2.

(c) With pure zinc & dil. H_2SO_4, some $CuSO_4$ has to be added.

(d) Anhydrous $CaCl_2$ is used to absorb moisture from evolved H_2.

(a) b, c d but not a

(b) a, c, d but not b

(c) a & b but not c & d

(d) a & c but not b & d

18. $Y \underset{H_2,Pd}{\overset{O_2}{\rightleftharpoons}} X + H_2O_2$

In the above reaction, X & Y are respectively

(a) 2 - ethyl anthraquinone & 2 - ethyl anthraquinol

(b) Marshall's acid & Caro's acid

(c) Barium peroxide & Barium oxide

(d) None of these

19. Hydrolysis of 1 mole of peroxo disulphuric acid gives -

(a) 2 moles of H_2SO_4

(b) 2 moles of Caro's acid (H_2SO_5)

(c) 1 mole of H_2SO_4 & 1 mole of H_2SO_5

(d) None of these

20. A metal 'X' does not react with cold water but evolves highly inflammable gas with steam. Then, 'X' can be -

(a) Na (b) Fe

(c) Ni (d) Cr

21. In industrial preparation of H_2 gas from water gas $(CO + H_2)$, which of the following statement is correct?

(a) CO is oxidized to CO_2 with steam in the presence of catalyst followed by absorption of CO_2 in water under pressure.

(b) CO & H_2 are fractionally separated using differences in their densities.

(c) CO is removed by absorption in aq. Na_3AsO_3 solution

(d) H_2 is removed through occlusion with Pd.

Properties of Hydrogen & Hydrides

22. If x & y are the boiling point & dielectric constant of water respectively & a & b are that of D_2O, then

(a) $x > a, y > b$ (b) $x > a, y > a$

(c) $x = a, y = b$ (d) $x < a, y > b$

23. H_2 will not reduce which of the following oxide?

(a) Al_2O_3

(b) CaO

(c) FeO

(d) None of these

24. The various types of hydrides & examples of each type are given below:-

Hydride Type	Examples
A) Electron deficient	i) Na^+H^-
B) Saline & Ionic hydride	ii) CH_4
C) Electron precise	iii) NH_3
D) interstitial	iv) B_2H_6
E) Electron rich	v) CrH

Choose the correct match among the following:-

(a) (A)- ii, (B)-iv, (C)-v, (D)-iii, E-i

(b) (A)- iv, (B)-i, (C)-ii, (D)-v, E-iii

(c) (A)- iv, (B)-v, (C)-i, (D)-iii, E-ii

(d) None of these

25. Which statement is true about ionic hydride?

(i) they are binary compounds of hydrogen & metals

(ii) they are crystalline in nature.

(iii) they are generally volatile.

(iv) Common examples include CH_4, SiH_4 etc.

(a) ii, iii, iv but not i & iv

(b) i & iv but not ii & iii

(b) i & ii but not iii & iv

(d) iii & iv but not i & ii

26. Dihydrogen forms two different compounds with oxygen. One of the compounds, 'A' is most stable while the other compound 'B' is highly unstable & slowly decomposes to form 'A' along with a colorless gas 'C'. 'B' decolourise acidic solution of $KMnO_4$ whereas 'A' does not show any such reaction. A, B & C are, respectively-

(a) H_2O, O_2, HO_2

(b) H_2O_2, H_2O, H_3O_2

(c) H_2O, H_2O_2, O_2

(d) O_2, H_2O, H_2O_2

27. An element 'A' reacts with hydrogen to form a compound 'B' which on treatment with H_2O, liberates gas. Then, A & B are respectively -

(a) Nitrogen and Ammonia

(b) Chlorine and Hydrogen chloride

(c) Calcium and Hydrolith

(d) Cobalt and Cobalt Hydrate

28. Hydrogen forms compounds when it -

(a) Looses an electron

(b) Gains an electron

(c) Shares an electron

(d) Looses, gains or shares an electron

29. Which of the following hydride is polymeric?

(a) BeH_2 (b) CaH_2

(c) NaH (d) LiH

30. The most abundant element in the universe is -

(a) Hydrogen (b) Aluminium

(c) Carbon (d) Helium

31. Ionic Hydrides

(a) do not conduct electricity in solid state.

(b) are good reducing agent at high temperature

(c) give basic solution on reaction with H_2O.

(d) have low melting point.

The correct statements are-

(a) both a & b (b) both c & d

(c) a, b & c (d) All of the above

32. In which of the following reactions, does hydrogen act as an oxidizing agent?

(a) $H_2 + F_2 \longrightarrow$ (b) $H_2 + SiCl_4 \longrightarrow$

(c) $Na + H_2 \longrightarrow$ (d) $CuO + H_2 \longrightarrow$

33. The term hydride gap refers to which region of periodic table?

(a) Groups 6 to 9

(b) Groups 7 to 9

(b) Groups 7 to 10

(d) Groups 5 to 7

Compounds of Hydrogen, Hardness of water & Hydrogen peroxide

34. When zeolite, which is hydrated sodium aluminium silicate, is treated with hard water, the Na^+ ions are exchanged with-

(a) H^+

(b) Ca^{2+}

(c) Mg^{2+}

(d) Both 2 & 3

35. The dipole moment of H_2O_2 is 2.1 D while that of H_2O is 1.84 D. But, H_2O is a better solvent than H_2O_2 because-

(a) Its dipole moment is less.

(b) It is less corrosive.

(c) H_2O gets ionized during chemical reaction.

(d) H_2O_2 gets decomposed during chemical reactions.

36. H_2O_2 acts as oxidizing agent in-

(a) Neutral & acidic medium

(b) Acidic or alkaline medium

(c) Alkaline or neutral medium

(d) Acidic as well as alkaline medium.

37. $K_4[Fe(CN)_6] + X \xrightarrow{Y} K_3[Fe(CN)_6]$, then-

(a) X is O_3, Y is H_2O (b) X is H_2O, Y is O_3

(c) X is O_2, Y is O_3 (d) X is O_3, Y is H_2O

38. If, one assumes linear structure instead of bent structure for water, then which one of the following properties cannot be explained?

(a) The formation of intermolecular H bonds in water.

(b) High boiling point of water.

(c) Solubility of polar compounds in water.

(d) Ability of water to form co-ordinate covalent bonds.

39. The gases absorbed by alkaline pyrogallol & CCl_4 are, respectively-

(a) O_2, O_3 (b) O_3, O_2

(c) O_2, N_2 (d) O_3, N_2

40. In which of the following reactions, H_2O_2 acts an oxidizing agent?

(a) $IO_4^- + H_2O_2 \longrightarrow IO_3^- + H_2O + O_2$

(b) $2I^- + H_2O_2 + 2H^+ \longrightarrow I_2 + 2H_2O$

(c) $Ag_2O + H_2O_2 \longrightarrow 2Ag + H_2O + O_2$

(d) $2MnO_4^- + 6H^+ \longrightarrow 2Mn^{2+} + 8H_2O + 5O_2$

41. Which one of the following on oxidation gives H_2O_2?

(a) 2 - ethyl anthraquinol

(b) 3 - ethyl anthraquinone

(c) Anthracene

(d) 4 - ethylanthracene

42. Select the correct statement.

(a) K_2MnO_4 on reaction with O_3 gives $KMnO_4$.

(b) Volume strength of 10N H_2O_2 is 56.

(c) H_2O_2 is obtained by electrolysis of dil. H_2SO_4.

(d) All are correct

43. The concept of hybridization is applicable to -

(a) CH_4 (b) NaH

(c) BeH_2 (d) Both 1 & 3

44. H_2O_2 in its reaction with KIO_4 & NH_2OH respectively, is acting as-

(a) reducing agent, oxidizing agent

(b) oxidising agent, oxidsing agent

(c) oxidising agent, reducing agent

(d) reducing agent, reducing agent

45. The reagent(s) used for softening the temporary hardness of water is/are-

(a) $Ca(OH)_2$ (b) Na_2CO_3

(c) NaOCl (d) All of them

46. On reacting calgon with $MgSO_4$, the product obtained has chemical formula -

(a) $Na_2[Na_4(PO_3)_6]$

(b) $Na_2[Mg_2(PO_4)_6]$

(c) $Na_2[Na_4(PO_4)_6]$

(d) $Na_2[Mg_2(PO_3)_6]$

47. Which of the following disproportionate when treated with H_2O?

(a) SO_3 (b) F_2

(c) N_2 (d) Cl_2

48. The salt responsible for permanent hardness of H_2O is:-

(a) Na_2SO_4 (b) $Mg(HCO_3)_2$

(c) NaCl (d) $MgCl_2$

49. Which of the following compound is a peroxide?

(a) KO_2

(b) BaO_2

(c) MnO_2

(d) NO_2

50. Match the entry in column - I with its corresponding entry in column - II.

Column - I	Column - II
(a) Clark's process	a. Electron deficient hydride
(b) Permutit	b. $Ca(OH)_2$
(c) Calgon	c. $Na_2Al_2Si_2O_8.XH_2O$
(d) BH_3	d. $Na_2[Na_4(PO_3)_6]$

(a) 1 - b , 2 - a , 3 - d, 4 - c

(b) 1 - b , 2 - c , 3 - d, 4 - a

(c) 1 - b , 2 - d , 3 - c, 4 - a

(d) 1 - c , 2 - b , 3 - d, 4 - a

51. There are three samples of H_2O_2 Labeled as 10 volume, 15 volume & 20 volume. If equal volume 'V' of each sample are mixed & then oxidized by 0.1 M $KMnO_4$ in acidic medium then the volume of $KMnO_4$ used is approximately-

(a) 16 V (b) 20 V

(c) 30 V (d) 40 V

52. Identify the incorrect statement about 5.6 volume of H_2O_2 (volume strength = 5.6)

(a) It is 1.7% weight by volume $\left(\frac{g}{ml} \times 100\right)$

(b) Normality of solution is 1N.

(c) Molarity of solution is 0.5M.

(d) None of these

53. 20 ml of a sample of H_2O_2 gives 400 ml of O_2 measured at STP. The H_2O_2 sample should be labeled as -

(a) 5 vol. H_2O_2

(b) dil. H_2O_2

(c) anhydrous H_2O_2

(d) 20 vol. H_2O_2

54. The mass of hydrogen peroxide present in 2 litres of its 5 M solution is '*x*' and the mass of oxygen liberated by decomposition of its 200 ml solution is '*y*', then 'X' & 'Y' RESPECTIVELY, are-

(a) 340 g, 16 g (b) 34 g, 32 g

(c) 680 g, 32 gm (d) 220 g, 15 g

55. Which of the following reactions are 'HYDRATION REACTION'S?

(a) $MgO + H_2O \longrightarrow$

(b) $AlCl_3 + H_2O \longrightarrow$

(c) $Ca_3N_2 + H_2O \longrightarrow$

(d) None of these

56. Which of the following compounds can be oxidized & reduced by H_2O_2 in acidic & basic medium respectively?

(a) $[Fe(CN)_6]^{3-}$ & $[Fe(CN)_6]^{4-}$

(b) $[Fe(CN)_6]^{4-}$ & $[Fe(CN)_6]^{3-}$

(c) $KMnO_4$ & K_2MnO_4

(d) $Fe_2(SO_4)_3$ & $FeSO_4$

57. Heavy water is prepared by prolonged electrolysis of ordinary water. The principle is that H^+ is discharged more readily than D^+ because of some reasons. The reason which cannot be co - related to this observation is that -

(a) H^+ has higher mobility than D^+.

(b) Discharge potential of H^+ is lower than that of D^+.

(c) The reaction, $2H \longrightarrow H_2$ occurs faster than $2D \longrightarrow D_2$.

(d) Mass of D is more than H & so it lies in lower layer of electrolysis liquid.

58. What is the product formed when a compound of third element of group - 2 & second element of group - 15 is reacted with water?

(a) 1mole of phosphinic acid

(b) 2 moles of phosphoric acid

(c) 2 moles of phosphine

(d) 1mole of P_2O_5

59. Which of the following is responsible to unite two ice cubes when they are pressed over each other?

(a) Ion dipole forces (b) Van der waals forces

(c) Covalent bonding (d) Hydrogen bond

60. The solubility of ionic compound is maximum in -

(a) D_2O (b) CCl_4

(c) H_2O (d) HOD

61. What is incorrect about H_2O_2?

(a) It can act as both oxidising & reducing agent

(b) The two OH bonds lie in the same plane

(c) It is a pale blue liquid

(d) It can be oxidized by O_3

62. Which of the following is not true?

(a) Ordinary H_2O is electrolysed more rapidly than D_2O.

(b) D_2O freezes at lower temperature than H_2O.

(c) Reaction b/w H_2 & Cl_2 is faster than b/w D_2 & Cl_2.

(d) Bond dissociation energy for D_2 is greater than that for H_2.

63. An inorganic substance liberates O_2 on heating & turns an acidic solution of KI, brown & reduces $KMnO_4$ (acidified) solution. The substance is-

(a) H_2O_2 (b) $NaNO_3$

(c) $CaCO_3$ (d) $Pb(NO_3)_2$

64. Washing soda removes -

(a) Only permanent hardness of water

(b) Only temporary hardness of water

(c) Both temporary and permanent hardness of water

(d) Odour from water.

65. Which one of the following oxides do not give an acid on reaction with H_2O?

(a) CO_2 (b) Cl_2O_7

(c) N_2O_5 (d) Na_2O

66. Water gas is a mixture of

(a) CO and H_2O (b) O_2 and CO

(c) CO and N_2 (d) CO and H_2

67. When 1 mole of PbS reacts completely with H_2O_2

(a) H_2 is liberated

(b) O_2 is liberated

(c) 4 moles of H_2O_2 are consumed

(d) Sulphur is converted to sulphite

68. Select the correct option about H_2O_2

(a) H_2O_2 reduces MnO_4^- both in acidic & basic media.

(b) H_2O_2 oxidises Fe^{2+} both in acidic & basic media.

(c) H_2O_2 liberates I_2 from acidified KI solution & reduces I_2 to I^- ions in basic medium.

(d) All are correct

69. H_2O_2 is 5.6 volume, then -

(a) It is 1.7% wt. by volume.

(b) It is 1 N

(c) It is 5.6 M

(d) Both 1 & 2

70. Choose the correct match

List I	List II
(A) 10 volume	(P) 5.358 N
(B) 20 volume	(Q) 3.036 % (w/v)
(C) 30 volume	(R) 3.4 gm H_2O_2/100ml soln.
(D) 11.2 volume	(S) 1.785M

(a) A - Q, B - S, C - P, D - R

(b) A - R, B - P, C - S, D - Q

(c) A - Q, B - P, C - S, D - R

(d) None of these

71. Which of the following reactions depict the acidic nature of hydrogen peroxide?

(a) $H_2O_2 + 2NaOH \longrightarrow Na_2O_2 + 2H_2O$

(b) $H_2O_2 + Ba(OH)_2 \longrightarrow BaO_2 + 2H_2O$

(c) $H_2O_2 + Na_2CO_3 \longrightarrow Na_2O_2 + CO_2 + H_2O$

(d) All of these

72. Which of the following reactions depict the oxidizing nature of H_2O_2?

(a) $2K_4[Fe(CN)_6] + H_2SO_4 + H_2O_2 \longrightarrow 2K_3[Fe(CN)_6] + K_2SO_4 + 2H_2O$

(b) $HCHO + H_2O_2 \longrightarrow HCOOH + H_2O$

(c) $K_2Cr_2O_7 + H_2SO_4 + 4H_2O_2 \longrightarrow 2CrO_5 + K_2SO_4$

(d) All of these

73. Select the reaction with incorrect products

(a) $2K_3[Fe(CN)_6] + 2KOH + H_2O_2 \longrightarrow 2K_4[Fe(CN)_6] + 2H_2O + O_2$

(b) $KIO_4 + H_2O_2 \longrightarrow KIO_3 + H_2O + O_2$

(c) $NaNO_2 + H_2O_2 \longrightarrow HNO_3 + H_2$

(d) Both 1 & 2

74. In the reaction,

$H_2O_2 + O_3 \longrightarrow H_2O + 2O_2$, H_2O_2 acts as -

(a) An acid

(b) An oxidising agent

(c) A reducing agent

(d) None of these

75. Calgon is

(a) Na_2CO_3

(b) Na_2SO_4

(c) $Na_2Al_2Si_2O_8.XH_2O$

(d) $Na_2[Na_4(PO_3)_6]$

76. Which of the following statement is correct about synthetic resin method?

(a) The cation exchanger consists of granular insoluble organic acid resin having giant molecules with $-SO_3H$ or $-COOH$ groups.

(b) The anion exchanger contains giant organic molecules with basic groups derived from amines like $RNH_3^+\overline{O}H$.

(c) $\overline{O}H$ ions, produces during anion exchange, neutralize the H^+ ions produced during cation exchange.

(d) All are correct

Answers

1. (a)	**2.** (d)	**3.** (a)	**4.** (c)	**5.** (d)	**6.** (d)	**7.** (d)	**8.** (d)	**9.** (a)	**10.** (d)
11. (b)	**12.** (b)	**13.** (b)	**14.** (c)	**15.** (b)	**16.** (d)	**17.** (b)	**18.** (a)	**19.** (c)	**20.** (b)
21. (a)	**22.** (d)	**23.** (d)	**24.** (b)	**25.** (c)	**26.** (c)	**27.** (c)	**28.** (d)	**29.** (a)	**30.** (a)
31. (c)	**32.** (d)	**33.** (b)	**34.** (d)	**35.** (d)	**36.** (d)	**37.** (a)	**38.** (c)	**39.** (a)	**40.** (b)
41. (a)	**42.** (d)	**43.** (d)	**44.** (a)	**45.** (d)	**46.** (d)	**47.** (d)	**48.** (d)	**49.** (b)	**50.** (b)
51. (a)	**52.** (d)	**53.** (d)	**54.** (a)	**55.** (d)	**56.** (b)	**57.** (d)	**58.** (c)	**59.** (d)	**60.** (c)
61. (b)	**62.** (d)	**63.** (a)	**64.** (c)	**65.** (d)	**66.** (d)	**67.** (c)	**68.** (d)	**69.** (d)	**70.** (a)
71. (d)	**72.** (d)	**73.** (c)	**74.** (c)	**75.** (d)	**76.** (d)				

Explanations

1. (a) $Zn + 2NaOH \longrightarrow Na_2ZnO_2 + H_2O \uparrow$

$Na_2ZnO_2 \longrightarrow 2Na^+ + [ZnO_2]^-$

5. (d) $2Na(Hg) + 2H_2O \longrightarrow 2NaOH + H_2 \uparrow$

To minimize the rate of reaction, alkali metals are used in the form of amalgams.

$Mg + 2H_2O \longrightarrow Mg(OH)_2 + H_2 \uparrow$

$3Fe + 4H_2O \longrightarrow Fe_3O_4 + 4H_2 \uparrow$

6. (d) $NaH + H_2O \longrightarrow NaOH + H_2 \uparrow$

$LiH + H_2O \longrightarrow LiOH + H_2 \uparrow$

$CaH_2 + 2H_2O \longrightarrow Ca(OH)_2 + 2H_2 \uparrow$
(hydrolith)

$Sn + NaOH \longrightarrow Na_2SnO_3 + H_2 \uparrow$

$Pb + NaOH \longrightarrow Na_2PbO_3 + H_2 \uparrow$

$Zn + NaOH \longrightarrow Na_2ZnO_2 + H_2 \uparrow$

$Al + NaOH \longrightarrow NaAlO_2 + H_2 \uparrow$

$F_2 + H_2O \longrightarrow HF + O_2 \uparrow$

7. (d) For the preparation of H_2 gas from granular zinc, conc. H_2SO_4, conc. HCl & conc. HNO_3 cannot be used.

$\underset{\text{(conc.)}}{Zn + H_2SO_4} \longrightarrow ZnSO_4 + H_2 \uparrow$

A part of this conc. H_2SO_4 reacts with H_2 produced & oxidises it to water.

$H_2SO_4 + H_2 \longrightarrow 2H_2O + SO_2 \uparrow$
(conc.)

⇒When conc. HCl is used, H_2 liberated by the action of conc. HCl on granular zinc will be impure as it contains fumes of volatile HCl.

Also, $ZnCl_2$ formed is insoluble in conc. HCl.

It forms a coating on granular zinc & the reaction stops after some time.

$Zn + 2HCl \longrightarrow ZnCl_2 + H_2 \uparrow$

HNO_3 acts both as an acid & as an oxidizing agent. The nascent H first formed reduces the HNO_3 into oxides of nitrogen.

$2HNO_3 + 2H \longrightarrow 2NO_2 + 2H_2O$

$2HNO_3 + 6H \longrightarrow 2NO + 4H_2O$

Most suitable acid is dilute H_2SO_4.

10. (d) $\underbrace{H_2 + CO}_{\text{water gas}} + H_2O \xrightarrow[FeCrO_4]{673K} CO_2 + 2H_2 \uparrow$

This is called water gas shift reaction.

The mixture of CO_2 & H_2 is passed through water under a P of 25 - 30 atm when CO_2 dissolves in water.

11. (b) Cu & Hg have less oxidation potential than H_2, so cannot react with dil. H_2SO_4 to evolve H_2. Although Pb reacts (has O.P. = + 0 13 v) with H_2SO_4 & evolves H_2 but $PbSO_4$ formed during the reaction forms a protective layer around unreacted Pb & prevent further reaction.

12. (b) $Zn + 2NaOH \xrightarrow{\Delta} Na_2ZnO_2 + H_2$

14. (c) Li^+H^-

$Li^+ \rightarrow$ smaller the size of cation, greater is the covalent character.

$8LiH + Al_2Cl_6 \longrightarrow 2LiAlH_4 + 6LiCl$

21. (a) $\underbrace{CO_{(g)} + H_{2(g)}}_{\text{Water gas}} + \underset{\text{steam}}{H_2O_{(g)}} \xrightarrow[FeCrO_4]{673K} CO_{2(g)} + 2H_{2(g)} \uparrow$

This chemical reaction in which CO of the water gas (or syn gas) react with steam to form & more dihydrogen is called water gas shift reaction.

26. (c) Compound 'A' = H_2O

Compound 'B' = H_2O_2

$H_2O_2 \longrightarrow H_2O + \frac{1}{2}O_2$

(colourless gas)

$H_2O_2 + KMnO_4 \longrightarrow H_2O + Mn^{2+}$

27. (c) $Ca + H_2 \longrightarrow CaH_2$

$CaH_2 + 2H_2O \longrightarrow Ca(OH)_2 + 2H_2 \uparrow$

31. (c) Ionic hydrides have high melting points.

33. (b) The elements of 7, 8 and 9 group do not form hydrides. This is actually known as HYDRIDE GAP.

34. (d) $Na_2Al_2Si_2O_8 \cdot XH_2O$

$\xrightarrow{Ca^{2+}} CaAl_2Si_2O_8 \cdot XH_2O + 2Na^+$

$\xrightarrow{Mg^{2+}} MgAl_2Si_2O_8 \cdot XH_2O + 2Na^+$

37. (a) O_3 is an oxidizing agent.

38. (c) Had water been linear, its dipole moment would have been zero.

40. (b)

41. (a)

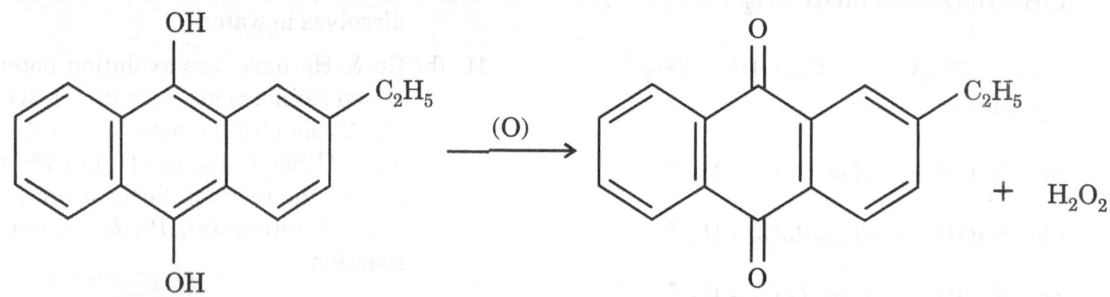

(2 - ethyl anthraquinone)

42. (d) Volume strength = $5.6 \times N$

$V.S. = 5.6 \times 10 = 56$

43. (d) NaH is an ionic hydride.

44. (a)

$+7 \leftarrow +5$

$KIO_4 + H_2O_2 \longrightarrow KIO_3 + H_2O + O_2$

(Reducing agent)

$-3 \leftarrow +5$

$\underline{N}H_4OH + 3H_2O_2 \longrightarrow H\underline{N}O_3 + 4H_2O$

(oxidising agent)

45. (d) Temporary hardness is due to bicarbonate of Ca & Mg.

$\rightarrow Ca(OH)_2$ is used to Clark's process.

$Ca(HCO_3)_2 + Ca(OH)_2 \longrightarrow 2CaCO_3 \uparrow + 2H_2O$

Na_2CO_3 process

$\rightarrow Ca(HCO_3)_2 + Na_2CO_3 \longrightarrow CaCO_3 \uparrow + 2NaHCO_3$

Also, $2\bar{O}Cl + 2H_2O \rightleftharpoons 2HOCl + 2\bar{O}H$

$Ca(HCO_3)_2 + 2\bar{O}H \longrightarrow CaCO_3 \uparrow + CO_3^{2-} + 2H_2O$

46. (d) $Na_2[Na_4(PO_3)_6] \xrightarrow[\text{(from hard water)}]{Mg^{2+}} Na_2[Mg_2(PO_3)_6] + 4Na^+$

47. (d) $Cl_2 + H_2O \longrightarrow HCl + HOCl$

51. (a) Volume Strength (V.S.) = $11.2 \times M$

$10 = 11.2 \times M$

$\frac{10}{11.2} = M$

$0.89M = M_1$

For 2nd Soln.

$V.S. = 11.2 \times M$

$\frac{15}{11.2} = M$

$1.34M = M_2$

For 3rd Soln.

$V.S. = 11.2 \times M$

$\frac{20}{11.2} = M$

$1.785M = M_3$

After mixing equal volumes 'V' of these solution,

$M_1V_1 + M_2V_2 + M_3V_3 = M_4V_4$

$0.89 \times V + 1.34 \times V + 1.785 \times V = M_4 \times 3_V$

$\frac{V(0.89 + 1.34 + 1.785)}{3V} = M_4$

$1.338M = M_4$

So, the final H_2O_2 solution obtained has an approximate molarity = 1.34 M

$MnO_4^- + H_2O_2 \longrightarrow O_2 + Mn^{2+}$

No. of gm eq. of MnO_4^- reacted = No. of gm eq. of H_2O_2 reacted

$\underset{(N\ V)}{0.1 \times 5 \times V} = \underset{(N\ V)}{1.34 \times 2 \times 3}$

$V = \frac{8.4}{0.5} = 16V$

52. (d) 5.6 volume H_2O_2 means 1ml of H_2O_2 liberates 5.6ml of O_2 at STP.

$H_2O_{2(l)} \longrightarrow H_2O_{(l)} + \frac{1}{2}O_{2(g)}$

So, 1ml H_2O_2 liberates O_2 = 5.6ml

100 ml H_2O_2 liberates $O_2 = 5.6 \times 100$

= 560 ml

Vol. of O_2 obtained at STP = 560 ml

No. of moles of O_2 obtained at STP

$= \frac{560}{22400} = 0.025$

So, mole of H_2O_2 that might have reacted

$= 0.025 \times 2 = 0.050$

So, 100 ml H_2O_2 had 0.050 moles of H_2O_2

Or

100 ml H_2O_2 had 1.7 gm (0.050×34) of H_2O_2.

So, H_2O_2 solution is 1.7% wt. by volume.

↱ Volume Strength $= 5.6 \times N$

$5.6 = 5.6 \times N$

$N = 1N$

↱ Volume Strength $= 11.2 \times M$

$\frac{5.6}{11.2} = M$

$0.5M = M$

53. (d) Volume strength refers to the volume of O_2 liberated by 1 vol. of H_2O_2.

So, 20 ml H_2O_2 liberates = 400 ml

1 ml H_2O_2 liberates $= O_2 = \frac{400}{20} = 20ml$

So, it is 20V H_2O_2.

54. (a) 1 L of 5M H_2O_2 contains

$H_2O_2 = 34 \times 5 = 170g$

2L will contain $H_2O_2 = 170 \times 2 = 340g$

200 ml of 5M solution will contain

$H_2O_2 = \frac{340}{2000} \times 200 = 34g$

$2H_2O_2 \longrightarrow 2H_2O + O_2$

68gm H_2O_2 gives O_2 = 32g

34gm H_2O_2 gives $O_2 = \frac{32}{68} \times 34 = 16g$

55. (d) All these are hydrolysis reactions.

$MgO + H_2O \longrightarrow Mg(OH)_2$

$$AlCl_3 + 3H_2O \longrightarrow Al(OH)_3 + 3HCl$$

$$Ca_3N_2 + 6H_2O \longrightarrow 3Ca(OH)_2 + 2NH_3$$

56. (b) (a) $2[Fe(CN)_6]^{4-} + H_2O_2 + 2H^+ \longrightarrow 2[Fe(CN)_6]^{3-} + 2H_2O$

(b) $2[Fe(CN)_6]^{3-} + 2\overline{O}H + H_2O_2 \longrightarrow 2[Fe(CN)_6]^{4-} + 2H_2O + O_2$

58. (c) $Ca_3P_2 + 6H_2O \longrightarrow 3Ca(OH)_2 + 2PH_3 \uparrow$

60. (c) H_2O has a very high dielectric constant.

61. (b) H_2O has open book like structure in which both the OH lie in different planes.

63. (a) $2H_2O_2 \xrightarrow{\Delta} 2H_2O + O_2$

$$2KI + H_2O_2 \xrightarrow{\text{Acidic}} 2KOH + \underset{\text{(brown)}}{I_2}$$

$$H_2O_2 + KMnO_4 \longrightarrow O_2 + Mn^{2+}$$

64. (c) It removes both temporary & permanent hardness.

$$CaCl_2 + Na_2CO_3 \longrightarrow CaCO_3 \downarrow + 2NaCl$$

$$CaSO_4 + Na_2CO_3 \longrightarrow CaCO_3 \downarrow + Na_2SO_4$$

$$Ca(HCO_3)_2 + Na_2CO_3 \longrightarrow CaCO_3 \downarrow + 2NaHCO_3$$

65. (d) $CO_2 + H_2O \longrightarrow H_2CO_3$

$$Cl_2O_7 + H_2O \longrightarrow 2HClO_4$$

$$N_2O_5 + H_2O \longrightarrow 2HNO_3$$

$$Na_2O + H_2O \longrightarrow 2NaOH$$

66. (d) Water gas is 1 : 1 mixture of CO and H_2.

As this mixture is used for the synthesis of methanol and a number of hydrocarbons, it is also known as synthesis gas.

67. (c) $PbS + 4H_2O_2 \longrightarrow PbSO_4 + 4H_2O$

69. (d) V.S. = $5.6 \times N$

71. (d) H_2O_2 is showing acidic nature in all three reactions.

73. (c) H_2O_2 oxidises nitrites to nitrates

74. (c) Here O_2 is an oxidizing agent where as H_2O_2 is a reducing agent.

75. (d) The complex salt of metaphosphoric acid, sodium hexanetaphosphate $(NaPO_3)_6$, is known as calgon. It is represented as $Na_2[Na_4(PO_3)_6]$.

76. (d) Cation exchanger

$$R-\overset{\overset{\displaystyle O}{\|}}{C}-O-H + Ca^{2+} \longrightarrow (RCOO)_2Ca + 2H^+$$

Anion exchanger

$$RNH_3^+\overline{O}H + Cl^- \longrightarrow RNH_3^+Cl^- + \overline{O}H$$

$$H^+ + \overline{O}H \longrightarrow H_2O$$

Smart Notes

Smart Notes

11

CHAPTER

s-Block Elements

1. Choose the correct option

   (a) Na exists in elemental form in earth's crust

   (a) Li is stored in paraffin wax.

   (c) Mg does not react with cold water.

   (d) $Na_2 CO_3$ is water soluble.

   (a) T T F T (b) F F F F

   (c) F F T T (d) F T T T

2. Which of the following on heating liberates Oxygen?

   (a) $Na_2 CO_3$ (b) $Li_2 CO_3$

   (c) $LiNO_3$ (d) LiOH

3. Which of the following metals on burning in moist air does not liberate NH_3?

   (a) Li (b) Mg

   (c) Ca (d) Rb

4. The alkali metals which form normal oxide, peroxide as well as superoxides are

   (a) K, Rb (b) Li, Na

   (c) Cb, Li (d) None of these

5. Prefix 'alkali' for alkali metals denotes

   (a) metallic nature (b) Non-metallic nature

   (c) Lustre (d) Ashes of plants

6. Which of the following alkali metal does not form alum?

   (a) Li (b) K

   (c) Rb (d) Na

7. Select the incorrect order

   (a) (Solid) Rb > K > Na > Li (Atomic radius)

   (b) Li > Na > K > Rb > Cs (Ionisation enthalpy)

   (c) Li > Na > k > Rb > Cs (Heat of atomization)

   (d) None of these

8. Select the correct reaction

   (a) $BeCl_2 + 2H_2O \longrightarrow Be(OH)_2 + 2HCl \uparrow$

   (b) $MgCl_2.6H_2O \xrightarrow{\Delta} MgO + 2HCl + 5H_2O$

   (c) $MgCl_2.6H_2O \xrightarrow[650\,k]{\text{dry HCl gas}} MgCl_2 + 6H_2O$

   (d) All are correct

9. Select the correct statement

   (a) In alkali metals, density increases as we go down the group but K has lower density than Na.

   (b) Melting & Boiling points of alkali metals decreases down the group.

   (c) All alkali metals give flame test

   (d) All are correct

10. Which one of the following energy is not considered while determining the oxidation potential?

(a) Sublimation energy

(b) Ionisation energy

(c) Hydration energy

(d) Lattice energy

11. Select the correct property

(a) Lithium - best reducing agent in aqueous medium

(b) KO_2 - used in submarines

(c) CsOH - maximum basic among all alkali metal hydroxides

(d) All are correct

12. Which one of the following is thermally unstable?

(a) NaOH (b) KOH

(c) LiOH (d) RbOH

13. Which of the following is colored?

(a) Oxides

(b) Peroxides

(c) Superoxides

(d) Both oxides & peroxides

14. Which of the following upon heating liberates NO_2?

(a) $LiNO_3$ (b) KNO_3

(c) $NaNO_3$ (d) Both (b) & (c)

15. Which of the following orders of solubility in H_2O is/are correct?

(a) $CsF > RbF > KF > NaF > LiF$

(b) $Ba(OH_2) > Sr(OH)_2 > Ca(OH)_2 > Mg(OH)_2 > Be(OH)_2$

(c) $BeCO_3 > MgCO_3 > CaCO_3 > SrCO_3 > BaCO_3$

(d) $BeSO_4 > MgSO_4 > CaSO_4 > SrSO_4 > BaSO_4$

(a) Only 1 & 2

(b) Only 2, 3 & 4

(c) Only 3 & 4

(d) all orders are correct

16. The low solubility of LiF & CsI in H_2O is because of high ________ energy & low ________ energy, respectively.

(a) lattice, hydration

(b) lattice, lattice

(c) hydration, hydration

(d) hydration, lattice

17. Select the incorrect option.

(1) Lithium reacts with nitrogen to form lithium nitride whereas all other alkali metals do not form nitrides.

(2) LiCl crystallizes as $LiCl.\ 2H_2O$ whereas halides of other alkali metal do not form hydrates.

(3) $Li + H-C\equiv C-H \longrightarrow$ No reaction

(4) $2Na + H-C\equiv C-H \longrightarrow Na^+C^- \equiv C^-\ Na^+ + H_2$

(5) Lithium forms amide on heating with NH_3

(a) All are correct

(b) only 1, 2 & 3 are correct

(c) 1, 2, 3 & 4 only correct

(d) both 2 & 4 are correct

18. Which of the following statements indicate common properties of lithium & magnesium?

(a) Both Li & Mg on burning form monoxides

(b) Carbonates of Li & Mg decompose to liberate CO_2

(c) Both react directly with nitrogen

(d) All the properties are common to lithium & Magnesium

19. The 'INCORRECT' Statement about the solubility of alkali metal in liquid NH_3 is

(a) The ammoniated electrons are responsible for the blue colour of solution & paramagnetic behavior

(b) The electrical conductivity is due to ammoniated cations $\left[M(NH_3)_x\right]^+$ as well as ammoniated electrons $\left[e^-(NH_3)_y\right]^-$.

(c) If the concentration of solution is above 3M, solution is bronze colored & then diamagnetic in nature. Also, the conductivity decreases $\because$ ammoniated cations are bound by free unpaired electrons.

(d) None of these

20. NaOH is manufactured by electrolysis of aq. NaCl solution. The cell used is called Castner Kellner cell. Which of the following statements is/are correct for Castner kellners cell?

(a) Outer compartment has graphite anode & Hg cathode

(b) Reaction in outer compartment are

At Anode : $2Cl^- \longrightarrow Cl_2 + 2e^-$

AT Cathode : $Na^+ + e^- \longrightarrow Na$

$Na + Hg \longrightarrow Na(Hg)$

Sodium amalgam.

(c) Central compartment has Hg anode & Fe cathode.

At Cathode : $2H^+ + 2e^- \longrightarrow H_2$

(in central Compartment)

At Anode : $\bar{O}H \longrightarrow OH + e^-$

(in central Compartment)

$Na(Hg) + OH \longrightarrow NaOH + Hg$

(d) All are correct

21. Which of the following reactions are not involved in Solvay's process for the manufacture of Na_2CO_3?

1. $NH_3 + H_2O + CO_2 \longrightarrow NH_4^+ HCO_3^-$
2. $NH_4^+HCO_3^- + NaCl \longrightarrow \overset{+}{Na}HCO_3^- + NH_4^+Cl^-$
3. $2NaHCO_3 \xrightarrow{\Delta} Na_2CO_3 + H_2O + CO_2 \uparrow$
4. $2NH_4Cl + Ca(OH)_2 \longrightarrow 2NH_3 + CaCl_2 + H_2O$
5. $CaCO_3 + H_2O + CO_2 \longrightarrow Ca(HCO_3)_2$

(a) 1 & 2 only (b) 3 & 4 only

(c) 3, 4 & 5 only (d) 5 only

22. Which of the following statements is the for all the alkali metals?

(a) Their nitrates decompose on heating to give NO_2 & O_2

(b) Their carbonates decompose on heating to give CO_2& metals oxide

(c) They react with oxygen to give mainly MO_2

(d) They react with halogens to give halides of formule M^+X^-.

23. Match the reactions given in column - I with the nature/characteristics of the products given is column-II

	Column-I		Column-II
(a)	$Na_2SO_4 + C + CaCO_3 \xrightarrow{\Delta}$	(p)	One of product has sp^2 hybridization of central atom.
(b)	$NaCl + NH_4HCO_3 \xrightarrow{\Delta}$	(q)	One of the products has sp^2 hybridization of central atom
(c)	$Na_2CO + Ca(OH_2) \xrightarrow{\Delta}$	(r)	One of the products is insoluble as precipitate
(d)	$KOH + NO (2:4$ by mole ratio)	(s)	One of the products is a neutral oxide

(a) $(A \rightarrow p, r, s)$; $(B \rightarrow p, q, s)$;$(C \rightarrow p)$; $(D \rightarrow p, q, s)$

(b) $(B \rightarrow p, s, q)$; $(A \rightarrow q, p, s)$;$(C \rightarrow s)$; $(D \rightarrow p, q)$

(c) $(A \rightarrow s, r, q)$; $(B \rightarrow p)$;$(C \rightarrow p, q, r)$; $(D \rightarrow s, p, q)$

(d) None of these

24. Consider the following statements

Statement-1 :

Concentrated solution of an alkali metal in liquid NH_3 is bronze colored & diamagnetic.

Statement-2 :

Increasing stability of peroxide or superoxide of alkaline earth metal down the group is due to stabilization of large anions by large cations through lattice energy effects.

Statement-3 :

$Be(OH)_2$ react with acids & alkali both to form $[Be(OH_2)_4]Cl_2$ & $[Be(OH)_4]^{2-}$ respectively

Arrange in correct order of true/false

(a) T T T (b) T F F

(c) FFF (d) FFT

25. When CO_2 is bubbled into an aqueous solution of Na_2CO_3, which of the following is formed?

(a) NaOH (b) $NaHCO_3$

(c) H_2O (d) HCOONa

26. Some of the orders are mentioned below. How many of these are 'WRONG'?

(a) $CsBr > RbBr > KBr > NaBr > LiBr$

(Ionic character)

(b) $BaCO_3 > CaCO_3 > MgCO_3 > BeCO_3$

(Thermal stability)

(c) $Ba > Sr > Ca > Mg > Be$

(Reactivity towards air)

(d) $CsOH > RbOH > KOH > NaOH > LiOH$

(Basic nature of hydroxides)

(e) $CsOH > RbOH > KOH > NaOH > LiOH$

(Thermal stability of hydroxides)

(a) 3 of these are wrong

(b) 2 of these are wrong

(c) 4 of these are wrong

(d) None of these are wrong

27. Select the incorrect reaction

(a) $NH_4Cl + NaOH \longrightarrow NH_3 + NaCl + H_2O$

(b) $Na_2SO_3 + S \xrightarrow{\Delta} Na_2S_2O_3$

(c) $2Na_2S_2O_3 + I_2 \longrightarrow Na_2S_4O_6 + 2NaI$

(d) None of these is incorrect

28. Consider the following statements

(a) Among alkali metal halides, cesium fluoride is most ionic is nature.

(b) KOH is a better absorber of CO_2 than NaOH

(c) KO_2 is a better oxidant than K_2O_2

(d) All alkali metals form hydrates.

Assign True/False to these statements

(a) TTTT (b) TTTF

(c) FTTT (d) TTFF

29. Which one of the following is liberated which the blue solutions of alkali metals in liquid NH_3 decompose very slowly?

(a) NH_3 (b) N_3H

(c) H_2 (d) N_2

30. Which one of these gives HCHO on heating?

(a) HCOONa (b) HCOOK

(c) HCOOLi (d) All of these

31. A metal 'M' reacts with N_2 to give a compound A (metal nitride). 'A' on heating at high temperature gives back 'M' & 'A' on reacting with H_2O gives a gas B. 'B' turns Copper sulphate solution blue on passing through it. 'M' and 'B' can be-

(a) Al & NH_3 (b) Li & NH_3

(c) Na & NH_3 (d) Mg & NH_3

32. Pick out the statement(s) which is (are) not true about the diagonal relationship of Li & Mg

(a) Polarising power of Li^+ & Mg^{2+} are almost same.

(a) Like lithium Mg decomposes water very fast

(c) LiCl & $MgCl_2$ are deliquescent.

(d) Like Li, Mg does not form solid bicarbonates.

(a) 1 & 2 (b) 2 & 3

(c) Only 2 (d) Only 1

33. $NaHCO_3$ & $KHCO_3$ belong to same category of compounds. Which of the following statement are true about them?

(a) $NaHCO_3$ is less soluble than $KHCO_3$.

(a) $NaHCO_3$ has infinite chain of hydrogen bonding whereas $KHCO_3$ forms dimeric H - bonding in solid state.

(c) Both can exist independently in solid state.

(d) All are correct.

34. The pair of compounds which cannot exist together in solution is-

(a) $NaHCO_3$ & NaOH (b) NaOH & KOH

(c) Na_2CO_3 & NaOH (d) $NaHCO_3$ & NaCl

35. The alkali metal with lowest heat of atomization is -

(a) Li (b) Na

(c) K (d) Cs

36. Na and K react with water much more vigorously than Li, because-

(a) These have high hydration energies as compared to lithium

(b) These have higher melting point than lithium

(c) These have lower melting point than lithium

(d) These have lower hydration energy than lithium

37. Which of the following statements is not true for alkali metals?

(a) Alkali metals dissolve in NH_3 & give blue solution.

(b) Lithium is the strongest reducing agent in aq. Medium.

(c) Na_2CO_3 is thermally stable.

(d) Density of K is more than that of Na.

38. A gas X is passed through water to form a saturated solution. The aqueous solution on treatment with $AgNO_3$ gives a white precipitate. The saturated aqueous solution also dissolves magnesium ribbon with evolution of a colourless gas Y. Identify X & Y.

(a) $X = CO_2, Y = Cl_2$

(b) $X = Cl_2, Y = CO_2$

(c) $X = Cl_2, Y = H_2$

(d) $X = H_2, Y = Cl_2$

39. Which of the following have peroxy linkage?

(a) BaO_2 (b) Na_2O_2

(c) Na_2O (d) Both (a) & (b)

40. Select the correct statement.

(a) Cs is most metallic among group-I elements.

(b) Li has maximum oxidation potential due to very high hydration energy

(c) LiCl is more covalent than NaCl

(d) All are correct

41. Which of the following are NOT 'points of similarities' between lithium & magnesium?

(a) The oxides, LiO_2 & MgO do not combine with excess oxygen to give any superoxide

(b) Solid hydrogen carbonates are not formed by lithium & magnesium

(c) Both LiCl & $MgCl_2$ are deliquescent

(d) None of these

42. Sodium is not present in -

(a) Glauber's salt (b) Soda ash

(c) Chile salt petre (d) Salt petre

43. $Na_2CO_3 + H_2O + CO_2 \longrightarrow (A)$.

White crystalline powder (A) on reaction with phenolphthalein gives -

(a) Pink colour (b) Yellow colour

(c) No Colour (d) Black Colour

44. The correct increasing order of densities of alkali metals is

(a) $Li < Na < K < Rb < Cs$

(b) $Li > Na > K > Rb > Cs$

(c) $Cs > Rb > Na > K > Li$

(d) None of these

45. In the manufacture of NaOH by Castner Kellner cell, by product obtained is -

(a) O_3 (b) Cl_2

(c) O_2 (d) Na_2CO_3

46. Which of the following is the correct reaction which occurs at cathode of outer compartment is Castner Kellner process?

(a) $2H^+ + 2e^- \longrightarrow H_2$

(b) $Na^+ + e^- \longrightarrow Na \xrightarrow{Hg} Na(Hg)$

(c) $\bar{O}H \longrightarrow OH + e^-$

(d) $2Cl^- \longrightarrow Cl_2 + 2e^-$

47. Identify the mismatch

(a) Chile saltpeter - $NaNO_3$

(b) Nitre - KNO_3

(c) Potash alum - $K_2SO_4.Al_2(SO_4)_3.24H_2O$

(d) None of these

48. Lithium is strongest reducing agent among alkali metals due to high-

(a) Electron affinity (b) hydration energy

(c) lattice energy (d) ionisation energy

49. Which alkali metal gives least wavelength of light emitted in a flame?

(a) Li (b) Na

(c) Cs (d) Rb

50. Metal 'A' reduces silica converting itself into B. 'B' absorbs moisture & converts into C. When C is heated with the reduction product of silica, a gas is liberated. Then, A, B, C & the gas are, respectively-

(a) Na, $Na_2O, NaOH, H_2 \uparrow$

(b) Na, Na_2O, Na_2CO_3, $O_2 \uparrow$

(c) Na, Na_2O, Na_2CO_3, $CO_2 \uparrow$

(d) Na, Na_2O_2, NaOH, $H_2 \uparrow$

51. Which salt of potassium contains anions with angular shape?

(a) KI_3 (b) KN_3

(c) K_2S_3 (d) K_2O_2

52. A compound of sodium does not give CO_2 when heated but it gives CO_2 when treated with dilute acids.The compound is found to contain 37.1 percent Na & 1.6 percent H_2 by wt. The compound in discussion is

(a) $Na_2CO_3 \cdot 10H_2O$ (b) $Na_2CO_3 \cdot H_2O$

(c) $NaHCO_3 \cdot 5H_2O$ (d) $NaHCO_3 \cdot 10H_2O$

53. Which of the following statements are correct?

(A) $NaHCO_3$ is used as mild antiseptic.

(B) liquid sodium metal is used as coolant in fast breeder nuclear reactor

(C) BeO is used as refractory material

(D) Cs & K are used as electrodes in photoelectric cells

(a) both C & D (b) B, C & D

(c) Both A & C (d) All of these are correct

54. In Castner kellner cell, Na^+ is discharged instead of H^+ ions because

(a) Hg reduces the mobility of Na^+

(b) the discharge potential of Na^+ is lower than that of H^+ ions at Hg cathodes

(c) Hg directly combines with H^+ ions

(d) All of these

55. LiCl > NaCl, holds good for-

(a) Covalent character

(b) Solubility in ether

(c) conduction in fused state

(d) All of these

56. All the product are same when A & B separately react with H_2O. A & B are respectively

(a) Ca, CaO

(b) Na, NaH

(c) NaOH, Na_2CO_3

(d) Na, k

Group II

57. Arrange the following in decreasing order of their hydration energies

$Be^{2+}, Mg^{2+}, Ca^{2+}, Sr^{2+}, Ba^{2+}$

(a) $Be^{2+} > Mg^{2+} > Ca^{2+} > Sr^{2+} > Ba^{2+}$

(b) $Ba^{2+} > Sr^{2+} > Ca^{2+} > Mg^{2+} > Be^{2+}$

(c) $Be^{2+} > Ca^{2+} > Sr^{2+} > Mg^{2+} > Ba^{2+}$

(d) $Ba^{2+} > Sr^{2+} > Mg^{2+} > Ca^{2+} > Be^{2+}$

58. Select the correct order.

(a) $Li^+ > Na^+ > K^+ > Rb^+ > Cs^+$

(Hydration Energy)

(b) $Ba^{2+}_{(aq.)} > Sr^{2+}_{(aq.)} > Ca^{2+}_{(aq.)} > Mg^{2+}_{(aq.)} >$

$Be^{2+}_{(aq.)}$ (Ionic mobility)

(c) Be > Mg > Ca > Sr > Ba (Ionisation energy)

(d) All are correct

59. Choose the 'INCORRECT' option

(a) $\underset{\text{(group II)}}{IE_1} > \underset{\text{(group.I)}}{IE_1}$

(b) $\underset{\text{(group I)}}{IE_1} > \underset{\text{(group.II)}}{IE_2}$

(c) Be is not regarded as true alkaline earth metal.

(d) None of these

60. Which of the following reactions depict the acidic & basic nature of $Be(OH)_2$, respectively?

(a) $Be(OH)_2 + 2\bar{O}H \longrightarrow [Be(OH)_4]^{2-}$ (Beryllate ion)

$Be(OH)_2 + 2HCl + 2H_2O \longrightarrow [Be(H_2O)_4]Cl_2$

(b) $Be(OH)_2 + 2HCl + 2H_2O \longrightarrow [Be(H_2O)_4]Cl_2$,

$Be(OH)_2 + 2\bar{O}H \longrightarrow [Be(OH)_4]^{2-}$

(c) Both of these

(d) None of these

61. Correct order of thermal stability of gp. II carbonates is

(a) $BeCO_3 > MgCO_3 > CaCO_3 > SrCO_3 > BaCO_3$

(b) $BaCO_3 > SrCO_3 > CaCO_3 > MgCO_3 > BeCO_3$

(c) $BaCO_3 = SrCO_3 > CaCO_3 = MgCO_3 > BeCO_3$

(d) $SrCO_3 > CaCO_3 > MgCO_3 > BeCO_3 > BaCO_3$

62. Select the incorrect reactions among the following.

1. $2Mg(NO_3)_2 \xrightarrow{\Delta} 2MgO + 4NO_2 + O_2$
2. $MgCO_3 + 2HNO_3 \longrightarrow Mg(NO_3)_2 + CO_2 + H_2O$
3. $Al_4C_3 + 12H_2O \longrightarrow 4Al(OH)_3 + 3CH_4$
4. $Be_2C + 4H_2O \longrightarrow 2Be(OH)_2 + CH_4$
5. $(NH_4)_2BeF_4 \xrightarrow{\Delta} BeF_2 + 2NH_4F$
6. $BeO + C + Cl_2 \xrightarrow{600-800k} BeO + COCl_2$

(a) Only 1, 2 and 3 are incorrect

(b) Only 4, 5 & 6 are incorrect

(c) Only 3 & 4 are incorrect

(d) Only 6 is correct

63. Which of the following reaction is a correct representation for an alkaline earth metal (M) dissolved in liquid NH_3?

(a) $M + (x+y)NH_3 \longrightarrow [M(NH_3)_x]^+ + [e(NH_3)_y]^-$

(b) $M + (x+y)NH_3 \longrightarrow [M(NH_3)_x]^{2+} + 2[e(NH_3)_y]^-$

(c) $M + (x+y)NH_3 \longrightarrow [M(NH_3)_x]^{3+} + 3[e(NH_3)_y]^-$

(d) Both (a) & (c)

64. When Mg burns in air, it produces

(a) MgO, MgO_2 (b) $MgO_2, MgCO_3$

(c) MgO Only (d) MgO & Mg_3N_2

65. The Gp. II metal which does not directly combine with hydrogen is

(a) Be (b) Ca

(c) Sr (d) Ba

66. Select the correct option.

(a) BeH_2 & MgH_2 are covalent & polymeric.

(b) Be & Mg fail flame test.

(c) CaH_2 on reacting with H_2O produces H_2.

(d) All are correct

67. Which of the following on thermal decomposition yields a basic as well as an acidic oxide?

(a) NH_4NO_3 (b) $NaNO_3$

(c) $KClO_3$ (d) $CaCO_3$

68. One mole of Mg_3N_2 on reaction with excess H_2O gives

(a) 2 moles of HNO_3 (b) 2 moles of NH_3

(c) 1 mole of NH_3 (d) 1 mole of HNO_3

69. $BeCl_2$ exists as a polymer, monomer & dimer in ___, ___ &___ states, respectively (above 1200 K)

(a) Solid, vapour, liquid

(b) Solid, solid, solid

(c) Liquid, vapour, solid

(d) Solid, liquid, liquid

70. An element A of gp. II gives brick red colour in Bunsen flame. (A) burns in N_2 atmosphere to give (B), which gets hydrolysed to produce gas (C) and an alkaline solution (D). The solution (D) on exposure to air produces a thin solid layer (E) on the surface.

Then (A), (B), (C), (D) & (E) are respectively.

(a) $Ca, Ca_3N_2, NH_3, Ca(OH)_2, CaCO_3$

(b) $Mg, Mg_3N_2, NH_3, Mg(OH)_2, MgCO_3$

(c) $Mg, Mg_3N_2, O_2, MgSO_4, MgCO_3$

(d) $Ca, Ca_3N_2, N_2O_5, Ca(OH)_2, CaCO_3$

71. Identify the mismatch

Compound		Common Name
(a) $MgSO_4 \cdot 7H_2O$	$\rightarrow$	Epsom Salt
(b) CaO	$\rightarrow$	Quick lime
(c) Suspension of $Mg(OH)_2$ in water	$\rightarrow$	Milk of Magnesia
(d) None of these		

72. Which of the following statements is incorrect?

(a) $LiHCO_3$ does not exist in the solid state.

(b) KO_2 is paramagnetic

(c) Solubility of $Ba(OH)_2$ in water is more than the solubility of $Mg(OH)_2$.

(d) Li_2CO_3 is more stable than Rb_2CO_3

73. When water is added to compound (A) (of Gp. II element), solution B is formed. When CO_2 is passed through this solution, milkiness appear due to (C). Identify (A), (B) & (C), respectively

(a) $CaO, Ca(OH)_2, CaCO_3$

(b) $Ca(OH)_2, CaCO_3, CaSO_4$

(c) $Ca_3N_2, Ca(OH)_2, CaSO_4$

(d) None of these

74. Identify the mismatch

Compound	Common Name
(a) Plaster of Paris	$CaSO_4 \cdot \frac{1}{2}H_2O$
(b) Gypsum	$CaSO_4 \cdot 2H_2O$
(c) Dead burnt plaster	Anhydrous $CaSO_4$
(d) None of these is a mismatch	

75. A metal M forms soluble sulphate & water insoluble hydroxide $M(OH)_2$. Its oxides is MO & it is amphoteric & possesses high melting point. The metal M is

(a) Be (b) Mg

(c) Ca (d) Sr

76. Magnesium forms Mg^{2+} & not Mg^{+} which of the following is a correct reason regarding this?

(a) Magnesium (II) is insoluble in water.

(b) Commonly higher oxidation states are preferred by metals

(c) Ionic radius of Mg (II) is smaller than that of Mg (I)

(d) High hydration energy & high lattice energy

77. The sum of the number of H_2O molecules in gypsum & plaster of paris is

(a) 3

(b) $3/2$

(c) $5/2$

(d) $7/2$

78. Select the correct order against the property mentioned

(a) $BeCO_3 > MgCO_3 > CaCO_3 > BaCO_3$ (Solubility)

(b) $Ba(OH)_2 > Ca(OH)_2 > Mg(OH)_2 > Be(OH)_2$ (Solubility)

(c) $Cs > Rb > K > Na > Li$ (Reactivity with H_2O)

(d) All are correct

79. Beryllium has less negative value of reduction potentials compared to other alkaline earth metals due to

(a) The smaller hydration energy of Be^{2+}

(b) The large value of enthalpy of atomization of Be metal.

(c) Large value of ionization energy of Be metal

(d) Both (b) & (c) are correct

80. Crude common salt becomes damp on keeping in air because

(a) It is hygroscopic in nature.

(b) It contains $MgCl_2$ & $CaCl_2$ as impurities which are deliquescent

(c) It releases Cl_2

(d) Both (a) & (b)

81. A metal carbide on hydrolysis gives an acidic gas which is cylindrically symmetrical. Reaction of carbide with nitrogen gas gives a nitrogenous fertilizer which is more suitable to the plants. The carbide is-

(a) acetylide of Mg

(b) methanide of Be

(c) allylide of Ca

(d) acetylide of Ca

82. Select the true statement(s).

(a) Be & Al, both become passive on reaction with HNO_3 due to formation of oxide layer.

(b) Chlorides of both Be & Al are lewis acids.

(c) Hydroxides of both Be & Al are soluble in alkali as well as in acid.

(d) All are correct.

83. A substance absorbs CO_2 & violently reacts with water. The substance is-

(a) $CaCO_3$ (b) CaO

(c) H_2SO_4 (d) ZnO

84. Property of all the alkaline earth metals that increases with their atomic number is-

(a) ionization energy

(b) solubility of their hydroxides

(c) solubility of their sulphate

(d) electronegativity

85. A chemical 'A' is used for preparation of washing soda to recover NH_3. When CO_2 is bubbled through an aqueous solution 'A', the solution turns milky. What is the chemical formula of 'A'?

(a) $Ca(HCO_3)_2$ (b) CaO

(c) $Ca(OH)_2$ (d) $CaCO_3$

86. Select the incorrect statement.

(a) In Solvays process, the by - product is $CaCl_2$.

(b) In Castner kellner process, the Hg layer acts as anode in central compartment.

(c) Castner kellner process is used to prepare mainly NaCl.

(d) Both 1 & 2

87. A metal M readily forms a water soluble sulphate MSO_4, water insoluble $M(OH)_2$ & oxide MO. The hydroxide is soluble in NaOH. The metal M is -

(a) Mg (b) Be

(c) Ca (d) Sr

88. Correct order of thermal stability is -

(a) $MgCO_3 > BeCO_3 > CaCO_3 > SrCO_3$

(b) $BeCO_3 > MgCO_3 > CaCO_3 > SrCO_3$

(c) $SrCO_3 > CaCO_3 > MgCO_3 > BeCO_3$

(d) None of these

89. Which is/are not correct configuration of s-block elements?

(a) $(Ar)3d^{10}4s^1$ (b) $(Ar)3d^{10}4s^2$

(c) $(Ar)4s^1$ (d) Both (a) & (b)

90. A metal M readily forms a water soluble MSO_4, water insoluble hydroxide $M(OH)_2$ & oxide MO. The oxide & hydroxide are soluble in NaOH. The metal 'M' is

(a) Be (b) Mg

(c) Ca (d) Sr

91. Which of the following salts on heating gives a mixture of two gases?

(a) $Mg(NO_3)_2$

(b) $NaNO_3$

(c) KNO_3

(d) $RbNO_3$

92. The electrolysis of molten KCl· $MgCl_2$· $6H_2O$ gives-
(a) Mg only (b) K only
(c) K & Mg only (d) Mg & Cl_2

93. The correct formula for Epsom salt is -
(a) $Mg(OH)_2$ (b) $MgSO_4$
(c) $MgSO_4 \cdot 7H_2O$ (d) $BaSO_4$

94. Which of the following is correct wrt beryllium?
(a) It is rendered passive by nitric acid
(b) It forms Be_2C
(c) Its hydride is e^- deficient & polymeric
(d) all are true

95. Which of the following is not the correct order of property indicated?
(a) $BeCO_3 < MgCO_3 < CaCO_3$ (thermal stability)
(b) $BeCl_2 > LiCl > NaCl$ (Covalent character)
(c) LiOH > NaOH > KOH > RbOH > CsOH (Solubility)
(d) LiF > NaF > KF > RbF (melting point)

96. Consider the following structures for hydrated gp. II metal ions-

$x = [Be(H_2O)_n]^{2+}$

$y = [Mg(H_2O)_n]^{2+}$

$z = [Ca(H_2O)_n]^{2+}$

What is the correct order of size of GP. II metal ions?
(a) $x > y > z$ (b) $y > x > z$
(c) $x = y > z$ (d) $z > x > y$

97. Which one of the following tests/reagents cannot be used to distinguish between the pair of species?
(a) Ca & Mg: flame test
(b) Cs_2SO_4 & $BaSO_4$: Water
(c) $NaHCO_3$ & Na_2CO_3 : Heat
(d) KNO_3 & $CsNO_3$: Heat

98. Which of the following does not exist?
(a) BeF_6^{2-} (b) AlF_6^{3-}
(c) BeF_4^{2-} (d) BeF_2

99. A compound x has the following characteristics
(a) It is used as a disinfectant
(b) It is used to prepare bleaching powder
(c) its suspension in water is called milk of lime
The compound x is-
(a) $Ca(OH)_2$ (b) CaO
(c) $CaCO_3$ (d) $CaSO_4 \cdot 2H_2O$

100. The oxidation state of most electronegative elements in the products of reaction of BaO_2 with dil. H_2SO_4 are
(a) 0 & 1 (b) 1 & 2
(c) 2 & 0 (d) 2 & +1

101. An alkaline metal gives a salt with chlorine which is sparingly soluble in water at room temperature, but fairly soluble in boiling water It also forms a sulphate whose mixture with a sulphide of a transition metal is called 'lithopone' & is used as a white pigment. The alkaline earth metal is
(a) Ca (b) Mg
(c) Sr (d) Ba

102. Two metal A & B belong to the same group of the periodic table. Metal A forms an insoluble oxide but a soluble sulphate. Metal B forms a soluble oxide but an insoluble sulphate. Both metal A & B hydroxides which are soluble in alkalis. A & B are
(a) Ba & Mg (b) Na & K
(c) Mg & Ba (d) K & Rb

103. $CaSO_4 \cdot 2H_2O \xrightarrow{200^\circ C} (B) \xrightarrow[\text{heating}]{\text{Strong}} (C)$

$\downarrow$

(A)

A, B & C, respectively are-
(a) $CaSO_4 \cdot \frac{1}{2}H_2O, CaSO_4, CaO$
(b) CaO, $CaSO_4, Ca(OH)_2$
(c) $CaSO_4$, $Ca(OH)_2$, CaO
(d) None of these

104. A solid compound X on heating, gives CO_2 gas and a residue which when mixed with H_2O forms Y. On passing an excess of CO_2 through aqueous solution of X, a clear solution Z is obtained. On boiling Z, the compound X is re-formed. The compounds X, Y & Z are respectively.

(a) K_2CO_3, $CaCO_3$, $Ca(HCO_3)_2$

(b) $CaCO_3$, $Ca(OH)_2$, $Ca(HCO_3)_2$

(c) $CaCO_3$, $Ca(HCO_3)_2$, $Ca(OH)_2$

(d) $CaCO_3$, Na_2CO_3, $Ca(OH)_2$

105. On heating a mixture of one mole each of, Li_2CO_3, $MgCO_3$ & Na_2CO_3, the volume of CO_2 evolved under STP condition will be-

(a) 67 22 L (b) 33 6 L

(c) 44 8 L (d) 22 4 L

106. $Mg \xrightarrow{Air} x + y$

$$x \xrightarrow{H_2O} \underset{\text{(color less gas)}}{z} \xrightarrow{H_2O} \text{solution} \xrightarrow{CuSO_4} A$$

In the above sequence of reactions, the substances X, Y, Z & A are respectively.

(a) Mg_3N_2, MgO, NH_3 $[Cu(NH_3)_4]SO_4$

(b) Mg_3N_2, MgO, NH_3, $MgSO_4$

(c) $Mg(NO_3)_2$, MgO_2, H_2O_2, $CuSO_4 \cdot 5H_2O$

(d) None of these

107. One mole of a substance 'A' on reacting with excess of H_2O, gives two moles of a readily combustible gas & an alkaline solution. The alkaline solution obtained gives white turbidity with SO_2. The substance 'A' is

(a) CaH_2

(b) NaH

(c) $Ca(OH)_2$

(d) Na_2SO_4

Answers

1. (d) **2.** (c) **3.** (d) **4.** (a) **5.** (d) **6.** (a) **7.** (d) **8.** (d) **9.** (d) **10.** (d)

11. (d) **12.** (c) **13.** (c) **14.** (a) **15.** (d) **16.** (a) **17.** (c) **18.** (d) **19.** (d) **20.** (d)

21. (d) **22.** (d) **23.** (a) **24.** (a) **25.** (b) **26.** (d) **27.** (d) **28.** (a) **29.** (c) **30.** (c)

31. (a) **32.** (c) **33.** (d) **34.** (a) **35.** (d) **36.** (c) **37.** (d) **38.** (c) **39.** (d) **40.** (d)

41. (d) **42.** (d) **43.** (a) **44.** (c) **45.** (b) **46.** (b) **47.** (d) **48.** (b) **49.** (c) **50.** (a)

51. (a) **52.** (b) **53.** (d) **54.** (b) **55.** (d) **56.** (b) **57.** (a) **58.** (d) **59.** (d) **60.** (a)

61. (b) **62.** (d) **63.** (b) **64.** (*) **65.** (a) **66.** (d) **67.** (d) **68.** (b) **69.** (a) **70.** (a)

71. (d) **72.** (d) **73.** (a) **74.** (d) **75.** (a) **76.** (d) **77.** (c) **78.** (d) **79.** (d) **80.** (b)

81. (d) **82.** (d) **83.** (b) **84.** (b) **85.** (c) **86.** (c) **87.** (b) **88.** (c) **89.** (d) **90.** (a)

91. (a) **92.** (d) **93.** (c) **94.** (d) **95.** (c) **96.** (a) **97.** (d) **98.** (a) **99.** (a) **100.** (b)

101. (d) **102.** (c) **103.** (a) **104.** (a) **105.** (c) **106.** (a) **107.** (a)

Explanations

2. (c) $LiNO_3 \xrightarrow{\Delta} Li_2O + NO_2 \uparrow$

3. (d) $Li + N_2 \longrightarrow Li_3N$

$Li_3N + H_2O \longrightarrow LiOH + NH_3 \uparrow$

$Mg + N_2 \longrightarrow Mg_3N_2$

$Mg_3N_2 + H_2O \longrightarrow Mg(OH)_2 + NH_3 \uparrow$

$Ca + N_2 \longrightarrow Ca_3N_2$

$Ca_3N_2 + H_2O \longrightarrow Ca(OH)_2 + NH_3 \uparrow$

In group-I, Li is the only alkali metal which reacts directly with N_2.

5. (d) Alkali metals word is derived from Arabic word al-qis meaning plant ashes. The ashes of plants are rich in carbonates of Na & k.

6. (a) Small sized Li cannot accommodate six water molecules around it

8. (d) Hydrolysis of $BeCl_2$ occurs

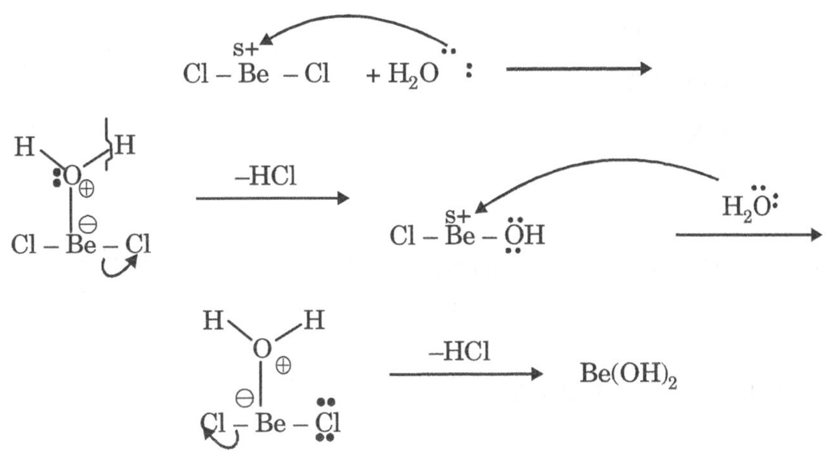

\# $MgCl_2.6H_2O \xrightarrow{\Delta} MgO + 2HCl + 5H_2O$

Here, initially water gets removed & then undergoes hydrolysis

\# $MgCl_2.6H_2O \xrightarrow[650\,k]{dry\,HCl\,gas} MgCl_2 + 6H_2\ddot{O}$

Here, hydrolysis does not occur owing to acidic medium.

9. (d) Density generally ↑ ses down gp. For alkali metals due to large ↑ se in atomic masses as compared to their volumes.

B.pt decreases down the gp. As the metallic bonding becomes weaker when we go down the group.

10. (d) (a) $M_{(s)} \xrightarrow{\text{Sublimation energy}} M_{(g)}$

(b) $M_{(g)} \xrightarrow{I.E.} M^+_{(g)}$

(c) $M^+_{(g)} \xrightarrow[\text{Energy}]{\text{Hydration}} M^+_{(aq)}$

11. (d) Lithium is best reducing agent in aq. med. due to very high hydration enegy.

\# $2KO_2 + 2H_2O \longrightarrow 2KOH_{(aq)} + II_2O_{2(aq.)} + O_{2(g)}$

$KO_2 + 2CO_{2(g)} \longrightarrow 2K_2CO_3 + 3O_{2(g)}$

12. (c) $LiOH \xrightarrow{\Delta} Li_2O + H_2O$

13. (c) Superoxides are colored as they are paramagnetic

14. (a) $4LiNO_3 \longrightarrow 2Li_2O + LiNO_2\uparrow + O_2\uparrow$

15. (d) For alkali metal florides & gp. II hydroxides, the solubility increases does the group because the decrease in lattice energy is more than decrease in hydration energy.

For group II carbonates & sulphates, solubility goes on decreasing down group. It is because decrease in lattice energy on moving down the group is less than decrease in hydration energy.

17. (c) (5) Lithium form imide (Li_2NH) on heating with while other alkali metals form amide of formula .

18. (d) (a) $4\,Li + O_2 \longrightarrow 2Li_2O$, $2Mg + O_2 \longrightarrow 2MgO$ $\longrightarrow$ Monoxides are formed

(b) $Li_2CO_3 \longrightarrow Li_2O + CO_2 \uparrow$

$MgCO_3 \longrightarrow MgO + CO_2 \uparrow$

(c) $Li + N_2 \longrightarrow Li_3N$

$Mg + N_2 \longrightarrow Mg_3N_2$

21. (d) Reactions 1, 2 & 3 are required to prepare Na_2CO_3.

Reaction '4' is used to regenerate NH_3 .

NH_4Cl, produced in reaction '2' is reacted with lime water to regenerate NH_3.

Reaction '5' has no significance in Solvay's process.

22. (d) $NaNO_3 \xrightarrow[\text{heating}]{\text{strong}} NaNO_2 + \frac{1}{2}O_2 \uparrow$

All alkal metal nitrates on strong heating give corresponding nitrite $+O_2$ except for $LiNO_3$

$LiNO_3 \xrightarrow{\Delta} Li_2O + NO_2 + O_2$

→ Alkali metal carbonates except lithium carbonate are thermally stable

$Li_2CO_3 \xrightarrow{\Delta} Li_2O + CO_2 \uparrow$

$Na_2CO_3 \xrightarrow{\Delta} \text{No effect}$

23. (a) (a) $Na_2SO_4 + 4C + CaCO_3 \xrightarrow{\Delta} \underset{\text{C in } CO_3^{2-} \text{ is} (sp^2)}{Na_2CO_3} + CaS \downarrow + \underset{\text{(CO is a neutral oxide)}}{4CO}$

(b) $NaCl + \overset{+}{NH_4}\, HCO_3^- \longrightarrow NaHCO_3(sp^2) + NH_4Cl(sp^3)$

(c) $Na_2CO_3 + Ca(OH)_2 \longrightarrow CaCO_3 \downarrow (sp^2) + NaOH$

(d) $2KOH + 4NO \rightarrow 2KNO_2(sp^2) + N_2O$ (neutrual oxide) $+ H_2O(sp^3)$

25. (b) $Na_2CO_3 + H_2O + CO_2 \longrightarrow 2NaHCO_3$

30. (c) $2HCOOLi \longrightarrow Li_2CO_3 + HCHO$

$2HCOONa \xrightarrow{\Delta} H_2 + \underset{\text{COONa}}{\overset{|}{\text{COONa}}}$

34. (a) $NaHCO_3$ is an acidic salt. So, it will react with base.

35. (d) Metallic bonding is weakest in Cs

38. (c) $Cl_2 + H_2O \longrightarrow HCl + O_2$

$AgNO_3 + HCl \longrightarrow \underset{\text{(White ppt.)}}{AgCl \downarrow} + HNO_3$

39. (d) BaO_2 is $Ba^{2+}O_2^{2-}$

Na_2O_2 is Na_2O_2

Both are peroxides

42. (d) Salt petre (potassium nitrate) is KNO_3.

43. (a) $Na_2CO_3 + H_2O + CO_2 \longrightarrow 2NaHCO_3$

Now, $NaHCO_3$ is an acidic salt, so it gives pink colour with phenolphthalein

44. (c) $\underset{\text{(Density)}}{\rho} = \frac{\text{Mass}}{\text{Volume}}$

There is exception in case of Na & K. As we move from Na to K, the increase in volume exceeds the increase in mass

45. (b) Brine soln. is electrolyzed

At cathode $Na^+ + e^- \longrightarrow Na$

$Na + Hg \longrightarrow Na(Hg)$

At anode $Cl^- \longrightarrow \frac{1}{2}Cl_2 + e^-$

49. (c) It is because energy emitted from de-excitation of e⁻ will be maximum in cesium.

50. (a) $\underset{A}{4Na} + SiO_2 \longrightarrow \underset{B}{2Na_2O} + Si$

$\underset{B}{Na_2O} + H_2O \longrightarrow \underset{C}{2NaOH}$

(reduction product of silica, SiO_2 will be Si)

$2NaOH + Si + H_2O \longrightarrow Na_2SiO_3 + H_2 \uparrow$

51. (a) (a) KI_3

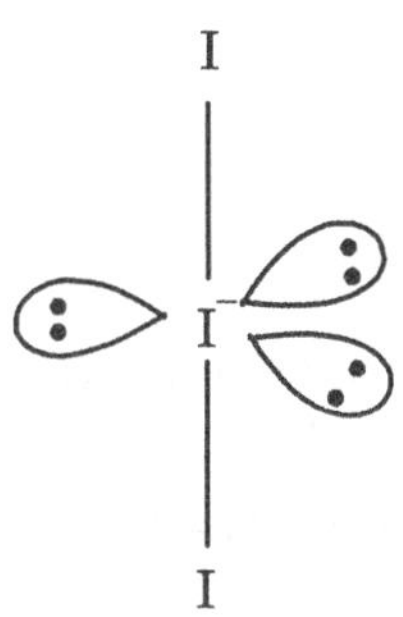

ANION- $\underset{(Linear)}{I_3^-}$

(b) KN_3

ANION N_3^-

$N = N \rightleftharpoons N^{\ominus}$
(Linear)

(c) K_2S_3

ANION $\rightarrow S_3^{2-}$

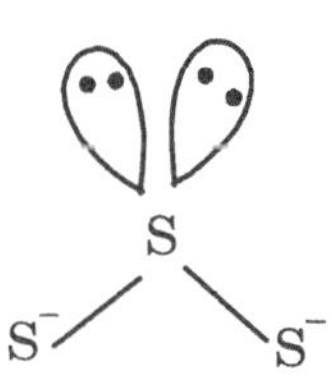

(d) K_2O_2

ANION $\rightarrow O_2^{2-}$

$^{\ominus}:\ddot{\underset{..}{O}}-\ddot{\underset{..}{O}}:^{\ominus}$

52. (b) $Na_2CO_3 \cdot H_2O$

% age of hydrogen by wt.

$= \frac{2}{124} \times 100 = 1.6\%$

%age age of sodium by wt.

$= \frac{46}{124} \times 100 = 37.1\%$

56. (b) $2Na + 2H_2O \longrightarrow 2NaOH + 2H_2 \uparrow$

$NaH + H_2O \longrightarrow NaOH + H_2 \uparrow$

57. (a) Greater are charge density, greater is hydration energy.

58. (d) Greater the hydration energy of an ion $\longrightarrow$ more hydrated is the ion $\longrightarrow$ more bulky is an ion $\longrightarrow$ lesser is its mobility.

59. (d)

IE ↑ Ses $\longrightarrow$

Li	Be
Na	Mg
K	Ca
Rb	Sr
Cs	Ba
Fr	Ra

$\underset{(group.\ II)}{IE_1} > \underset{(group.\ I)}{IE_1}$

\# $\underset{(group.\ I)}{IE_1} > \underset{(group.\ II)}{IE_2}$

1) $\underset{(g)}{Na} \xrightarrow{IE_1} \underset{(g)}{Na^+} + e^-$
$\downarrow$ $1s^2 2s^2 2p^6 3s^1$; $\downarrow$ $1s^2 2s^2 2p^6$

2) $\underset{(g)}{Na^+} \xrightarrow{IE_2} \underset{(g)}{Na^{2+}} + e^-$
$\downarrow$ $1s^2 2s^2 2p^6$

(Noble gas configuration)

1) $\underset{(g)}{Mg} \xrightarrow{IE_1} \underset{(g)}{Mg^+} + e^-$
$\downarrow$ $1s^2 2s^2 2p^6 3s^2$; $\downarrow$ $1s^2 2s^2 2p^6 3s^1$

2) $\underset{(g)}{Mg^+} \xrightarrow{IE_2} \underset{(g)}{Mg^{2+}} + e^-$
$\downarrow$ $1s^2 2s^2 2p^6 3s^1$

(No Noble gas configuration)

\# Be is not regarded as true alkaline earth metal because its oxide & hydroxide are amphoteric.

60. (a) $Be(OH)_2 + 2\bar{O}H \longrightarrow [Be(OH)_4]^{2-}$

$\downarrow$

(Lewis acid)

HO – Be – OH

Be(4) : $1s^2 2s^2$ (G.S.)

$1s^2 2s^1 2p^1$ (E.S.)

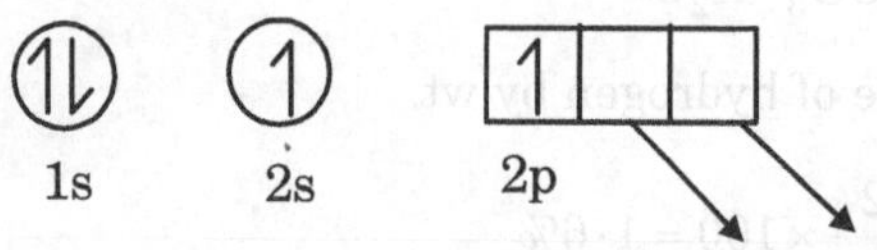

VACANT 2p-orbitals of Be

$$HO-\overset{s+}{Be}-OH + 2\bar{O}H \longrightarrow [Br(OH)_4]^{2-}$$

$$\# \ Be(OH)_2 + 2HCl + 2H_2O \longrightarrow [Be(H_2O)_4]Cl_2$$

(Arrhenius Base)

61. (b) Bigger anions are stabilized by bigger cations

62. (d) $BeO + C + Cl_2 \xrightarrow{600-800K} BeCl_2 + CO$

63. (b) Gp. II elements show + 2 Oxidation state.

64. $2Mg + O_2 \longrightarrow 2MgO$

$3Mg + N_2 \longrightarrow Mg_3N_2$

65. (a) Be reacts with O_2 (air) which results in the formation of the layer of BeO over Be. So, Be does not directly combine with H_2.

66. (d) 1.

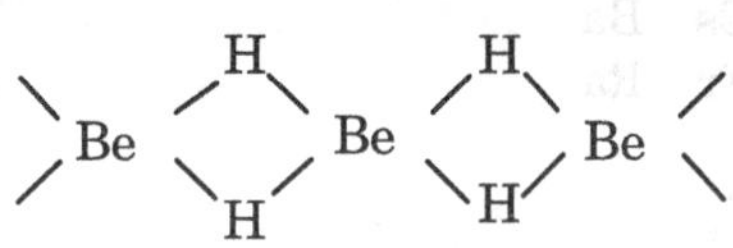

Each Be forms 4 banana (3c - 2e) bonds.

2. Size of Be & Mg is small, so electronic excitation is difficult in Be & Mg

$\# \ CaH_2 + 2H_2O \longrightarrow Ca(OH)_2 + 2H_2 \uparrow$

67. (d) $CaCO_3 \xrightarrow{\Delta} \underset{\text{(Basic oxide)}}{CaO} + \underset{\text{(Acidic oxide)}}{CO_2}$

68. (b) $Mg_3N_2 + 6H_2O \longrightarrow 3Mg(OH)_2 + 2NH_3$

69. (a)

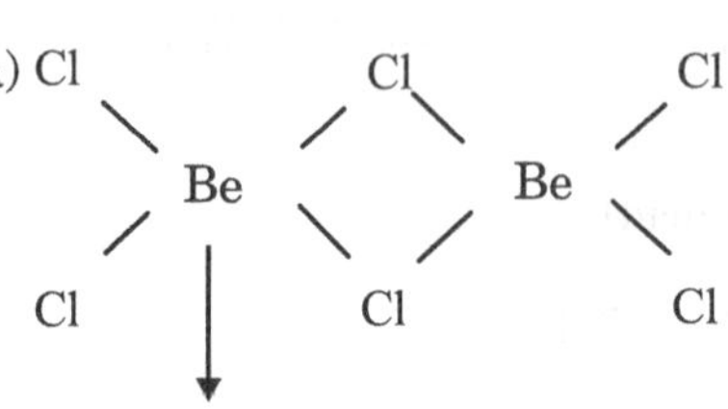

sp^3 hybridised
(Polymeric form)

Cl – Be – Cl
(monomer)

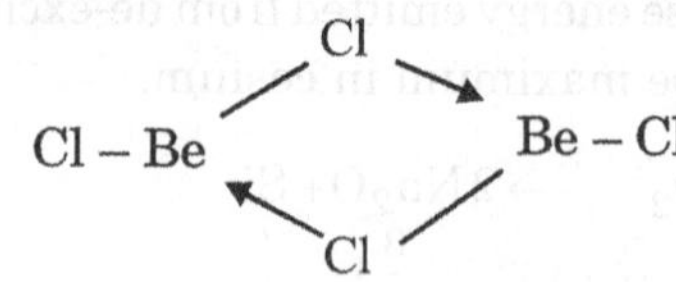

(Dimeric form) sp^2 hyb. Be

In the vapour state, $BeCl_2$ exists as a monomer with linear structure above 1200 K. But below 1200 K, It exist dimeric structure even in vapour state.

70. (a) Brick red coloration is shown by Ca.

So, (A) is Ca.

$\underset{(A)}{3Ca} + N_2 \longrightarrow \underset{(B)}{Ca_3N_2}$

$Ca_3N_2 + H_2O \longrightarrow \underset{(D)}{3Ca(OH)_2} + \underset{(C)}{2NH_3}$

$\underset{(D)}{Ca(OH)_2} + \underset{(Air)}{CO_2} \longrightarrow CaCO_3 + H_2O$

72. (d) Rb_2CO_3 is more stable than Li_2CO_3 ($\because$ Rb^+ is bigger)

73. (a) $\underset{(A)}{CaO} + H_2O \longrightarrow \underset{(B)}{Ca(OH)_2}$

$Ca(OH)_2 + CO_2 \longrightarrow \underset{(C)}{CaCO_3} + H_2O$

75. (a) BeO is amphoteric & possesses high M.pt.

76. (d) High hydration energy & high lattice energy more than compensates the sum of 1st & 2nd Ionisation Mg^{2+} energies in .

77. (c) $CaSO_4 \cdot 2H_2O$ (Gypsum)

$CaSO_4 \cdot \frac{1}{2}H_2O$ (P.O.P)

81. (d) $CaC_2 + 2H_2O \longrightarrow Ca(OH)_2 + C_2H_2 \uparrow$

$CaC_2 + N_2 \longrightarrow \underbrace{CaCN_2 + C}_{\text{NITROLIM}}$

83. (b) $CaO + CO_2 \longrightarrow CaCO_3$

$CaO + H_2O \longrightarrow Ca(OH)_2$; $\Delta H = -ve$

86. (c) Castner kellner shell is used to manufacture NaOH.

87. (b) $Be(OH)_2 + 2\bar{O}H \longrightarrow [Be(OH)_4]^{2-}$ Beryllate ion

88. (c) Bigger anions (CO_3^{2-}) are stabilized by bigger cations.

89. (d) Both (a) & (b) are d-block elements.

91. (a) $Mg(NO_3)_2 \longrightarrow MgO + NO_2 \uparrow + O_2 \uparrow$

92. (d) $MgCl_2 \longrightarrow Mg^{2+} + 2Cl^-$

$Mg^{2+} \downarrow Mg$ (Cathode); $2Cl^- \downarrow Cl_2$ (Anode)

KCl is unaffected by applied voltage.

96. (a) Because of higher charge density on Be^{2+}, it has the maximum hydration energy & hence, the largest size.

97. (d) Both KNO_3 & $CsNO_3$ on being heated give corresponding nitrites & O_2. So, they can't be distinguished merely by heating.

98. (a) Be cannot expand its octet & cannot show covlency of 6.

100. (b) $BaO_2 + H_2SO_4 \longrightarrow BaSO_4 + H_2O_2$

In $BaSO_4$, oxidation state of O is –2

In H_2O_2, oxidation state of O is 1.

101. (d) $BaCl_2$ is sparingly soluble in water at room temperature. But fairly soluble in boiling water.

$(BaSO_4 + ZnS)$ is called lithopone.

102. (c) MgO is insoluble but $MgSO_4$ is soluble BaO is soluble but $BaSO_4$ is insoluble.

103. (a) $CaSO_4 \cdot 2H_2O \xrightarrow{200^\circ C} (B) \xrightarrow[\text{heating}]{\text{Strong}} (C)$

$\downarrow$ (A)

104. (a) $\underset{(X)}{CaCO_3} \xrightarrow{\Delta} CaO + CO_2 \uparrow_{(g)}$

$CaO + H_2O \longrightarrow \underset{(Y)}{Ca(OH)_2}$ soluble

$\underset{(X)}{CaCO_3} + H_2O + CO_2 \longrightarrow \underset{(Z)}{Ca(HCO_3)_2}$

$\underset{(Z)}{Ca(HCO_3)_2} \longrightarrow \underset{(X)}{CaCO_3} + H_2O + CO_2$

105. (c) $Li_2CO_3 \xrightarrow{\Delta} Li_2O + CO_2 \uparrow$

$MgCO_3 \xrightarrow{\Delta} MgO + CO_2 \uparrow$

$Na_2CO_3 \xrightarrow{\Delta}$ No effect

Total 2 moles of CO_2 are evolved. So, value of CO_2 evolved at STP = $2 \times 22 \cdot 4$

= 44 8L

106. (a) $Mg \xrightarrow{\text{Air}} \underset{(y)}{MgO} + \underset{(x)}{Mg_3N_2}$

$\underset{(x)}{Mg_3N_2} + H_2O \longrightarrow MgO + \underset{(z)}{NH_3} \uparrow$

$\underset{(z)}{NH_3} \xrightarrow{H_2O} NH_4OH \xrightarrow{CuSO_4}$

$[Cu(NH_3)_4]SO_4$

107. (a) $CaH_2 + 2H_2O \longrightarrow Ca(OH)_2 + 2H_2 \uparrow$,
$Ca(OH)_2$ ↓ alkaline solution; H_2 ↓ readily combustible gas

$Ca(OH)_2 + SO_2 \longrightarrow CaSO_3 + H_2O$

$Ca(OH)_2 + SO_2 \longrightarrow CaSO_3 + H_2O$

Smart Notes

12

CHAPTER

Some p-Block Elements

1. The correct order of Atomic radii of Gp. 13 elements is

(a) Tl > In > Al > Ga > B

(b) Tl > In > Ga > Al > B

(c) Tl > Ga > In > Al > B

(d) Tl > Ga > Al > B > In

2. Which one of the following is incorrect?

(a) B > Tl > Ga > Al > In (Ionization Energy)

(b) Al > Ga > In > Tl (Reducing power)

(c) $Tl^{3+} < Tl^{+}$ (Stability of oxdn. State)

(d) None of these

3. Some statements are given regarding Gp.13 elements. The number of correct statements are.

1. The stability of monovalent ions follows the order - $Tl^{+} > In^{+} > Ga^{+}$

2. $Ga^{3+} > Ga^{+}$ (stability)

3. $In^{3+} > In^{+}$ (stability)

4. $Tl^{+} > Tl^{3+}$ (stability)

5. $Tl^{+} > In^{+}$ (Reducing power)

6. $Tl^{3+} > In^{3+}$ (oxidizing power)

(a) 3 statements/orders are correct

(b) 5 orders/statements are correct

(c) 2 orders are correct

(d) All are correct.

4. Following two reactions are given,

(1) $Al(OH)_3 + NaOH \longrightarrow Na^{+}[Al(OH)_4]^{-}$

(2) $2\,Al(OH)_3 + 3H_2SO_4 \longrightarrow Al_2(SO_4)_3 + 6H_2O$

Select the correct option wrt these reactions.

(a) In Reaction 1, $Al(OH)_3$ acts as Bronsted acid whereas in reaction 2, $Al(OH)_3$ acts as Bronsted base.

(b) In Reaction 1, $Al(OH)_3$ acts as lewis acid whereas in reaction 2, $Al(OH)_3$ acts as Arrhenius base.

(c) In Reaction 1, $Al(OH)_3$ acts as lewis base whereas in reaction 2, $Al(OH)_3$ acts as lewis acid.

(d) None is correct.

5. Select the correct option among the following

(a) Monohalides of Ga & In undergo disproportionation.

(b) B & Al directly combine with N_2 to form corresponding nitrides.

(c) $(BN)_n$ is called Inorganic graphite

(d) All are correct

6. Oxide & hydroxide of which of the following Gp.13 element is amphoteric

(a) B (b) Al

(c) In (d) Tl

7. Which one of the following is not correct for allotropic forms of C?

(a) C60 molecule has a shape like soccer ball & is called Buckminster fullerene

(b) Graphite fibers embedded in plastic form high strength composites

(c) In graphite, C is sp^2 hybridised .

(d) In diamond, C is sp^3 hybridised & paramagnetic.

8. Select the 'INCORRECT REACTION.'

(a) $2Al + 6HCl \longrightarrow 2AlCl_3 + 3H_2 \uparrow$

(b) $2Al + 2NaOH + 6H_2O \longrightarrow 2Na^+[Al(OH)_4]^- + 3H_2 \uparrow$

(c) $AlN + H_2O \longrightarrow Al(OH)_3 + NH_3$

(d) None of these is incorrect

9. The correct order of lewis acidic strength of trihalides of Boron is

(a) $BF_3 > BCl_3 > BBr_3$

(b) $BCl_3 > BBr_3 > BF_3$

(c) $BBr_3 > BCl_3 > BF_3$

(d) $BCl_3 > BBr_3 = BF_3$

10. Actual (theoretical) B - F bond lengths is 149 pm. But in BF_3 the observed bond length is

(a) 160 pm

(b) 330 pm

(c) 170 pm

(d) 130 pm

11. Which of the following reactions are correct?

(a) $2NaBH_4 + I_2 \longrightarrow B_2H_6 + 2NaI + H_2$

(b) $2BF_3 + 6NaH \xrightarrow{450K} B_2H_6 + 6NaF$

(c) $4BF_3 + 3LiAlH_4 \longrightarrow 2B_2H_6 + 3LiF + 3AlF_3$

(d) All are correct

12. Which one of the following is NOT TRUE for structure of Diborane?

(a) Both the B atoms are sp^3 hybridized.

(b) There are 2 banana bonds (3c - 2e) & four 2c - 2e, bonds

(c) It is an electron deficient compound.

(d) None of these

13. Select the correct match.

Reaction	Products
1. $B_2H_6 + O_2 \longrightarrow$	(W) $Na^+(BH_4)^-$
2. $NaH + B_2H_6 \xrightarrow{ether}$	(X) $B_2O_3 + H_2O$
3. $B_2H_6 + H_2O \longrightarrow$	(Y) $2BH_3 \cdot NMe_3$
4. $B_2H_6 + 2NMe_3 \longrightarrow$	(Z) $H_3BO_3 + H_2$

(a) 1 – X, 2 – W, 3 – Z, 4 – Y

(b) 2 – X, 3 – Z, 4 – W, 1 – Y

(c) 2 – X, 2 – Y, 3 – Z, 4 – W

(d) 4 – W, 3 – Y, 2 – X, 1 – Z

14. Consider the following reaction

$$3B_2H_6 + 6NH_3 \xrightarrow[temp]{low} A \xrightarrow{473k} B$$

A & B in above reaction are, respectively

(a) $[BH_2(NH_3)_2]^+[BH_4]^-$ & $B_3N_3H_6$

(b) $B_3N_3H_6$ & B_2O_3

(c) $[BH_2(NH_3)_2]^+(BH_4)^-$ & BN

(d) None of these

15. Select the correct statement about hydrolysis of Borax.

(a) Partial hydrolysis of borax yields sodium hydroxide & tetraboric acid

(b) Complete hydrolysis of borax gives sodium hydroxide & orthoboric acid

(c) In both the above circumstances, the aqueous solution of borax is alkaline

(d) All are correct.

16. Which of these is 'INCORRECT' w.r.t BORAX BEAD TEST?

(a) $Na_2B_4O_7 \xrightarrow{heat} 2NaBO_2 + B_2O_3$

(b) $CoO + B_2O_3 \xrightarrow{\Delta} \underset{\text{(blue)}}{\underset{\text{Cobalt metaborate}}{Co(BO_2)_2}}$

(c) $Cr_2O_3 + 3B_2O_3 \xrightarrow{\Delta} \underset{\text{(green)}}{\underset{\text{Chromium metaborate}}{Cr(BO_2)_3}}$

(d) $NiO + B_2O_3 \longrightarrow \underset{\text{(blue)}}{\underset{\text{Nickel metaborate}}{Ni(BO_2)_2}}$

17. Which of the following Boron Compounds can be used to prepare

(a) $Na_2B_4O_7$ (b) BCl_3

(c) BN (d) All

18. In aqueous solution, H_3BO_3 behaves as

(a) Arrhenius acid (b) Lewis acid

(c) Bronsted acid (d) Protonic acid

19. Select the correct use of respective Compounds of Boron.

(a) Borax is used in the preparation of medicinal soaps due to its antiseptic properties.

(b) Aq. Solution of boric acid called boric lotion is used as mild antiseptic.

(c) B is extremely hard refractory solid & is used in making bullet proof vest.

(d) All uses are correct.

20. Which one of the following is partially hydrolysed by water?

(a) BF_3 (b) BCl_3

(c) BBr_3 (d) Both 2 & 3

21. Borazine is considerably more reactive than benzene. Select the correct reaction.

(a) $B_3N_3H_6 + 3HCl \longrightarrow B_3N_3H_9Cl_3$

(b) $B_3N_3H_6 + 3HCl \longrightarrow B_3N_3Cl_3 + \frac{9}{2}H_2 \uparrow$

(c) $B_3N_3H_6 + HCl \longrightarrow$ No Reaction

(d) None of these

22. Correct statement about Borazine is

(a) It is aromatic

(b) It is isoelectronic with organic benzene

(c) It is more reactive than organic benzene because of its polar nature

(d) All are correct

23. If H_b & H_T are the bridged hydrogen & terminal hydrogen respectively in the structure of diborane, then which statement is incorrect about this compound?

(a) Hybridization of Boron in the molecule is sp^3

(b) Bond angle H_bBH_b is greater than H_tBH_t

(c) Bond length of B.......H_b is greater than $B-H_t$.

(d) Bond angle of H_bBH_b is lesser than H_tBH_t

24. Boron $+ A \longrightarrow B \xrightarrow{H_2O} C + D$

Compound 'C' has a layered structure & Compound 'D' on reaction with diborane forms a compound E which is called 'Inorganic benzene'. The Compounds A & C respectively are -

(a) O_2 & $(BN)_x$ (b) N_2 & $B(OH)_3$

(c) H_2 & B_2H_6 (d) Cl_2 & H_3BO_3

25. (a) $Al_4C_3 + H_2O \longrightarrow \underset{\text{(ppt)}}{X} + \underset{\text{(gas)}}{Y}$

(b) $X + NaOH \longrightarrow Z$

In the above reactions

(I) Z is a co-ordinate complex

(II) Reducing power of Y is greater than similar compound formed by lead.

(III) X acts as Lewis acid in second reaction

(IV) Hybrid state of central atom in Al_4C_3 & $Y_{(gas)}$ is same.

Correct statements are -

(a) I, III, IV but not II (b) II & IV only

(c) both I & III only (d) All are correct.

26. $x + 3LiAlH_4 \longrightarrow A + LiF + 3AlF_3$

$y + 3LiAlH_4 \longrightarrow B + LiCl + 3AlCl_3$

In the above reactions, x & y are boron compounds. Then,

(a) x is more acidic than y.

(b) x & y both are electron rich compounds

(c) Bond order in x is more than that in Y.

(d) x can be easily hydrolyzed as compared to y.

27. The ratio of number of bonds in an aromatic compound obtained by treating 2 mole NH_3 & 1 mole B_2H_6 at high temperature is 'x'. This ratio is 'y' in the gaseous compound obtained by acidification of metal carbonate. Then, $x + y$ is -

(a) 4 (b) 8

(c) 5 (d) 3

28. Which of the following is incorrect order of the property mentioned along?

(a) $Al < Ga < In < Tl$ (stability of +1 oxidation state)

(b) $Al < Ga < In < Tl$ (stability of + 3 oxidation state)

(c) $AlX_3 > GaX_3 > InX_3 > TlX_3$ (increasing Lewis acid character)

(d) $Ga < Al < In < Tl$ (increasing atomic radius)

29. B – F bond length is smallest in -

(a) BF_4^- (b) $BF_3 \leftarrow CO$

(c) $BF_3 \leftarrow NMe_3$ (d) BF_3

30. Among the following the metal protected by a layer of its own oxide is -

(a) Al (b) B

(c) C (d) Ga

31. Select the correct match

Column - I	Column-II
(A) Orthoboric acid	(P) sp^2 hybridized Boron
(B) Diborane	(Q) Ring having $6\pi e^-$ delocalized
(C) Borazine	(R) Reacts with H_2O
(D) Boron nitride	(S) Bond having partial double bond character
	(T) sp^3 hybridized Boron

(a) (A) → P,R,S ; (B) → R, T; (C) → P, Q, R, S ; (D) – P, Q, R, S

(b) (A) → R, S, T ; (B) → Q, T; (C) – R, S, Q ; (D) – Q, R, T

(c) (A) → R, S, T ; (B) → Q, T; (C) – Q, S, T ; (D) – Q, R, T

(d) None of these

32. Consider the following statements

(1) Boron forms covalent compounds

(2) Trihalides of boron are electron deficient

(3) Boron can show maximum covalency of 6

Of these, the correct statements are -

(a) only 1 & 2 (b) only 2 & 3

(c) only 3 (d) All are correct

33. Which one of the following is NOT correct?

(a) Anhydrous $AlCl_3$ is a covalent compound.

(b) Hydrated $AlCl_3$ is ionic in nature.

(c) Aluminium reacts with NaOH to give H_2 gas

(d) Boron does not give H_2 gas with NaOH

34. Gallium & indium forms dihalides as follows -

(1) $GaCl_3 + Ga \rightarrow 2GaCl_2$

(2) $In + HCl_{(gas)} \rightarrow 2InCl_2$

The oxidation states of Ga & In in these dihalides are -

(a) +1 & +3 (b) +2 only

(c) +3 & +3 (d) +4 & +4

35. Select the correct match

Oxide	Nature
(a) $B_2O_3 \longrightarrow$	Acidic oxide
(b) $SiO_2 \longrightarrow$	acidic oxide
(c) $Al_2O_3 \longrightarrow$	amphoteric oxide

(d) None of these is incorrect

36. Select the incorrect reaction.

(a) $B_2H_6 + NH_3 \xrightarrow[\text{NH}_3 \text{ low temp.}]{\text{excess}} B_2H_6 \cdot 2NH_3$

(b) $B_2H_6 + NH_3 \xrightarrow[\text{NH}_3 \text{ high temp.}]{\text{excess}} (BN)_n$

(c) $B_2H_6 + NH_3 \xrightarrow[\text{NH}_3 \text{ higher temperature}]{\text{ratio } 2NH_3 : 1B_2H_6} B_3N_3H_6$

(d) None of these is incorrect

37. When H_3BO_3 is heated above 300K, it gives -

(a) Metaboric acid (b) Tatraboric acid

(c) Borax (d) Borazine

38. A coloured salt of cobalt is subjected to borax bead test. The transparent borax bead turns blue. From the given reaction, identify the one which does not take place during the process.

(a) $4NaBO_2 + CO_2 \longrightarrow Na_2B_4O_7 + Na_2CO_3$

(b) $Na_2B_4O_7 \longrightarrow 2NaBO_2 + B_2O_3$

(c) $Na_2[B_4O_5(OH)_4] . 8H_2O \longrightarrow Na_2B_4O_7 + 10H_2O$

(d) $CoO + B_2O_3 \longrightarrow \underset{\text{(blue)}}{Co(BO_2)_2}$

39. Diborane forms an adduct with -

(a) PF_3 (b) BF_3

(c) H^+ (d) NH_4^+

40. Aluminium is more reactive than iron but still aluminium is less easily corroded than iron because -

(a) Aluminium is a noble metal

(b) Oxygen forms a protective oxide layer

(c) Iron turns passive

(d) None of these

41. Boric acid is a very weak acid but in the presence of certain organic compounds, it acts as a strong acid. Which one of the following organic compounds can affect such change?

(a) Formic acid

(b) Glycol (Cis - 1, 2 - diol)

(c) Acetic acid

(d) Ethylene

42. In B_2H_6, the type of overlapping of orbitals involved in banana bond formation is -

(a) $sp^3 - s$ (b) $sp^2 - s - sp^2$

(c) $sp^3 - s - sp^3$ (d) None of these

43. Which of the following are true about diborane ?

(a) it has two bridging hydrogens & four terminal hydrogen.

(b) When methylated, the product is $(CH_3)_4B_2H_2$.

(c) The bridging hydrogens are in a plane perpendicular to the rest.

(d) All of the above

44. Which of the following statements is false?

(a) Small amines such NH_3,CH_3NH_2 & $(CH_3)_2NH$ as give unsymmetrical cleavage of diborane.

(b) Boron hydrides are readily hydrolysed.

(c) Diborane is an e^- deficient compound.

(d) None of these

G.p. 14

45. Identify the correct order wrt the property mentioned

(a) C > Si > Ge > Pb> Sn (Ionisation Energy)

(b) C > Si > Ge > Sn > Pb (Stability of + 4 oxidation state)

(c) Pb > Sn > Ge > Si > C (Stability of +2 oxidation state)

(d) All orders are correct

46. Tendency to catenate in Gp 14 follows the order $C >> Si > Ge \approx Sn$; C has maximum tendency to Catenate because

(a) IE of C is minimum

(b) IE of C is maximum

(c) C is most abundant element

(d) C-C bond enthalpy is maximum

47. Nitrogen in trisilylamine is sp^3 hybridized where as Nitrogen in trimethylamine is sp^3 hybridized because

(a) Nitrogen is smaller in size as compared to silicon

(b) Silicon is smaller is size as compared to Nitrogen

(c) There is $p\pi$ - $d\pi$ back bonding in trisilylamine whereas it is absent in trimethylamine.

(d) None of these

48. Select the correct statement

(a) $(CH_3)_3$ N is a stronger base than $(SiH_3)_3$ N

(b) $(SiF_6)^{2-}$ exists but $(CF_6)^{2-}$ does not

(c) $(SiF_6)^{2-}$ is known but $(SiCl_6)^{2-}$ is not.

(d) All are correct

49. Silicon cannot form graphite type of an allotrope because

(a) It is smaller in size

(b) of its high ionization energy

(c) of its inability to form $p\pi - p\pi$ bonding

(d) It cannot catenate

50. Oxalic acid $(H_2C_2O_4)$ $\xrightarrow[H_2SO_4]{conc.}$ X + Y. X is a neutral oxide whereas Y can absorb in

(a) H_2SO_4 (b) $CuCl_2$

(c) Cu_2Cl_2 (d) KOH

51. Which oxide of C has 'NO' overall dipole moment?

(a) CO (b) C_3O_2

(c) CO_2 (d) Both (b) & (c)

52. A colourless gas which burns with blue flame & reduces CuO to Cu is

(a) N_2 (b) CO

(c) CO_2 (d) NO_2

53. CO_2 is gas whereas SiO_2 is a solid because

(a) C is more electronegative than Si.

(b) C can Catenate more than Si.

(c) C is larger than Si.

(d) C can form $p\pi$ - $p\pi$ bonds & exist as discrete molecule held by weak Vander Waal's forces. Si can't form $p\pi$ - $p\pi$ bond with itself. To satisfy its tetra valency it has to form 4 Sigma bonds & hence polymeric network.

54. Silicones are synthetic organosilicon compounds containing repeated R_2SiO units held by Si – O – Si linkages. Select the TRUE STATEMENT(S) about Silicones.

(a) Linear polymer is formed by R_2SiCl_2

(b) Cross linked polymer is formed by $RSiCl_3$

(c) R_3SiCl is used to block chain length

(d) All are correct

55. Pyrosilicates have the general formula

(a) SiO_4^{4-} (b) $Si_2O_7^{6-}$

(c) $(SiO_2)_n$ (d) $\left[(Si_4O_{11})^{6-}\right]_n$

56. Which of the following tetrachlorides can't be hydrolysed by water?

(a) $SiCl_4$ (b) CCl_4

(c) Both of these (d) None of these

57. The correct formulae for water gas & producer gas are, respectively

(a) $CO + H_2$, $CO + N_2$

(b) $CO + N_2$, $CO + H_2$

(c) $CO + CH_3OH$, $CO + CH_3COOH$

(d) $CO + Cl_2$, $CO + O_2$

58. In diamond & graphite, the hybridization of C are, respectively

(a) sp, sp (b) sp, sp^2

(c) sp^2, sp^3 (d) sp^3, sp^2

59. Which one of the following does not exist?

(a) $PbCl_4$ (b) $PbBr_4$

(c) PbI_4 (d) Both (a) & (b)

60. A certain compound, when burnt, give three oxides. The first turned lime water milky, the second turned anhydrous $CuSO_4$ blue & the third formed an amphoteric compound. The elements present in the compound are -

(a) B, H & C (b) C, B & Si

(c) C, H & Al (d) C, B & Al

61. A mineral consists of three 'SiO_4' tetrahedra that shares corners with adjacent tetrahedral. The mineral also contains Ca^{2+} ions, Cu^{2+} ions & H_2O molecules in the ratio 1 : 1: 1. The complete formula for the mineral is -

(a) $CaCuSi_3O_{10} \cdot H_2O$

(b) $CaCu_2Si_3O_{10} \cdot 2H_2O$

(c) $Ca_2CuSi_2O_5 \cdot H_2O$

(d) $Ca_2Cu_2Si_3O_{10} \cdot 2H_2O$

62. Match the carbides in Column -I with their hydrolysis products in Column-II.

Column-I	Column-II
(a) SiC	p. Hydrocarbon with a $C \equiv C$ is formed
(b) Al_4C_3	q. Unsaturated hydrocarbon is produced
(c) CaC_2	r. No hydrocarbon is produced
(d) Mg_2C_3	s. Saturated hydrocarbon is produced.

(a) $a-r; b-p; c-q,s; d-p,s$

(b) $a-r; b-s; c-p,q; d-p,q$

(c) $a-q,s; b-p,r; c-p; d-r$

(d) $a-p,s; b-q,r; c-s; d-r$

63. Carbon forms an anion containing one σ & 2π bonds, then the anion is

(a) C^{4-} (b) C_2^{2-}

(c) C_3^{4-} (d) C_3^{-}

64. An oxide 'x' of element of gp. 14 has the following properties -

(a) It has a three dimensional network solids.

(b) It has a very high melting point

(c) Kieselgur is an amorphous form of this oxide

(d) It forms water glass on combining with NaOH

Oxide 'x' is -

(a) CO (b) SiO_2

(c) CO_2 (d) SnO

65. In which of the following compounds vacant orbitals participate in hybridization?

(I) B_2H_6 (II) $N(SiH_3)_3$

(III) BF_4^-

(a) I only (b) both I & II

(c) I, II & III (d) both I & III

66. Which of the following is a consequence of inert pair effect?

(a) $SnCl_2$ acts as reducing agent.

(b) $SnCl_4$ acts as oxidizing agent.

(c) $TlCl_3$ acts as oxidizing agent.

(d) PbO_2 is an oxidant

(a) a, c & b (b) only d

(c) a, b, c and d (d) both c & d

67. Which of the following is a consequence of inert pair effect?

(a) $SnCl_2$ acts as a reducing agent

(b) $SnCl_4$ acts as an oxidizing agent.

(c) SnO_2 is amphoteric

(d) PbO_2 is an oxidant

(e) CCl_2 is unstable but $PbCl_2$ is stable.

(a) a, d, e (b) d, e

(c) a, b, c, d, e (d) a, b, c

68. HCOOH on being dehydrated with conc. H_2SO_4 gives -

(a) CO_2 (b) CO

(c) C_3O_2 (d) Both (a) & (c)

69. Carbon sub oxide (C_3O_2) can be prepared by dehydrating 'x' with P_4O_{10}. 'x', is -

(a) Maloic acid (b) Fumaric acid

(c) Malonic acid (d) Salicylic acid

70. Silicones repel water due to -

(a) the presence of alkyl groups pointed towards surface

(b) strong Si – O – Si bonds

(c) less surface area

(d) high magnitude of Vander waal's forces

71. N_2 is passed through heated CaC_2. Then, which of following options is 'INCORRECT' for the product formed?

(a) Hybridisation of C is sp in cyanamide ion.

(b) Urea is an intermediate formed during the hydrolysis of above product

(c) Hydrolysis of the above product gives rise to N_2 gas.

(d) Both (a) & (b)

72. A Scarlet red compound (X) on treatment with conc. HNO_3 gives compound (Y) & (Z). (Z) with HCl produces a chloride compound (A) which can also be produced by treating (X) with conc. HCl. Compound X, Z & A will be respectively -

(a) $Mn_3O_4, MnO_2, MnCl_2$

(b) $Pb_3O_4, Pb(NO_3)_2, PbCl_2$

(c) $Fe_3O_4, Fe_2O_3, FeCl_2$

(d) $Fe_3O_4, FeO, FeCl_3$

73. $Si_2O_7^{6-}$ anion is obtained when -

(a) No oxygen of SiO_4 tetrahedron is shared with another SiO_4 tetrahedron.

(b) One oxygen of SiO_4 tetrahedron is shared with another SiO_4 tetrahedron.

(c) two oxygen of SiO_4 tetrahedron are shared with another SiO_4 tetrahedron.

(d) three or all four oxygen of a tetrahedron are shared with other SiO_4 tetrahedron

74. $C(OH)_4$ is unstable while $Si(OH)_4$ is relatively stable because -

(a) Carbon can form stronger 2pπ-2pπ bonds with oxygen by eliminating H_2O whereas Silicon, due to its bigger size will try to form 3pπ-2pπ bond with O which are relatively unstable

(b) Silicon is more electro – ve than C.

(c) Si - Si bond is stronger than C - C bond.

(d) Both (b) & (c)

75. $(Me)_2SiCl_2$ On hydrolysis will produce -

(a) $Me_2Si = O$ (b) $Me_2SiCl(OH)$

(c) Me_3SiCl (d) $-\!\!\left[O-(Me)_2Si-O\right]_n\!\!-$

76. The species present in solution when CO_2 is dissolved in water are -

(a) $CO_2, H_2CO_3, HCO_3^-, CO_3^{2-}$

(b) H_2CO_3, CO_3^{2-} only

(c) CO_3^{2-} & HCO_3^- only

(d) CO_2, H_2CO_3 only

77. Select the correct statement (s).

(a) CO can be prepared by heating HCOOH with Conc. H_2SO_4.

(b) CO_2 is anhydride of H_2CO_3.

(c) SiCis called carborundum.

(d) All are correct

78. Choose the correct match.

Column I	Column II
(1) Ortho silicate	(p) $(SiO_3)_n^{2n-}$
(2) Chain silicate	(q) $(Si_2O_5)_5^{10-}$
(3) Pyro silicate	(r) SiO_4^{4-}
(4) Cyclic silicate	(s) $(Si_4O_{11})_n^{6n-}$
	(t) $Si_2O_7{}^{6-}$

(a) 1 - r , 2 - p , s , 3 - t , 4 - p

(b) 1 - t , 2 - p , 3 - s , 4 - r

(c) 1 - p , 2 - p , 3 - s , t , 4 - r

(d) None of these

79. Select the correct statement among the following

(a) Water gas has more calorific value than producer gas.

(b) SiO_2 is a linear molecule like CO_2.

(c) $SiCl_6^{2-}$ is known but SiF_6^{2-} is not known.

(d) Both 2 & 3 are correct.

80. Which of the following statements is/are incorrect?

(a) $B_4O_7^{2-}$ on hydrolysis with acid/ H_2O yields $B(OH)_3$.

(b) SiO_4^{4-} on hydrolysis with acid/ H_2O yields $Si_2O_7^{2-}$.

(c) $MeSiCl_3$ on hydrolysis & then condensation gives a complex cross linked polymer of silicones.

(d) CO_2 is an acidic oxide.

81. Select the incorrect reaction among the following wrt preparation of CO.

(a) $H_2C_2O_4 \xrightarrow[H_2SO_4]{conc.} CO_{(g)} + CO_{2(g)} + H_2O$

(b) $K_4[Fe(CN)_6] + 6H_2SO_4 + 6H_2O \xrightarrow{\Delta}$ $2K_2SO_4 + FeSO_4 + 3(NH_4)_2SO_4 + 6CO$

(c) $C_{(s)} + CO_{2(g)} \longrightarrow 2CO_{(g)}$

(d) None of these

82. Which of the following is/are correct about CO & CO_2?

(a) CO reduces I_2O_5 to I_2.

(b) CO_2 is absorbed by $Ca(OH)_2$

(c) CO_2 reacts with sodium metaborate to form borax & sodium carbonate

(d) All are correct

83. Select the incorrect match

(a) $PbO \longrightarrow$ Amphoteric oxide

(b) $PbO_2 \longrightarrow$ Oxidising agent

(c) $CO_2 \longrightarrow$ Oxidising agent

(d) $SnCl_4.8H_2O \longrightarrow$ Butter of tin

Answers

1. (a)	**2.** (d)	**3.** (b)	**4.** (b)	**5.** (d)	**6.** (b)	**7.** (d)	**8.** (d)	**9.** (c)	**10.** (d)
11. (d)	**12.** (d)	**13.** (a)	**14.** (a)	**15.** (d)	**16.** (d)	**17.** (d)	**18.** (b)	**19.** (d)	**20.** (a)
21. (a)	**22.** (d)	**23.** (b)	**24.** (b)	**25.** (c)	**26.** (c)	**27.** (c)	**28.** (b)	**29.** (d)	**30.** (a)
31. (a)	**32.** (a)	**33.** (d)	**34.** (a)	**35.** (d)	**36.** (d)	**37.** (a)	**38.** (a)	**39.** (a)	**40.** (b)
41. (b)	**42.** (c)	**43.** (d)	**44.** (d)	**45.** (d)	**46.** (d)	**47.** (c)	**48.** (d)	**49.** (c)	**50.** (d)
51. (d)	**52.** (b)	**53.** (d)	**54.** (d)	**55.** (b)	**56.** (b)	**57.** (a)	**58.** (d)	**59.** (c)	**60.** (c)
61. (d)	**62.** (b)	**63.** (b)	**64.** (b)	**65.** (d)	**66.** (d)	**67.** (b)	**68.** (b)	**69.** (c)	**70.** (a)
71. (c)	**72.** (b)	**73.** (b)	**74.** (a)	**75.** (d)	**76.** (a)	**77.** (d)	**78.** (a)	**79.** (a)	**80.** (b)
81. (d)	**82.** (d)	**83.** (d)							

Explanations

1. (a) Poor screening effect of d electrons will decrease the size of Ga.

3. (b) $In^+ > Tl^+$ (Reducing power) [∵ In is more stable in +3]

Tl is more stable in +1, due to inert pair effect.

4. (b)

$$Al(OH)_3 + Na^+\overline{OH} \longrightarrow Na^+[Al(OH)_4^-]$$

Lewis acid (Base)

5. (d) $3\,GaX \longrightarrow 2Ga + Ga^{3+} + 3X^-$

$3\,InX \longrightarrow 2In + In^{3+} + 3X^-$

(Ga & In are not stable in +1 hence tend to disproportionate)

6. (b) Al_2O_3 & $Al(OH)_3$, both are amphoteric.

$$2Al(OH)_3 + 3H_2SO_4 \longrightarrow Al_2(SO_4)_3 + 6H_2O$$

↓

Arrhenius base

9. (c) In BF_3, B has a vacant 2p-orbital.

2pπ - 2pπ BACK BONDING OCCURS]

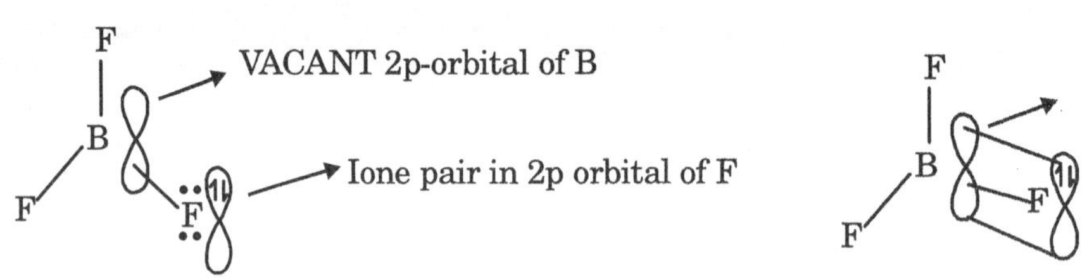

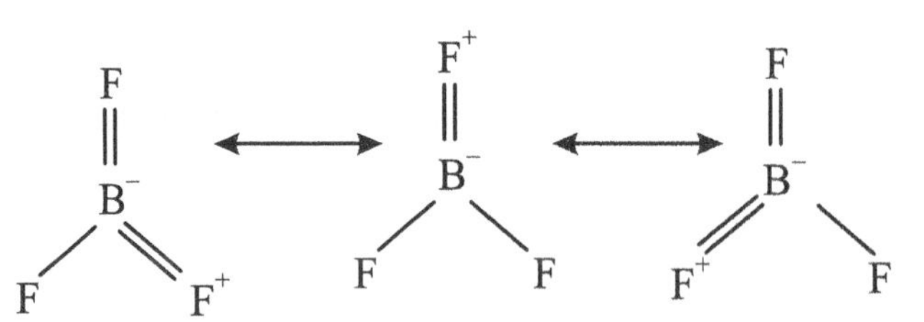

In BF_3, there is effective 2pπ - 2pπ back bonding. In $BaCl_3$, back bonding is less effective (2pπ - 3pπ) So, in BF_3, B is almost satisfied, so it is a weaker lewis acid.

10. (d) In BF_3, due to 2pπ - 2pπ back bonding, there is partial double bond character in B - F bond & its bond length is decreased.

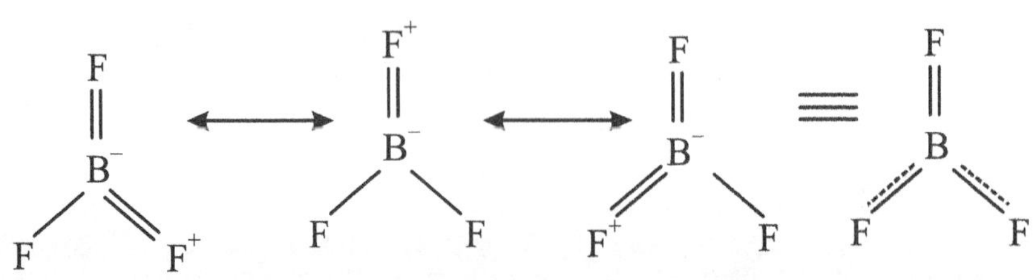

12. (d)

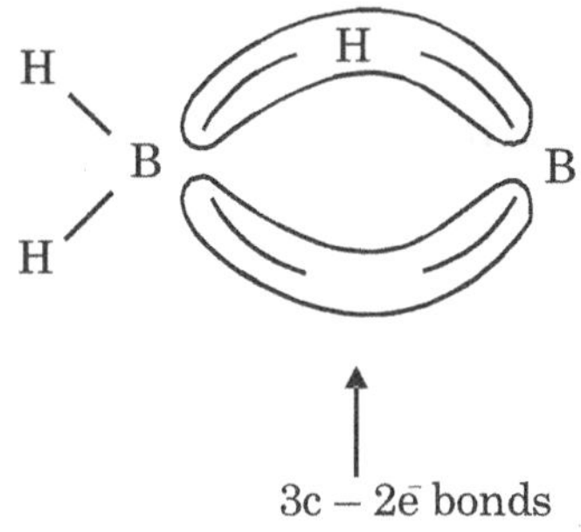

14. (a)

$$3B_2H_6 + 6NH_3 \xrightarrow[\text{temp}]{\text{low}} 3[BH_2(NH_3)_2]^+(BH_4)^-$$

$$\downarrow$$

$$2B_3N_3H_6 + 12H_2 \uparrow$$

15. (d) Borax is a Salt of strong base and weak acid, so it undergoes anionic hydrolysis.

$$Na_2B_4O_7 + 2H_2O \longrightarrow 2NaOH + H_2B_4O_7$$

$$H_2B_4O_7 + 5H_2O \longrightarrow 4H_3BO_3$$

16. (d) $Ni(BO_2)_2$ is brown.

17. (d) $Na_2B_4O_7 + 2HCl + 5H_2O \longrightarrow 4H_3BO_3 + 2NaCl$

$$Na_2B_4O_7 + H_2SO_4 + 5H_2O \longrightarrow 4H_3BO_3 + Na_2SO_4$$

$$BCl_3 + 3H_2O \longrightarrow H_3BO_3 + 3HCl$$

$$BN + 3H_2O \longrightarrow H_3BO_3 + NH_3$$

18. (b) $B(OH)_3 + H_2O \longrightarrow (B(OH)_4)^- + H^+$

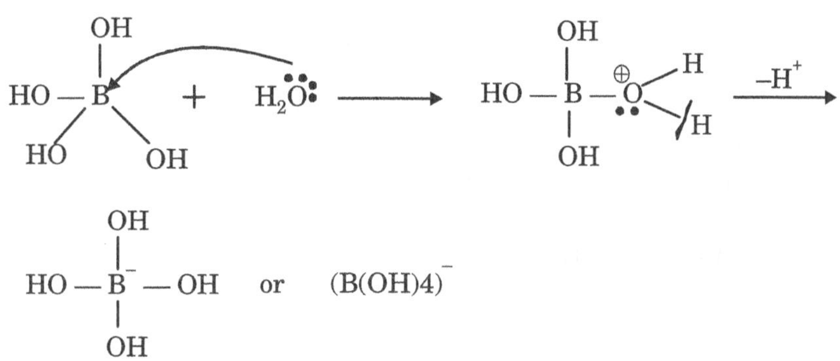

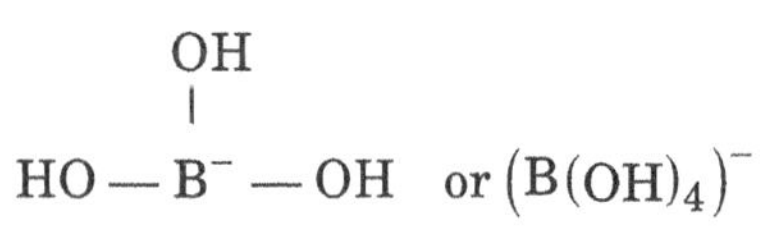

20. (a) Trihalides of Boron are completely hydrolysed except BF_3.

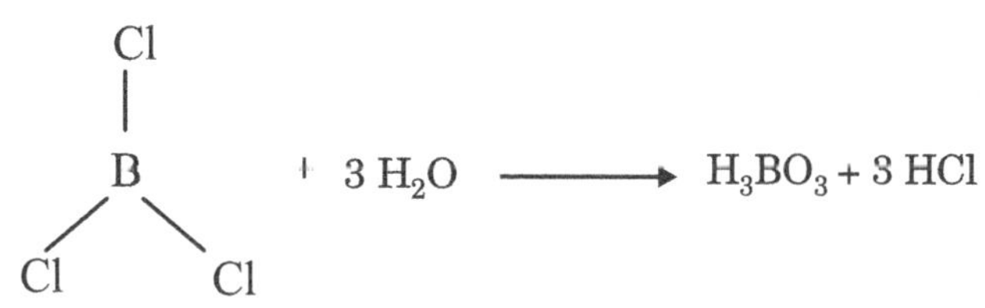

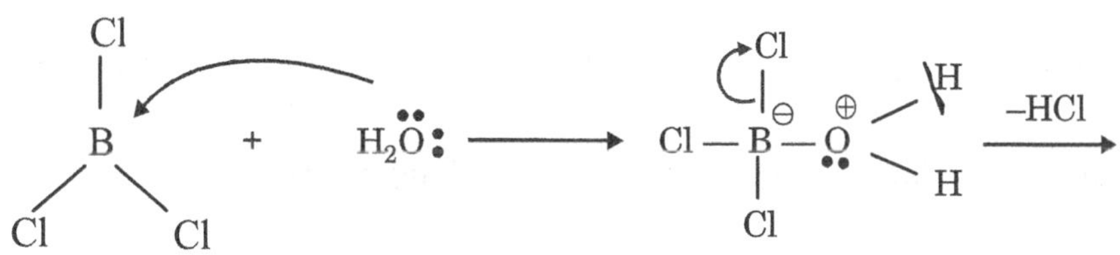

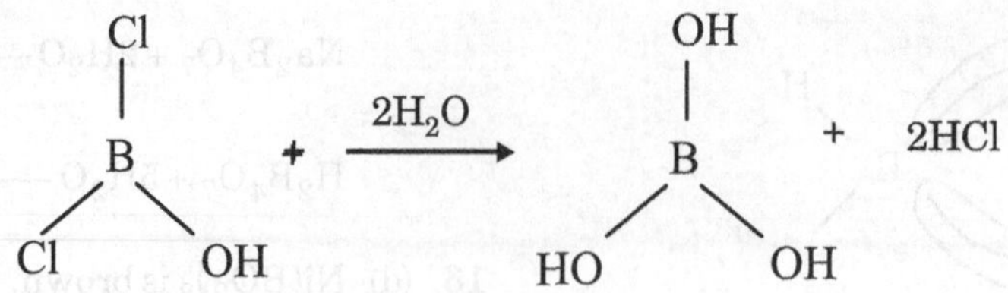

BF_3 hydrolysis incompletely & forms fluoroborates.
This is because HF first formed reacts with H_3BO_3.

$$\begin{aligned} 4BF_3 + 12H_2O &\longrightarrow 4H_3BO_3 + 12HF \\ 12HF + 3H_3BO_3 &\longrightarrow 3H^+ + 3(BF_4)^- + 9H_2O \\ \hline 4BF_3 + 3H_2O &\longrightarrow H_3BO_3 + 3H^+ + 3(BF_4)^- \end{aligned}$$

21. (a)

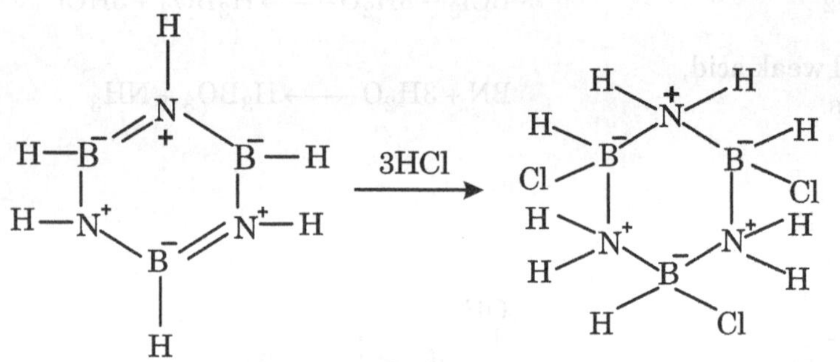

24. (b) Boron + $\underset{(A)}{N_2} \longrightarrow \underset{(B)}{BN} \xrightarrow{H_2O} \underset{(C)}{H_3BO_3} + \underset{(D)}{NH_3}$

25. (c) (1) $Al_4C_3 + H_2O \longrightarrow \underset{\text{ppt. (X)}}{Al(OH)_3} + \underset{(Y)}{CH_{4(gas)}}$

(2) $Al(OH)_3 + \overset{+}{Na}\overset{-}{OH} \longrightarrow \underset{(Z)}{Na^+[Al(OH)_4]^-}$

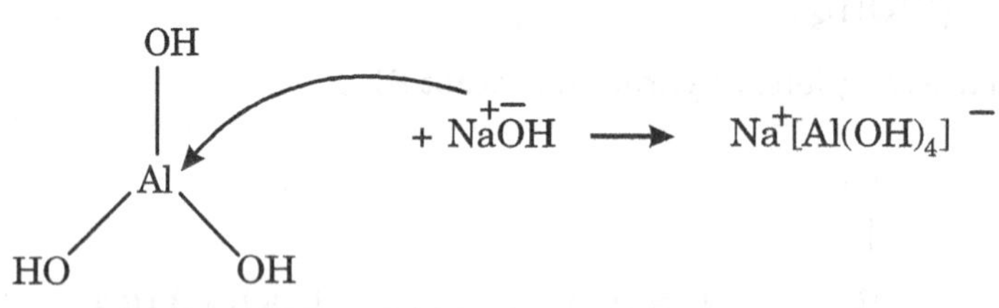

(Lewis acid)

(Al has vacant p orbitals)

$\longrightarrow$ Reducing power of $Y_{(CH_4)}$ is less than PbH_4 because of greater bond strength of C – H bond

$\longrightarrow$ Al_4C_3 is an ionic carbide $Al^{3+}C^{4-}$. So, no chance of any hybridization

26. (c) x is & Y is BCl_3. Both the trihalides are electron deficient

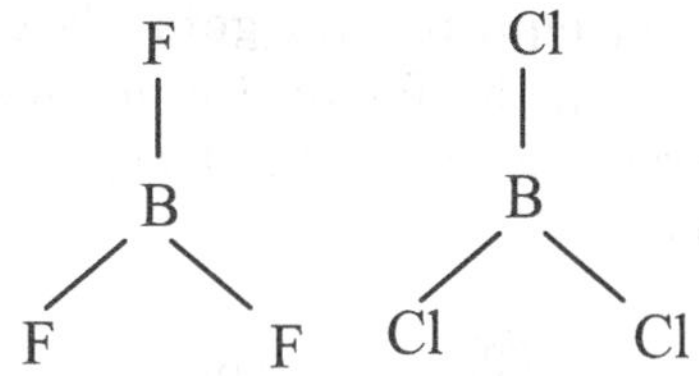

(y) BCl_3 is more acidic than $BF_3(x)$ because of less effective $2p\pi - 3p\pi$ back bonding in BCl_3.

Because of effective $2p\pi - 3p\pi$ back bonding in BF_3, the B – F bond order will be increased

⟶ Because of partial double bond character in B – F bond the bond strength is enhanced & hence hydrolysis becomes tougher

27. (c) NH_3 & B_2H_6 react at high temperature in ratio, 2 : 1 to form borazine ($B_3N_3H_6$)

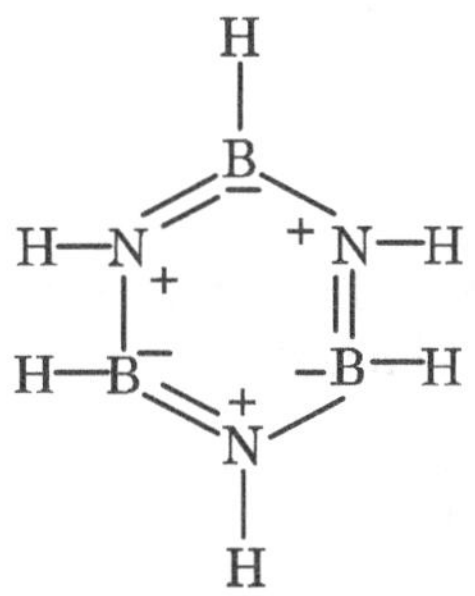

$$x = \frac{\text{No. of } \sigma \text{ bonds}}{\text{No. of } \pi \text{ bonds}} = \frac{12}{3} = 4$$

So, $x = 4$

⟶ Acidification of a metal carbonate

$$CaCO_3 + 2HCl \longrightarrow CaCl_2 + H_2O + \underset{\text{(gas)}}{CO_2} \uparrow$$

Gaseous compound is CO_2

$O = C = O$

$$y = \frac{\text{No. of } \sigma \text{ bonds}}{\text{No. of } \pi \text{ bonds}} = \frac{2}{2} = 1$$

$$\boxed{\therefore x + y = 5}$$

28. (b) Al^{3+} is more stable

Tl^+ is stable.

$Al^{3+} > Ga^{3+} > In^{3+} > Tl^{3+}$ (actual order)

29. (d) In $F - \underset{F}{\overset{F}{B}} \leftarrow F^-$, $BF_3 \leftarrow CO$ & $BF_3 \leftarrow NH_3$,

B now has 'NO VACANT ORBITAL' for 'BACK BONDING'. In BF_3, boron has a vacant p-orbital due to which $2p\pi$- $2p\pi$ back bonding occurs & B - F bond length is reduced.

30. (a) Al_2O_3 forms a protective layer.

31. (a) Orthoboric acid is H_3BO_3

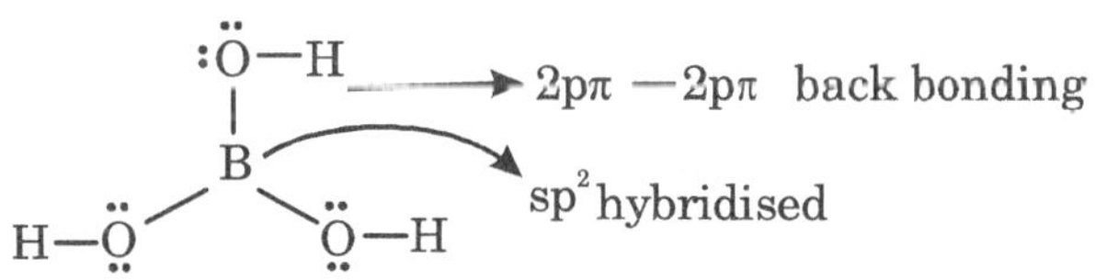

here. So, partial double bond character

$$B(OH)_3 + H_2O \rightleftharpoons [B(OH)_4]^- + H^+$$

Dibroane is B_2H_6

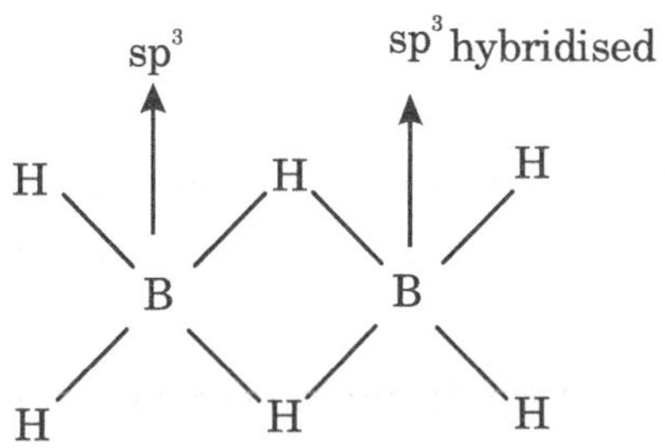

Borazine is -

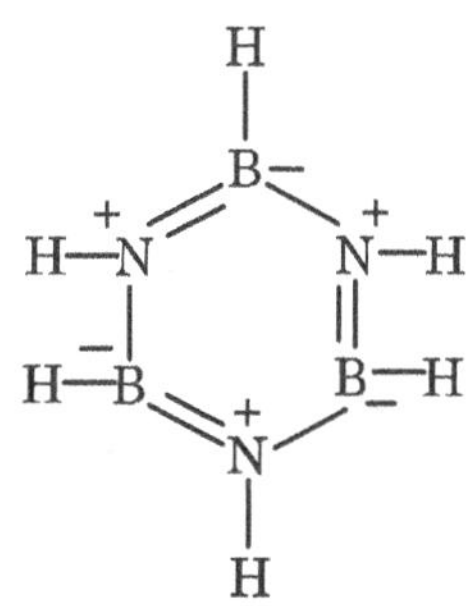

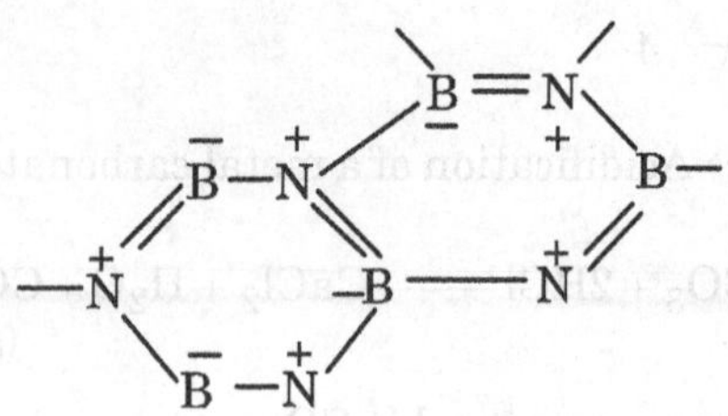

33. (d) (a)

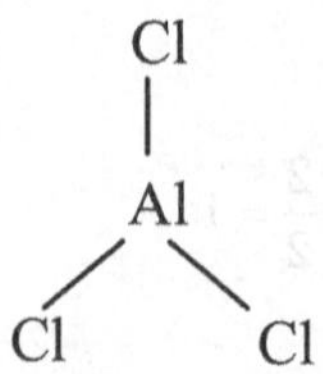

(Anhydrous)

(b) Hydrated $AlCl_3$ exists as $[Al(H_2O)_6]^{3+} + 3Cl^-$

So, it is ionic

(c) $2M + 2NaOH + 6H_2O \rightarrow 2Na^+(Al(OH)_4)^-$

$+ 3H_2 \uparrow$

(d) $2B + 6NaOH \rightarrow 2Na_3BO_3 + 3H_2$

34. (a) These are more properly written as $Ga^+(GaCl_4)^-$ & $In^+(InCl_4)^-$ & here these are in +1 & +3 oxidation states.

37. (a) $H_3BO_3 \xrightarrow[370K]{\Delta} HBO_2$

39. (a) B_2H_6 is a lewis acid which will form adduct with a lewis base (PF_3) having lone pair.

41. (b) $B(OH)_3 + 2H_2O \rightleftharpoons [B(OH)_4]^- + H_3O^+$ ____(1)

To enhance the acidic property of ortho boric acid, the added compound must be a C is - alcohol.

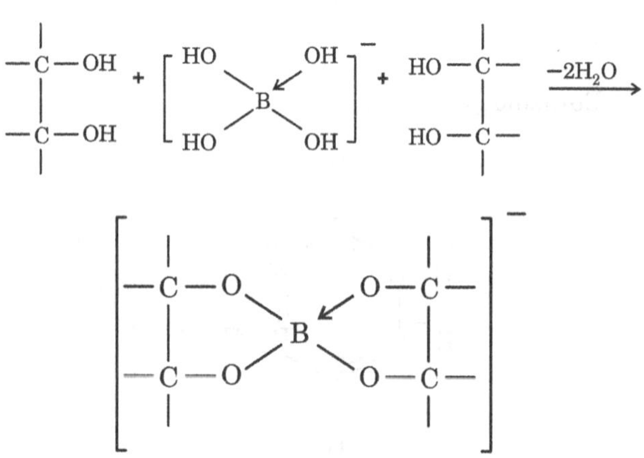

The Cis alcohol forms a stable complex with $(B(OH)_4)^-$ & thus removing it from reaction 1. So, equilibrium gets disturbed hence favoring the forward reactions & thereby increasing the acidic strength of orthoboric acid.

43. (d)

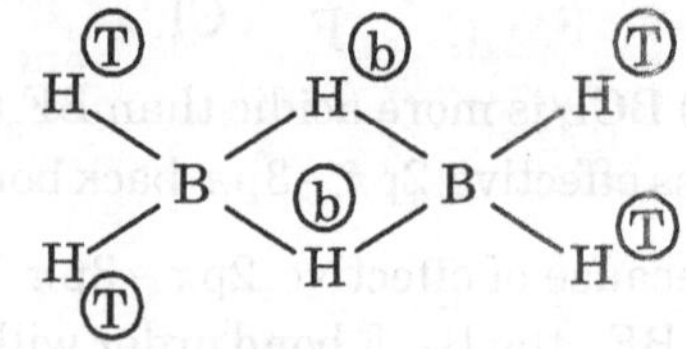

Ⓣ— Terminal H's

ⓑ— Bridging H's

44. (d)

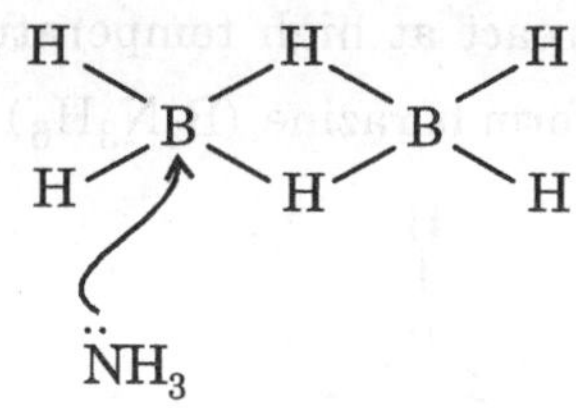

45. (d) I.E. of Pb > IE. of Sn (∵ of increased effective nuclear charge)

47. (c)

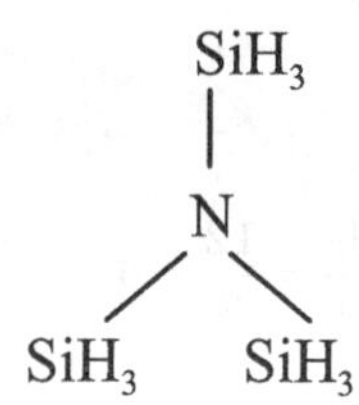

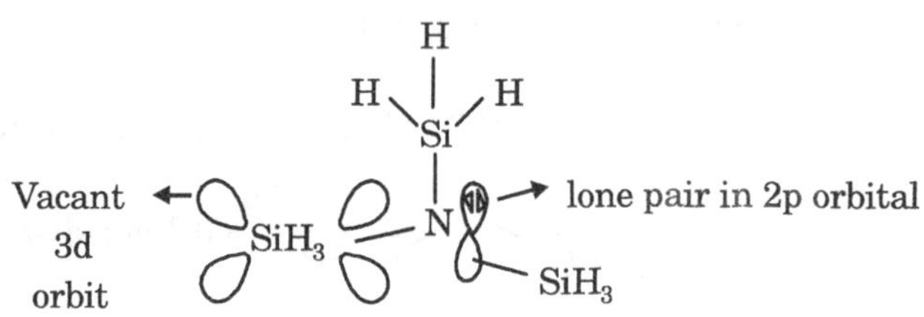

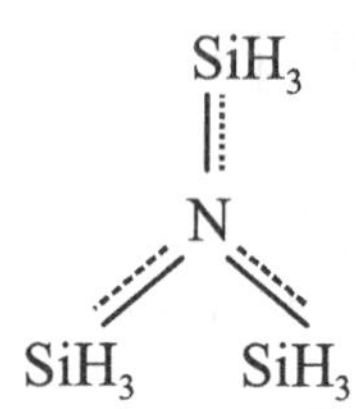

48. (d) There is pπ - dπ back bonding in $(SiH_3)_3N$, so lone pair on N gets delocalized, hence it is a weaker bases

In $(CH_3)_3N$, the lone pair is not delocalized. So, it is a stronger base.

$(CF_6)^{2-}$ does not exist because C lacks vacant d-orbitals.

$(SiF_6)^{2-}$ is known but $(SiCl_6)^{2-}$ is not six large chloride ions cannot be accommodated around Si^{4+}

50. (d) $H_2C_2O_4 \xrightarrow[H_2SO_4]{conc.} \underset{\text{Neutral}}{CO} + \underset{\text{Acidic}}{CO_2}$

CO_2 is absorbed in KOH

51. (d) $\overset{\ominus}{\ddot{C}} \equiv \overset{\oplus}{\ddot{O}}$

$O=C=C=C=O, O=C=O$

52. (b) $CuO + CO \longrightarrow Cu + CO_2 \uparrow$

54. (d) 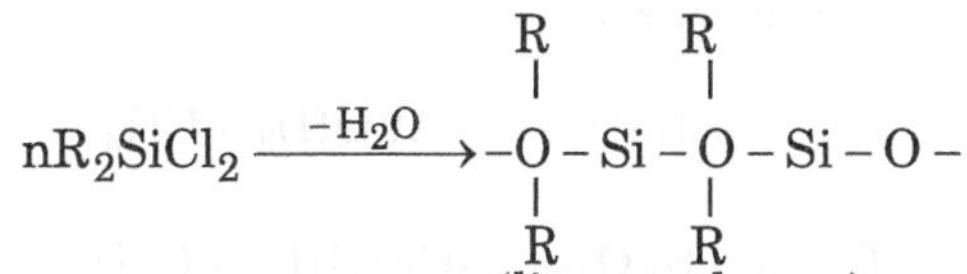

(linear polymer)

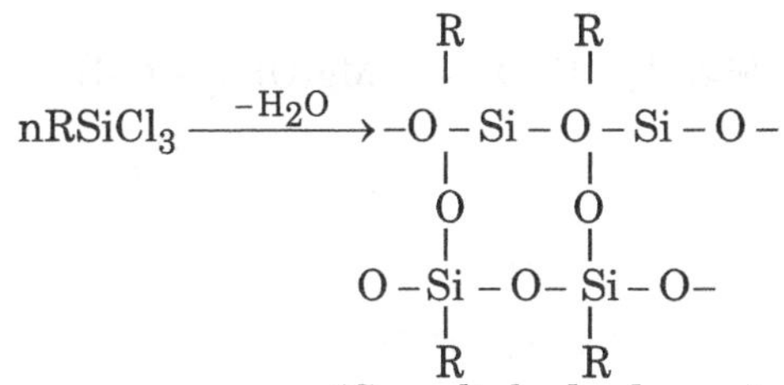

(Cross linked polymer)

56. (b)

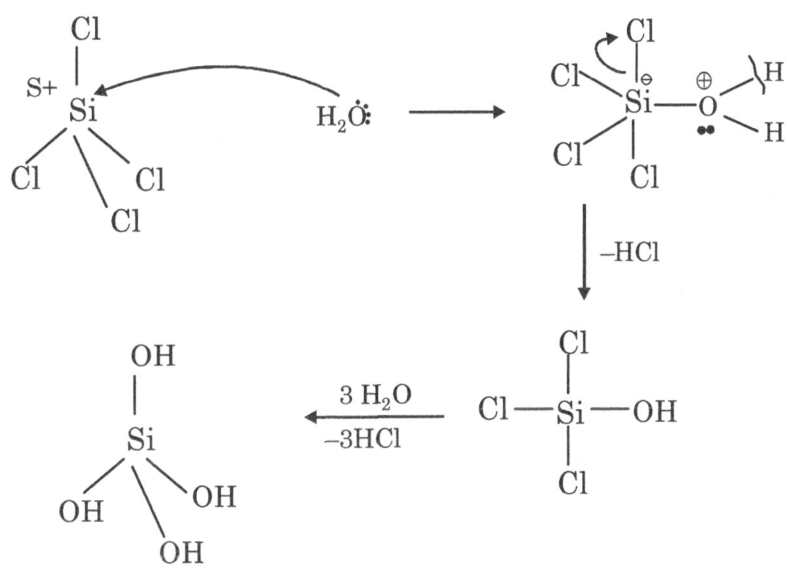

(2) C lacks vacant d-orbitals

59. (c) The non-existence of PbI_4, is probably due to strong oxidizing power of Pb^{4+} & strong reducing power of I^-.

60. (c) CO_2 turns time water milky.

H_2O turns anhydrous $CuSO_4$ blue

Al_2O_3 & $Al(OH)_3$ are amphoteric

61. (d) In $Ca_2Cu_2Si_3O_{10} \cdot 2H_2O$

ratio of $Ca^{2+} : Cu^{2+} : H_2O = 2:2:2 = 1:1:1$

Also, it is neutral i.e. charge must be balanced

$Ca_2\,Cu_2Si_3$ (O_{10}) ⟶ Total anionic charge = 20 (i.e. – 20)

$Ca_2\,Cu_2Si_3$ ⟶ Total Cation charge = + 20

$2Ca^{2+} = +4$ $\quad$ $3Si^{4+} = +12$

$2Cu^{2+} = +4$ So, $+4+4+12 = +20$

62. (b) $SiC + H_2O \longrightarrow$ No reaction

(SiC is inert)

$Al_4C_3 + H_2O \longrightarrow Al(OH)_3 + CH_4$

$CaC_2 + H_2O \longrightarrow Ca(OH)_2 + C_2H_2$
$(H - C \equiv C - H)$

$Mg_2C_3 + H_2O \longrightarrow Mg(OH)_2 + C_3H_4$
$(CH_3 - C \equiv C - H)$

63. (b) $C_2^{2-} \longrightarrow \overset{\ominus}{\ddot{C}} \equiv C^{\cdot\cdot\ominus}$

66. (d) Tl is stable in +1, so $TiCl_3$ (where Tl is n +3) will act Oxidising agent & itself get reduced to +1.

Pb is stable in +2.

67. (b) Inert pair effect shows its impact in Pb.

68. (b) $HCOOH + H_2SO_4 \rightarrow CO + H_2O$

69. (c)

$$CH_2(COOH)_2 \xrightarrow[150^\circ C]{P_4O_{10}} C_3O_2 + 2H_2O$$

Malonic acid

71. (c)

(a) $:\overset{\ominus}{\ddot{N}} = C = \overset{\ominus}{\ddot{N}}:$ (cyanamide ion)

↓

sp-hybridised

(b) $CaC_2 + N_2 \xrightarrow[1378K]{1273-} \underbrace{CaCN_2 + C(graphite)}_{\text{Nitrolim}}$

$CaCN_2 + H_2O \longrightarrow CaO + H_2NCN$

$H_2NCN + H_2O \longrightarrow \underset{\text{(Urea)}}{H_2NCONH_2}$

$H_2NCONH_2 + H_2O \longrightarrow 2NH_3 + CO_2$

$CaO + CO_2 \longrightarrow CaCO_3$

$CaCN_2 + 3H_2O \longrightarrow CaCO_3 + 2NH_3$

72. (b) 'X' will be Pb_3O_4

$\underset{(X)}{Pb_3O_4} + 4HNO_3 \longrightarrow \underset{(Z)}{Pb(NO_3)_2} + \underset{(Y)}{PbO_2} + 2H_2O$

$\underset{(Z)}{Pb(NO_3)_2} + 2HCl \longrightarrow \underset{(A)}{PbCl_2} + 2HNO_3$

$\underset{(X)}{Pb_3O_4} + 8HCl \longrightarrow \underset{(A)}{3PbCl_2} + 4H_2O + Cl_2$

73. (b)

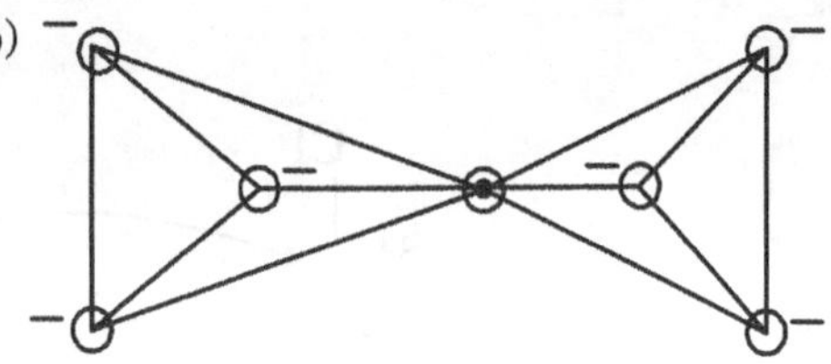

$Si_2O_7^{6-}$

76. (a) $CO_2 + H_2O \rightleftharpoons H_2CO_3$

$H_2CO_3 \rightleftharpoons H^+ + HCO_3^- \rightleftharpoons H^+ + CO_3^{2-}$

80. (b) $SiO_4^{4-} \xrightarrow{\text{Hydrolysis}} Si_2O_7^{6-}$

82. (d) (a) $5CO + I_2O_5 \longrightarrow I_2 + 5CO_2$

(b) $CO_2 + Ca(OH)_2 \longrightarrow CaCO_3 + H_2O$

(c) $4NaBO_2 + CO_2 \longrightarrow Na_2B_4O_7 + Na_2CO_3$

83. (d) $SnCl_4.5H_2O$ is known as butter of tin or oxymercurate of tin.

Smart Notes

Smart Notes

13 CHAPTER

Organic Chemistry-Some Basic Principles and Techniques (GOC)

Electronic effects

1. Which of the following is the correct order of -I effect?

(a) $-F > -Cl > -Br > -I$

(b) $-\overset{\oplus}{N}R_3 > -C \equiv C-H > -CH=CH_2$

(c) $-OR > -OH > -NH_2$

(d) All are correct

2. Which of the following is are correct representation of inductive effect?

(a) 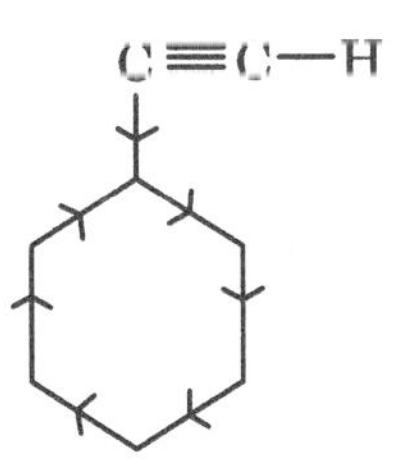

(b) 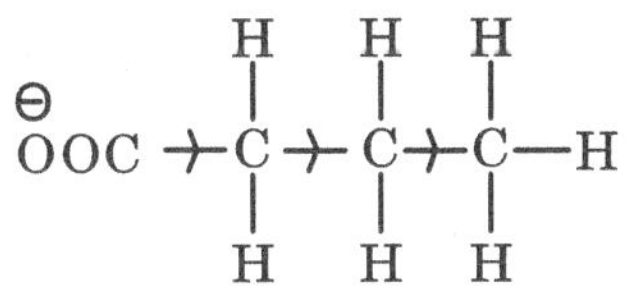

(c) 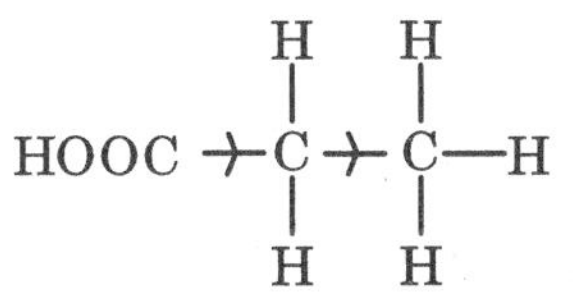

(d) Both (a) & (c)

3. Arrange the following in correct order of their acidic strengths.

(I) $CH_2=CH-CH_2-\overset{O}{\overset{\|}{C}}-O-H$,

(II) $CH_3-CH_2-CH_2-\overset{O}{\overset{\|}{C}}-O-H$

(III) $H-C\equiv C-CH_2-\overset{O}{\overset{\|}{C}}-O-H$

(a) I > II > III (b) III > I > II

(c) III > I > II (d) III > II > I

4. Which of the following orders can be explained on the basis of inductive effect?

(a) Acidic strength:- $O_2N-CH_2-COOH > F-CH_2-COOH > H_3CO-CH_2-COOH > CH_3CH_2COOH$

(b) Dipole moment:- $CH_3NO_2 > CH_3COOH > CH_3F > CH_3OH$

(c) Stabilities of carbocations:-

$(CH_3)_3C^{\oplus} > (CH_3)_2\overset{\oplus}{C}H > CH_3\overset{\oplus}{C}H_2$

(d) Stabilities of Carbanion:-

$\overset{\ominus}{C}H_3 > CH_3-\overset{\ominus}{C}H_2 > CH_3-\overset{\ominus}{C}H-CH_3 > CH_3-\overset{\ominus}{C}(CH_3)-CH_3$

5. Select the correct order of stabilities of resonating structures.

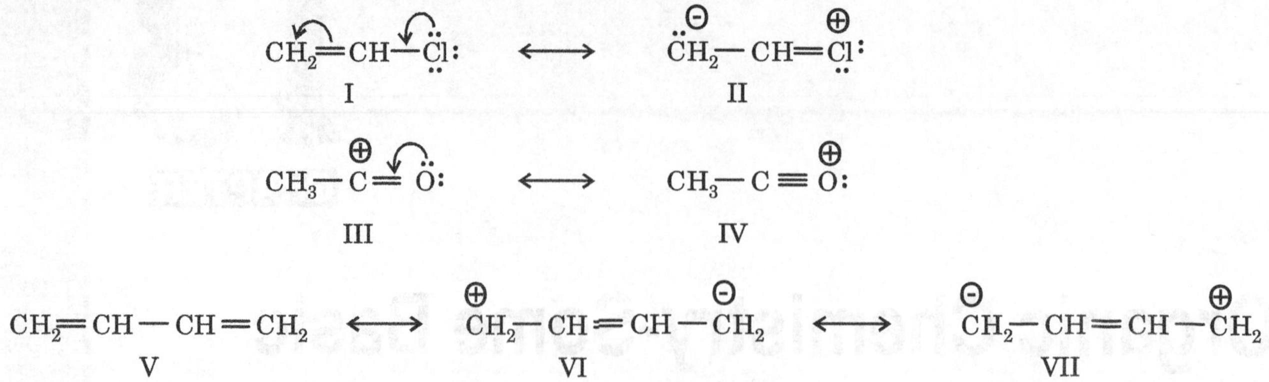

(a) II > I (b) III > IV (c) V > VI > VII (d) None of these

6. Select the correct statements wrt the structures given below-

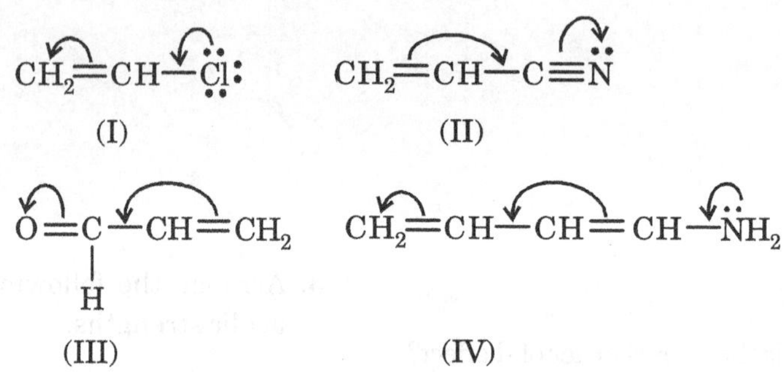

(a) Structures I& IV exhibit + M effects of - Cl& $-NH_2$ groups respectively.

(b) Structure III shows -M effect wrt - CHO group.

(c) Structure II shows - M effect wrt - CN group.

(d) All are correct.

7. Mesomeric effect occurs in which of the following?

(1) $CH_2 = CH - C \equiv C - H$ (2) $CH_2 = CH - \ddot{O} - CH_3$ (3) 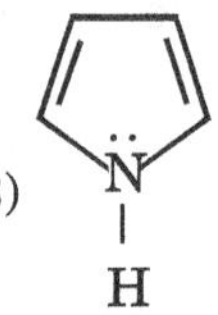

(a) Only 1 & 2 (b) Only 1 (c) Only 2 & 3 (d) All of the above

8. Which of the following orders are NOT explained by hyper conjugation?

(a) Relative stabilities of carbocations

$$(CH_3)_3C^{\oplus} > (CH_3)_2CH^{\oplus} > CH_3-\overset{\oplus}{C}H_2$$

(b) Relative stabilities of free radicals

$$(CH_3)_3C^{\bullet} > (CH_3)_2CH^{\bullet} > CH_3-\dot{C}H_2$$

(c) Relative stabilities of carbanions

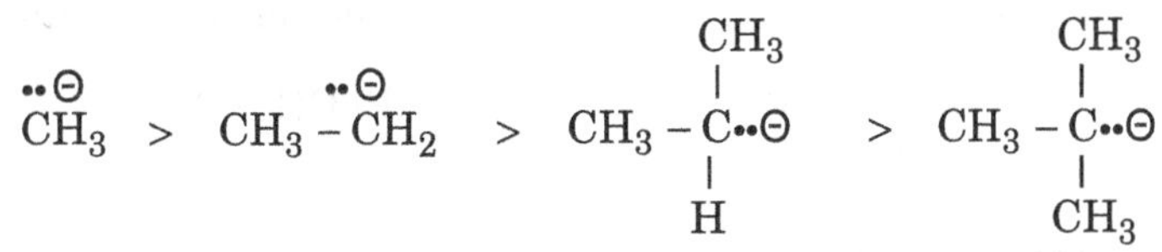

(d) Relative stabilities of alkenes

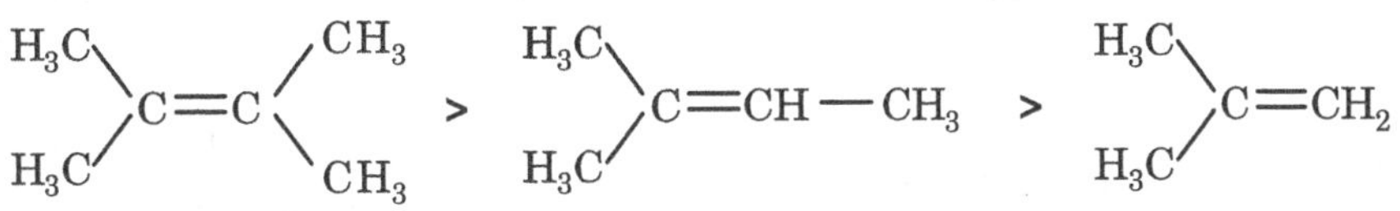

9. In propene, due to hyper conjugation-

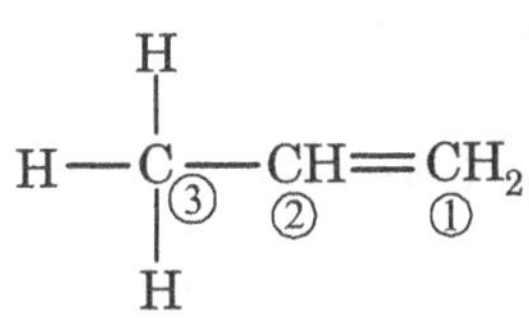

(a) $C_1 - C_2$ bond lengths is more than expected.

(b) $C_2 - C_3$ bond length is less than expected.

(c) Separation of charges occurs.

(d) All are correct.

10. Select the Incorrect statement.

(a) pyrrole is more basic than pyridine.

(b) methanamine is more basic than aniline.

(c) hybridisation of carbanion may be sp^3, sp^2 & sp.

(d) All are incorrect.

11. How many Primary (1°) carbon atoms are present in 4 - ethyl - 2, 3 - dimethyl hexane?

(a) 1 (b) 3

(c) 5 (d) 6

12. Which of tho following aro O, P dirootoro but deactivators?

(a) – Cl (b) – N = O, –S = O

(c) $- CH = CH_2$ (d) All of these

13. Which of the following is/are planar?

(a) Methyl carbocation

(b) Methyl free radical

(c) Methyl carbanion

(d) Both (a) & (b)

14. Which of the following groups exert significant +M as well as - I effect ?

(a) $-\overset{\oplus}{S}(CH_3)_2$ group (b) $-\overset{\oplus}{N}(CH_3)_3$ group

(c) –ONO group (d) $-NO_2$ group

15. Conjugate base of which compound is most stable-

(a) $CH_3 - CH_2 - NO_2$ (b) $CH_3 - NO_2$

(c) CH_3CH_2Cl (d) CH_3CH_2Br

16. Which one of the following compounds can be deprotonated by $\bar{O}H$ fastest?

(a) HCOOH, pka = 3.8 (b) H_2S, pka = 7

(c) Toluene, pka = 41 (d) CH_3NH_2, pka = 40

17. Identify the correct acidic strength order in the following compound.

(1) 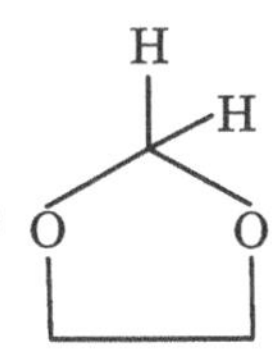

(2) 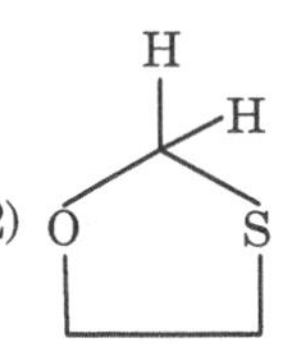

(3) 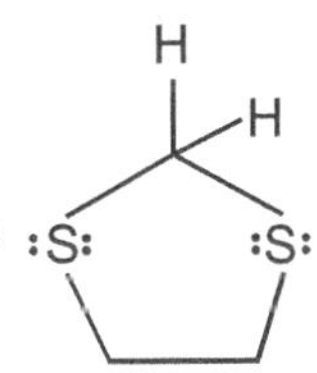

(a) 1 > 2 > 3 (b) 3 > 2 > 1

(c) 1 = 2 = 3 (d) 3 > 1 > 2

18. Which of the following is most basic?

(a) 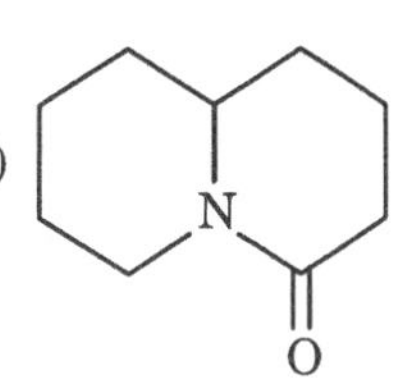

(b) 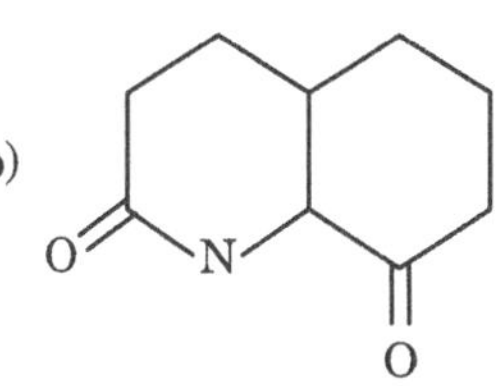

(c) 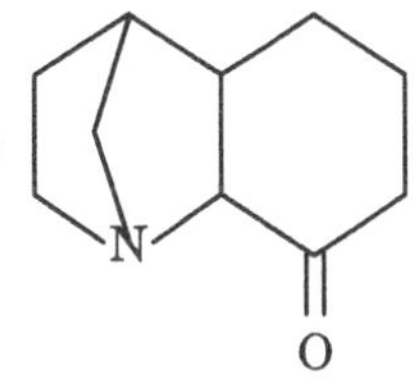

(d) 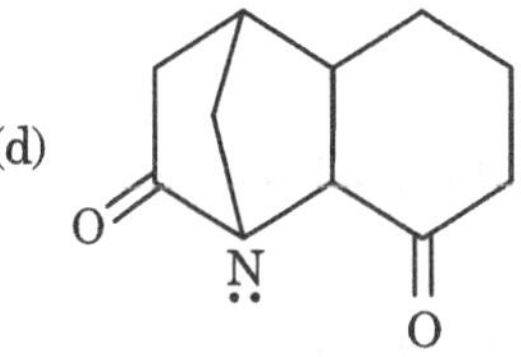

19. Which of the following compounds will have net dipole moment?

(a) 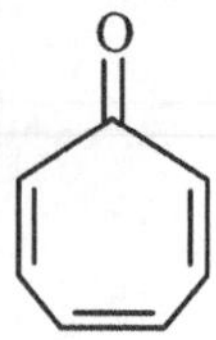 (b) 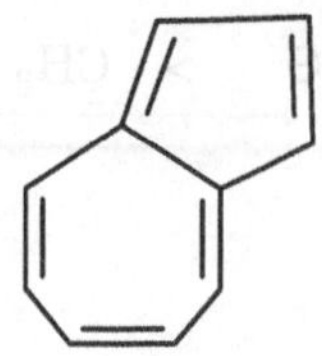

(c) 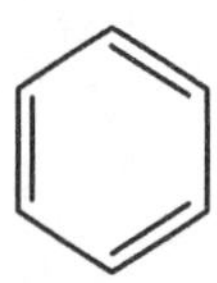 (d) Both (a) & (b)

20. Which of the following order for 'ACIDIC STRENGTH' is correct?

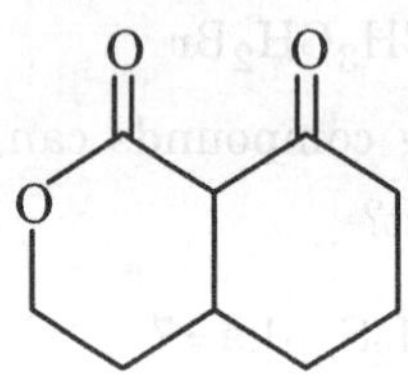
(P)

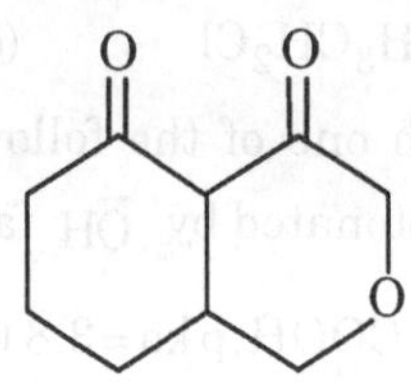
(Q)

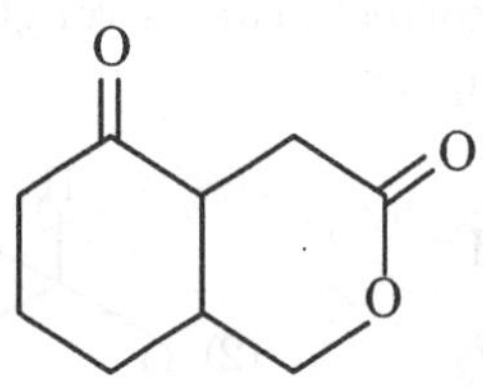
(R)

(a) P > Q > R (b) Q > P > R

(c) R > Q > P (d) R = Q > P

21. Arrange the following resonating structure in the decreasing order of their stability -

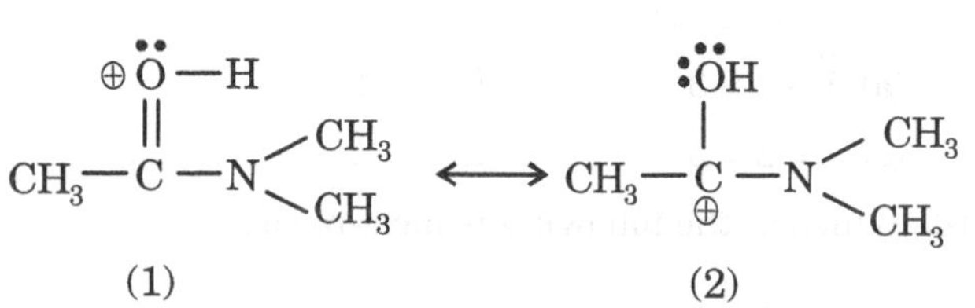
(1) (2)

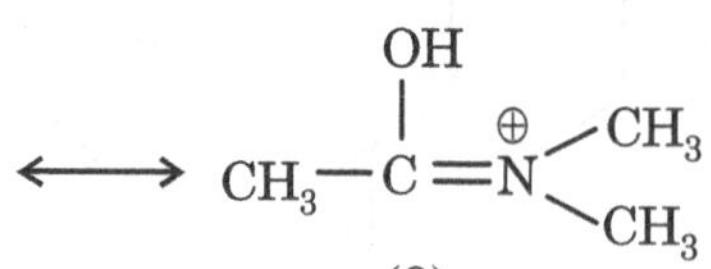
(3)

(a) 1 > 2 > 3

(b) 2 > 1 > 3

(c) 3 = 1 > 2

(d) 3 > 1 > 2

22. Arrange the following molecules in the decreasing order of C – N bond length.

(P) $CH_2 = NH$ (Q) $CH_3 - CH_2 - NH_2$

(R) 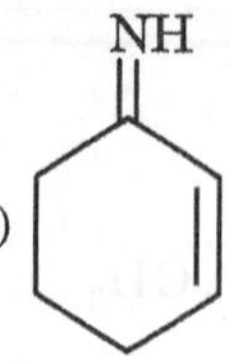 (S) 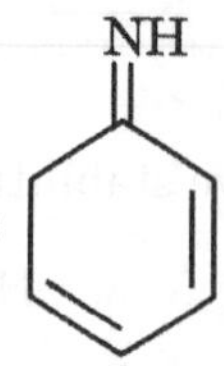

(a) Q > S > R > P (b) P > Q > R > S

(c) Q > S > R = P (d) None of these

23. Choose the correct order of bond length from the molecule given below

$$CH_2 \overset{x}{=} CH \overset{z}{-} CH \overset{y}{=} CH_2$$

(a) $x > y > z$ (b) $z > x = y$

(c) $x = z > y$ (d) $y > x > z$

24. Which is most acidic?

(a) $ClCH_2COOH$ (b) $O_2N - CH_2COOH$

(c) CH_3COOH (d) CH_3CH_2COOH

25. The correct order of ease of abstraction of hydrogen by $NaNH_2$ (base) in the following compound is –

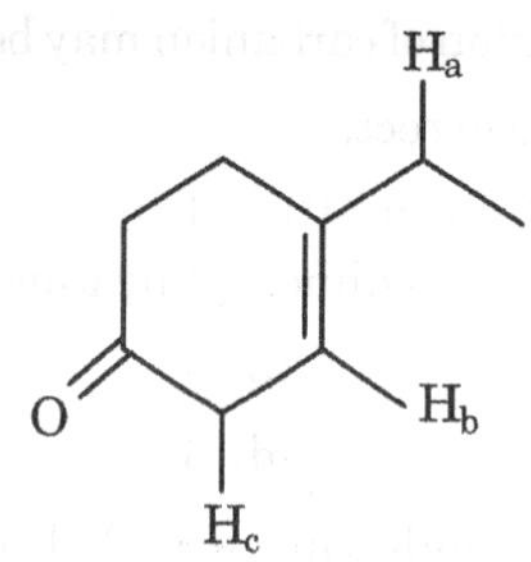

(a) $H_c > H_a > H_b$ (b) $H_a > H_b > H_c$

(c) $H_a > H_c > H_b$ (d) None of these

26. Which of the following has minimum bond length of C = C?

(a) 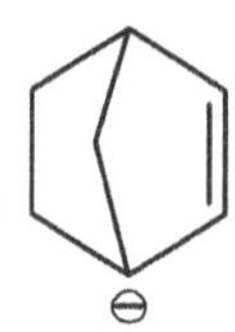 (b) 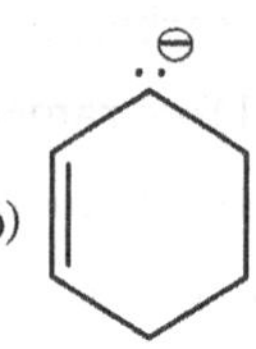

(c) 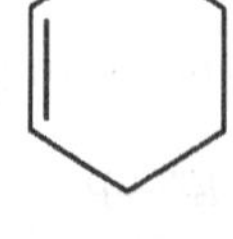 (d) 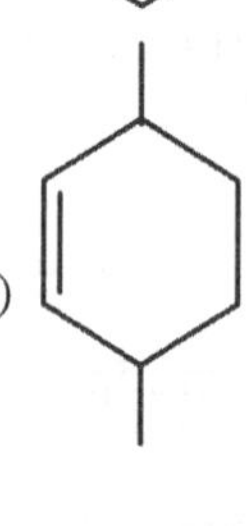

27. Consider the following resonating structures-

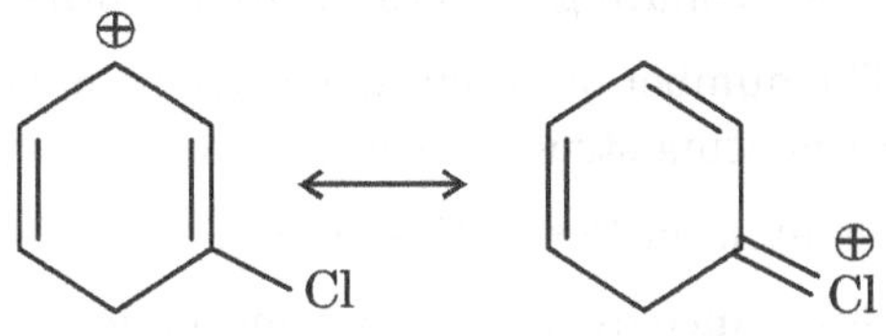

I II

(a) Structure I is more stable than II.

(b) Structure II is more stable than I.

(c) Both structures are equally stable.

(d) None of these

28. Which one of the following has longest C = C bond length?

(a) $CH_2 = C = CH_2$

(b) $CH_3 - CH = CH_2$

(c) $CH_3 - C(CH_3)_2 - CH = CH_2$

(d) $CH_3 - C(CH_3) = CH_2$

29. Select the incorrect order-

(a) $CH_3 - NH_2 > CH_2 = \ddot{N}H > CH_3 - C \equiv \ddot{N}$ (Basic strength)

(b)

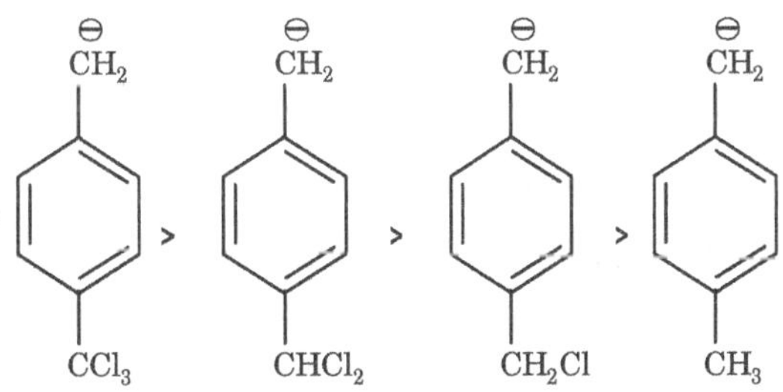

(stability of carbanions)

(c)

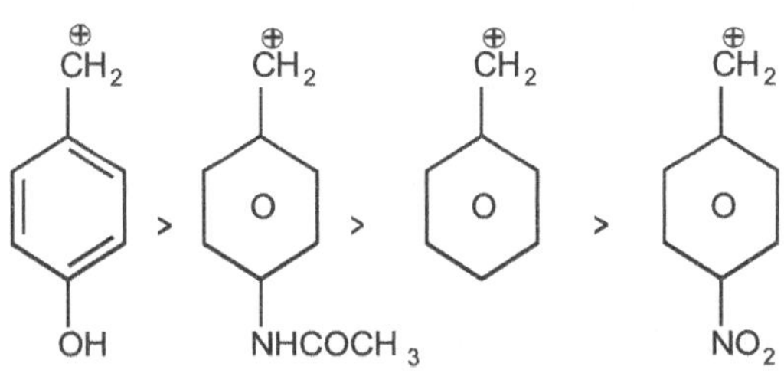

(stability of carbocations)

(d) None of these

30. The most unlikely representation of resonating structure of p - nitrophenoxide is -

(a)

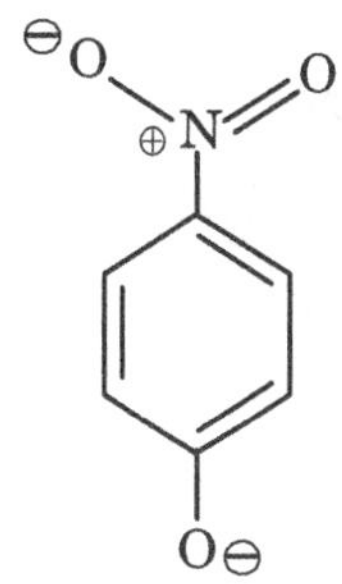

(b)

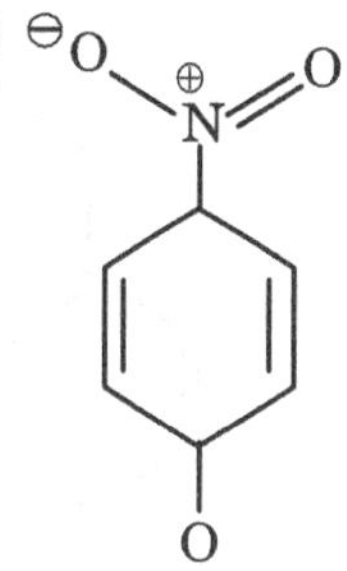

(c)

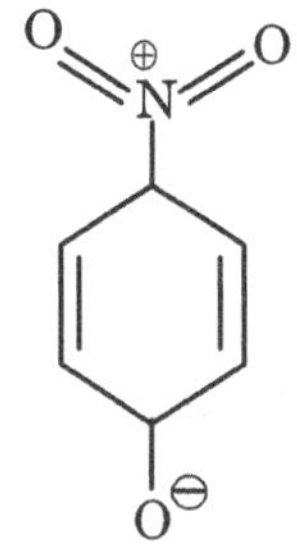

(d)

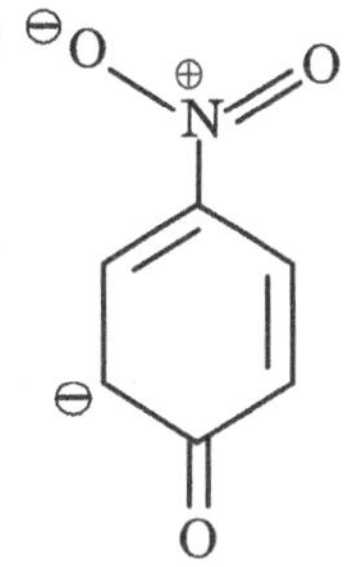

31.

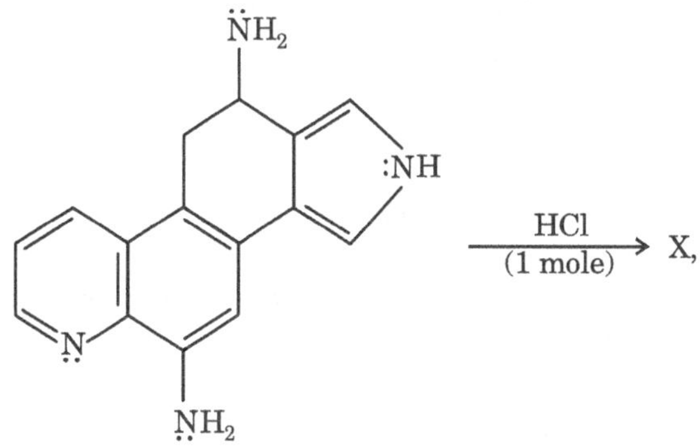

X will be –

(a)

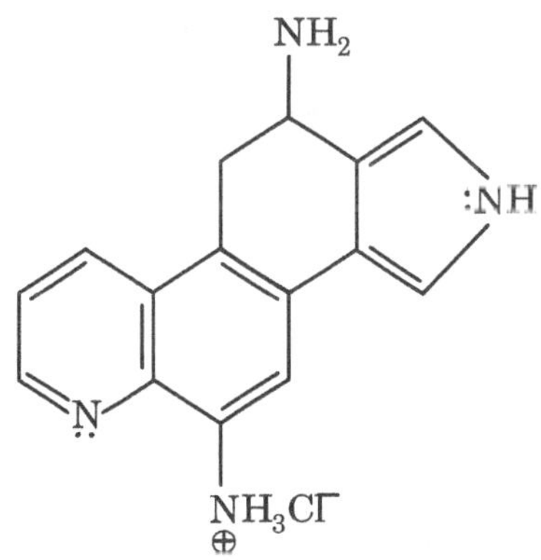

(b)

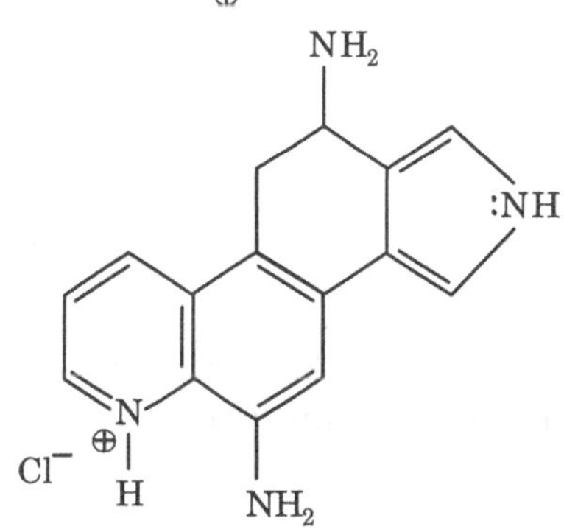

(c) 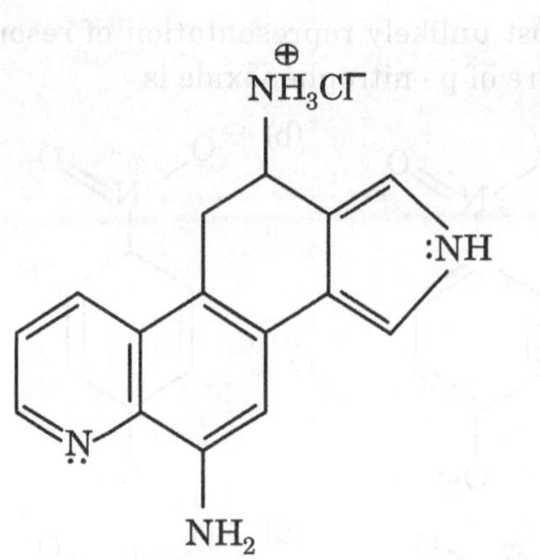

(d) 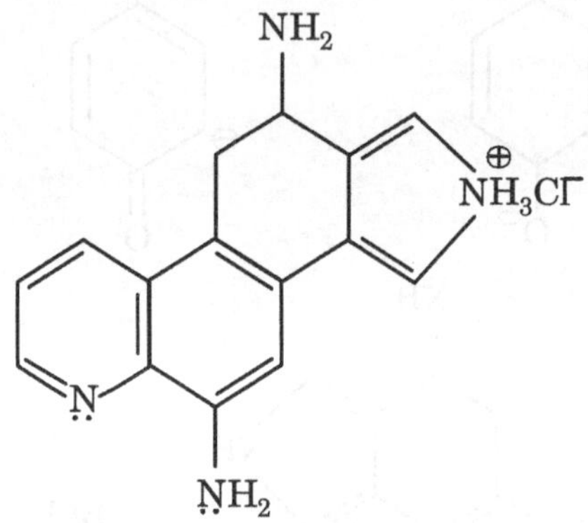

32. Which of the following compound is Non – aromatic?

(a) 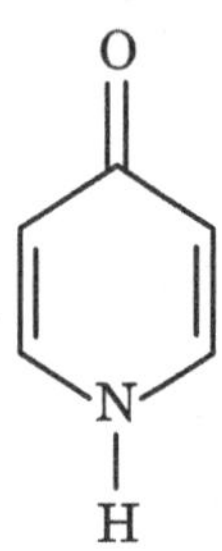 (b) 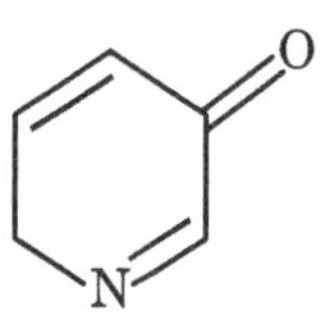

(c) 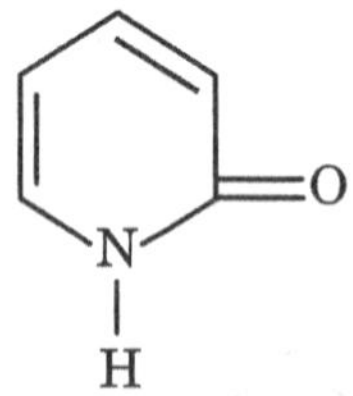 (d) 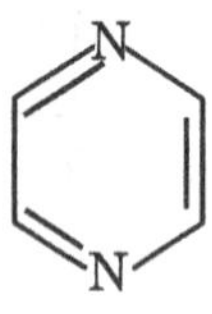

33. Which of the following has highest inductive effect?

(a) CH_3 (b) $-CH_2-CH_3$

(c) $-CH(CH_3)_2$ (d) $(CH_3)_3C-$

34. What is not true about resonance?

(a) The resonating structures are hypothetical.

(b) The number of unpaired electrons in various resonating structures are same.

(c) No atom moves in Resonance.

(d) Resonance hybrid is most energetic

35. In the reaction of C_6H_5Y, the major product (> 60%) is meta isomer, so the group Y is –

(a) – COOH (b) $-NH_2$

(c) – OH (d) – Cl

36. The correct stability order of the following resonating structures is-

(1) $H_2C=\overset{\oplus}{N}=\ddot{N}:^{\ominus}$ (2) $H_2\overset{\oplus}{C}-\ddot{N}=\ddot{N}:^{\ominus}$

(3) $H_2\overset{\ominus}{\ddot{C}}-\overset{\oplus}{N}\equiv \ddot{N}$ (4) $H_2\overset{\ominus}{\ddot{C}}-\ddot{N}=\overset{\oplus}{N}:$

(a) 1 > 2 > 3 > 4 (b) 1 > 3 > 2 > 4

(c) 2 > 1 > 3 > 4 (d) 2 > 1 > 3 > 4

37. What is the correct increasing order of bond lengths of the bonds indicated as I, II, III & IV in the following compound?

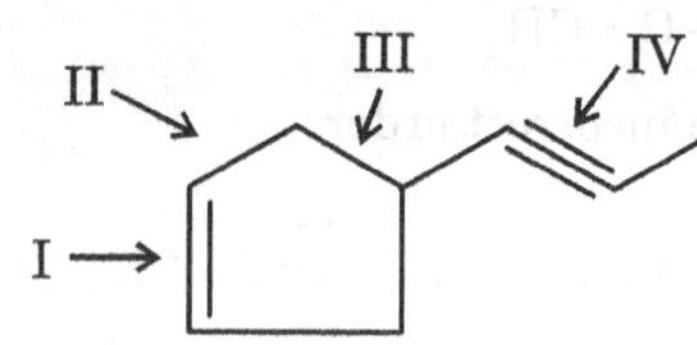

(a) 3 > 2 > 1 > 4 (b) 2 > 1 > 3 > 4

(c) 1 > 2 > 3 > 4 (d) 1 = 2 > 3 = 4

38. Most acidic H is present in-

(a) $CH_3-\overset{O}{\overset{\|}{C}}-H$

(b) $CH_3-\overset{O}{\overset{\|}{C}}-CH_2-\overset{O}{\overset{\|}{C}}-OCH_3$

(c) $(CH_3CO)_3C-H$

(d) $(CH_3)_3COH$

39. Which of the following is maximum acidic?

(a) $R-\overset{O}{\overset{\|}{C}}-O-H$ (b) $R-\underset{O}{\underset{\|}{\overset{O}{\overset{\|}{S}}}}-O-H$

(c) ROH (d) RCH_2OH

40. The benzene ring which is most & least activated for electrophllic substitution reaction are, respectively

(1) 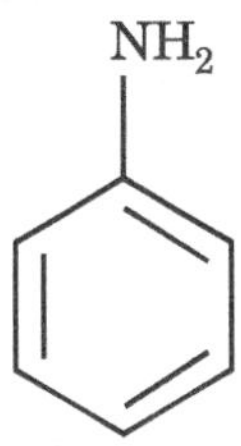

(2) 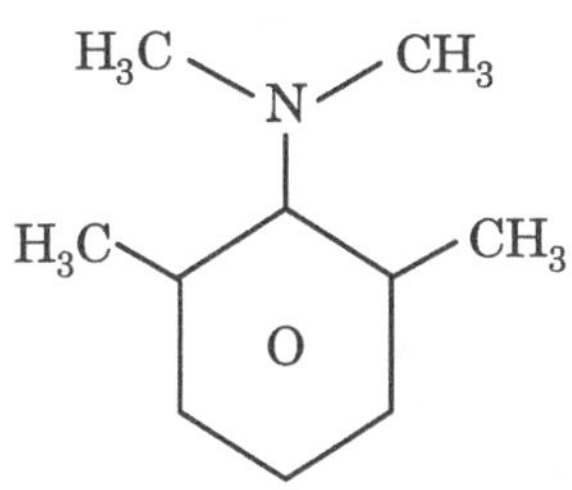

(3) 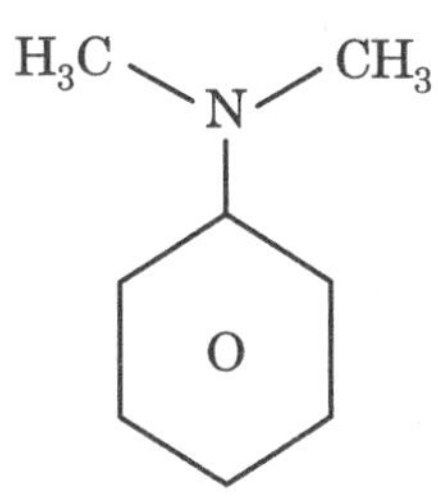

(4) 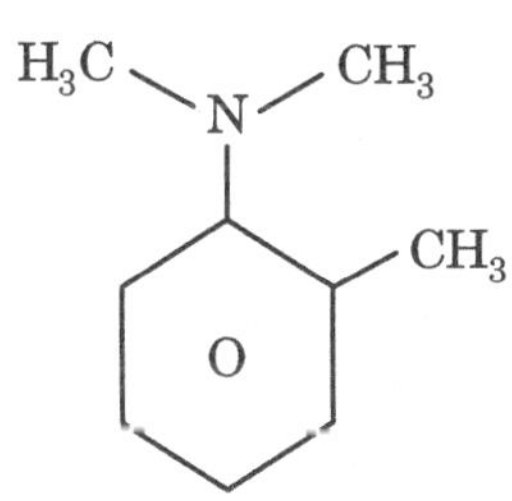

(a) 1, 2 (b) 2, 3
(c) 2, 4 (d) 4, 2

41. Which of the following is the strongest base?

(a) 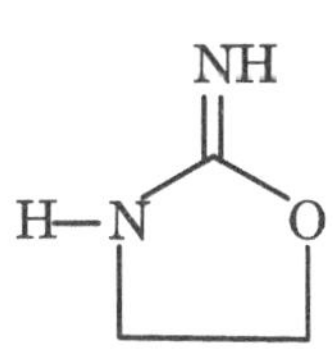
(b) 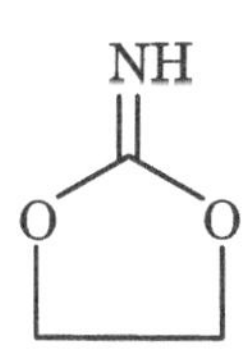
(c) 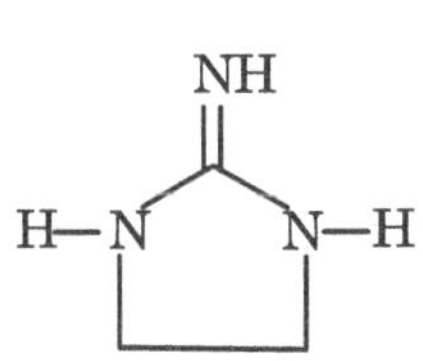
(d) 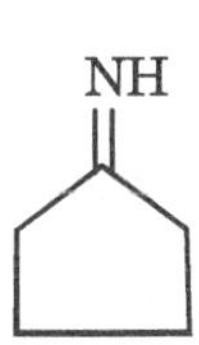

42. Which of the following is non - aromatic ?

(a) 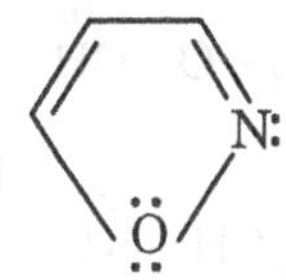
(b) 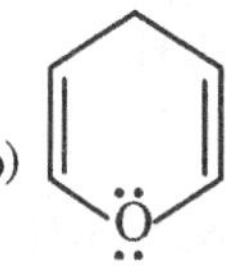
(c) 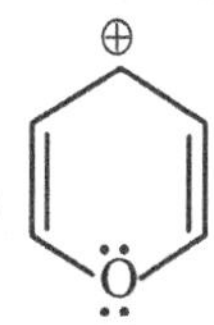
(d) None of these

Reaction Intermediates

43. The most & least stable carbocations among the following are respectively -

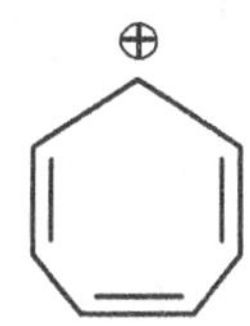
I

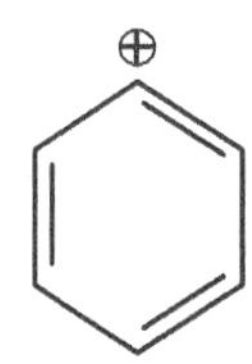
II

III

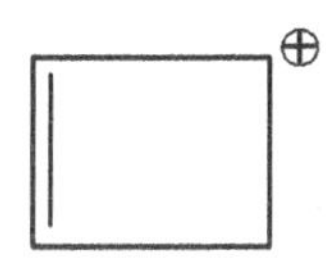
IV

(a) I & III (b) II & IV
(c) I & IV (d) III & IV

44. Which of following are aromatic in nature?

(a) 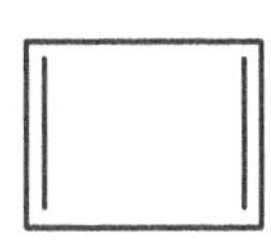
(b) 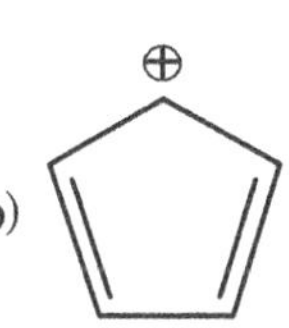
(c) 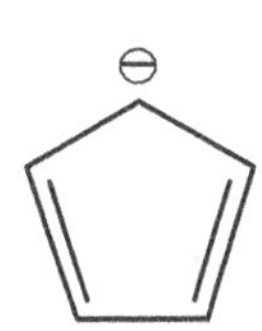
(d) 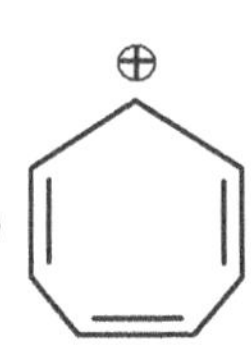

(a) Only b (b) a & b
(c) c & d (d) a, b, c, d

45. Most stable carbocation is

(a) $(CH_3)_2\overset{\oplus}{C}H$ (b) $CH_3-O-\overset{\oplus}{C}H_2$

(c) $CH_3-NH-\overset{\oplus}{C}H_2$ (d) $CH_3CH_2\overset{\oplus}{C}H_2$

46. Which of the following is not aromatic ?

(a) 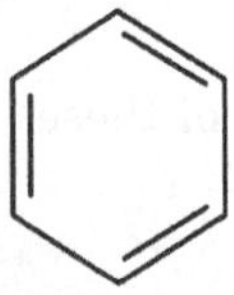 (b) 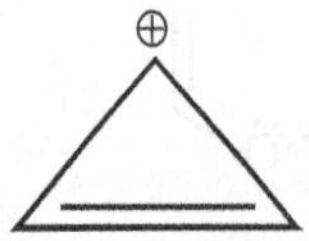

(c) 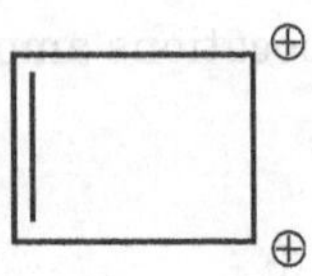 (d) 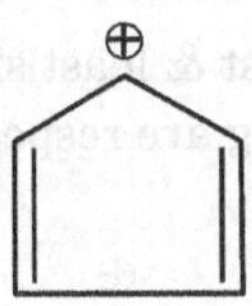

47. Most basic out of the following is -

(a) 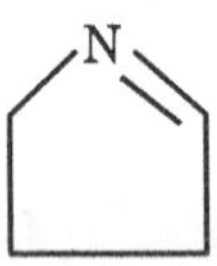 (b) 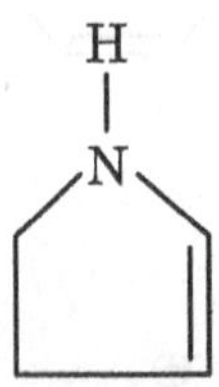

(c) 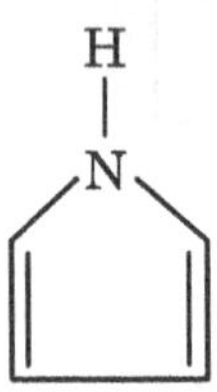 (d) 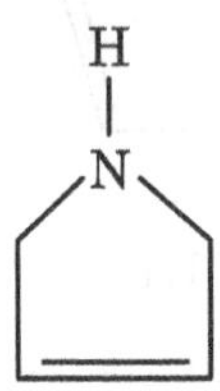

48. Select the correct order against the property mentioned.

(a) 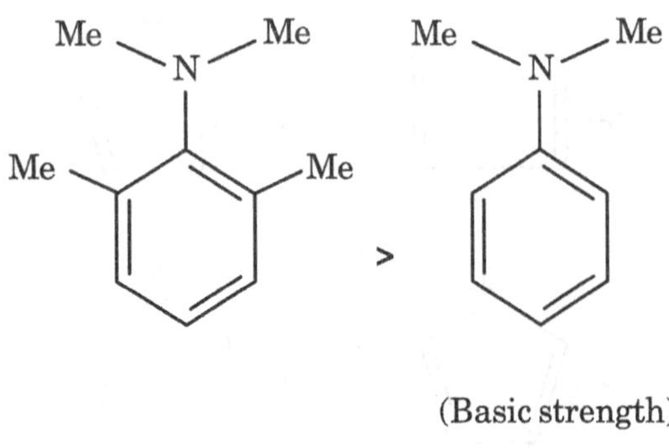

(Basic strength)

(b) $H-C\equiv C^{\ominus}$ > $CH_2=\overset{\ominus}{C}H$ > $\overset{\ominus}{C}H_3$

(stability)

(c) 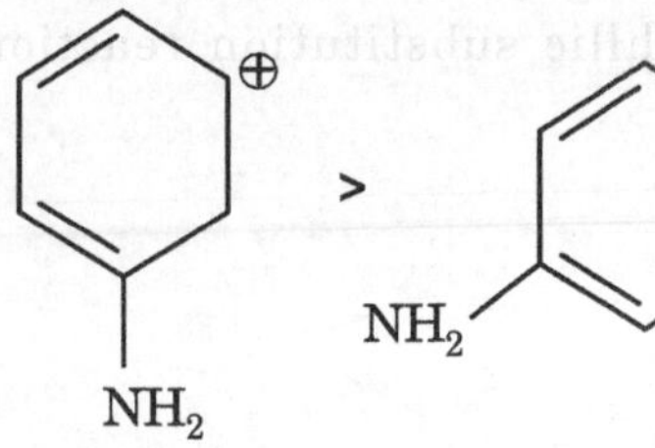 (stability)

(d) All are correct

49. In which pair, second ion is more stable than first?

(a) 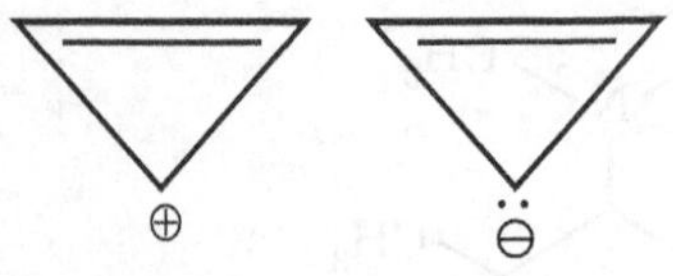

(b) 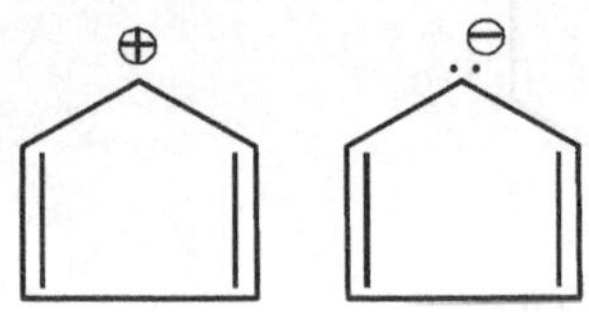

(c) 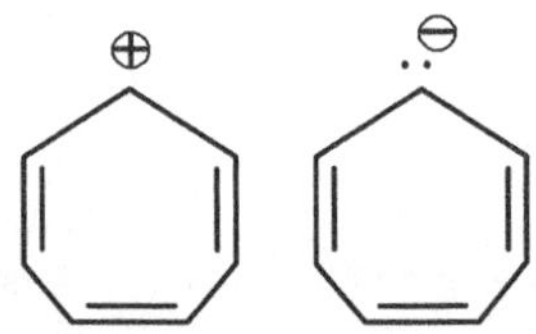

(d) 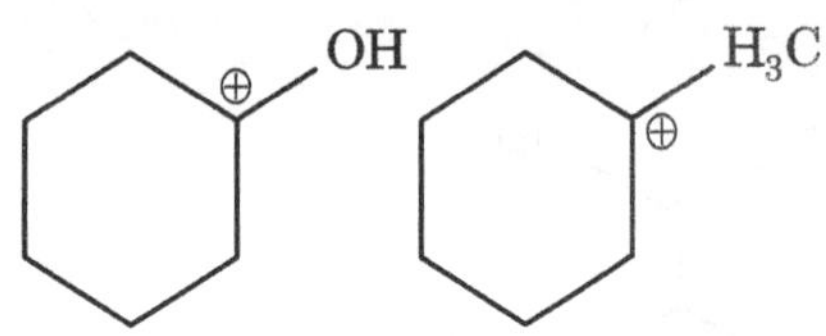

50. In which of the following, resonance occurs to stabilize the intermediate (carbocation or a carbanion) (Ignore hyper conjugation if possible)

(a) 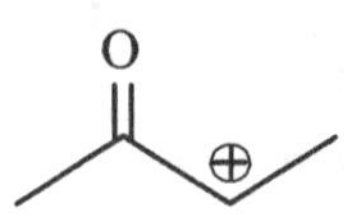

(b) 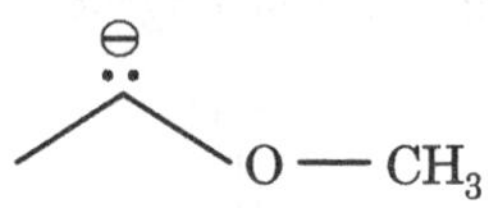

(c) 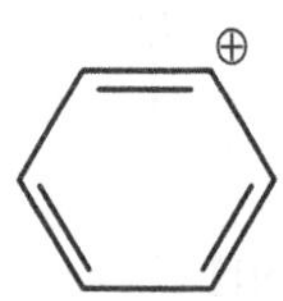

(d) 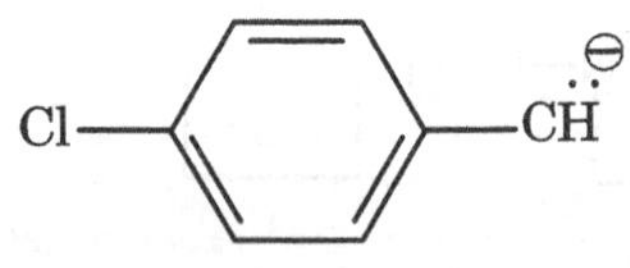

51. The correct stability order of following carbocation is -

(a) 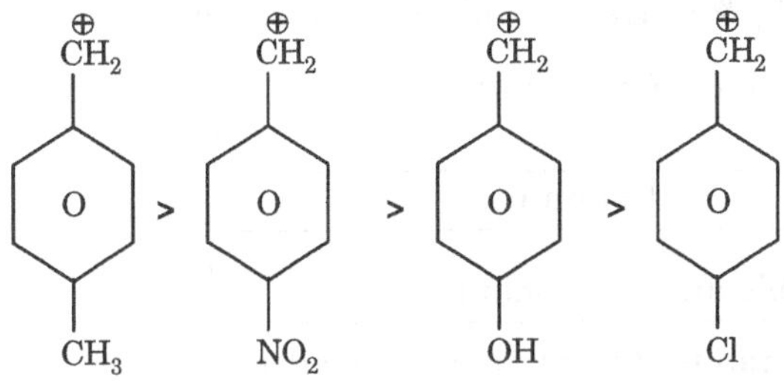

(b) 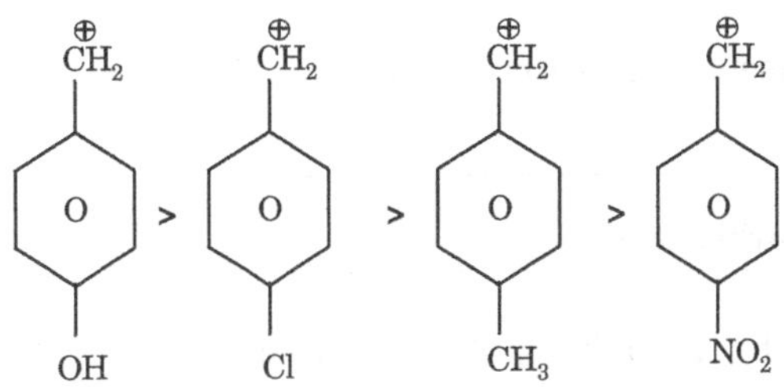

(c) 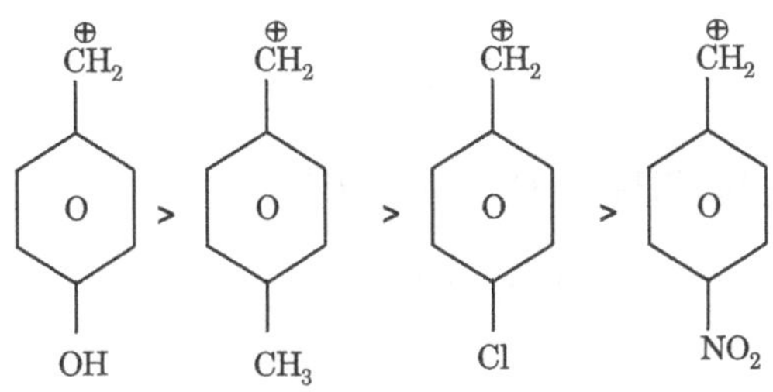

(d) None of these

52. Which of the following resonating structures of 1-methoxy - 1, 3 - butadiene is least stable?

(a) $CH_2-CH=CH-CH=\overset{\oplus}{\underset{..}{O}}-CH_3$

(b) $CH_2-CH-\overset{\ominus}{\ddot{C}H}-CH=\overset{\oplus}{\underset{..}{O}}-CH_3$

(c) $\overset{\ominus}{\ddot{C}H_2}-\overset{\oplus}{C}H-CH=CH-\ddot{\underset{..}{O}}-CH_3$

(d) $CH_2=CH-\overset{\ominus}{\ddot{C}H}-\overset{\oplus}{C}H-\ddot{\underset{..}{O}}-CH_3$

53. Which one of the following carbocations will not re-arrange ?

(a) 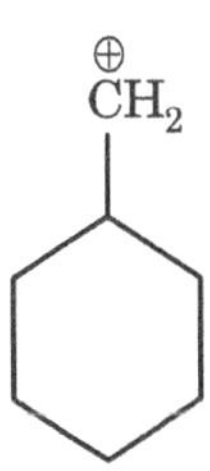

(b) 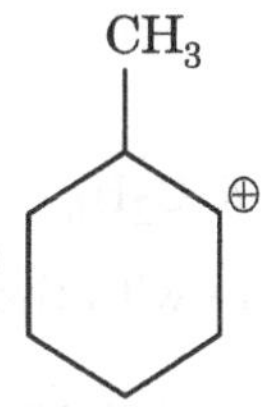

(c) $CH_3-\underset{}{\overset{CH_3}{\overset{|}{C}H}}-\overset{\oplus}{C}H-CH_3$

(d) 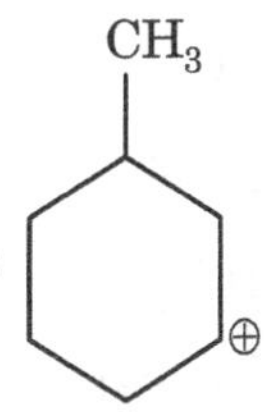

54. Which of the following is most stable?

(a) 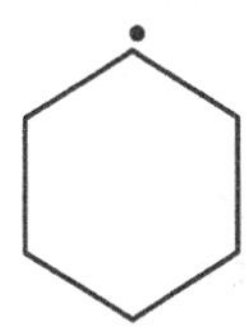

(b) 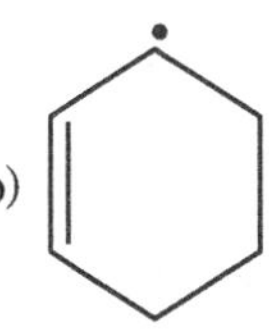

(c) 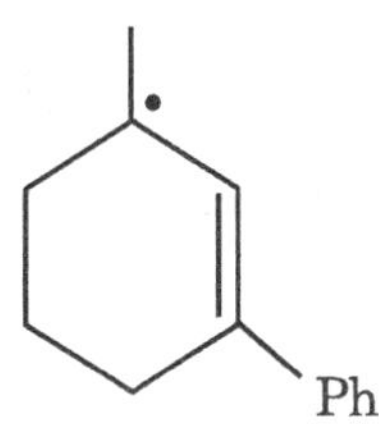

(d) 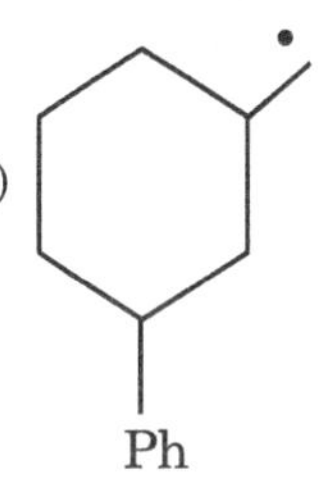

55. 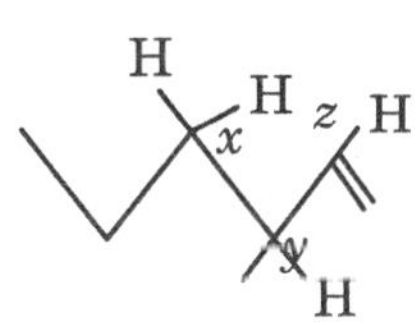

The decreasing order of ease of homolysis of C – H bond x, y & z is -

(a) $y > x > z$ (b) $z > x > y$

(c) $z > y > x$ (d) $y > z > x$

56. In which case [C – O] bond length is minimum

(a) 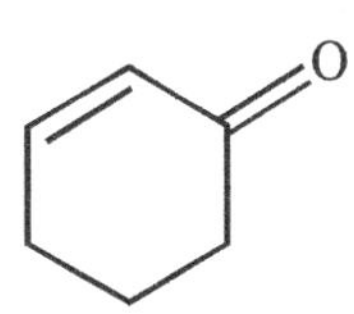

(b) 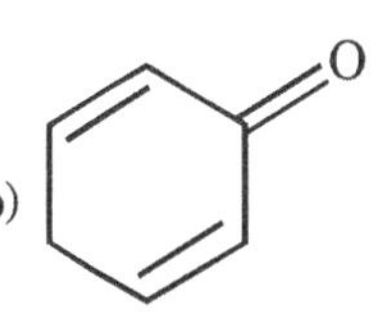

(c) 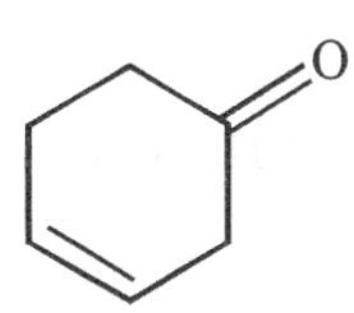

(d) 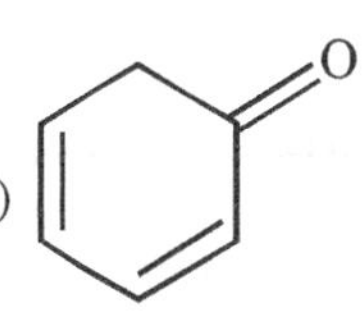

57. Heterolysis of propane gives-

(a) $\dot{C}H_3$ & $\dot{C}_2H_5$ (b) $\overset{\oplus}{C}H_3$ & $\overset{\ominus}{C}_2H_5$ (c) $\overset{\ominus}{C}H_3$ & $\overset{\oplus}{C}_2H_5$ (d) None of these

58. Match the ions given in Column I with their nature given in Column - II.

Column I	Column II
(a) $CH_3-O-\overset{\oplus}{C}H-CH_3$	(a) Stable due to resonance
(b) $F_3C^{\oplus}$	(b) Destablised due to inductive effect
(c) $CH_3-\overset{CH_3}{\underset{CH_3}{C}}\,{:}^{\ominus}$	(c) Stabilised by hyper conjugation
(d) $CH_3-\overset{\oplus}{C}H-CH_3$	(d) a 2° carbocation

(a) 1 - a , b, c ;2 - a, b ; 3 - b, 4 - c, d (b) 1 - a , b ; 2 - a, b, c ; 3 - b, c; 4 - c, d

(c) 1 - c , d ; 2 - b, a ; 3 - a, b, c; 4 - d (d) None of these

59. Arrange the following carbanions in the decreasing order of their stabilities-

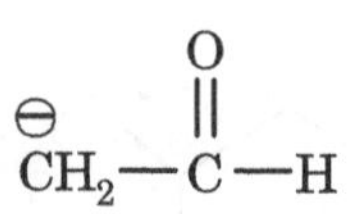
(I)

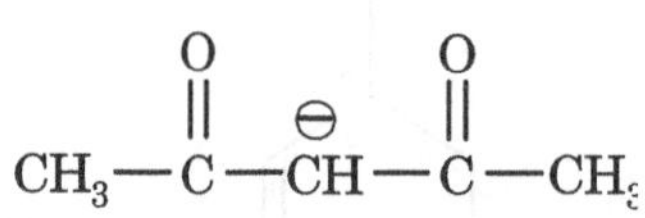
(II)

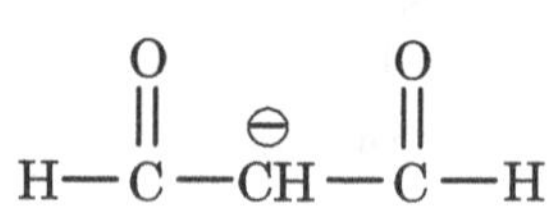
(III)

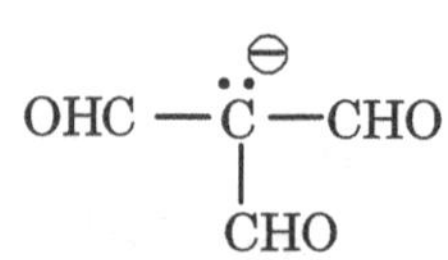
(IV)

(a) IV > III > II > I (b) III > II > IV > I

(c) IV = III > II > I (d) I > II > III > IV

60. The following carbocation rearranges to-

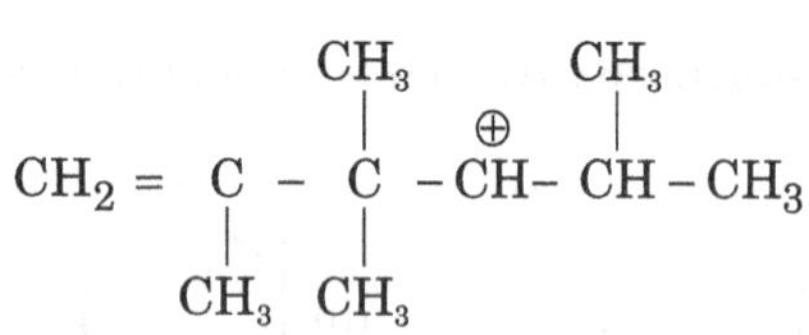

(a) 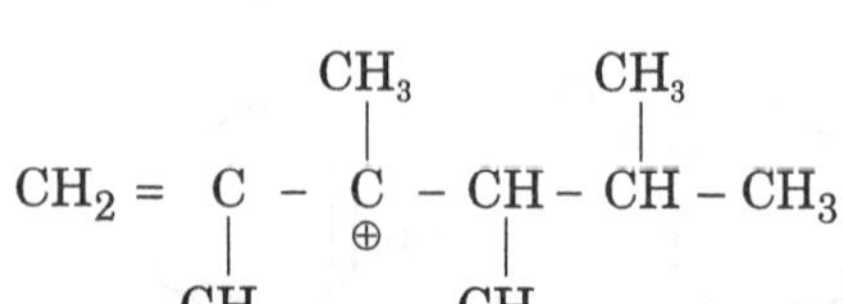

(b) 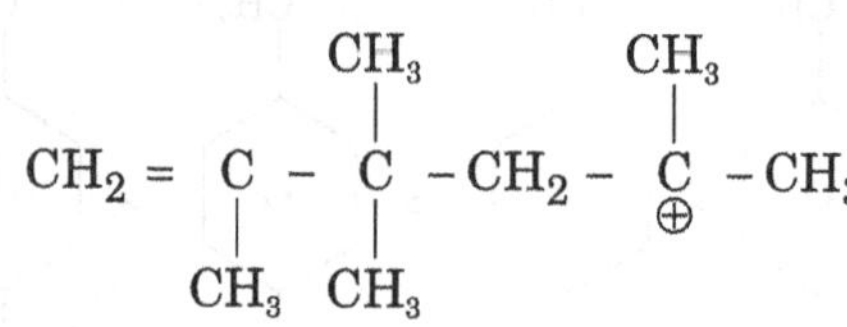

(c) 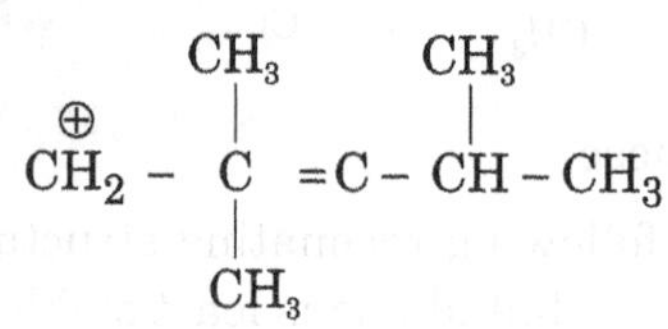

(d) None of these

61. Arrange the following carbocations in the decreasing order of their stability -

(1) 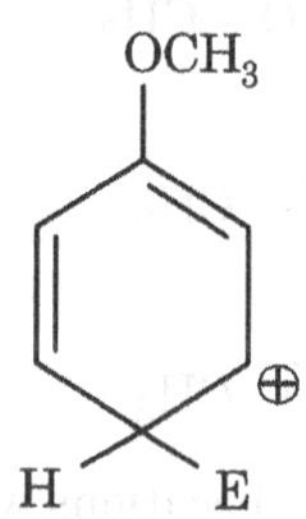

(2) 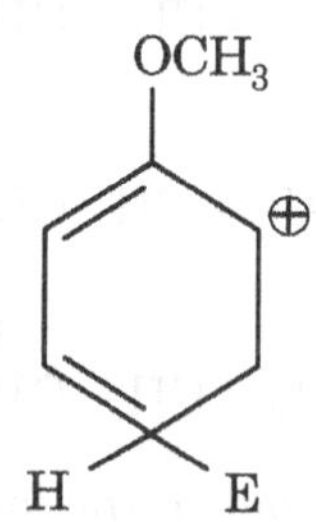

(3) 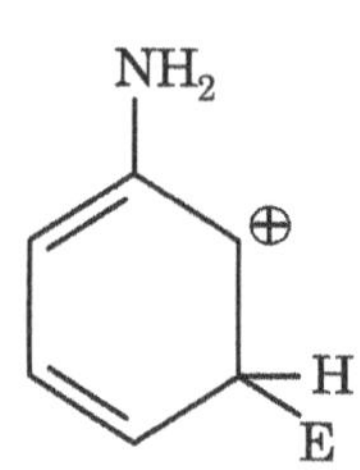

(4) 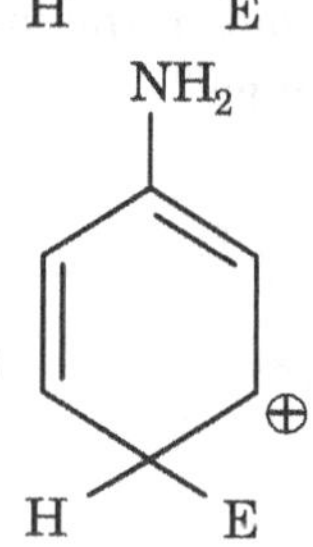

(a) 4 > 1 > 3 > 2 (b) 1 > 2 > 3 > 4

(c) 3 > 2 > 4 > 1 (d) 4 > 1 = 3 > 2

62. The correct order of stabilities of following carbocations is-

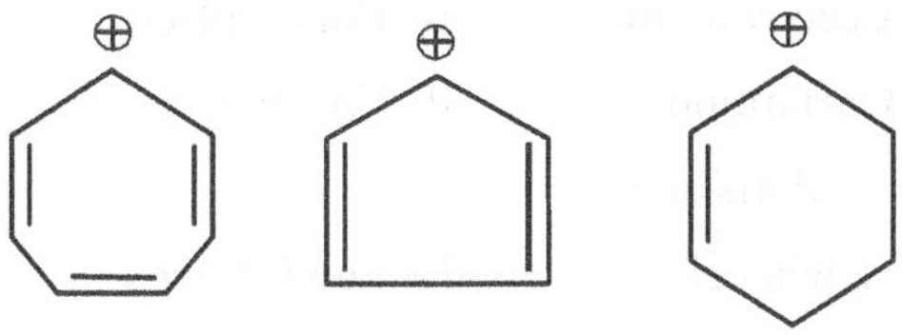

1 2 3

(a) 1 > 2 > 3 (b) 1 > 3 > 2

(c) 3 > 1 > 2 (d) 2 > 1 > 3

63. Select the correct order wrt the property mentioned along.

(a) 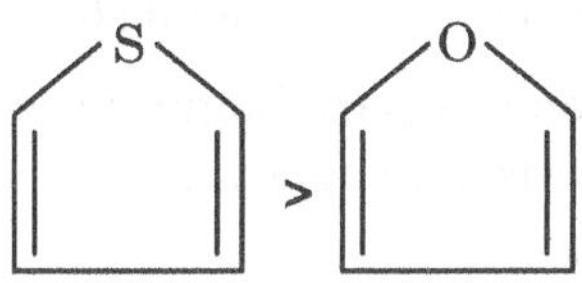 (Resonance energy)

(b) 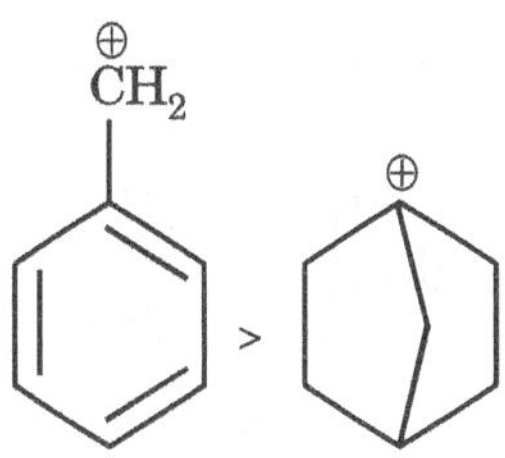

(Stability of carbocation)

(c) 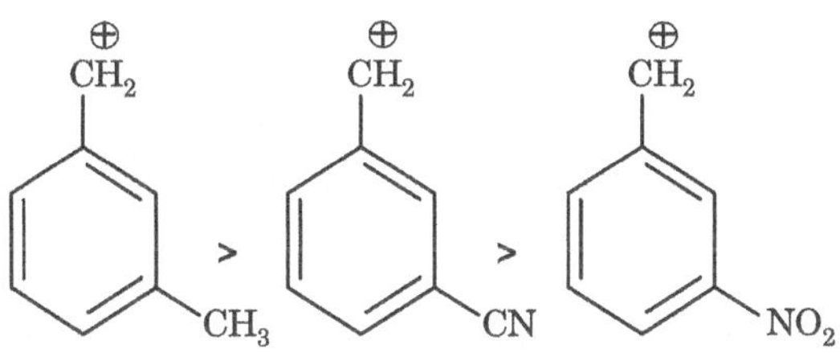

(d) All order are correct

64. Select the most stable carbocation

(a) $\overset{\oplus}{C}H_3$ (b) $CH_3-\overset{\oplus}{C}=O$

(c) $CH_3-\overset{\oplus}{C}=NH$ (d) $\overset{\oplus}{C}H_2-CH_2-Cl$

65. In which case, conjugate acid is stabilized by resonance?

(a) 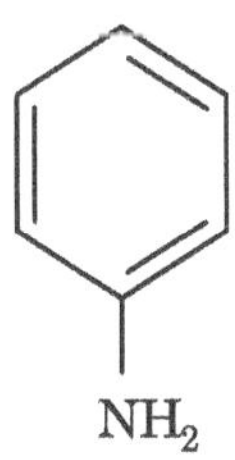 (b) 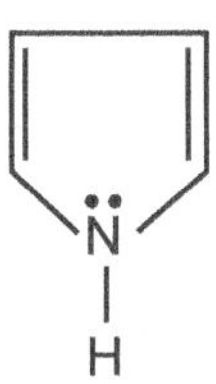

(c) 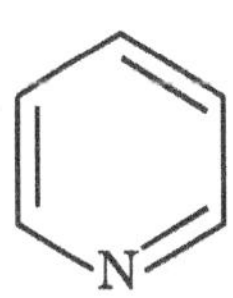 (d) Both (a) & (b)

66. Out of the following structures, the ones with highest & lowest energy, are respectively

(a) 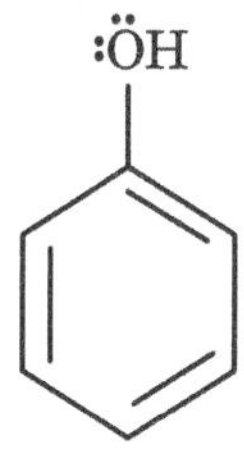 (b) 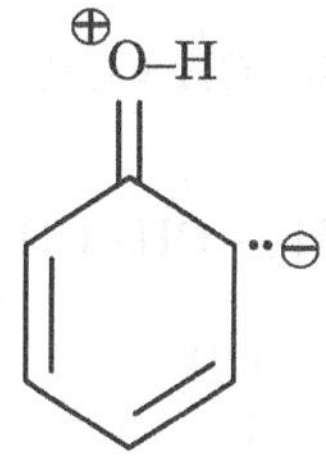

(c) 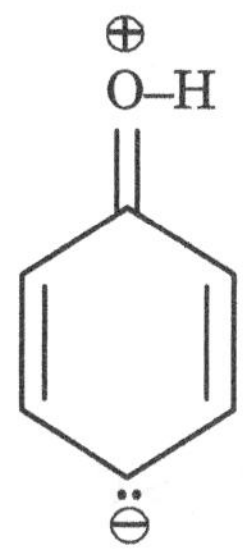 (d) 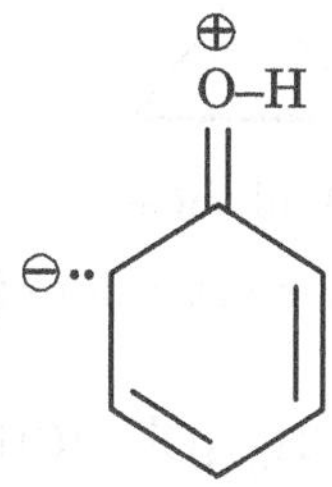

(a) a & b (b) a & c

(c) c & a (d) b & a

67. In which case, resonance is not possible?

(a) $C_6H_5-\overset{\oplus}{C}H-CH_3$

(b) $CH_3-CH=CH-O^{\ominus}$

(c) $CH_2=CH-\overset{\oplus}{N}H_3$

(d) $\overset{\ominus}{C}H_3-CH=CH-\overset{\oplus}{C}H-CH_3$

68. The correct set of stability order is

(a) $CH_2=\overset{\oplus}{C}H > CH_3-\overset{\oplus}{C}H_2$

(b) $CH_2=\overset{\ominus}{C}H > CH_3-\overset{\ominus}{C}H_2$

(c) $CH_2=\overset{\ominus}{C}H > H-C\equiv C^{\ominus}$

(d) $CH_2=\dot{C}H > CH_3-\dot{C}H_2$

69. Identify the most stable carbocation

(a) 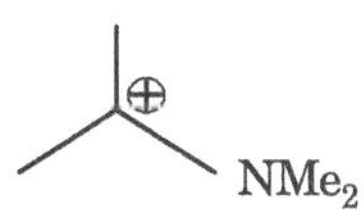

(b) 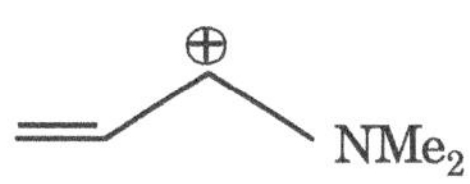

(c) 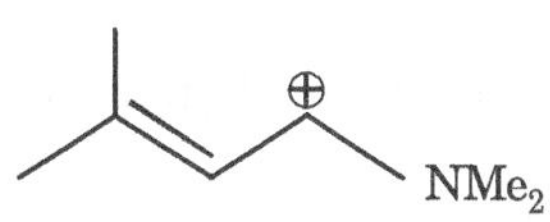

(d) 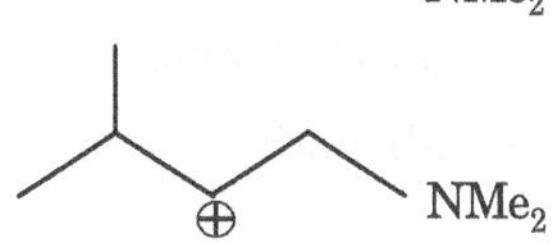

70. The carbocation stabilized by mesomeric effect is-

(a) $Ph-CH=CH-\overset{\oplus}{C}H_2$

(b) $CH_3-\overset{\oplus}{C}H-CH_3$

(c) 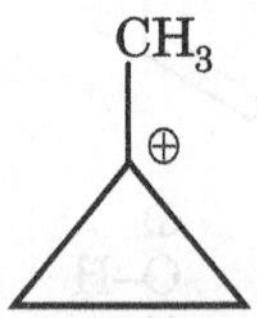

(d) All of these

71. The correct order of stability is-

(a) $(CH_3)_3\dot{C} > (CH_3)_2\dot{C}H > \dot{C}H_3 > C_6H_5\dot{C}H_2$

(b) $C_6H_5\overset{\oplus}{C}H_2 > \overset{\oplus}{C}H_3 > (CH_3)_2\overset{\oplus}{C}H > (CH_3)_3\overset{\oplus}{C}$

(c) $C_6H_5\overset{\ominus}{C}H_2 > \overset{\ominus}{C}H_3 > (CH_3)_2\overset{\ominus}{C}H > (CH_3)_3\overset{\ominus}{C}$

(d) All of these

72. Select the incorrect order of stability of carbocations.

(a) 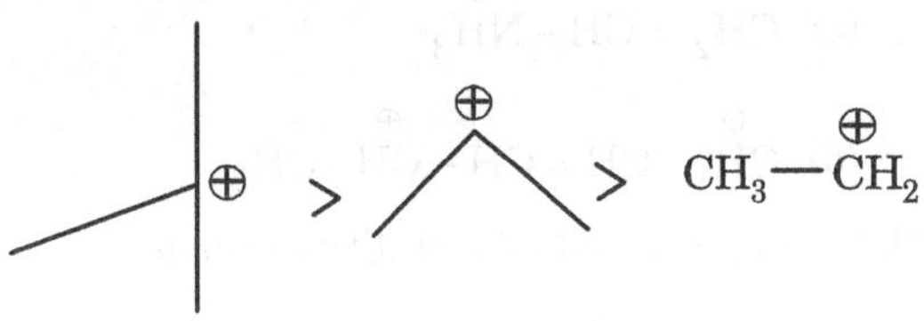

(b) $Ph-\overset{\oplus}{C}H-CH=CH_2 > Ph-CH=CH-\overset{\oplus}{C}H_2$

$> Ph-CH=\overset{\oplus}{C}-CH_3$

(c) 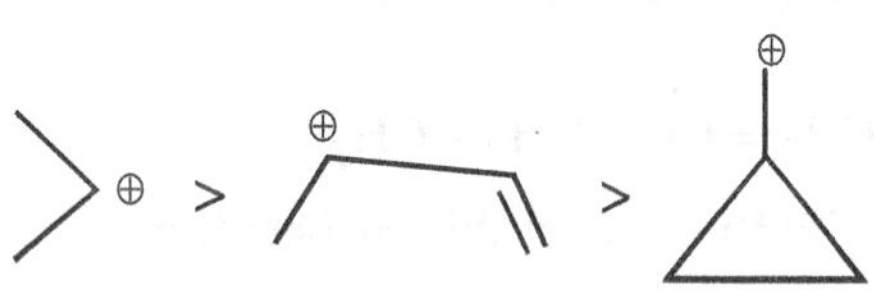

(d) 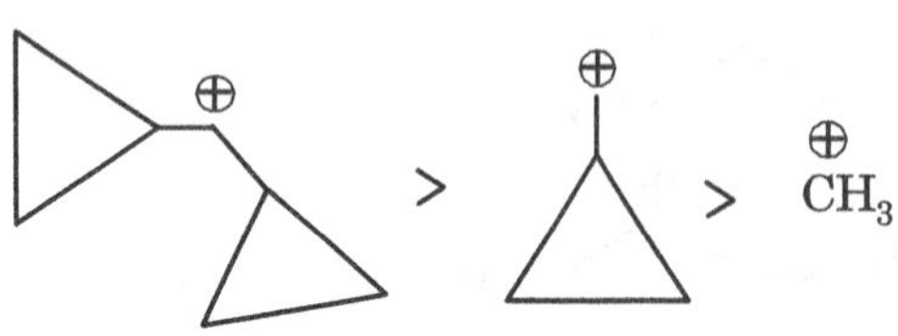

73. Which of the following intermediate is formed by homolytic fission?

(a) Carbocation (b) Carbanion

(c) Free radical (d) Both (a) & (b)

74. The reaction intermediate where C has a complete octet, is-

(a) Free radical (b) Carbocation

(c) Carbanion (d) Carbene

75. Carbanions are-

(a) Lewis acids (b) Lewis bases

(c) Arrehenius acids (d) Arrehnius bases

76. Which of the following is correct wrt carbenes?

(a) Singlet carbene is bent where triplet carbene is linear.

(b) In Singlet carbene, C is in excited state whereas in triplet carbene, C is in ground state.

(c) Singlet carbene is paramagnetic whereas triplet carbane is diamagnetic.

(d) All are correct.

77. Which one of the following is least stable?

(a) 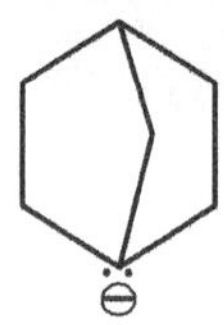

(b) 

(c) $H-C\equiv C^{\bullet\bullet\ominus}$

(d) 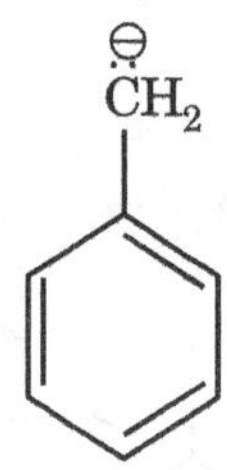

78. Arrange the following free radicals in in correct order of stability

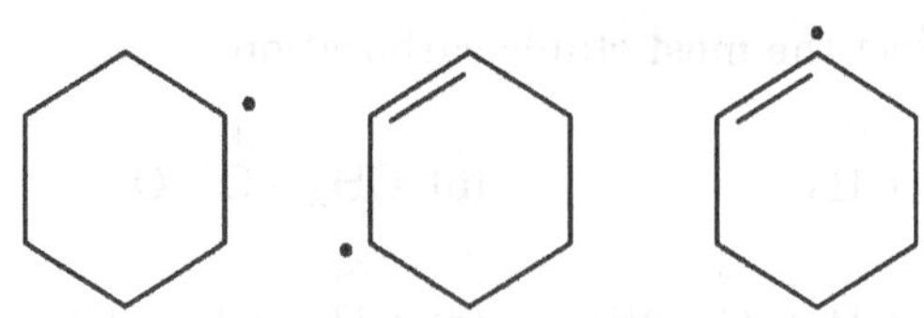

(I) (II) (III)

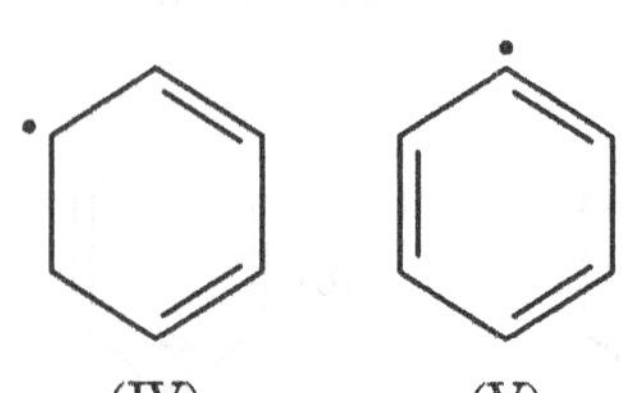

(IV) (V)

(a) IV > II > I > III > V

(b) IV > II = I > III > V

(c) IV > IV > III > II > I

(d) None of these

79. Which of the following example has both the resonating structures contributing equally to the resonance hybrid?

(a) 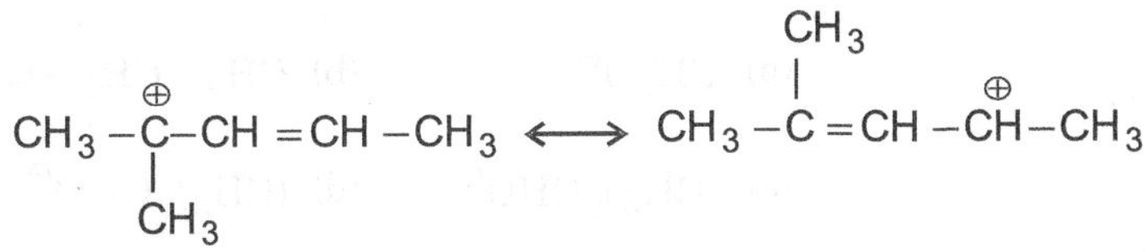

(b) 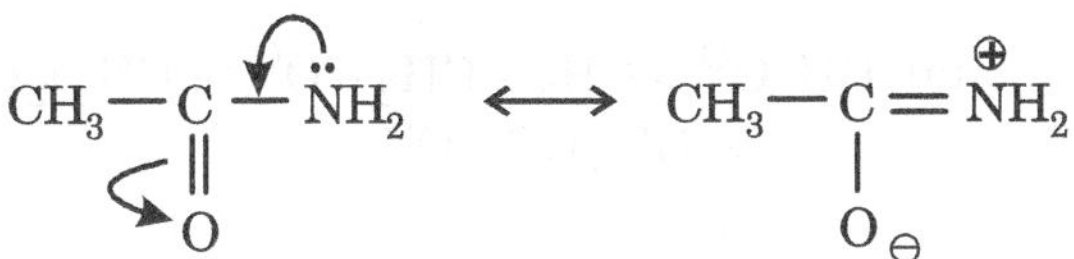

(c) 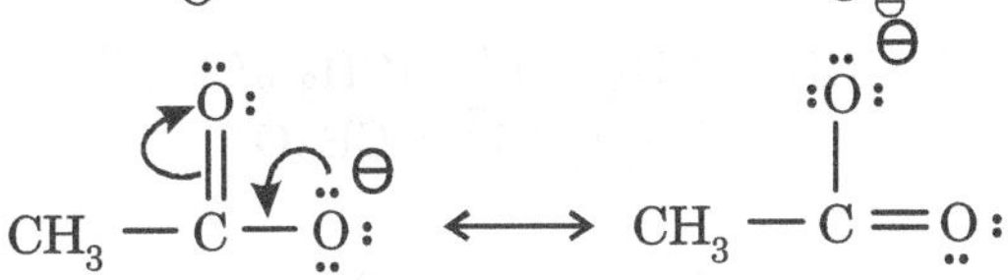

(d) All of the above

80. Select the incorrect order.

(a) 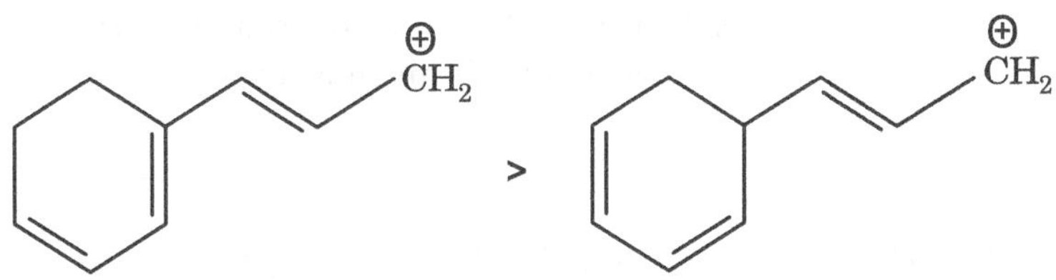

(stability of carbocation)

(b) 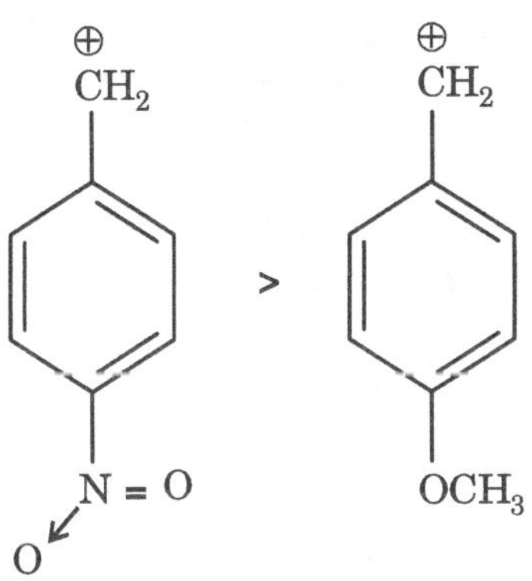 (stability of carbocation)

(c) 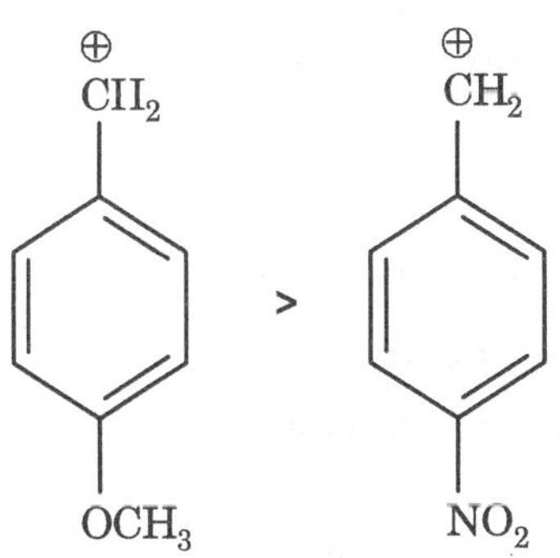 (stability of carbocation)

(d) Both (a) & (b) are incorrect

81. The most unstable carbocation is-

(a) $CH_3-\overset{\oplus}{C}H_2$ (b) $\overset{\oplus}{C}H_2-CHO$

(c) $\overset{\oplus}{C}H_2-OCH_3$ (d) $CH_3-\overset{\oplus}{\underset{CH_3}{C}}-H$

82. In which of the following carbocations, Ist is more stable than IInd?

(a)

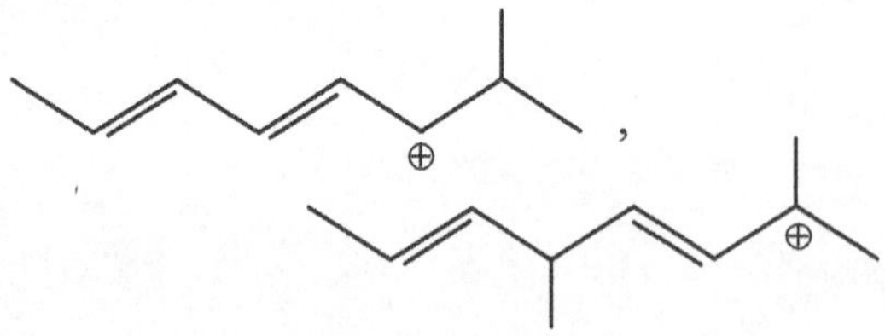

(b) $CH_3-O-\overset{\oplus}{C}H-CH_3$,

$CH_3-\overset{\oplus}{\underset{CH_3}{C}}-CH_2-CH_3$

(c)

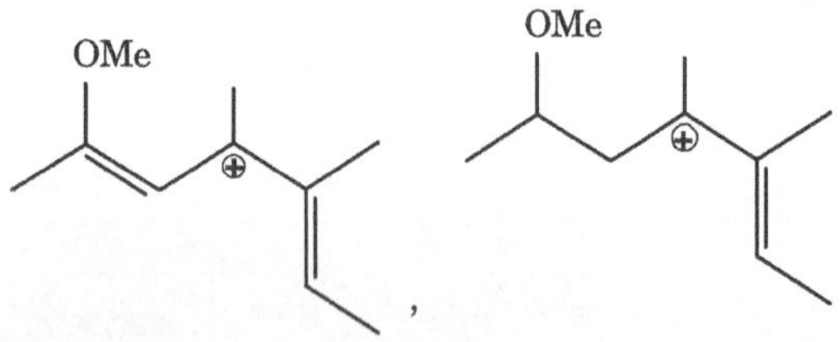

(d) All of these

Nomenclature, Electrophiles & Nucleophiles

83. IUPAC name of the following compound is -

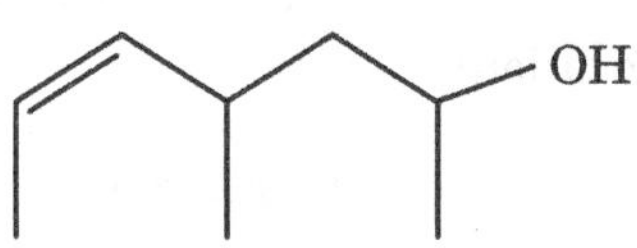

(a) 1, 3 - Dimethyl hex - 4 - en - 1 - ol

(b) 4, 6 - Dimethyl hex - 2 - en - 6 - ol

(c) 4 - Methyl hept - 2 - en - 6 - ol

(d) 4 - Methyl hept - 5 - en - 2 - ol

84. Which of the following series contains only nucleophiles ?

(a) $NH_3, H_2O, ^-CN, I^-$

(b) $AlCl_3, NH_3, H_2O, I^-$

(c) $AlCl_3, BF_3, H_2O, NH_3$

(d) $AlCl_3, BF_3, NO_2^{\oplus}, NH_3$

85. Arrange the following nucleophiles in decreasing order of reactivity towards S_N2

(a) $CH_3O^{\ominus}$ (b) $CH_3-CH_2-O^{\ominus}$

(c) $(CH_3)_2CHO^{\ominus}$ (d) $(CH_3)_3C-O^{\ominus}$

(a) $CH_3O^{\ominus} > CH_3-CH_2-O^{\ominus} > (CH_3)_2CHO^{\ominus} > (CH_3)_3C-O^{\ominus}$

(b) $CH_3-CH_2-O^{\ominus} > (CH_3)_2CHO^{\ominus} > (CH_3)_3CO^{\ominus} > CH_3O^{\ominus}$

(c) $CH_3O^{\ominus} > (CH_3)_3CO^{\ominus} > (CH_3)_2CHO^{\ominus} > CH_3CH_2O^{\ominus}$

(d) None of these

86. Leaving ability in decreasing order of some nucleophiles is given below -

$N_2 > CF_3-\overset{\ominus}{SO_3} >$

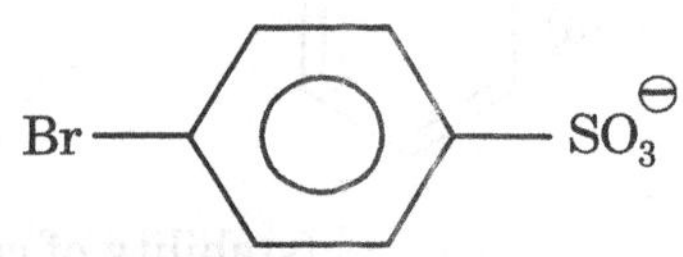

$>$

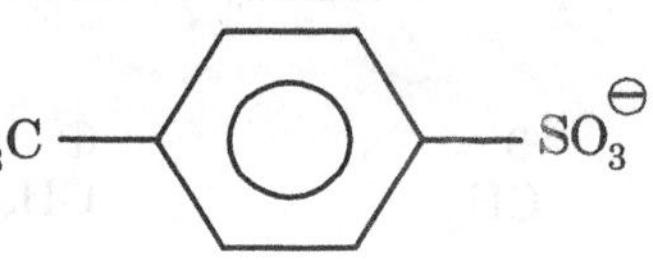

Which of the following statements is/are correct about the leaving groups?

(a) The leaving group should be stable after leaving with the bonding pairs of electrons

(b) Good leaving groups should be weak bases.

(c) Strong bases are good leaving group.

(d) Both (a) & (b) are correct.

87. The IUPAC name of smallest saturated ether which is optically active is-

(a) 2 - Methoxy - 2 - Methyl propane

(b) 2 - Methoxy butane

(c) 3 - Methoxy - 3- Methyl hexane

(d) 3 - Methoxy butane

88. Which halide ion is the best nucleophile in DMSO solution?

(a) F^- (b) Cl^-

(c) Br^- (d) I^-

89. In which of the following pairs, the first one is more nucleophillic than the second?

(a) $(CH_3)_2\overset{\ominus}{N}, \overset{\ominus}{N}H_2$

(b) $C_6H_5O^{\ominus}, (CH_3)_2CHO^{\ominus}$

(c) $CH_3O^{\ominus}, (CH_3)_3CO^{\ominus}$

(d) Both (a) & (b)

90. Which of the following are not nucleophillic?

(a) NH_4^+

(b) PF_6^-

(c) BF_4^-

(d) All of these

91. Select the 'INCORRECT IUPAC NAME'

	Molecule	IUPAC Name
(a)	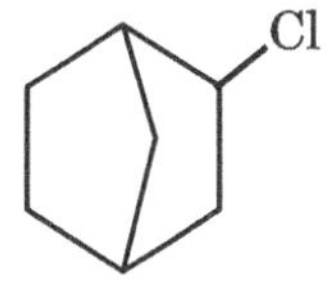	→ 2- Chloro bicycb [2.2.1] heptane
(b)	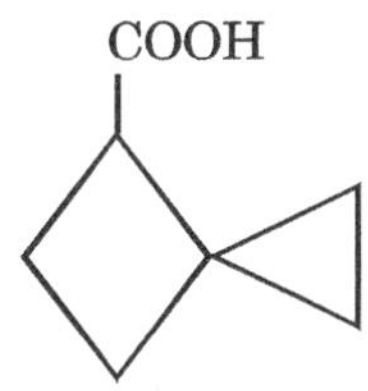	→ Spiro [2 . 3] hexane- 4 - Carboxylic acid
(c)	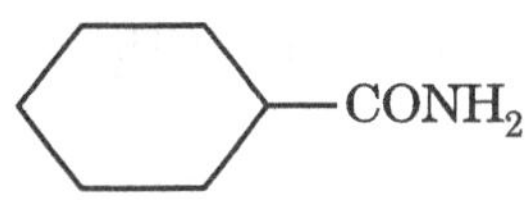	→ Cyclohexane Carboxamide

(d) All are correct

92. Select the correct match.

Column - I (Molecules)	Column - II (IUPAC Names)
(a) $Cl-\overset{O}{\overset{\parallel}{C}}-CH_2-\overset{O}{\overset{\parallel}{C}}-CH_2-\overset{O}{\overset{\parallel}{C}}-H$	Cyclohexane - 1, 2 - dicarboxylic anhydride
(b) 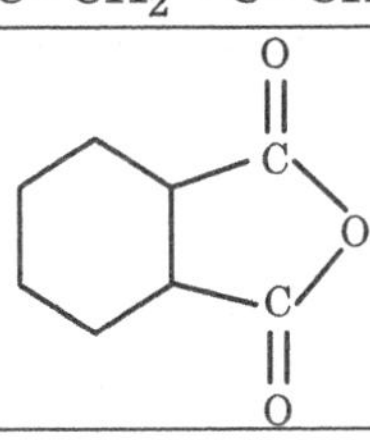	1 - Cyclobutyl pentane
(c) 	Cyclopropane Carboxylic acid
(d) 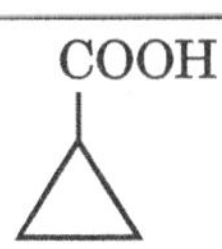	3, 5 - dioxopenanoyl chloride

(a) 1 - d; 2 - a; 3 - b; 4 - c (b) 1 - c; 2 - d; 3 - b; 4 - a (c) 1 - a; 2 - b; 3 - c; 4 - d (d) None of these

93. Select the correct IUPAC name of the following compound

$$NC-CH_2-CH_2-\overset{O}{\overset{\parallel}{C}}-OH$$

(a) 4- Cyanopropanoic acid

(b) 3 - Cyanopropanoic acid

(c) 3 - Carboxypropanenitile

(d) 4 - Carboxy propane Cyanide

94. The IUPAC name of acetylacetone is-

(a) Pentane - 2, 4 - diol

(b) Pentane - 2, 4 - dione

(c) Butane - 2, 4 - diol

(d) Butane - 2, 4 - dione

95. The correct IUPAC name of 2 - ethylpent-3-yne

(a) 4 - Methyl hex - 2 - yne

(b) 4 - ethyl pentyne

(c) 3 - ethyl pentyne

(d) None of these

96. Select the 'Incorrect' Hybridisation?

(a) $CH_2 = CH - \overset{\bullet\bullet\ominus}{C}H_2$ ($\uparrow$ sp^2)　(b) $CH_3 - \overset{\oplus}{C}H_2$ ($\uparrow$ sp^2)

(c) $H - C \equiv C^{\oplus}$ ($\uparrow$ sp)　(d) None of these

97. Of the following compounds which one has a 'WRONG' IUPAC Name?

(1) $CH_3 - CH_2 - CH_2 - \overset{O}{\overset{||}{C}} - O - CH_2 - CH_3$
ethyl butanoate

(2) $CH_3 - \underset{CH_3}{\underset{|}{CH}} - \overset{}{\underset{O}{\underset{||}{C}}} - \underset{CH_3}{\underset{|}{CH}} - CH_3$
2 – 4, – dimethyl – 3 – pentanone

(3) $H_3C - \underset{OH}{\underset{|}{CH}} - \overset{CH_3}{\overset{|}{CH}} - CH_3$
2 – Methyl – 3 – butanol

(4) $CH_3 - CH_2 - CH_2 - CHO$
butanal

(a) 1　(b) 2

(c) 3　(d) 4

98. Which of following an 'INCORRECT IUPAC Name?

(a) 1 - Ethoxy - 1-propanamine

(b) Cyclohexanoic acid

(c) 1 - Amino - 1 - Ethoxypropane

(d) Both (b) & (c)

99. The correct IUPAC name of 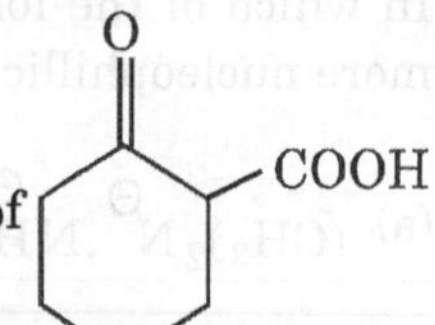 is–

(a) 2 - Oxocyclohexane carboxylic acid

(b) 2 - Carboxycyalohexanoic acid

(c) 2 - Ketocyclohexanoic acid

(d) 2 - Oxo - 1 - carboxy cyclohexane

100. Certain compounds with their corresponding common names are given. How many of these are 'WRONG' Common names?

$CH_3 - \overset{O}{\overset{||}{C}} - COOH$ – Pyruvic acid,

$CH_3 - CH = CH - COOH$ – Crotonic acid,

$CH_2 = CH - C \equiv C - H$ – Vinyl acetylene,

$C_6H_5 - CH = CH - CHO$ – Cinamaldehyde

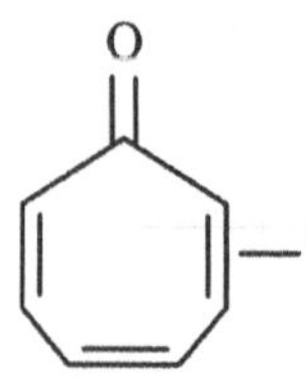 — Tropone , 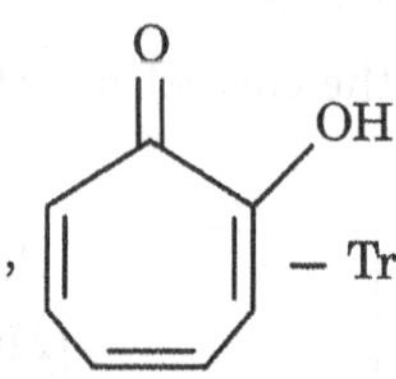 – Tropolone ,

(a) 3　(b) 5

(c) 6　(d) None of these

101. The IUPAC Name of $(C_2H_5)_2\ NCH_2\ \underset{Cl}{\underset{|}{CH}}COOH$ is–

(a) 2 - Chloro - 4 - N - ethylpentanoic acid

(b) 2 - Chloro - 3 - (N, N - diethylamino) - propanoic acid

(c) 2 - Chloro - 2- Carboxy - N- ethyl methane

(d) Both 1 & 3 are correct

102. Which of the following molecules have been correctly named as per IUPAC nomenclature?

(a) $CH_3 - \overset{H}{\overset{|}{\underset{OH}{\underset{|}{C}}}} - COOH$
2-Hydroxy propanoic acid

(b) 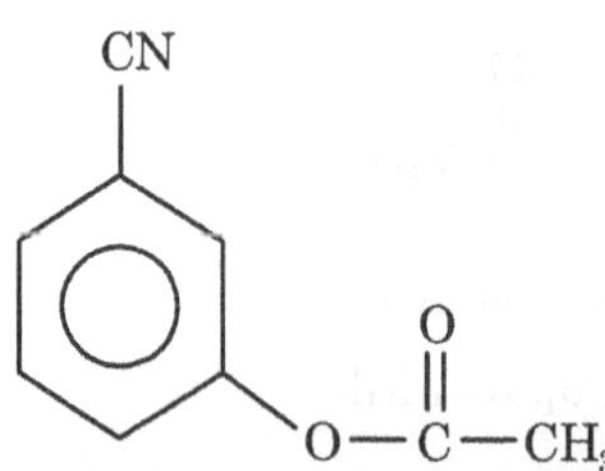
3- Cyanophenyl ethanoate

(c) 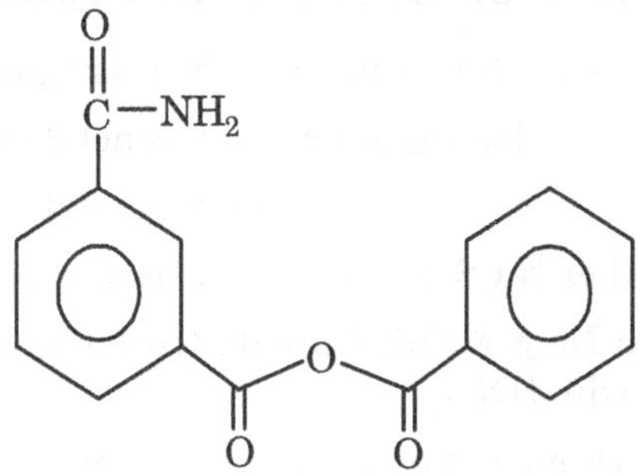

Benzene Carboxylic 3 - Carbamoyl benzene Carboxylic anhydride

(d) All of these are correct

103. The IUPAC name of the compound $CH_2 = CH - CH(CH_3)_2$ is -

(a) Propyl ethene

(b) 2 - ethenyl propane

(c) 3 - methyl but - 1 - ene

(d) 2 - methyl but - 3 - ene

104. The correct IUPAC name of the compound given below is -

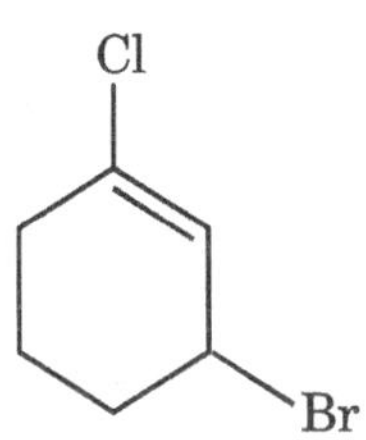

(a) 2 - bromo - 6 - chlorocyclohex - 1 - ene

(b) 6 - bromo - 2 - chlorocyclohex - 1 - ene

(c) 1 - bromo - 3 - chlorocyclohex - 1 - ene

(d) 3 - bromo - 1 - chlorocyclohex - 1 - ene

105. The correct IUPAC name of

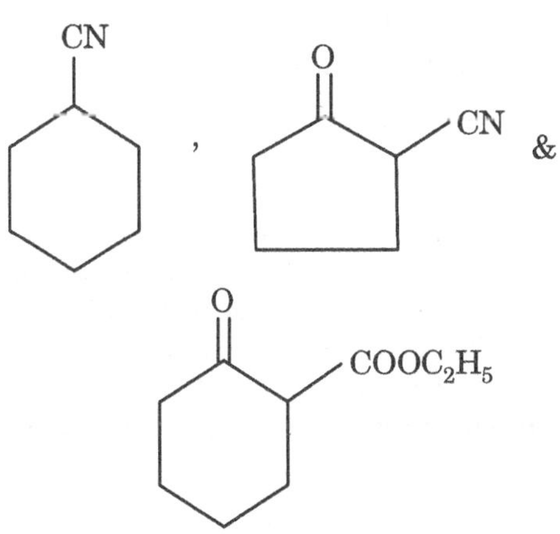

are, respectively -

(a) Cyclohexane cyanide, 2 - ketocyclopentane cyanide & ethyl 2 - ketocyclohexanoate

(b) Cyclohexane cyanide, 2 - ketocyclopentane cyanide & ethyl - 2 ketocyclohexanoate

(c) Cyclohexane carbonitrile, 2 - oxocyclopentane - 1 - carbonitrile& ethyl (2 - oxocyclohexane) - 1 - carboxylate

(d) None of these

106. The IUPAC name of

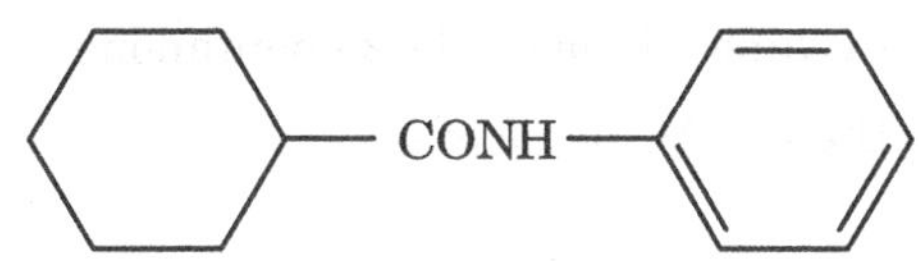

is –

(a) N - cyclohexylbenzamide

(b) N - phenyl - n - cyclohexylmethanamide

(c) N - phenyl Cyclohexane Carboxamide

(d) N - Cyclohexyl - N - phenyl Methanamide

107. Which one of the following is sec - butyl phenyl vinyl methane?

(a) 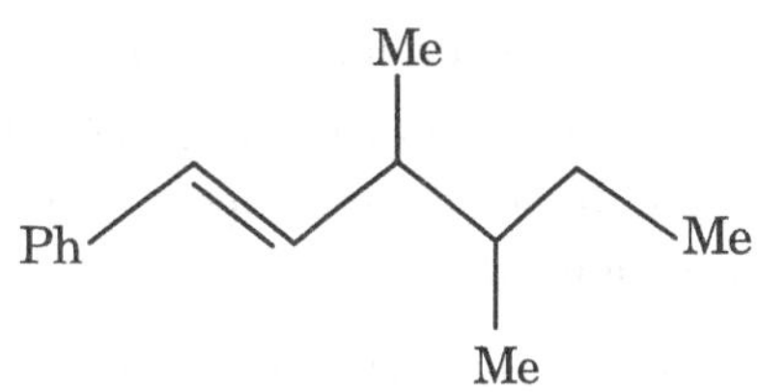

(b) 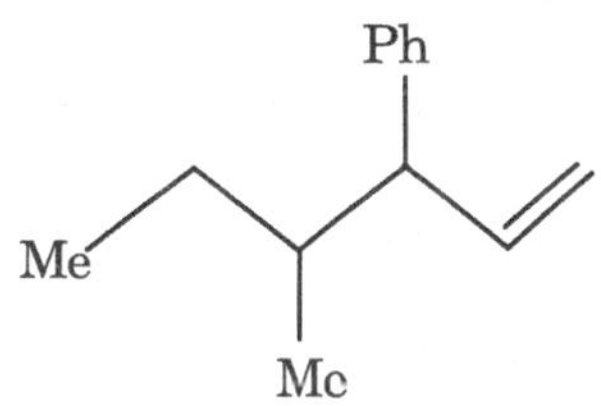

(c) 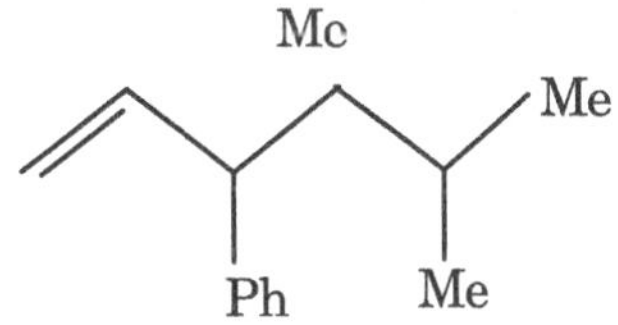

(d) None of these

108. The correct IUPAC name of the molecule given below is -

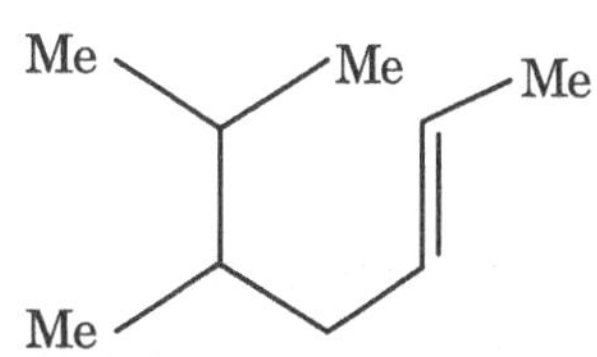

(a) 5, 6 - dimethyl hept - 2 - ene

(b) 2, 3 - dimethyl hept - 5 - ene

(c) 5, 6 - dimethyl hept - 3 - ene

(d) 5 - isopropyl hex - 2 - ene

109. The correct IUPAC name of the compound 'X' is -

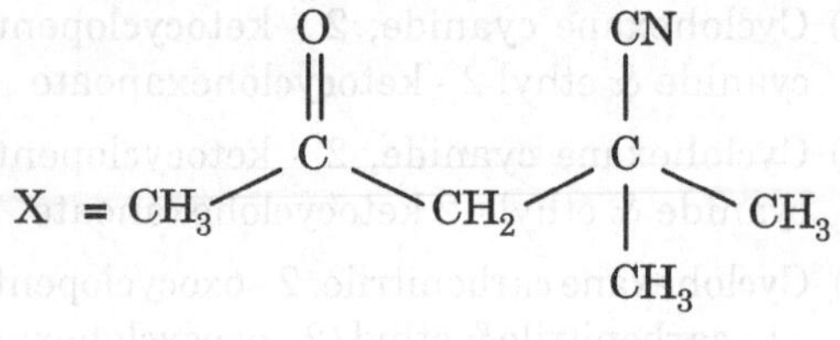

(a) 4 - Cyano - 4 - methyl oxopentane

(b) 2 - Cyano - 2 - methyl - 4 - oxopentane

(c) 2, 2 - dimethyl - 4 - oxopentanenitrile

(d) 4 - Cyano - 4 - methyl - 2 - pentanone

110. Give the IUPAC name of -

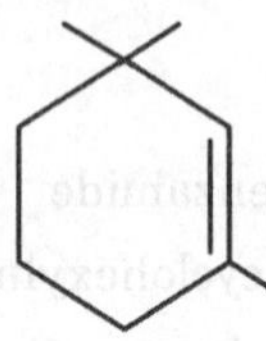

(a) 1, 1, 3 - Trimethylcyclohex - 2 - ene

(b) 1, 3, 3 - Trimethylcyclohex - 1 - ene

(c) 1, 1, 5 - Trimethylcyclohex - 5 - ene

(d) 2, 6, 6 - Trimethylcyclohex - 1 - ene

111. The correct decreasing order of priority for the functional groups of organic compounds in IUPAC system of nomenclature is -

(a) $-CONH_2, -CHO, -SO_3H, -COOH$

(b) $-COOH, -SO_3H, -CONH_2, -CHO$

(c) $-SO_3H, -COOH, -CONH_2, -CHO$

(d) $-CHO, -COOH, -SO_3H, -CONH_2$

112. The IUPAC name for the compound

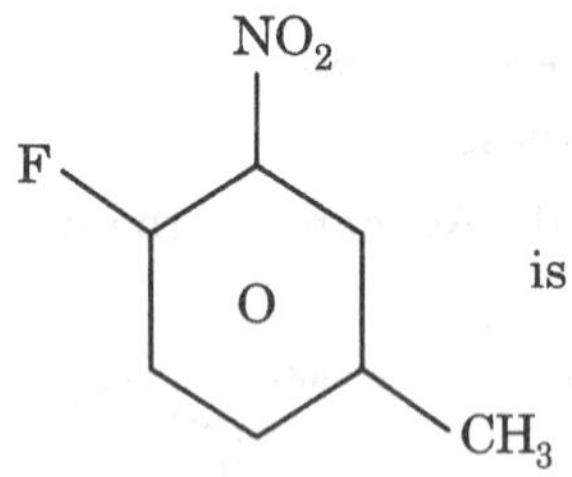

is -

(a) 1 - fluoro - 4 - methyl - 2 - nitrobenzene

(b) 4 - fluoro - 1 - methyl - 3 - nitrobenzene

(c) 4 - methyl - 1 -fluoro - 2 - nitrobenzene

(d) 2 - fluoro - 5 - methyl - 1 - nitrobenzene

Purification of Organic Compounds

113. During estimation of halogens by lassaignes test,

(a) NaCN is destroyed but Na_2S is not destroyed

(b) Na_2S is destroyed but NaCN is not destroyed

(c) Neither Na_2S nor NaCN are destroyed

(d) Both Na_2S &NaCN are destroyed on boiling with dil. HNO_3

114. In the Duma's method for the estimation for nitrogen, 0.0237 grams of an organic compound gave 2.21ml of nitrogen at 754.32 mm of Hg pressure at 18°C (Aq. Tension at 18°C is 15.4 mm of Hg) Therefore, the percentage of nitrogen in the compound is -

(a) 20.67 % (b) 10.6 %

(c) 11.2 % (d) 13.9 %

115. Kjeldahl's method is not applied for the estimation of nitrogen in -

(a) CH_3NO_2

(b) 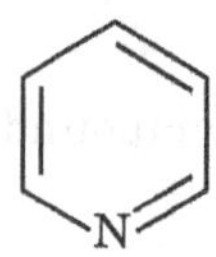

(c) 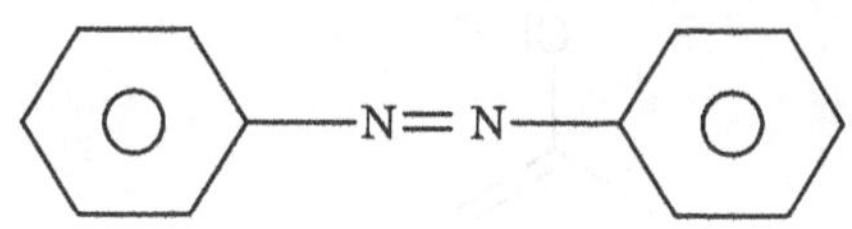

(d) All of these

116. Presence of blood red colour shows the presence of-

(a) Nitrogen

(b) Sulphur

(c) Nitrogen &Sulphur together

(d) Halogen

117. A mixture of approximately 95% ethanol & 50% H_2O can be separated by -

(a) Fractional distillation

(b) Vacuum distillation

(c) Azeotropic distillation

(d) Steam distillation

118. For the estimation of nitrogen, 1.4g of an organic compound was digested by Kjeldahl method & the evolved ammonia was absorbed in 60ml of M/10 H_2SO_4. The unreacted acid required 20ml of M/10 NaOH for complete Neutralization. The percentage of nitrogen in the compound is -

(a) 5% (b) 6%

(c) 10% (d) 3.5%

119. In lassaigne's test, the organic compound is first fused with sodium metal. The sodium metal is used because-

(1) The melting point of sodium is low

(2) Sodium reacts with elements of organic compound to form inorganic ionic salts.

(3) Sodium salts are water soluble

(4) Sodium does not react with H_2O

(a) Only 1 is correct

(b) Both 3 and 4 are correct

(c) 1, 2 & 3 only are correct

(d) All are correct

120. Which of the following statements about the compound P, Q, R & S given below is correct ?

(P) $H_2N-NH_3^+Br^-$

(Q) 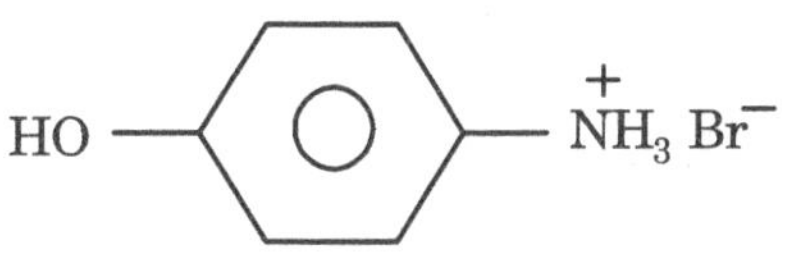

(R) 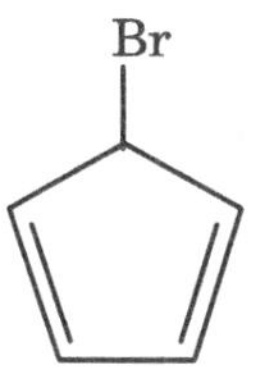

(S) 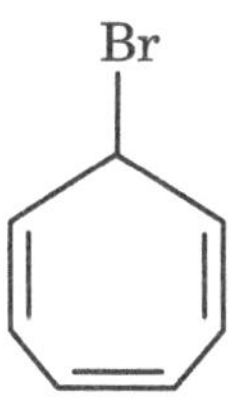

(a) Sodium fusion extract of the compounds P & Q give Prussian blue colour with $FeSO_4$.

(b) All the four compounds (P, Q, R & S) give pale yellow precipitate with $AgNO_3$.

(c) Only P & Q gives pale yellow ppt. with $AgNO_3$.

(d) Only P, Q & S gives pale yellow ppt. with $AgNO_3$.

121. In Kjeldahl's method for the estimation of nitrogen, K_2SO_4 & $CuSO_4$ are used , their function is -

(1) K_2SO_4 raises the b.pt. & ensures complete reaction.

(2) $CuSO_4$ acts as catalyst

(3) K_2SO_4 acts as catalyst & $CuSO_4$ raises the b.pt.

(a) Only 3 is correct

(b) Both 1 & 2 are correct

(c) Only 2 is correct

(d) All are correct

122. Which of the following will not give lassaignes' test for nitrogen?

(a) $C_6H_5NHNH_2$ (b) NH_2CONH_2

(c) $PhN = NPh$ (d) NH_2NH_2

123. Which of the following will give blood red colour while doing Lassaignes' test for nitrogen

(a) 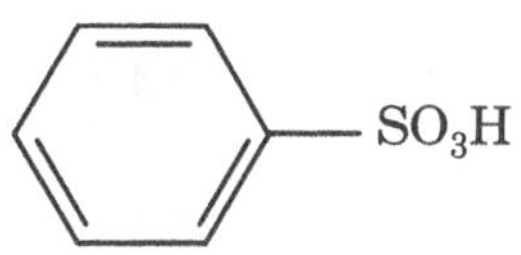

(b) $H_2N-\overset{\overset{O}{||}}{C}-NH_2$

(c) $NH_4^+NO_3^-$

(d) 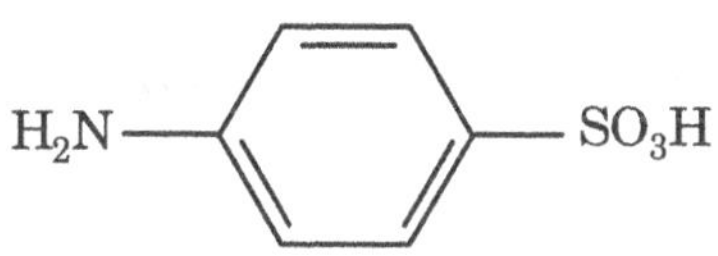

124. In Carius method of estimation of halogens, 250 mg of an organic compound gave 141 mg of AgBr. The % age of bromine in the compound is (at. of Ag = 108, Br = 80)

(a) 48 (b) 60

(c) 24 (d) 36

Answers

1. (d)	2. (b)	3. (b)	4. (d)	5. (d)	6. (d)	7. (d)	8. (c)	9. (d)	10. (a)
11. (c)	12. (d)	13. (d)	14. (c)	15. (b)	16. (a)	17. (b)	18. (c)	19. (d)	20. (b)
21. (d)	22. (a)	23. (b)	24. (b)	25. (a)	26. (a)	27. (b)	28. (d)	29. (d)	30. (c)
31. (c)	32. (b)	33. (d)	34. (d)	35. (a)	36. (b)	37. (a)	38. (c)	39. (b)	40. (a)
41. (c)	42. (b)	43. (a)	44. (c)	45. (c)	46. (d)	47. (d)	48. (d)	49. (b)	50. (d)
51. (c)	52. (c)	53. (d)	54. (c)	55. (a)	56. (c)	57. (c)	58. (a)	59. (a)	60. (a)
61. (a)	62. (b)	63. (d)	64. (c)	65. (c)	66. (c)	67. (c)	68. (b)	69. (c)	70. (a)
71. (c)	72. (c)	73. (c)	74. (c)	75. (b)	76. (a)	77. (a)	78. (a)	79. (c)	80. (d)
81. (b)	82. (d)	83. (d)	84. (a)	85. (a)	86. (d)	87. (b)	88. (a)	89. (c)	90. (d)
91. (d)	92. (a)	93. (b)	94. (b)	95. (a)	96. (d)	97. (c)	98. (d)	99. (a)	100. (d)
101. (b)	102. (d)	103. (c)	104. (d)	105. (c)	106. (c)	107. (b)	108. (a)	109. (c)	110. (b)
111. (b)	112. (a)	113. (d)	114. (b)	115. (d)	116. (c)	117. (c)	118. (c)	119. (c)	120. (d)
121. (b)	122. (d)	123. (d)	124. (c)						

Explanations

3. (b) $\underset{\text{sp}}{H-C\equiv C}-CH_2-\overset{\overset{O}{\parallel}}{C}-O-H$, $\underset{\text{sp}^2}{CH_2=CH}-CH_2-\overset{\overset{O}{\parallel}}{C}-O-H$, $CH_3-\underset{\text{sp}^3}{CH_2}-CH_2-\overset{\overset{O}{\parallel}}{C}-OH$

Acidic strength $\propto -I$ effect

$-I$ effect : sp > sp^2 > sp^3

5. (d) Structure I is more stable than II because I is uncharged. IV is more stable than III because of more no. of covalent bonds and also, in structure IV, all the atoms (C & O) have complete octet.

V > VI = VII (Stability orders)

No charge separation in structure V. Structures VI & VII are indistinguishable.

7. (d) 

(only one pie bond participates)

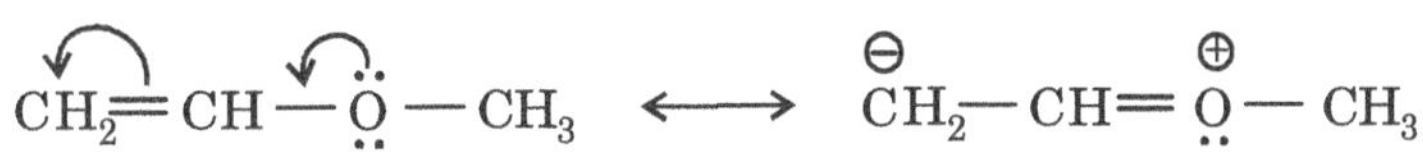

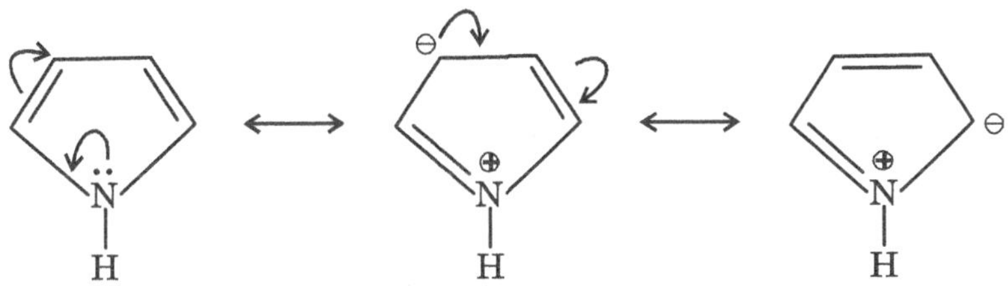

8. (c) Relative stabilities of carbocations is explained on the basis of hyper conjugation & + I effect.

Relative stabilities of free radicals is also explained on the basis of hyper conjugation & + I.

⇒ NO HYPERCONJUGATION OCCURS IN CARBANIONS

⇒ Relative stabilities of alkenes can be explained by hyper conjugation.

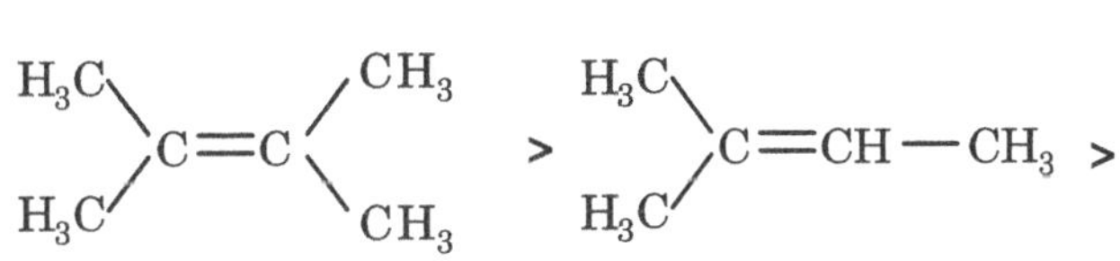

(12∝–H) (9∝ H)

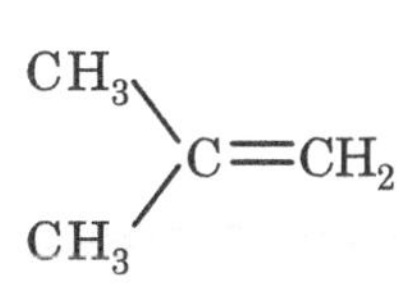

(6∝ H)

9. (d)

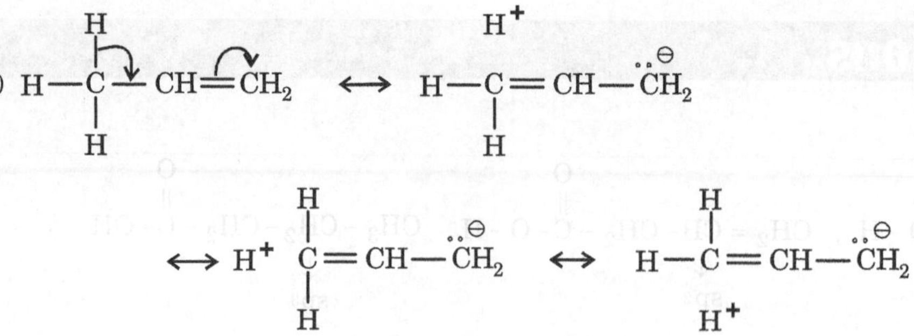

10. (a)

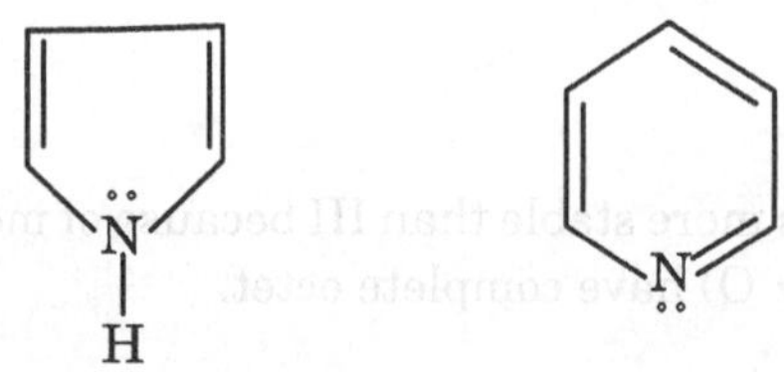

(I) Pyrrole (II) Pyridine

In pyridine, the lone pair resides in sp^2 hypridised orbital which is present in the plane of molecule. The p-orbitals forming the π – bonds are present perpendicular to the plane of molecule. So, lone pair cannot participate in resonance. Hence, it is more basic.

→ In pyrrole, the lone pair is present in unhybridised p- orbital & can participate in resonance. So, it is less basic.

→

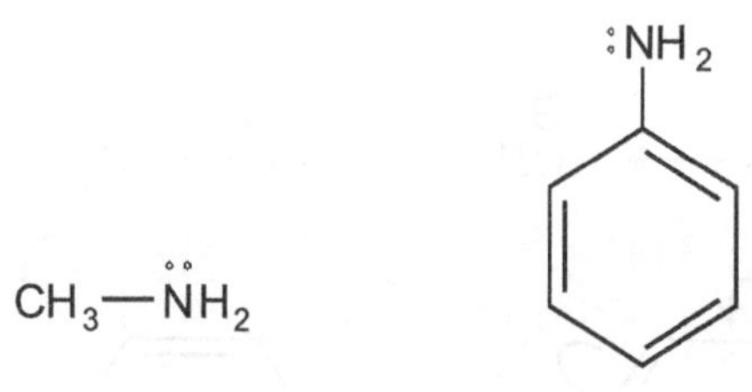

(Methanamine) (Aniline)

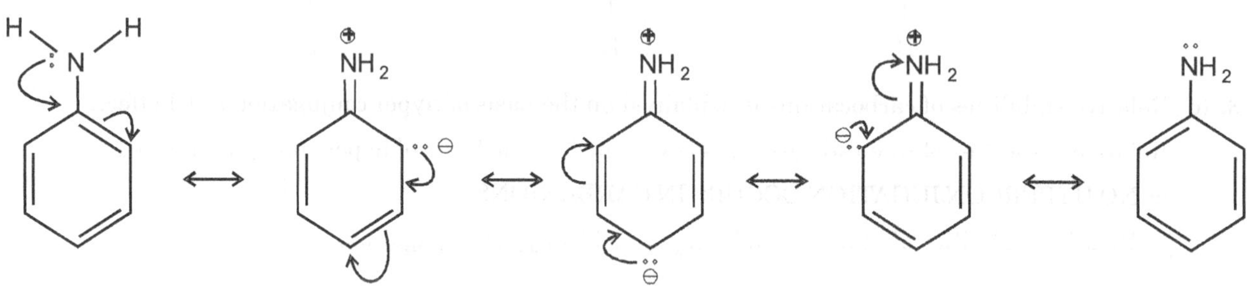

In methanamine, the lone pair on N is localized. It is 'Not Delocalised'. So, it is more basic than Aniline.

$\Rightarrow \overset{\ominus}{C}H_3 \rightarrow sp^3$ hybridised carbanion

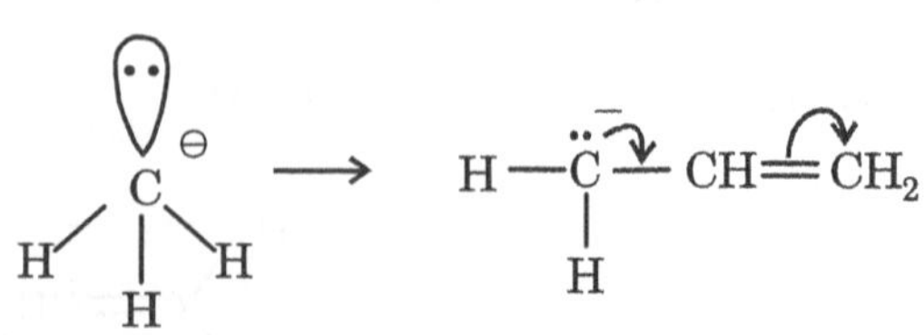

the carbanion which participates in mesomeric effect is sp^2 hybridised.

→ In $H-C\equiv\overset{\ominus}{C}$, the carbanion is sphyridised.

11. (c) 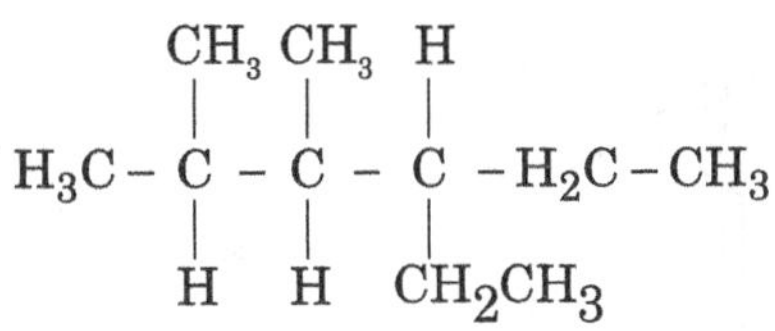

15. (b) 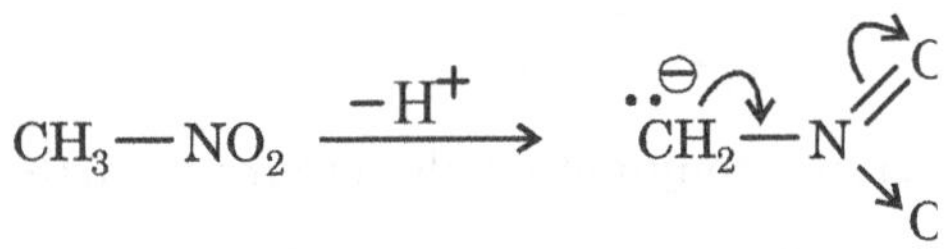

(Stabilized by resonance)

16. (a) Lesser the pka, more is the acidic strength.

17. (b) 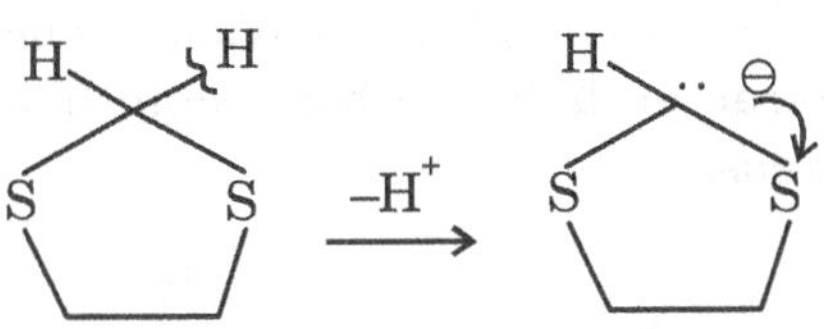

S has vacant 3d - orbitals i.e. d - orbital resonance is possible with two S atoms.

18. (c) In option 1

l.p. of N is involved in resonance

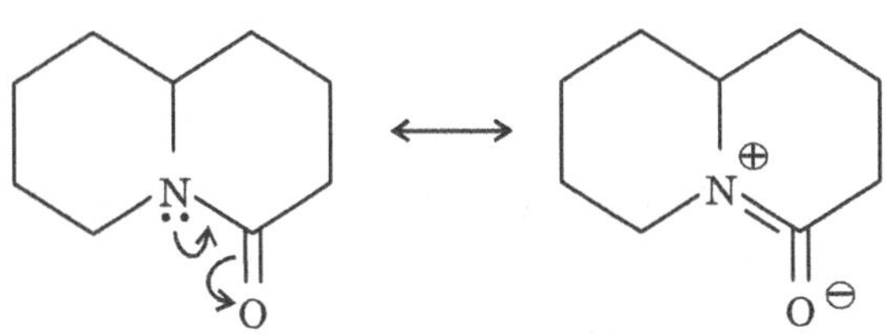

Similarly, l.p. of N is involved in resonance in option 2

However, in option 3 & 4 the l.p. of N are not involved in resonance but in option 4, $-\overset{O}{\overset{\|}{C}}-$ exerts good - I effect (due to its close proximity to N).

19. (d) 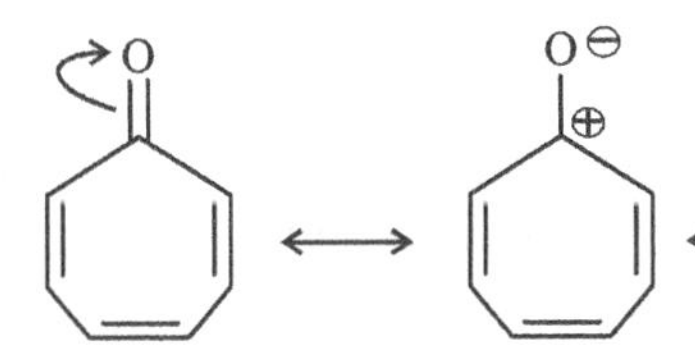

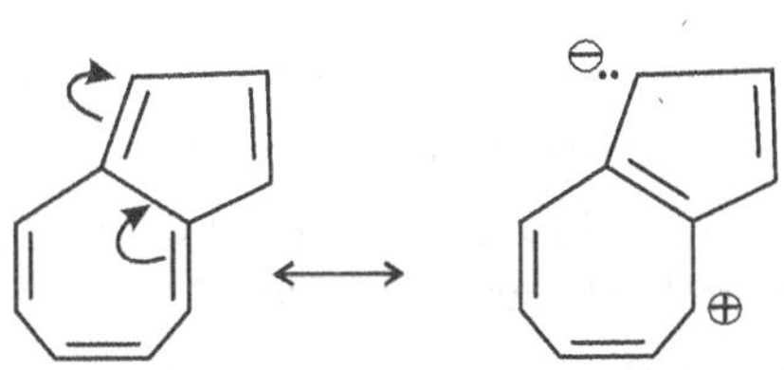

Both rings becomes aromatic

20. (b) SolutionIn Q, the conjugate base is surely stabilized by resonance.

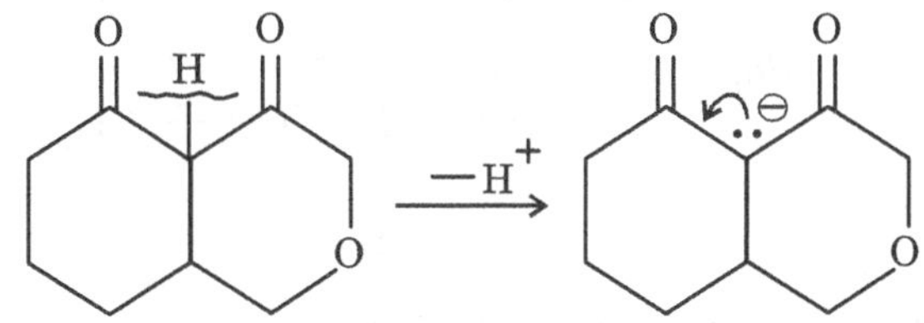

(Resonance stabilized with both ~C = O gps.)

In P,

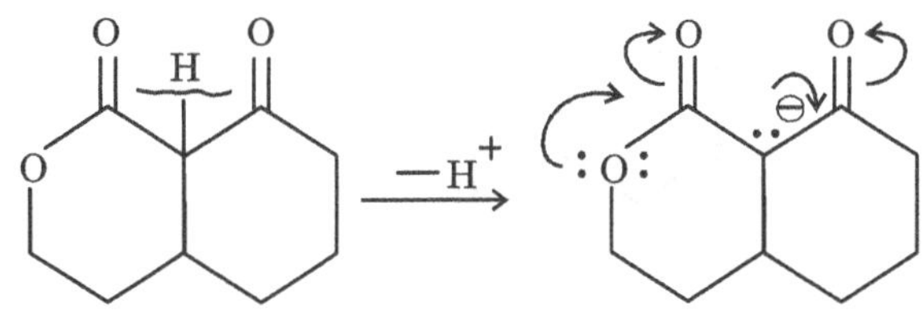

(The - ve charge is delocalized from two ends but resonance can also occur from O on the left).

21. (d) Structure I & 3 have max. number of covalent bonds & also , octet is complete.

But in 3, + ve charge is resting on 'N' & in 1, + ve charge is on 'O'. So, 3 is more stable than 1 followed by 2nd structure.

22. (a) In Q, pure single bond in C – N.

So, bond length is maximum.

In S, extended conjugation is there

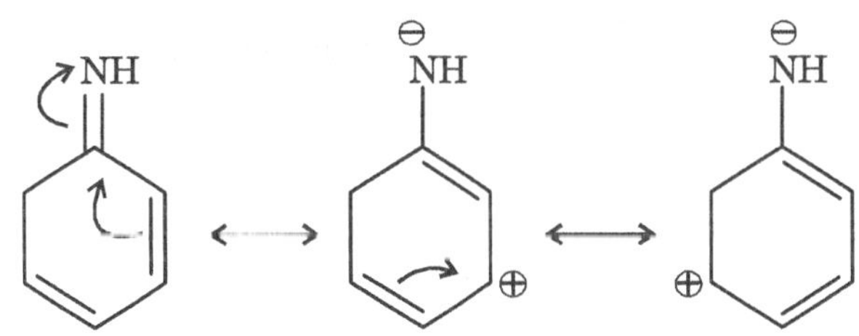

So, single bond character develops in otherwise doubly bonded C & N.

In R, there is lesser resonance.

In P, there is pure double bond character in C N bond.

23. (b) In buta – 1, 3 – diene,

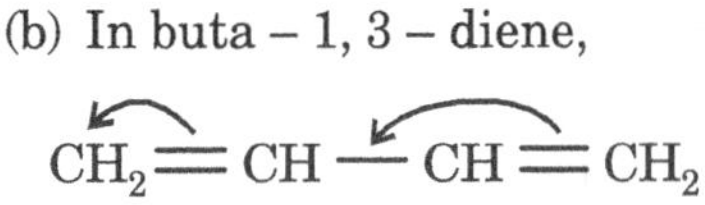

I

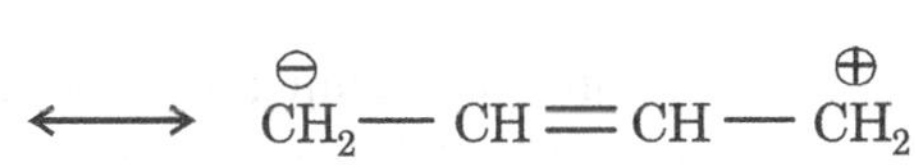

II

Resonating structure I is more stable than II, so it is more contributing towards the resonance hybrid.

So, $\overset{x}{CH_2{=}}\!-\!\overset{z}{CH{-}}\!-\!\overset{y}{CH{=}}\!-\!CH_2$

$$z > x = y$$

24. (b) → Acidic S. ∝ – I effect

25. (a)

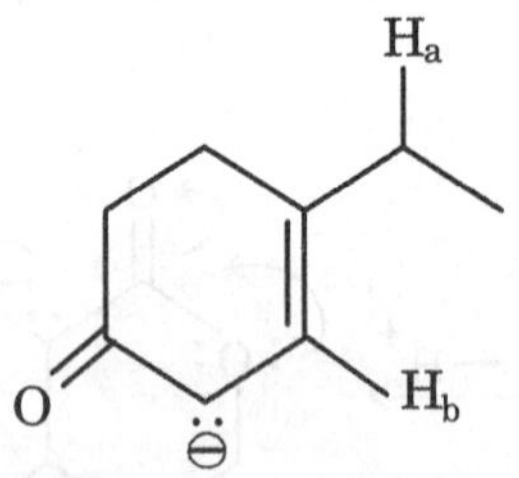

(after abstraction of Hc, conjugate base is stabilized with resonance from two ends)

26. (a)

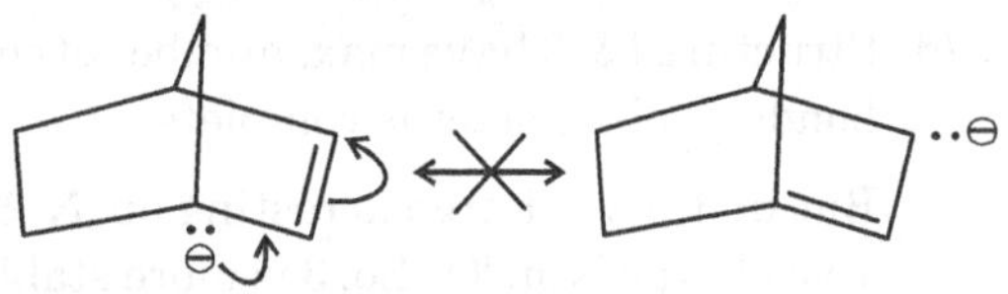

It is in violation of Bredt's rule.

Also, 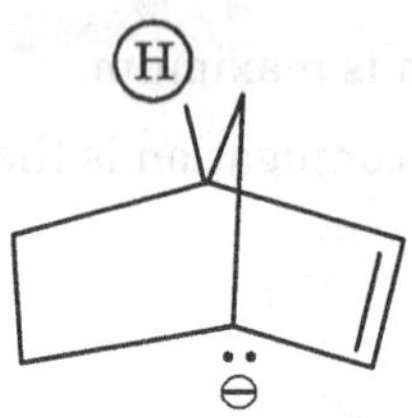 the encircled ∝ – H

can not be utilized in hyper conjugation, again due to Bredt's rule.

So, the double will have pure double bond character only & hence, the least bond length.

In option 2, the carbanion is stabilized by resonance thus introducing single bond character in the double bond.

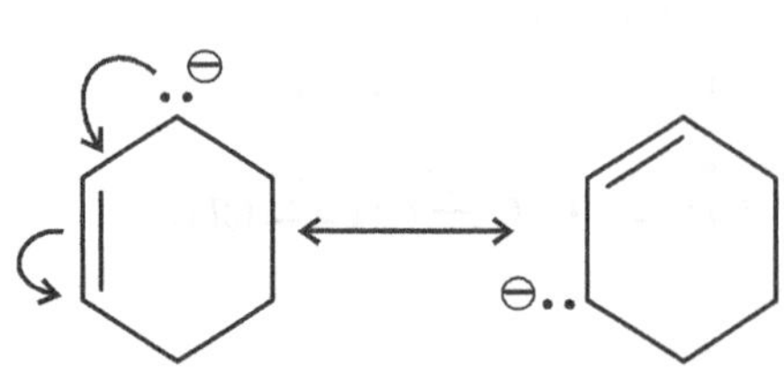

In third option hyper conjugation can occur, introducing single bond character in the double bond.

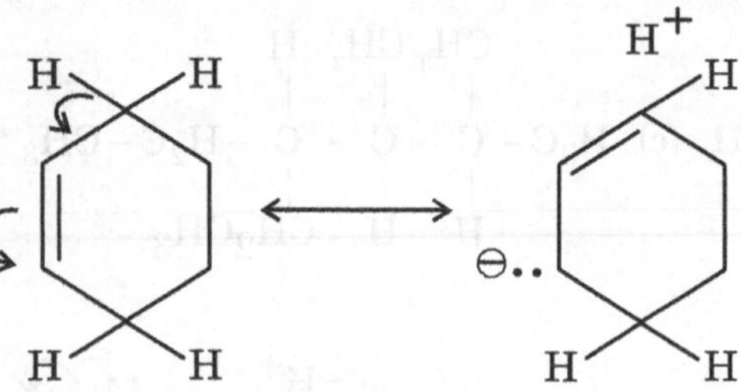

→ Hyper conjugation also occur in fourth option.

27. (b) More no. of covalent bonds impart more stability.

28. (d) In option 4, max.hyper conjugation is possible. So, C = C bond will have maximum single bond Character & hence bond length will get increased.

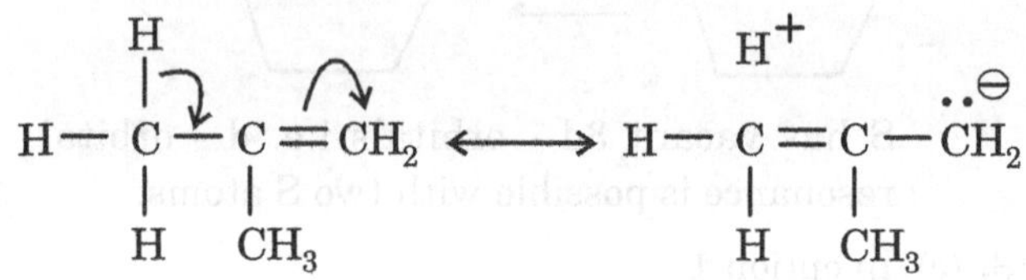

29. (d) In option 1, decide on the basis of hybridization. In $CH_3-\underset{\underset{sp}{\downarrow}}{C}\equiv N$, the sp hybridized C reduces e^- density on N Thus reducing its basic strength .

In Option 2,

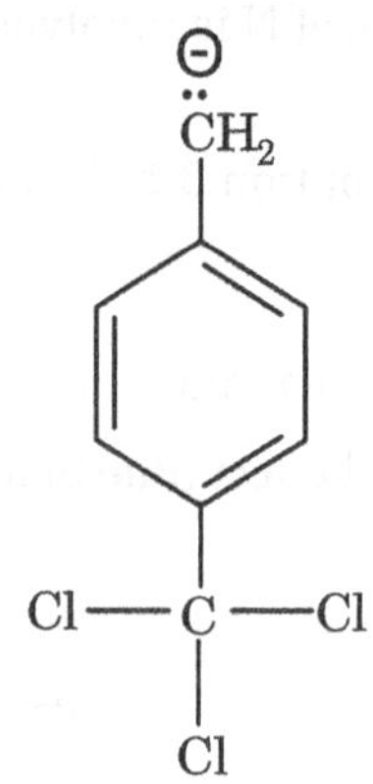

Carbanionis maximumstable due to maximum reverse hyperconjugationof $-CCl_3$ gp.

In Option 3, look for the stability of carbocations on the basis of +M effect.

30. (c) 'N' cannot expand its octet.

31. (c) H^+ will attack at most basic N.

32. (b) 2nd molecule violates Huckel's rule & is non-planar due to sp^3 hybridized C.

33. (d) Greater the no. of C – H bonds in the alkyl group, greater will be + I effect.

35. (a)

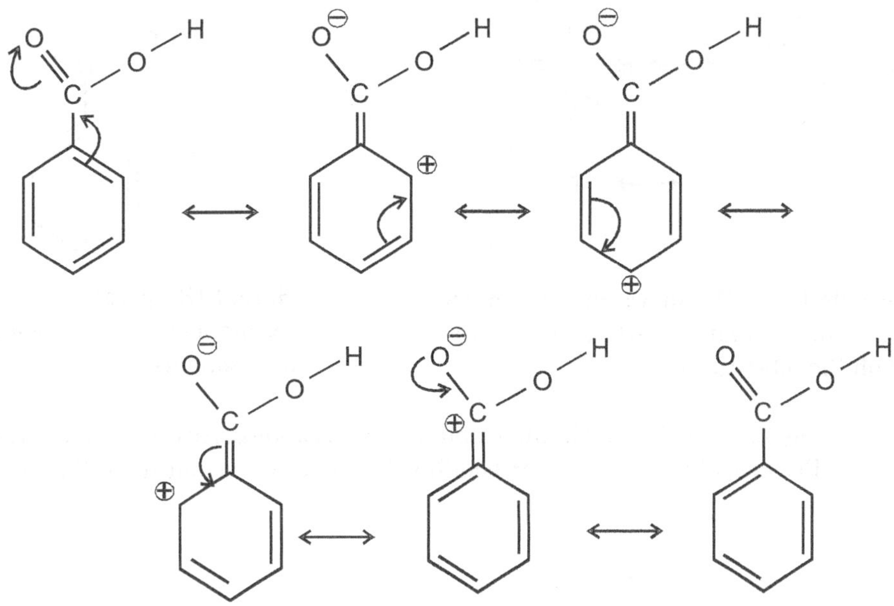

Due to – M effect in C_6H_5COOH, the ortho & para positions become e– deficient & So, it will direct E^+ at meta position.

36. (b) In structure I & III, the no. of covalent bonds are greater than in II & IV. Also, octet rule is followed in I & III whereas it is violated in II (wrt C) & IV (wrt terminal N). So, structure I & III are more stable than II & IV. Out ofI& III, I is more stable because – ve charge rests on N (which is more electro – ve). Out of II & IV, II is more stable because – ve charge rests on N, which is more electro – ve.

So, final stability order is 1 > 3 > 2 > 4.

37. (a) Formation of bond I $\longrightarrow sp^2 - sp^2$ overlap

Formation of bond II $\longrightarrow sp^2 - sp^2$ overlap

Formation of bond III $\longrightarrow sp^2 - sp^2$ overlap

Formation of bond IV $\longrightarrow sp^2 - sp^2$ overlap

38. (c)

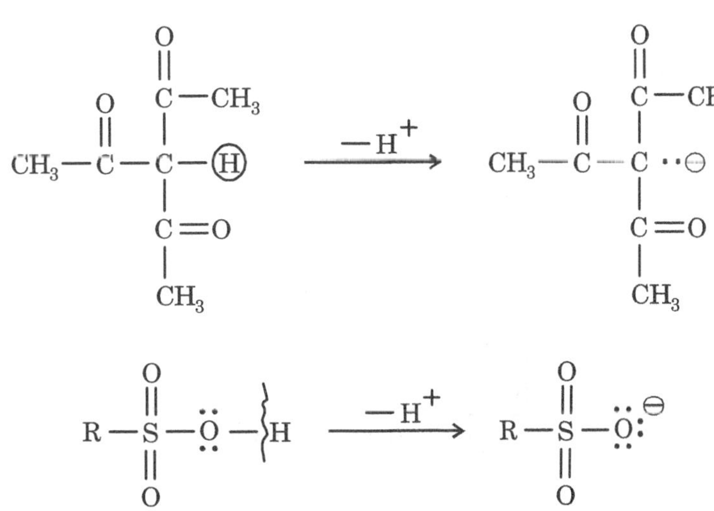

So,

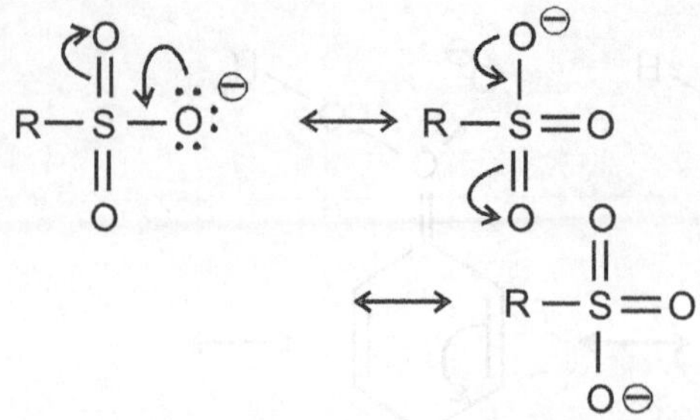

In molecule (2)

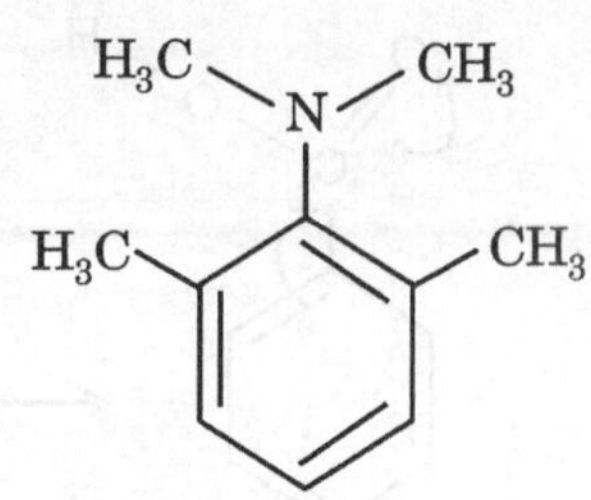

Here, SIR operates (i. e. steric inhibition to resonance) & hence lone pair is not involved in resonance.

40. (a) In molecule (1), $-NH_2$ shows maximum +R effect, so ring is maximum activated towards electrophillic substitution.

41. (c) Because its conjugate acid is maximum stabilized by resonance. Also, equivalent resonating structures are formed. Besides, effective 2pπ - 2pπ bonding, less electro – ve nature of N than O stabilizes it more.

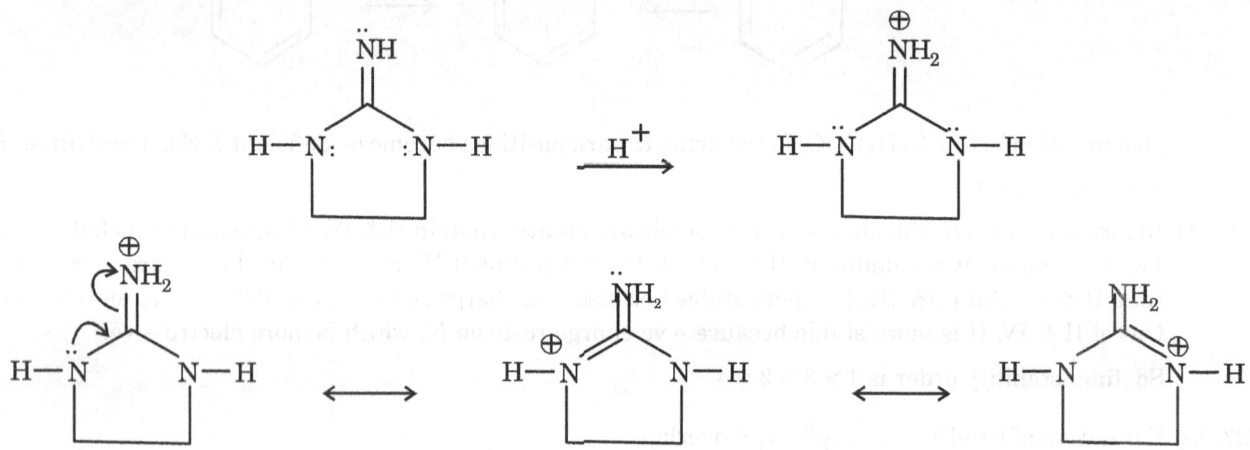

42. (b) (a)

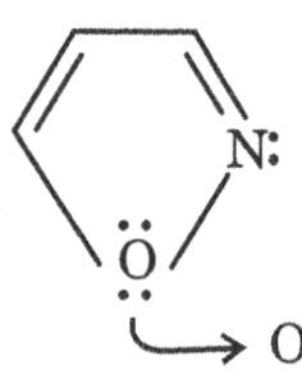

$\rightarrow$ l . p. Not involved in Resonance

One of the l. p. involved in resonance
PLANAR, 6 p e^- , Aromatic

(b)

$\rightarrow$ sp^3 hybridised

□e^- = 6 , Not Completely delocalised

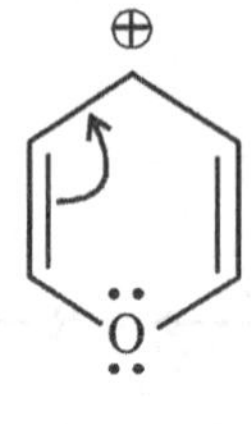

6□e^-, Aromatic

43. (a) 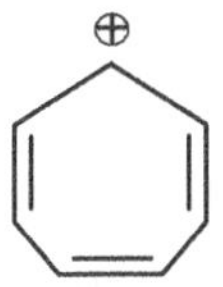

Tropyllium Carbocation (Aromatic, highly stable)

Cyclopentadienyl Carbocation (Anti-aromatic, highly unstable)

44. (c) Cyclic, planar compounds with $(4n + 2)\ \pi e^-$ are aromatic.

45. (c) Greater + M effect. (2p – 2p overlap & lesser electro – ve nature of N in comparison to O).

46. (d) It is not aromatic as it violates Huckel's rule it is rather anti-aromatic & highly unstable.

47. (d) No resonance & lone pair of electrons on sp^3 hybridized N.

48. (d) 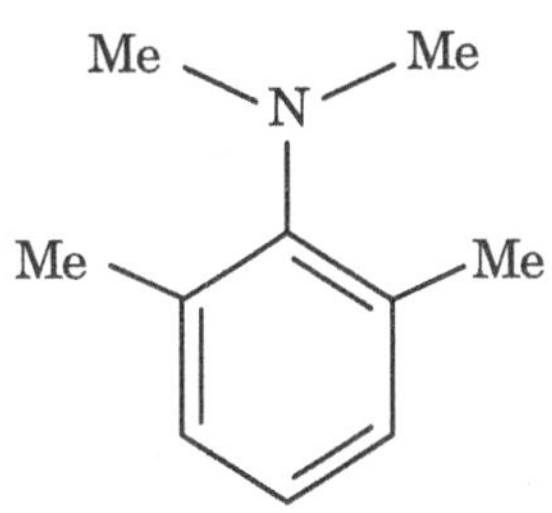

Here, SIR (steric inhibition to resonance) operates due to which lone pair on 'N' is not involved in resonance. Hence, more basic,

For option 2 –

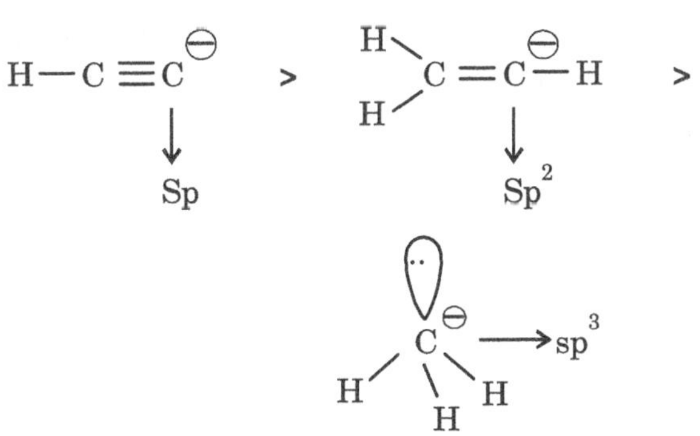

(–ve charge resting on sp hybridized C i.e. most electro – ve C is most stable)

For option 3 –

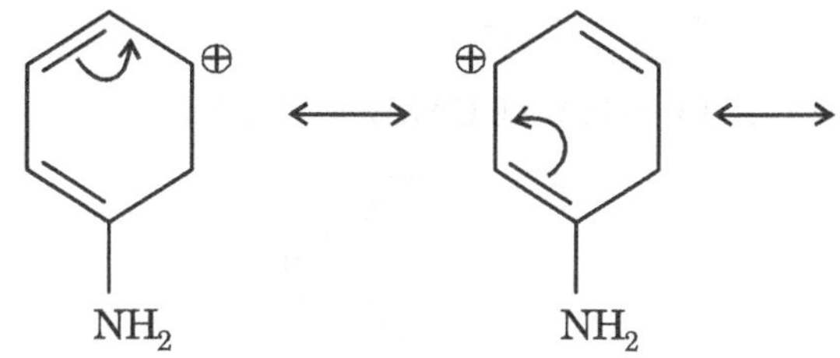

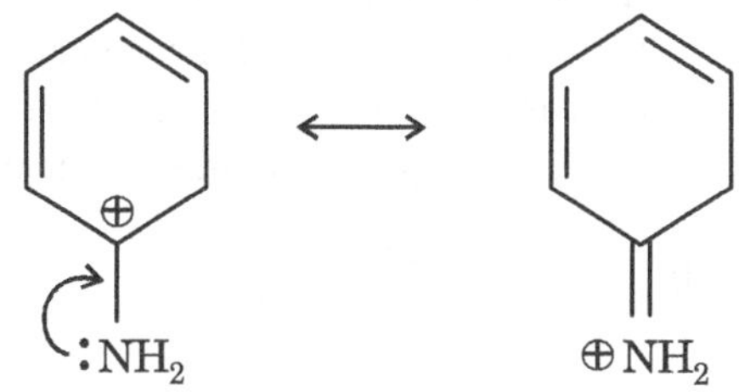

(More resonating structure because of extended resonance)

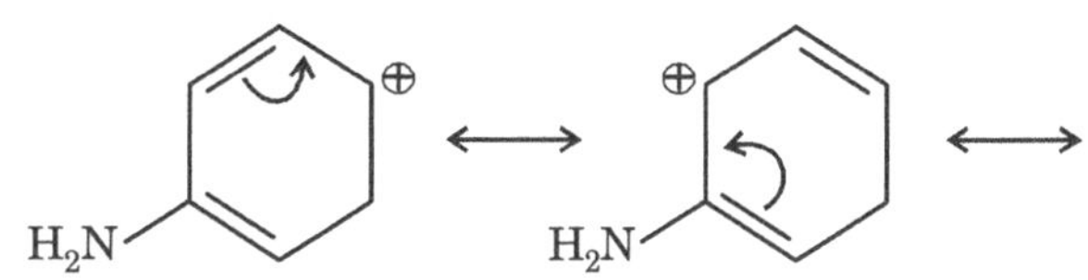

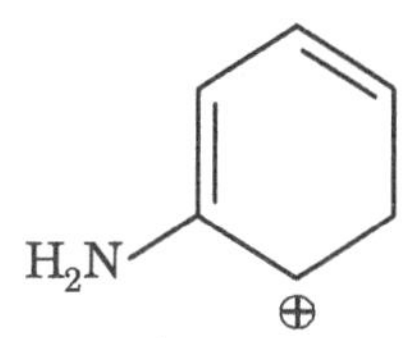

49. (b) In option (b) –

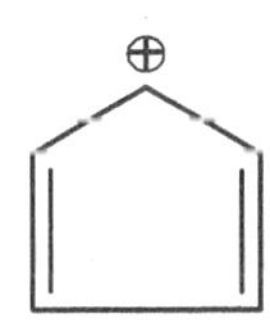 ⟶ Anti Aromatic

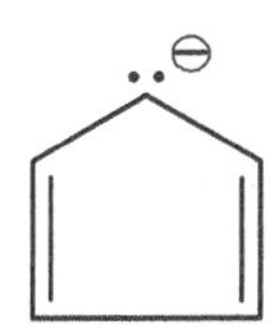 ⟶ $6\pi e^-$ (Aromatic)

50. (d) 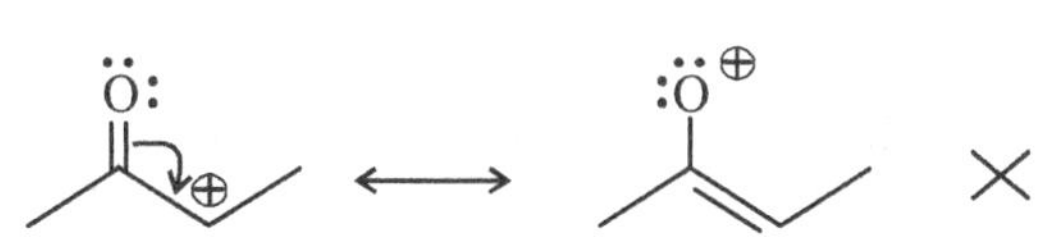

(It is prohibited because octet of O is compromised) & cloud will not move away from electro – ve element).

(b) 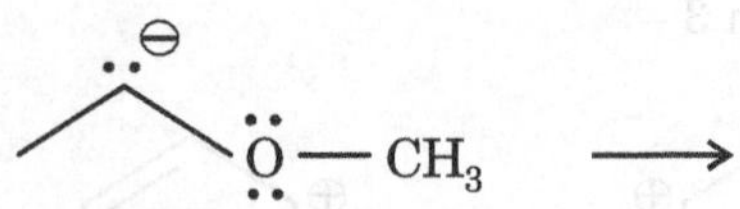

(c) In, PHENYL CARBOCATION,

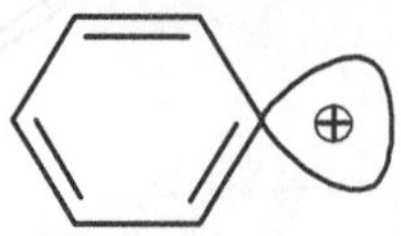

the vacant orbital lies in the plane of ring & hence the π e– cloud of the ring cannot stabilize the carbocation whereas the p orbitals forming π bonds are perpendicular to the plane of ring.

(d) 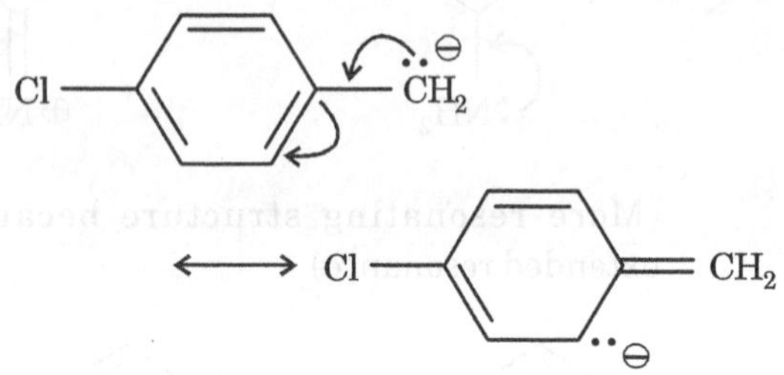

51. (c) 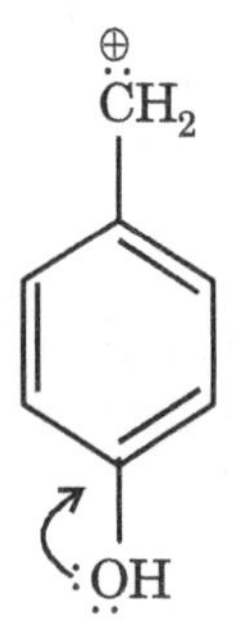 (Greatest stability due to + M)

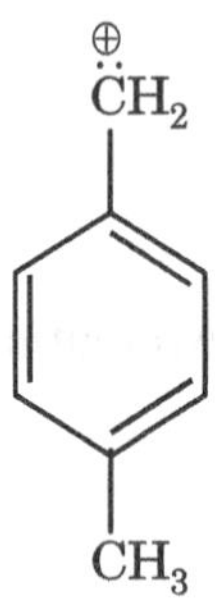 (Stable due to + I &hyperconjugation)

52. (c) In the given resonating structures of 1 – methoxy – 1, 3 – butadiene, structures in option a & b are more stable because octet for all the C atoms is complete. The structure in option d is more stable than structure C because +ve charge on electron deficient C is stabilized by the presence of pair of electrons on adjacent O atom & by negatively charged adjacent C.

54. (c) Extensive delocalisation of odd e– is there.

55. (a) 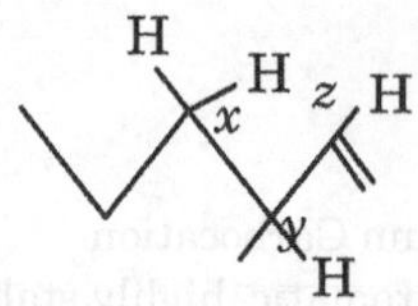

On cleaving 'y' homolytically, the free radical formed will be maximum stable (by Mesomeric & by hyper conjugation). So, its ease of cleavage is maximum.

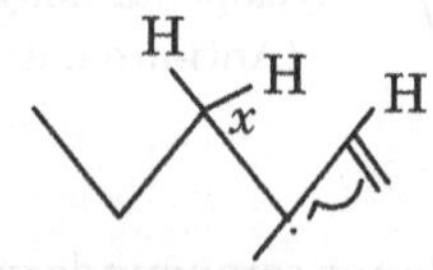

56. (c) In option (a), (b) & (d) resonance is possible, thus introducing single bond character in otherwise (–C = O) double bond.

57. (c) 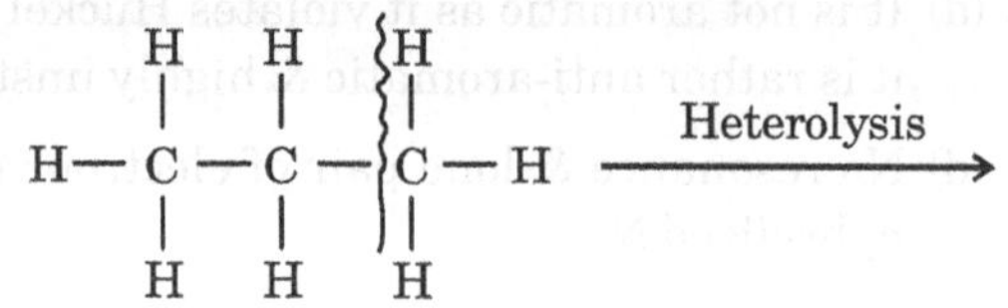

$\overset{..\ominus}{CH_3—CH_2} + \overset{\oplus}{CH_3}$

Set I

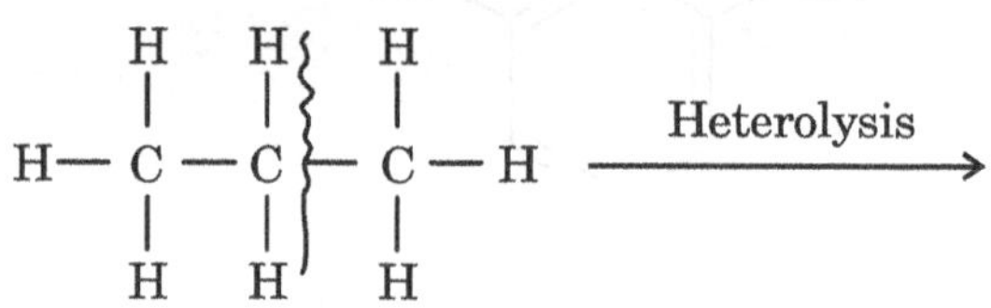

$CH_3—\overset{\oplus}{CH_2} + \overset{\ominus}{CH_3}$

Set II

Now, Out of set I & set II, set II will give major product on account of greater stabilities of carbocation &carbanions formed.

58. (a)

(stabilised by Resonance)

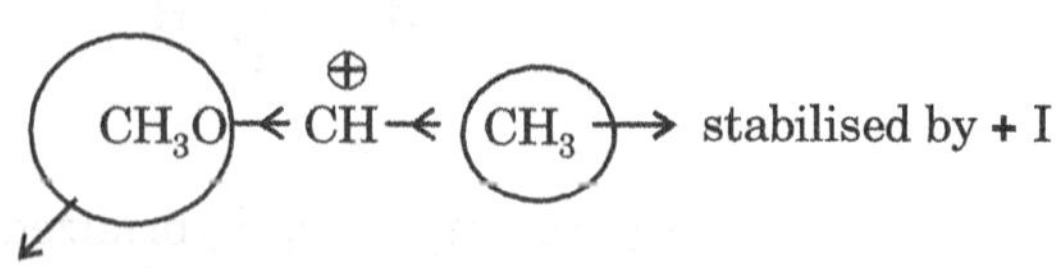

destabilised by - I

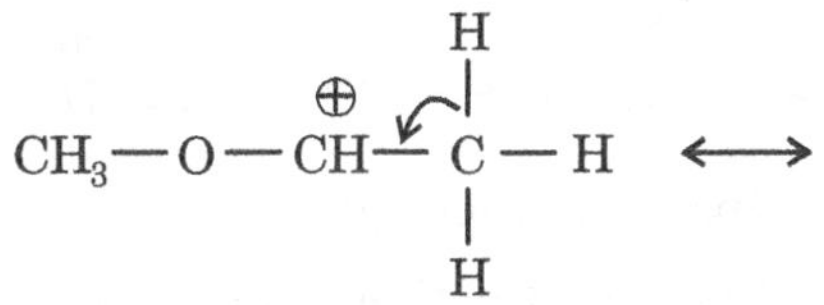

(stabilised by hyperconjugation)

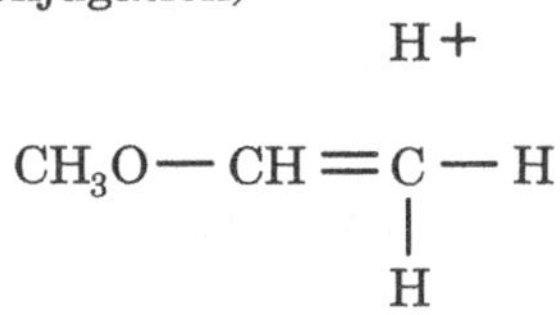

(b)

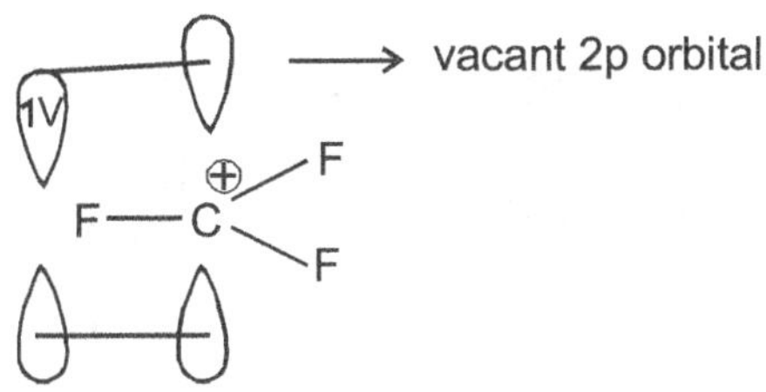

$2p\pi - 2p\pi$

back bonding (resonance stabilization)

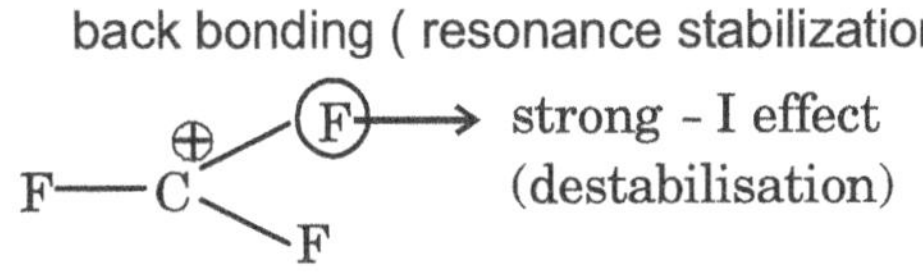

59. (a)

(IV)

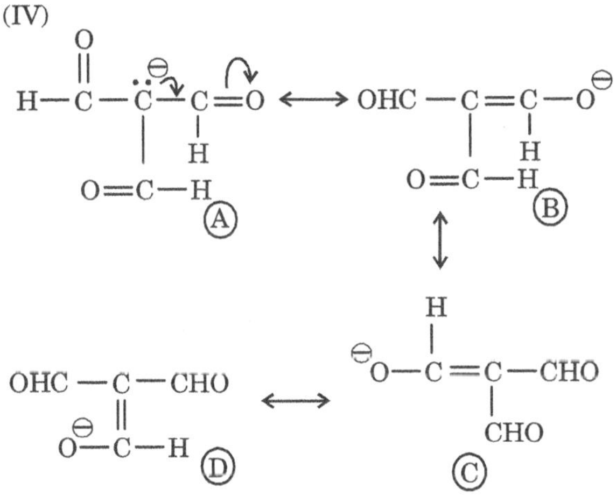

Total 4 resonating structures, of which B, C & D are equally contributing.

In II & III

(II)

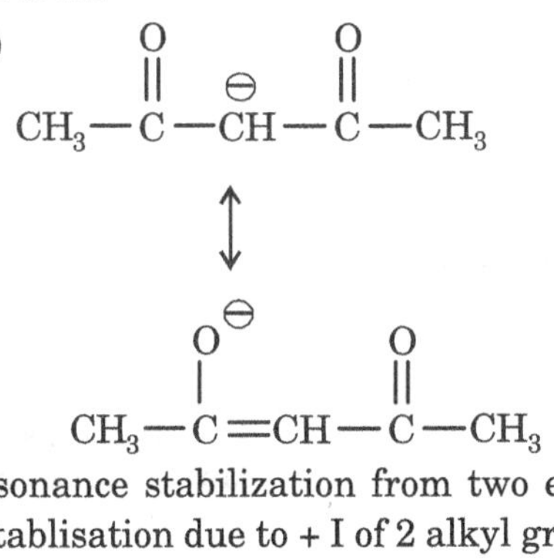

(Resonance stabilization from two ends but destablisation due to + I of 2 alkyl groups.)

(III)

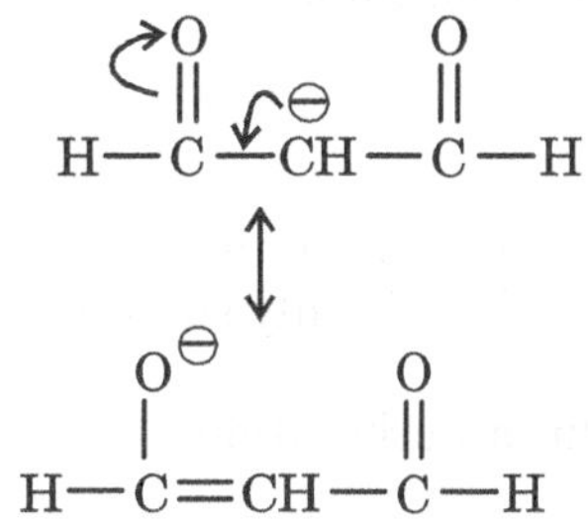

(Resonance stabilization, No destabilization by + I).

So, II is less stable than III.

In IV,

(Resonance only from one end)

60. (a)

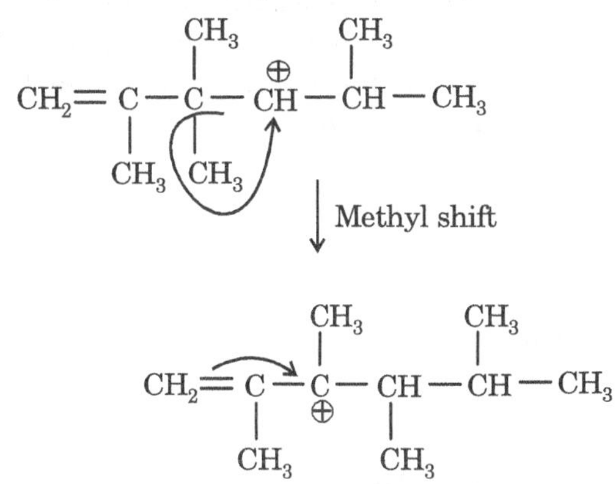

61. (a) In 4th carbocation resonance wrt both π bonds & + M of $-NH_2$ stabilize the carbocation.

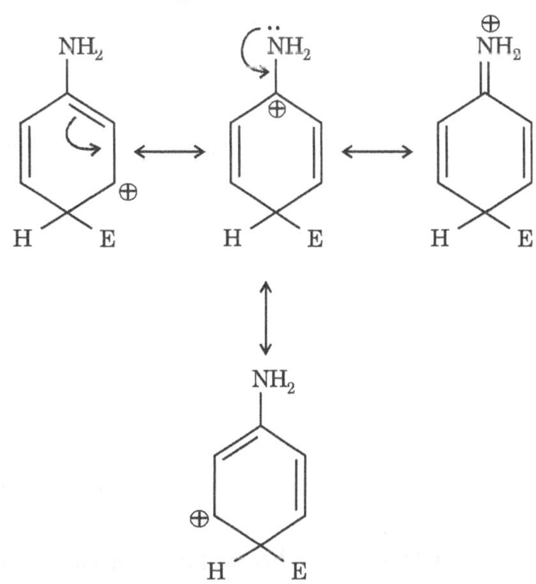

+ M of $-NH_2$ > + M of $- OCH_3$

So, 3 is more stable than 1.

Out of 2 & 4, $-OCH_3$ has more – I than $-NH_2$. So, 2nd is least stable.

62. (b)

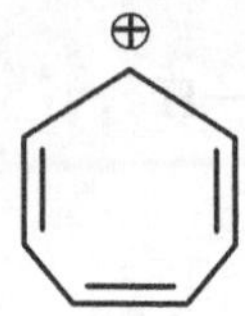

Aromatic (highly stable)

Tropyllium Carbocation

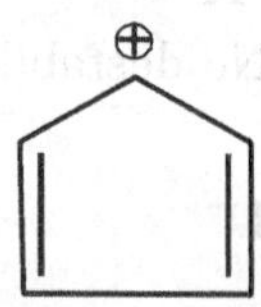

Aromatic (highly stable)

63. (d) (a)

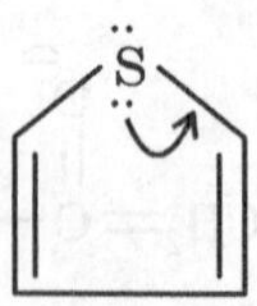

S is less electronegative than O, so it has a greater electron releasing capacity, hence aromatic character, more stable & more resonance energy.

(b)

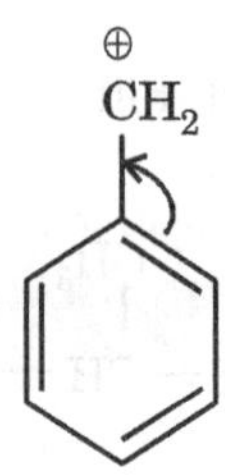

(stabilized by resonance)

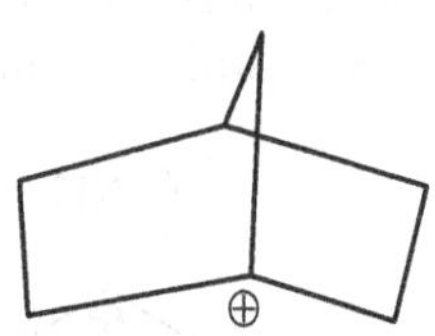

(Bridge head carbocation, not stabilized by hyper conjugation, violation of Bredt's rule

(c)

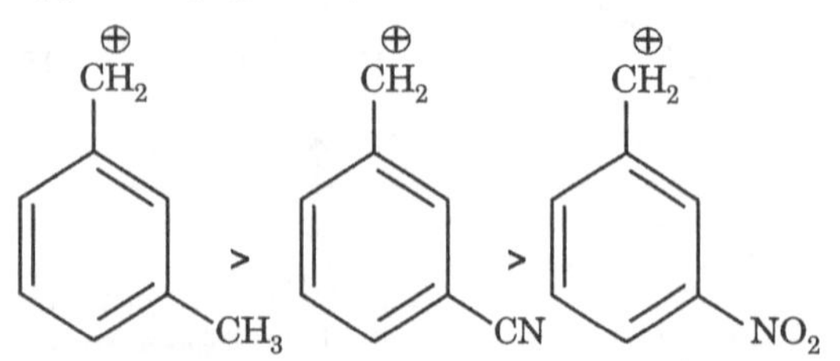

All the three groups i.e. $-CH_3$, $-CN$ and NO_2 are at meta positions, so no resonance advantage to the carbocation.

$-CH_3$ exerts + I & makes carbocation stable.

$-CN$ & $-NO_2$ exerts – I. & make carbocation unstable.

64. (c) $CH_3-\overset{\oplus}{C}=\ddot{N}-H \longleftrightarrow CH_3-C\equiv\overset{\oplus}{N}-H$

(2pπ – 2pπ overlap)

Here, there is effective overlap and less electronegative nature of N as compared to O.

$CH_3-\overset{\oplus}{C}=\ddot{O}: \longleftrightarrow CH_3-C\equiv\overset{\oplus}{O}:$

65. (c)

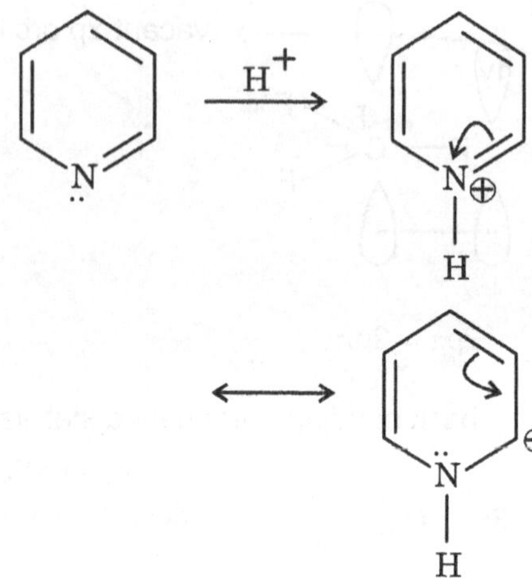

66. (c)

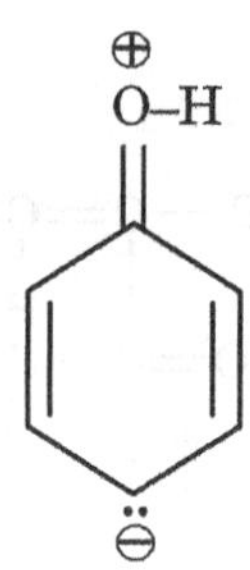

Max. unstable (opposite charges are farthest apart)

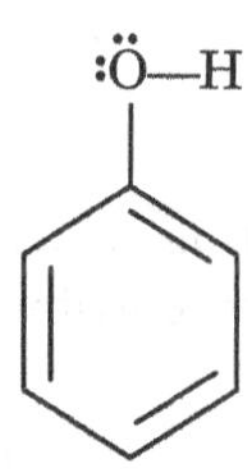

Max. stable (neutral molecule)

67. (c)

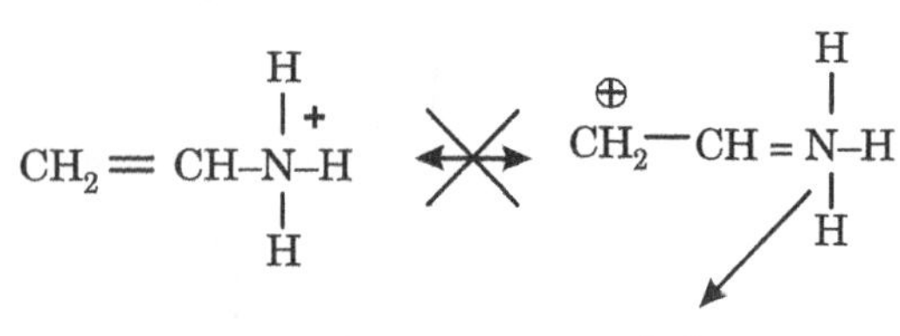

(N can't show expanded octet)

68. (b) 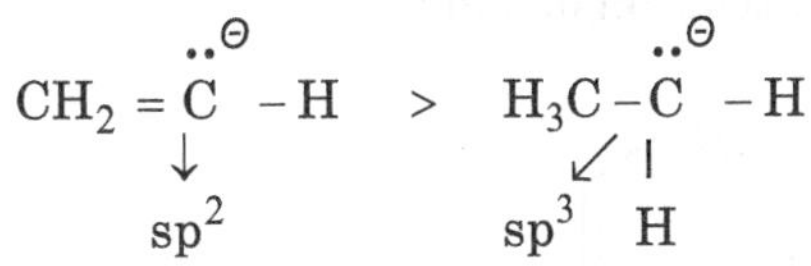

–ve charge on more electronegative C will be more stable.

69. (c) 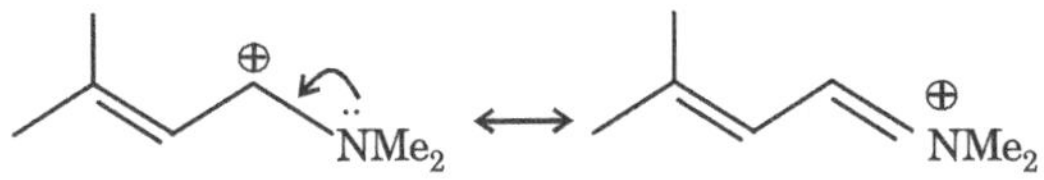

(Resonance from π bond & l. p. of N)

Besides, + I of two $-CH_3$ groups.

73. (c) $A - A \xrightarrow{hv} \dot{A} + \dot{A}$

$(A\bullet\{\bullet A)$

Carbocations & Carbanions are formed by heterolytic clearage.

74. (c) 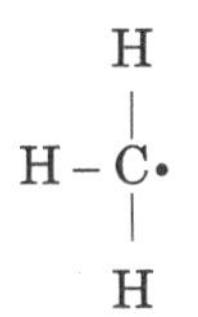 7 Valence e^- (after bonding)

(Methyl free radical)

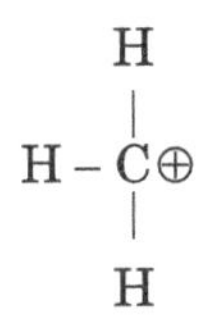 6 Valence e^- (after bonding)

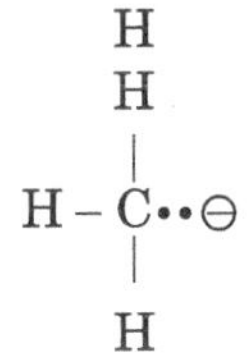 8 Valence e^- (after bonding)

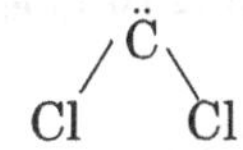 6 Valence e^- (after bonding)

(Dichlorocarbene)

75. (b) Carbanions possess a lone pair & they seek a proton or some other positive centre to which they can donate their election pair

76. (a) SINGLET CARBENE

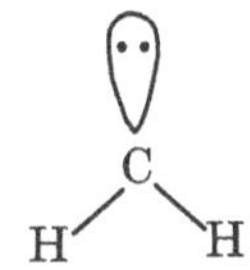

C (6) :- $1s^2\ 2s^2\ 2p^2$

G.S. → 

C is sp² hybridised here & is in ground state. Its shape is Bent.

Triplet Carbene

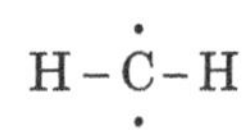

C (6) :- $1s^2\ 2s^2\ 2p^2$

G.S. 

Excited State 

C is sp hybridized here & is in excited state. Shape- linear

77. (a) 2 & 4 are stabilised by resonance

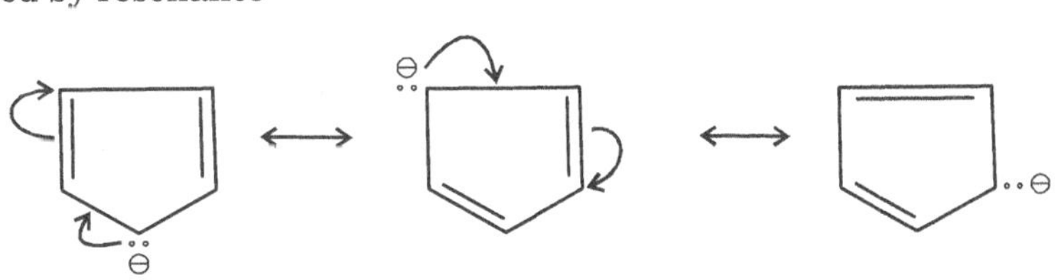

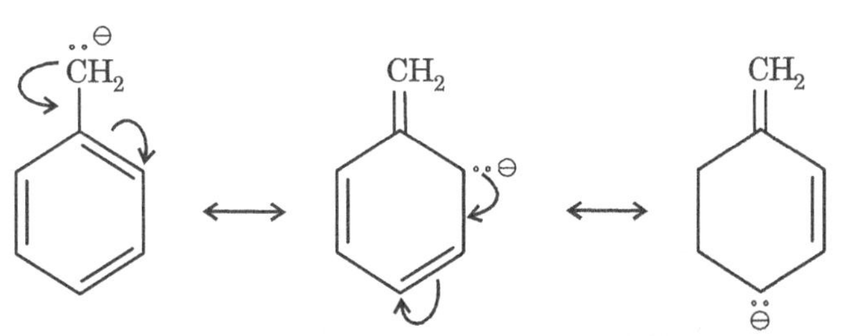

$H-C\equiv \underset{\underset{sp}{\uparrow}}{C}^{\bullet\bullet\ominus}$ → Here C is sp hypridised & –ve. Charge rests on more electro -veC thus enhancing its stability.

78. (a) Maximum stable free radical is IV because of extensive delocatisation of odd e-

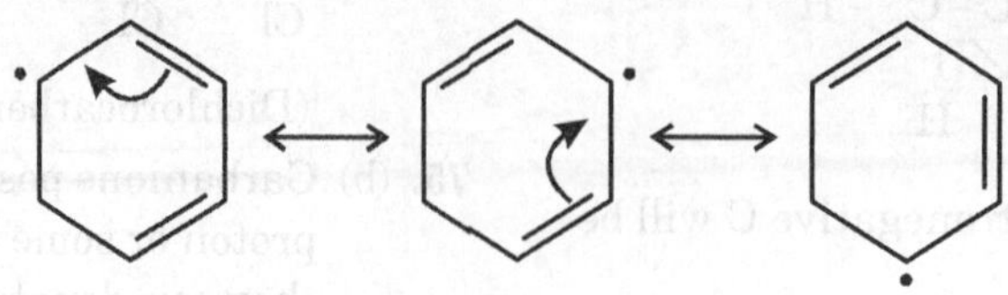

In structure II, the delocalisation of odd e– is less.

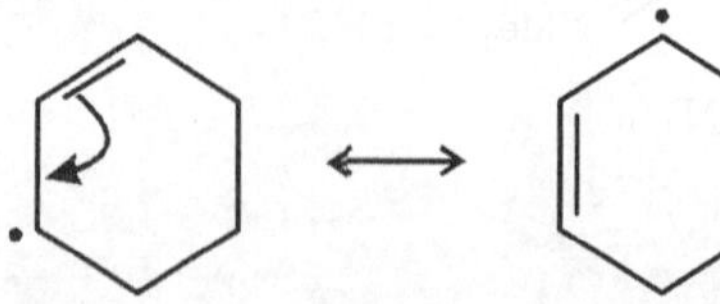

In III & V, the free radical is not delocalized & is surrounded by sp^2 hyb. C's (more extensively in V), So, they are less stable.

79. (c) In option 3, both the resonating structures are indistinguishable.

80. (d) Solution In option 1

Extensive delocalization of +ve charge.

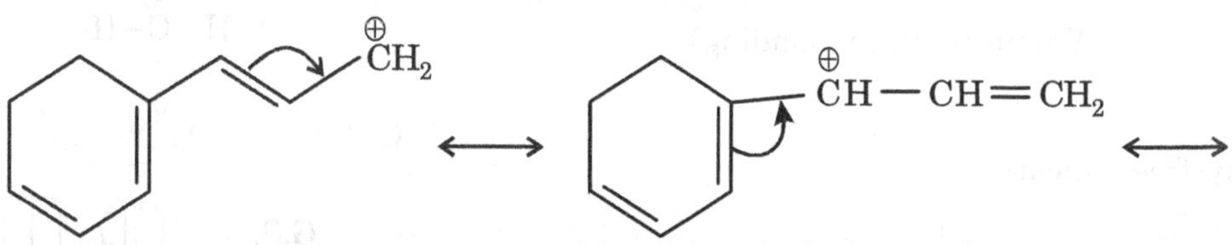

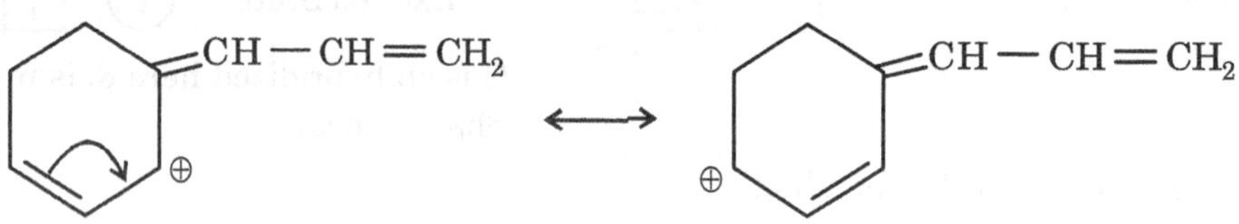

But in

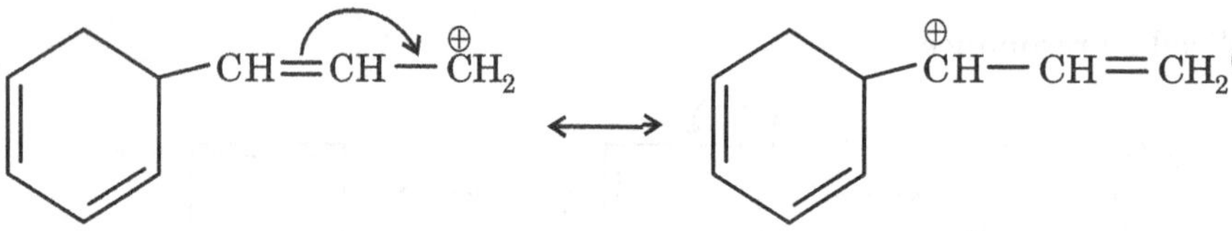

In option 2nd,

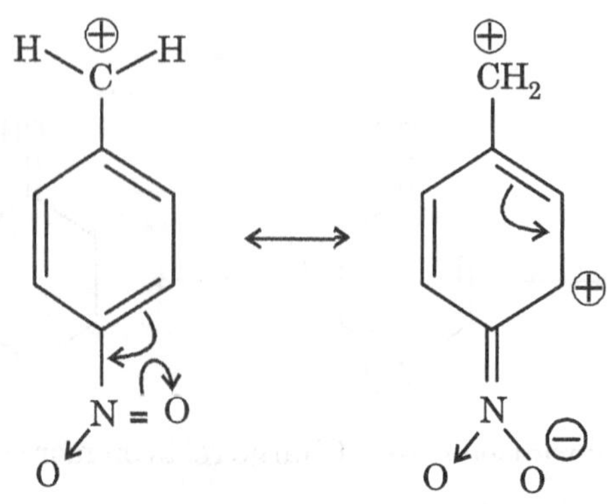

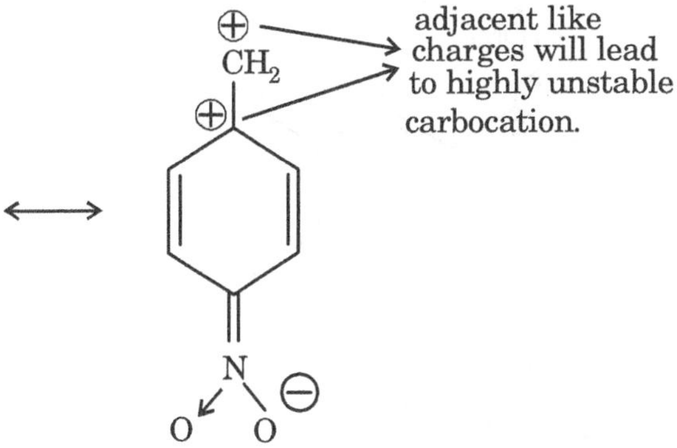

(–NO_2 exerts – M & – I)

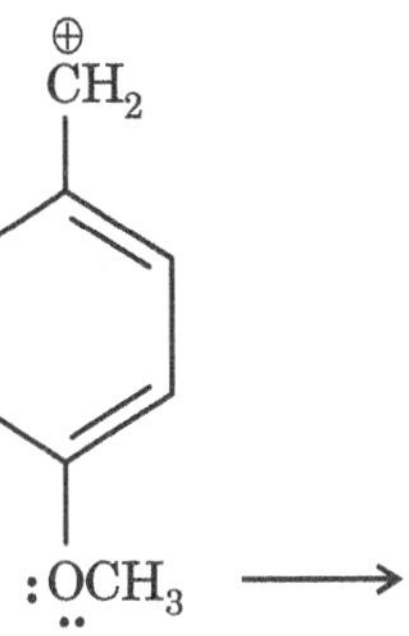

(- OCH_3) exerts + M & –I. However, its + M dominates over its –I.

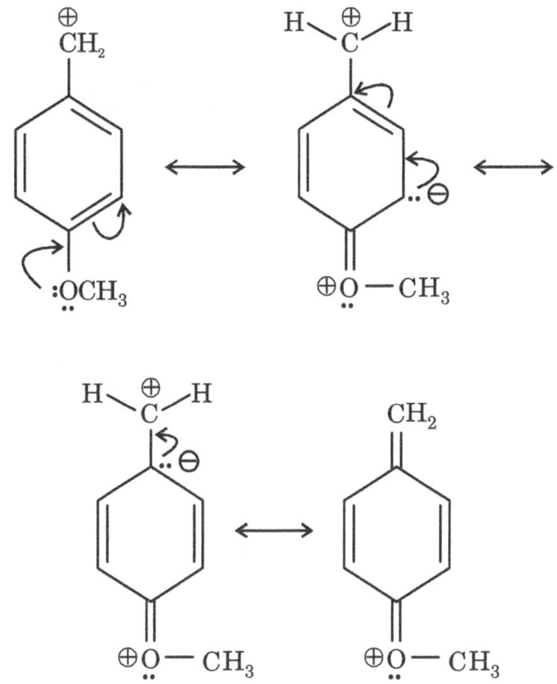

(+ve charge of carbocation is neutralized)

81. (b) Among the given carbocations, 2nd is most unstable.

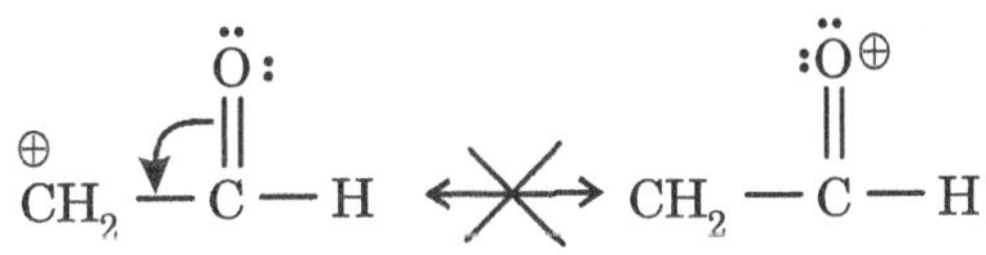

It looks stabilised by mesomeric effect but actually, it is not. Once π bond moves away from O, its octet is compromised. So, resonance cannot occur. Rather - CHO exerts its - I effect.

82. (d)

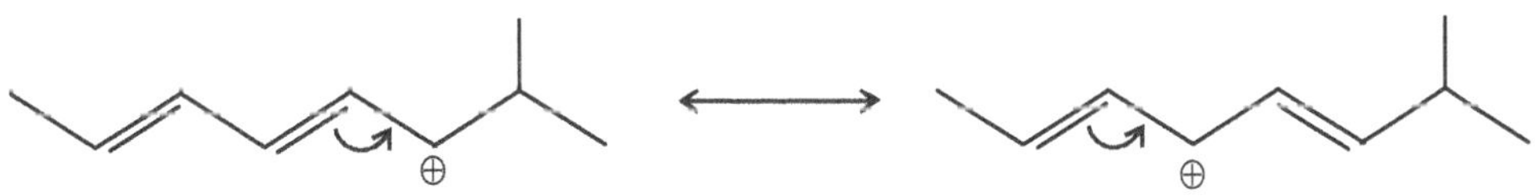

(Continuos resonance)

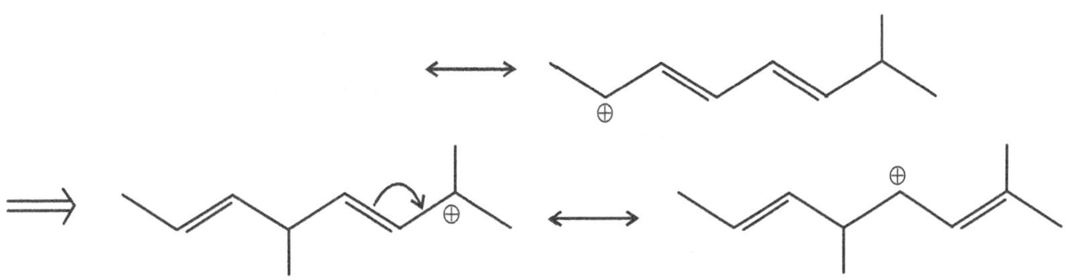

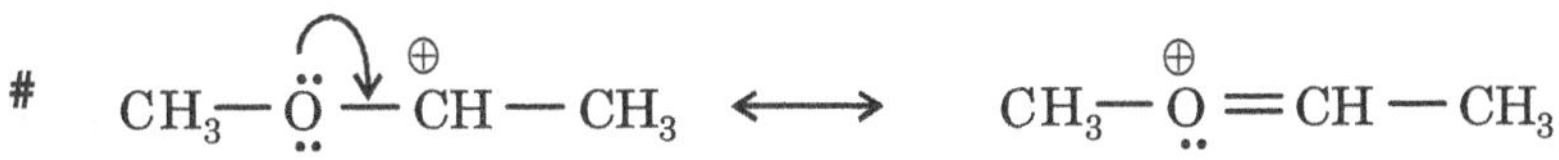

I II

(Resonance, dispersal of +ve charge, octet complete in structure II)

#

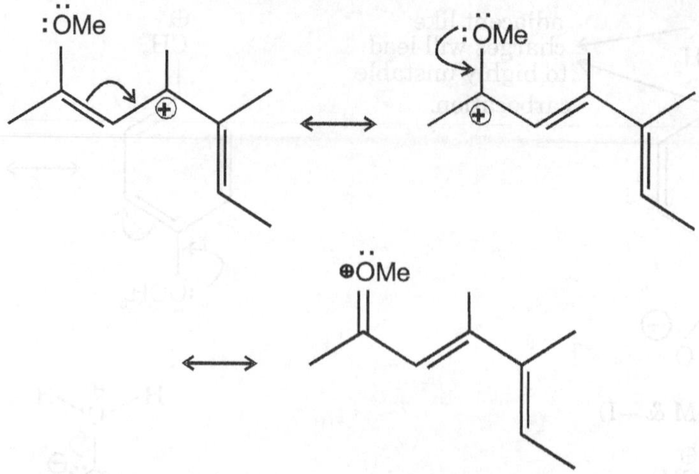

Also,

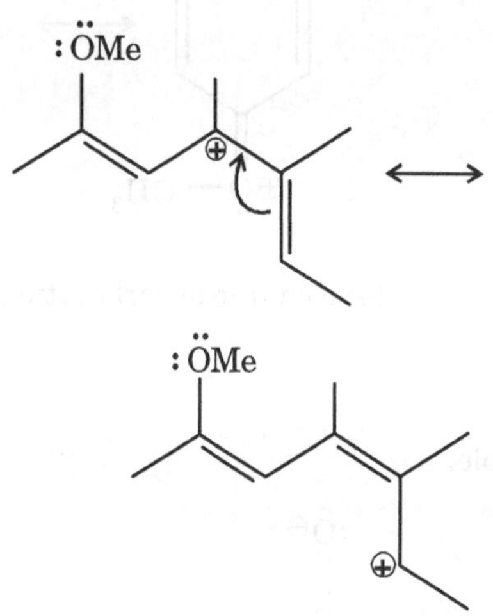

85. (a) Steric hindrance reduces nucleophillicity.

87. (b) $CH_3-O-\overset{\displaystyle H}{\underset{\displaystyle CH_2CH_3}{C}}-CH_3$

89. (c) Out of $CH_3O^{\ominus}$ & $CH_3-\overset{\displaystyle CH_3}{\underset{\displaystyle CH_3}{C}}-O^{\ominus}$, $CH_3O^{\ominus}$

is a stronger nucleophile than ter-butoxide

$(CH_3-\overset{\displaystyle CH_3}{\underset{\displaystyle CH_3}{C}}-O^{\ominus})$.

Steric factors reduce nucleophillicity of ter – butoxide.

94. (b)

$$\underset{1}{CH_3}-\underset{2}{\overset{O}{\overset{||}{C}}}-\underset{3}{CH_2}-\underset{4}{\overset{O}{\overset{||}{C}}}-\underset{5}{CH_3}$$

(Acetyl acetone)

IUPAC Name ⟶ Pentane – 2, 4 – dione

95. (a)

$$H_3C-\overset{\displaystyle H}{\underset{\displaystyle {}^{5}CH_2\overset{6}{C}H_3}{{}^{4}C}}-\overset{3}{C}\equiv\overset{2}{C}-\overset{1}{C}H_3$$

4 – Methyl hex – 2 – yne

96. (d) In allylic carbanions

(a)

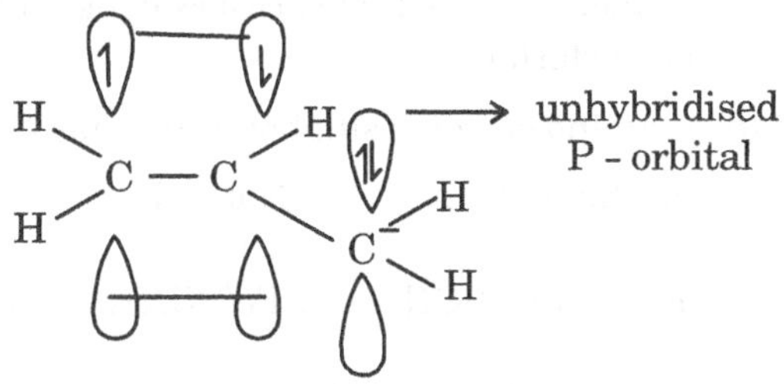

The Carbanions which participate in resonance are sp^2 hybridised.

(b)

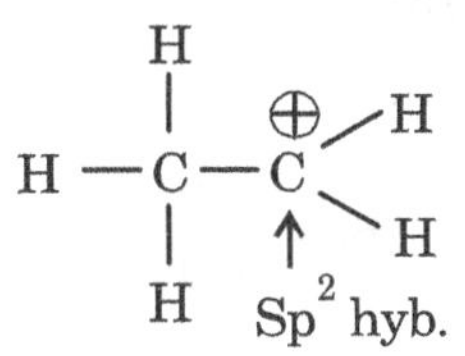

(c) $H-C\equiv \overset{\oplus}{C}$

97. (c)

$\overset{1}{H_3C}-\overset{2}{C}H(OH)-\overset{3}{C}H(CH_3)-\overset{4}{C}H_3$

3 – Methyl butan – 2 – ol

98. (d)

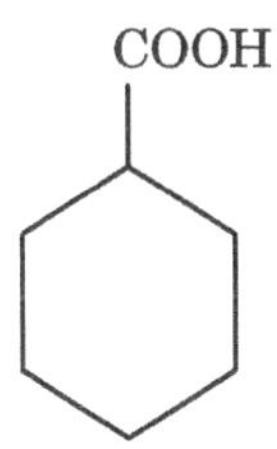

→ Cyclohexane carboxylic acid

$CH_3CH_2O-\overset{1}{C}H(NH_2)-\overset{2}{C}H_2-\overset{3}{C}H_3$ › 1– Ethoxy propanamine

103. (c) $\overset{1}{C}H_2=\overset{2}{C}H-\overset{3}{C}H(CH_3)-\overset{4}{C}H_3$

104. (d)

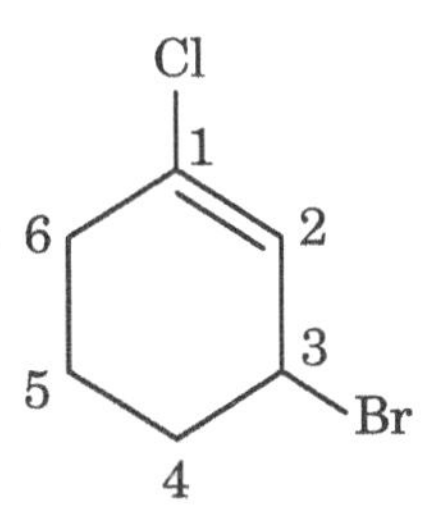

3 – bromo – 1 – chlorocyclohex – 1 – ene

113. (d) $NaCN+HNO_3 \longrightarrow NaNO_3+HCN\uparrow$;

$Na_2S+2HNO_3 \longrightarrow 2NaNO_3+H_2S\uparrow$

114. (b) $P_{wet\,gas}=754.32$ mm of Hg

Aq. Tension 15.4 mm of Hg

$P_{dry\,gas}=754.32-15.4=738.92$ mm of Hg

Apply, $PV=nRT$

$\frac{738.92}{760}\text{atm}\times\frac{2.21}{1000}$

$=\frac{w}{28}\times0.0821\times291$

$w=0.025\,g$

Percentage of nitrogen in compound

$=\frac{0.025}{0.0237}\times100=10.6$

118. (c)

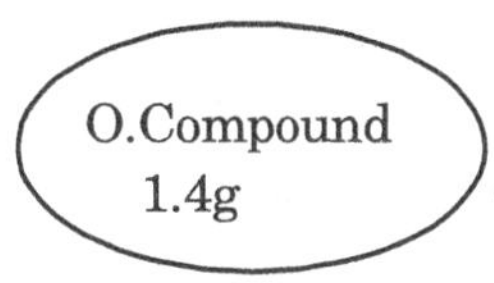

$\longrightarrow NH_3 + H_2SO_4$ (60ml, $\frac{1}{10}$M)

or $NH_3+H_2SO_4$ ($\frac{1}{5}$N, 60 m) $(\because N=M\times n\text{ factor})$ $N=\frac{1}{10}\times2=\frac{1}{5}N$

Meq. Of acid $=\frac{1}{5}\times60=12$ Meq.

(Milli equivalents)

$NaOH + H_2SO_4 \longrightarrow$

NaOH: 20ml, $\frac{1}{10}M(\text{or}\frac{1}{10}N)$

Meq.of NaOH $=20\times\frac{1}{10}=2$ Meq.

Meq.of H_2SO_4 neutralized by NaOH = 2 Meq.

(by law of equivalence)

$\therefore$ Meq.of H_2SO_4 reacted with $NH_3=10$

So, Meq. of NH_3 = 10

Gram equivalents of $NH_3 = \frac{10}{1000} = 10^{-2}$

No. of moles of NH_3 =

($\because$ No. of gm equivalents = no. of moles × n factor)

n – factor for NH_3 = 1

$\therefore$ No. of moles = No. of gmeqvts. of NH_3 = 10^{-2} of NH_3

1 mole of NH_3 has N = 14g

0.01 mole of NH_3 has N = 14 × 0.01 = 0.14g

Percentage of N in organic compound

$= \frac{0.14}{1.4} \times 100 = 10\%$

120. (d) Compound 'R' will not be able to release Br^- $\because$ the cyclopentadienyl anion will be anti aromatic (highly unstable)

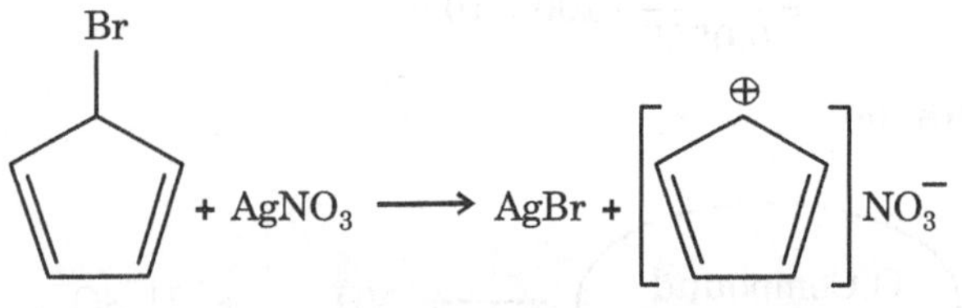

122. (d) Hydrazine (NH_2NH_2) Lacks C & So, NaCN cannot be formed & hence lassaignes' test can't be performed.

123. (d) *p*-amino benzene sulphonic acid has C, N & S So, Na + C + N + S → NaSCN

$$Fe^{3+} + 3NaSCN \longrightarrow Fe(SCN)_3 + 3Na^+$$

Ferric thiocyanate
(Blood red coloration)

124. (c) AgBr has M. mass 188g

188g AgBr has Br = 80g

141×10^{-3} AgBr has Br $= \frac{80}{800} \times 141 \times 10^{-3}$

= 0.06 g Br

% age of Br in Organic Compound

$= \frac{0.06}{250 \times 10^{-3}} \times 100 = 24\%$

Smart Notes

Smart Notes

14

CHAPTER

Isomerism

1. Match the following.

Column – I	Column – II
(a) $CH_3 - CH_2 - CH_2 - CH_3$ & $CH_3CH(CH_3)CH_3$	(i) Position isomers
(b) $CH_3 - CH_2 - O - CH_2CH_3$ & $CH_3OCH_2CH_2CH_3$	(ii) Ring chain isomers
(c) 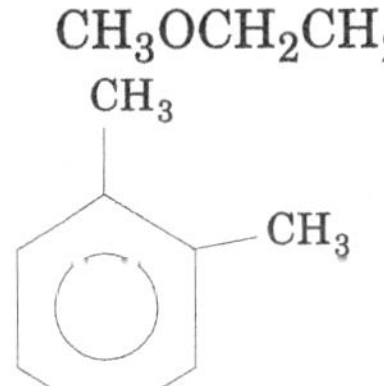 & 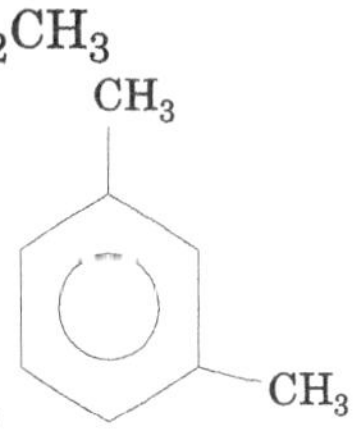	(iii) Chain isomers
(d) $CH_3 - CH = CH_2$ & 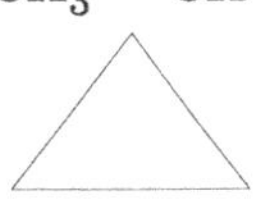	(iv) metamers

(a) a - iii, b - iv, c - i, d - ii (b) a - iii, b - ii, c - iv, d - i

(c) a - ii, b - iv, c - i, d - iii (d) a - iii, b - iv, c - ii, d - i

2. Which is / are incorrect relation for the following pair of compounds?

(a) 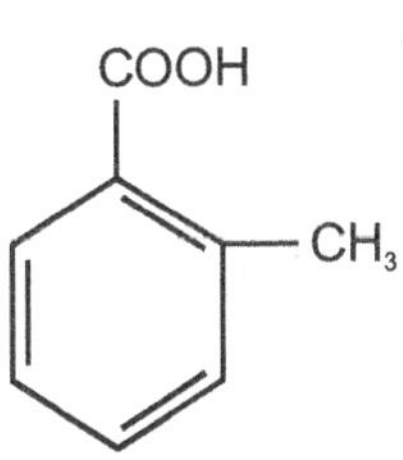 & 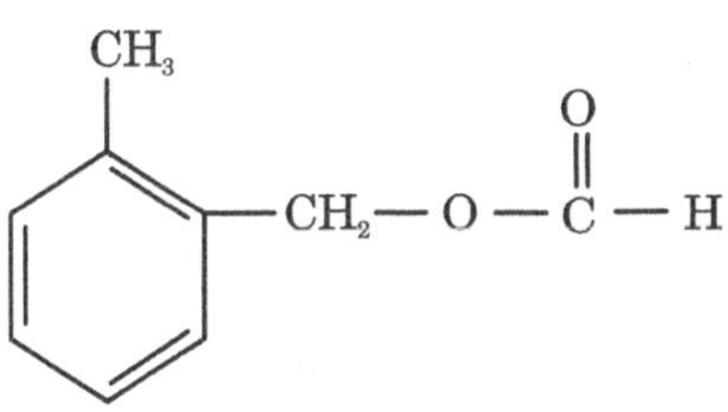 Functional isomers

(b) 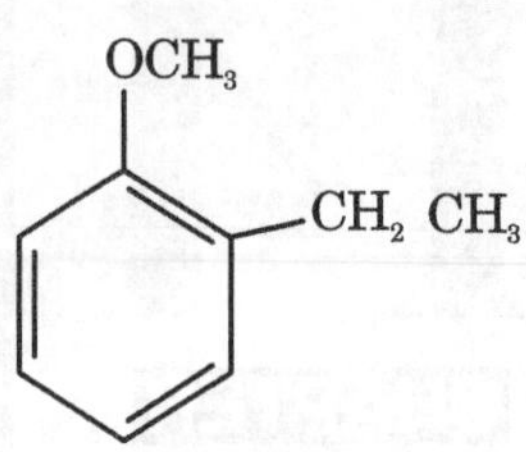 & 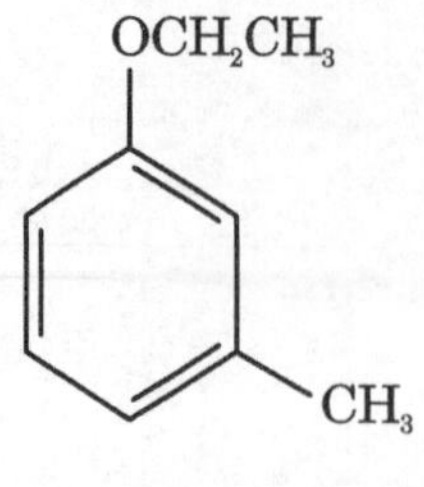

Metamers

(c) $CH_3CH_2CH_2OH$ & $CH_3OCH_2CH_3$

Metamers

(d) 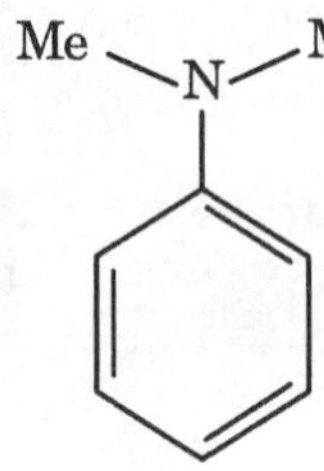 & 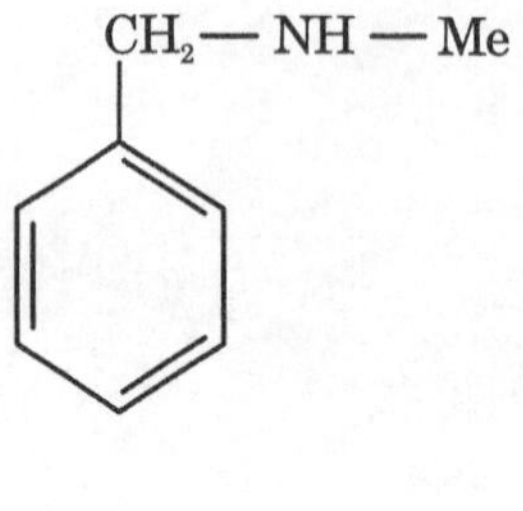

Functional isomers

3. The number of 2° amines possible with the formula $C_5H_{13}N$ are -

(a) 6 (b) 8

(c) 7 (d) 5

4. Total cyclic isomers of C_6H_{12} are - (excluding stereoisomers)

(a) 12 (b) 14

(c) 10 (d) 8

5. The compounds $CH_3-CH_2-NHCHO$ & $CH_3-\underset{NH_2}{\underset{|}{CH}}-CHO$ are

(a) Tautomers

(b) Position isomers

(c) Functional isomers

(d) Chain isomers

6. Which of the following pair is the chain isomers?

(a) 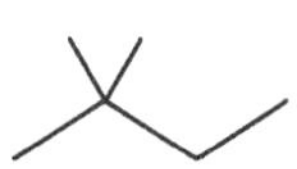 , 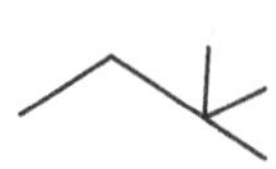

(b) 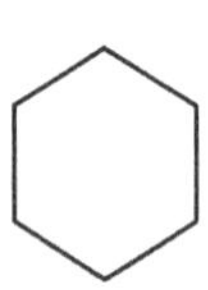 , 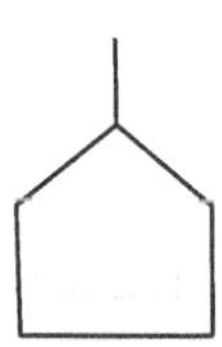

(c) 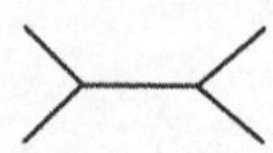 , 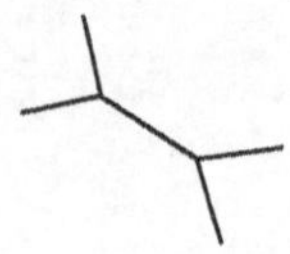

(d) 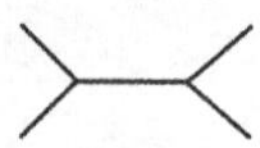 , 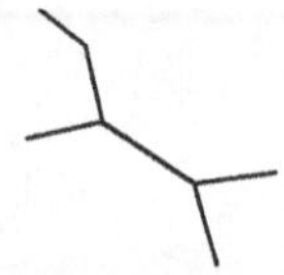

7. Which of the following is a pair of metamers?

(a) $CH_3-CH_2-\overset{O}{\overset{||}{C}}-CH_2-CH_3$ & $CH_3-\overset{O}{\overset{||}{C}}-CH_2-CH_2-CH_3$

(b) 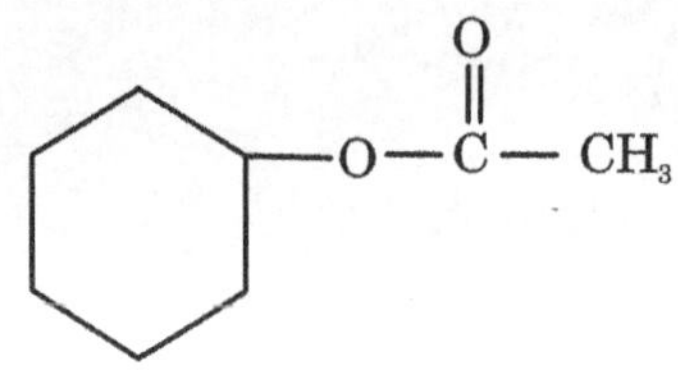 & 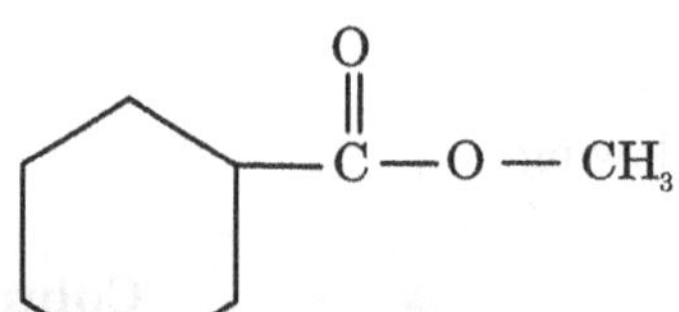

(c) $CH_3-CH_2-\overset{O}{\overset{||}{C}}-O-H$ & $CH_3-\overset{O}{\overset{||}{C}}-O-CH_3$

(d) Both (a) & (b)

8. Which of the following is incorrect-?

(a) $CH_3CH_2CH_2COCH_3$ & $CH_3CH_2COCH_2CH_3$ are metamers.

(b) $CH_3-CH_2-C\equiv C-H$ & $CH_3-C\equiv C-CH_3$ are position isomers.

(c) $CH_3-CH_2-CH_2-NH_2$ & $CH_3-\underset{NH_2}{\underset{|}{CH}}-CH_3$ tautomers.

(d) CH_3CH_2OH & $(CH_3)_2O$ are functional isomers.

Tautomerism

9. Which of the following has the highest % of enol content?

(a) $CH_3-\overset{O}{\overset{\|}{C}}-O-CH_3$

(b) $CH_3-\overset{O}{\overset{\|}{C}}-CH_2-\overset{O}{\overset{\|}{C}}-OCH_3$

(c) $C_6H_5-\overset{O}{\overset{\|}{C}}-CH_2-\overset{O}{\overset{\|}{C}}-OCH_3$

(d) $CH_3-\overset{O}{\overset{\|}{C}}-CH_2-\overset{O}{\overset{\|}{C}}-CH_3$

10. Which of the following will not show tautomerism?

(a) $CH_3-\overset{O}{\overset{\|}{C}}-CH_2-\overset{O}{\overset{\|}{C}}-OC_2H_5$

(b) 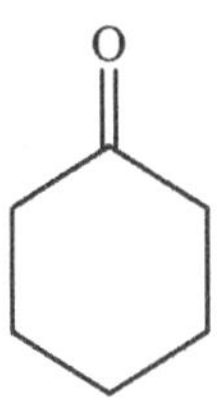

(c) 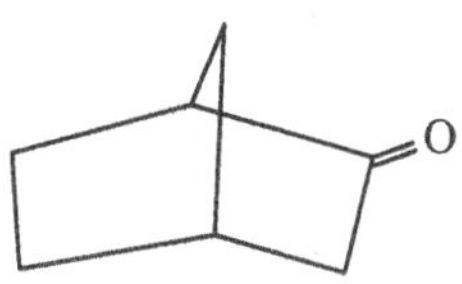

(d) 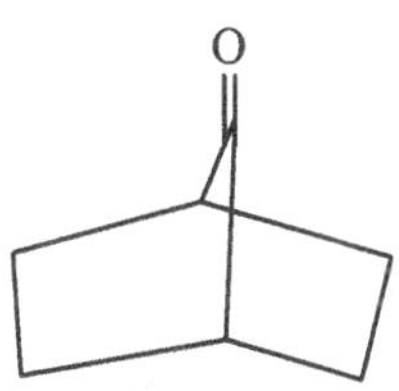

11. Tautomerism is possible in -

(a) $C_6H_5-\overset{O}{\overset{\|}{C}}-H$

(b) $C_6H_5-\overset{O}{\overset{\|}{C}}-C_6H_5$

(c) $CH_3-\underset{CH_3}{\overset{CH_3}{\overset{|}{\underset{|}{C}}}}-\overset{O}{\overset{\|}{C}}-C_6H_5$

(d) $C_6H_5-\overset{O}{\overset{\|}{C}}-CH_2-CH_3$

12. Which of the following is correct tautomeric form of

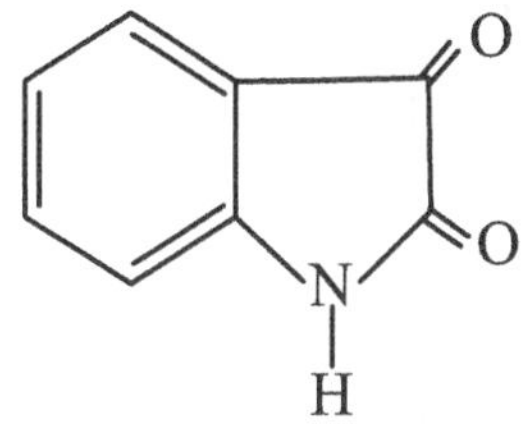 ?

(a) 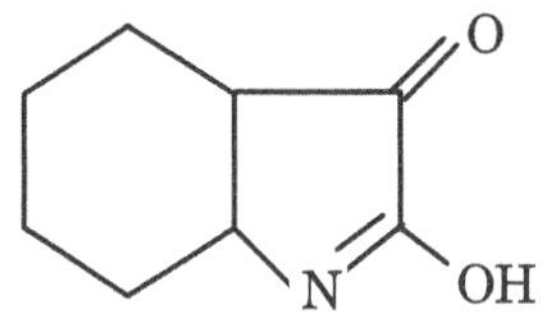

(b) 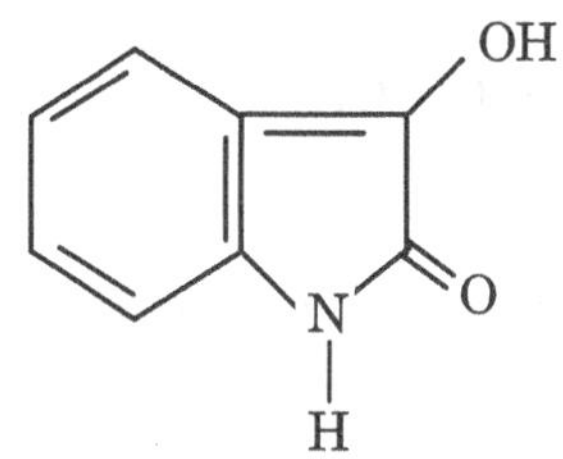

(c) 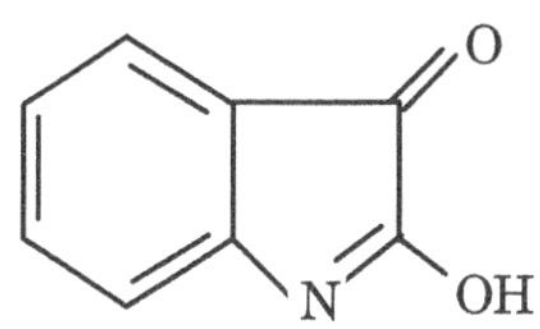

(d) None of these

13. Which of the following molecules can exhibit tautomerism ?

(a) 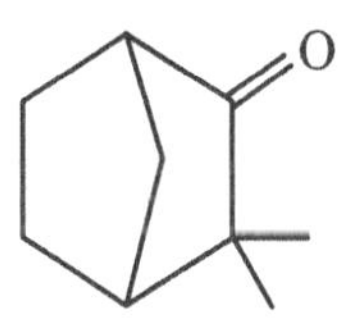 (b) 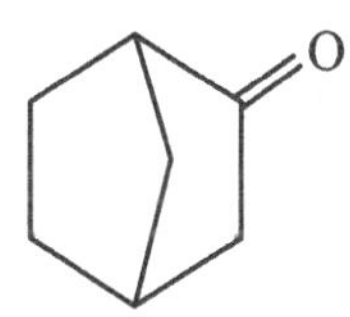

(c) 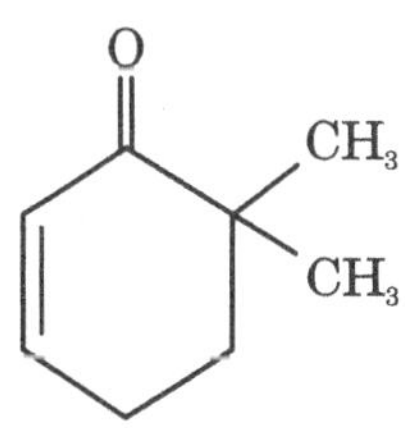 (d) 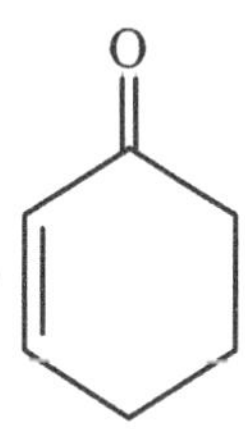

(e) 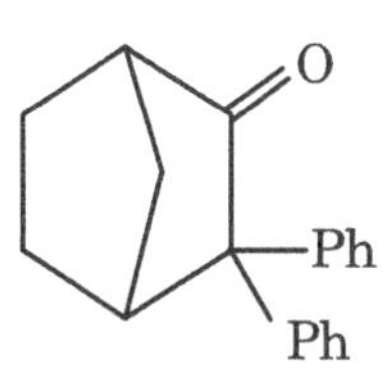

(a) a, b & d only (b) b, c & d only

(c) c, d & e only (d) None of these

14. The most stable enolic form of 2, 4 - pentanedione is -

(a) $CH_2=\underset{\displaystyle OH}{\underset{|}{C}}-CH_2-\overset{\displaystyle O}{\overset{||}{C}}-CH_3$

(b) $CH_3-\underset{\displaystyle OH}{\underset{|}{C}}=CH-\overset{\displaystyle O}{\overset{||}{C}}-CH_3$

(c) $CH_2=\underset{\displaystyle OH}{\underset{|}{C}}-\underset{\displaystyle OH}{\underset{|}{C}}=CH_2$

(d) $CH_3-\underset{\displaystyle OH}{\underset{|}{C}}=C=\underset{\displaystyle OH}{\underset{|}{C}}-CH_3$

15. Which of the following compounds show tautomerism ?

(a) $H_2N-\overset{\displaystyle O}{\overset{||}{C}}-NH_2$

(b) $H_2N-\overset{\displaystyle S}{\overset{||}{C}}-NH_2$

(c) 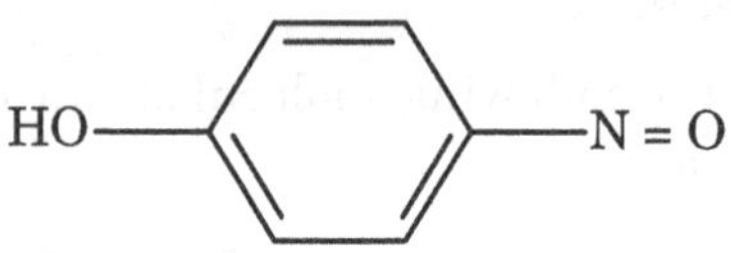

(d) All of these

16.

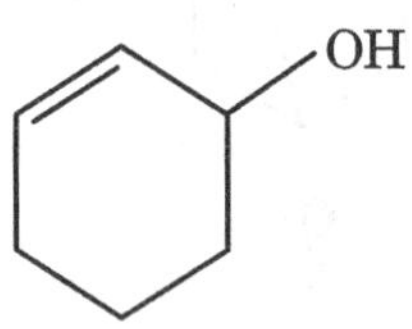

(I)

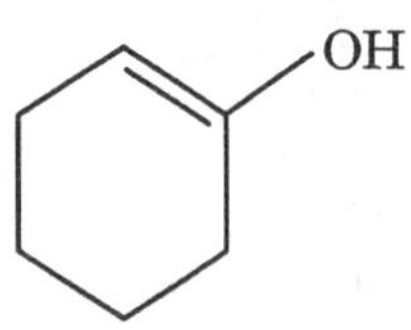

(II)

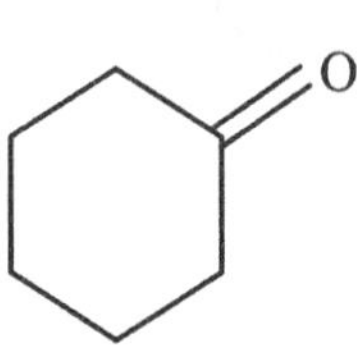

(III)

III is functional isomer but not tautomer of

(a) I only (b) II only

(c) I & II (d) None of these

17. Phenol is a tautomer of -

(a) 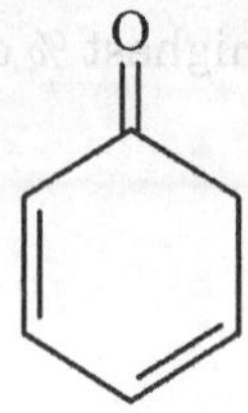

(b) 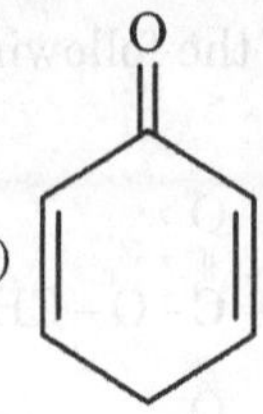

(c) 

(d) Both 1 & 2

18. Select the incorrect option.

(a) The enol form of 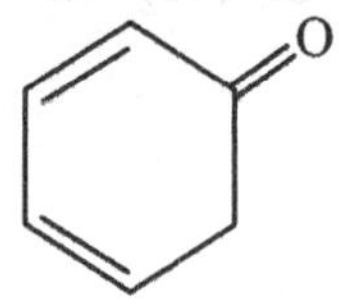 is aromatic.

(b) The molecule, 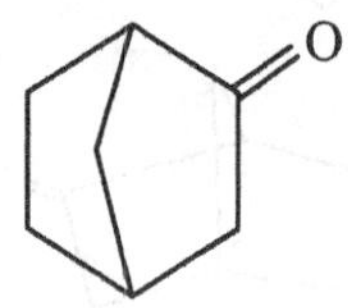 cannot tautomerise.

(c) Tautomerism is possible in 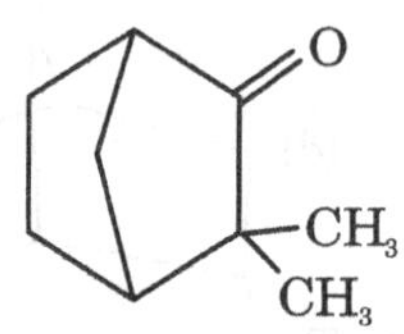.

(d) Both 2 & 3.

19. Among the following given compounds, the correct order of enol content is -

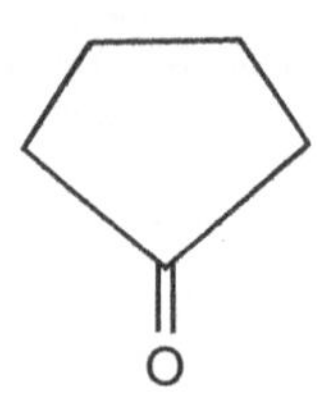

(I)

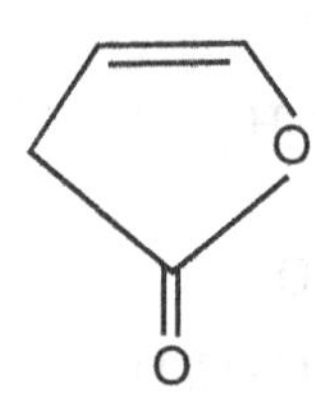

(II)

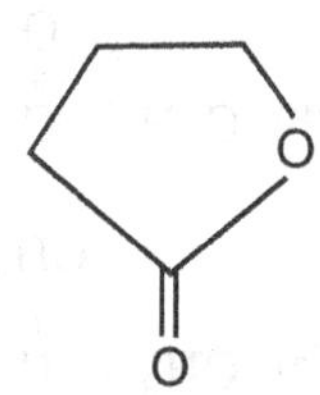

(III)

(a) I > II > III (b) III > II > I

(c) II > I > III (d) II > III > I

20. Which of the following has unstable enol form ?

(a) 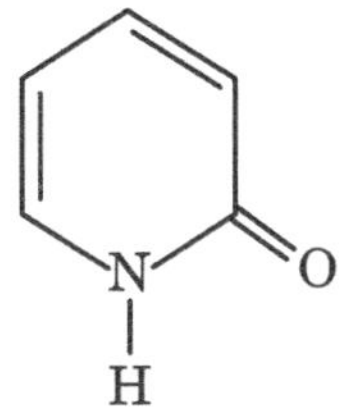

(b) 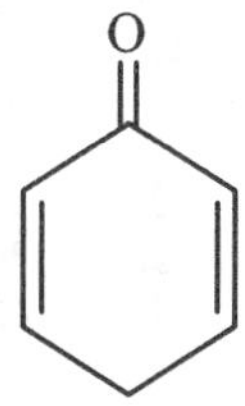

(c) 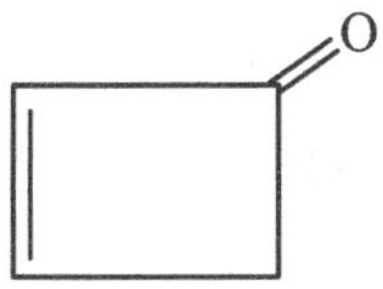

(d) Both 1 & 2

Optical Isomerism

21. Which of these are chiral molecules?

(a) 1, 3 - dichloropopa - 1 , 2 - diene

(b) 2 - hydroxy propanoic acid

(c) 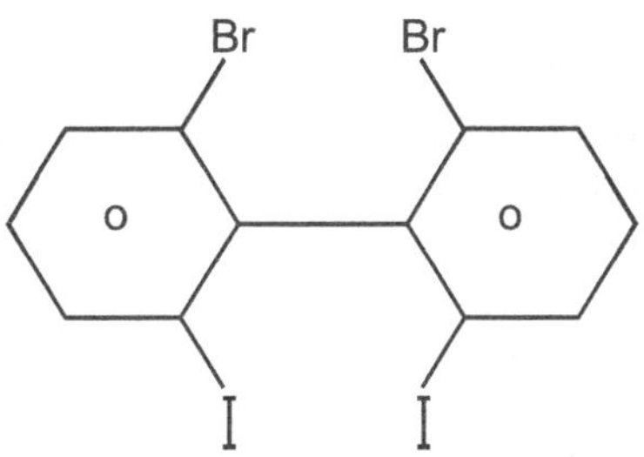

(a) a, b, & c (b) Both a & b

(c) Only b (d) Both b & c

22. Trans - 1- chloro - 4- bromo cyclohexane is-

(a) Chiral molecule because it has two chiral centres.

(b) Chiral molecule but it does not have any chiral center.

(c) Achiral because it has a centre of symmetry.

(d) Achiral because it has a plane of symmetry.

23. Which of the following is optically inactive?

(a) 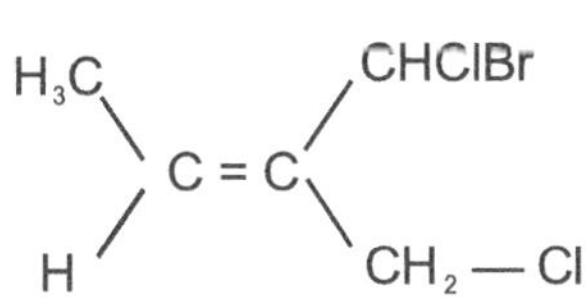

(b) 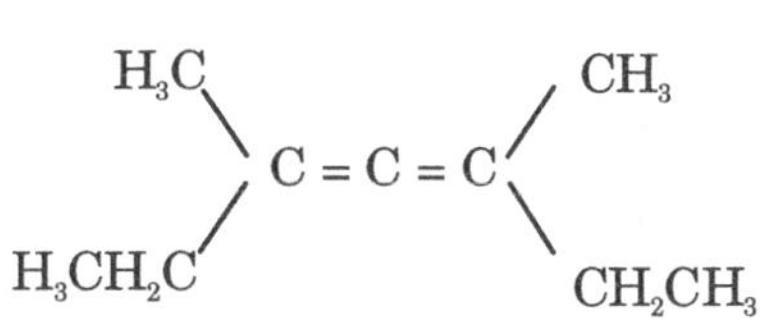

(c) 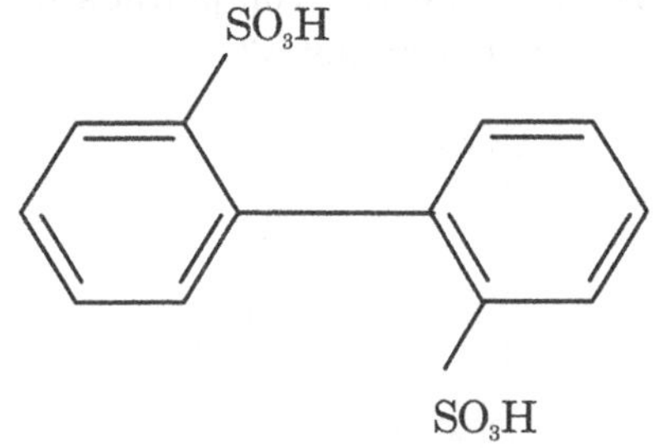

(d) 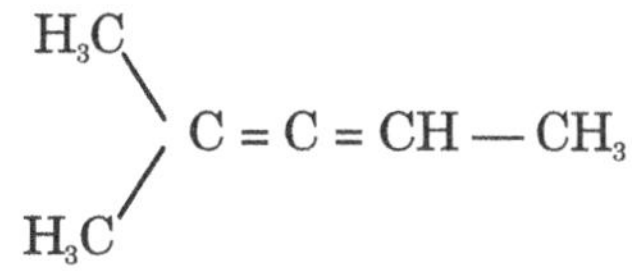

24. Which of the following compounds possess a chiral centre?

(a) 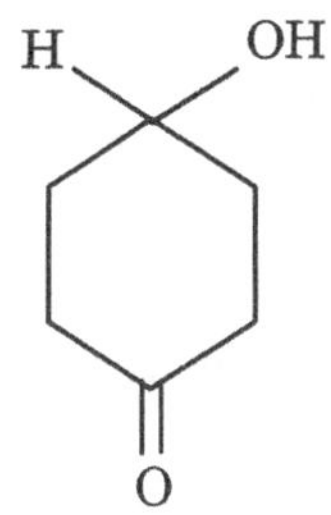

(b) 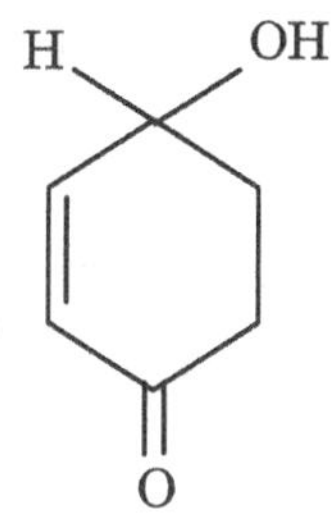

(c) 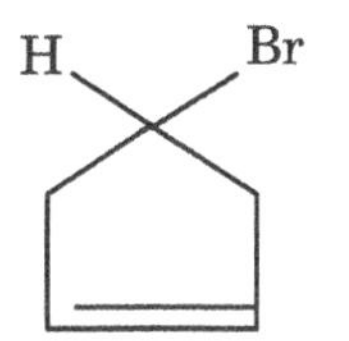

(d) 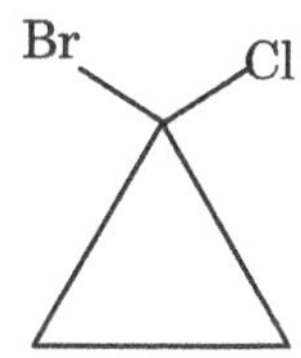

25. The number of stereogenic centres in the molecule-

$OHC-(CHCl)_2-CHO$ is / are

(a) 1 (b) 2

(c) 3 (d) 4

26. Which of the following has a centre of symmetry?

(a) 

(b) 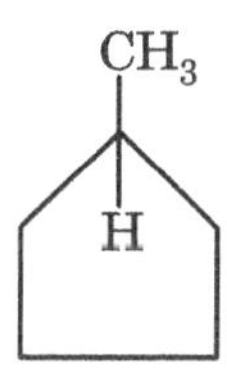

(c) 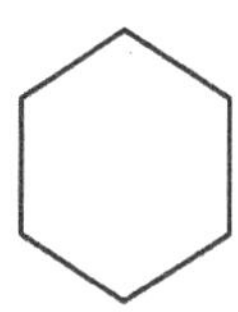

(d) Both 1 & 2

27. Which of the following compounds does not contain a plane of symmetry?

(a) 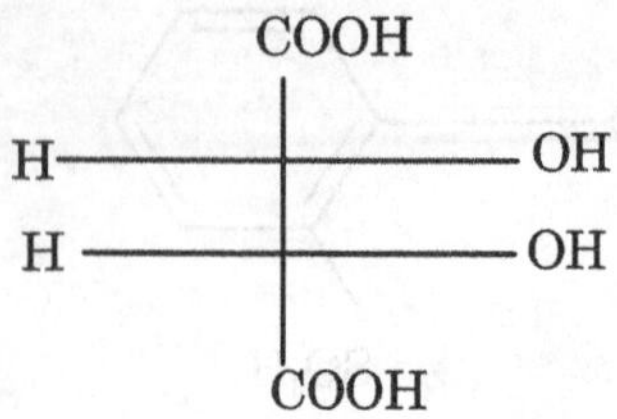

(b) 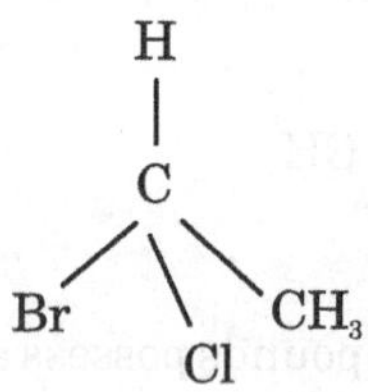

(c) 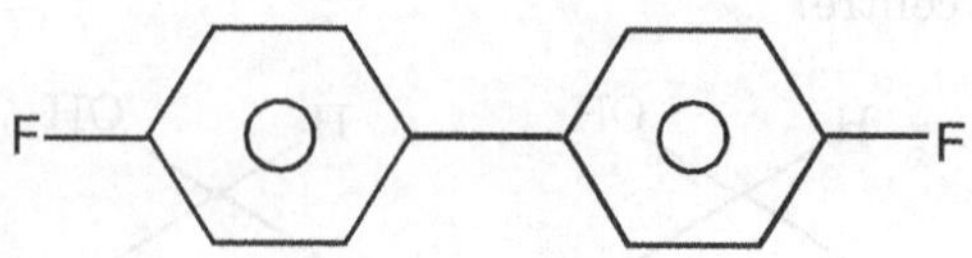

(d) $CH_3 - CH_3$

28. Which statement is correct about the following structures?

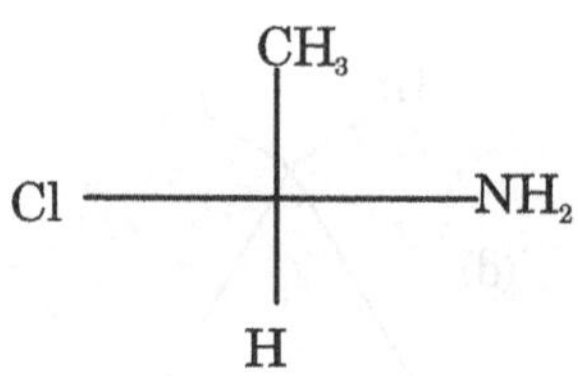

(I)

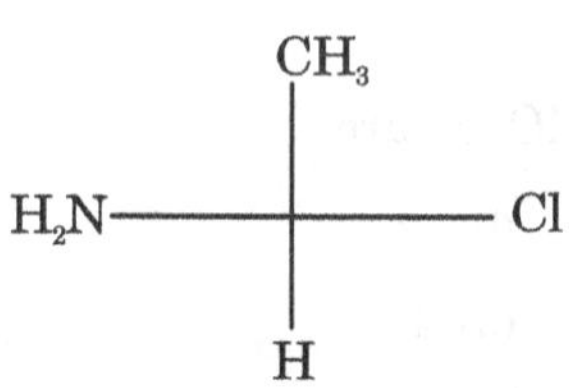

(II)

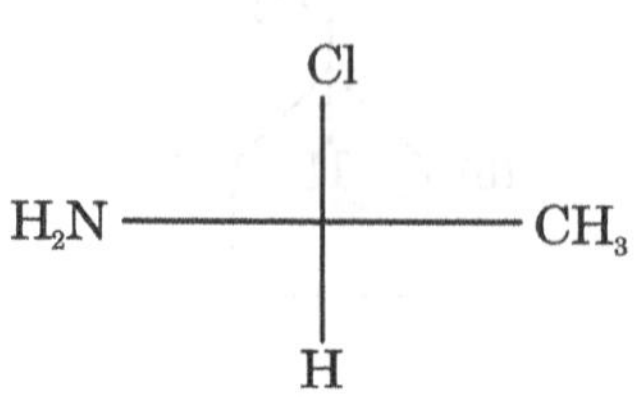

(III)

(a) I & II are enantiomers.

(b) II & III are enantiomers.

(c) I & III are identical

(d) All are correct

29. Which is chiral?

(a) 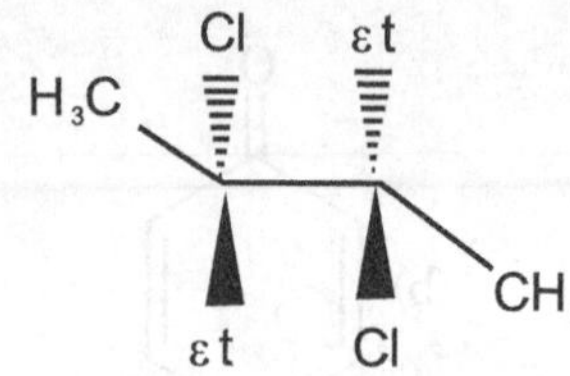

(b) 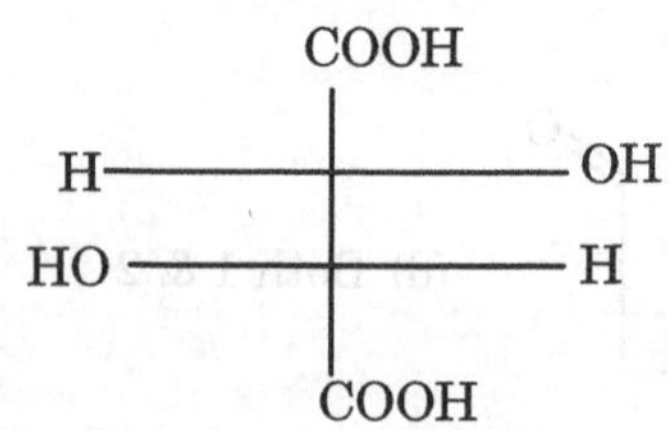

(c) 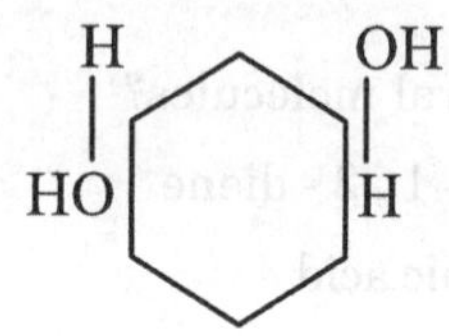

(d) All of these

30. Which of the following structure is not meso - 2, 3 - butanediol ?

(a) 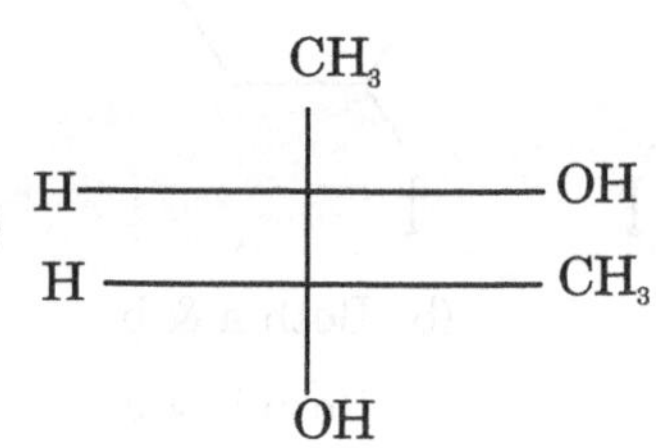

(b) 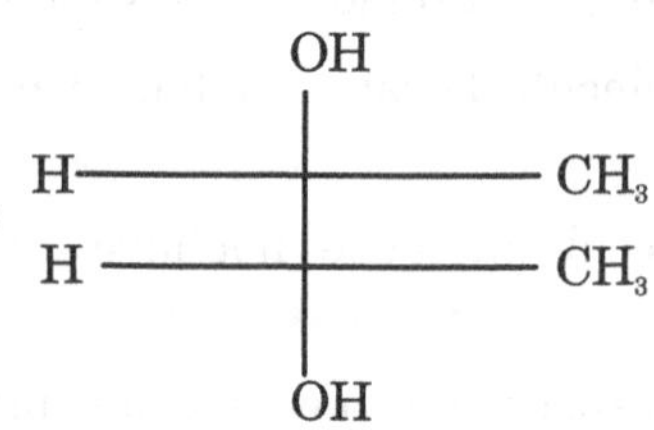

(c) 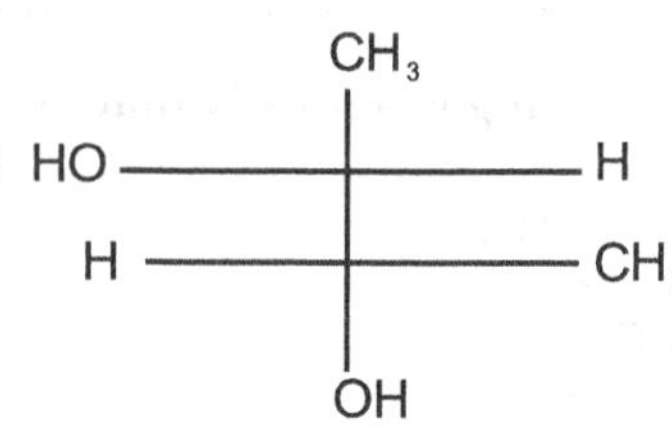

(d) 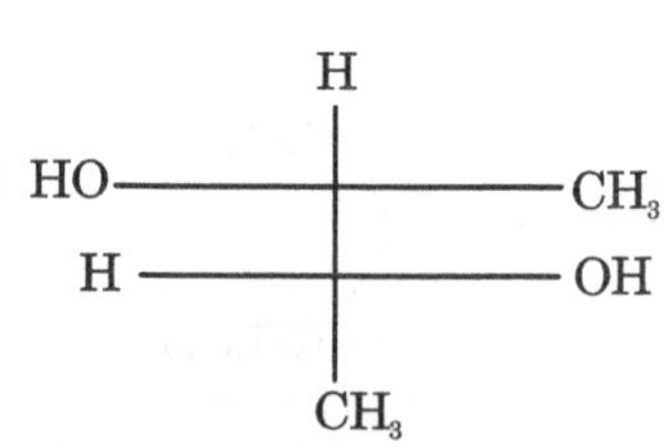

31. Which of the following compounds can be optically inactive?

(a) 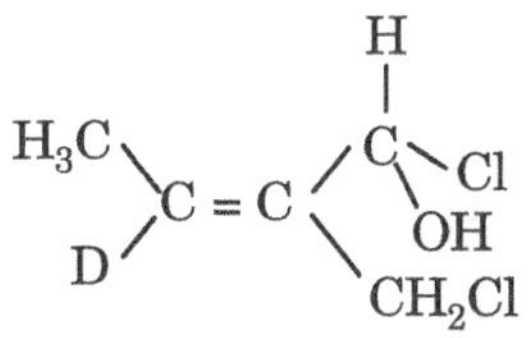

(b) 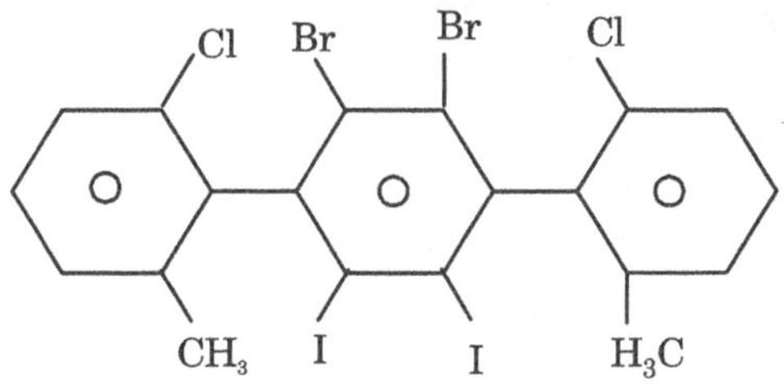

(c) 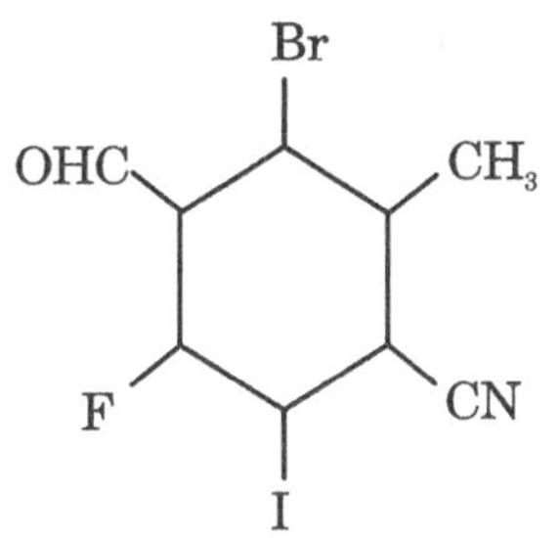

(d) 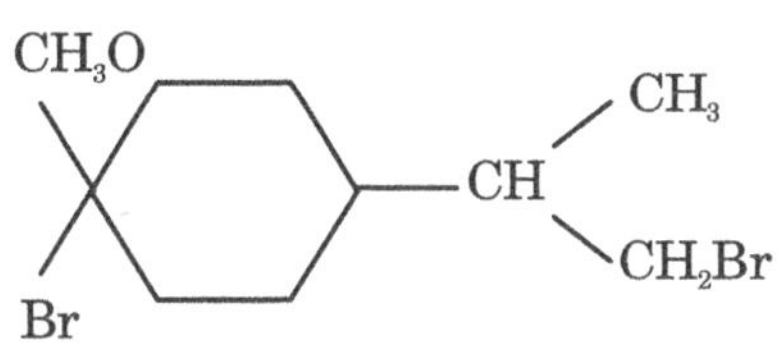

32. Which of the following is non-superimposable on its mirror image?

(a) 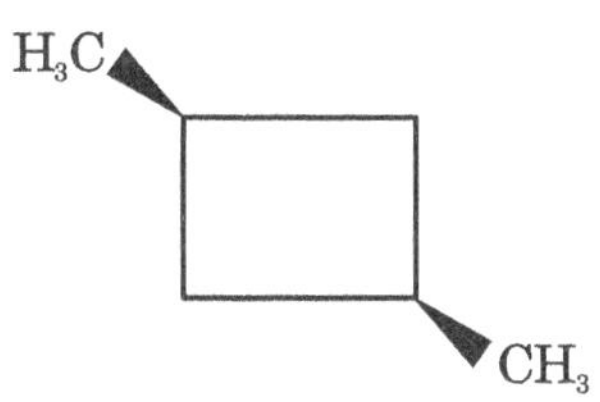

(b) 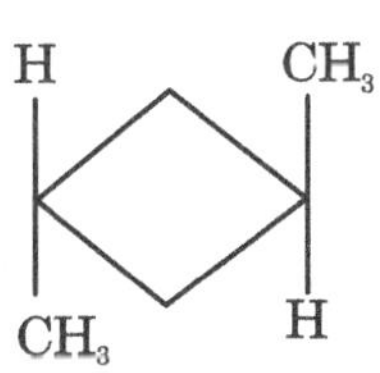

(c) 

(d) 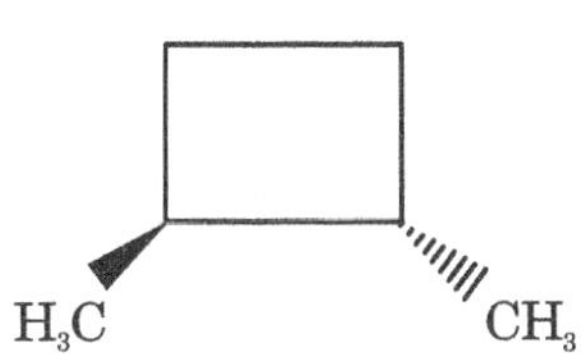

33. In the stereomer given below, after replacing H_Y with T (tritium) & H_X with D (Deuterium), structure M is obtained. But replacing H_Y with D & H_X with T (in original stereomer), structure N is obtained. M & N represent a pair of

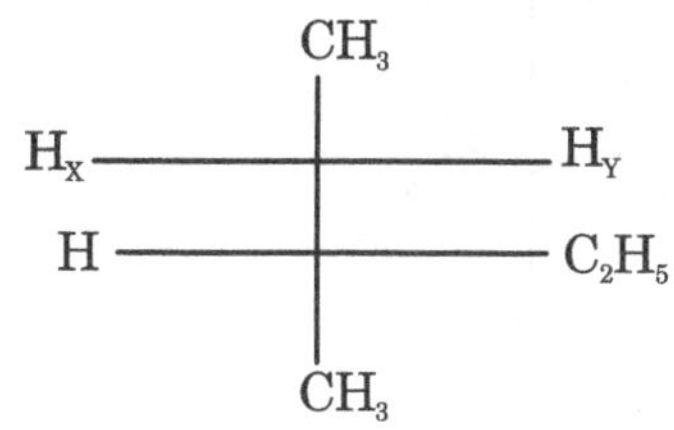

(a) enantiomers

(b) diastereomers

(c) E – 2 isomers

(d) Constitutional isomers

34. Identity the correct statement about the compound

$$CH_3(OH)CH-CH=CH-CH(OH)CH_3$$
(Y)

(a) If stereochemistry around the double bond in Y is Cis, the number of enantiomers possible for Y is 2.

(b) There are 4 meso stereomers possible for Y.

(c) Total no. of stereoimers possible for Y are 2.

(d) All stereoisomers of Y are optically active.

(a) only a, b & d (b) only a & d

(c) only c & d (d) only a

35. Which of the following statements are correct?

(a) The compound 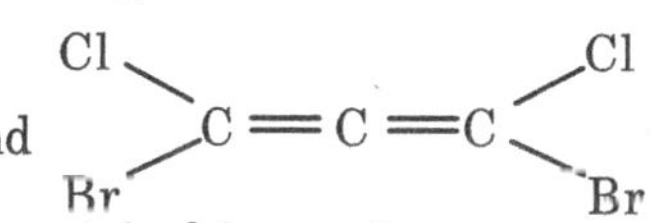 will not show geometrical isomerism.

(b) Enol content of 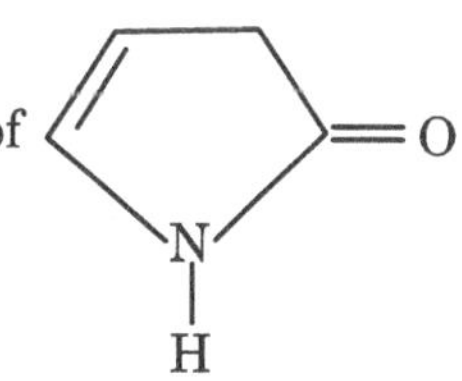 is greater than that of 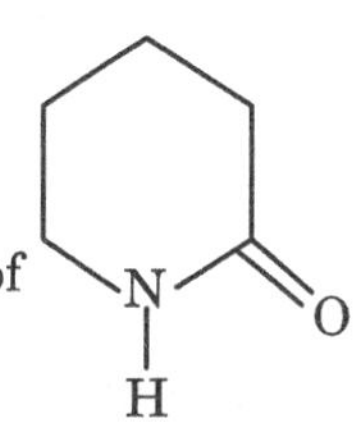

(c) Propene cannot show geometrical isomerism.

(d) All are correct

36. Which of the following conformation is most stable?

(a)

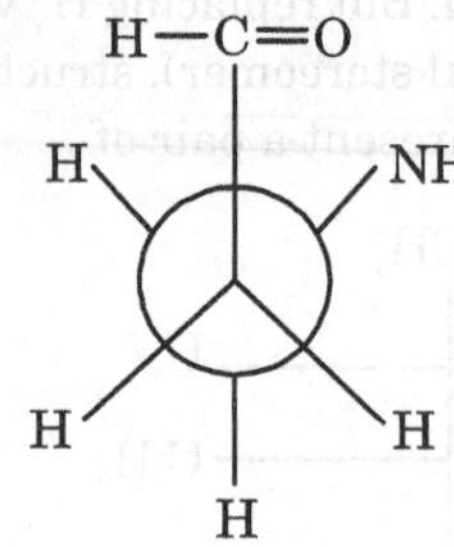

(b)

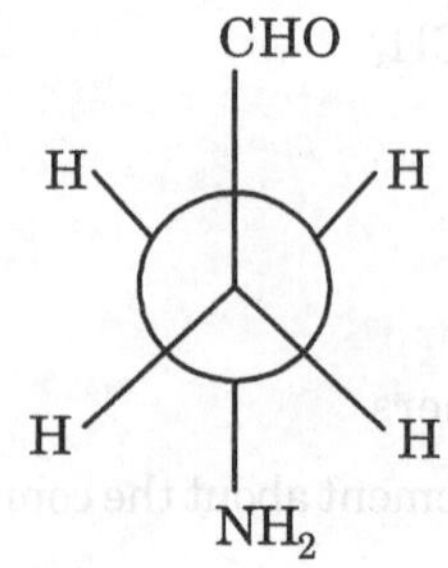

(c)

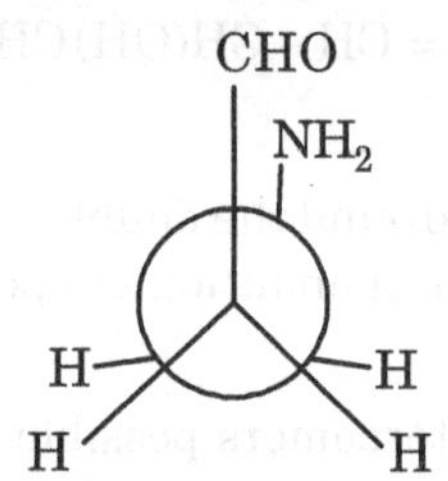

(d)

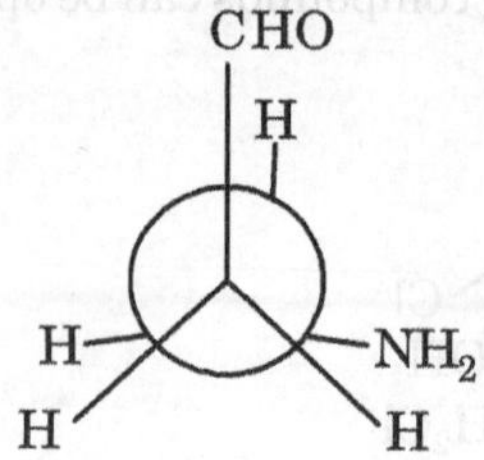

37. Which of the following shows geometrical isomerism ?

(a)

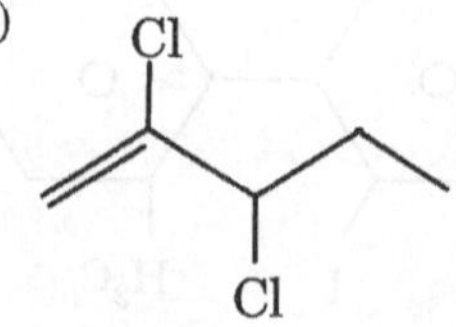

(b) $C_6H_5 - CH = NOH$

(c)

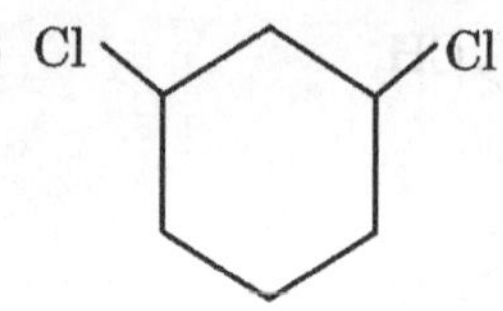

(d) Both 2 & 3

38. Consider the following conformations -

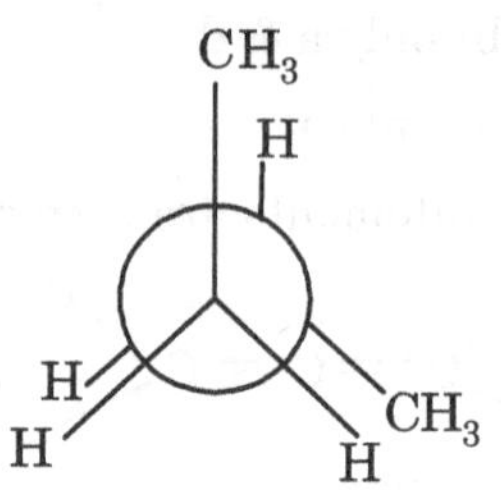

(a)

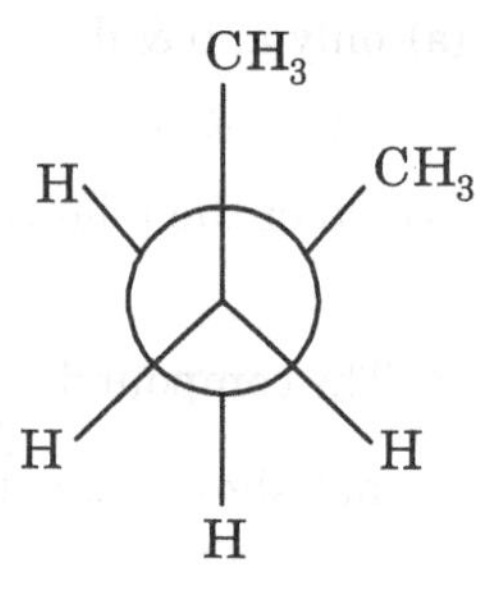

(b)

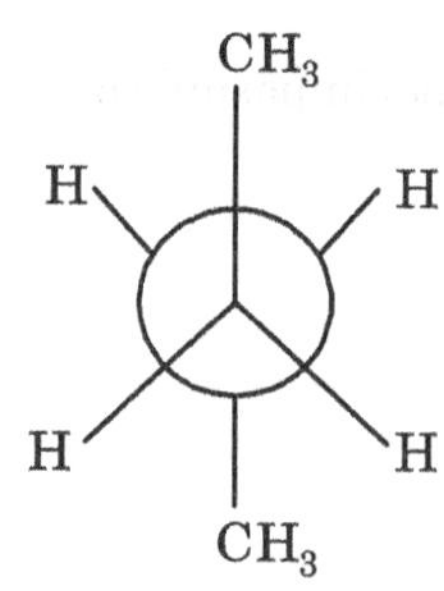

(c)

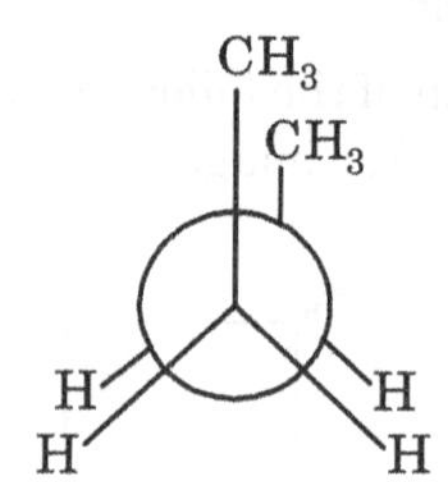

(d)

Relative stabilities of the above varies as -

(a) d > a > c > b (b) c > b > a > d (c) a > b > c > d (d) b > a > c > d

39. Identify the incorrect statement.

(a) Magnitude of torsional strain depends on the angle of rotation across C - C bond.

(b) Staggered form of ethane has least torsional strain.

(c) Eclipsed form of ethane has maximum torsional strain.

(d) Staggered form of ethane has maximum torsional strain.

40. In eclipsed & staggered conformation of ethane, bond angle and bond length are respectively -

(a) same, same (b) different, same (c) same, different (d) different, different

41. Which of the following compound can show geometrical isomerism ?

(a) 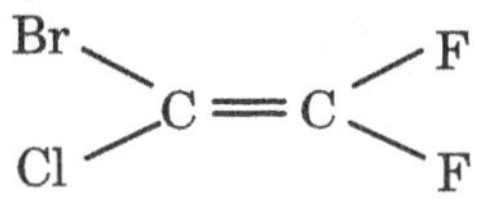

(b) 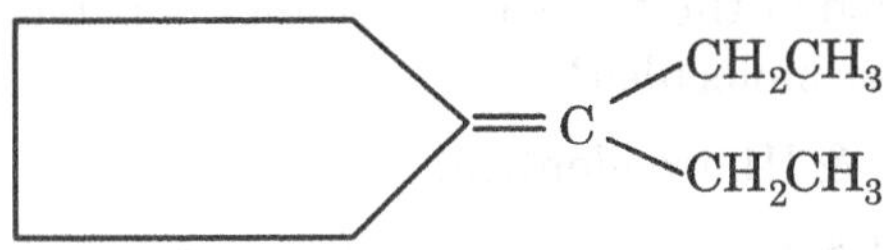

(c) 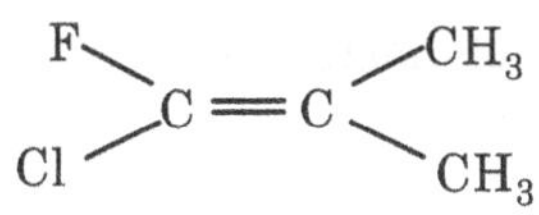

(d) 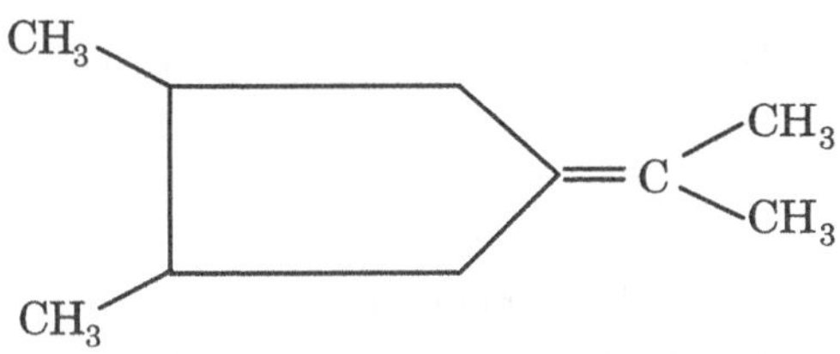

42. The most stable conformation of ethylene glycol is -

(a) Anti (b) Gauche

(c) Partially eclipsed (d) fully eclipsed

43. The geometrical isomerism is shown by -

(a) 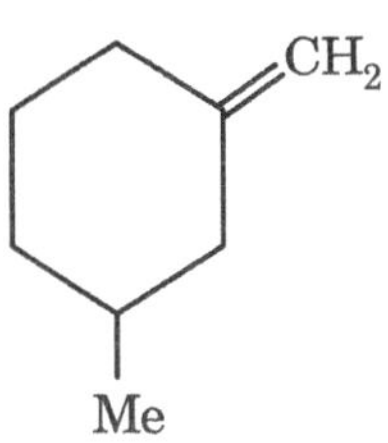

(b) 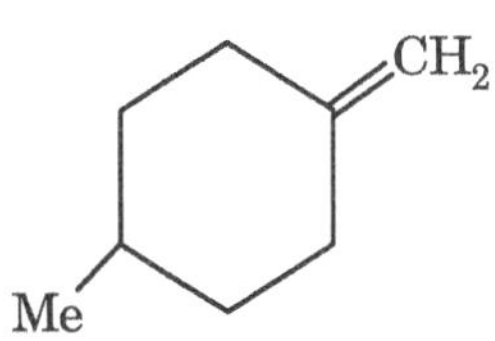

(c) 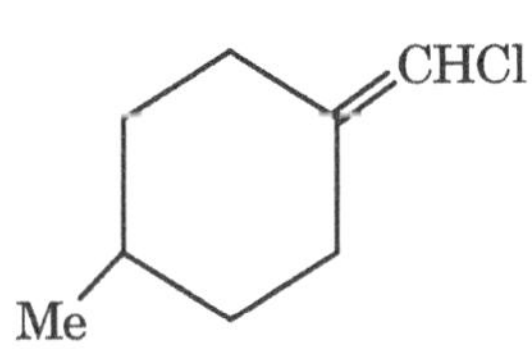

(d) 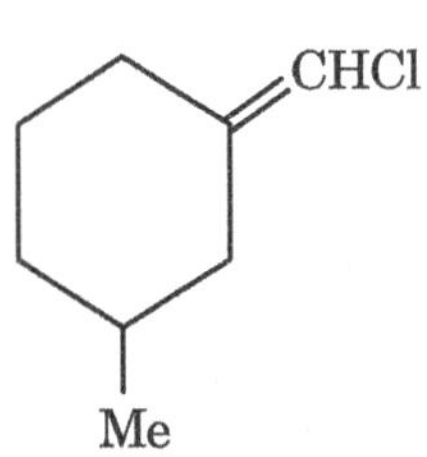

44. Which of the following can show geometrical isomerism ?

(a) 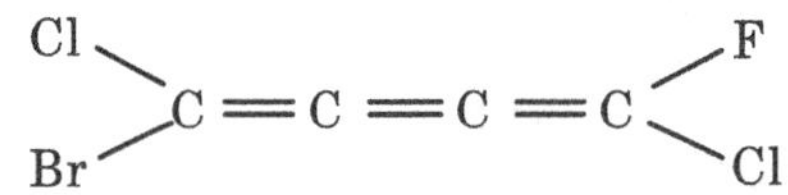

(b) 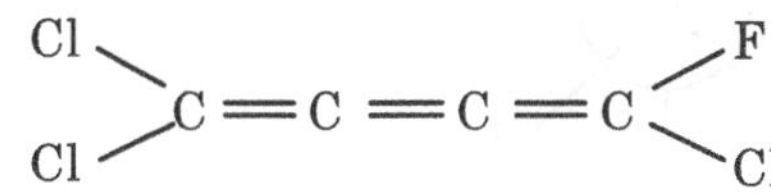

(c) 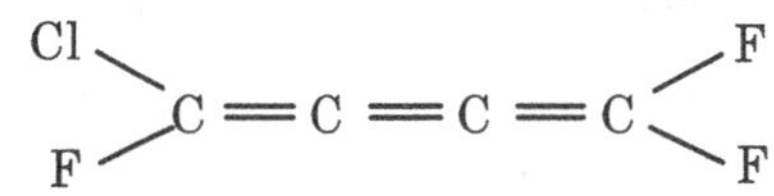

(d) 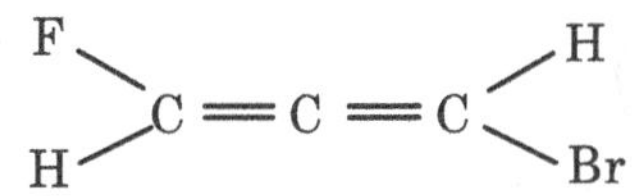

45. Plane of symmetry is not present in -

(a) 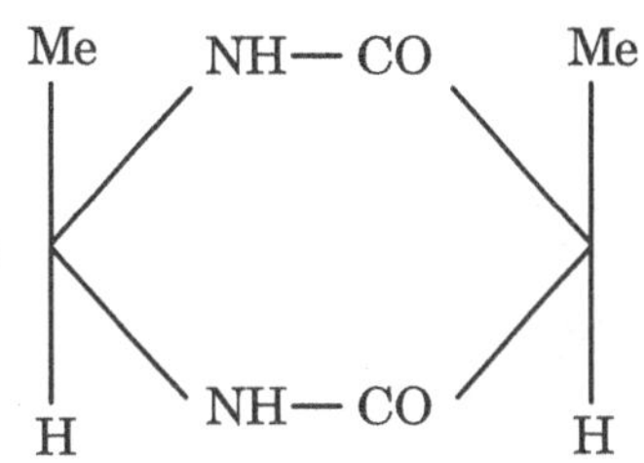

(b) 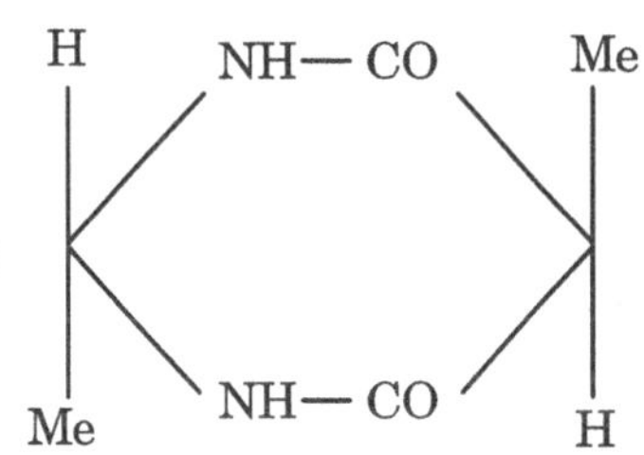

(c) 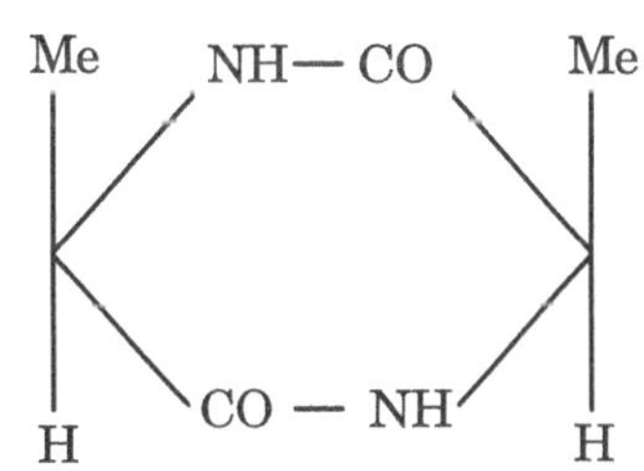

(d) All of these

46. 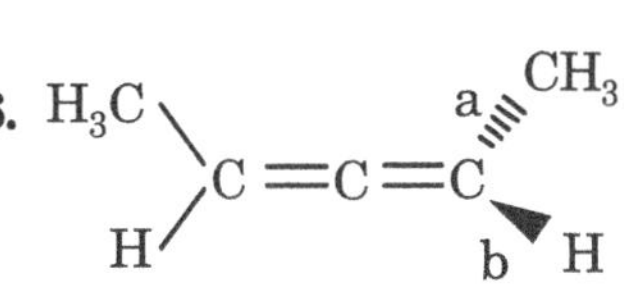

The molecule obtained by interchanging the groups at bond a & b in the above structure is related to the above structure as a -

(a) Diastereomer (b) Enantiomer

(c) Identical (d) messo

47. Among the following Newmann projection of meso - 2 , 3 - butenediol are -

(P) 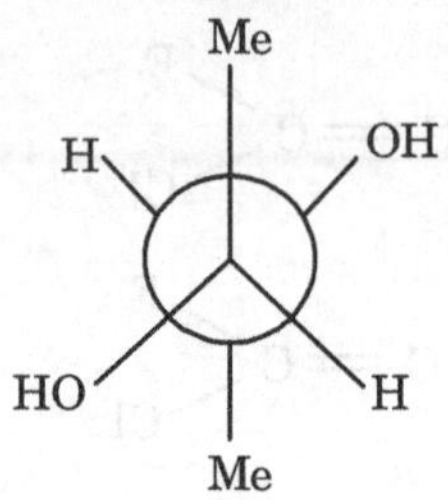

(Q) 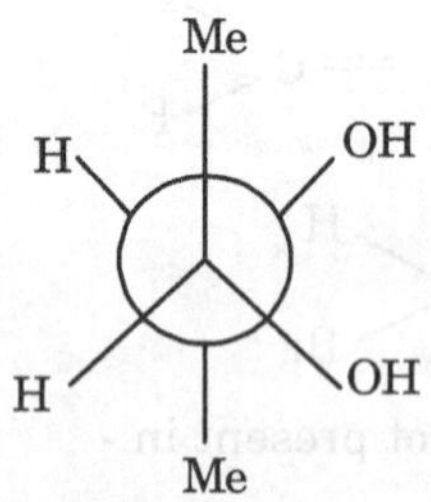

(R) 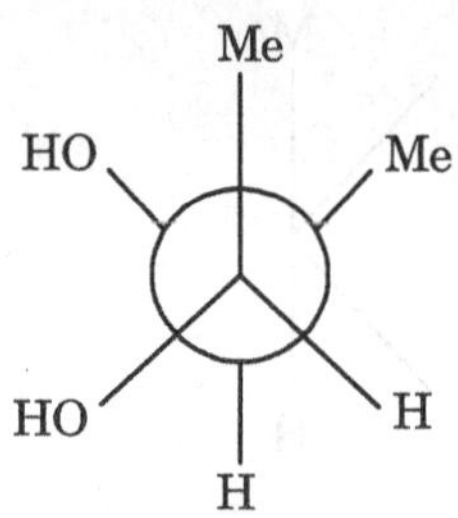

(S) 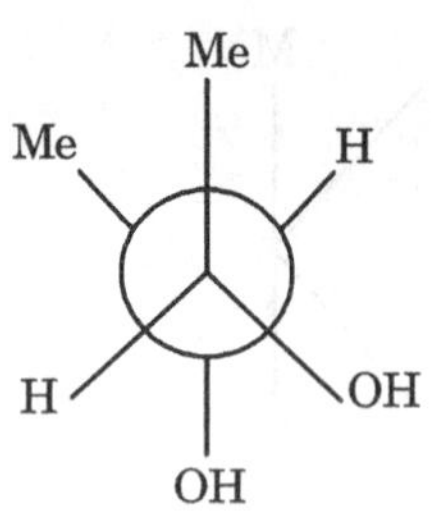

(a) P, Q (b) P, R

(c) R, S (d) Q, S

48. Ratio of pair of diastereomers & enantiomers of 1 - chloro - 2 - methyl cyclopropane is -

(a) 2 : 1 (b) 4 : 3

(c) 3 : 1 (d) 1 : 2

49. The pair of structure given below are -

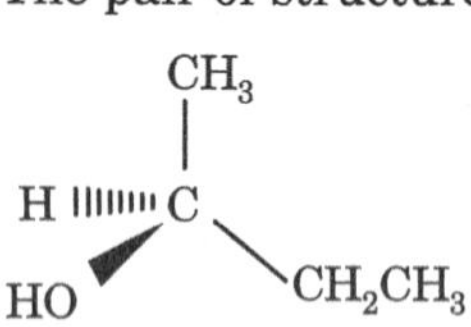

I

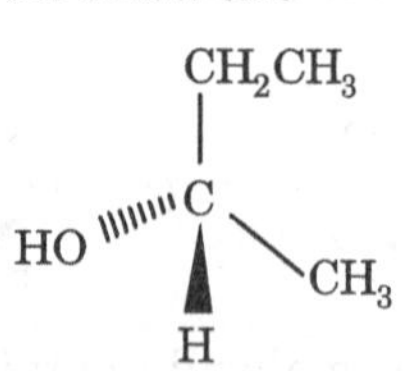

II

(a) Enantiomers (b) Diastereomers

(c) Identical (d) Position isomers

50. 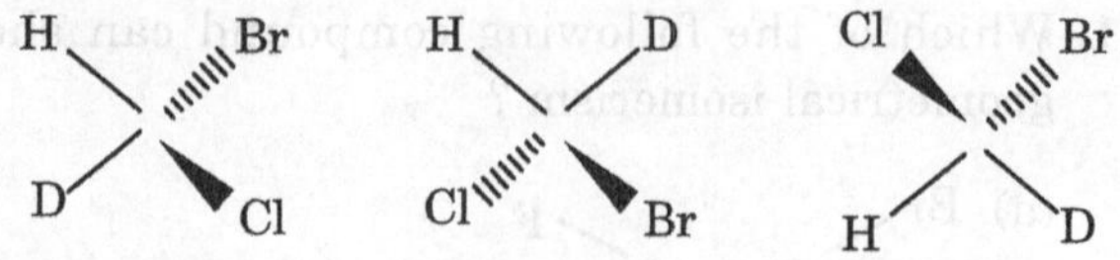

(I) (II) (III)

Which of the following statements is true about these molecules?

(a) I & III are identical

(b) I & II are a pair of enantiomers

(c) I & II are a pair of diastereomers.

(d) Both (a) & (b)

51. 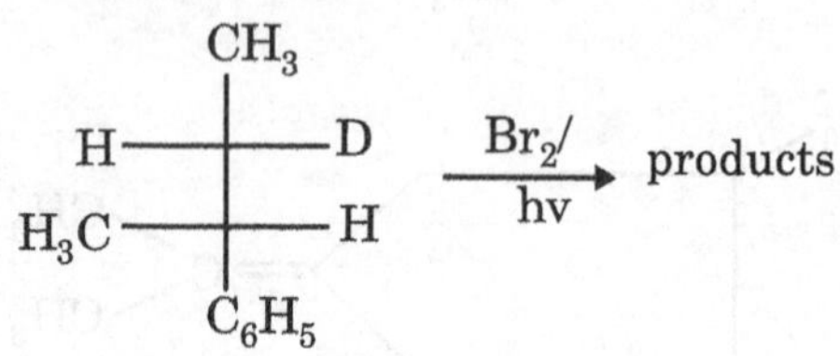

For the above monobromination reaction, the major product obtained will exist as

(a) pair of enantiomers

(b) pair of diastereomers

(c) a single stereoisomers

(d) meso form.

52. The compound which will not loose chirality when treated with $LiAlH_4$ is

(a) 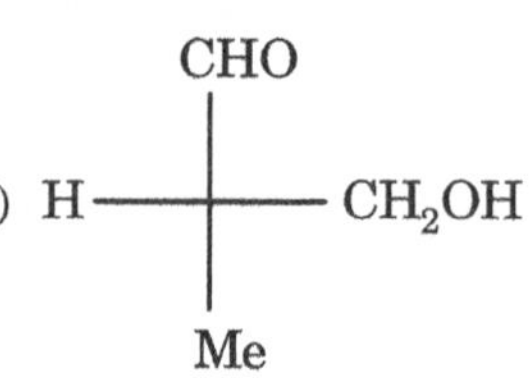

(b) 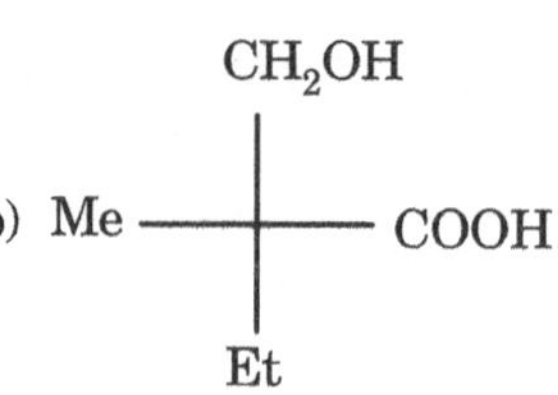

(c) 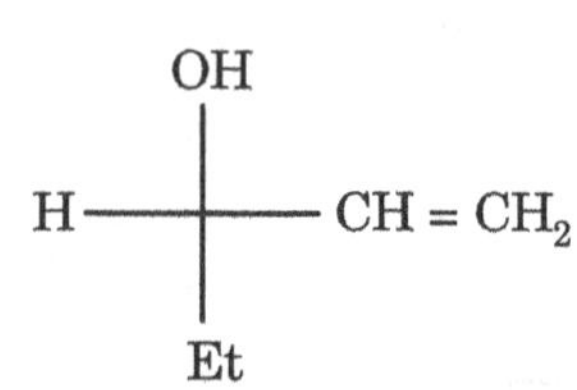

(d) 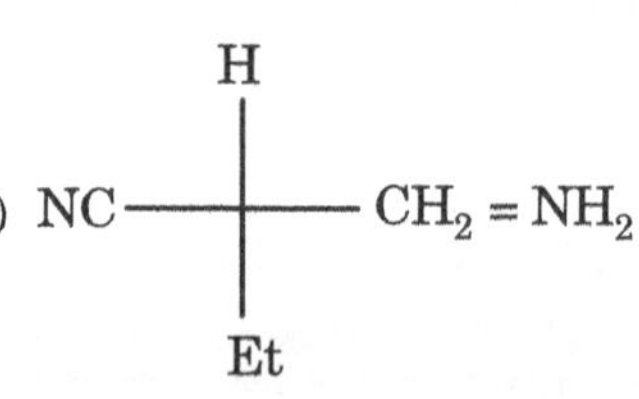

53. $\underset{\text{S - Laticacid}}{CH_3CH(OH)-COOH} + \underset{\text{2 - butarol (d \& l form)}}{CH_3-\overset{\overset{\displaystyle OH}{|}}{C}H-C_2H_5} \xrightarrow[H_2SO_4]{\text{conc.}} P$

Solution of P formed is

(a) optically inactive as 100% racemic mixture is obtained

(b) optically active as 100% pure enantiomer is obtained

(c) optically active as 100% inversion occurs

(d) optically active as products are non-mirror image sterisomers

54. Rank the following substituents in order of decreasing priority order according to CIP Rules

$-CH(CH_3)_2$ (1) $-CH_2Br$ (2)

$-CH_2CH_2Br$ (3)

(a) 2 > 3 > 1 (b) 1 > 3 > 2

(c) 3 > 1 > 2 (d) 2 ? 1 ? 3

55. The absolute configurations of the two chiral C's in the following molecule are

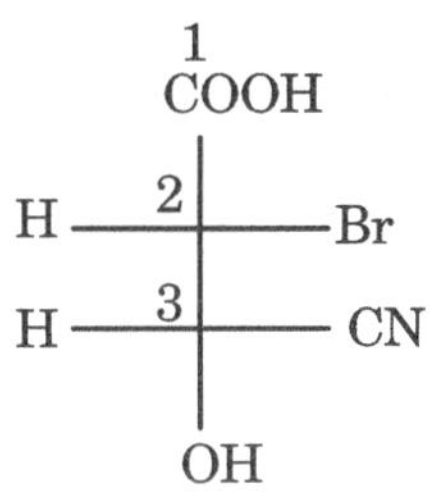

(a) 2R, 3S (b) 2R, 3R

(c) 2S, 3S (d) 2S, 3R

56. Which of the following can show geometrical isomerism?

(a)

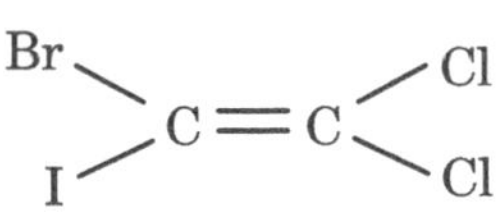

(b)

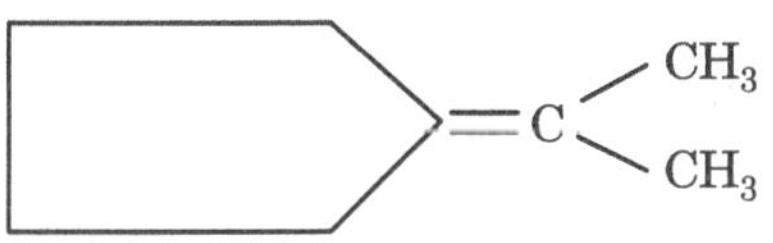

(c)

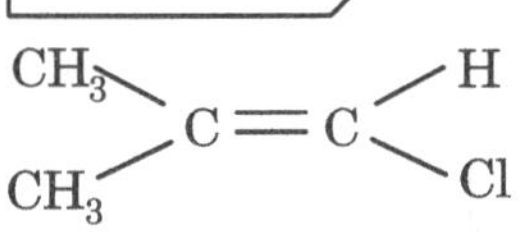

(d)

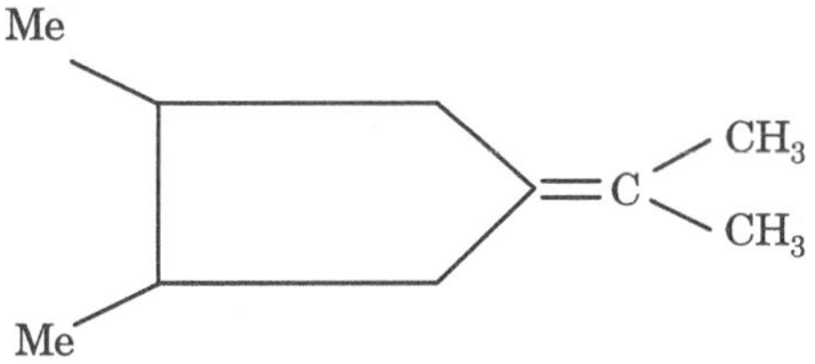

57. Which of the following has plane of symmetry as well as centre of symmetry?

(a)

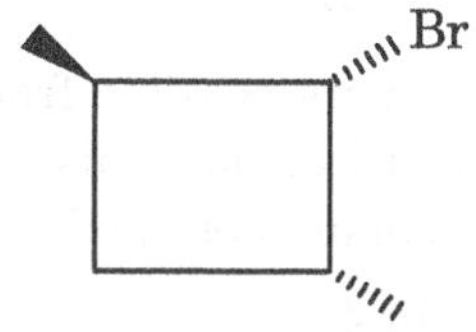

(b)

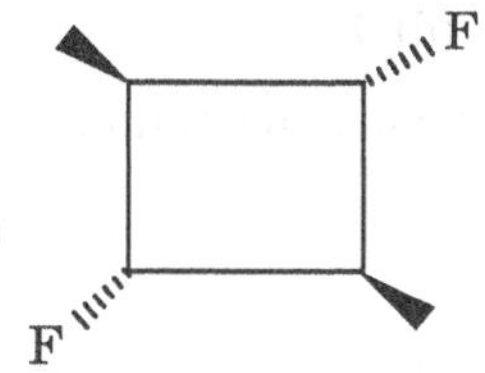

(c)

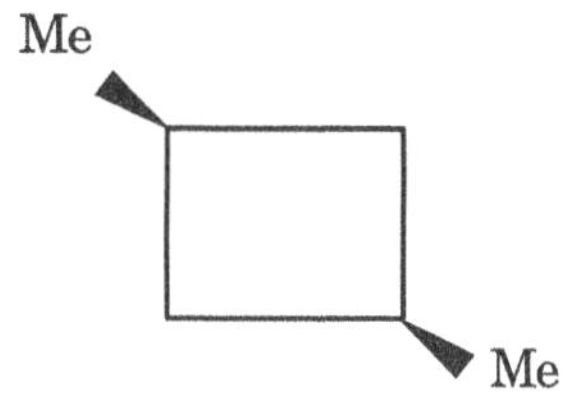

(d)

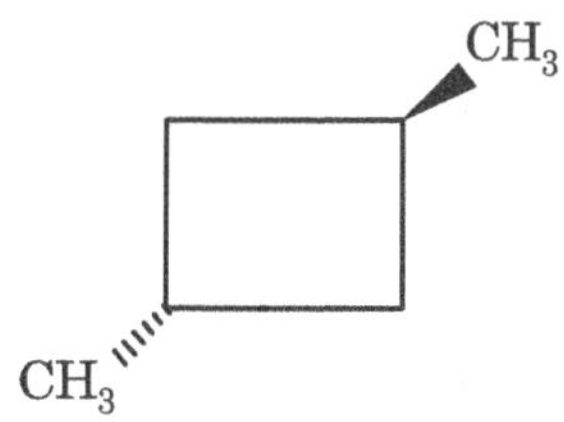

58. Identity the true statement (s) about the following.

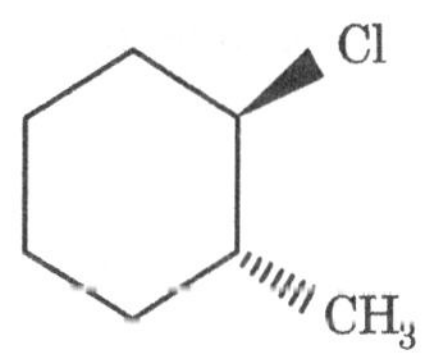

(W)

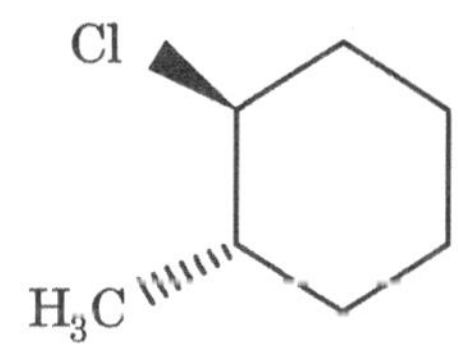

(X)

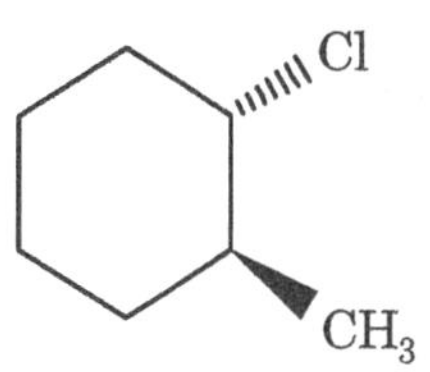

(Y)

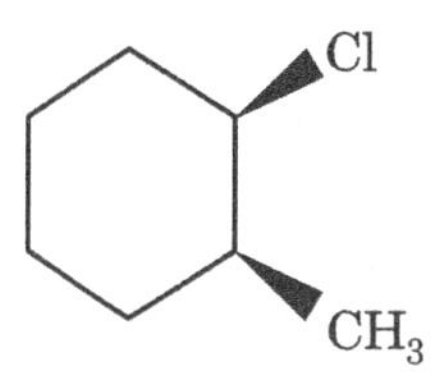

(Z)

(a) w, x & y have same physical properties.

(b) x, y & z have same physical properties.

(c) w, x & z have same physical properties.

(d) w, y & z have same physical properties.

59. $CH_3-\underset{OH}{\underset{|}{CH}}-CH_2-CH_2-\underset{Br}{\underset{|}{CH}}-CH_3$

Mixture of all stereoisomers possible from this structure is subjected to fractional distillation. Then , how many fraction are obtained.

(a) 2 (b) 4

(c) 6 (d) 7

60. Which of the following will show tautomerism?

(a) 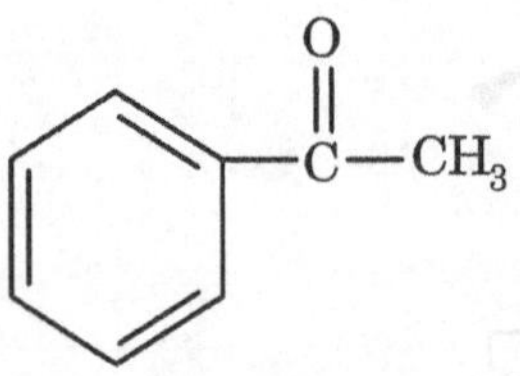

(b) 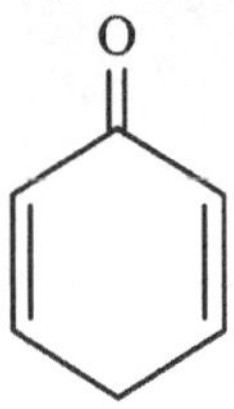

(c) $CH_3-\overset{O}{\overset{||}{C}}-CH_2-\overset{O}{\overset{||}{C}}-CH_3$

(d) All

61. The correct statement(s) concerning the structures P, Q, R & S is/are-

(P) 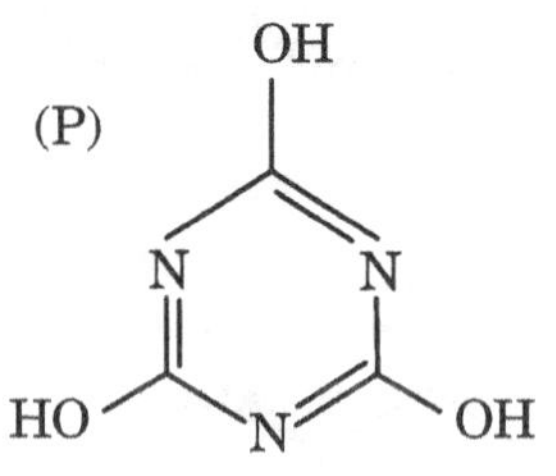

(Q) 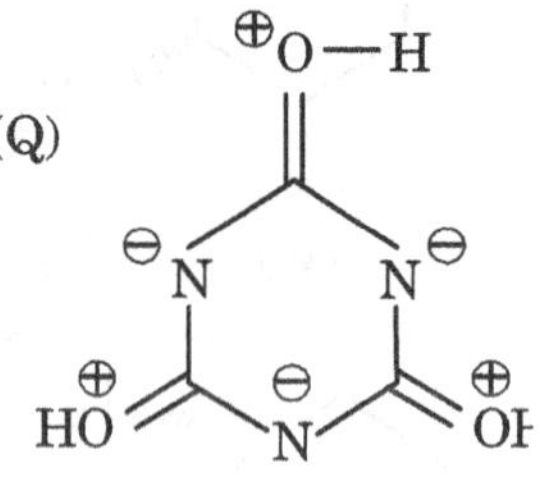

(R) 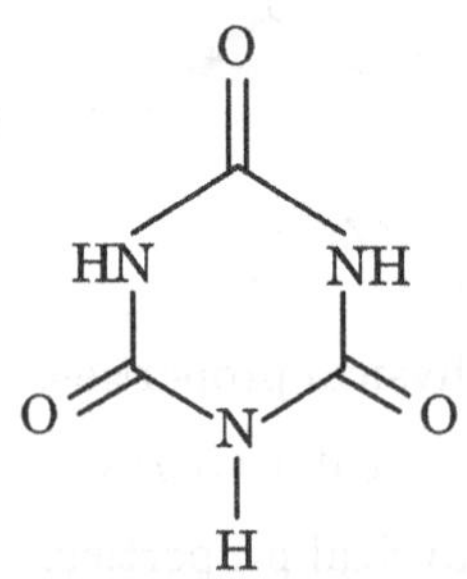

(S) 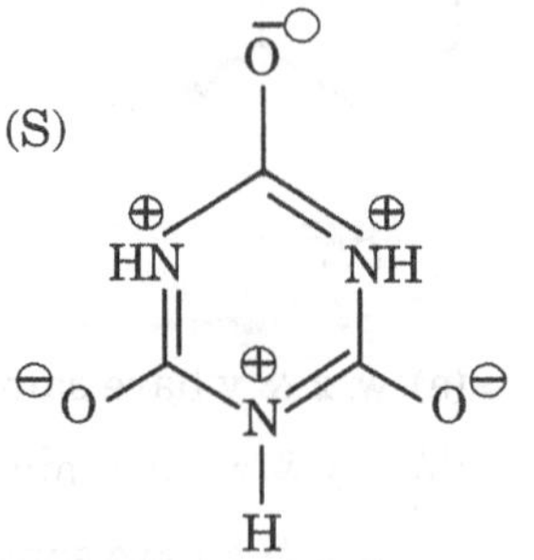

(1) Q & S are tautomers

(2) R & S are resonating structures.

(3) P & R are tautomers.

(4) P & Q are resonating structures.

(a) Only 1 is correct

(b) Only 2 & 3 are correct

(c) Only 2, 3 & 4 are correct

(d) None is correct

62. Which of following can show geometrical isomerism?

(a) 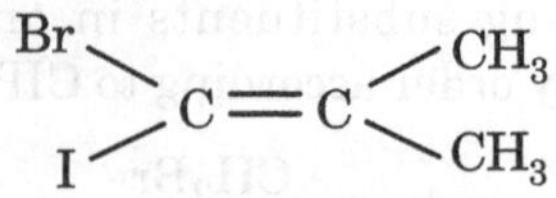

(b) 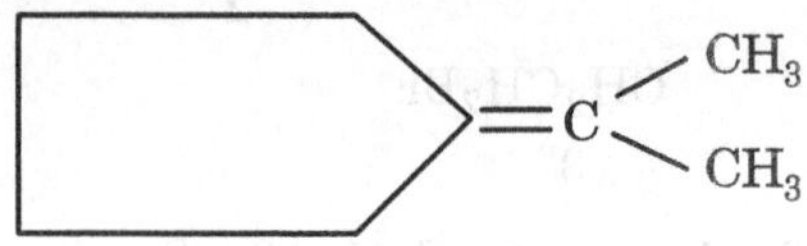

(c) 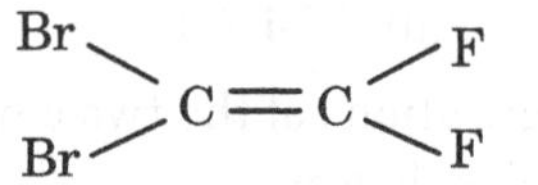

(d) 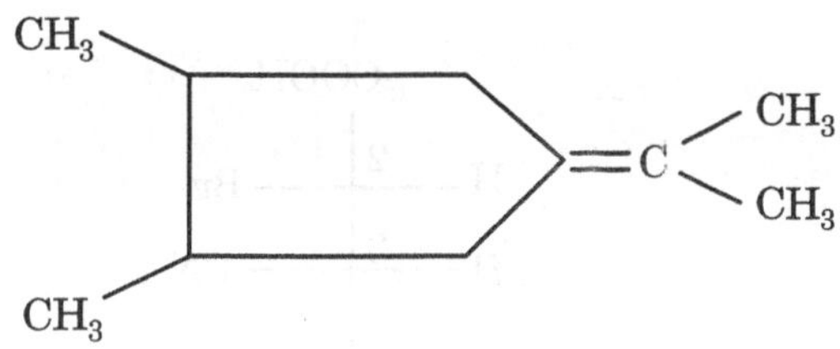

63. If optical rotation produced by

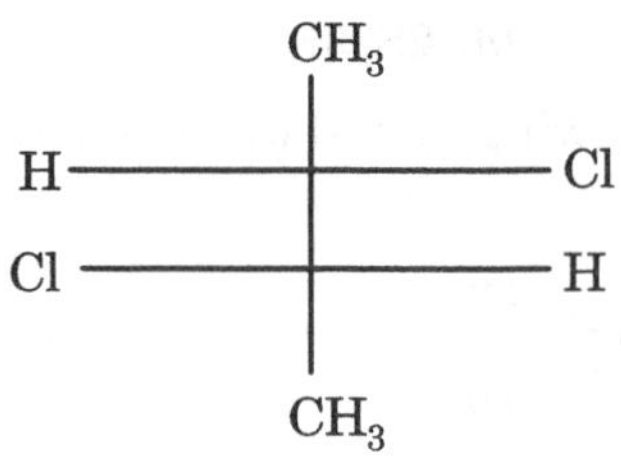

is +36°, then that produced by

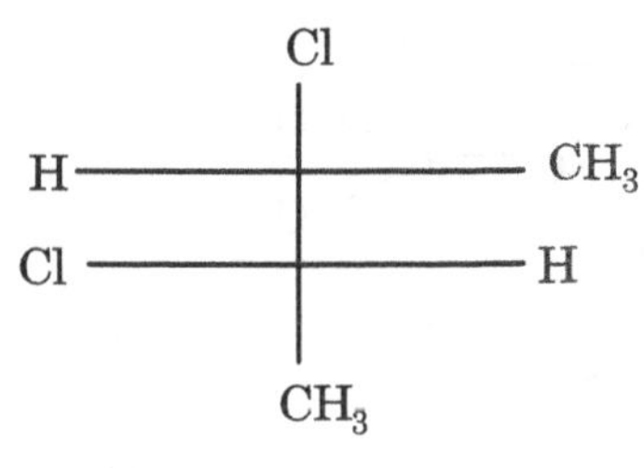

is ,

(a) + 36° (b) - 36°

(c) 0 (d) + 42°

64. A compound shows no optical rotation in a given solvent. Then, which is correct?

(a) It may be a racemic mixture.

(b) It may be a meso compound.

(c) It may not have a chiral centre

(d) All of these are possible

65. 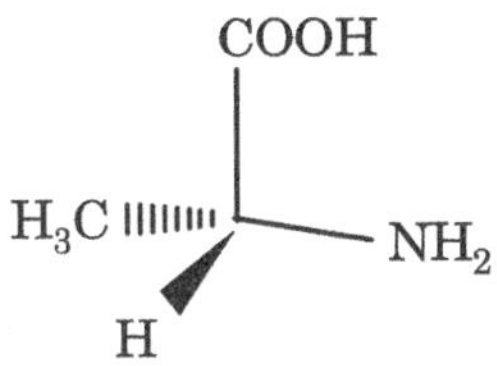

Assign D or L configuration

(a) D

(b) L

(c) Cannot be assigned D or L

(d) Depends on direction of rotation of light

66. 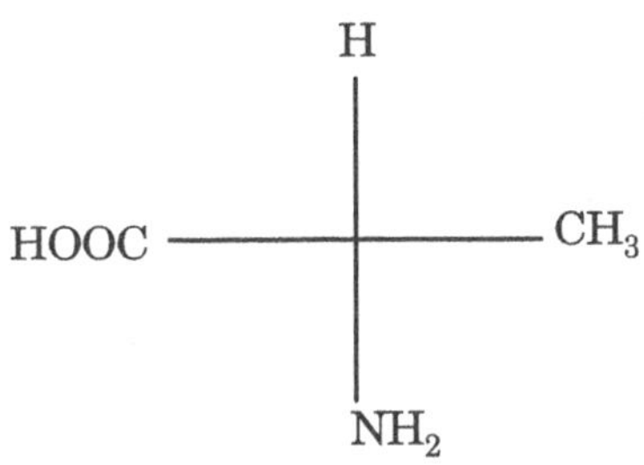

The configuration of above compound is-

(a) S (b) D

(c) L (d) Meso

67. The no. of metamers possible for the molecular formula $C_4H_{10}O$, are-

(a) 3 (b) 2

(c) 4 (d) 1

68. The cyclic structural isomers possible for is/are

(a) 2 (b) 3

(c) 4 (d) 1

69. The compound which will not show geometrical isomerism is/are

(a) Propane (b) But - 1 - ene

(c)  (d) All of these

70. The pair of compounds which are isomers to each other, are-

(a) 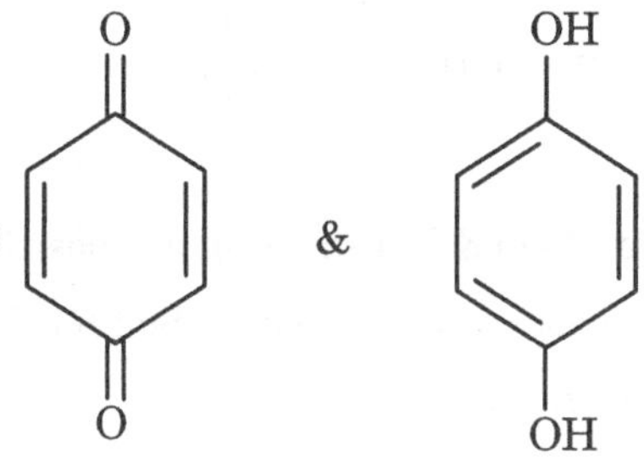

(b) 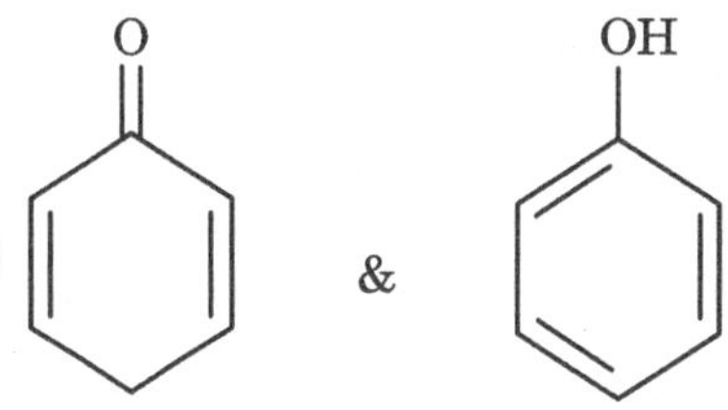

(c) 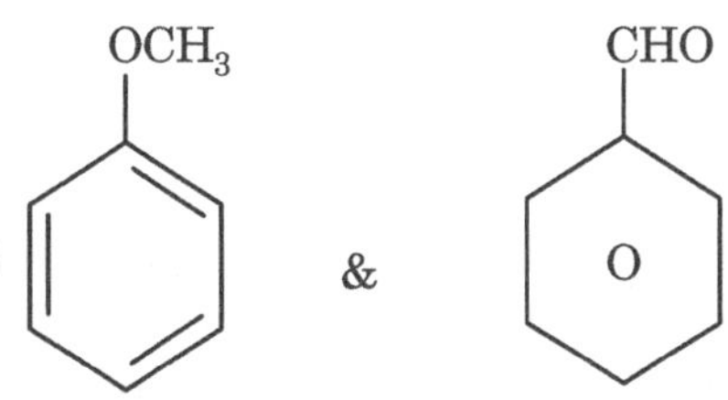

(d) All of these

71. Which of the following is not observed in alkenes?

(a) Chain isomerism

(b) Geometrical isomerism

(c) Metamerism

(d) Position isomerism

72. Consider the following three molecules-

(I) $CH_3-CH_2-CH_2-CH_2-OH$

(II) $CH_3-CH_2-\underset{\displaystyle OH}{\underset{|}{CH}}-CH_3$

(III) $CH_3-\overset{\displaystyle OH}{\overset{|}{\underset{\displaystyle CH_3}{\underset{|}{C}}}}-CH_3$

Among these three, IIIrd is the chain isomer of -

(a) I only

(b) II only

(c) Both I & II

(d) None of these

73. Select the incorrect statement

(a) $CH_2 = CH = CH = CH_2$ & 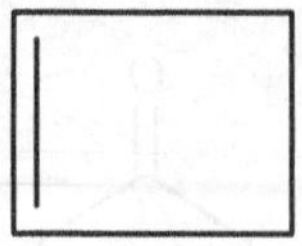 are ring chain isomers.

(b) Butane - 1- ol & butan - 2- ol are position isomers.

(c) Metamerism develops when the function group is divalent.

(d) None of these

74. How many molecules among the following exhibit 'TAUTOMERISM?

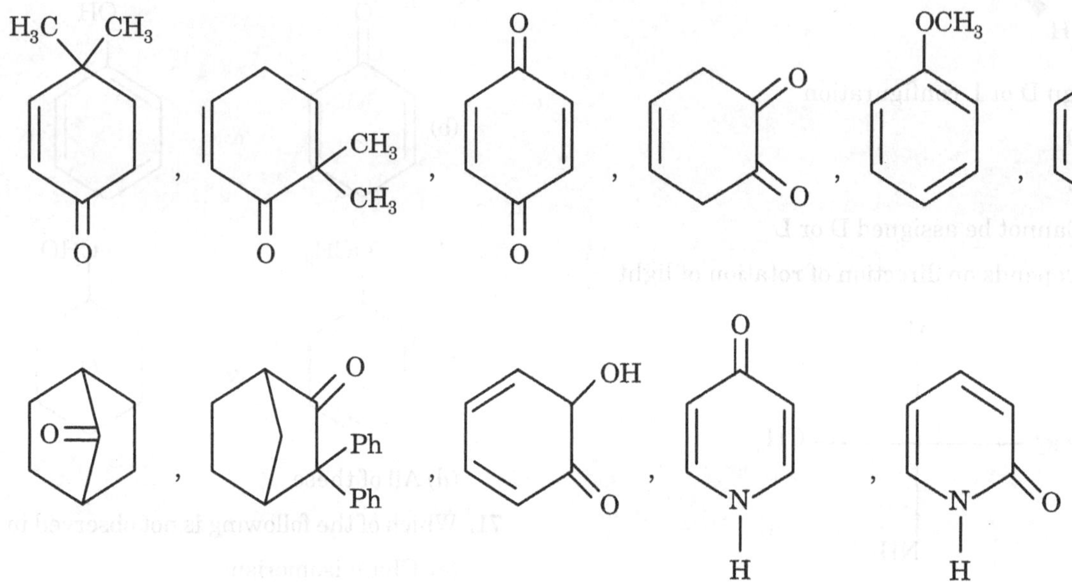

(a) 5 (b) 2 (c) 6 (d) 7

75. The absolute configuration of the following molecule is-

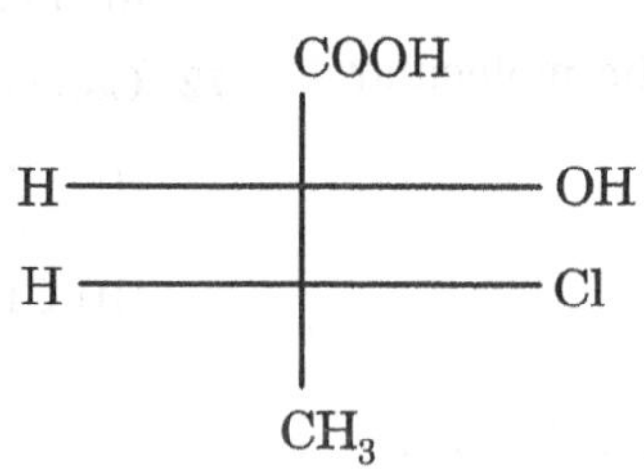

(a) 2R, 3S (b) 2S, 3R (c) 2S, 3S (d) 2R, 3R

Answers

1. (a)	**2.** (c)	**3.** (a)	**4.** (a)	**5.** (c)	**6.** (b)	**7.** (d)	**8.** (c)	**9.** (d)	**10.** (d)
11. (d)	**12.** (c)	**13.** (b)	**14.** (b)	**15.** (d)	**16.** (a)	**17.** (d)	**18.** (d)	**19.** (c)	**20.** (c)
21. (a)	**22.** (d)	**23.** (d)	**24.** (b)	**25.** (b)	**26.** (*)	**27.** (b)	**28.** (d)	**29.** (b)	**30.** (a)
31. (b)	**32.** (d)	**33.** (b)	**34.** (d)	**35.** (d)	**36.** (a)	**37.** (d)	**38.** (b)	**39.** (d)	**40.** (a)
41. (d)	**42.** (b)	**43.** (d)	**44.** (a)	**45.** (c)	**46.** (b)	**47.** (b)	**48.** (a)	**49.** (c)	**50.** (d)
51. (b)	**52.** (c)	**53.** (d)	**54.** (d)	**55.** (a)	**56.** (d)	**57.** (d)	**58.** (a)	**59.** (a)	**60.** (d)
61. (c)	**62.** (d)	**63.** (c)	**64.** (d)	**65.** (a)	**66.** (b)	**67.** (a)	**68.** (b)	**69.** (d)	**70.** (b)
71. (c)	**72.** (c)	**73.** (d)	**74.** (d)	**75.** (b)					

Explanations

2. (c) Option 3rd are functional isomers because there molecular formula is same but functional groups are different.

3. (a) $CH_3-NH-CH_2-CH_2-CH_2-CH_3$, $CH_3-NH-CH(CH_3)-CH_2-CH_3$,

$CH_3-NH-CH_2-CH(CH_3)-CH_3$, $CH_3-NH-C(CH_3)_2-CH_3$;

$CH_3-CH_2-NH-CH_2CH_2CH_3$ & $CH_3-CH_2-NH-CH(CH_3)-CH_3$

4. (a)

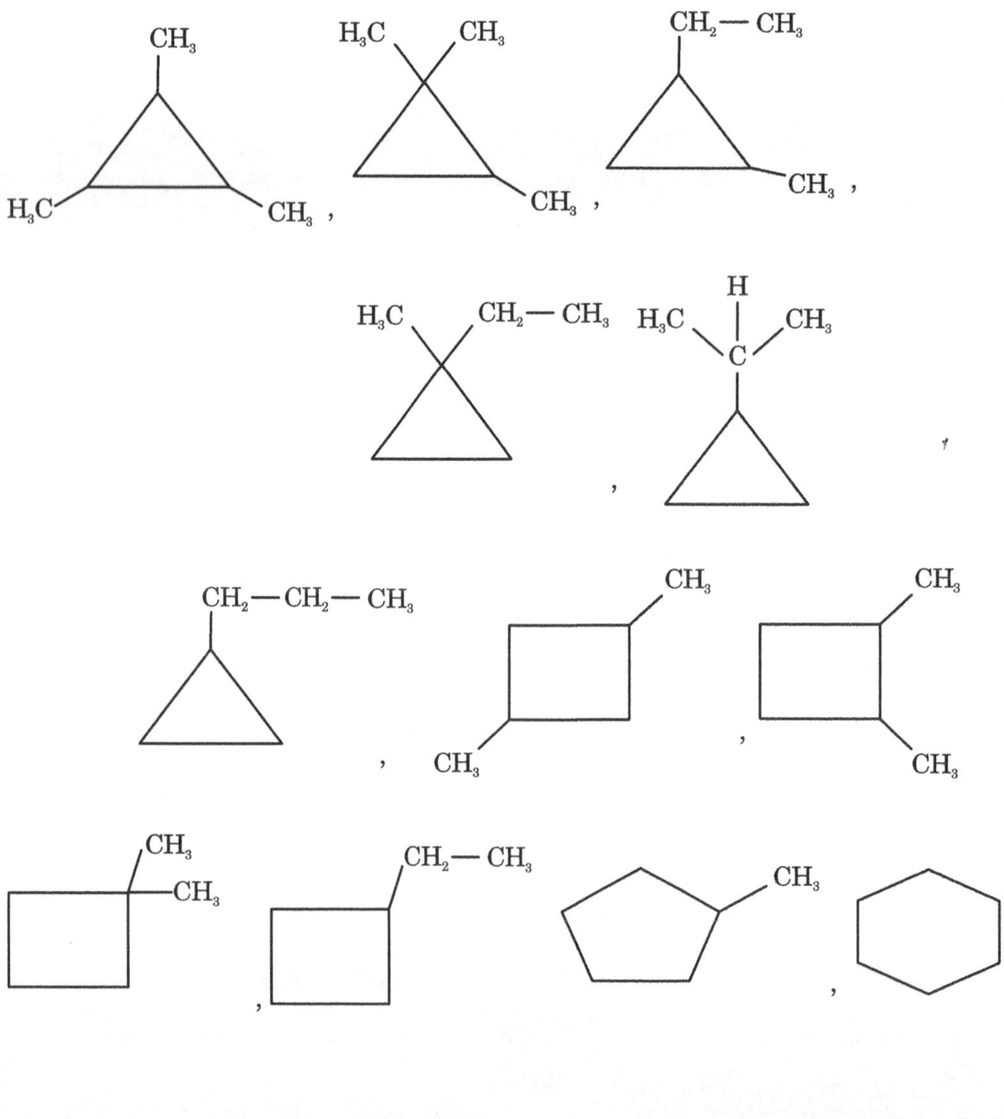

5. (c) $CH_3-CH_2-NHCHO$

$$CH_3-CH_2-\underset{\underset{H}{|}}{N}-\overset{\overset{O}{||}}{C}-H$$

It is a substituted amide while the structure

$CH_3-\underset{\underset{NH_2}{|}}{CH}-CHO$ is an amino aldehyde.

6. (b)

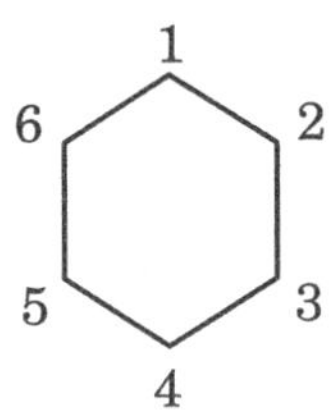

Cyclohexane

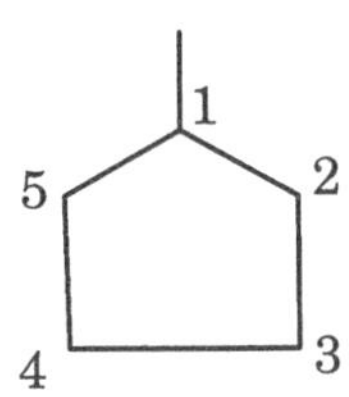

Methyl Cyclopanaten

7. (d) Metamers belong to same functional groups.
Option 3rd are functional isomers.

8. (c) $CH_3-CH_2-CH_2-NH_2$ &
(Propan - 1 - amine)

$CH_3-\underset{\underset{NH_2}{|}}{CH}-CH_3$ are position isomers.

(propan - 2 - amine)

9. (d) Option '4' has maximum acidic $\propto -$H's i.e. maximum active methylene group.

10. (d)

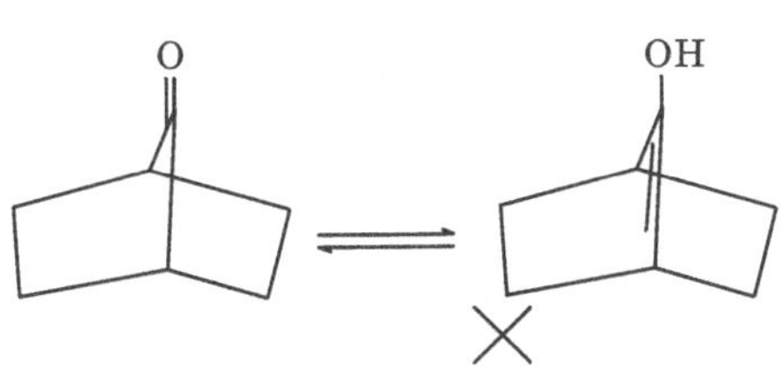

(It violates BREDT'S Rule)

11. (d) $C_6H_5-\overset{\overset{O}{||}}{C}-\underset{\underset{H}{|}}{\overset{\overset{H}{|}}{C}}-CH_3$

This molecule has $\propto -$H.

13. (b)

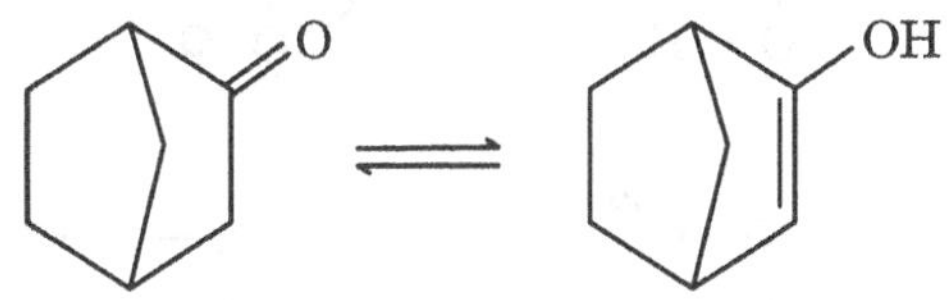

(c)

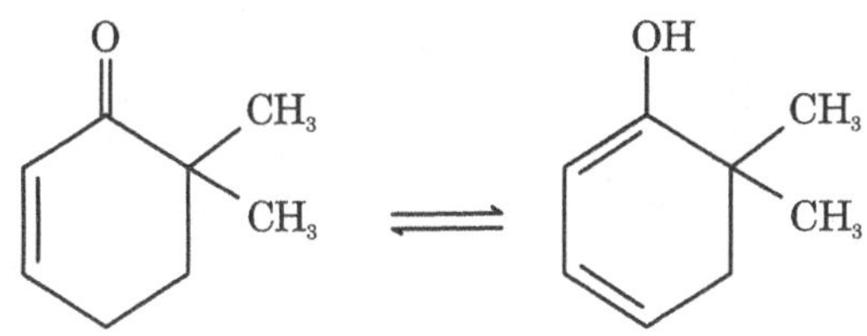

(d)

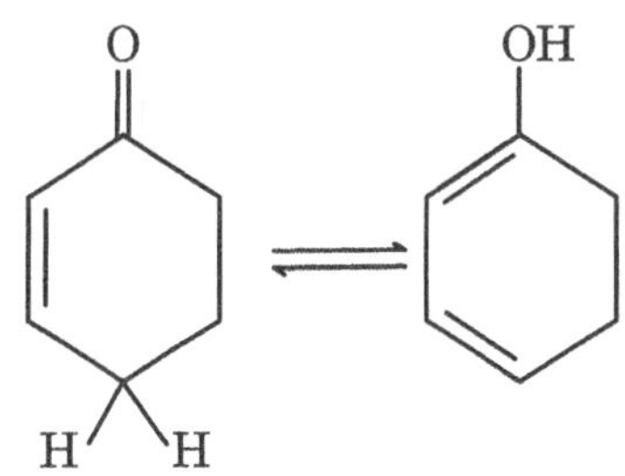

or

↑
enolizable H's

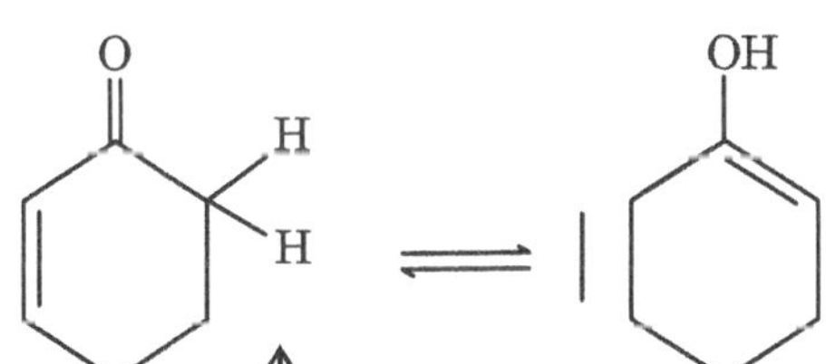

↑
enolizable H's

15. (d) (a) $H_2N-\overset{\overset{O}{||}}{C}-NH_2 \rightleftharpoons HN=\overset{\overset{HO}{|}}{C}-NH_2$

(b) $H_2N-\overset{\overset{S}{||}}{C}-NH_2 \rightleftharpoons HN=\overset{\overset{SH}{|}}{C}-NH_2$

(c)

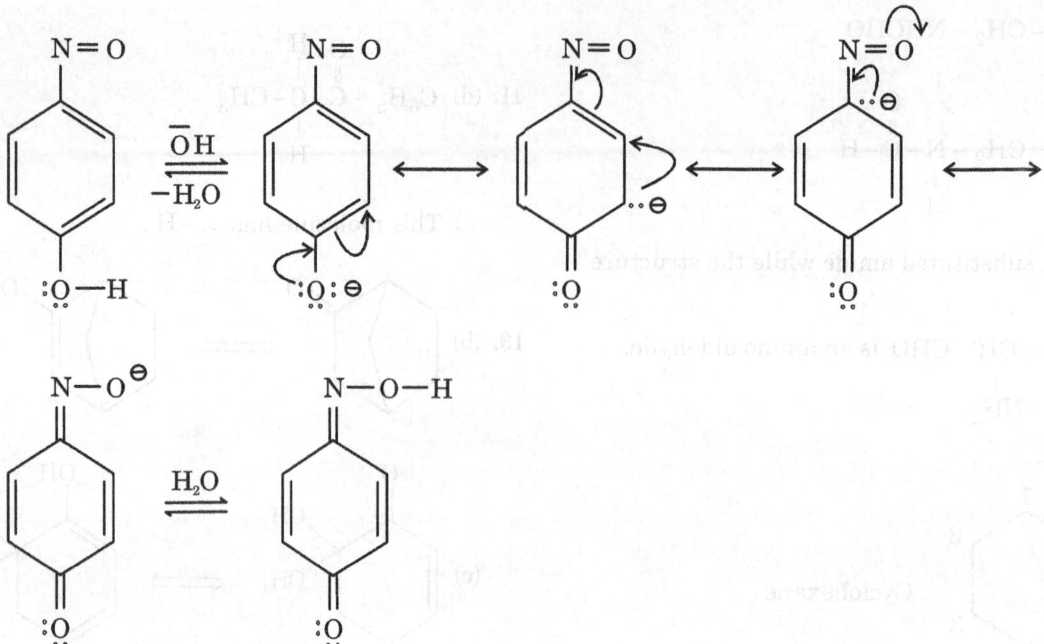

16. (a)

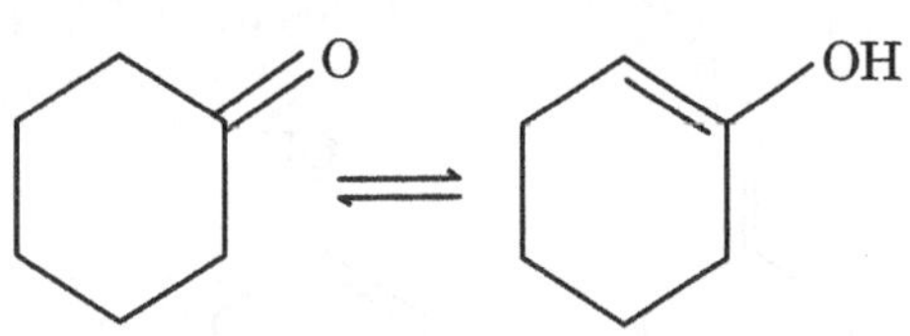

(enol form where -OH os bonded to doubly bonded C)

Or

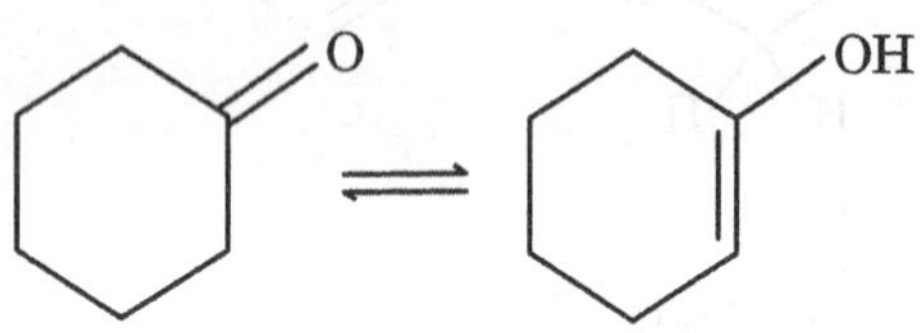

(enol form where -OH os bonded to doubly bonded C)

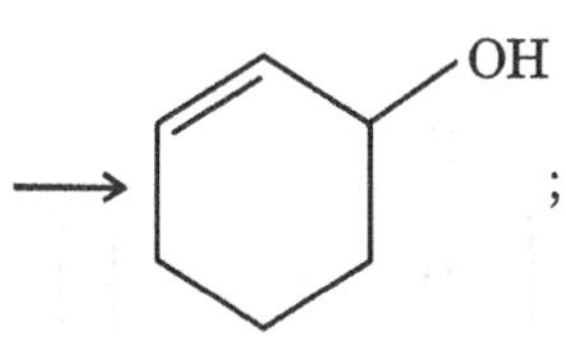

; it is not enol form.

(I)

17. (d) (a)

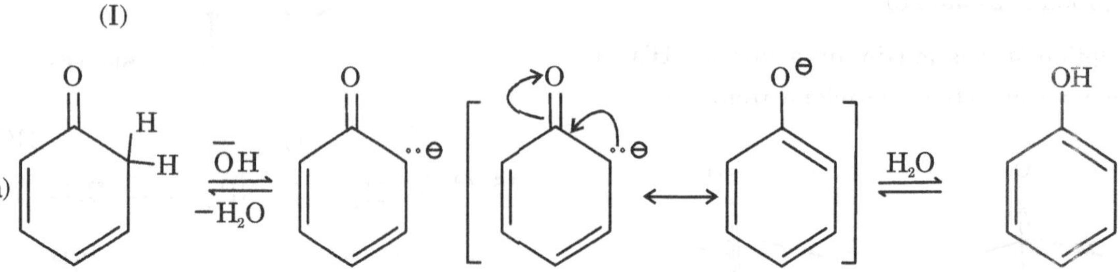

(b)

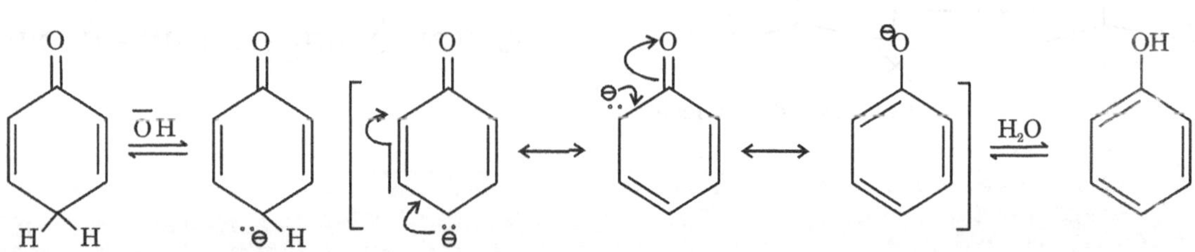

18. (d) (a)

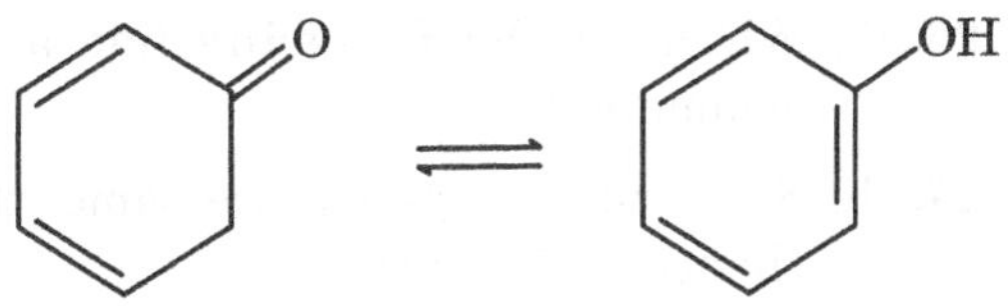

(b)

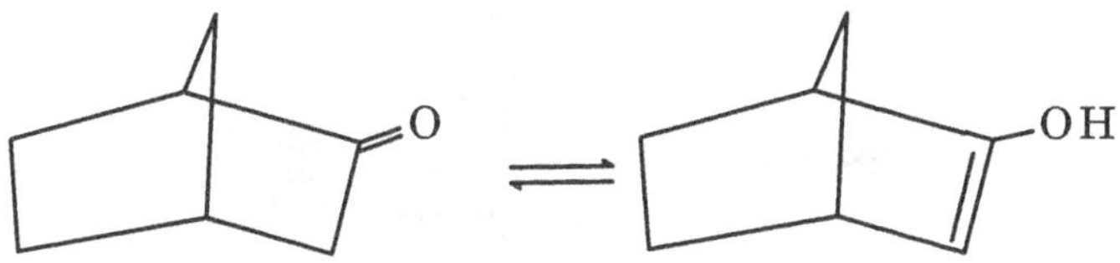

(c)

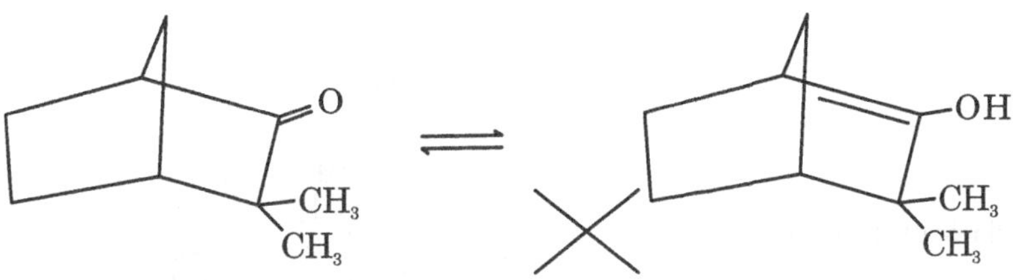

(BREDT'S RULE IS VIOLATED)

19. (c) Enol form of II will be aromatic

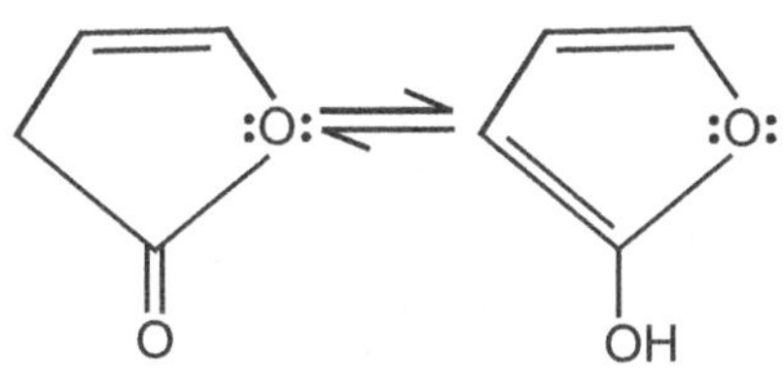

(Aromatic)

Among I and III, I is ketone having more acidic H's for enolization. So, enol content of I > III.

20. (c)

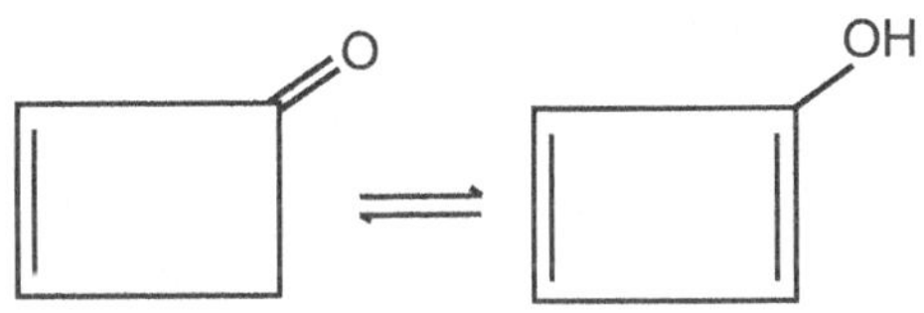

(ANTI -AROMATIC)

21. (a) (a)

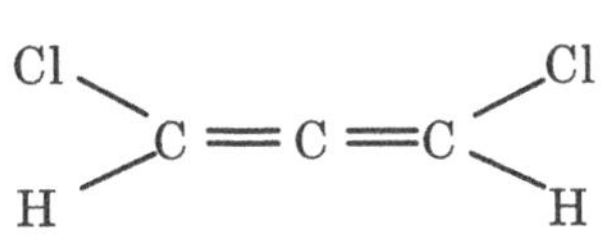

1, 3 - dichloropropa - 1, 2 - diene

(It is an alkene& the groups on terminal C's will not be planar, hence no symmetry)

(b)

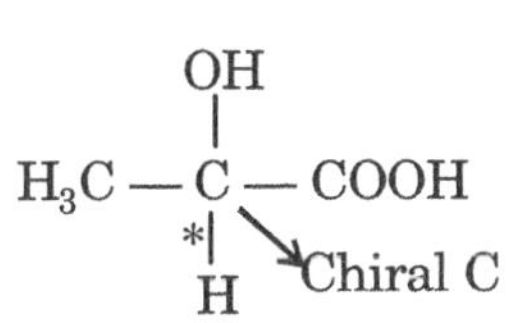

(c)

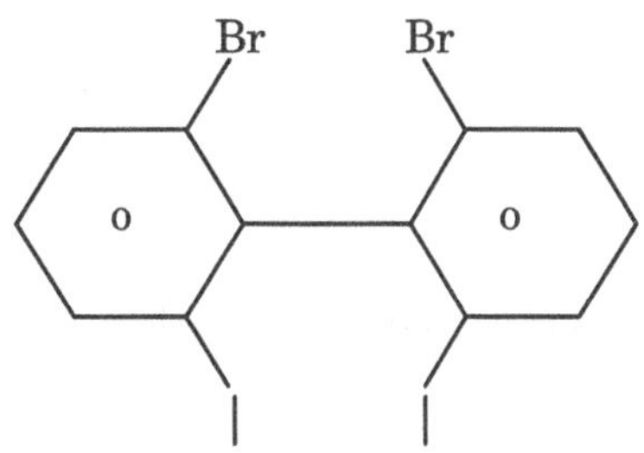

Ortho-tetrasubstituted biphenyl will loose planarity due to repulsions, hence no symmetry.

22. (d)

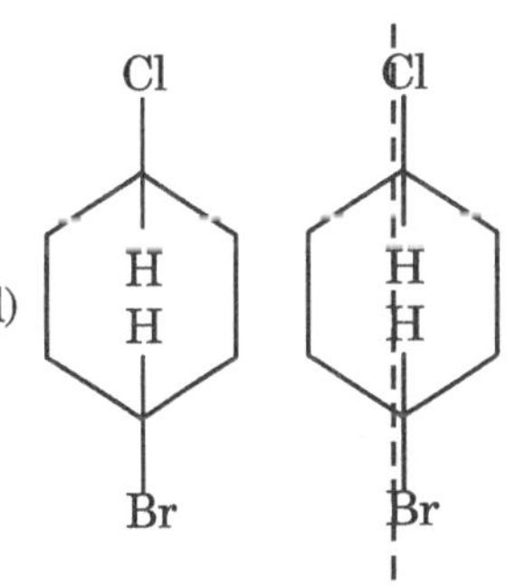

23. (d) Option 1 has a chiral C marked '*'

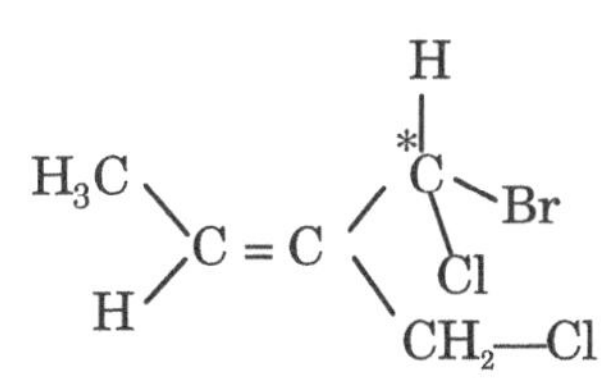

Optically active

Option 2 is an allene with even no. of double bonds and different gps. attached at terminal C 's, so will show optical activity.

In option 3, $-SO_3OH$ gps. Are bulky & hence will cause rotation in biphenyls. So, rings loose plamarity and hence, the molecule is chiral.

In option 4, the left terminal C of allene (double bonded) has 2 methyl gps. Attached to it. Hence, optically inactive.

24. (b) 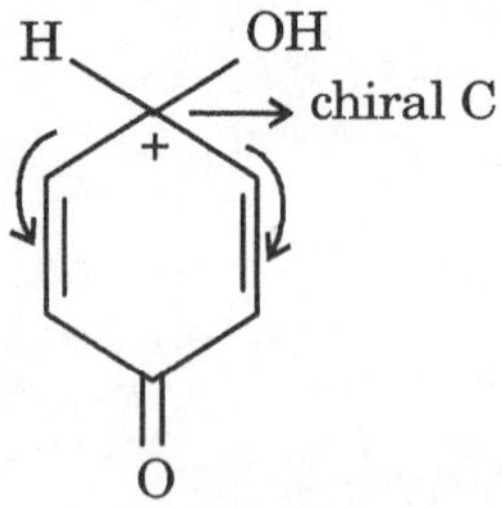

The 'C' marked '*' has –H, –OH & on moving

left there is –CH 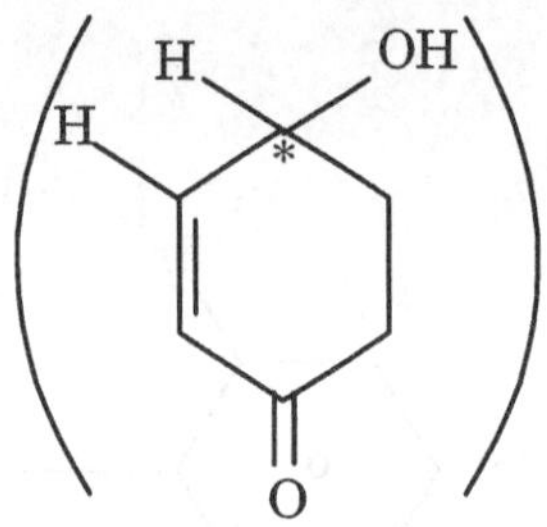 &

on moving right, $-CH_2$ is there 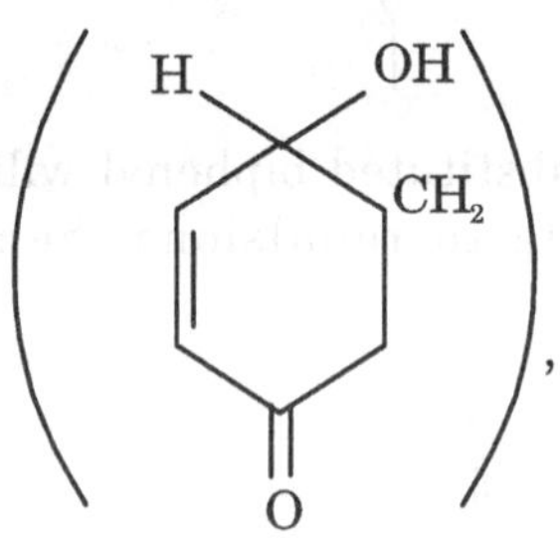, So, all the 4 groups are different. Hence, C is chiral.

25. (b) 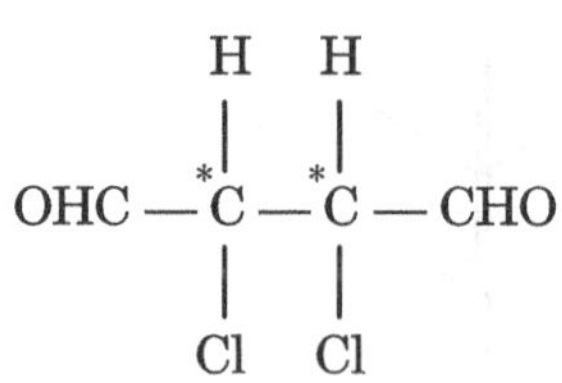

26. (*) Which of the following has a centre of symmetry?

27. (b) Since no Two groups are same, there is no chance of any POS.

28. (d) 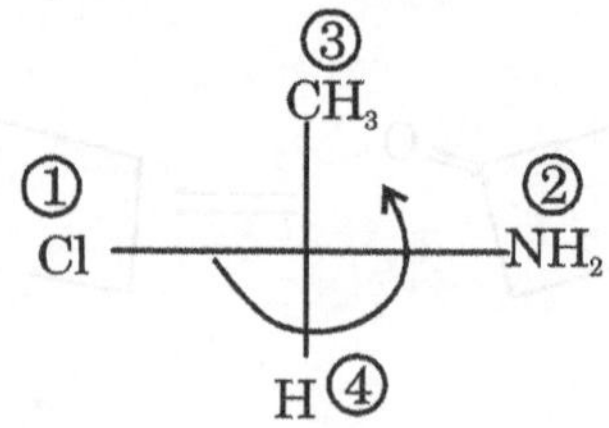

(I)

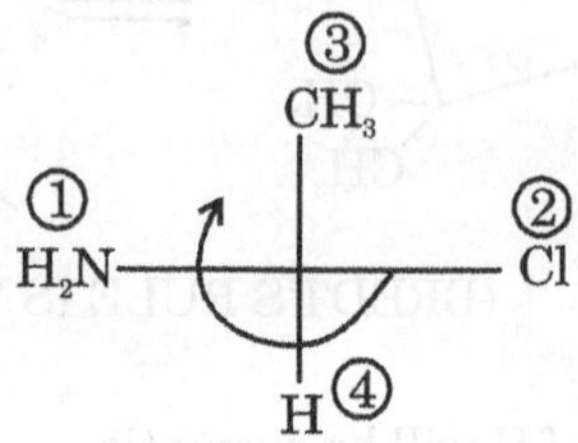

(II)

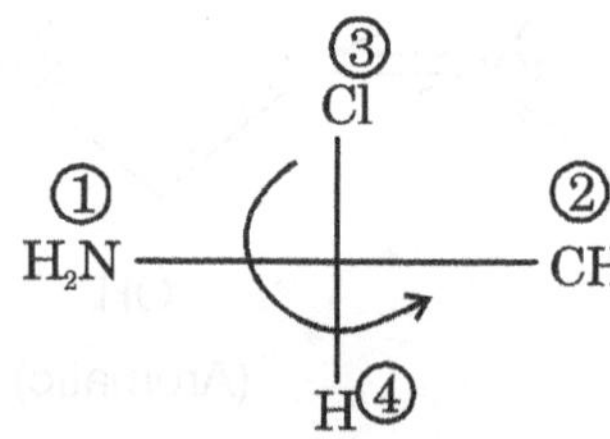

(III)

'S' - Configuration 'R' - Configuration

'S' - Configuration

(Priorities set on basis of CIP Rules)

So, I & II are enantiomers & I & III are identical. Also, II & III are enantiomers.

29. (b) Molecule 1 has centre of symmetry while molecule (C) has plane of symmetry.

30. (a) 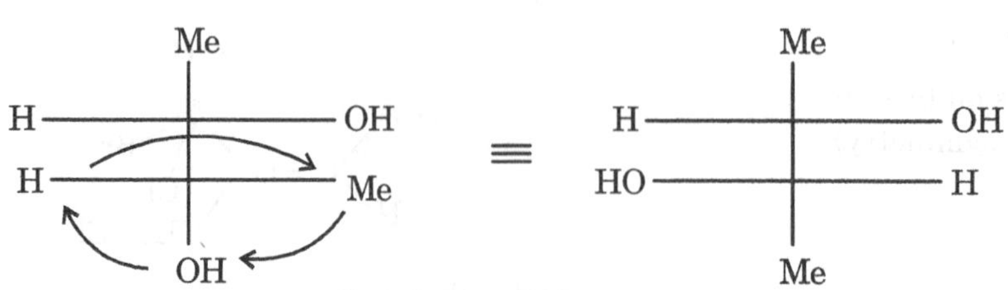

It is not meso.

31. (b) (a) 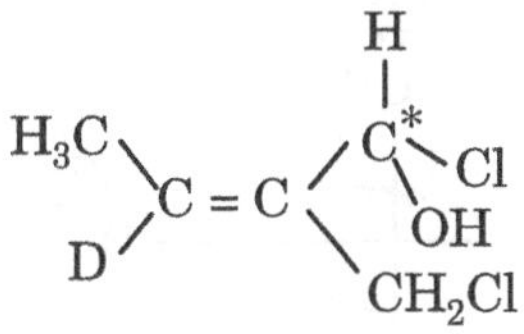

Chiral C, optically active

(b) The central ring will lie in plane while the rest will rotate. POS can be spotted.

(c) 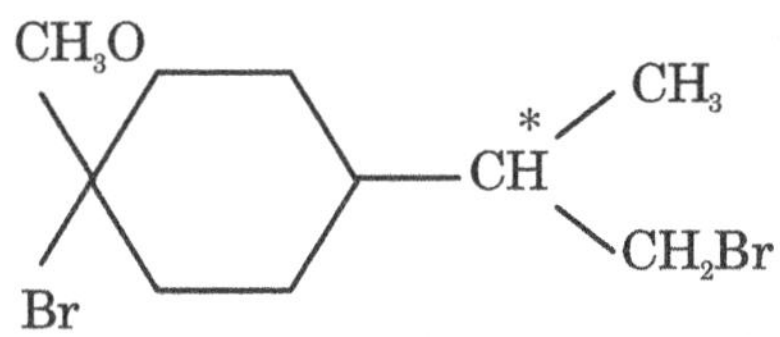

Chiral, optically active

32. (d) Objects (or compounds) which have plane of symmetry, will have their mirror images superimposable on them.

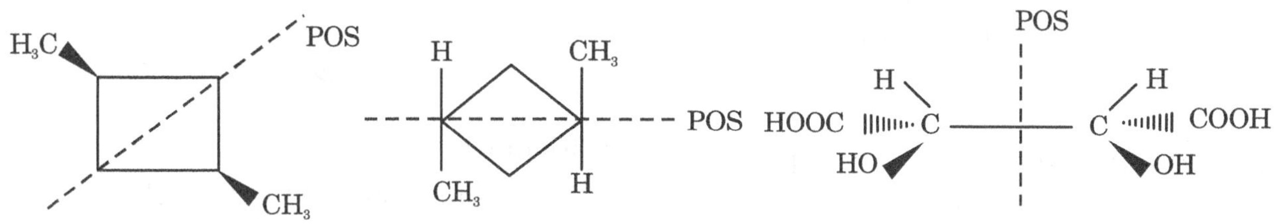

33. (b) Replacing H_Y with T & H_X with D

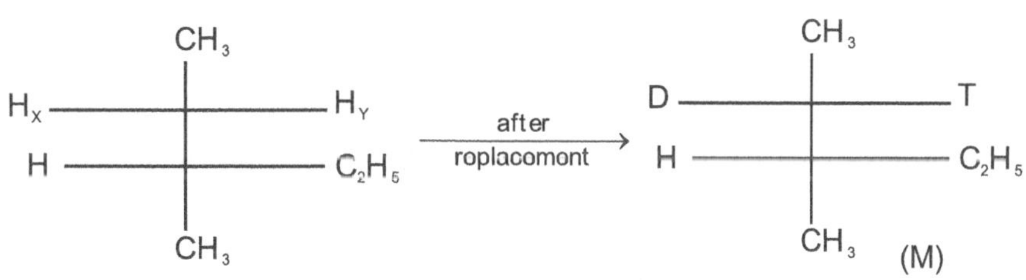

Now,

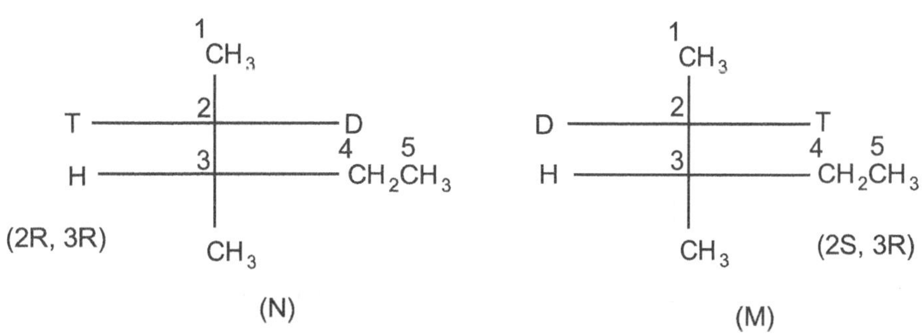

In N
At C-2

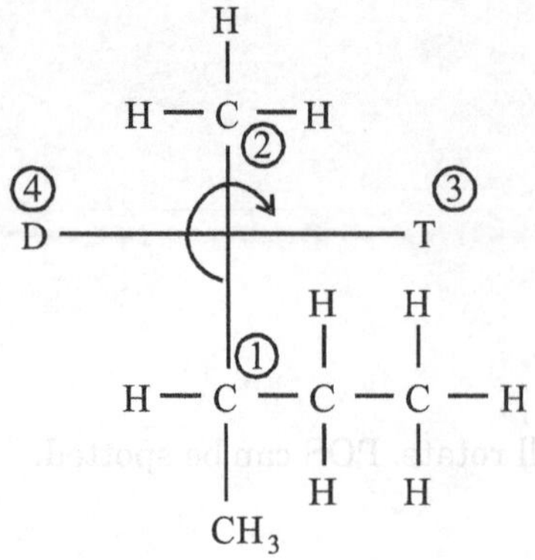

In M
At C-2

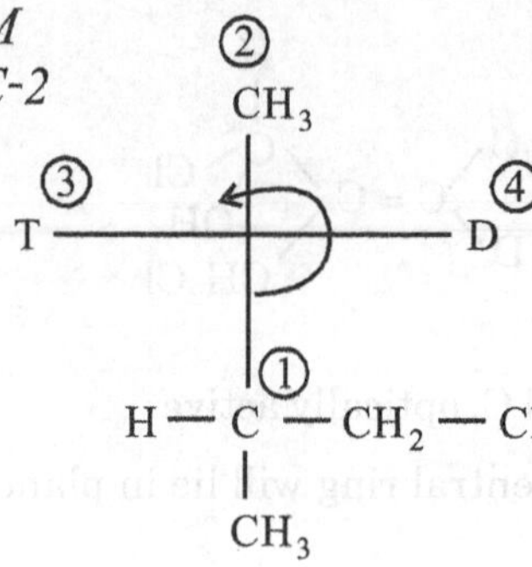

At C-3

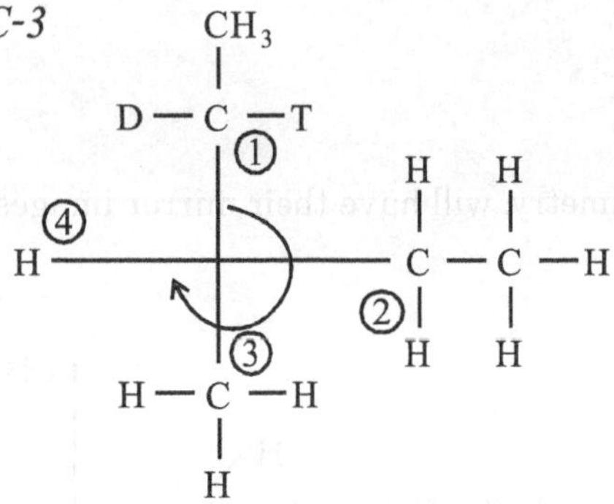

At C-3

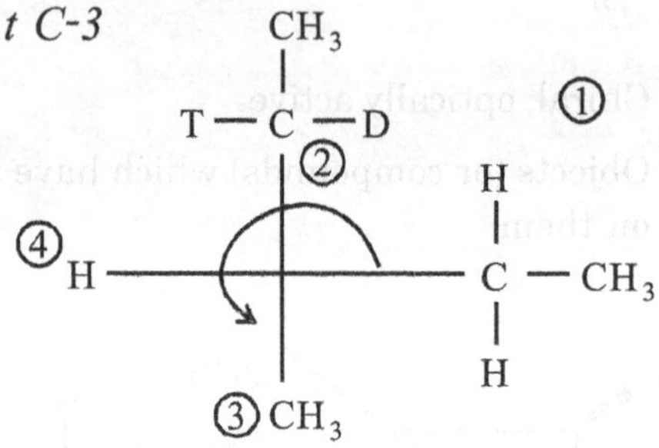

34. (d)

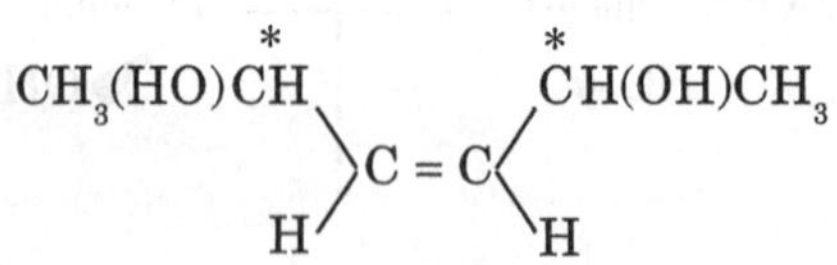

- 2 chiral C's
- No. of enantiomers $= 2^{n-1} = 2^{2-1} = 2^1 = 2$
- No. of meso forms $= 2^{\frac{n}{2}-1} = 2^{\frac{2}{2}-1} = 2^0 = 1$

35. (d)

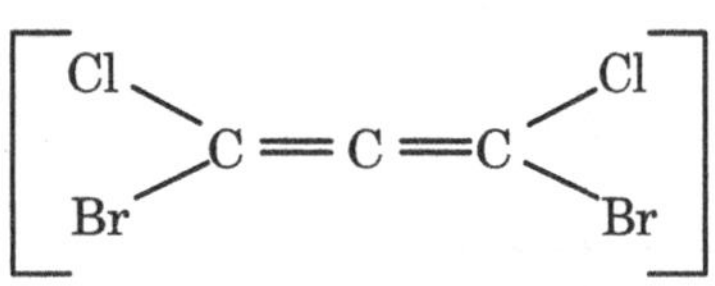

The terminal 4 groups i.e. Cl & Br are not in same plane. So, can't show geometrical isomerism.

In molecule, 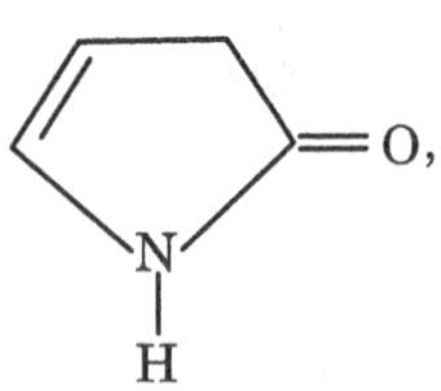 the enol content is higher because its enol form is aromatic.

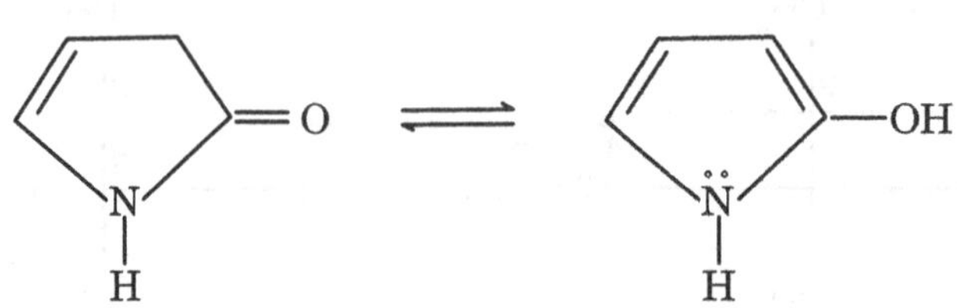

(AROMATIC)

⇒ Propene cannot show geometrical isomerism because the two groups (i.e. H's) attached to doubly bonded C are same.

$CH_3 - CH = CH_2$

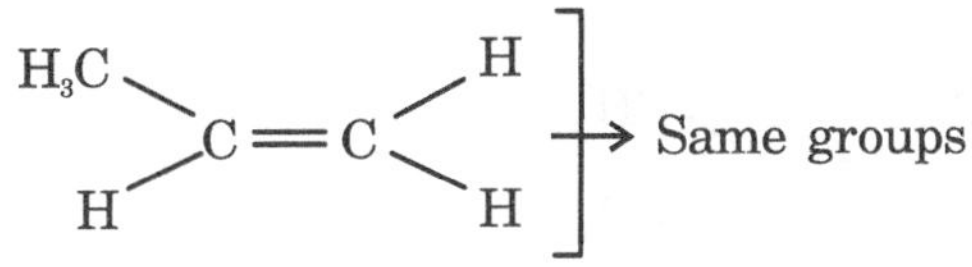

36. (a)

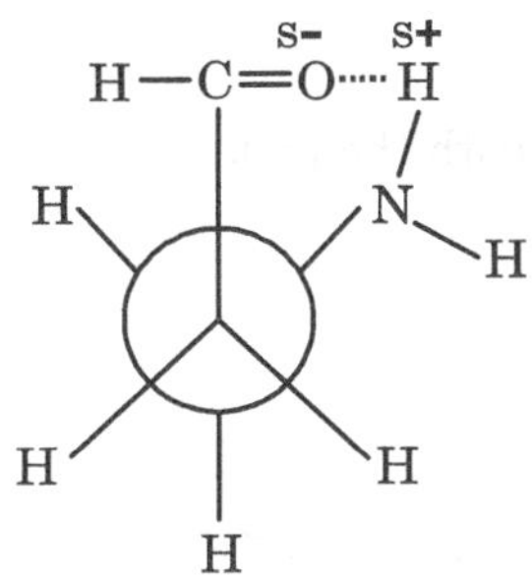

Max. stable conformation due to intra molecular H - BONDING.

37. (d) So.2nd molecule is an addoxime.

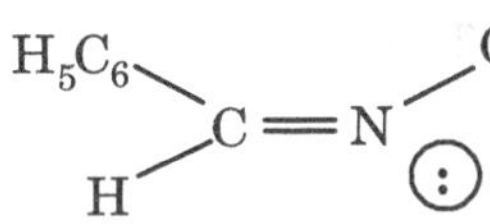

ANTI - FORM

(restricted rotation about C = N double bond and two different substituents on each C & N)

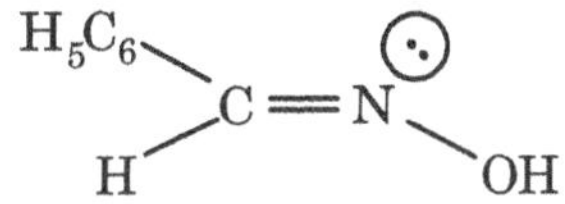

SYN - FORM

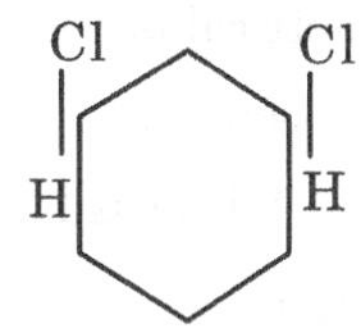

Cis - 1, 2 - dichloro cyclohexane

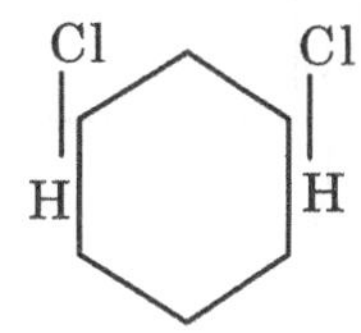

Trans - 1, 2 - dichloro cyclohexane

38. (b) (Min. steric strain)

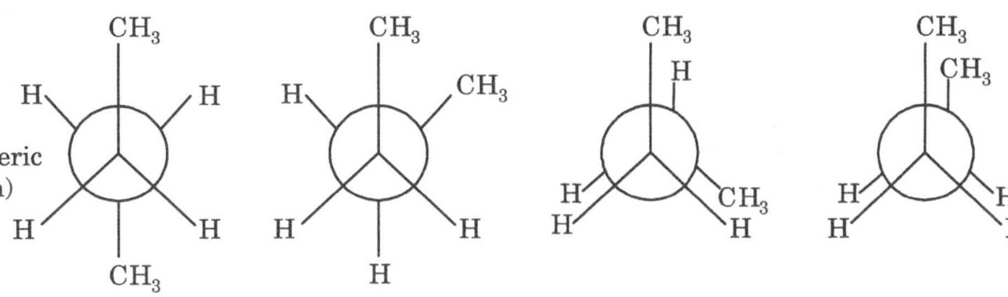

(Max. steric strain)

(c)	(b)	(a)	(d)
Anti form (Dihedral angle b/w methyl gps. = 180°)	Gauche form (Dihedral angle b/w methyl gps. = 60°)	Partially eclipsed (Dihedral angle b/w methyl gps., θ = 120°)	Fully eclipsed (Dihedral angle b/w methyl gps., θ = 0°)

39. (d)

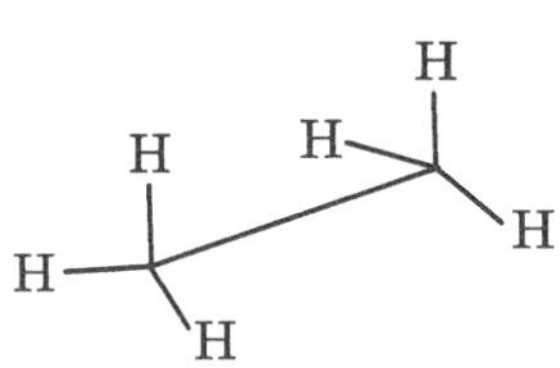

Eclipsed form

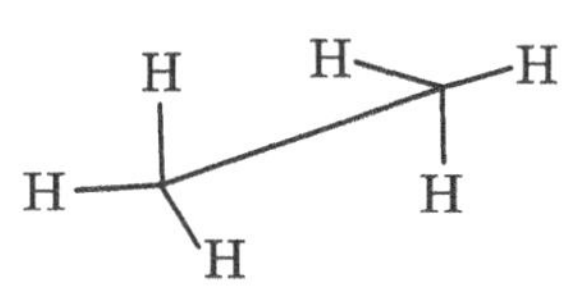

Staggered form

Staggered form has least torsional strain

40. (a) In various conformations, dihedral angle differ but bond angles and bond lengths will remain same.

41. (d)

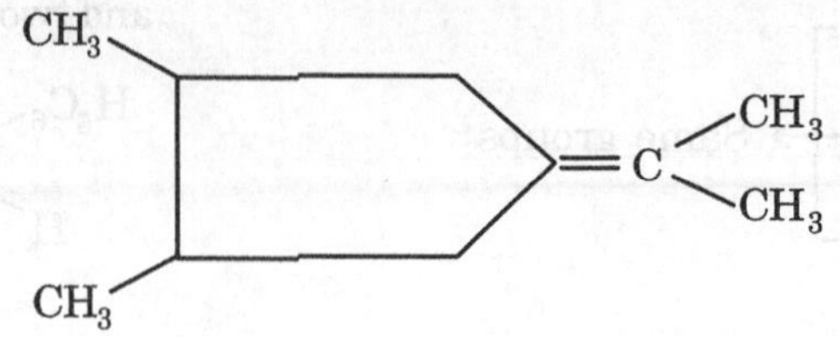

here, geometrical isomerism is not possible across double bond because both the gps. ($-CH_3$) are same.

However, G. I. is possible wrt Ring.

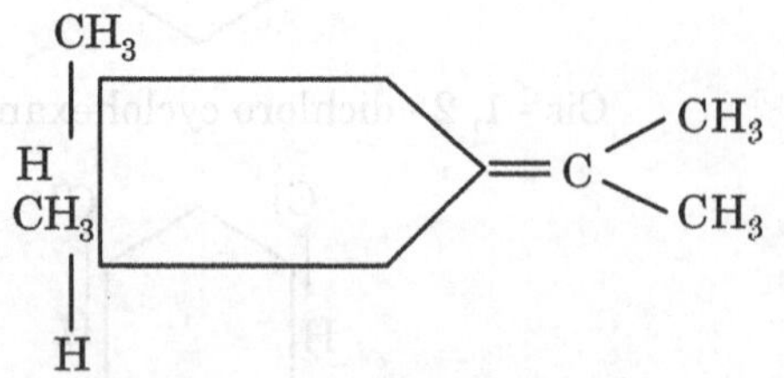

(Cis)

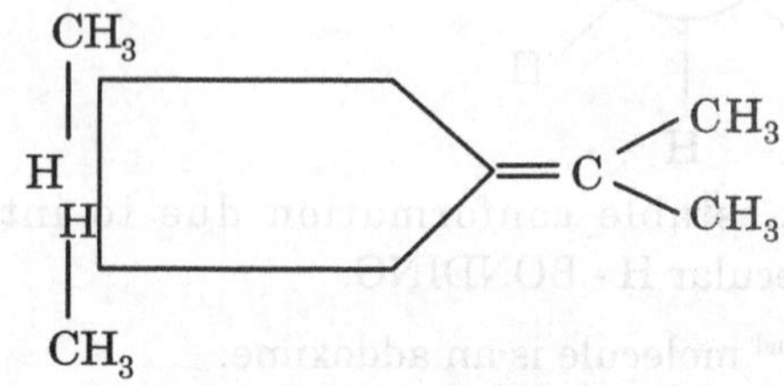

(Trans)

42. (b) Ethylene glycol is CH_2-OH | CH_2-OH

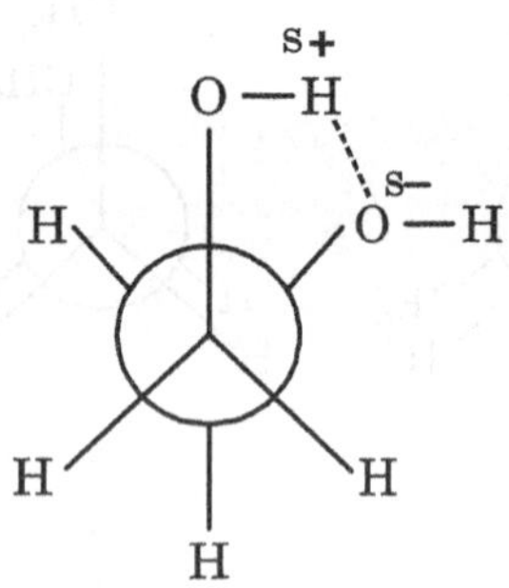

Gauche form more stable because of intra molecular H—bonding.

43. (d) In option d,

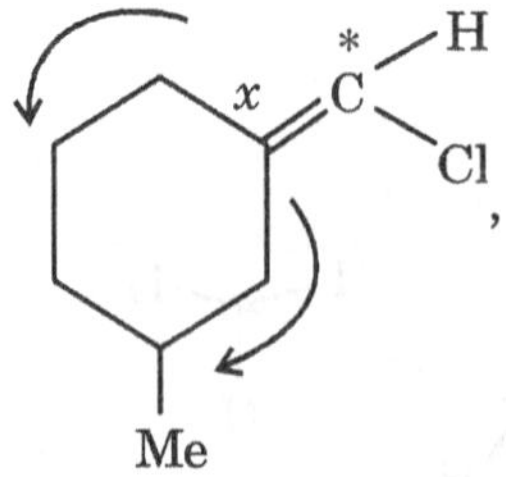

, doubly bonded C (*) has 2 different groups i. e. H & Cl.

Also, the doubly bonded C in the ring (marked x) has different groups if we trace across the ring.

44. (a) In option 1,

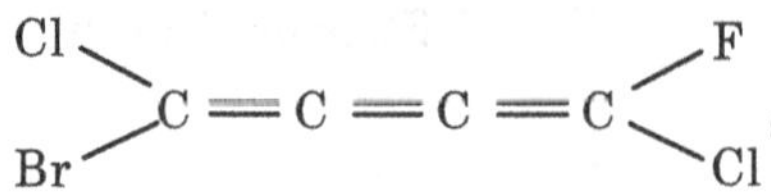

, molecule is planar & all the 4 extreme groups all in same plane.

Also, the groups bonded to doubly bonded C's are different.

45. (c) (a) 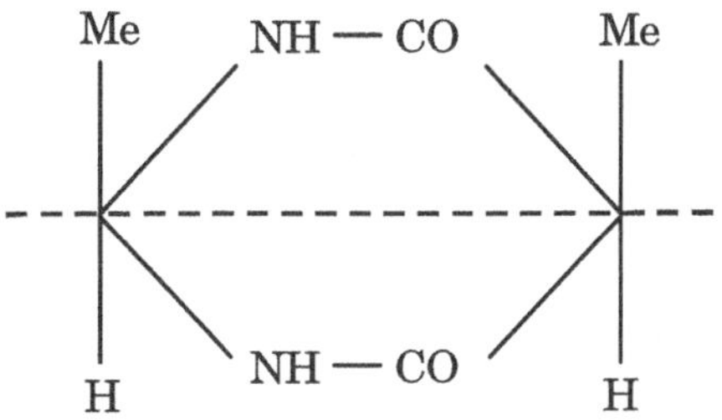

POS

(b) 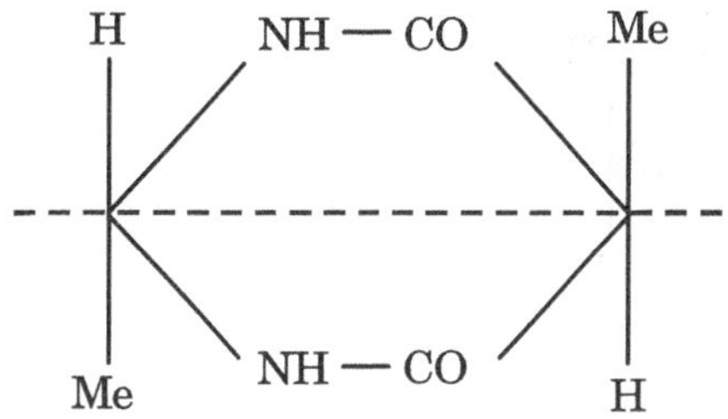

POS

46. (b)

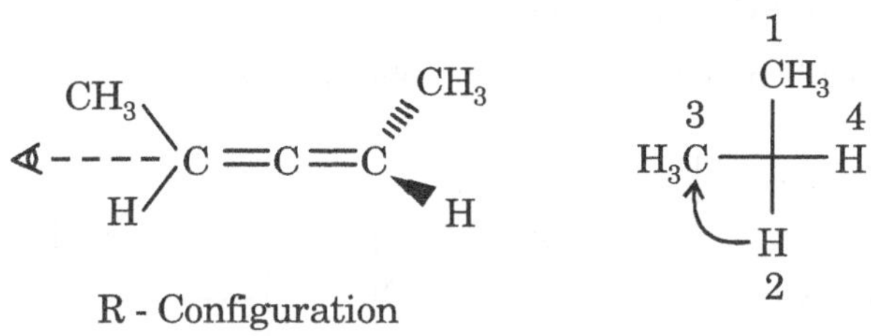

R - Configuration

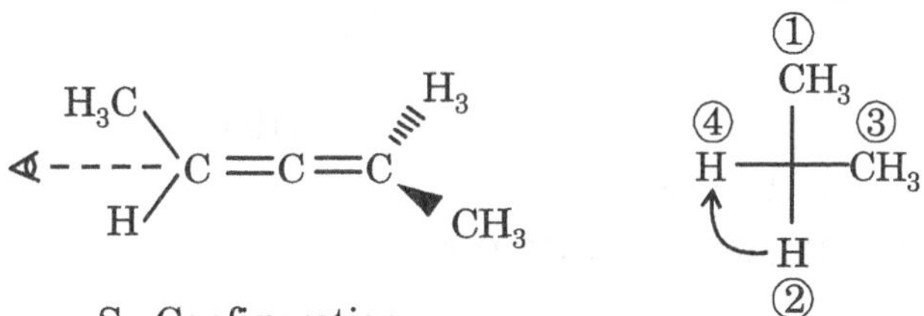

S - Configuration

Here, the R & S configurations are determined by precedences of the groups attached to the axial section when viewed along that axis.

→ Front plane is given priority over other.

→ Final structure is given priority from 2 to 3.

47. (b) Performs rotations & check for plane of symmetry

(P).

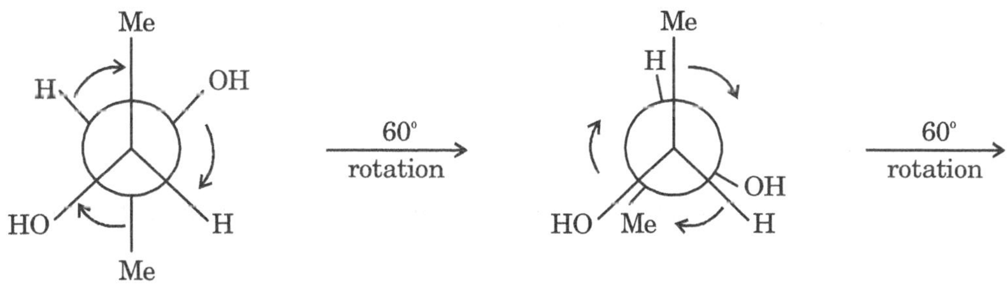

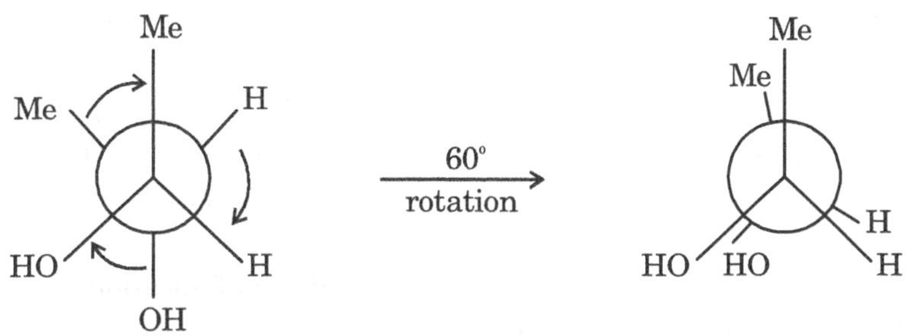

It will have plane of symmetry Similarly for R.

48. (a) 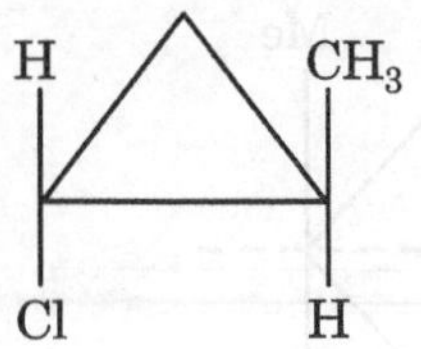

Trans - 1 - Chloro - 2 - methyl cyclopropane 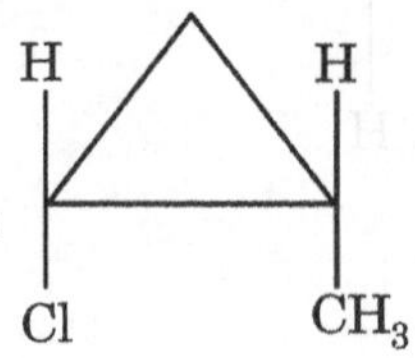

Cis - 1 - Chloro - 2 - methyl cyclopropane

Since, both are optically active, enantiomers obtained = 4

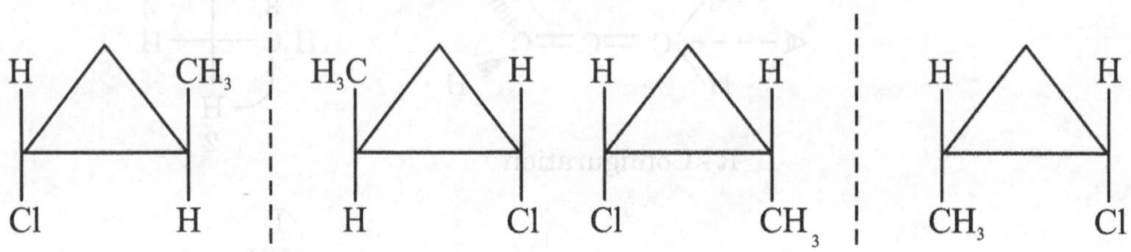

(I) (II) (III) (IV)

Enantiomers : - I & II, III & IV

Diastereomers : - I & III, I & IV, II & III, II & I

Pair of diastereomers = 4

Pairs of enantiomers = 2

Ratio = 2 : 1

49. (c) For wedge & dash, if lowest priority group is on dash, directly look for $\frac{R}{S}$ configuration.

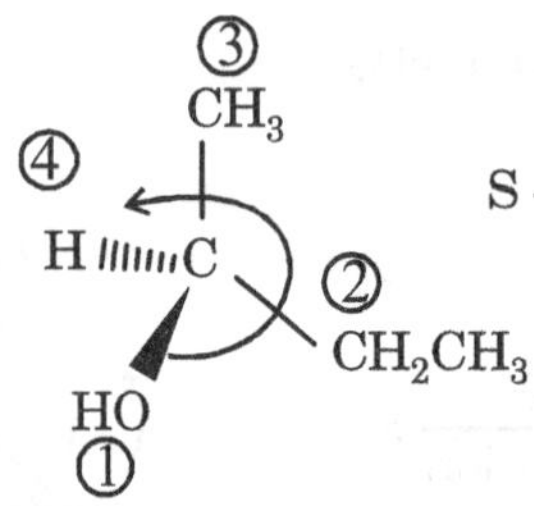

S - Configuration

→ (For wedge & dash), if lowest priority group is not on dash, manipulate & bring it on dash, find the configuration & reverse it.

For structure II -

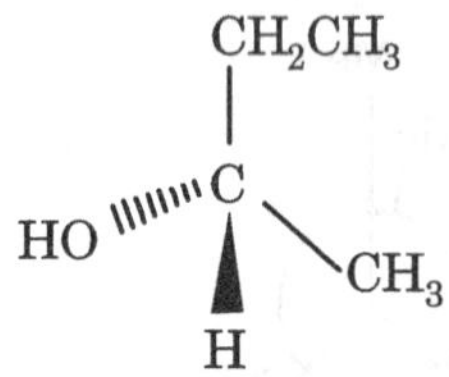 After manipulation, 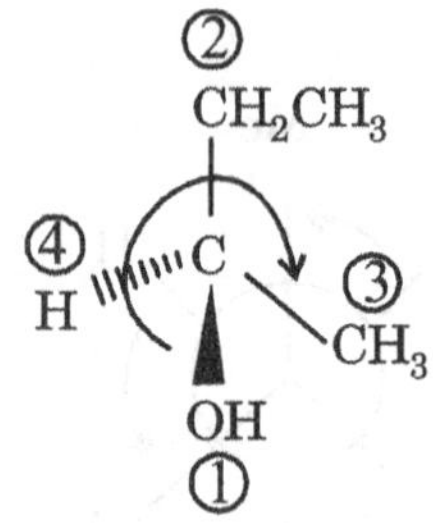

R - Configuration

It means that actual molecule has S - configuration.

So, I & II are identical molecules.

50. (d) 1.

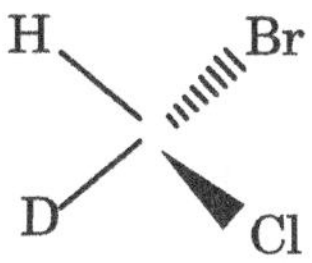

Check $\frac{R}{S}$ here. Because the lowest priority group is not on dash (| | | | | | | |), manipulate the molecule & bring it on dash.

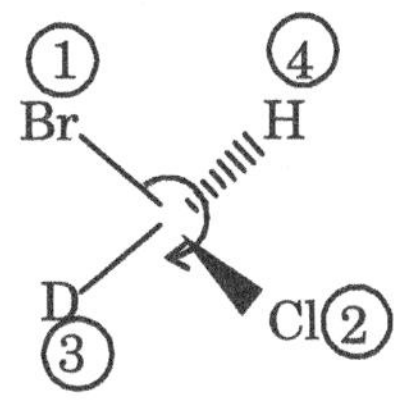

R - configuration

The original molecule has S-configuration.

2.

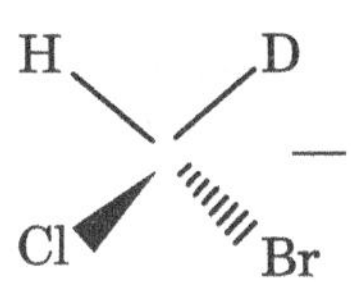

— (II)

Manipulate the groups & bring the lowest priority group on dash.

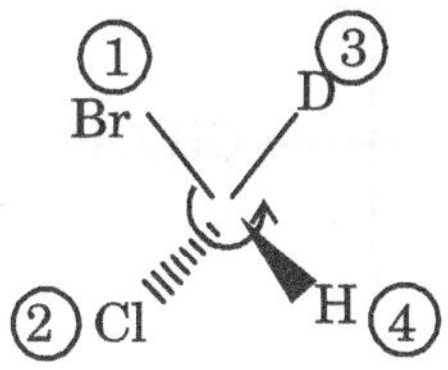

S - configuration

So, the original molecule has R-configuration

3.

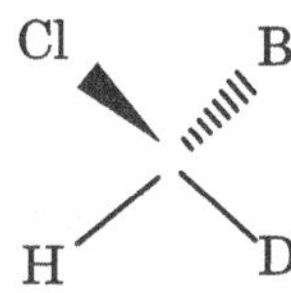

Since the lowest priority group is not on dash (| | | | | | | |), manipulate & bring it on dash.

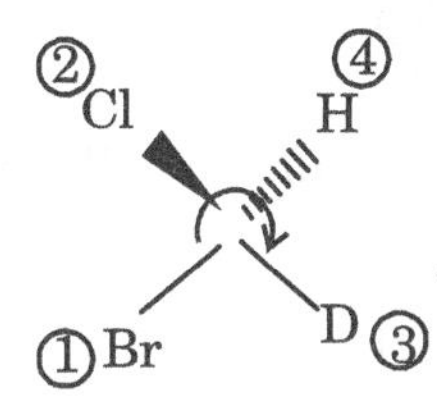

this molecule has R-configuration, so original molecule will have S-configuration.

51. (b)

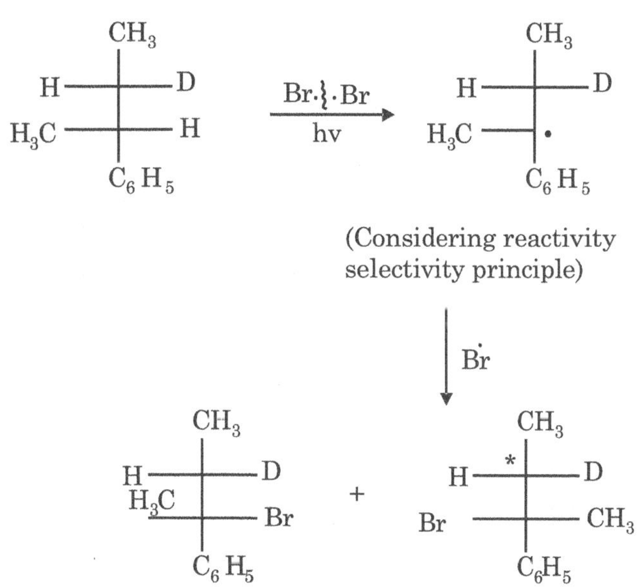

A single chiral C marked '*' already exists. In addition to that, another chiral centre will get created & we'll get a pair of diasteromers.

52. (c) (a)

(Chirality lost)

(b) 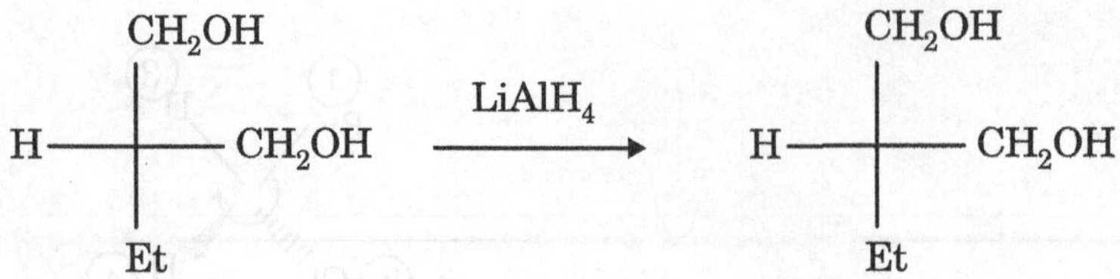

(Chirality lost)

(c) 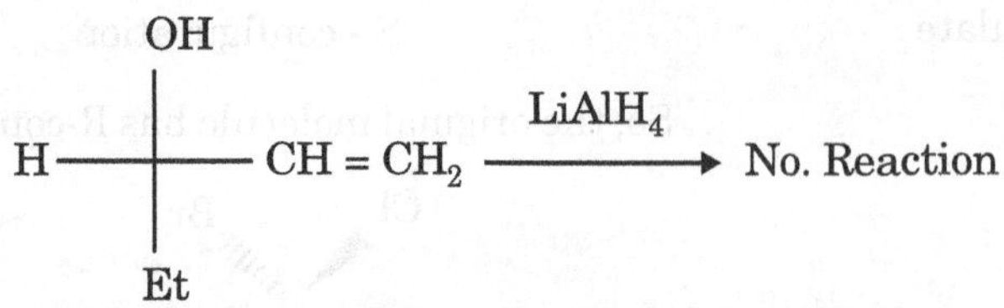

Chirality is retained

53. (d) 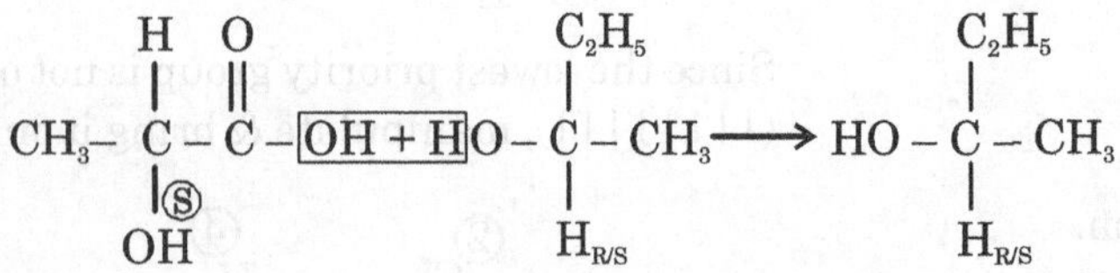

Products obtained will have configuration S, R & S, S (which are diastereomers)

54. (d) 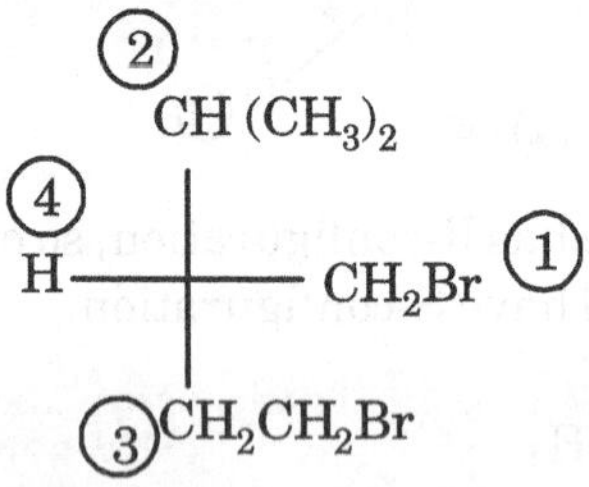

55. (a) At C – 2

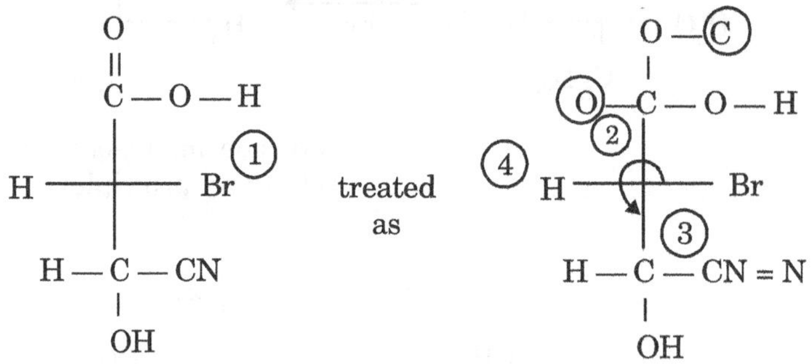

R' at C – 2

At C–3

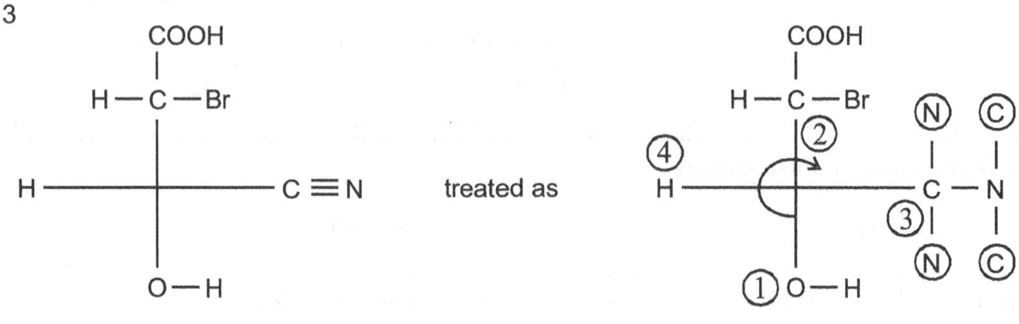

'S' at C–3

56. (d) In option 4, G. I. is possible wrt Ring.

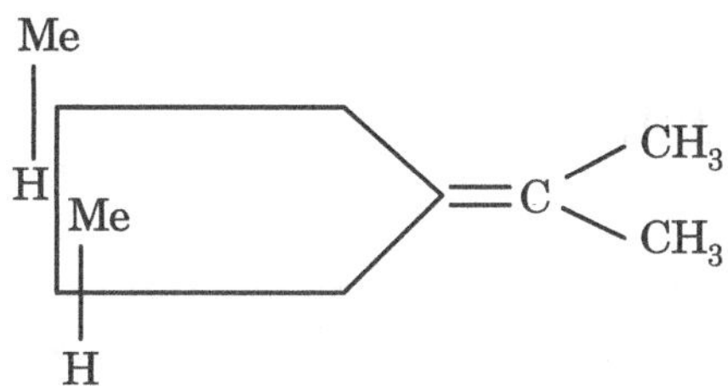

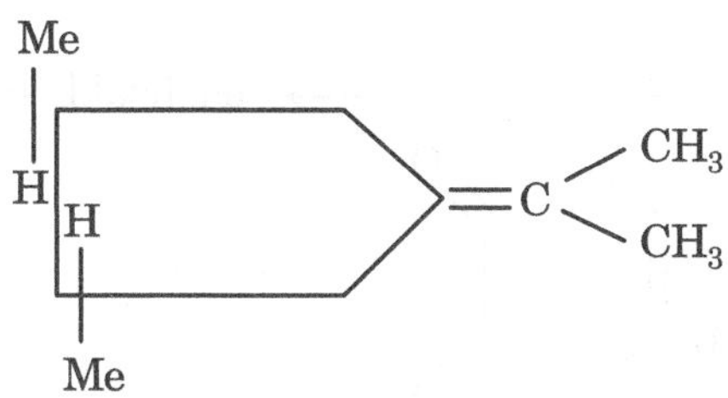

57. (d)

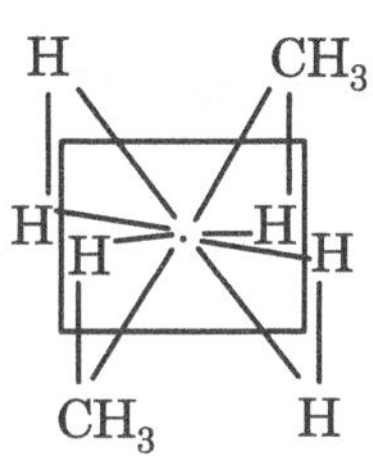

COS

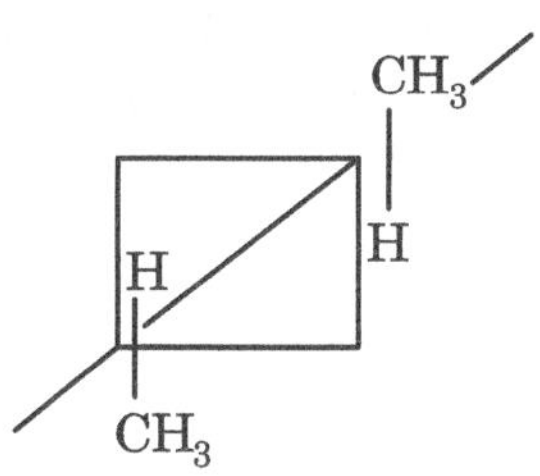

POS

58. (a) Enantiomers have same physical properties.

59. (a) Diastereomers have different physical properties.

62. (d) For a molecule to show geometrical isomerism, both the groups attached to double bonded C should be different.

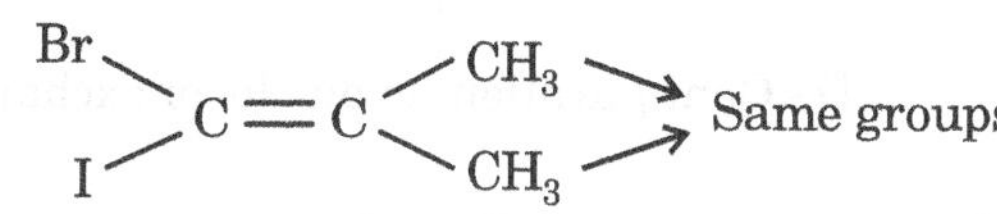

Geometrical isomerism is possible in fourth option wrt the ring.

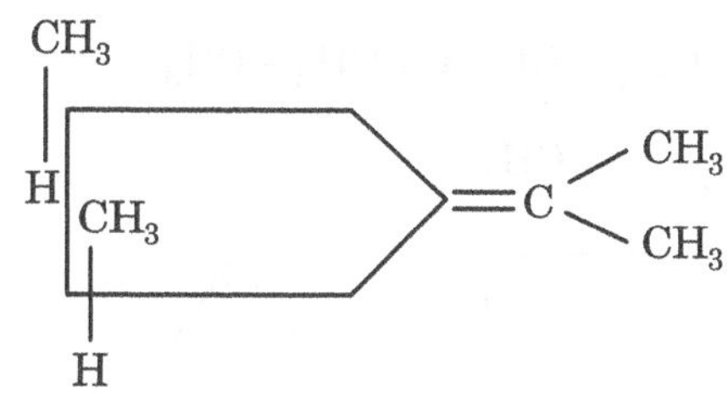

Cis

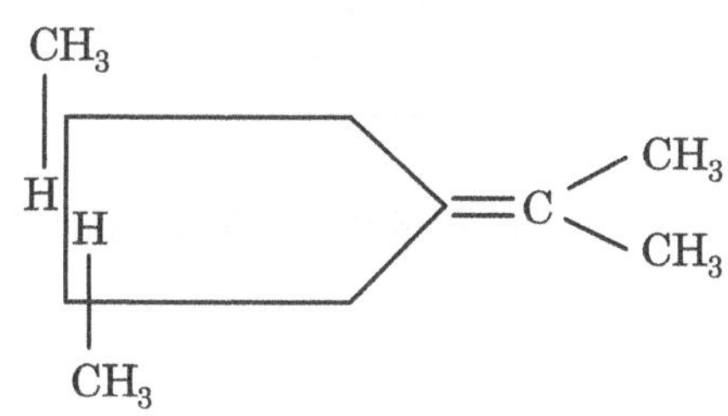

Trans

63. (c)

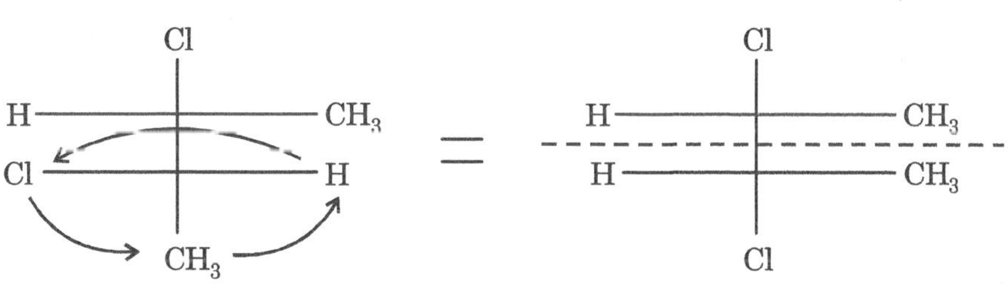

Meso

65. (a)

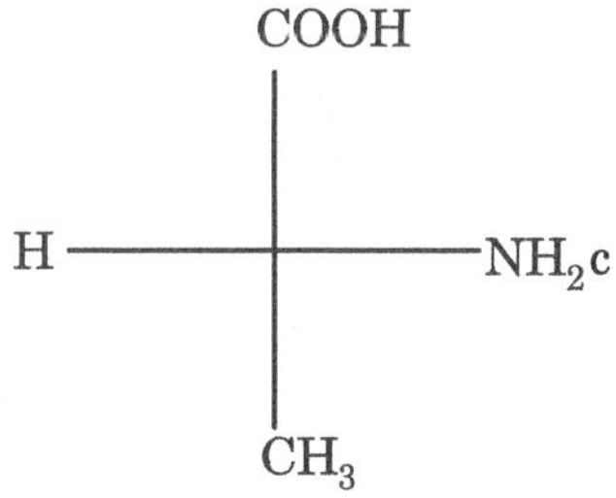

- NH_2 on Rt. Side, it is D configuration.

66. (b) 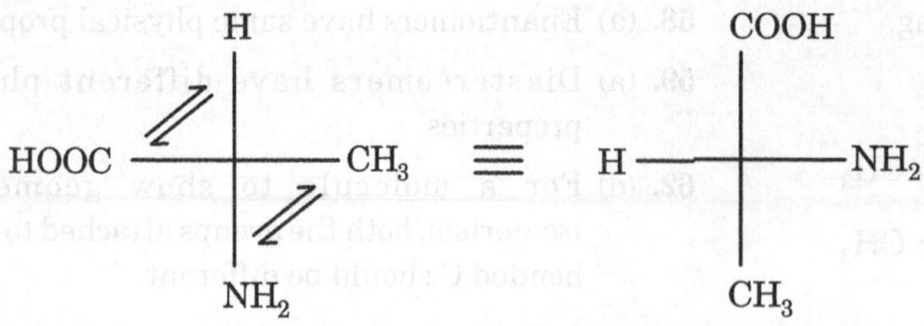

D - Configuration (even No. of exchanges are allowed)

67. (a) $CH_3-CH_2-CH_2-O-CH_3$

$CH_3-CH_2-O-CH_2-CH_3$

$CH_3-CH(CH_3)-O-CH_2-CH_3$

68. (b) 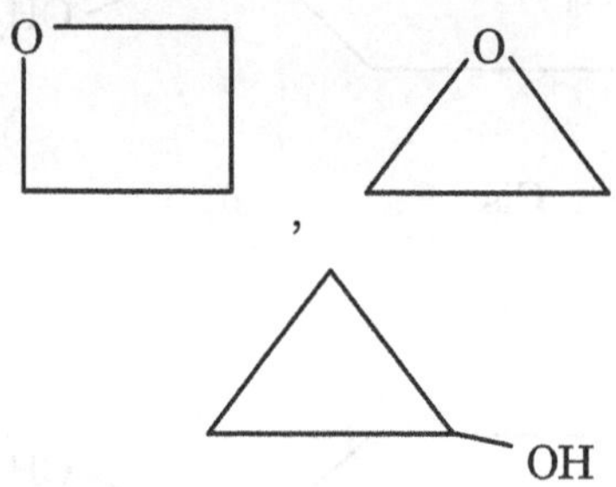

72. (c) Chain isomers differ in longest chain.

74. (d) (1) 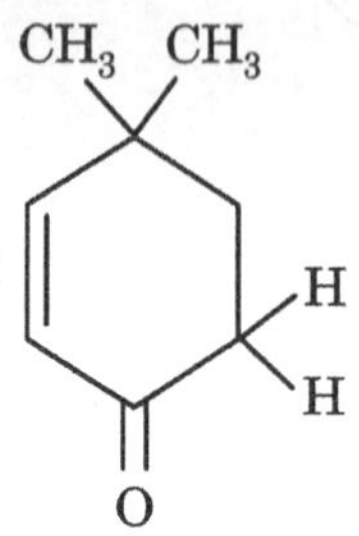 ⟶ enolizable H's

(2) 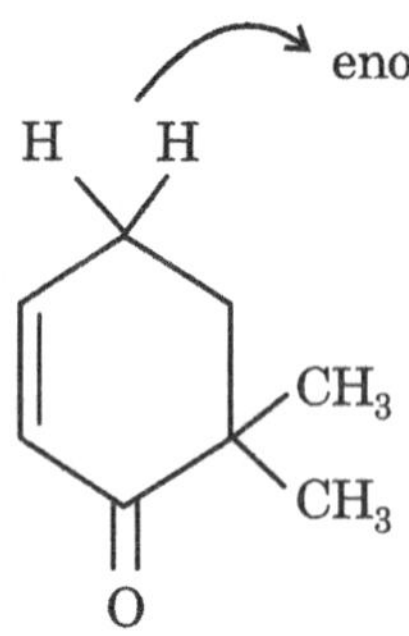

(3) 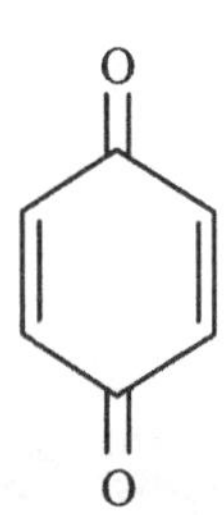 ⟶ 'NO' enolizable H's (will not show tautomerism)

(4) 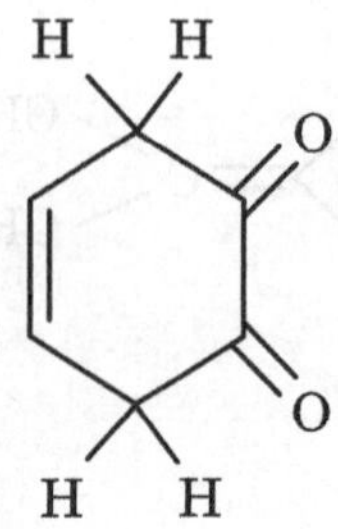 ⟶ enolizable H's ⟶ enolizable H's

(5) 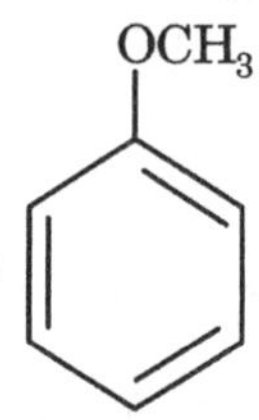 ⟶ NO 'enolizable H', will not show tautomerism.

(6) 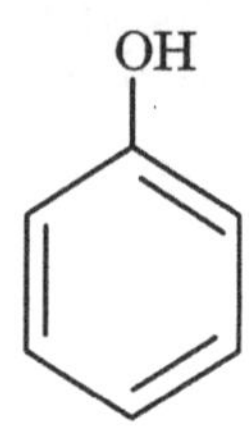 ⟶ already an 'enol' form' because — OH is bonded to double bonded C.

(7) 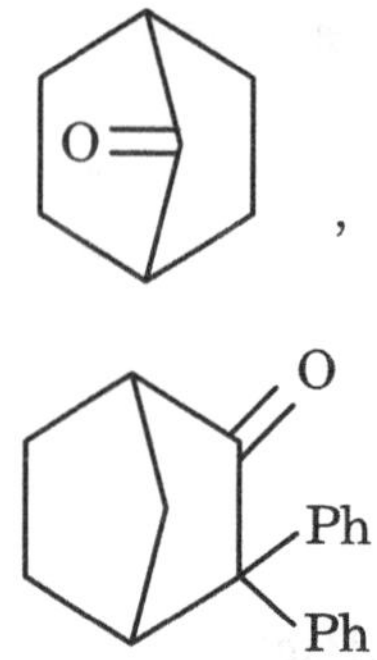 ⟶ Violation of 'BREDT'S RULE'

(8) 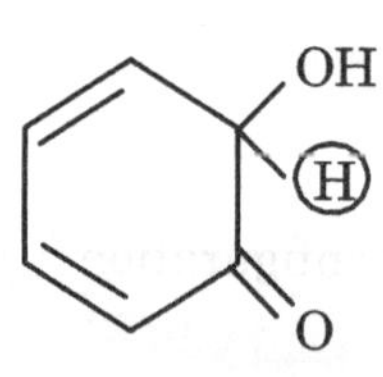 ⟶ enolizable H

(9) 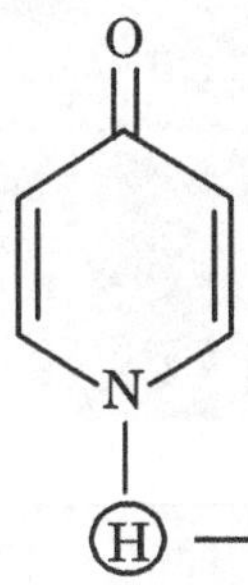 ⟶ enolizable H

(10) 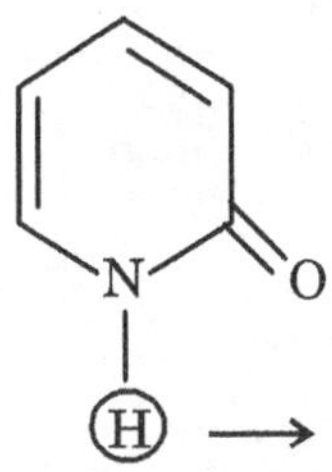 ⟶ enolizable H

75. (b) 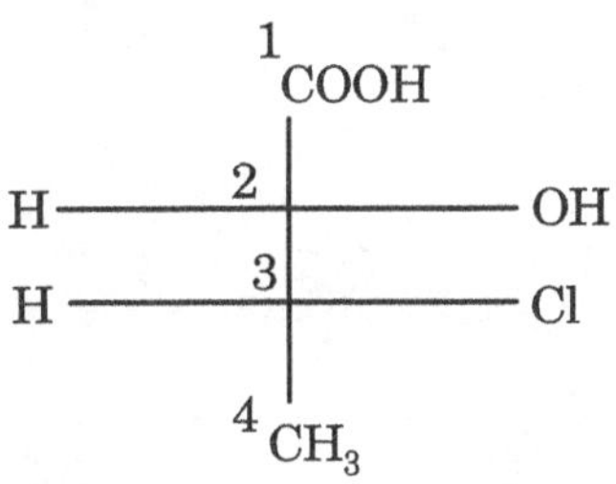

At C - 2

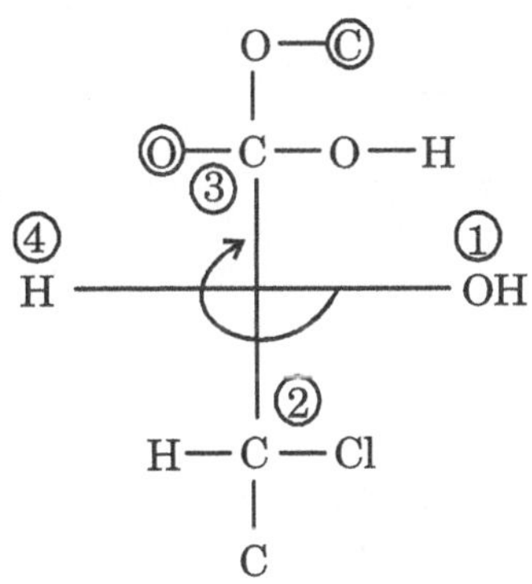

It is coming to be 'R' but since lowest priority group is on horizontal line, it will actually be 'S'.

So, 2S

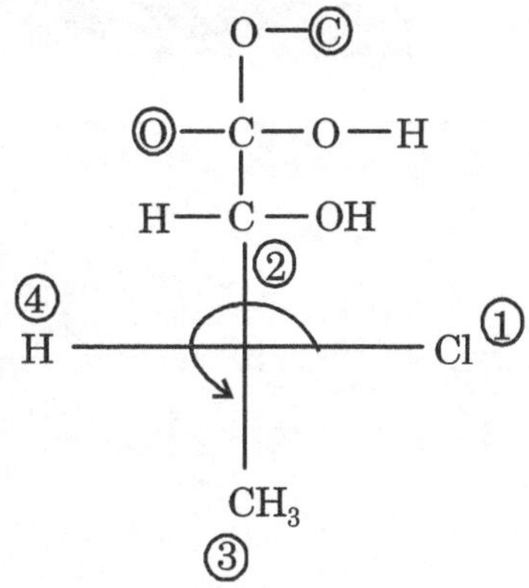

It is coming 'S' but since lowest priority gp. Is on horizontal line, it is actually 'R'.

So, 3R

Smart Notes

CHAPTER 15

Hydrocarbons

Alkanes (Preparation, Physical & Chemical Properties)

1. Which of the following alkanes can be synthesized by Wurtz reaction in good yield?

   (a) $(CH_3)_2CH - CH_2 - CH(CH_3)_2$

   (b) $(CH_3)_2CH - CH_2 - CH_2 - CH(CH_3)_2$

   (c) $CH_3 - CH_2 - C(CH_3)_2CH_2 - CH_3$

   (d) $(CH_3)_3C - CH_2 \;\; CH_2 \;\; CH_3$

2. Formation of free radical takes place with the absorption of minimum energy in the case of-

   (a) $CH_3 - (CH_2)_2 - CH_2\,Br$

   (b) $(CH_3)_2CHCH_2Br$

   (c) $CH_3 - \underset{\displaystyle Br}{\underset{|}{CH}} - CH_2 - CH_3$

   (d) $(CH_3)_3CBr$

3. Which of the following reagents do not give an alkane when reacted with Grignard's reagent?

   (a) $CH_3 - CH_2 - C \equiv C - H$

   (b) $CH_3\,CH_2\,OH$

   (c) $CH_3 - CH_2 - \underset{\displaystyle H}{\underset{|}{N}} - H$

   (d) $CH_3\,OCH_2\,CH_3$

4. An alkane having molecular weight 86 gives two mono chlorinated products. The alkane is-

   (a) 2, 3- Di methyl butane

   (b) 2, 2 - Di methyl propane

   (c) n – butane

   (d) both 2 & 3

5. $Ph - CH_2 - \overset{\displaystyle CH_3}{\overset{|}{CH}} - CH_3 \xrightarrow[\text{(Monochlorination)}]{Cl_2/hv}$

   Which statement is / are correct about photochemical chlorination of the above compound?

   (a) The major product will have chiral C atom but is optically inactive.

   (b) The intermediate free radical of the major product is resonance stabilised.

   (c) The intermediate free radical is 3° for major product.

   (d) Both 1 & 2 are correct.

6. Identify the chain propagation of chlorination of methane

(a) $\dot{C}H_3 + \dot{C}l \longrightarrow CH_3Cl$

(b) $CH_3Cl + \dot{C}l \longrightarrow \dot{C}H_2Cl + HCl$

(c) $Cl - Cl \xrightarrow{hv} \dot{C}l + \dot{C}l$

(d) $\dot{C}l + \dot{C}l \longrightarrow Cl_2$

7. Ethane cannot be obtained by-

(a) Heating methyl bromide with Na metal in dry ether.

(b) Hydrogenation of ethane

(c) Sodium acetate on Kolbe's electrolysis.

(d) Hydrolysis of Al_4C_3.

8. 1 - Bromo - 3 - Chlorocyclobutane will react with 2 moles of Na in ether producing–

(a) 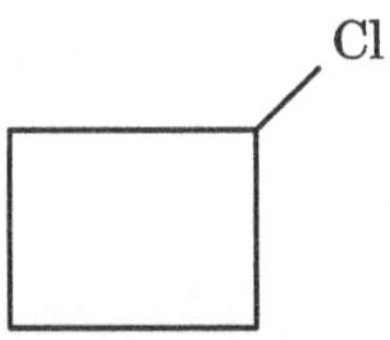

(b) 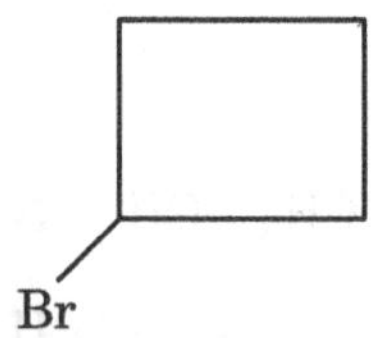

(c) 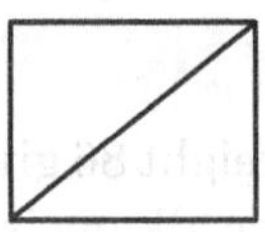

(d) None of these

9. Which of the following is not obtained on Wurtz reaction of C_2H_5Cl?

(a) C_4H_{10} (b) C_2H_4

(c) C_2H_6 (a) C_3H_8

10. Which of the following represent Sabatier Senderen's reaction?

(a) $R - X \xrightarrow{Zn, HCl}$

(b) $R - X \xrightarrow[ether]{Na, dry}$

(c) $R - X \xrightarrow{Zn, dry\ ether}$

(d) $R - CH = CH_2 \xrightarrow{Ni, 523-573k}$

11. Which of the following is not obtained as a product on electrolysis of potassium propionate?

(a) C_4H_{10} (b) $C_2H_5\ COOC_2H_5$

(c) C_2H_4 (d) $C_2H_5\ COOH$

12. Which sodium salt on heating with soda lime gives propane?

(a) $CH_3 - CH_2 - \overset{O}{\overset{||}{C}} - O^-Na^+$

(b) $CH_3 - CH_2 - CH_2 - \overset{O}{\overset{||}{C}} - O^-Na^+$

(c) $(CH_3)_2\,CH - \overset{O}{\overset{||}{C}} - O^-Na^+$

(d) Both (b) & (c)

13. Process of decarboxylation by soda lime gives alkanes with

(a) Some number of C atoms

(b) One C less than in the parent compound

(c) C atoms more than in parent compound

(d) None of these

14. Which of the following on hydrolysis gives methane?

(a) Al_4C_3 (b) Be_2C

(c) CaC_2 (d) both (a) & (b)

15. Which of the following gas cannot be prepared by Kolbe's electrolysis?

(a) C_2H_6 (b) C_4H1_0

(c) CH_4 (d) Both (a) & (c)

16. An alkane (M. wt 72) forms only one monochlorinated product. Its formula is–

(a) $(CH_3)_4C$ (b) $CH_3(CH_2)_3CH_3$

(c) $(CH_3)_2CHCH_2CH_3$ (d) 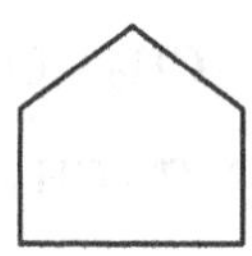

17. Which of the following compounds should undergo chlorination faster than the remaining three?

(a) n-pentane $(CH_3 - CH_2 - CH_2 - CH_2 - CH_3)$

(b) neopentane $CH_3 - \underset{CH_3}{\overset{CH_3}{\underset{|}{\overset{|}{C}}}} - CH_3$

(c) Isopentane $CH_3 - \underset{CH_3}{\underset{|}{C}} - CH_2 - CH_3$

(d) n-butane $(CH_3 - CH_2 - CH_2 - CH_3)$

18. Characteristic reactions of alkanes are-

(a) addition reactions

(b) Substitution reactions

(c) polymerization reactions

(d) elimination reaction

19. The alkanes that yields two isomeric monobromo derivatives is- (excluding stereoisomers)

(a) Neopentane (b) Ethane

(c) Methane (d) Propane

20. Which of the following is correct sequence of steps in halogenation of an alkane?

(a) Propagation, initiation, termination

(b) Initation, propagation, termination

(c) Termination, intitation, propagation

(d) None of these

21. The highest boiling point is expected for-

(a) n-octane

(b) n-butane

(c) 2, 2, 3, 3- tetramethyl butane

(d) propane

22. Identify the incorrect statement.

(a) addition of hydrogen to alkenes take place in presence of a catalyst.

(b) stability of alkenes is inversely proportional to heat of hydrogenation

(c) acid catalyzed dehydration of 3° alcohols occurs via free radical formation.

(d) dehydrohalogenation of RCl with alc. KOH yields highly substituted alkene as a major product

23. The No. of Chiral compounds possible on mono-chlorination of 2 - methyl butane is -

(a) 2 (b) 4

(c) 6 (d) 8

24. The total isomers formed by dichlorination of cyclohexane will be-

(a) 4 (b) 9

(c) 7 (d) 11

25. Corey House reaction involves

(a) Formation of an organometallic compound.

(b) Nucleophilic substitution reaction

(c) Nucleophilic addition reaction

(d) Both (a) & (b)

26. Hydrolysis of an ester gives a carboxylic acid, sodium salt of which on Kolbe's electrolysis yields ethane. Then , the ester is –

(a) Methyl ethanoate

(b) Methyl methanoate

(c) Ethyl methanoate

(d) Methyl propanoate

27. Consider the following reaction

$$\text{Hydrocarbon (A)} \xrightarrow{H_2/Ni} 2-\text{Methyl pentane} \xrightarrow[h\nu]{Cl_2} \underset{(X)}{\text{optically active Monochloro products}}$$

$$\text{All Mono chlorinated products} \xrightarrow[\text{distillation}]{\text{fractional}} \underset{(Y)}{\text{No. of fraction obtained}}$$

Value of X & Y, respectively is-

(a) 5, 6 (b) 6, 5

(c) 4, 5 (d) 5, 4

28. Total number of product (except stereoisomers) obtained on dichlorination of methyl cyclo-propane are–

(a) 3 (b) 6

(c) 4 (d) 7

29.

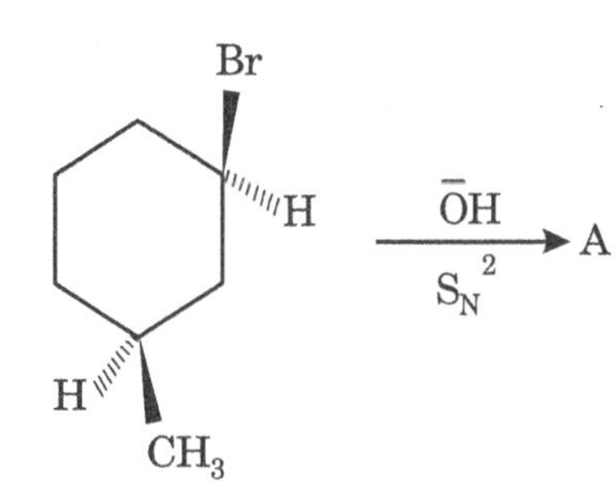

(1R, 3S) Cis 1-Bromo-3-Methyl Cyclohexane

The product A formed in the above reaction is

(a) (1R, 3S) - Cis - 3 - methyl cyclohexanol
(b) (1S, 3R) - Cis - 3 - methyl cyclohexanol
(c) (1S, 3S), - Trans - 3 - methyl cyclohexanol
(d) (1R, 3R) - Trans - 3 - methyl cyclohexanol

30. 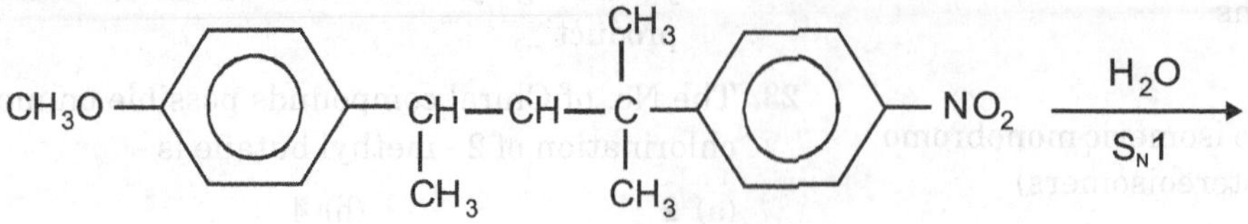

The major product formed in the above reaction is–

(a) 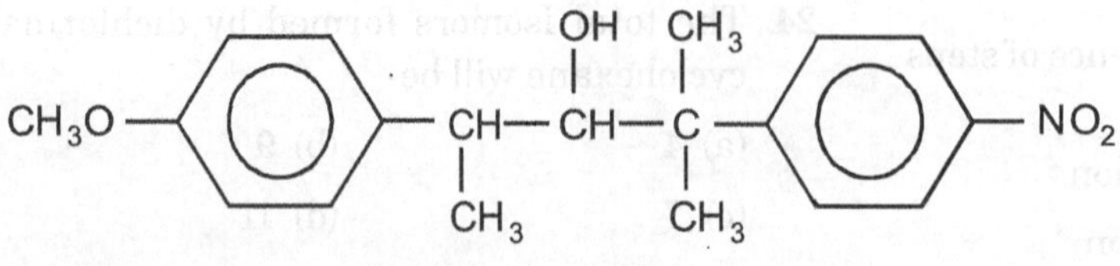

(b) 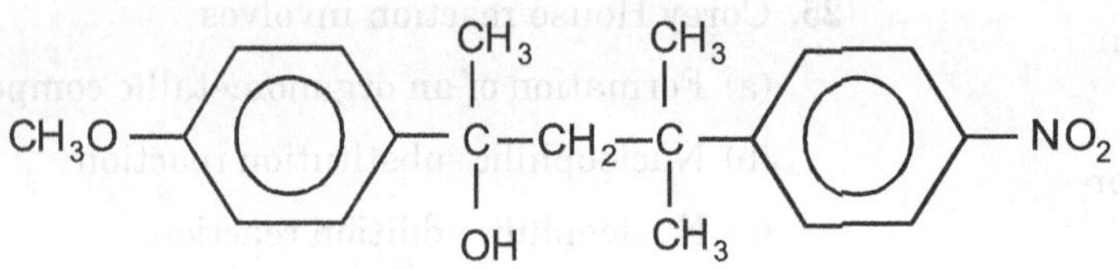

(c) 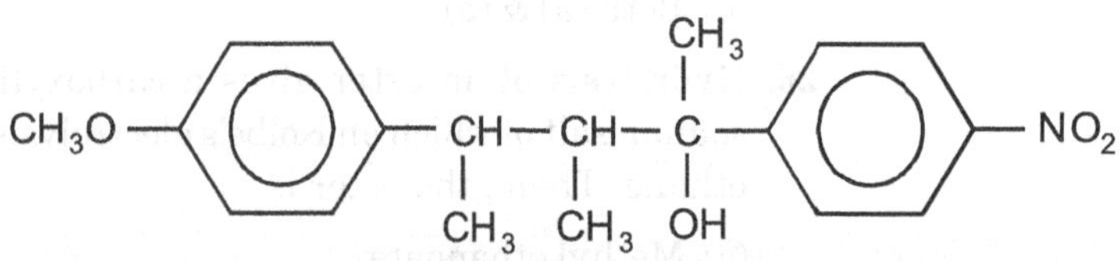

(d) Both (a) & (c)

31. In the following reaction, the correct product is

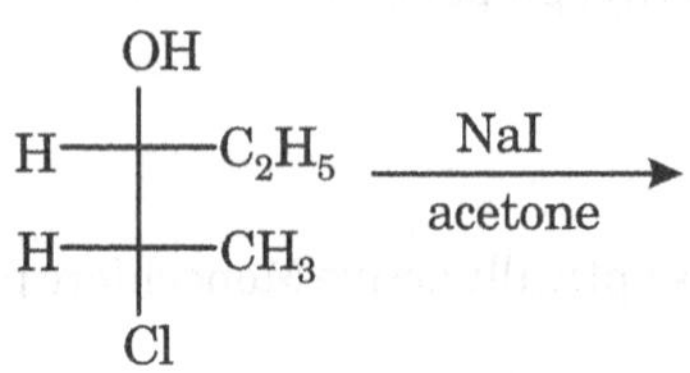

(a) 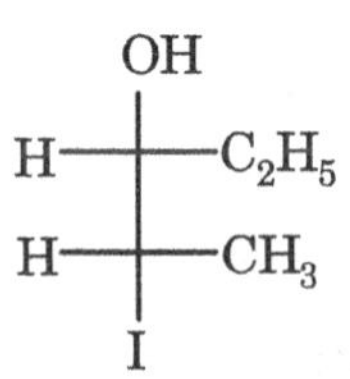
(b) 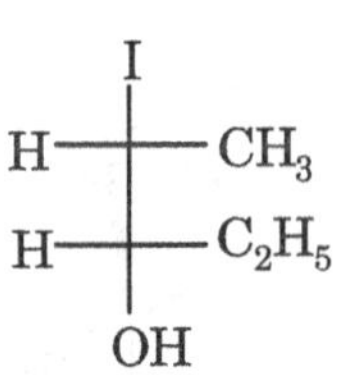

(c) 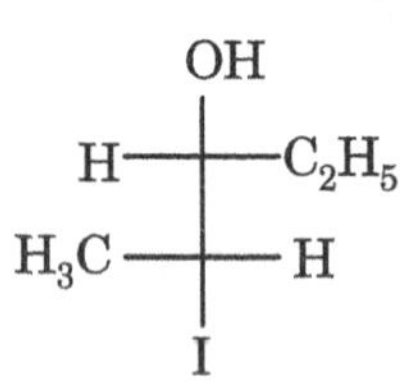
(d) 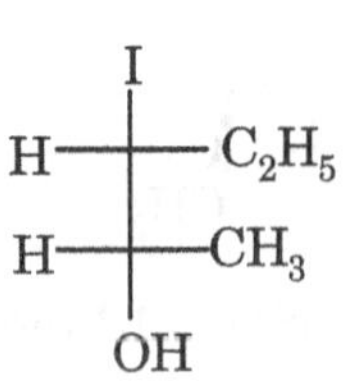

32. Which of the following will give racemic mixture on undergoing $S_N{}^1$?

(a) $CH_3-CH(C_2H_5)-Br$

(b) $CH_3-C(Br)(C_2H_5)-CH_3$

(c) $CH_3-CH(C_2H_5)-CH_2Br$

(a) Only a
(b) a, b & c only
(c) b & c only
(d) a & c only

33. Identify the correct reactivity order for the S_N1 reaction.

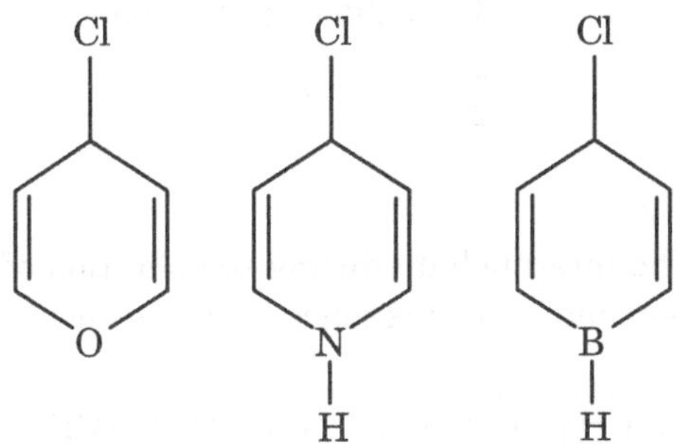

(1) (2) (3)

(a) $2 > 1 > 3$ (b) $2 = 1 > 3$

(c) $1 > 2 > 3$ (d) $3 > 2 > 1$

34.

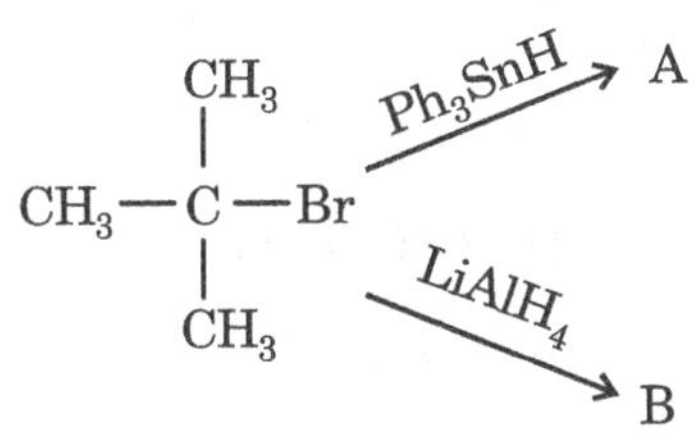

Which of the following statements is/are true about A & B?

(a) A is an alkene but B is an alcohol.

(b) A is an alkane but B is an alkene.

(c) A is an alcohol but B is an acid

(d) None of these

35.

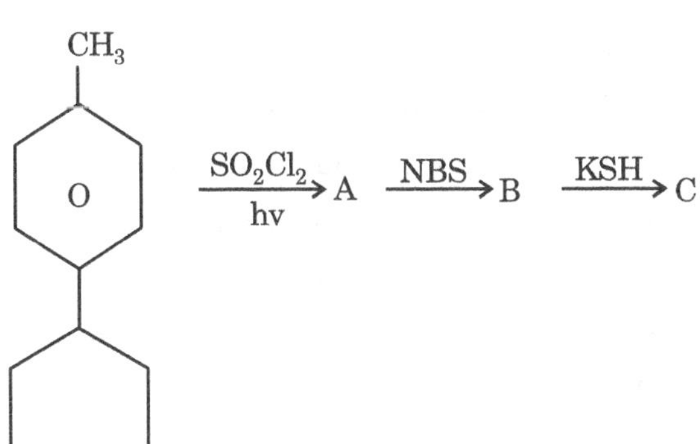

The product 'C'

(a)

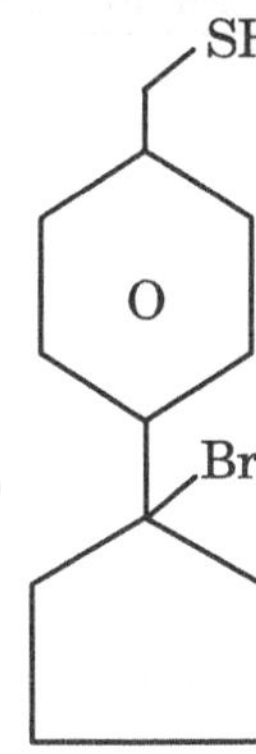

(b)

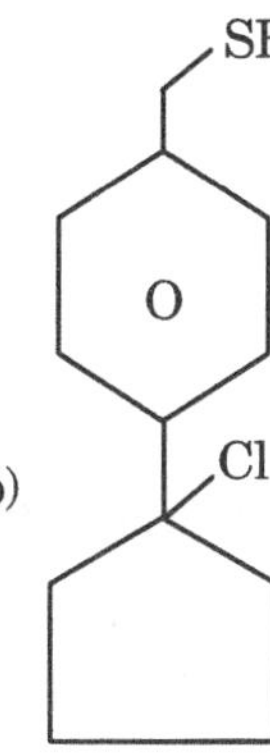

(c)

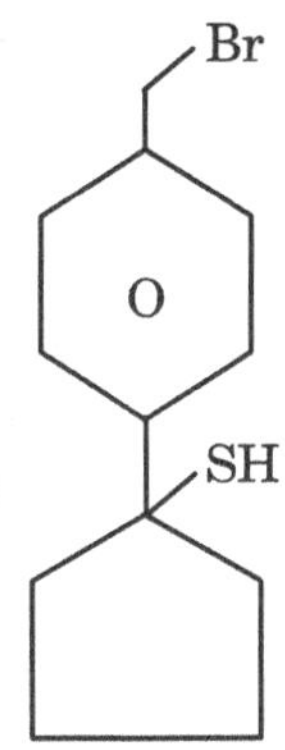

(d)

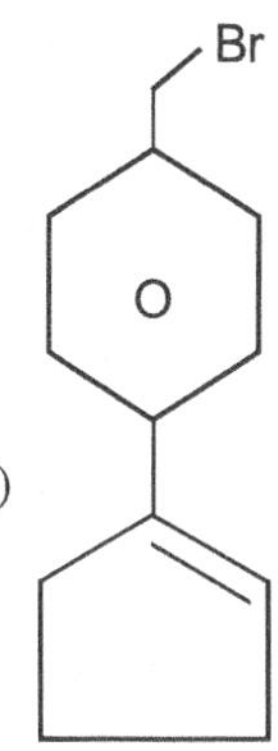

36. The major products formed in the following reactions are

(1) $(C_2H_5)_3N + CH_3-CH_2-Br \longrightarrow (A)$

(2) $(C_2H_5)_3N + CH_3-\underset{\underset{I}{|}}{CH}-CH_2-CH_3 \longrightarrow (B)$

A & B are –

(a) $A \longrightarrow CH_2 = CH_2$;

$B \longrightarrow CH_3-CH=CH-CH_3$

(b) $A \longrightarrow (CH_3-CH_2)_4\overset{+}{N}\overset{-}{Br}$;

$B \longrightarrow CH_3-\underset{\underset{N^+(C_2H_5)_3\ I^-}{|}}{CH}-CH_2-CH_3$

(c) $A \longrightarrow (CH_3CH_2)_4\overset{+}{N}\overset{-}{Br}$;

$B \longrightarrow CH_2 = CH-CH_2CH_3$

(d) None of these

37. 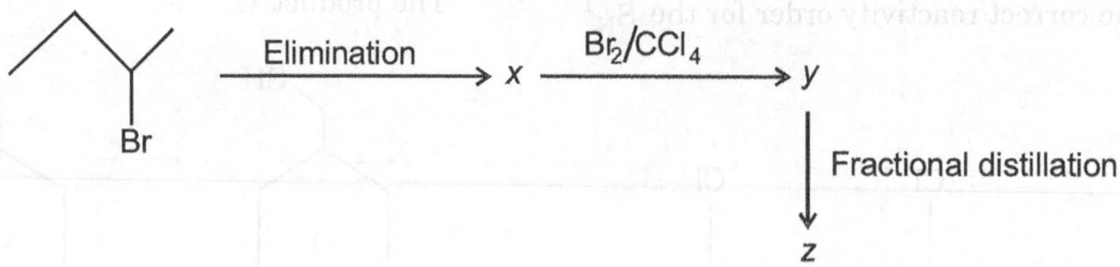

x, y & z are, respectively -

(x & y are the no. of possible products including stereomers & Z is the no. of fractions)

(a) $x = 5, y = 3, z = 5$ (b) $x = 3, y = 5, z = 3$

(c) $x = 2, y = 3, z = 3$ (d) $x = 3, y = 4, z = 3$

Alkenes

38. The final product 'X' is–

$$CH_3-CH=CH-CH_3 \xrightarrow{NBS} A \xrightarrow[KOH]{alc.} X$$

(a) $CH_2=CH-CH=CH_2$

(b) $CH_3-CH=CH-CH_2$ (with Br on the terminal CH_2)

(c) $CH_3-C\equiv C-CH_3$

(d) $CH_2-CH_2-CH=CH_2$ (with Br on the first CH_2)

39. Consider the following orders & select the 'INCORRECT ONE?'

(a) 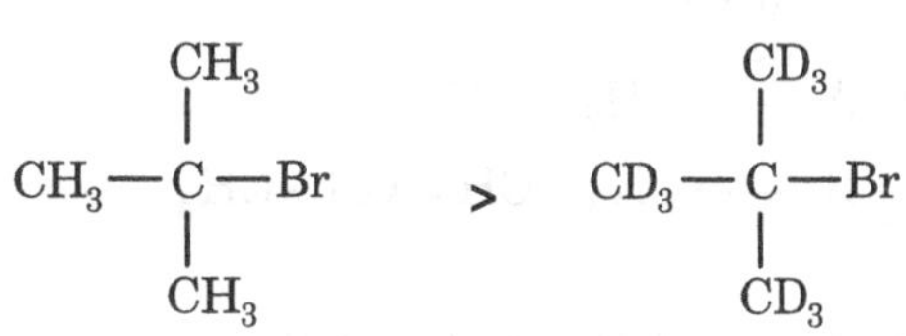

(Rate of E_2 reaction)

(b) 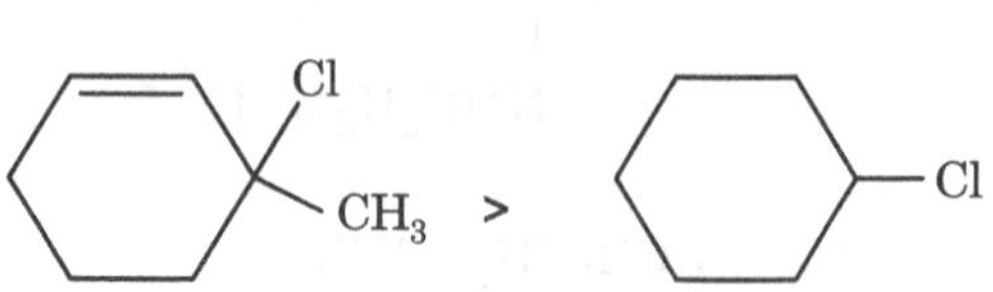

(Rate of SN^1)

(c) $CH_2=CH-CH_2-Br > CH_3-CH_2-CH_2-Br$ (Rate of SN^2)

(d) None of these is incorrect

40. The intermediate during the addition of HCl to propene in the presence of organic peroxide is–

(a) $CH_3-\dot{C}H-CH_3$ (b) $CH_3\overset{\oplus}{C}HCH_3$

(c) $CH_3CH_2\dot{C}H_2$ (d) $CH_3CH_2\overset{\oplus}{C}H_2$

41. $CH_3-CH_2-CH=CH_2 \xrightarrow[\text{(high T)}]{Cl_2,\ 500^\circ C} A \xrightarrow[KOH]{alc.} B$, the compound B is –

(a) $CH_3-CH=CH-CH_2$

(b) $CH_2=CH-CH=CH_2$

(c) $CH_2=CH-CH_2-CH_3$

(d) $CH_2=C=CH-CH_3$

42. Consider the following reaction -

$$CH_3-CH_2-\underset{F}{\overset{H}{C}}-CH_3 \xrightarrow{Base} \underset{\text{(Major)}}{A} + \underset{\text{(Minor)}}{B}$$

What is true for A?

(a) It is formed by E_2 mechanism.

(b) It can't show geometrical isomerism.

(c) It is formed by SN^1 mechanism.

(d) None of these are true.

43. Select the correct order of reactivity for E_2 with alcoholic KOH.

(a) 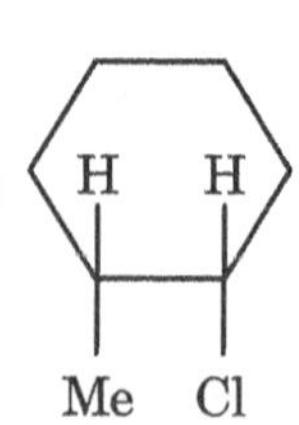

(b) 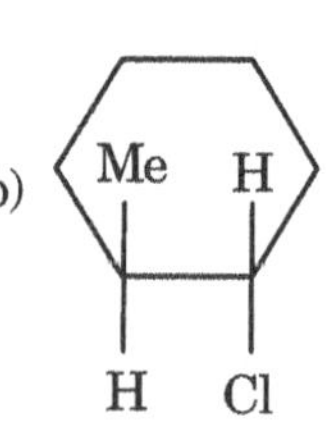

(c) 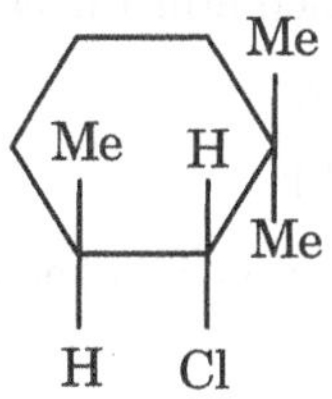

(a) a > b > c (b) b > c > a

(c) c > b > a (d) a = b = c

44. Which of the following cannot undergo E_2 reaction with alcoholic KOH?

(a) 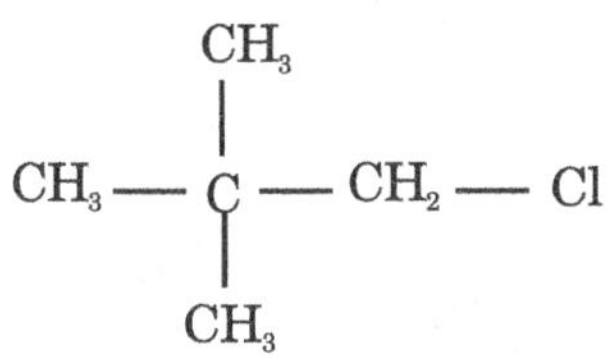

(b) 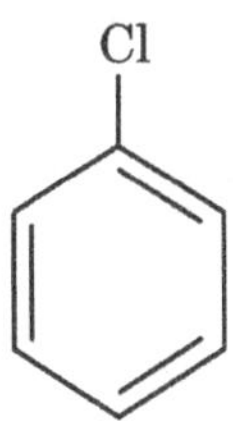

(c) 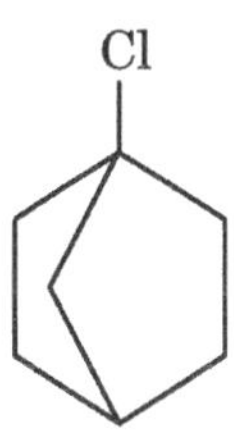

(d) All of these

45. 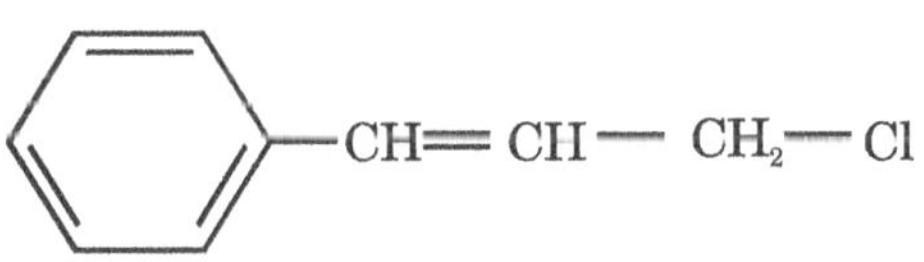

The no. of stereomers formed in solvolysis of given compound in water will be -

(a) 2 (b) 3

(c) 4 (d) 1

46. 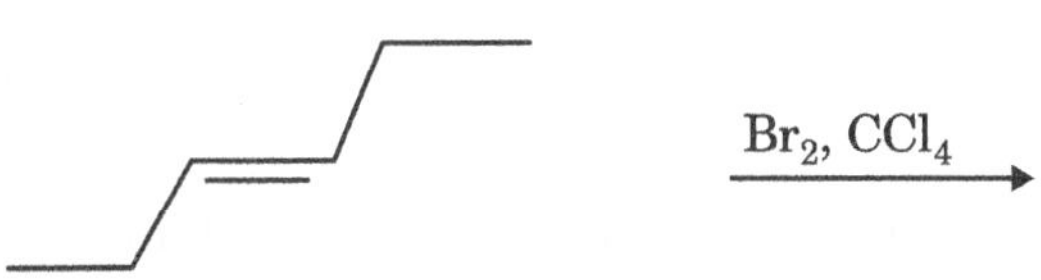

The major products in the above reaction is/are

1. 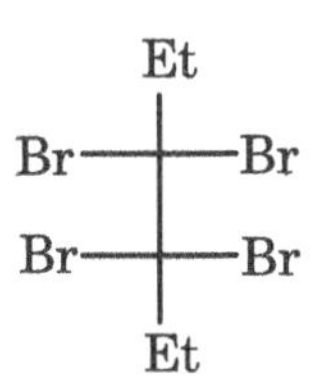

2. 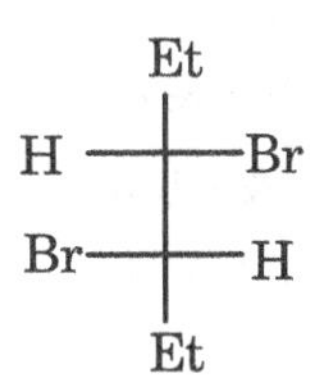

3. 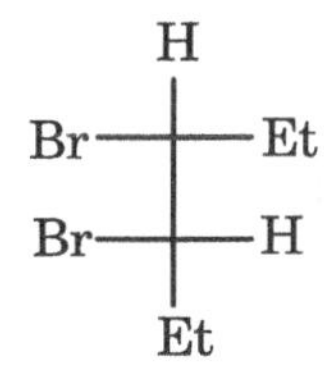

4. 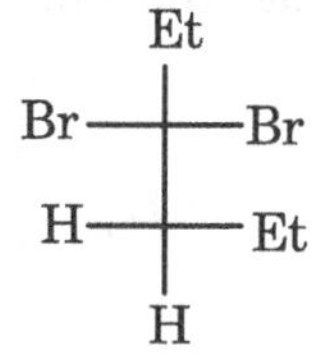

(a) 2 & 3 only (b) 1, 2 & 3

(c) 4 only (d) 2, 3 & 4

47. 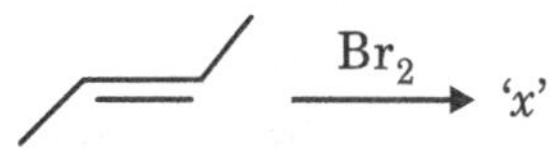

'x' is

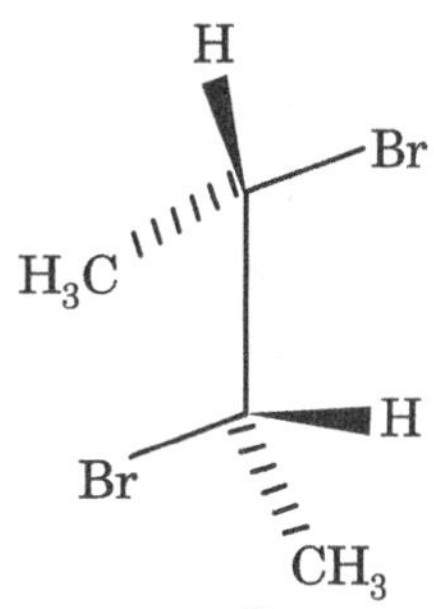

(I)

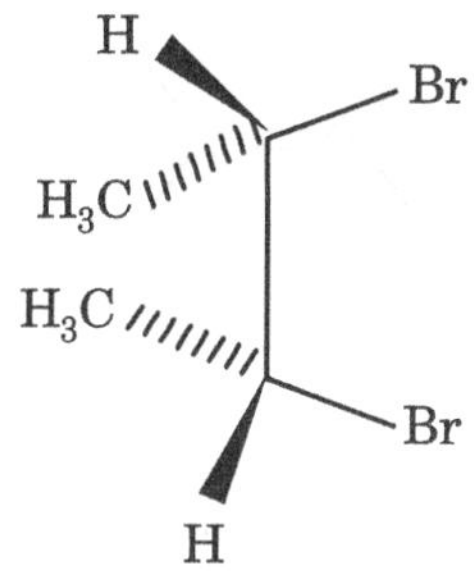

(II)

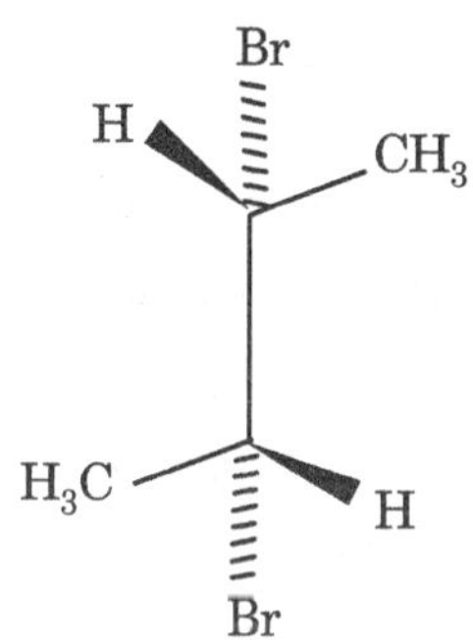

(III)

(a) I & III (b) II & III

(c) II only (d) III only

48. 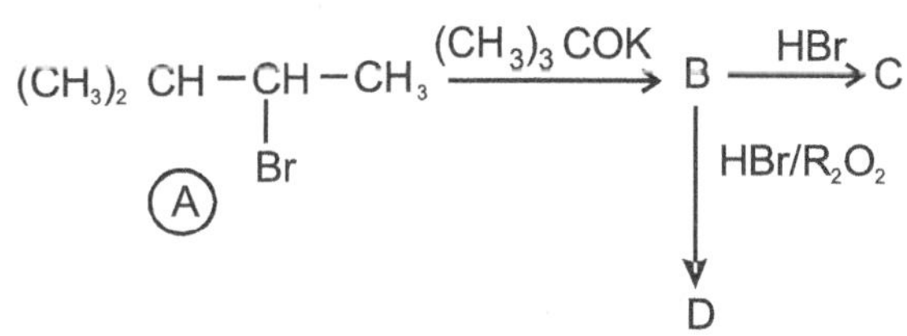

The correct order of S_N2 reactivity for A, C & D will be

(a) D > A > C (b) D > C > A

(c) C > A > D (d) A > C > D

49. Consider the following reaction-

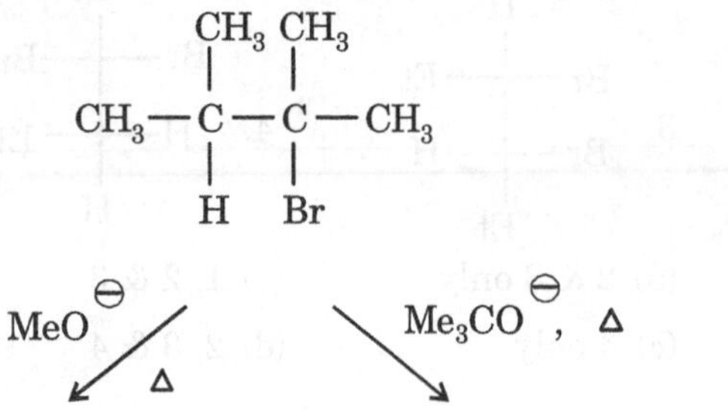

A (Major product) B (Major product)

A &B, respectively are-

(a) 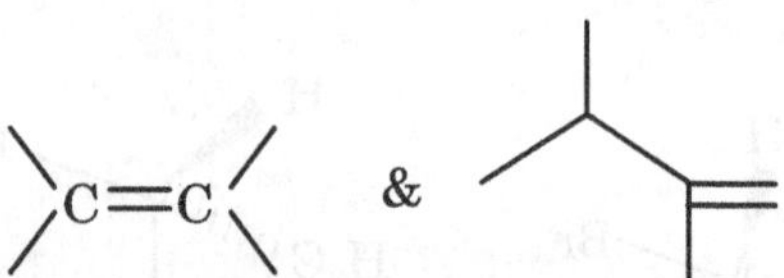

(b) 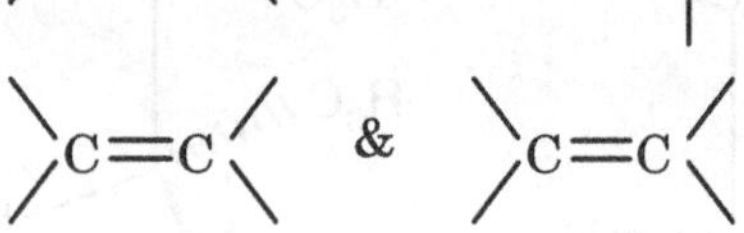

(c) 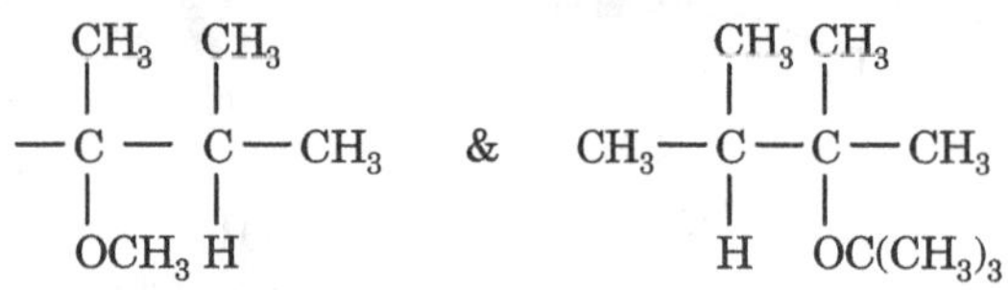

(d) None of these

50. 2 - Phenyl propene on acid catalyzed hydration gives -

(a) 2 - Phenyl - 2 - propanol

(b) 2 - Phenyl - 1 - propanol

(c) 3 - Phenyl - 1 - propanol

(d) 1 - Phenyl - 2 - propanol

51. For the following reactions –

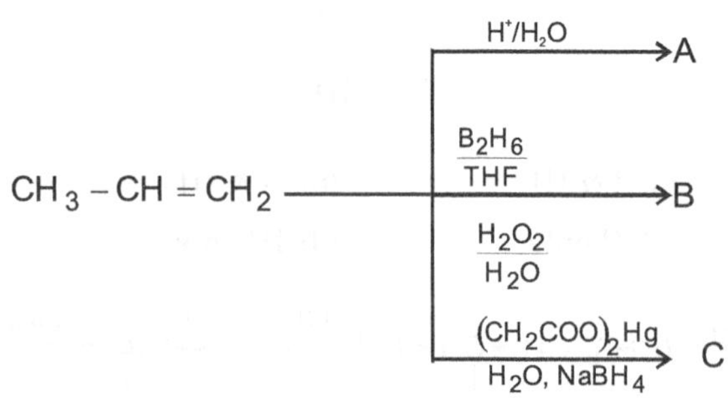

The incorrect is / are

(a) A, B & C are alcohols

(b) A & B are structural isomers

(c) A & C are identical

(d) Formation of A & C involves formation of transition state while that of B involves formation of a carbocation

52. H – I does not show Kharasch effect because -

(a) H – I bond is very strong.

(b) H — I bond fails to yield I°

(c) I° generated adds on to double bond according to Markovnikov's rule

(d) I° combines to form I_2 molecule

53. For which of the following reaction, ΔG is highly (maximum) negative?

(a) 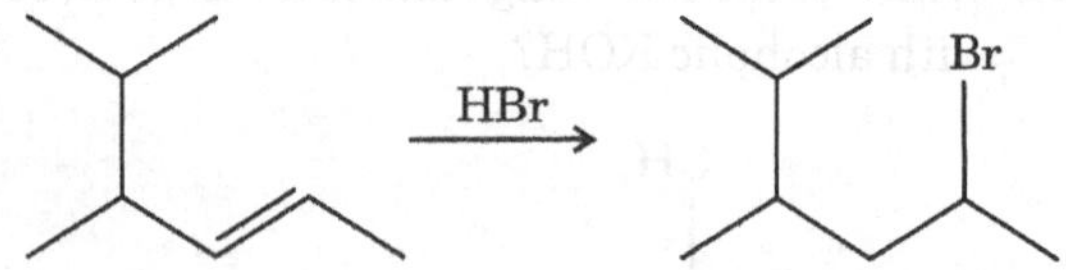

(b) 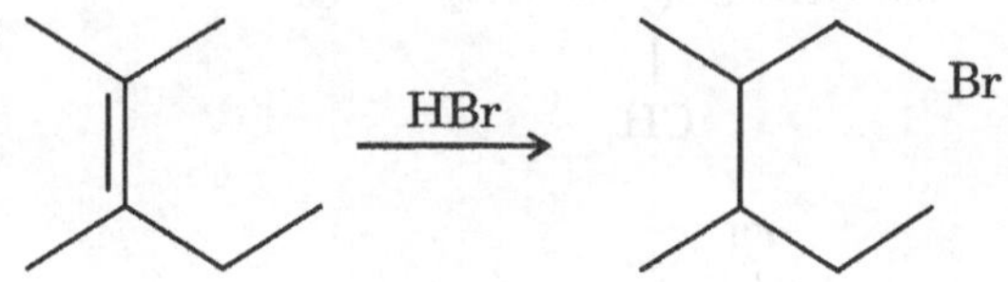

(c) 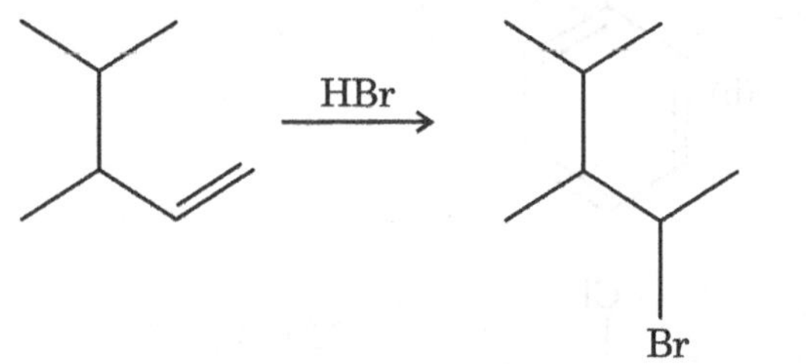

(d) 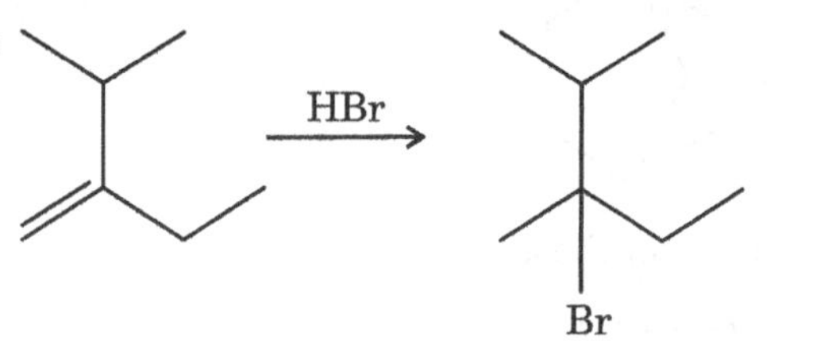

54. What is not true regarding the following reaction?

$$CH_3-CH=CH-Ph \xrightarrow[R_2O_2]{HCl}$$

(a) The product is formed via carbocation formation.

(b) The product mixture contains 5 stereomers.

(c) The major product formed is a racemic mixture.

(d) The product is formed via formation of a free radical.

(a) Both c & d

(b) Both a & c

(c) a, b & c but not d

(d) Both b & d

55. Select the incorrect match

(a) 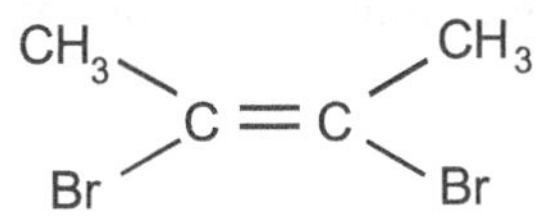 $\xrightarrow[\text{Pd/C}]{H_2}$ Meso Compound

(b) 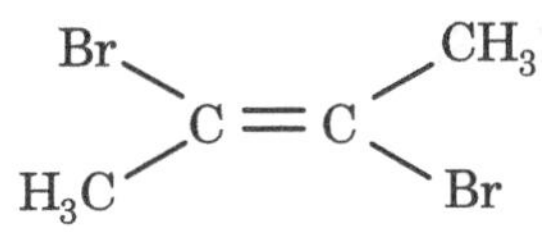 $\xrightarrow[\text{Wilkinson's Catalyst}]{H_2\ +}$ Racemic Mixture

(c) 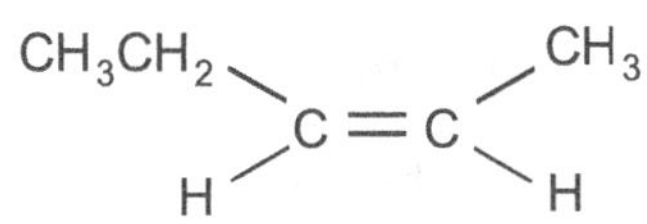 $\xrightarrow[\text{Syn addition}]{H_2/Pt}$ Racemic Mixture

(d) None of these

56. In the presence of peroxide, HCl & HI do not give Anti - Markownikoff's addition to an alkene because –

(a) Both the steps are endothermic in both the cases

(b) Both the steps are exothermic in both the cases

(c) One of the steps is endothermic in both the cases

(d) None of these

57. Rate of Dehydration when the following given compounds are treated with conc. H_2SO_4 is –

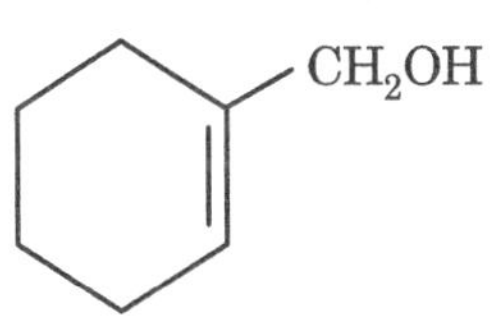

(P)

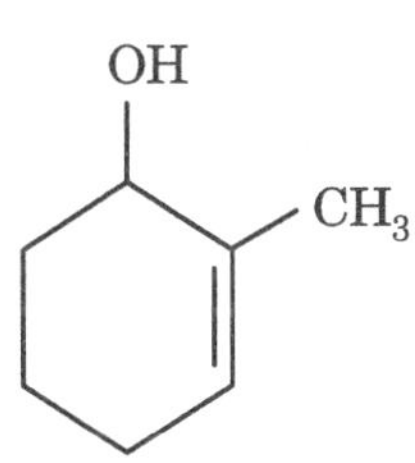

(Q)

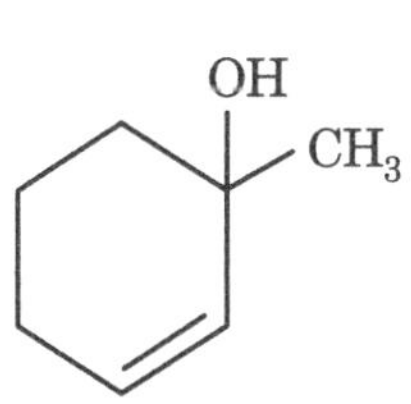

(R)

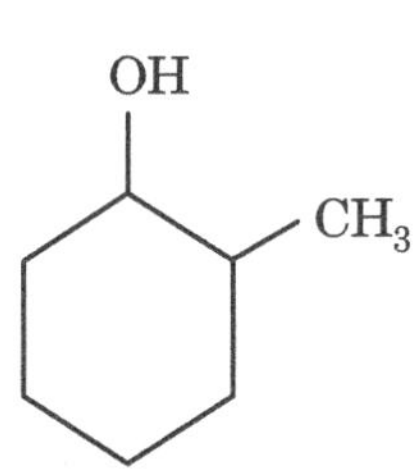

(S)

(a) P > Q > R > S (b) Q > P > R > S

(c) R > Q > P > S (d) R > Q > S > P

58. Trans - 2 - butene&trans - 2 - pentene on treatment with Br_2 in the presence of CCl_4 gives, respectively

(a) Racemic mixture &meso compound

(b) Meso compound and racemic mixture

(c) Meso compound and Meso compound

(d) Both (a) & (b)

59. $CH_3-CH_2-Cl \xrightarrow[\text{KOH}]{\text{alc.}} CH_2=CH_2$, the most probable mechanism for this reaction is –

(a) E_1 (b) S_N

(c) E_2 (d) E_{1cb}

60. $CH_3-CH=CH_2$ is reacted with some reagents. Results are accordingly shown below. Match these from list - I to list - II.

	List – I		List – II
P	$\frac{KMnO_4}{H^+}$	1	Propane – 1, 2 – diol
Q	$\frac{KMnO_4}{OH}$	2	Propane – 1 – ol
R	$O_3\left(\frac{H_2O}{Zn}\right)$	3	Propane – 2 – ol
S	$\frac{B_2H_6}{NaOH} / H_2O_2$	4	$CH_3COOH + CO_2$
T	$\frac{H_2O}{H^+}$	5	$CH_3CHO + HCHO$

	P	Q	R	S	T
(a)	1	4	5	3	2
(b)	1	4	5	2	3
(c)	4	1	5	2	3
(d)	4	1	2	3	4

61. How many compounds are more reactive than ethene towards electrophilic addition?

$CH_3-CH=CH_2$, $H-\overset{O}{\overset{||}{C}}-CH=CH_2$,

$Cl-CH_2=CH_2$, $CH_3-O-CH=CH_2$

$H-C\equiv C-H$, $Ph-CH=CH_2$

(a) 6 (b) 7

(c) 3 (d) 4

62. The major product of dehydration of following alcohol is -

$CH_3-\underset{CH_3}{\underset{|}{CH}}-CH_2-CH_2OH \xrightarrow[\Delta]{H^+}$ Alkene

(a) $CH_3=\underset{CH_3}{\underset{|}{C}}-CH-CH_3$

(b) $CH_2=\underset{CH_3}{\underset{|}{C}}-CH_2-CH_3$

(c) $CH_3-\underset{CH_3}{\underset{|}{CH}}-CH=CH_2$

(d) 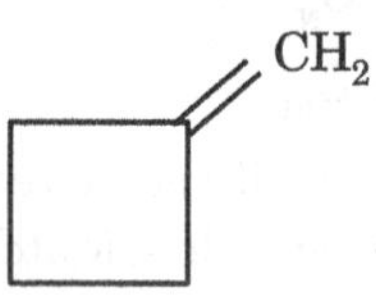

63. The major product of the following reactions is -

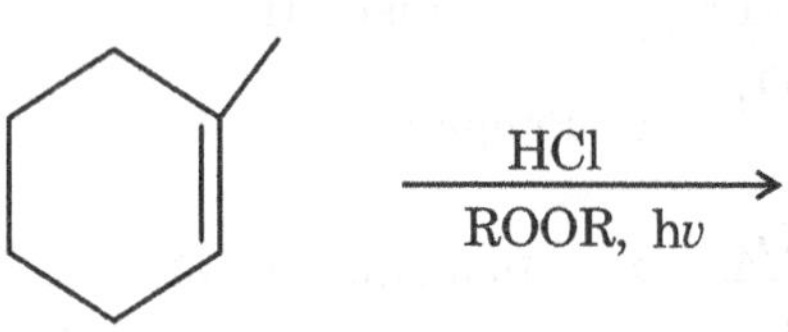

(a) 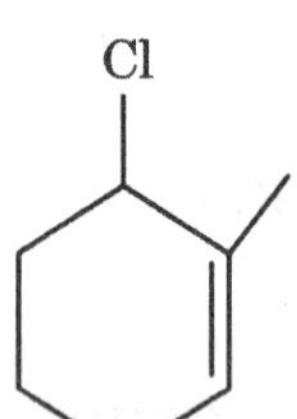 (b) 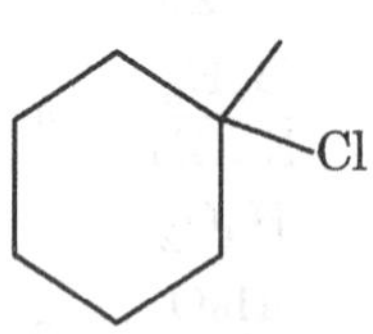

(c) 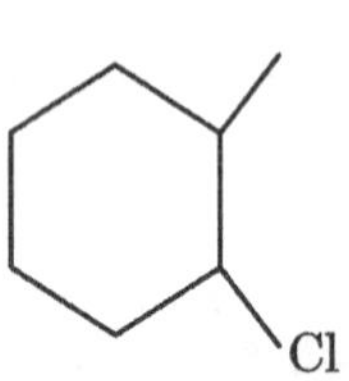 (d) 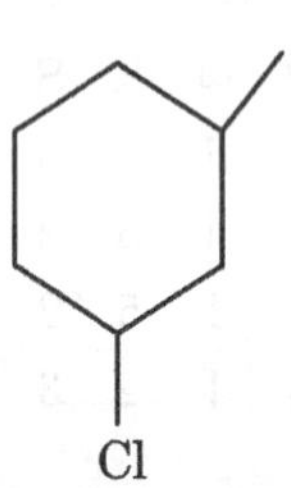

64. Arrange the following alkyl halides in decreasing order of rate of β-elimination reaction with alc. KOH (E_2).

(a) $CH_3-\underset{CH_3}{\underset{|}{\overset{H}{\overset{|}{C}}}}-CH_2Br$

(b) CH_3-CH_2-Br

(c) $CH_3-CH_2-CH_2-Br$

(a) 1 > 2 > 3 (b) 3 > 2 > 1

(c) 2 > 3 > 1 (d) 1 > 3 > 2

65. 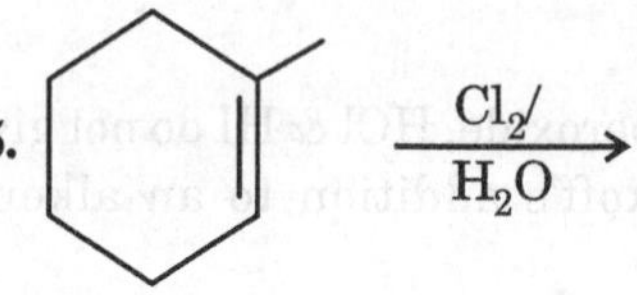 Major product in the reaction is -

(a) 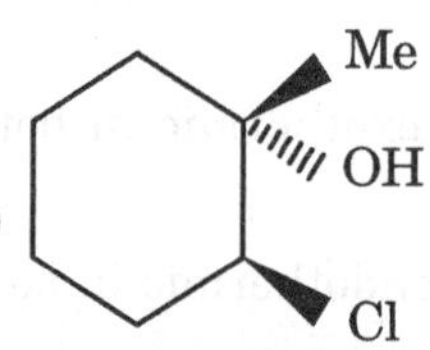 (b) 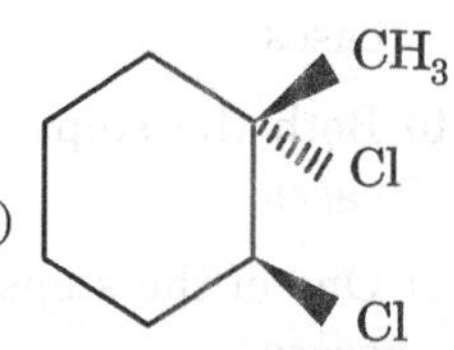

(c) 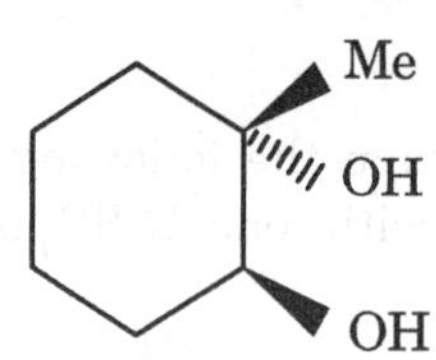 (d) 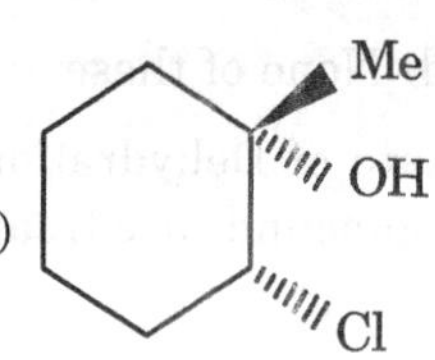

66. Which of the following reactions result in the formation of a pair of diastereomers?

(a) 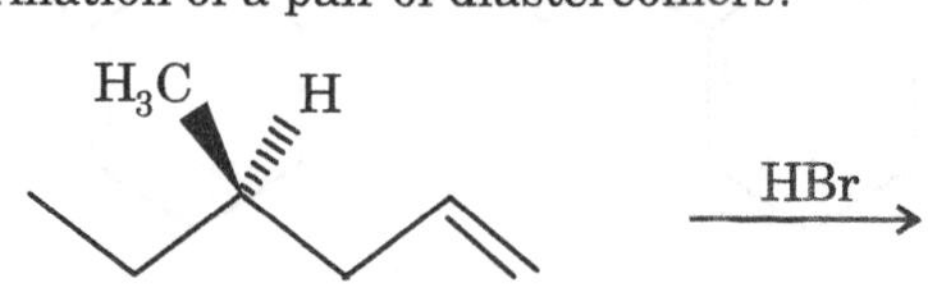

(b) 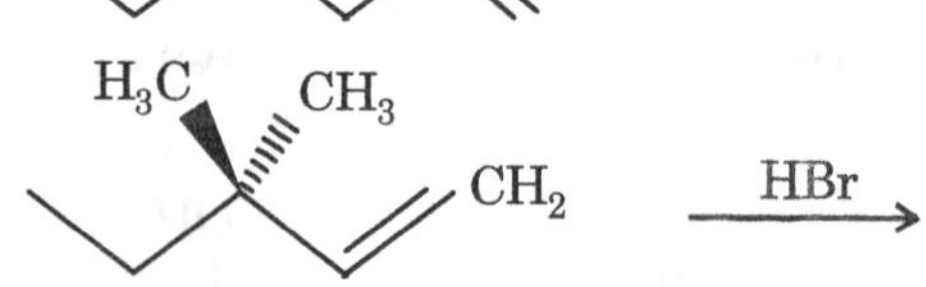

(c) 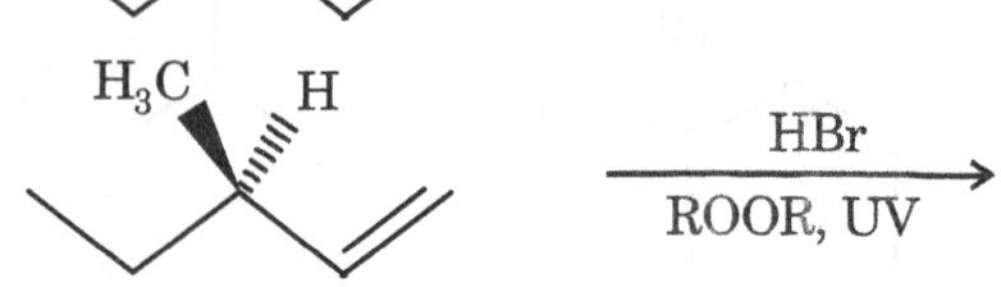

(d) Both (b) & (c)

67. 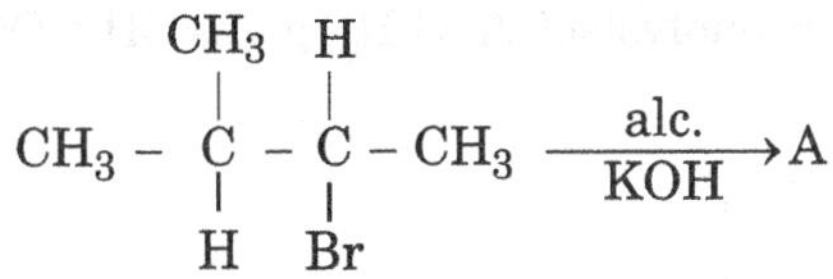

(Major product). A wil be

(a) $(CH_3)_2C = CH - CH_3$

(b) $(CH_3)_2CH - CH{=}CH_2$

(c) $CH_2{=}C(CH_3) - CH_2 - CH_3$

(d) $(CH_3)_2CH - C \equiv C - H$

68. The product obtained on the reduction of but - 2 - yne in the presence of Lindlar's Catalyst is-

(a) trans but - 2 - ene

(b) Cis but - 2 - ene

(c) Trans but - 2 - ene&Cis but - 2 - ene

(d) None of these

69. The possible products, in the order of decreasing yield, obtained from the reaction of 3 - bromo - 2, 3 - dimethylpentane with alc. KOH, are -

(a) $(CH_3)_2C = C(CH_3) - CH_2CH_3 > (CH_3)_2CHC(CH_3) = CH - CH_3 > (CH_3)_2CH - C(CH_2 - CH_3) = CH_2$

(b) $(CH_3)_2C = C(CH_3) - CH_2CH_3 < (CH_3)_2CHC(CH_3) = CH - CH_3 < (CH_3)_2CH - C(CH_2 - CH_3) = CH_2$

(c) $(CH_3)_2CH - C(CH_2 - CH_3) = CH_2 > (CH_3)_2C = C(CH_3) - CH_2CH_3 > (CH_3)_2CHC(CH_3) = CH - CH_3$

(d) None of these

70. Consider the following reaction-

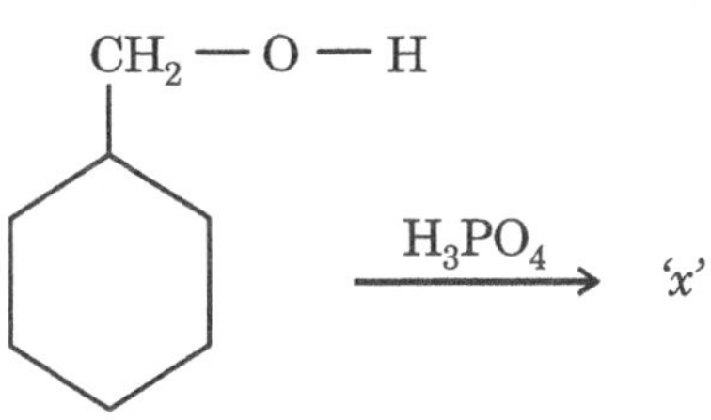

The major product 'x' is

(a) 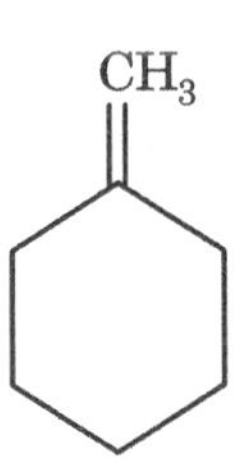

(b) 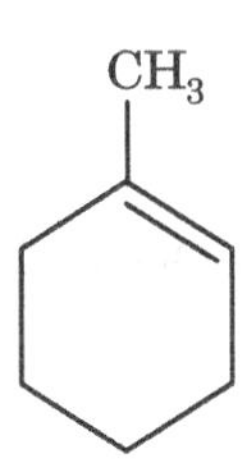

(c) 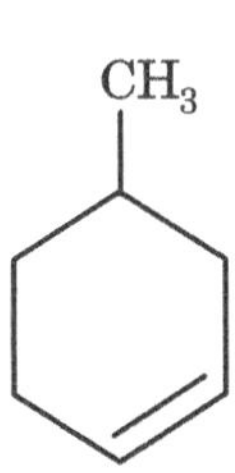

(d) 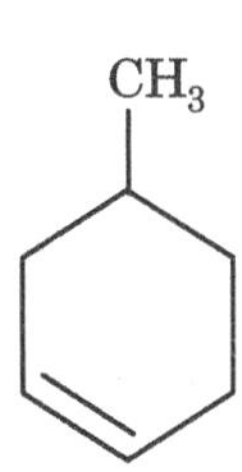

71. Select the correct statement (s)

(a) Heat of hydrogenation of hept - 1 - ene, hept - 3 - ene& 2 - methyl hex - 2 - ene follow the order hept - 1 - ene>hept - 3 - ene> 2 - methyl hex - 2 - ene.

(b) Case of Acid catalyzed dehydration of alcohols to alkenes follows the order 3° > 2° > 1°, based upon the stabilities of respective Carbocations.

(c) $\overset{\ominus}{O}H$ is a poor leaving group.

(d) All are correct.

72. Anti Markowinkoff's rule is not observed in -

(a) pent - 2 - ene (b) but - 2 - ene

(c) but - 1 - ene (d) propene

73. 1 – Butyne $\xrightarrow[\text{(Controlled reduction)}]{\text{lindlar's Catalyst}}$ A $\xrightarrow[\text{Lenzoyl peroxide}]{\text{HCl in}}$ B $\xrightarrow[\text{KOH}]{\text{alc.}}$ C

C, in the above reaction, is -

(a) 1 - butane (b) 2 - butene

(c) 1 - butyne (d) 2 - butyne

74. Reaction of HBr with propene in the presence of peroxide gives mainly -
(a) iso - propyl bromide
(b) allyl bromide
(c) n - propyl bromide
(d) 3 - bromopropane

75. The alkene that gives same product with HBr in the absence as well as in the presence of peroxide is-
(a) 1 - butane
(b) propene
(c) 2 - methyl propene
(d) 2 - butene

76. Addition of which of the following molecules to alkenes involve carbocation as an intermediate -
(a) HX (b) $\frac{H^+}{H_2O}$
(c) X_2 (halogen) (d) Both (a) & (b)

77. The order of reactivity of following alkenes when subjected to acid catalyzed hydration is -
(1) $(CH_3)_2CH = CH_2$
(2) $CH_3 - CH = CH_2$
(3) $CH_2 = CH_2$
(a) 1 > 3 > 2 (b) 1 > 2 > 3
(c) 3 > 2 > 1 (d) 2 > 1 > 3

78. The only alcohol that can be prepared by the indirect hydration of alkene is –
(a) propyl alcohol (b) ethyl alcohol
(c) isobutyl alcohol (d) methyl alcohol

79. Select the incorrect reaction with the corresponding product, shown –
(a) $CH_3 - CH = CH - CH_3 \xrightarrow[KMnO_4]{Cold} CH_3CH(OH)CH(OH)CH_3$
(b) $CH_3 - C \equiv C - CH_3 \xrightarrow[KMnO_4]{Cold} CH_3 - \overset{O}{\overset{\|}{C}} - \overset{O}{\overset{\|}{C}} - CH_3$
(c) $CH_3 - C \equiv CH \xrightarrow[KMnO_4]{hot} CH_3COOH + CO_2 \uparrow$
(d) None of these

80. Oxidative ozonolysisof A (C_4H_8) gives CH_3COOH only. 'A' is –
(a) 
(b) 
(c) $CH_3 - CH = CH - CH_3$
(d) $CH_3 - CH_2 - CH = CH_2$

81. One mole of alkene X on ozonolysis gave one mole of acetone. The IUPAC name of alkene 'X' is -
(a) 2 - methyl but - 1 - ene
(b) 2 - methyl but - 2 - ene
(c) but - 2 - ene
(d) but - 1 - ene

82. Select the reaction with 'INCORRECT' product ?
(a) $CH_3 - CH = CH - CH_3 \xrightarrow[KMnO_4]{hot} CH_3COOH + CH_3COOH$
(b) 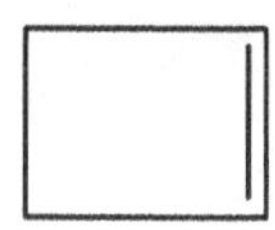 $\xrightarrow[KMnO_4]{hot}$ $\begin{array}{l} CH_2-COOH \\ | \\ CH_2-COOH \end{array}$
(c) 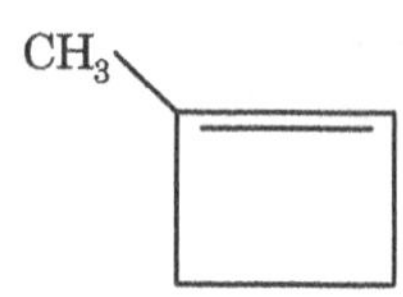 $\xrightarrow[KMnO_4]{hot}$ $\begin{array}{l} \quad\quad CH_3 \\ \quad\quad | \\ CH_2-C=O \\ | \\ CH_2-COOH \end{array}$
(d) None of these

83. Which will give an alkene on Kolbe's electrolysis?
(a) $CH_3COO^-Na^+$
(b) $CH_3 - CH_2 - COO^-Na^+$
(c) $\begin{array}{l} CH_3 - C - COO^-Na^+ \\ \quad\quad \| \\ CH_3 - C - COO^-Na^+ \end{array}$
(d) $\begin{array}{l} CH_2 - COO^-Na^+ \\ | \\ CH_2 - COO^-Na^+ \end{array}$

84. A hydrocarbon X adds one mole of hydrogen to give another hydrocarbon & decolorized bromine water. X reacts with hot $KMnO_4$ to give two moles of the same carboxylic acid. The structure 'X' is -

(a) $CH_3-CH=CH-CH_2CH_2CH_3$

(b) $CH_3-CH_2-CH=CH-CH_2CH_3$

(c) $CH_3-CH_2=CH_2-CH=CH-CH_3$

(d) $CH_2=CH-CH_2CH_2CH_3$

85. First step in acid catalyzed hydration of isobutene involves the attack of -

(a) $\dot{H}$ (b) H^+

(c) $\bar{O}H$ (d) H_2O

86. When 3, 3 - dimethyl - 2 - butanol is heated with conc. H_2SO_4, the major product obtained is -

(a) Cis & trans isomers of 2, 3 - dimethyl - 2 - butane

(b) 3, 3 - dimethyl - 1 - butane

(c) 2, 3 - dimethyl - 2 - butane

(d) 2, 3 - dimethyl - 1 - butane

87. The final product of the reaction given below -

$$CH_2=CH-\ddot{O}CH_3+HBr\xrightarrow[\text{temp.}]{\text{Room}}\text{'X'}$$

(a) $CH_3-\underset{\substack{|\\Br}}{CH}-OCH_3$

(b) $CH_2=CH-Br$

(c) CH_3OH

(d) Both (b) & (c)

88. 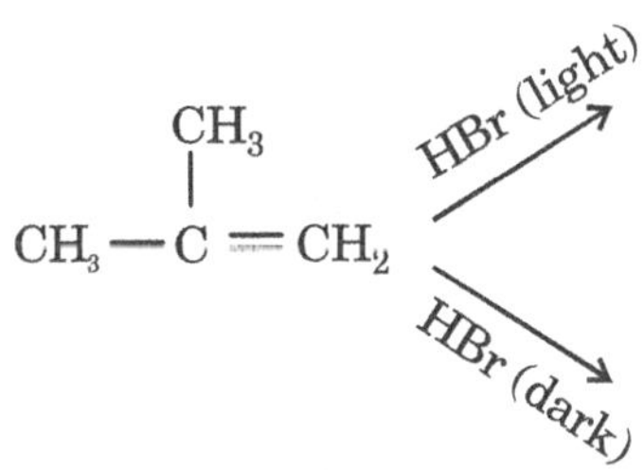

A & B are, respectively -

(a) 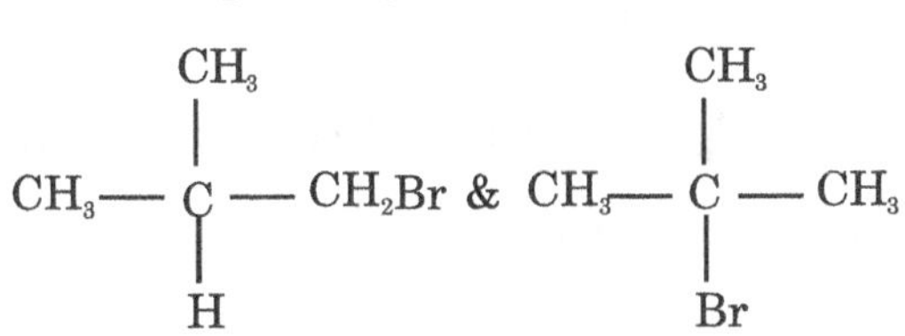

(b) 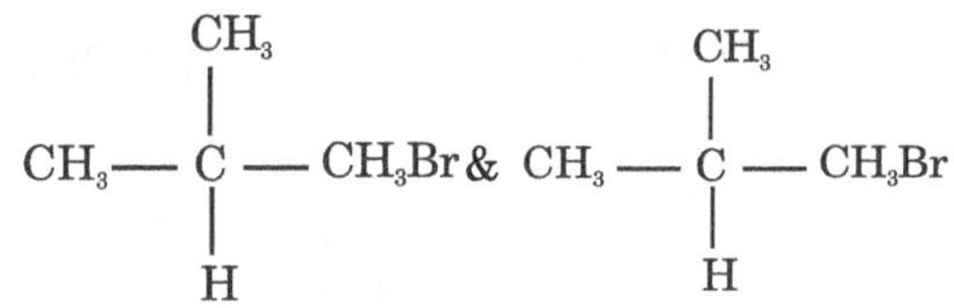

(c) 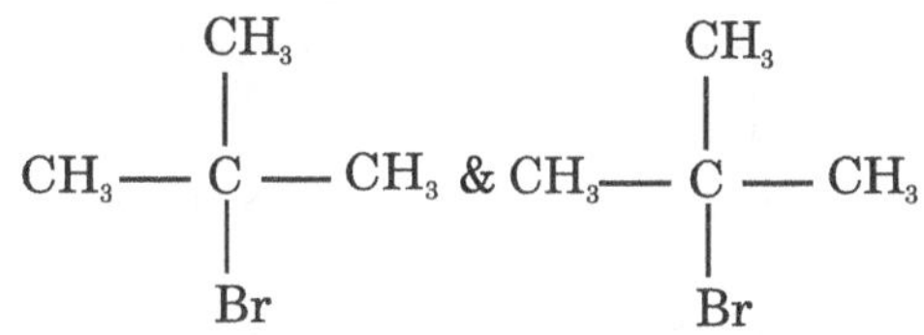

(d) 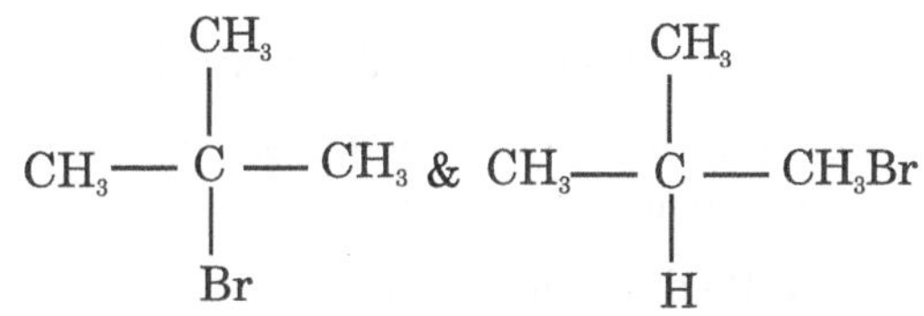

89. On reacting the following , the product formed is–

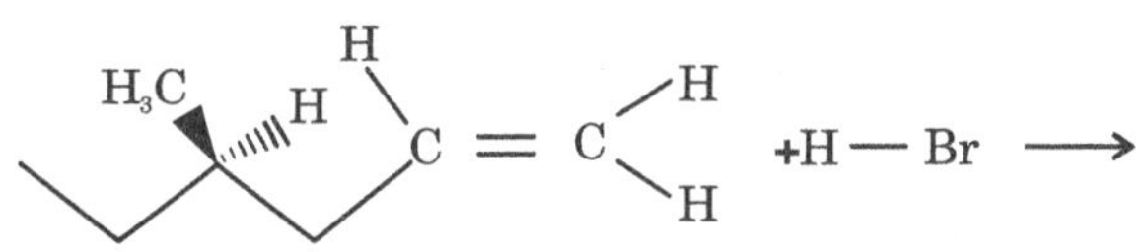

(a) Racemic mixture (b) Diastereomeric mixture

(c) Meso compound (d) None of these

90. On reacting butene with HBr, the major product formed is –

(a) (+) –2 – Bromobutane

(b) (–) – 2 – Bromobutane

(c) 1 – Bromobutane

(d) (±) 2 – Bromobutane

91. In which of the following cases, Markownikoff's rule is violated?

(a) $CH_3-NH-CH=CH_2 + H-Br \longrightarrow$

(b) $CH_3-S-CH=CH_2 + HBr \longrightarrow$

(c) $O_2N-CH=CH_2 + HBr \longrightarrow$

(d) Both (a) & (b)

92. Rate of reaction of following molecules towards electrophilic addition follows the order –

(1) 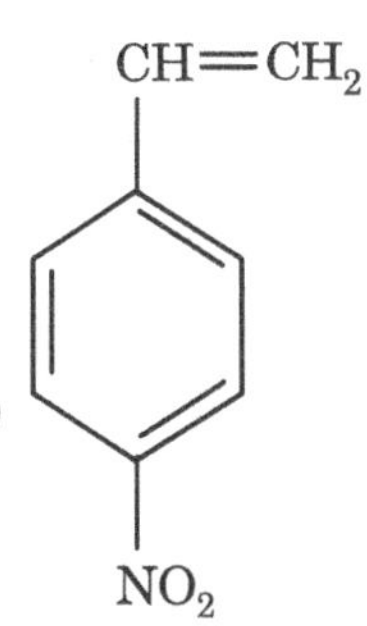

(2) 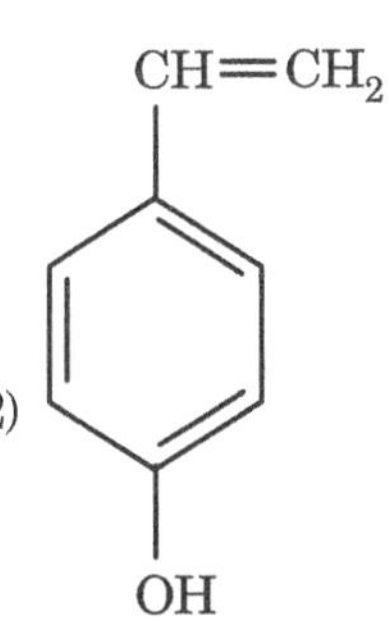

(3) 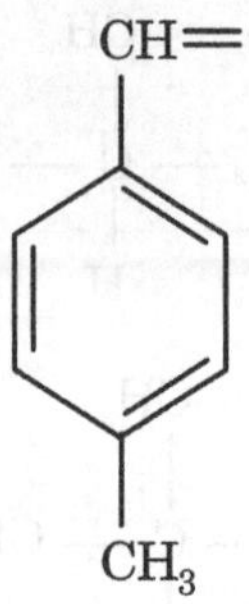 (4) 

(a) 4 > 2 > 3 > 1 (b) 2 > 3 > 4 > 1

(c) 1 > 2 > 3 > 4 (d) 3 > 2 > 4 > 1

93. Reaction of ethylene with Br_2 in water in the presence of NaCl gives –

(a) 1, 2 - dibromoethane

(b) 2 - bromoethanol

(c) 1 - bromo - 2 - chloroethane

(d) All of these

94. 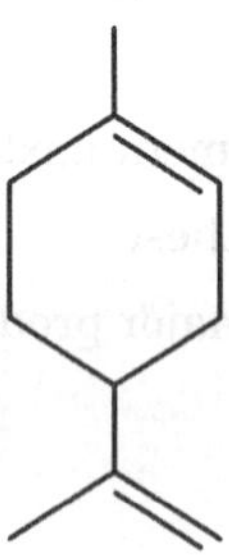 $\xrightarrow[\text{(1 mole)}]{H_2,\ Ni}$ 'X' is-

(a) 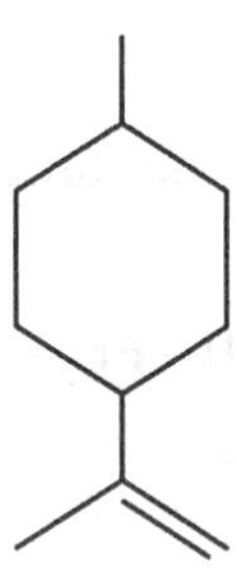 (b) 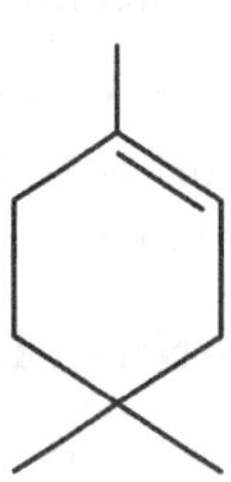

(c) 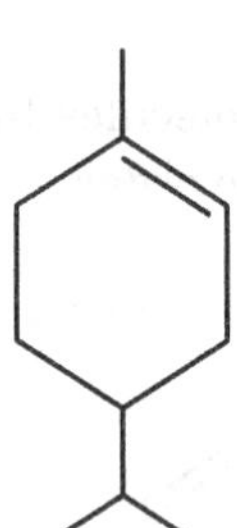 (d) Both (a) & (b)

95. Ease of acid catalyzed dehydration follows the order –

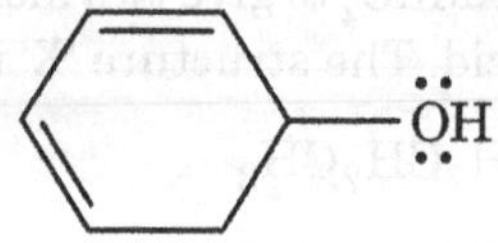
(1)

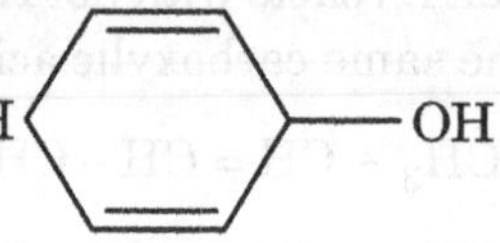
(2)

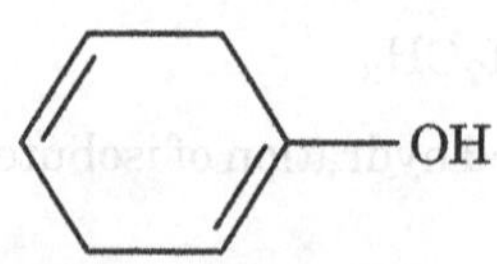
(3)

(a) 1 > 2 > 3 (b) 3 > 2 > 1

(c) 2 > 1 > 3 (d) 2 = 1 > 3

96. An alkene on ozonolysis yields only ethanal. There is an isomer of this , which on ozonolysis yields -

(a) Propanone & methanol

(b) Propanone & ethanol

(c) Ethanol & methanol

(d) Only propanone

97. Consider the following reaction–

$$A \xrightarrow[H_2O]{O_3} CH_3-\overset{\overset{O}{\|}}{C}-CH_3 + CH_3COOH + CH_3-\overset{\overset{O}{\|}}{C}-COOH$$

A can be –

(a) $CH_3-\underset{}{C}(=CH-CH_3)-CH=C(CH_3)_2$ i.e. $CH_3-\overset{\overset{CH-CH_3}{\|}}{C}-CH=C\langle^{CH_3}_{CH_3}$

(b) $CH_3-\overset{\overset{C(CH_3)_2}{\|}}{C}-CH=CH-CH_3$

(c) $CH_3-\overset{\overset{C(CH_3)_2}{\|}}{C}-CH=CH-CH_2-CH_3$

(d) Both (a) & (b)

98. Which of the following alcohols can't be prepared from an alkene?

(a) 

(b) 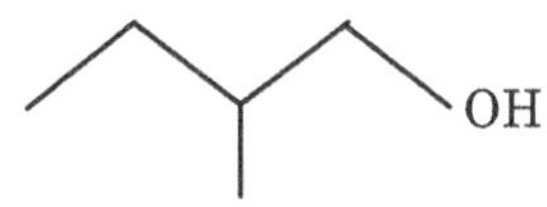

(c) 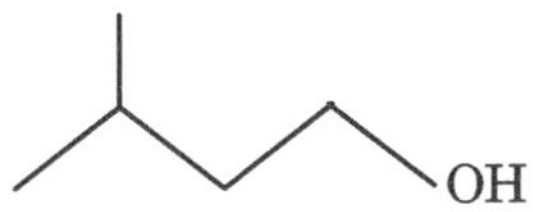

(d) 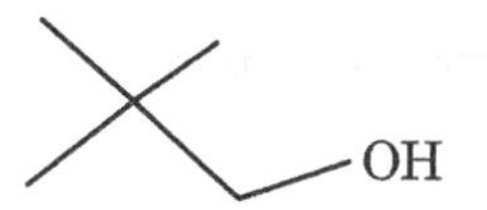

99. Which of the following orders of reactivity are correct towards acid catalyzed hydration of an alkene?

(a) 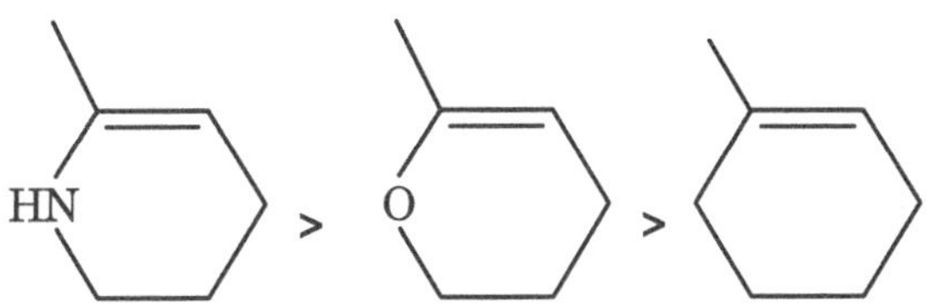

(b) $(H_3C)(CH_3O)C=CH_2$ > $(H_3C)(CH_3OCH_2)C=CH_2$

(c) $CH_3-CH=CH_2$ > $CH_2=CH_2$

(d) All are correct

100. Which of the following will give only one alkene on dehydrohalogenation? (including isomers)

(a) $CH_3-CH(Cl)-CH_2-CH_2-CH_3$

(b) $CH_3-CH_2-C(H)(Cl)-CH_2-CH_3$

(c) $CH_3-CH_2-CH_2-Cl$

(d) Both (a) & (b)

101. Consider the following reaction -

(R) – 3 – methoxy – butene $\xrightarrow[\text{(electrophillic addition)}]{\text{HBr}}$ A (Major)

(R) – 3 – methoxy – butene $\xrightarrow{\text{HBr}/R_2O_2}$

Which of the following is correct wrt above sequence of reactions?

(a) B is optically pure R isomer.

(b) A is a racemic mixture.

(c) The number of diastereomeric pairs in the major product A are 4.

(d) B does not have any stereo centre.

(a) a & b but not c & d

(b) a & c but not b & d

(c) b & c but not a & d

(d) b & d but not a & c

102. Select the correct option

(a) 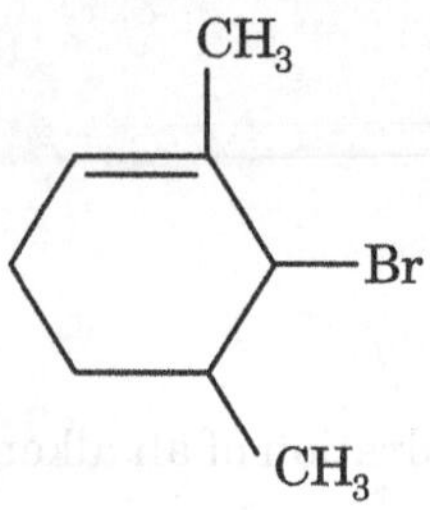 ⟶ Can undergo E_2 elimination.

(b) 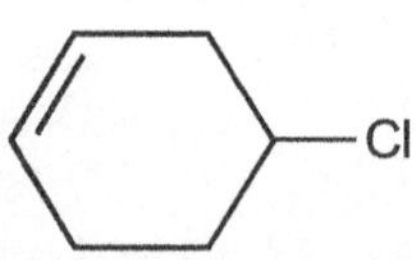 ⟶ loses chirality in both E_1 & E_2

(c) 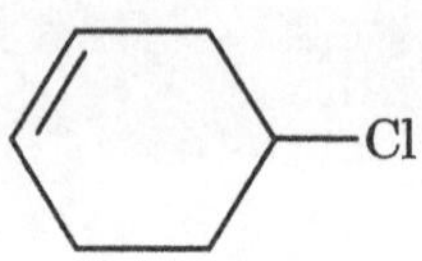 ⟶ produces same product in SN^1 & SN^2

(d) Both (a) & (c) are correct

Alkynes

103. Identify a reagent from the following list which can easily distinguish b/w 1 - butyne& 2 - butyne.

(a) $-Br_2$, CCl_4 (b) H_2, Lindlar's catalyst

(c) $HgSO_4$, dilute H_2SO_4 (d) Ammoniacal Cu_2Cl_2 Solution

104. $Ph-C\equiv C-CH_2CH_3 \xrightarrow{Na/liq.NH_3} A \xrightarrow{Br_2} B$

The final products 'B' is/are -

(a) Racemic mixture (b) Diastereomers

(c) Meso compound (d) Structural isomers

105. 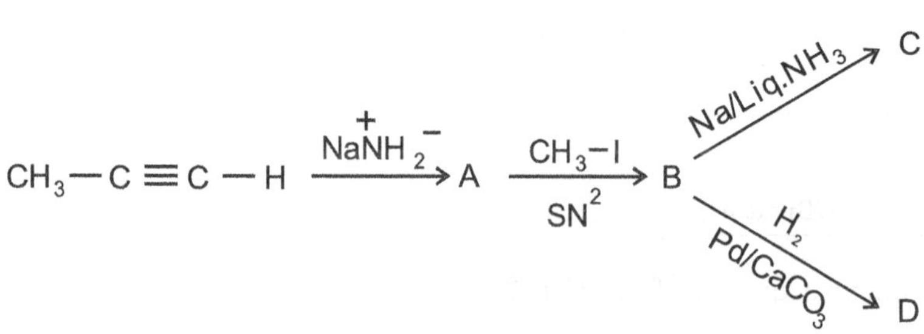

C & D represent -

(a) Enantiomer (b) Diastereomers

(c) Geometrical isomers (d) Meso compound

106. Identify the compound 'A' in the following sequence –

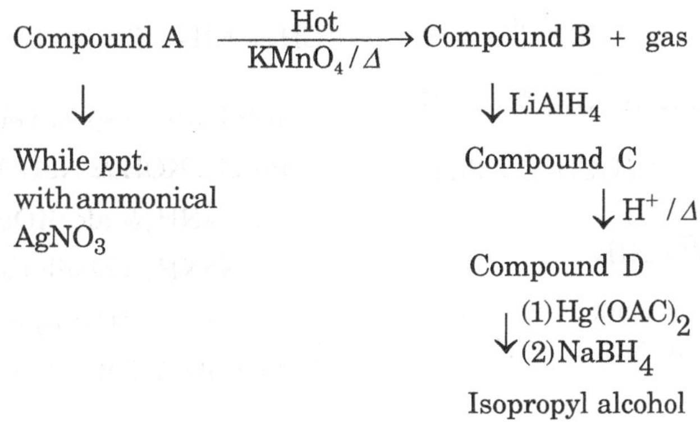

(a) $CH_3-CH_2-C\equiv C-H$ (b) $CH_3-CH=CH-CH_2-CH_3$

(c) $CH_3-CH_2-CH_3$ (d) $CH_2=CH-CH_3$

107. 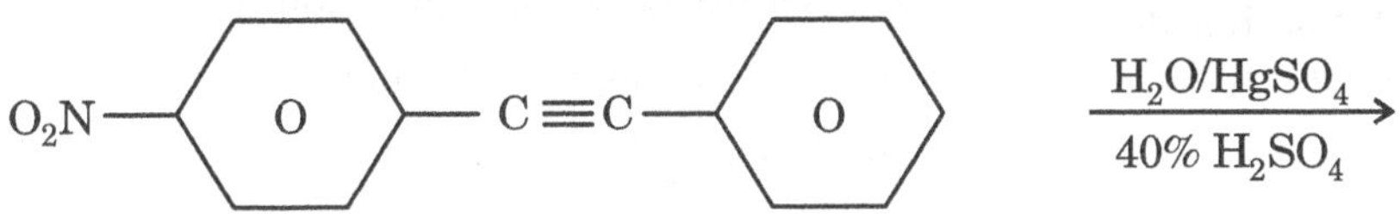 X,

Here, x is major product –

(a) 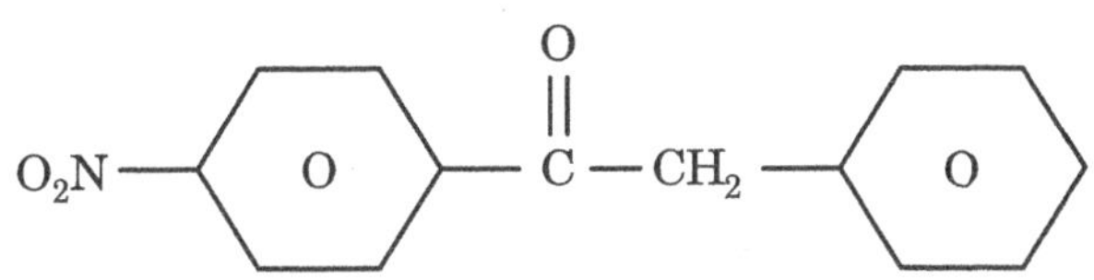

(b) 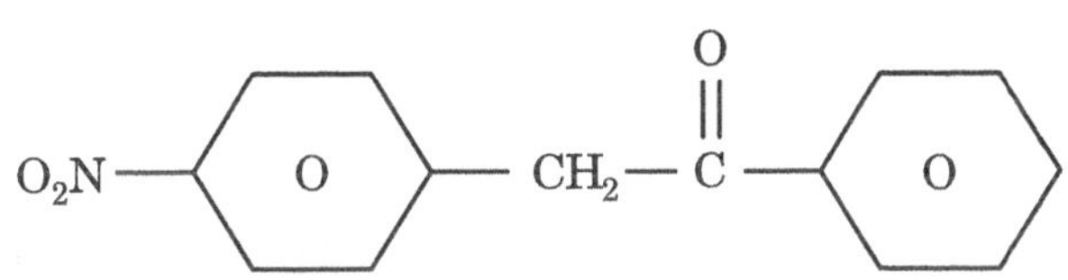

(c) 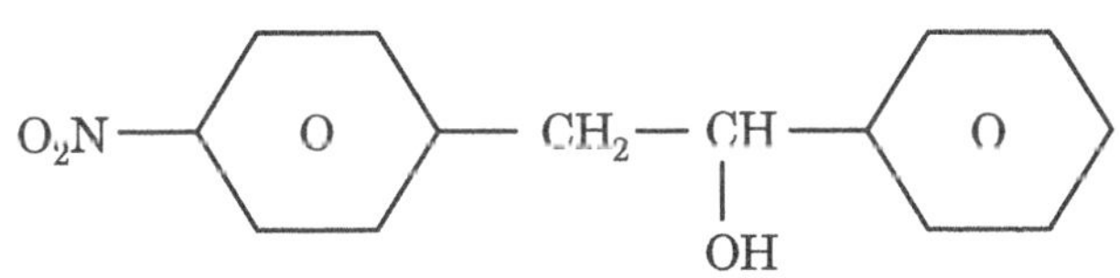

(d) None of these

108. Which of the following reaction will yield an alkyne?

(a) 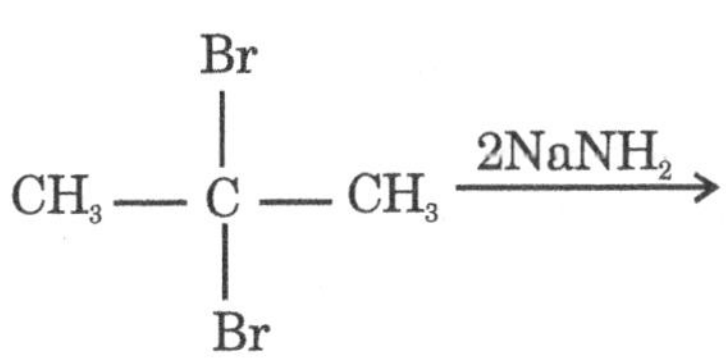

(b) $Cl_2CH-CHCl_2+Zn \xrightarrow{alc.}$

(c) $CaC_2+H_2O \longrightarrow$ (d) All of these

109. Which of these will not react with acetylene?

(a) NaOH (b) Ammoniacal $AgNO_3$

(c) Na (d) HCl

110. Which acid base reaction will give negligible yield?

(a) $H-C\equiv C-H+\overset{\ominus}{N}H_2 \rightleftharpoons H-C\equiv C^{\ominus}+NH_3$

(b) $H-C\equiv C-H+RO^{\ominus} \rightleftharpoons H-C\equiv C^{\ominus}+ROH$

(c) $CH_3COOH+RO^{\ominus} \rightleftharpoons CH_3COO^{\ominus}+ROH$

(d) $ROH+\overset{\ominus}{N}H_2 \rightleftharpoons RO^{\ominus}+NH_3$

111. Which of the following does not give a white precipitate with ammoniacal $AgNO_3$ Sol.?

(a) Propyne

(b) But - 1 - yne

(c) 3 - methyl but - 1 yne

(d) But - 2 - yne

112. A gas decolorizes alkaline $KMnO_4$ Sol. but does not give precipitate with silver nitrate. It is–

(a) CH_4 (b) C_2H_6

(c) C_2H_4 (d) C_2H_2

113. Consider the following reaction

$$CH_3-CH_2-C\equiv C-H \underset{B}{\overset{A}{\rightleftharpoons}} CH_3-C\equiv C-CH_3$$

A & B are, respectively-

(a) alc. KOH & $NaNH_2$

(b) $NaNH_2$ & alc. KOH

(c) $NaNH_2$ &Lindlar's Catalyst

(d) Lindlar's Catalyst & $NaNH_2$

114. Identify A & B in the following sequence-

$$CaC_2 + H_2O \longrightarrow A \xrightarrow[H_2SO_4]{HgSO_4} B$$

(a) A is ethylene, B is acetaldehyde

(b) A is acetylene, B is acetaldehyde

(c) A is ethane, B is ethanol

(d) A is acetylene, B is propionaldehyde

115. Consider the following reaction-

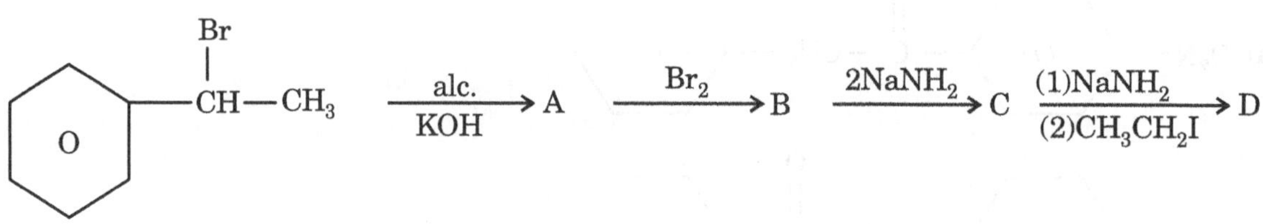

Which of the following would be D?

(a) 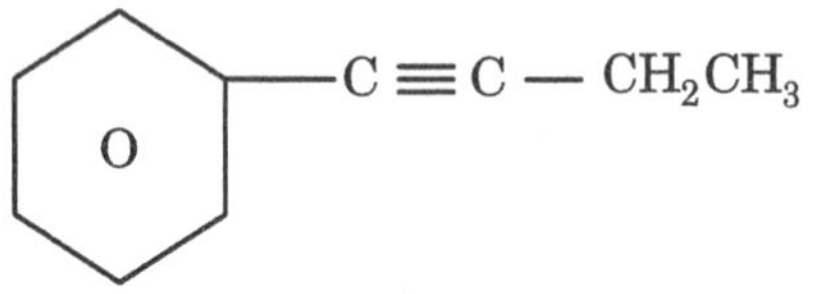

(b) 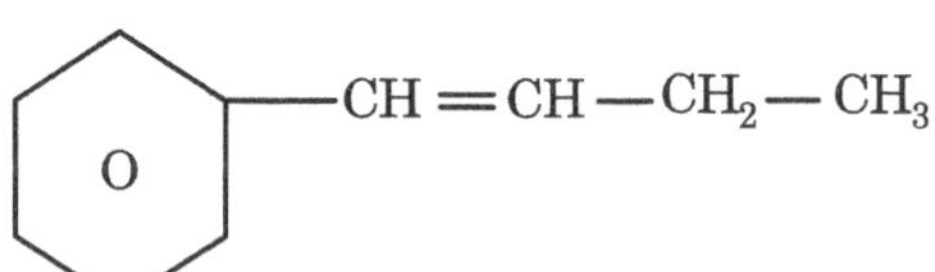

(c) 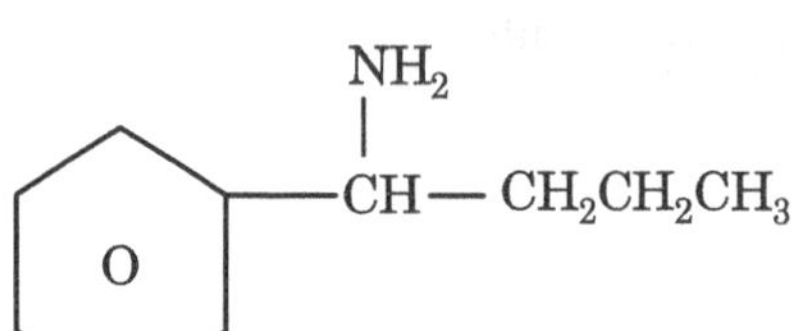

(d) 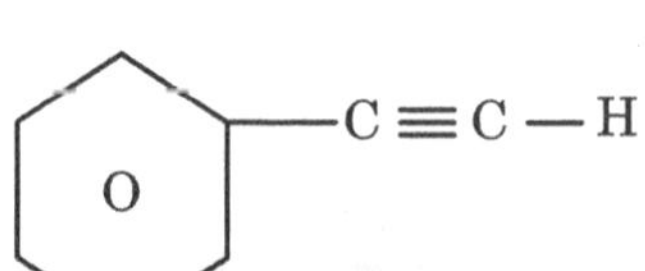

116. In the following sequence of reactions, the product (D) is–

$$H-C\equiv C-H \xrightarrow{HBr} A \xrightarrow{HBr} B \xrightarrow[KOH]{alc.} C \xrightarrow{NaNH_2} D$$

(a) Ethanol (b) Ethyne

(c) Ethanol (d) Ethene

117. The synthesis of pent - 2 - yne is best achieved by placing a chloroalkane into a mixture of sodium amide & an alkyne. The chloro alkane & alkyne respectively are–

(a) $CH_3-C\equiv C-H + CH_3CH_2CH_2-Cl$

(b) $CH_3-CH_2-C\equiv CH + CH_3Cl$

(c) $CH_3-C\equiv C-CH_3 + CH_3Cl$

(d) $CH_3-CH_2-C\equiv C-H + CH_3CH_2Cl$

Benzeme

118. Which of the reaction are correctly matched with their products?

(a) 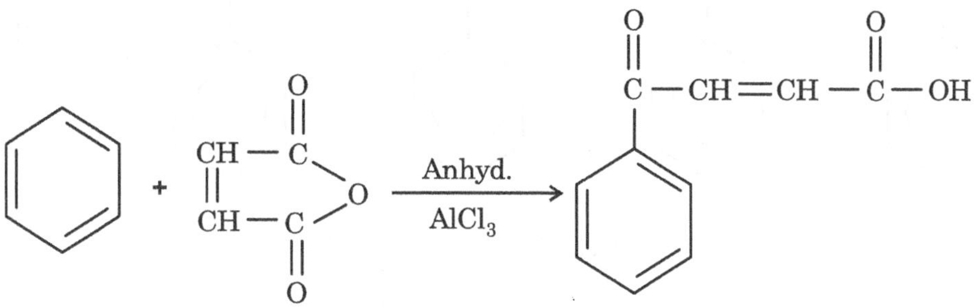

(Maleic anhydride)

(b) 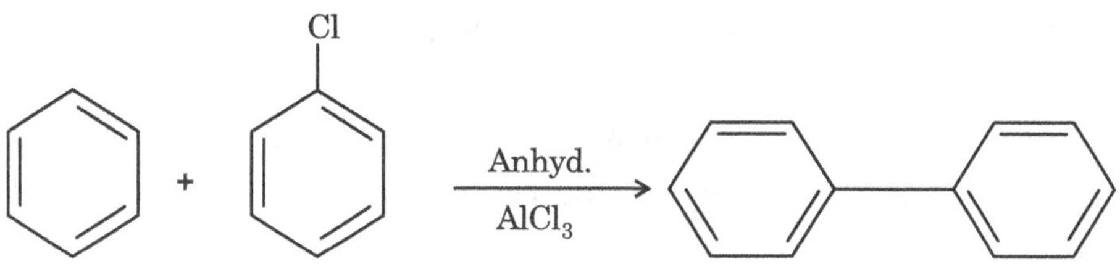

(c) 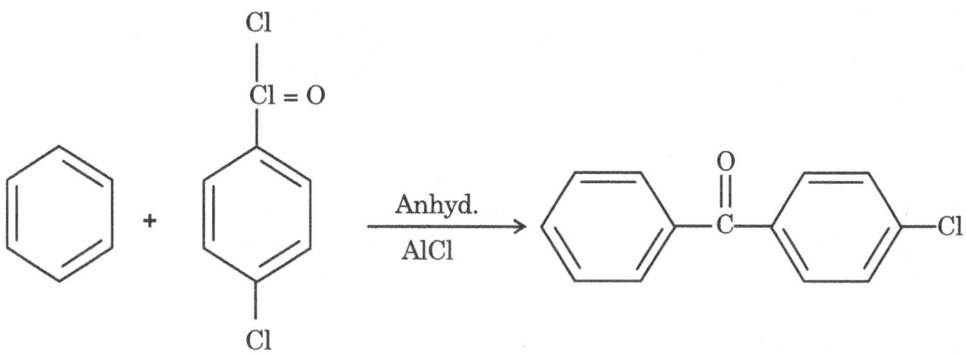

(d) Both (a) & (c) are correct

119. Which of the following can be used for nitration of benzene?

(a) $HNO_3 + H_2SO_4$ (nitrating mixture)

(b) CH_3COONO_2

(c) N_2O_5

(d) All of these

120. Which of the following molecules fail to give Friedal Craft's reaction?

(a) 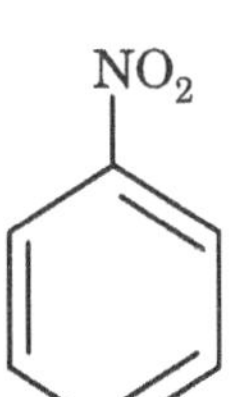

(b) 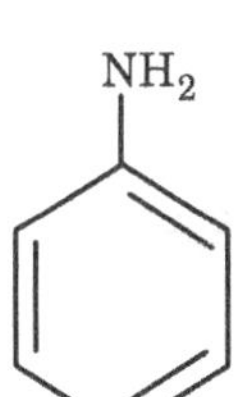

(c) 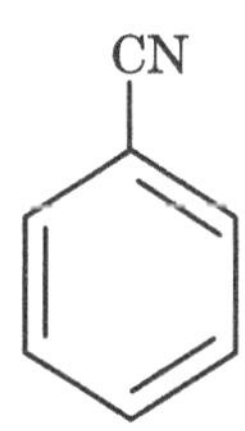

(d) All of these

121. Friedal Crafts alkylation of benzene can be performed by–

(a) $R-Cl + \text{Anhyd. } AlCl_3$

(b) $ROH/ROR + BF_3$

(c) $CH_3-CH=CH_2 \xrightarrow{H_2SO_4}$

(d) All of these

122. How many of the following are more reactive than benzene towards nitration?

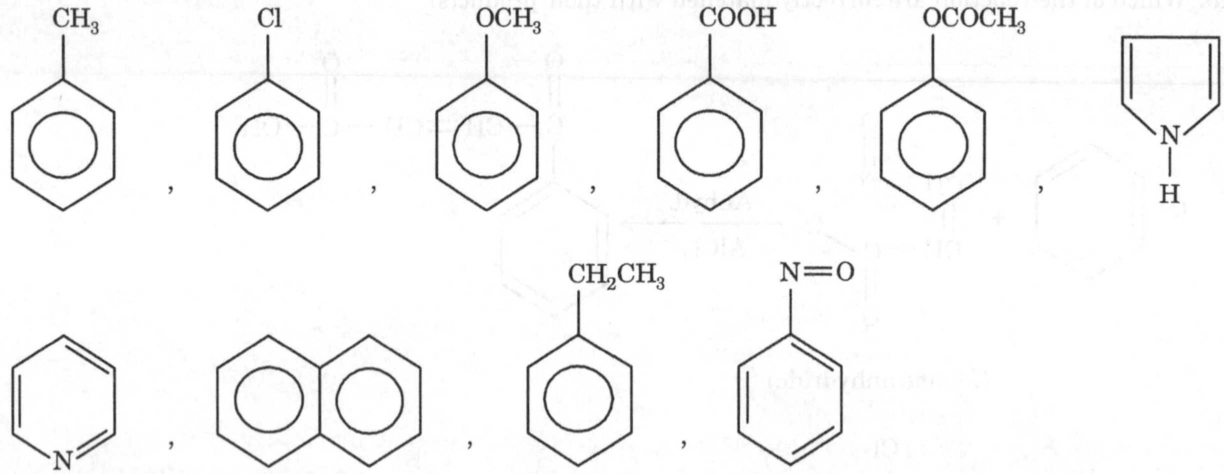

(a) 7 (b) 6 (c) 5 (d) 8

123. Rank the following in order of decreasing reactivity for nitration.

(i) 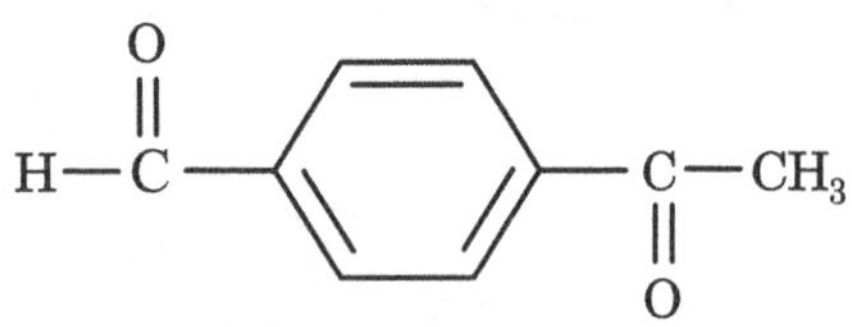

(ii) 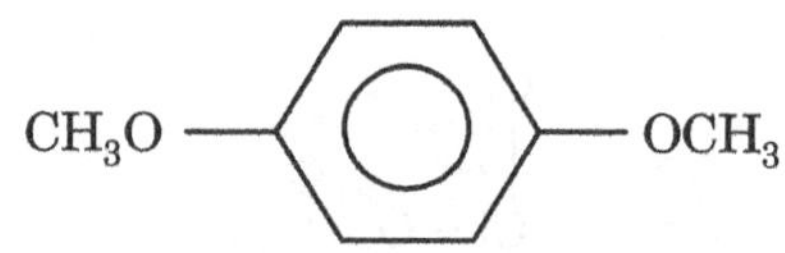

(iii) 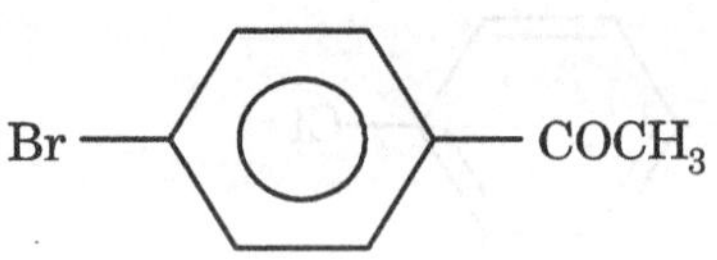

(iv) 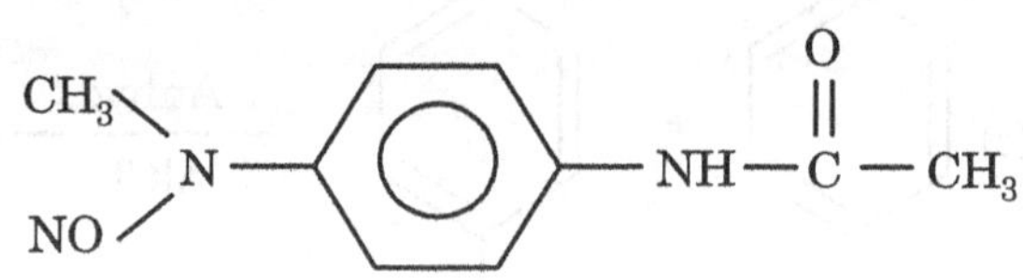

(a) ii > iv > iii > i (b) ii > iv > i > iii (c) ii = iv = i > iii (d) None of these

124. The major product 'Y' of the given reaction is -

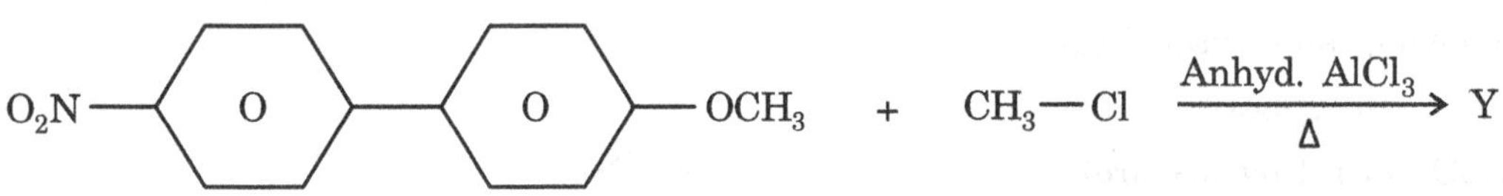

(a) 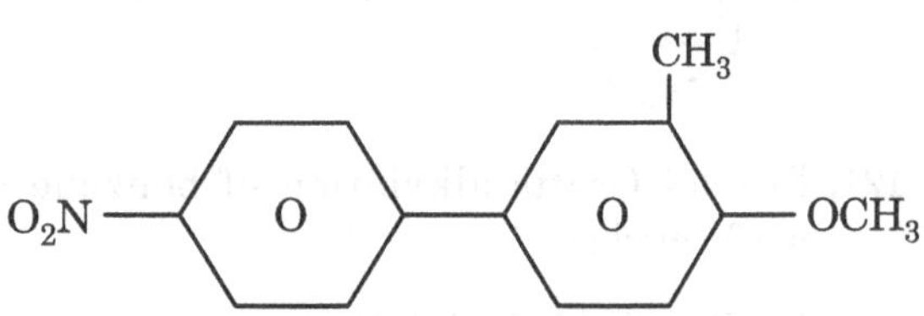

(b) 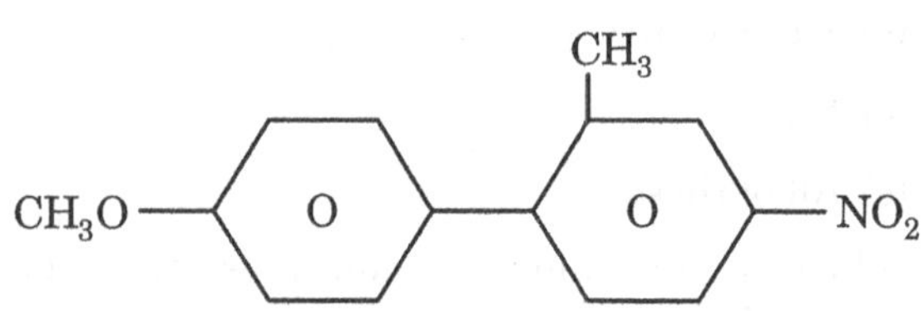

(c) 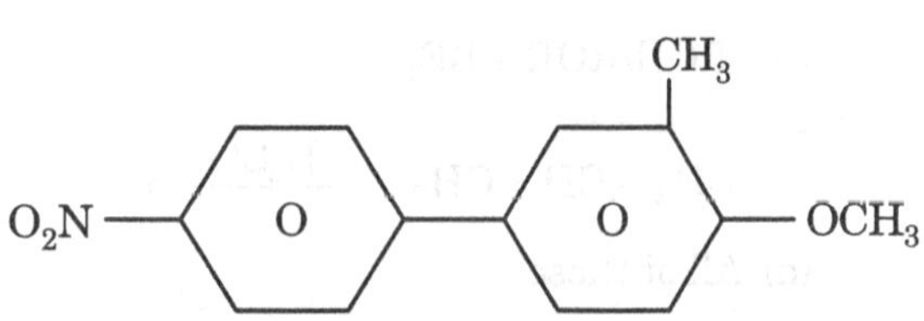

(d) None of these

125. Select the correct option.

(a) Nitration of the compound, 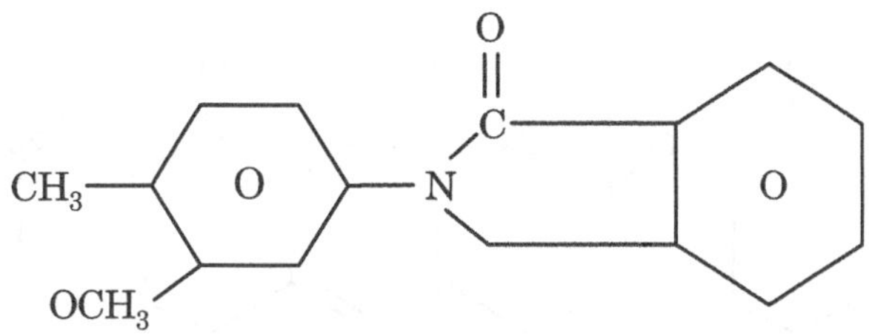 gives

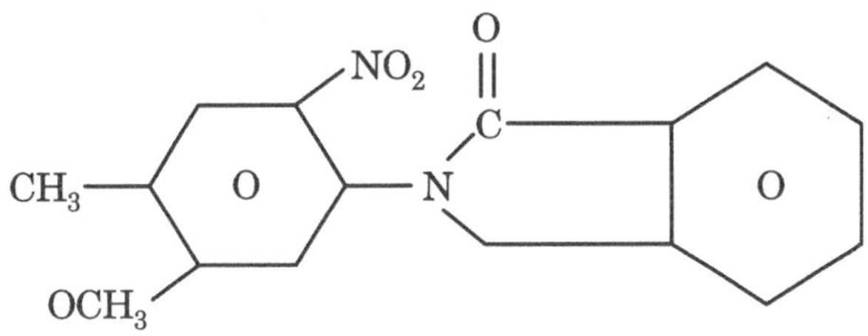

(b) In the compound, 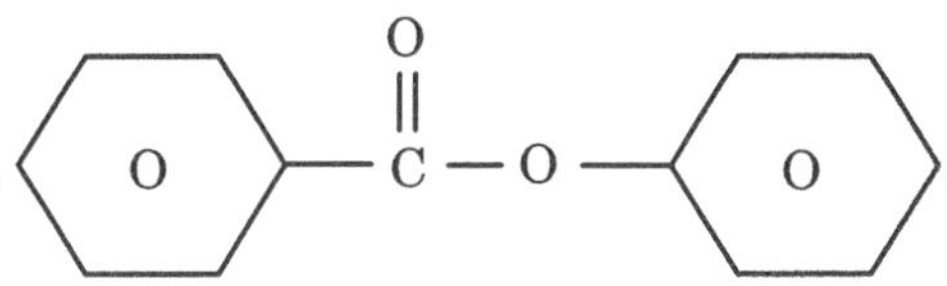 , nitration (electrophilic attack) occurs at II

(I) (II)

ring)

(c) Toluene is more reactive than benzene for electrophilic attack.

(d) All are correct.

126.

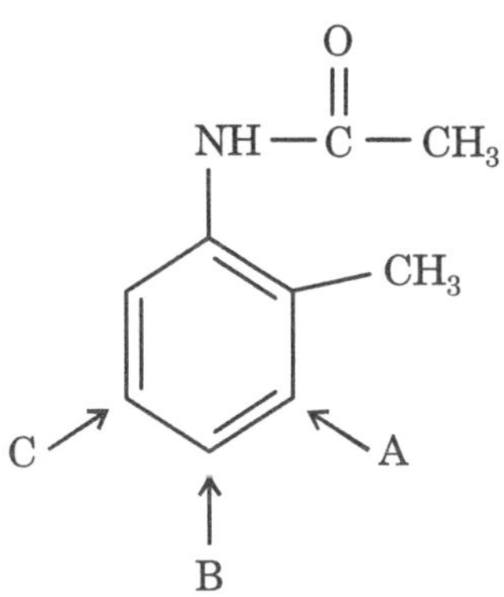

Identify the position where electrophilic aromatic substitution (EAS) is most likely to occur.

(a) A (b) B

(c) C (d) None of these

127. The major product formed in the reaction is -

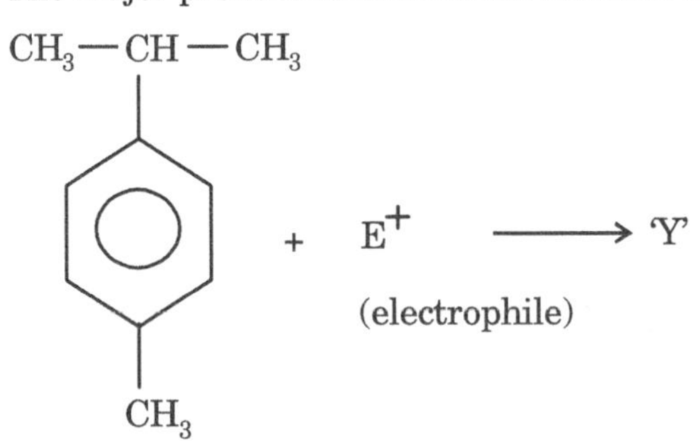

'Y' is –

(a) 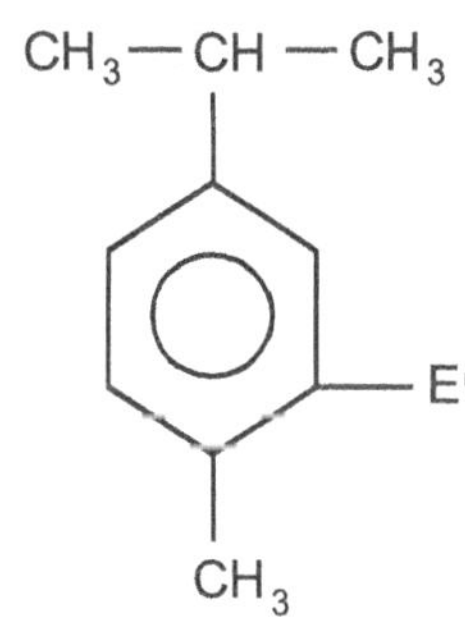 (b) 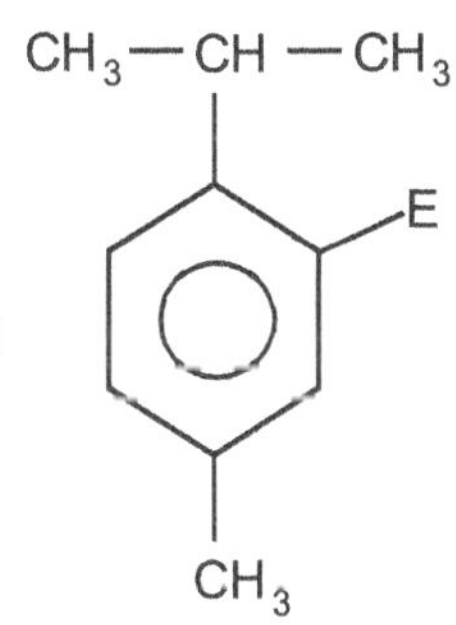

(c) 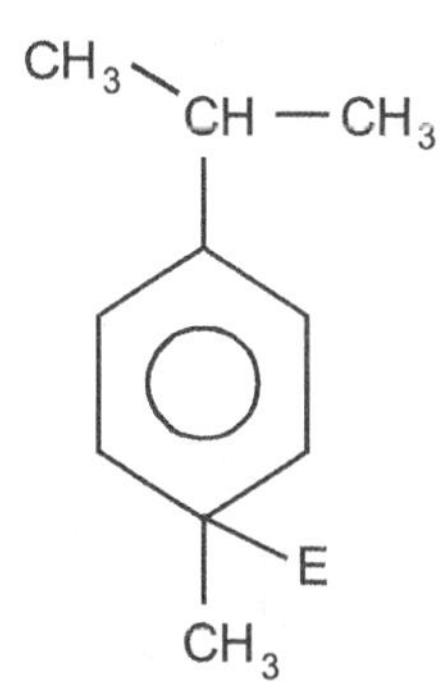 (d) None of these

128. Increasing order of rate of electrophilic substitution follows the order –

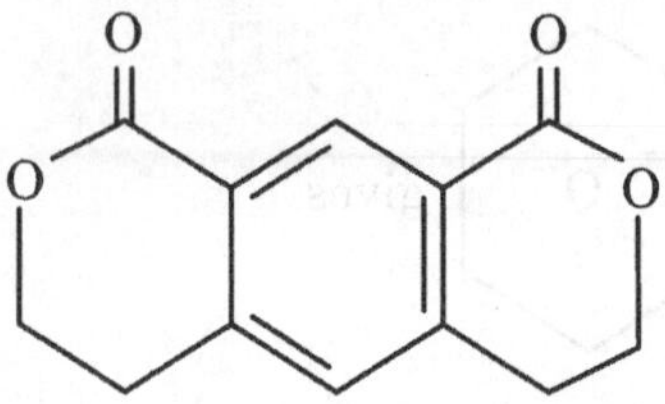

(I)

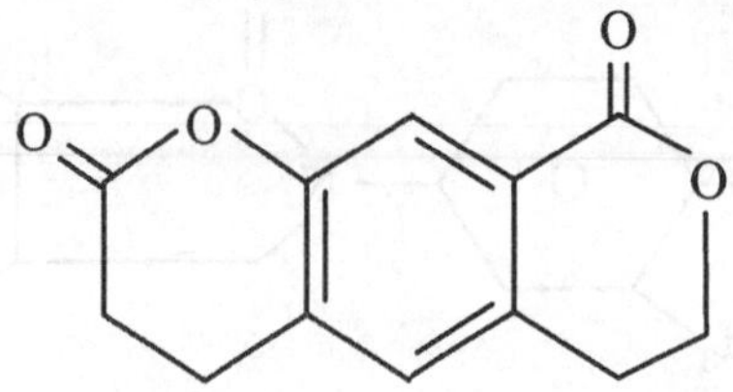

(II)

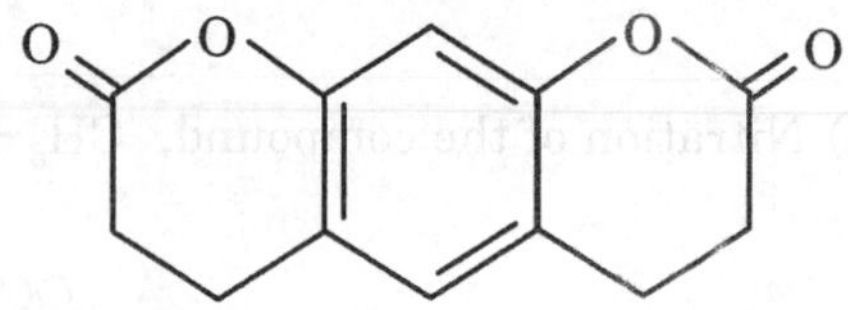

(III)

(a) I > II > III (b) I > III = II

(c) III > II > I (d) None of these

129. 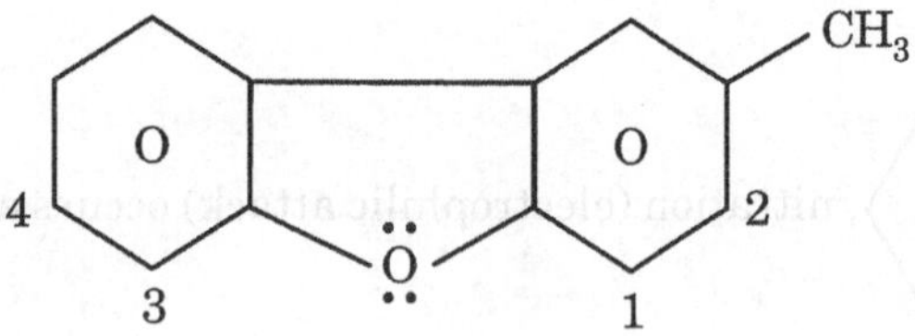

Identify the preferred position where electrophilic aromatic substitution takes place.

(a) 1 (b) 2

(c) 3 (d) 4

130. Which of the following order towards electrophilic aromatic substitution are correct?

(a) 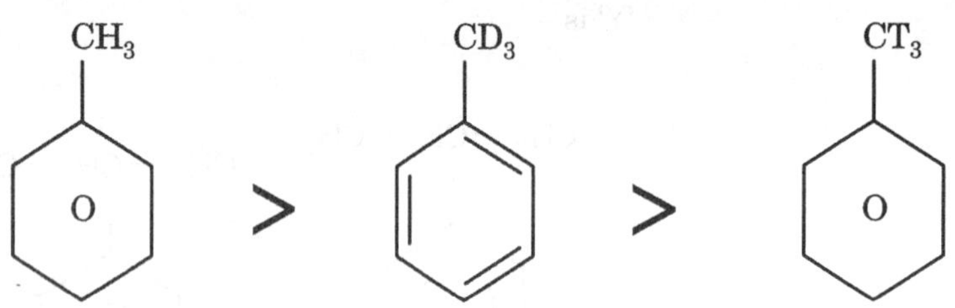

(b) 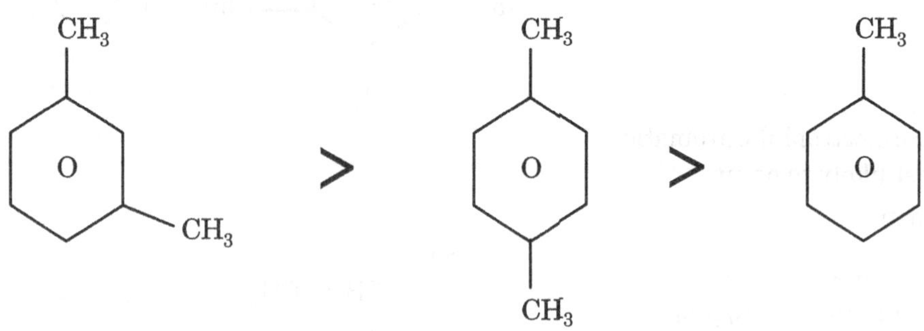

(c) 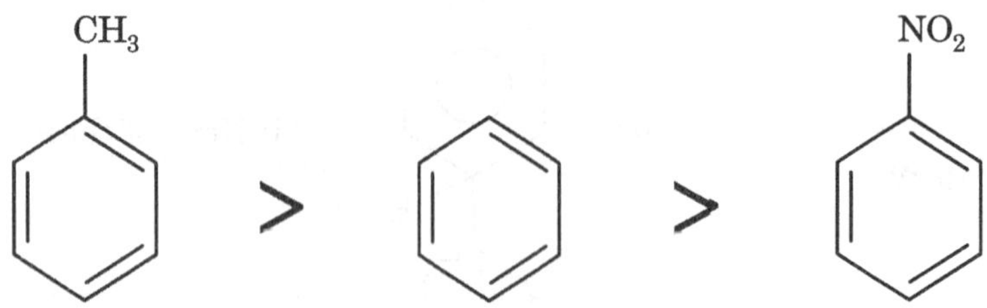

(d) All are correct

131. The major product of the following reaction is –

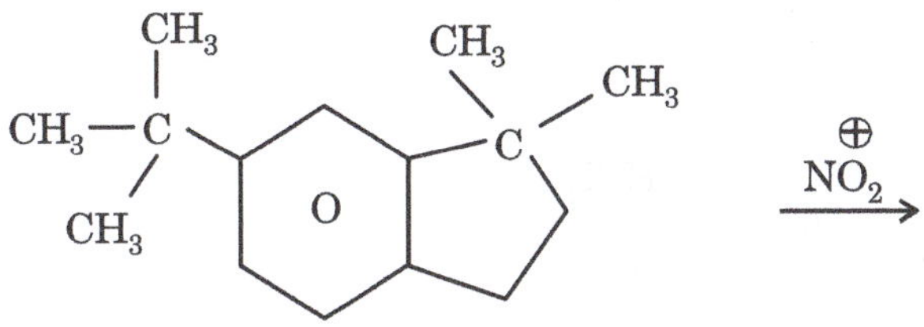

(a) 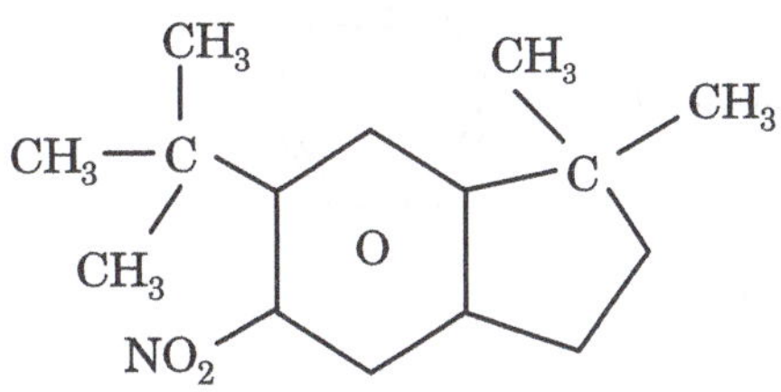

(b) 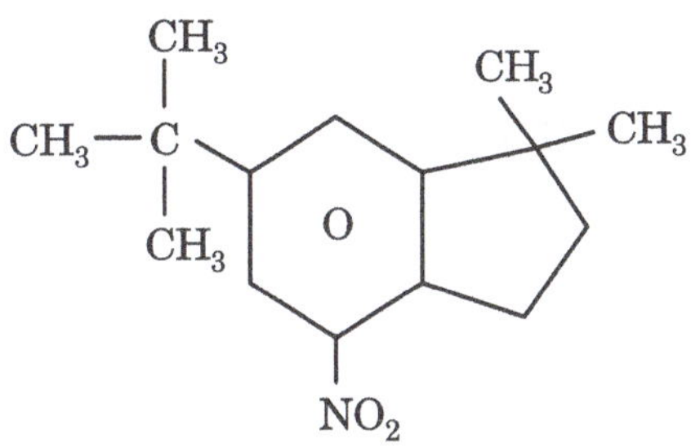

(c) 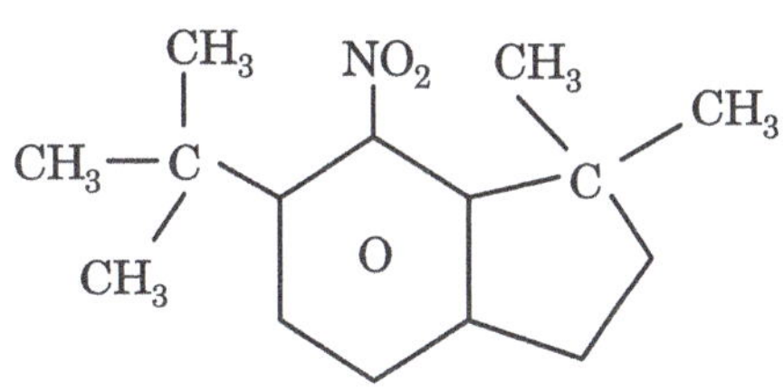

(d) None of these

132. The ring at which rate of bromination is fastest by using Br_2/Fe is –

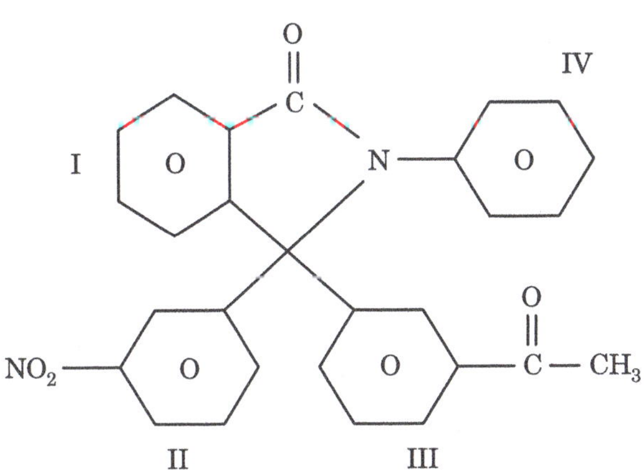

(a) I (b) II
(c) III (d) IV

133. O - Xylene on ozonolysis may give -

(a) $\underset{CHO}{\overset{CHO}{|}}$ only

(b) $CH_3-\overset{O}{\overset{||}{C}}-CHO$ only

(c) $\begin{matrix} CH_3-C=O \\ | \\ CH_3-C=O \end{matrix}$, $CH_3-\overset{O}{\overset{||}{C}}-CHO$ & $\underset{CHO}{\overset{CHO}{|}}$

(d) $CH_3-CH_2-\overset{O}{\overset{||}{C}}-CHO$ & $CH_3-\overset{O}{\overset{||}{C}}-CH_3$

134. $3H-C\equiv C-H \xrightarrow[\text{Fe tube}]{\text{Red hot}} X$

'X' is –

(a) 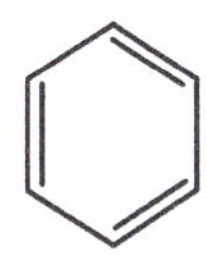

(b) 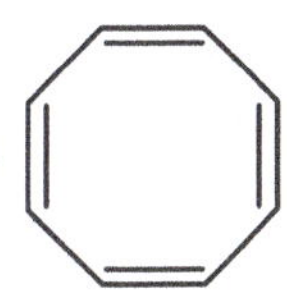

(c) 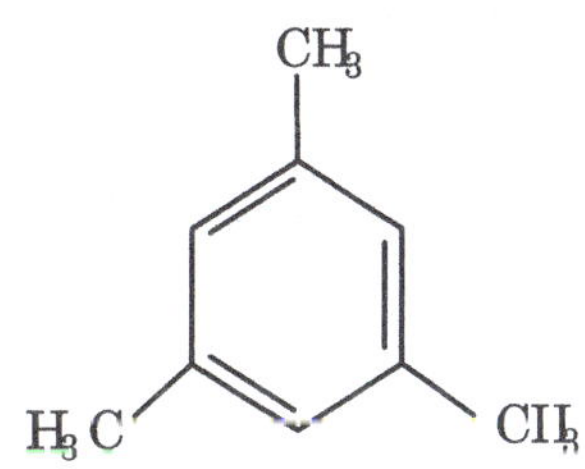

(d) 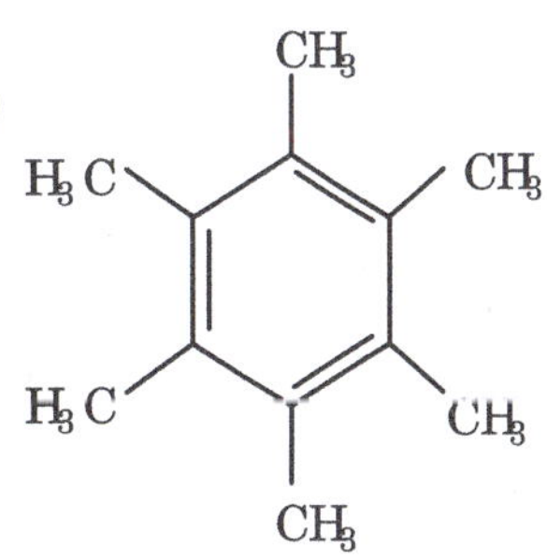

135. Friedal craft's reaction of benzene with isobutyl chloride produces -

(a) Isobutyl benzene

(b) ter - butyl benzene

(c) n - butyl benzene

(d) sec - butyl benzene

136. Which of the following structures correspond to the product expected, when excess of C_6H_6 reacts with CH_2Cl_2 in the presence of anhydrous $AlCl_3$?

(a) 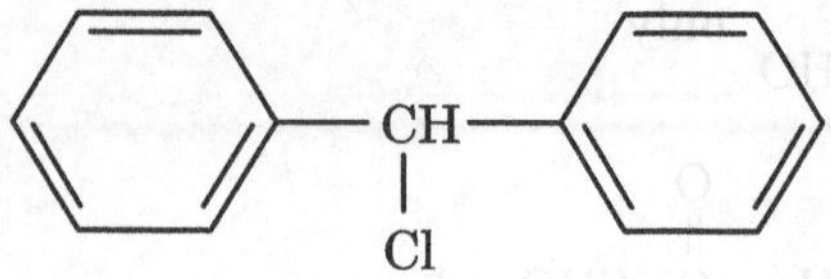

(b) 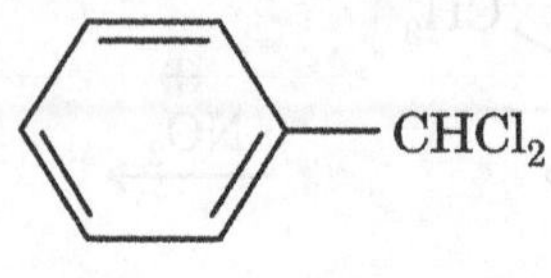

(c) 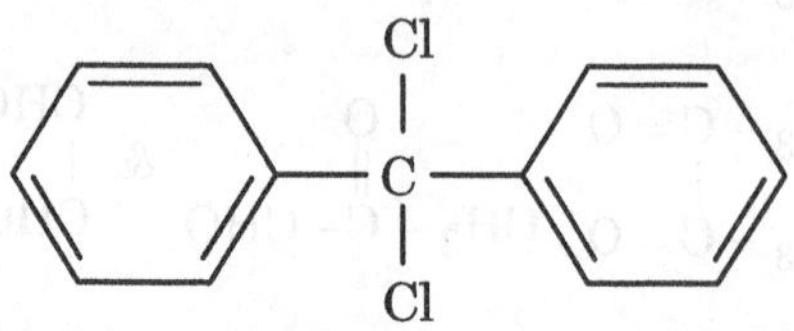

(d) 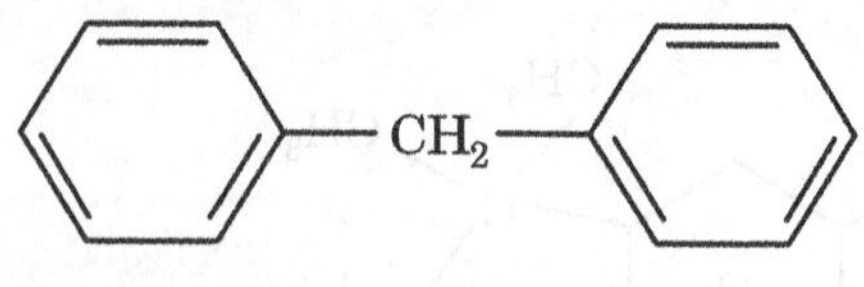

137. Select the correct reactions along with the corresponding products.

(a) 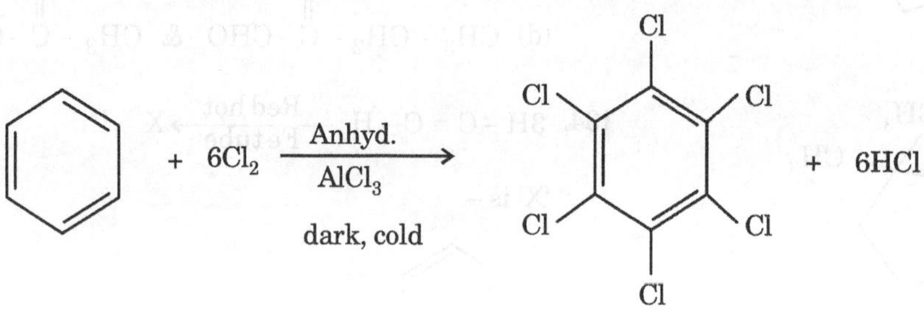

Hexachlorobenzene

(b) 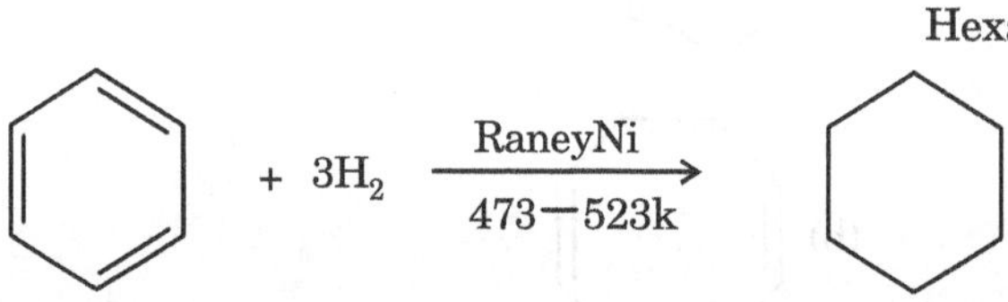

Cyclohexane

(c) 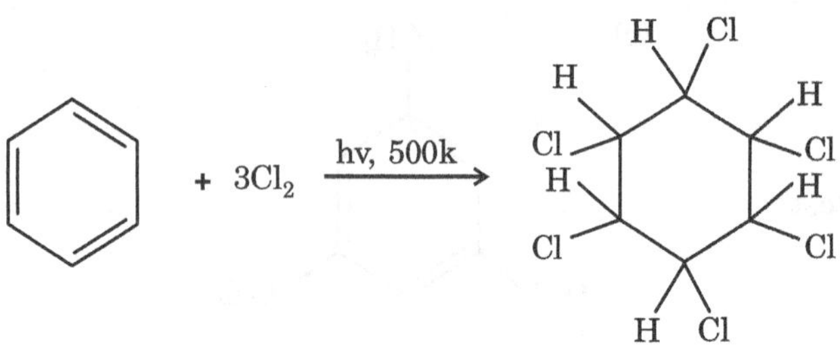

Benzene Hexachloride

(d) All are correct

138. 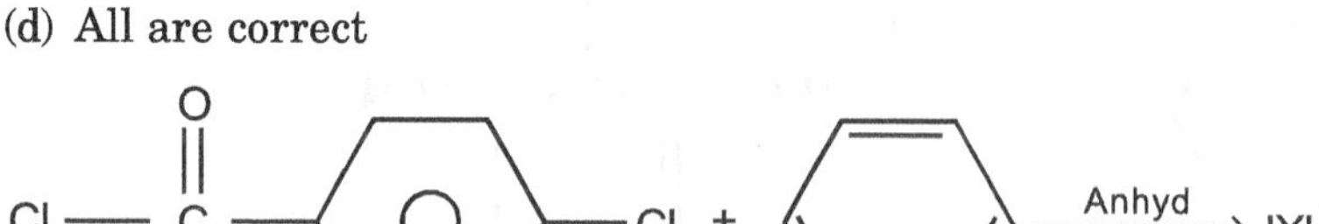

'X' is –

(a) 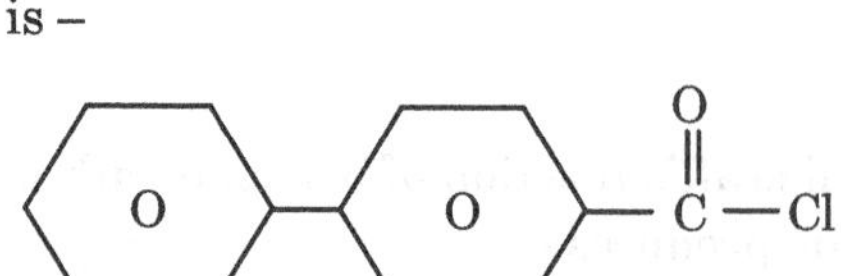

(b) 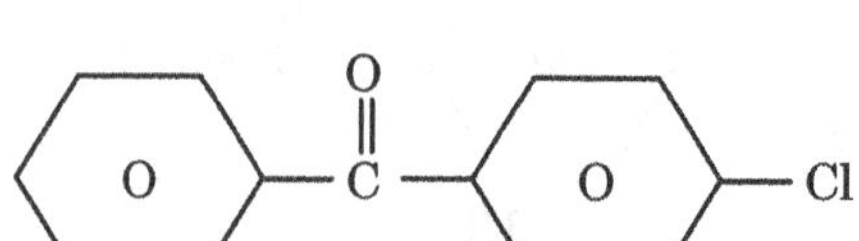

(c) 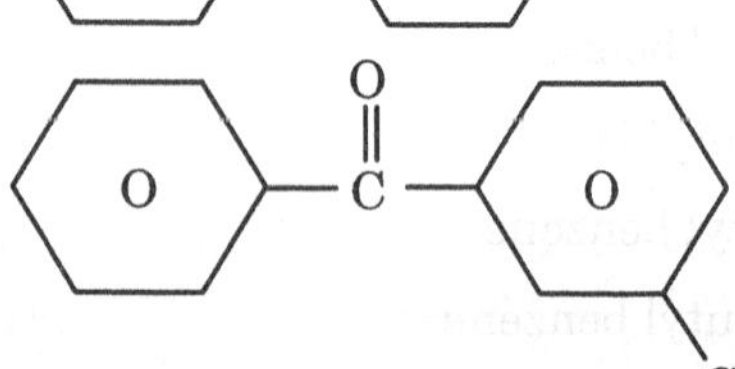

(d) 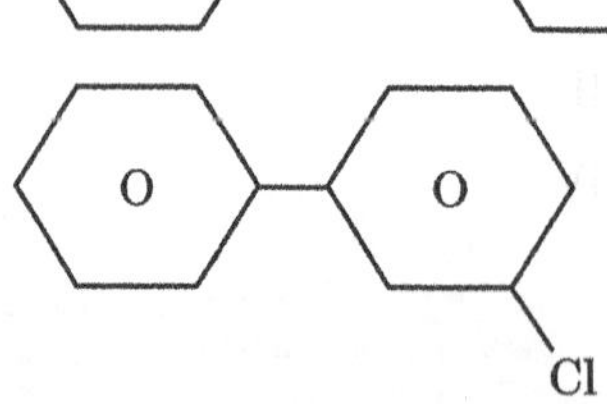

139. Friedal crafts acylation (i.e. introduction of $R-\overset{O}{\overset{\|}{C}}-$ group) can be brought about by which of the following reactions?

(a) 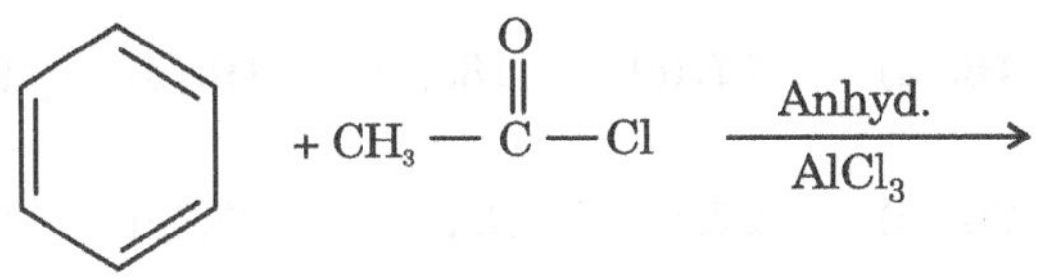

(b) 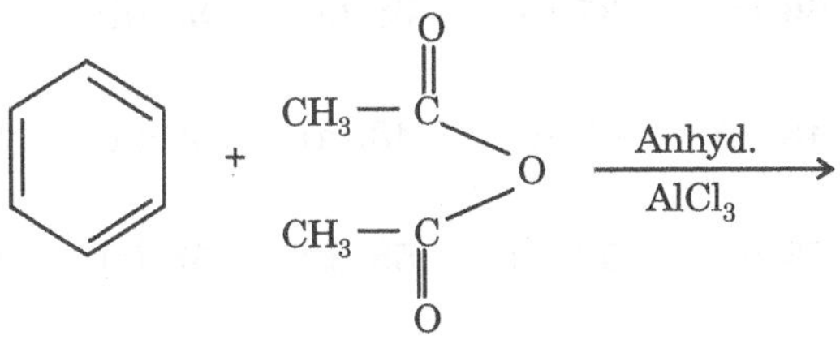

(c) 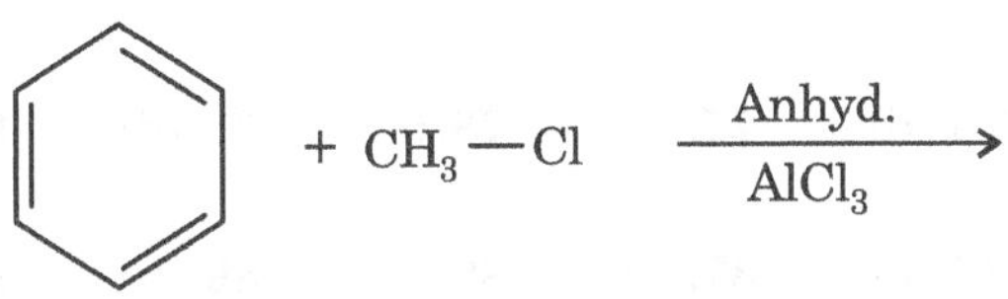

(d) Both (a) & (b)

140. The advantages of Friedal crafts acylation over alkylation are -

(a) The acyl group in the product withdraws electrons from the π system, making multiple substitutions harder.

(b) Rearrangements are also no longer a problem because the electrophile, acylium cation, is relatively stable.

(c) Both of these

(d) None of these

141. The amount of $AlCl_3$ used, is

(a) More in friedalcraft's alkylation than in acylation.

(b) More in friedalcrafts acylation than in alkylation

(c) Equal in both friedalcraft's alkylation & acylation

(d) None of these

142. Among the following, the compound that can be most readily sulphonated, is-

(a) Benzene (b) Nitrobenzene

(c) Chlorobenzene (d) toluene

143. Which of the following is major product for the mononitration of phenyl benzoate,

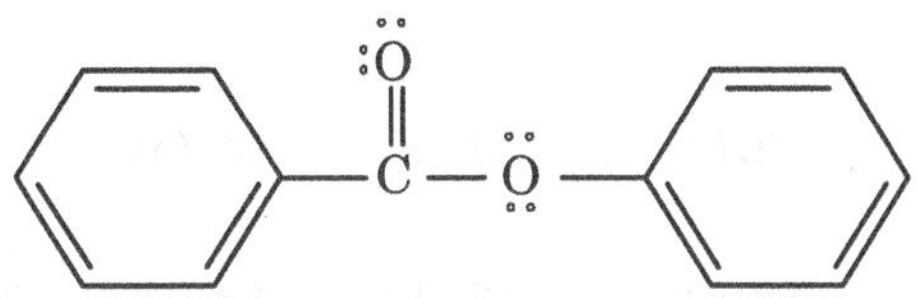

(a) 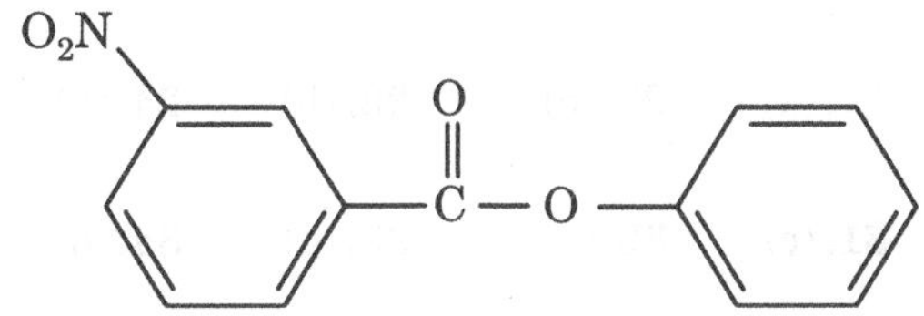

(b) 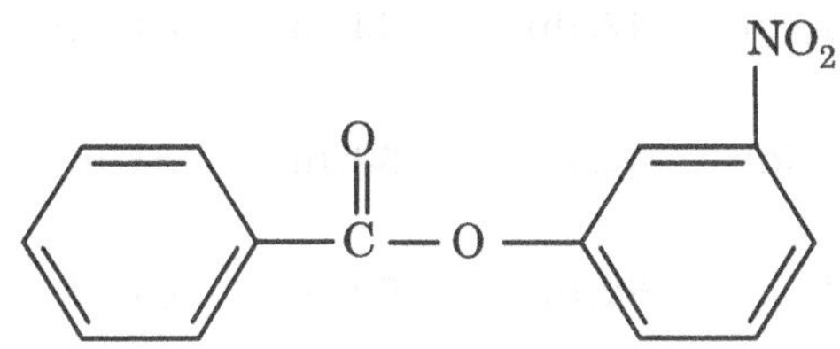

(c) 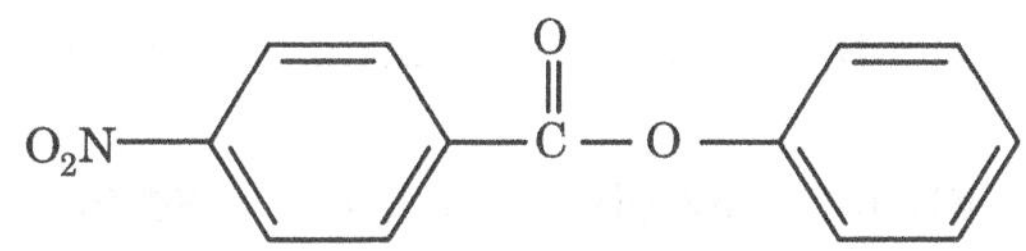

(d) 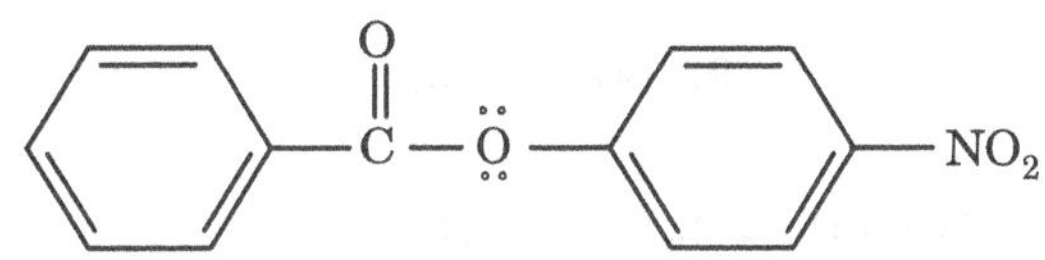

144. Benzene on reaction with ICl in the presence of anhydrous $AlCl_3$ gives –

(a) 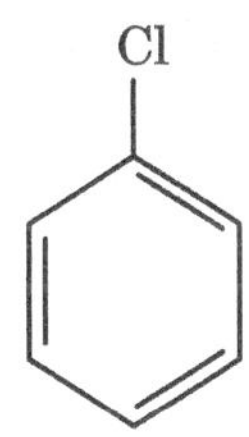

(b) 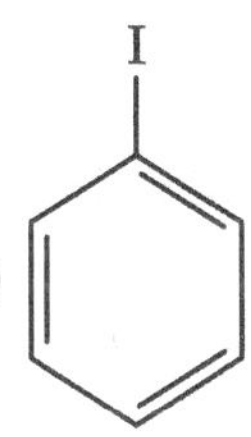

(c) 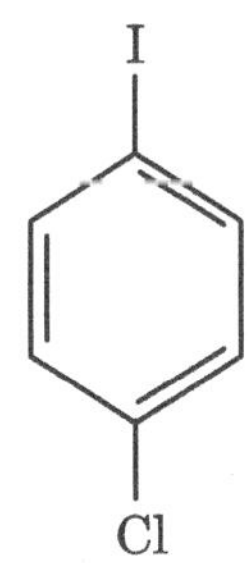

(d) 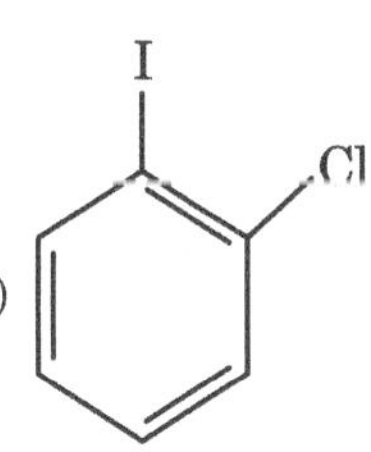

Answers

1. (b)	**2.** (d)	**3.** (d)	**4.** (a)	**5.** (d)	**6.** (b)	**7.** (d)	**8.** (c)	**9.** (d)	**10.** (d)
11. (d)	**12.** (d)	**13.** (b)	**14.** (d)	**15.** (c)	**16.** (a)	**17.** (c)	**18.** (b)	**19.** (d)	**20.** (b)
21. (a)	**22.** (c)	**23.** (b)	**24.** (b)	**25.** (d)	**26.** (a)	**27.** (b)	**28.** (b)	**29.** (c)	**30.** (b)
31. (c)	**32.** (a)	**33.** (a)	**34.** (b)	**35.** (b)	**36.** (c)	**37.** (b)	**38.** (a)	**39.** (d)	**40.** (b)
41. (b)	**42.** (b)	**43.** (a)	**44.** (d)	**45.** (c)	**46.** (c)	**47.** (c)	**48.** (a)	**49.** (a)	**50.** (a)
51. (d)	**52.** (d)	**53.** (d)	**54.** (d)	**55.** (d)	**56.** (d)	**57.** (c)	**58.** (b)	**59.** (c)	**60.** (c)
61. (c)	**62.** (a)	**63.** (b)	**64.** (d)	**65.** (a)	**66.** (a)	**67.** (a)	**68.** (b)	**69.** (a)	**70.** (b)
71. (d)	**72.** (b)	**73.** (b)	**74.** (c)	**75.** (d)	**76.** (d)	**77.** (b)	**78.** (b)	**79.** (d)	**80.** (c)
81. (b)	**82.** (d)	**83.** (d)	**84.** (b)	**85.** (b)	**86.** (c)	**87.** (c)	**88.** (a)	**89.** (b)	**90.** (d)
91. (c)	**92.** (a)	**93.** (d)	**94.** (c)	**95.** (a)	**96.** (a)	**97.** (d)	**98.** (d)	**99.** (d)	**100.** (c)
101. (a)	**102.** (b)	**103.** (d)	**104.** (a)	**105.** (b)	**106.** (a)	**107.** (b)	**108.** (d)	**109.** (a)	**110.** (b)
111. (d)	**112.** (c)	**113.** (a)	**114.** (b)	**115.** (a)	**116.** (b)	**117.** (b)	**118.** (d)	**119.** (d)	**120.** (d)
121. (d)	**122.** (b)	**123.** (a)	**124.** (c)	**125.** (d)	**126.** (b)	**127.** (a)	**128.** (c)	**129.** (a)	**130.** (d)
131. (b)	**132.** (d)	**133.** (c)	**134.** (a)	**135.** (b)	**136.** (d)	**137.** (d)	**138.** (b)	**139.** (d)	**140.** (c)
141. (b)	**142.** (d)	**143.** (d)	**144.** (b)						

Explanations

1. (b) Wurtzreaction is used to prepare alkanes with even no. of C atoms.

2. (d) 4th option is neopertyl bromide & it will readily give max. stable 3° free radical.

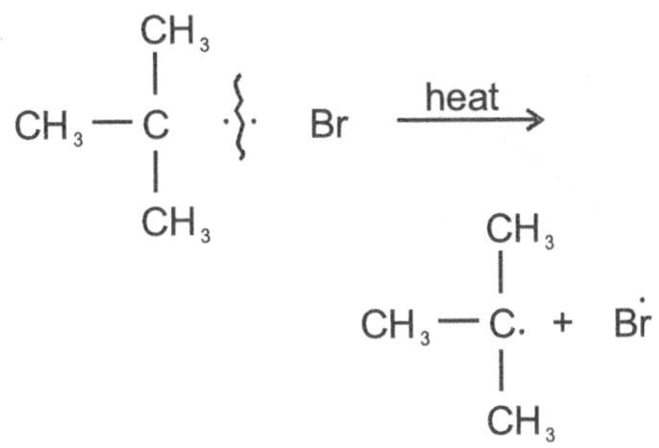

3. (d) $\overset{\delta-}{CH_3}-\overset{\delta+}{MgBr} \xrightarrow{\text{Acidic H}} CH_4$

Option (d) is an ether ($CH_3OCH_2CH_3$) which lacks an acidic H.

4. (a) 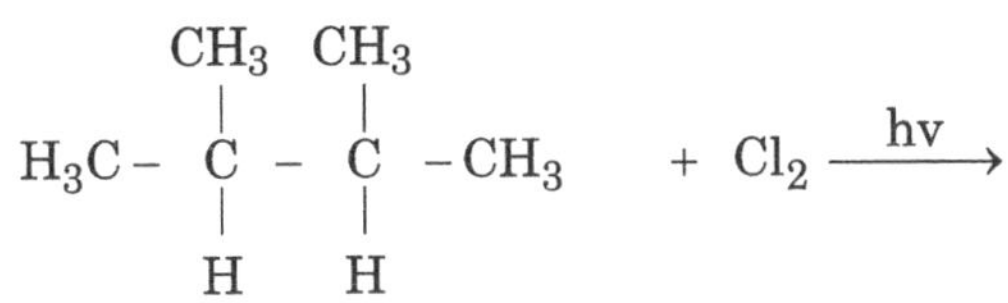

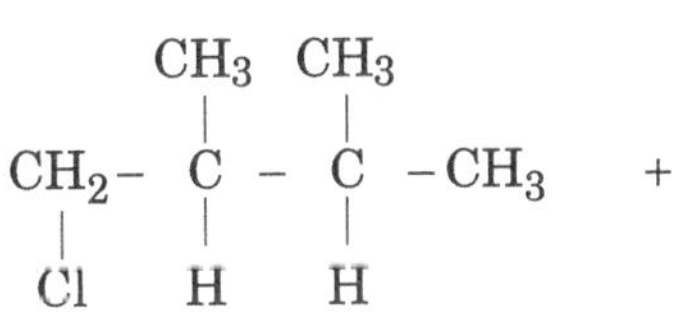

(1 – Chloro – 2, 3 – dimethylbutane)

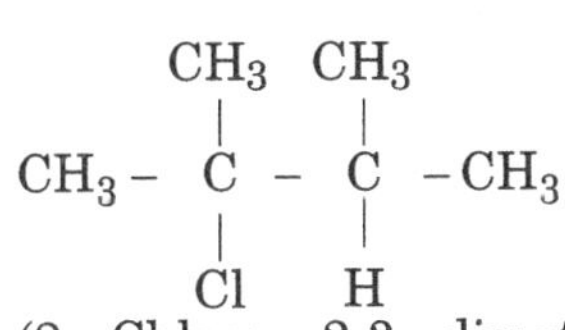

(2 – Chloro – 2,3 – dimethylbutane)

5. (d) 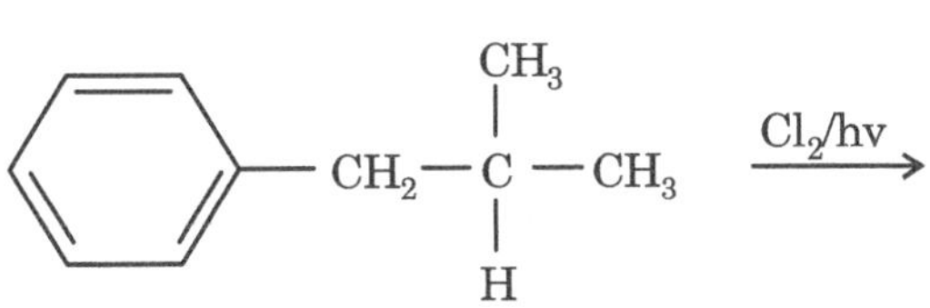

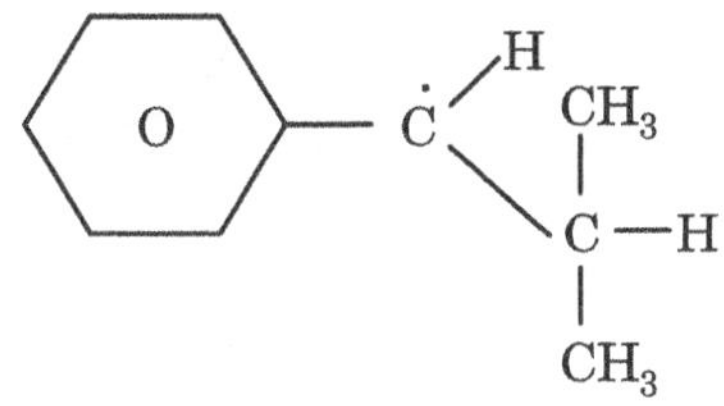

(Planar free radical, resonance stabilized & stabilized by hyper conjugation)

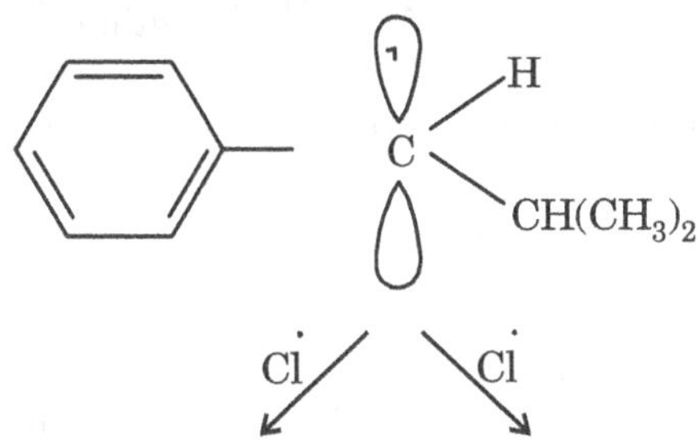

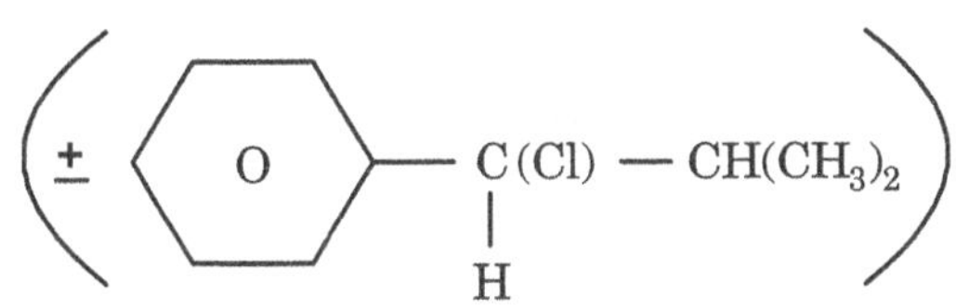

Racemic mixture is optically inactive.

6. (b) Option (a) & (d) are chain termination steps.

7. (d) $Al_4C_3 + H_2O \longrightarrow Al(OH)_3 + CH_4 \uparrow$

8. (c) 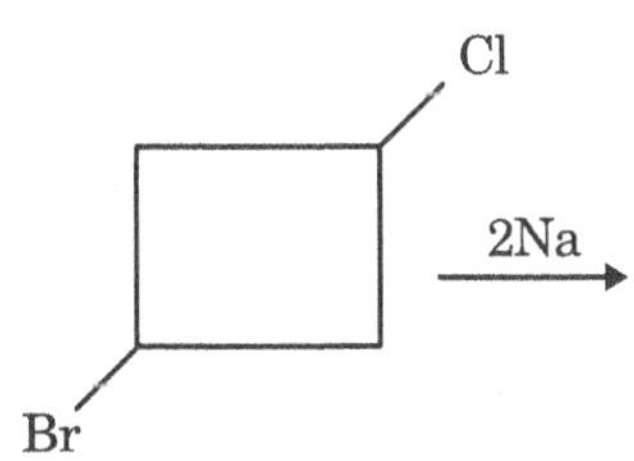

(1) $2Na \longrightarrow 2Na^+ + 2e^-$

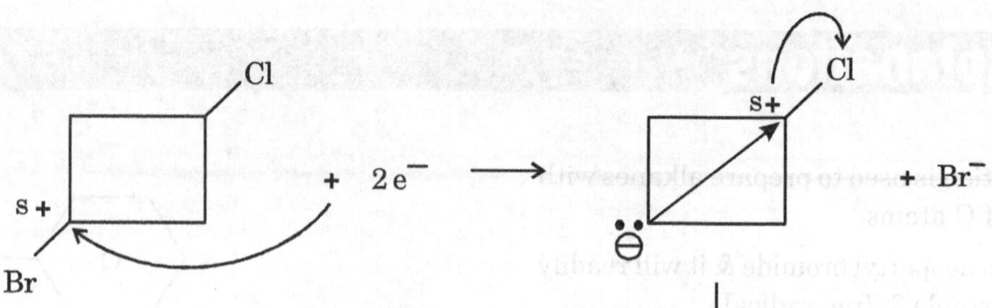

(2) (C—Br bond is weaker than C—Cl)

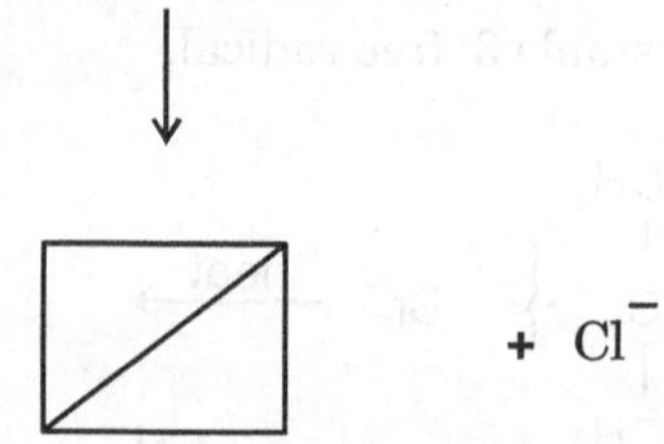

9. (d) $C_2H_5Cl + 2Na + C_2H_5Cl \xrightarrow[\text{ether}]{\text{dry}}$

(a) $\dot{Na} + CH_3—CH_2 \cdot \{ \cdot Cl \longrightarrow NaCl + CH_3—\dot{C}H_2$

(b) $CH_3—\dot{C}H_2 + CH_3—\dot{C}H_2 \xrightarrow{\text{dimerizes}} CH_3—CH_2—CH_2—CH_3$

(c) $H—C(H)(H)—\dot{C}H_2 + \dot{C}H_2—C(H)(H)—H \xrightarrow{\text{disproportionates}} CH_3—CH_3 + CH_2{=}CH_2$

11. (d) Kolbe's electrolysis

$$CH_3-CH_2-\overset{O}{\overset{\|}{C}}-O^{\ominus}K^+ \longrightarrow CH_3-CH_2-\overset{O}{\overset{\|}{C}}-\ddot{\underset{\cdot\cdot}{O}}{:}^{\ominus} + K^+$$

$$H_2O \rightleftharpoons H^+ + \overline{O}H$$

This reaction follows free radical mechanism

At Anode

(a) $CH_3—CH_2—\overset{O}{\overset{\|}{C}}—\ddot{\underset{\cdot\cdot}{O}}{:}^{\ominus} \xrightarrow{\text{oxdn.}} CH_3—CH_2—\overset{O}{\overset{\|}{C}}—\ddot{\underset{\cdot\cdot}{O}}\cdot + e^-$

(b) $CH_3—CH_2 \cdot \{ \cdot \overset{O}{\overset{\|}{C}}—\ddot{\underset{\cdot\cdot}{O}}\cdot \longrightarrow CH_3—\dot{C}H_2 + CO_2 \uparrow$

(c) $CH_3—\dot{C}H_2 + CH_3—\dot{C}H_2 \xrightarrow{\text{dimerizes}} CH_3—CH_2—CH_2—CH_3$

(d) $CH_3—\dot{C}H_2 + \dot{C}H_2—CH_3 \xrightarrow{\text{disproportionation}} CH_3—CH_3 + CH_2{=}CH_2$

(e) $CH_3—CH_2—\overset{O}{\overset{\|}{C}}—\ddot{\underset{\cdot\cdot}{O}}\cdot + \dot{C}H_2—CH_3 \longrightarrow CH_3—CH_2—\overset{O}{\overset{\|}{C}}—OCH_2CH_3$

12. (d) $CH_3CH_2CH_2-\overset{O}{\overset{\|}{C}}-O^-Na^+ + NaOH \xrightarrow{CaO} CH_3CH_2CH_3 + Na_2CO_3$

$$CH_3-\underset{CH_3}{\underset{|}{CH}}-\overset{O}{\overset{\|}{C}}-O^-Na^+ + NaOH \xrightarrow{CaO} CH_3CH_2CH_2 + Na_2CO_3$$

14. (d) $Al_4C_3 + H_2O \longrightarrow Al(OH)_3 + CH_4\uparrow \quad Al^{3+}\ C^{4-} + H_2O \longrightarrow Al(OH)_3 + CH_4\uparrow$

$$Be_2C + H_2O \longrightarrow Be(OH)_2 + CH_4\uparrow \quad Be^{2+}\ C^{4-} + H_2O \longrightarrow Be(OH)_2 + CH_4\uparrow$$

$$\underset{(Ca^{2+}C_2^{2-})}{CaC_2} + H_2O \longrightarrow Ca(OH)_2 + C_2H_2\uparrow$$

15. (c) Kolbe's electrolysis of $H-\overset{O}{\overset{\|}{C}}-O^-Na^+$ will give H_2 & that of $CH_3-\overset{O}{\overset{\|}{C}}-O^-Na^+$ will give C_2H_6 (following free radical mechanism)

16. (a) $\underset{(M.\ wt. = 72)}{CH_3-\overset{CH_3}{\overset{|}{\underset{CH_3}{\underset{|}{C}}}}-CH_3} + Cl-Cl \xrightarrow{hv} CH_3-\overset{CH_3}{\overset{|}{\underset{CH_3}{\underset{|}{C}}}}-CH_2Cl + HCl$

17. (c) Due to greater stability of 3° free radical, 3° H are maximum reactive & it is present in isopentane.

18. (b) Alkanes undergo free radical substitution reactions.

19. (d) $CH_3-CH_2-CH_3 + Br_2 \xrightarrow{hv} CH_3CH_2CH_2Br + CH_3-\underset{Br}{\underset{|}{CH}}-CH_3$

21. (a) Greater the molar mass, greater are the Vander Waals forces of attraction, greater is the boiling point.

22. (c) Acid catalyzed dehydration of 3° alcohols occurs via carbocation formation.

23. (b) $CH_3-\overset{CH_3}{\overset{|}{\underset{H}{\underset{|}{C}}}}-CH_2-CH_3 + Cl_2 \xrightarrow{H\nu} \underset{\text{(2 isomers d\&1)}}{\underset{Cl}{\underset{|}{CH_2}}-\overset{CH_3}{\overset{|}{\underset{H}{\underset{|}{C^*}}}}-CH_2-CH_3} + \underset{\text{(2 isomers d\&1)}}{CH_3-\overset{CH_3}{\overset{|}{\underset{H}{\underset{|}{C}}}}-\overset{H}{\overset{|}{\underset{Cl}{\underset{|}{{}^*C}}}}-CH_3}$

24. (b)

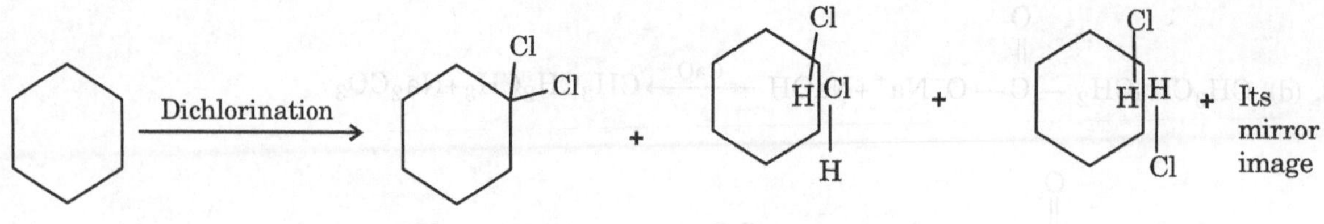

(1, 1 -Dichlorcyclohexane) (Cis -1, 2-dichloro cyclohexane) (Meso) (Trans-1, 2- dichloro cyclohexane)

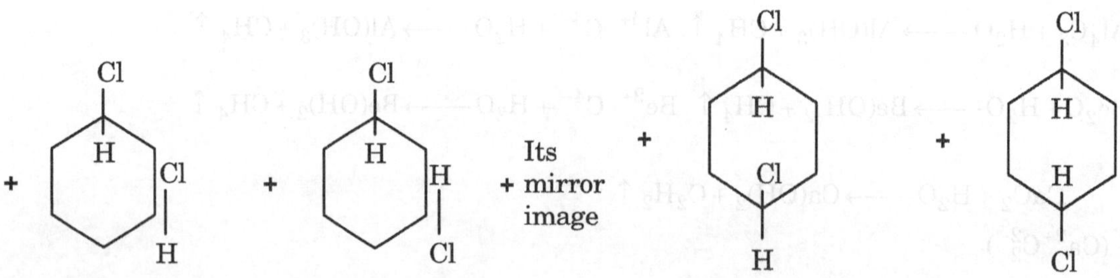

(Cis -1, 3-dichloro cyclohexane) (Meso) (Trans 1,3 dichloro cyclohexane) (Cis 1, 4-dichloro cyclohexane) No chiral C (Trans -1, 4 -dichloro cyclohexane) No chiral C

25. (d) Corey House reaction

Step 1 : involves formation of an organometallic compound.

$$R-X + 2Li \longrightarrow R-Li + LiX$$

Step 2 : this organometallic compound reacts with cuprous iodide to form alkyl lithium cuprate.

$$2R-Li + CuI \longrightarrow R_2CuLi + LiI$$

Step 3 : the dialkyl lithium cuprate (gilman's reagent) formed, reacts with alkyl halide to form higher alkane having both even & odd number of C atoms via nucleophillic substitution (SN^2)

$$R_2CuLi + R'X \longrightarrow R-R' + LiX + CuR$$

26. (a) $CH_3-\overset{O}{\overset{\|}{C}}-O-CH_3 \xrightarrow{H_3O^+} CH_3COOH + CH_3OH$

(Methyl ethanoate)

$$CH_3COOH \xrightarrow{NaOH} CH_3COONa \xrightarrow[\text{electrolysis}]{\text{Kolbe's}} C_2H_6$$

27. (b) $CH_3-\underset{H}{\overset{CH_3}{C}}-CH_2-CH_2-CH_3 + Cl_2 \xrightarrow{hv}$

$$\underset{Cl}{CH_2}-\underset{H}{\overset{CH_3}{\overset{*}{C}}}-CH_2CH_2CH_3 \ (d+l) + CH_3-\underset{Cl}{\overset{CH_3}{C}}-CH_2CH_2CH_3 +$$

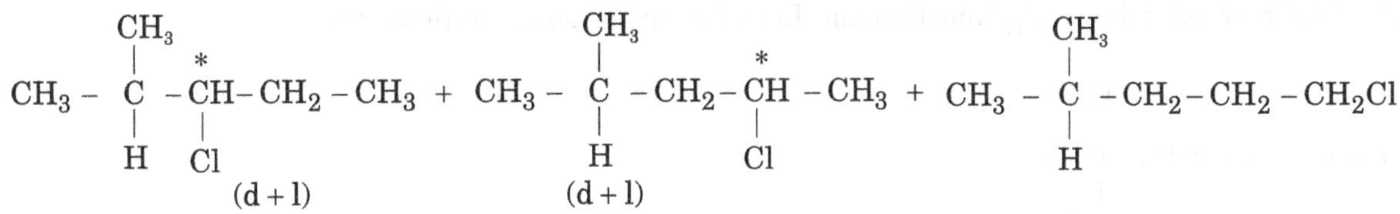

28. (b)

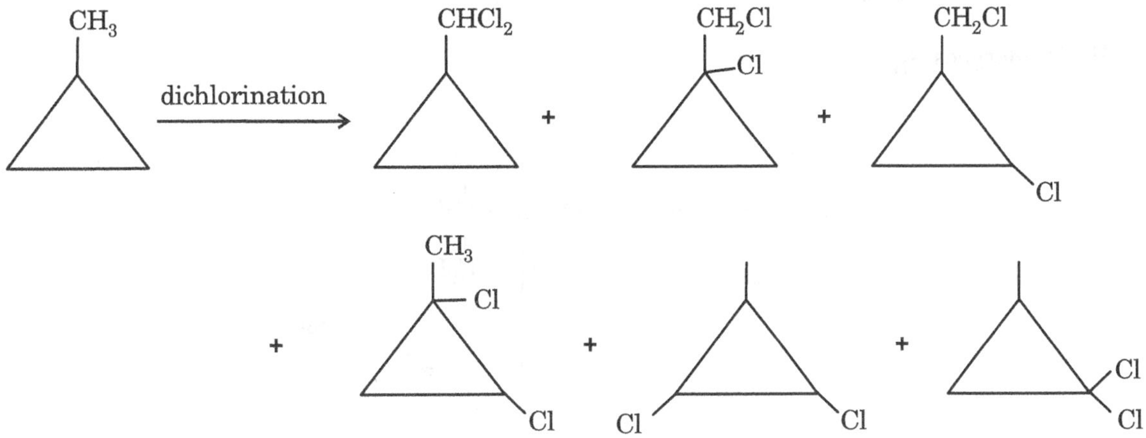

29. (c)

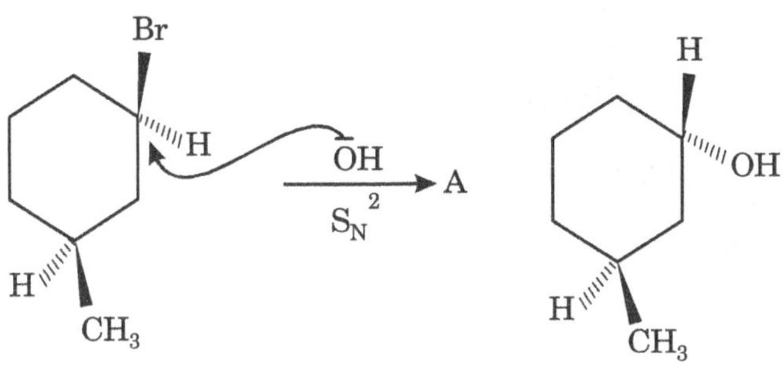

(1R, 3S) (1S, 3S)

30. (b)

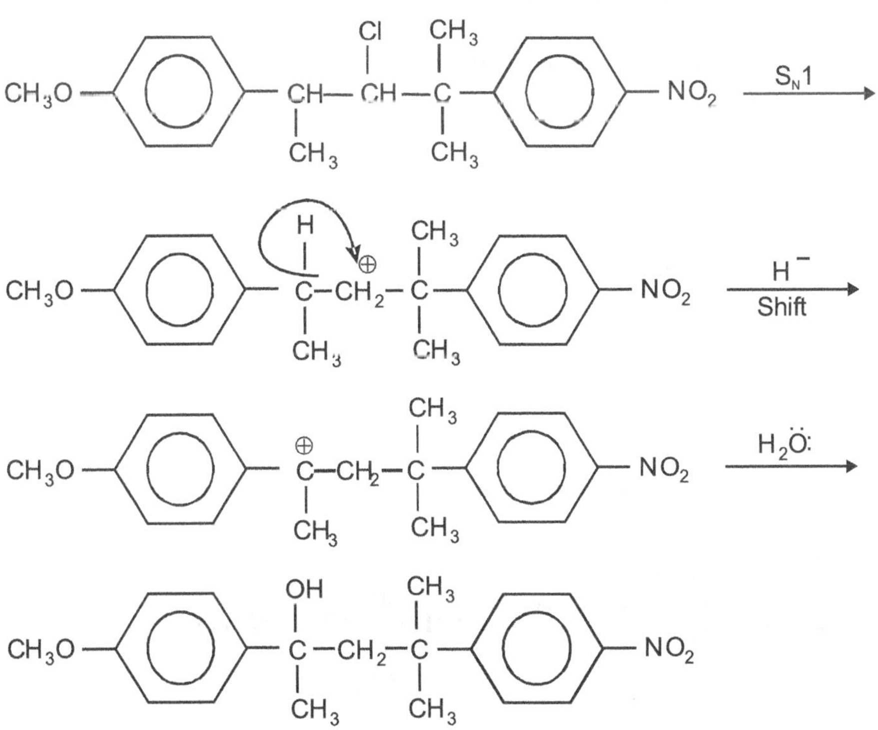

31. (c) The reaction follows S_N^2 mechanism. Look for inversion of configuration.

32. (a) In (a) i.e. 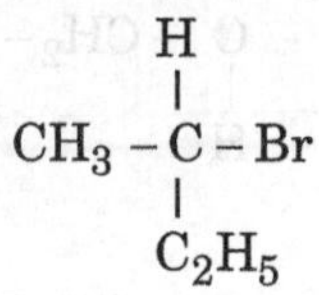

If 'a' undergoes S_N^1

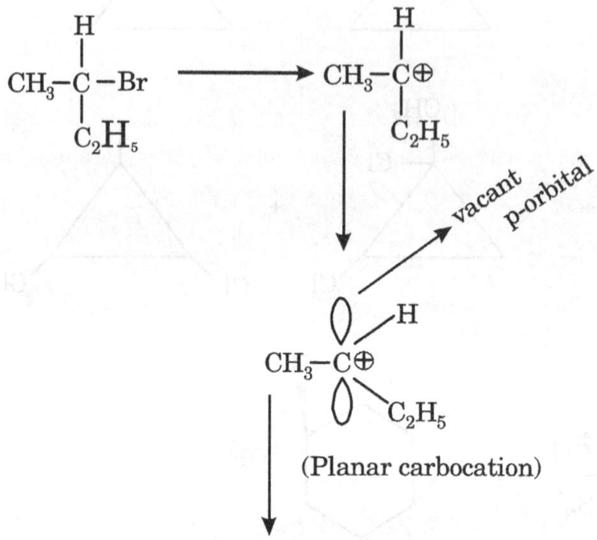

Chiral centre will be created

after Nucleophillic attack

& $Nu^{\ominus}$ attack is possible both

from above & below the plane,

so racemic mixture is obtained

(b) option b, lacks chiral C.

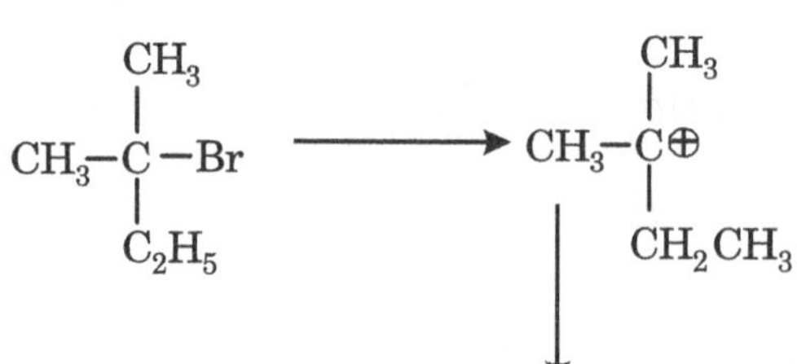

(No chiral centre or Carbon created)

(c) 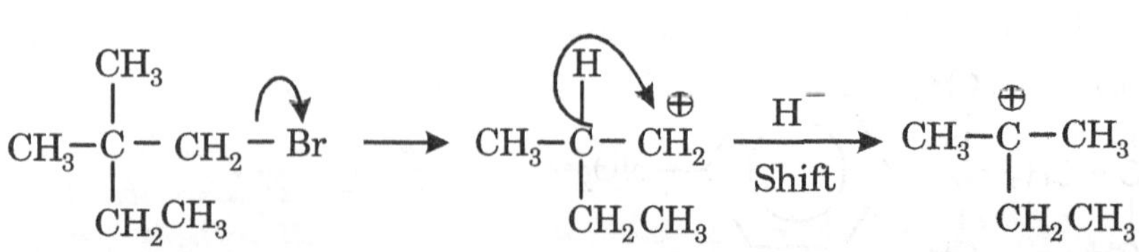 (No chiral C)

33. (a) (1) 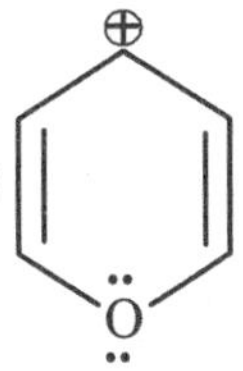 → Aromatic Carbocation ($6\pi\ e^-$, planar)

(2)  → more effective $2p\pi$ - $2p\pi$ overlap, lesser electro - *ve* nature of N than O.

(3) 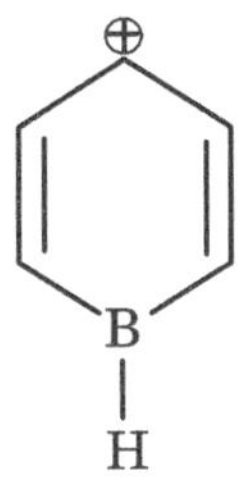 → ANTI AROMATIC, unstable

34. (b) 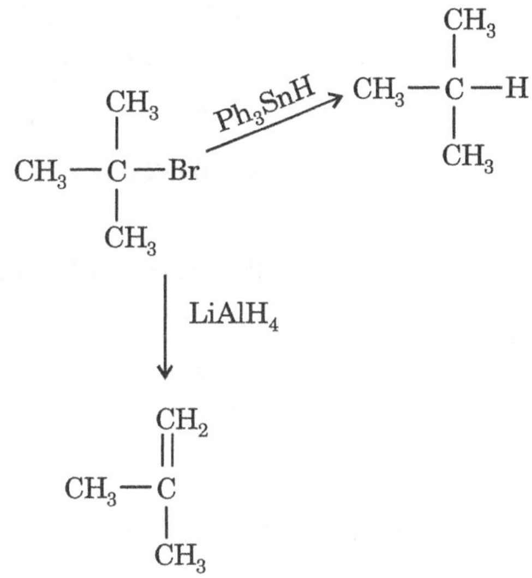

35. (b) 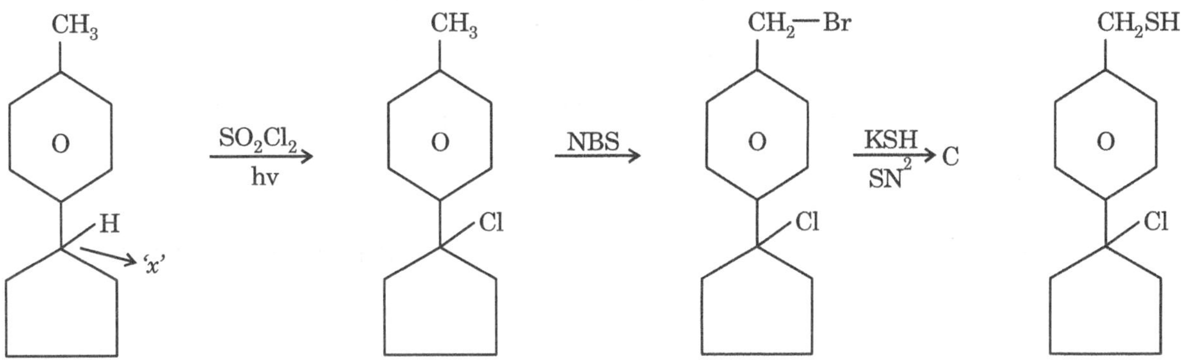

→ Free radical will be more stable at location 'X' (Resonance stabilized & 4 ∝ H for hyper conjugation)

(Free radical sub.)

36. (c) $CH_3-CH_2-Br+(CH_3CH_2)_3\ddot{N} \longrightarrow$ 1° halides cannot undergo E_1/S_N1 reactions & the base is not strong enough to carry out E_2. So, S_N2 is only option left.

$$\underset{\text{(Bulky base)}}{(C_2H_5)_3N} + CH_3-\underset{\substack{|\\I}}{CH}-CH_2-CH_3 \xrightarrow{\Delta} \underset{\text{(Hoffmann's Product)}}{CH_2=CH-CH_2-CH_3}$$

37. (b) 'x' is cis but-2-ene, trans but-2-ene & but-1-ene So, $x = 3$

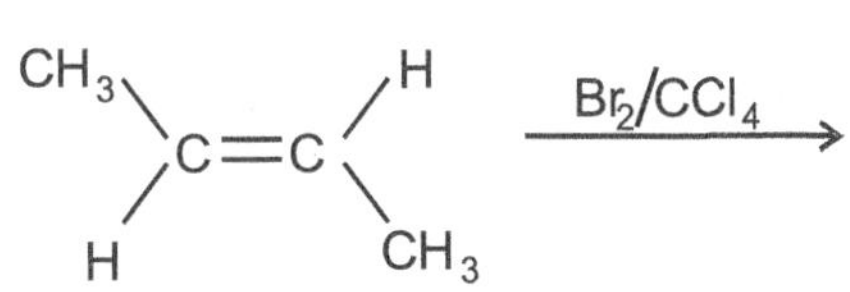 Meso-2, 3-dibromo butane.

Trans but-2-ene

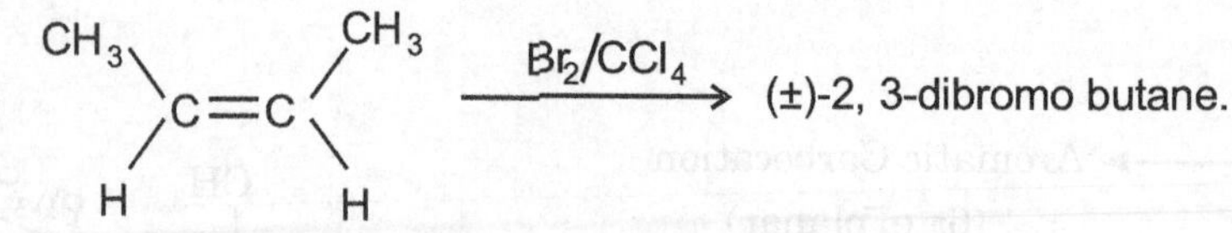

Cis but-2-ene

$$CH_3-CH_2-CH=CH_2 \xrightarrow{Br_2/CCl_4} (\pm)\ \text{1,2-dibromobutene}$$

$\therefore$ Y = 5

38. (a) $CH_3-CH=CH-CH_3$ + 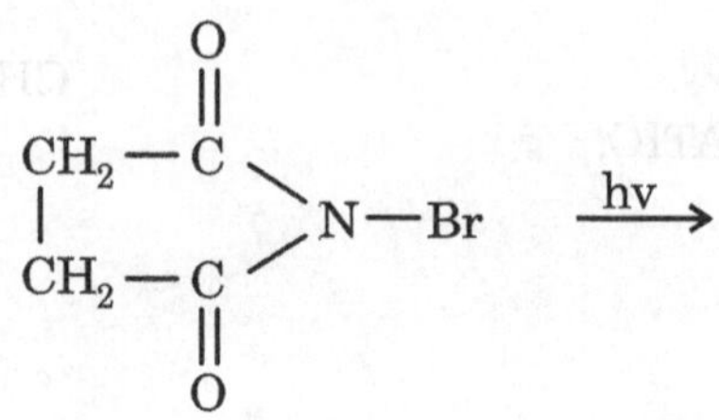

(Free radical sub. will occur)

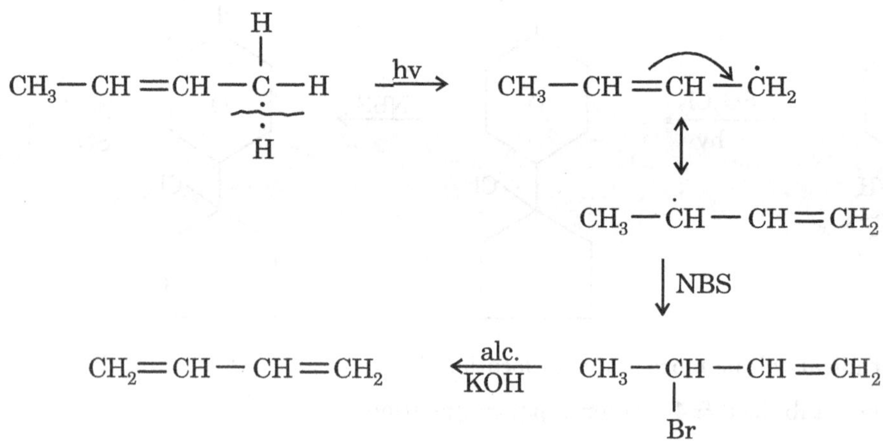

39. (d) (a) C – D bond is tough to break than C – H bond. So , rate of E_2 is lesser in $CD_3-C(CD_3)_2-Br$.

(b) 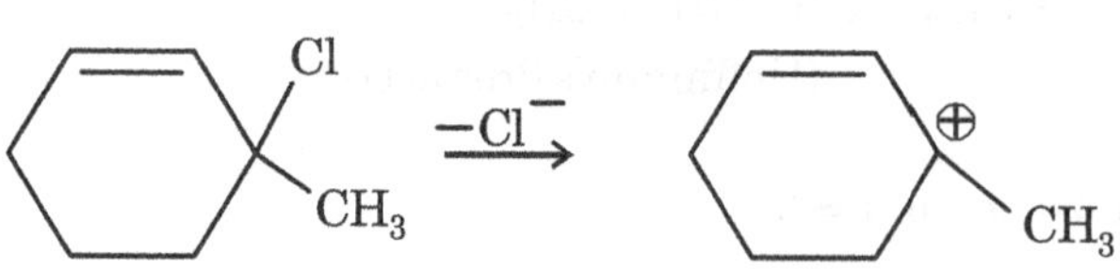

(Carbocation is resonance stabilised)

(c) $CH_2=CH-CH_2-Br$

(Transition state is stabilized by allylic delocalization of π electron cloud, activation energy is lowered & rate of SN^2 is enhanced).

40. (b) Peroxide effect operates only with HBr (via free radical mechanism).

41. (b) $CH_3-CH_2-CH=CH_2+Cl_2 \xrightarrow{\text{high T}} CH_3-\underset{\underset{Cl}{|}}{CH}-\underset{(A)}{CH=CH_2}$

(high T favors free radical substitution at allylic poisiton)

$\downarrow$ alc. KOH

$\underset{(B)}{CH_2=CH-CH=CH_2}$

$\downarrow$

Conjugated diene, stable product

42. (b) $CH_3-CH_2-\overset{\overset{H}{|}}{\underset{\underset{F}{|}}{C}}-CH_3 \xrightarrow{\text{Base}} \underset{\text{(Major A)}}{CH_3-CH_2-CH=CH_2} + \underset{\text{(Minor B)}}{CH_3-CH=CH-CH_3}$

43. (a) 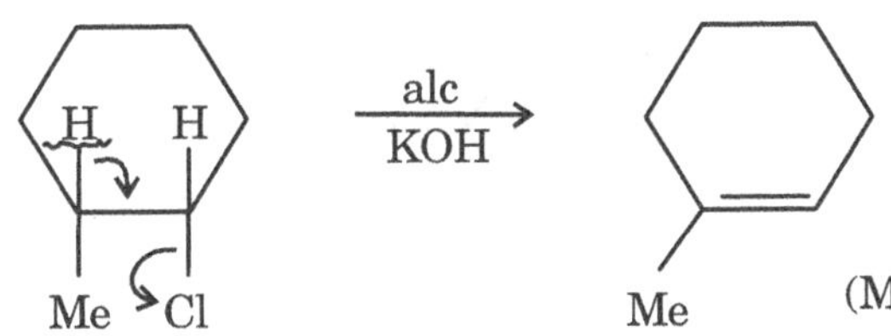

(Major product, stable alkene)

(E_2 elimination, anti-periplanar atoms are removed)

(b) 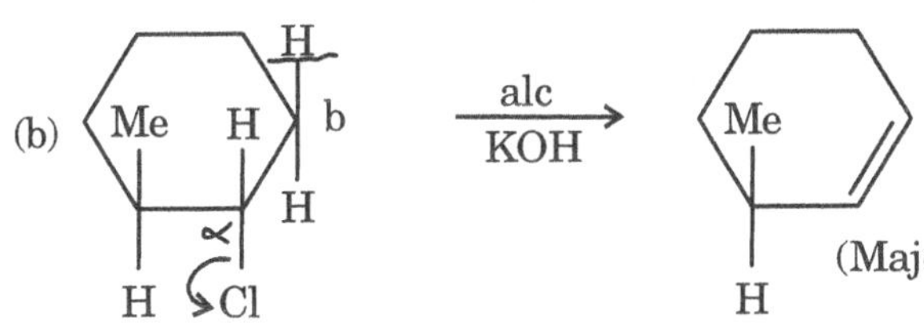

(Major product)

(c) 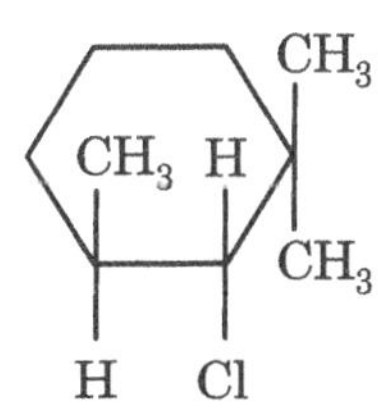 $\longrightarrow$ No E_2 reaction

(None of β H are anti - periplanarwrt Cl)

Maximum stable alkene is formed in case - 1, so, it will undergo E_2 readily.

44. (d) (a) $CH_3-\overset{\overset{CH_3}{|}}{\underset{\underset{CH_3}{|}}{C}}-CH_2-Cl \xrightarrow{E_2}$ Not possible, because neopentyl chloride lacks a β —H

(neopentyl chloride)

(b) 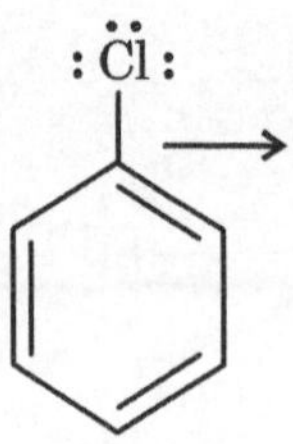
bond can't be broken with ease.
(partial double bond character)

(c) 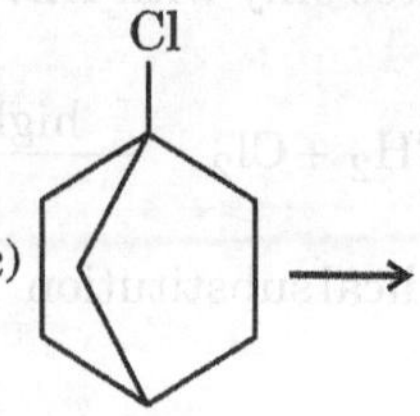

π bond can't be formed at bridge head location (violation of Bredt's rule)

45. (c) 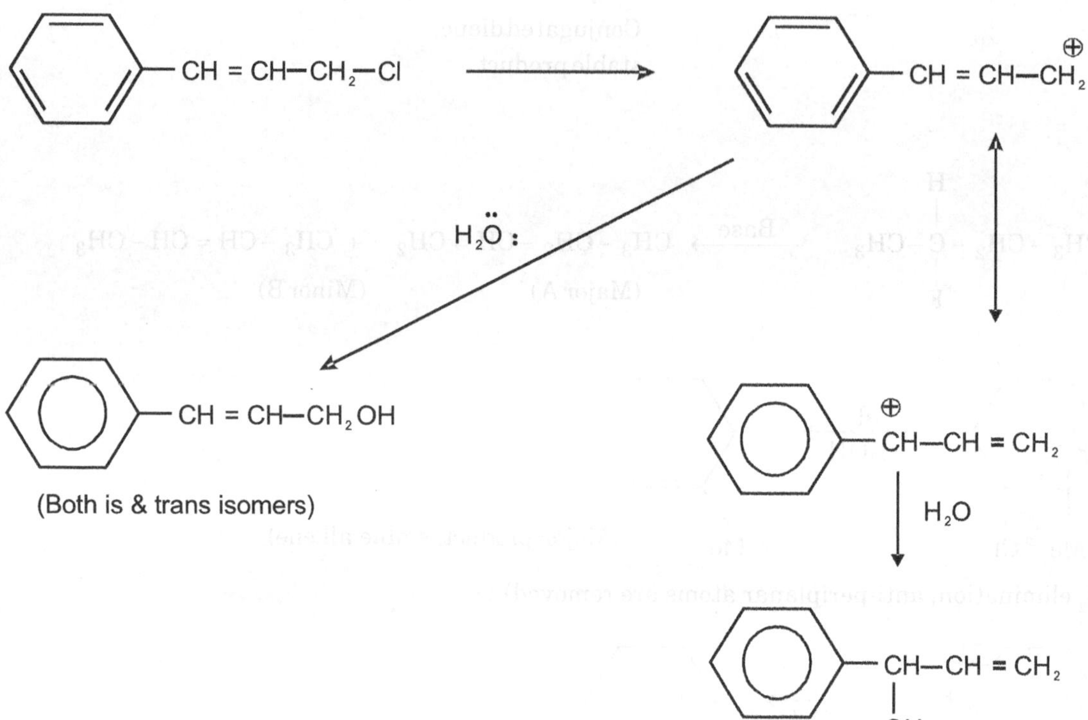

(Both is & trans isomers)

Total product = 4

(Chiral C, So 2 Stereomers)

46. (c) 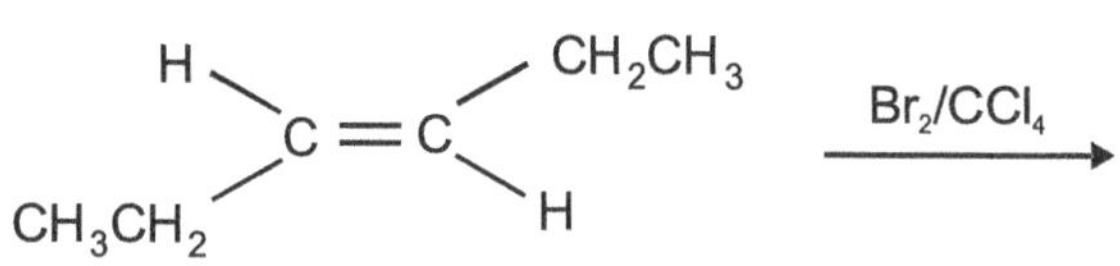

Br_2 show anti- addition.

Trans alkene + Anti - addition Meso product 4th molecule is

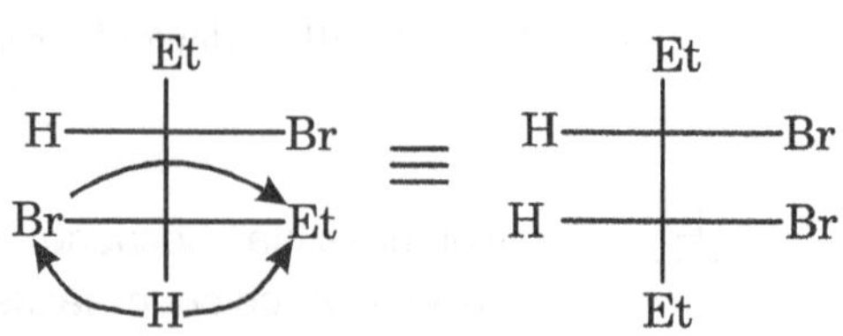

47. (c) Trans-but-2-ene + $Br_2 \xrightarrow[\text{addition}]{\text{Anti-}}$ meso compound In Meso compound, the 2 chiral C's will have opposite absolute configurations. If one chiral C will have R configuration, other one will have S configuration.

48. (a)

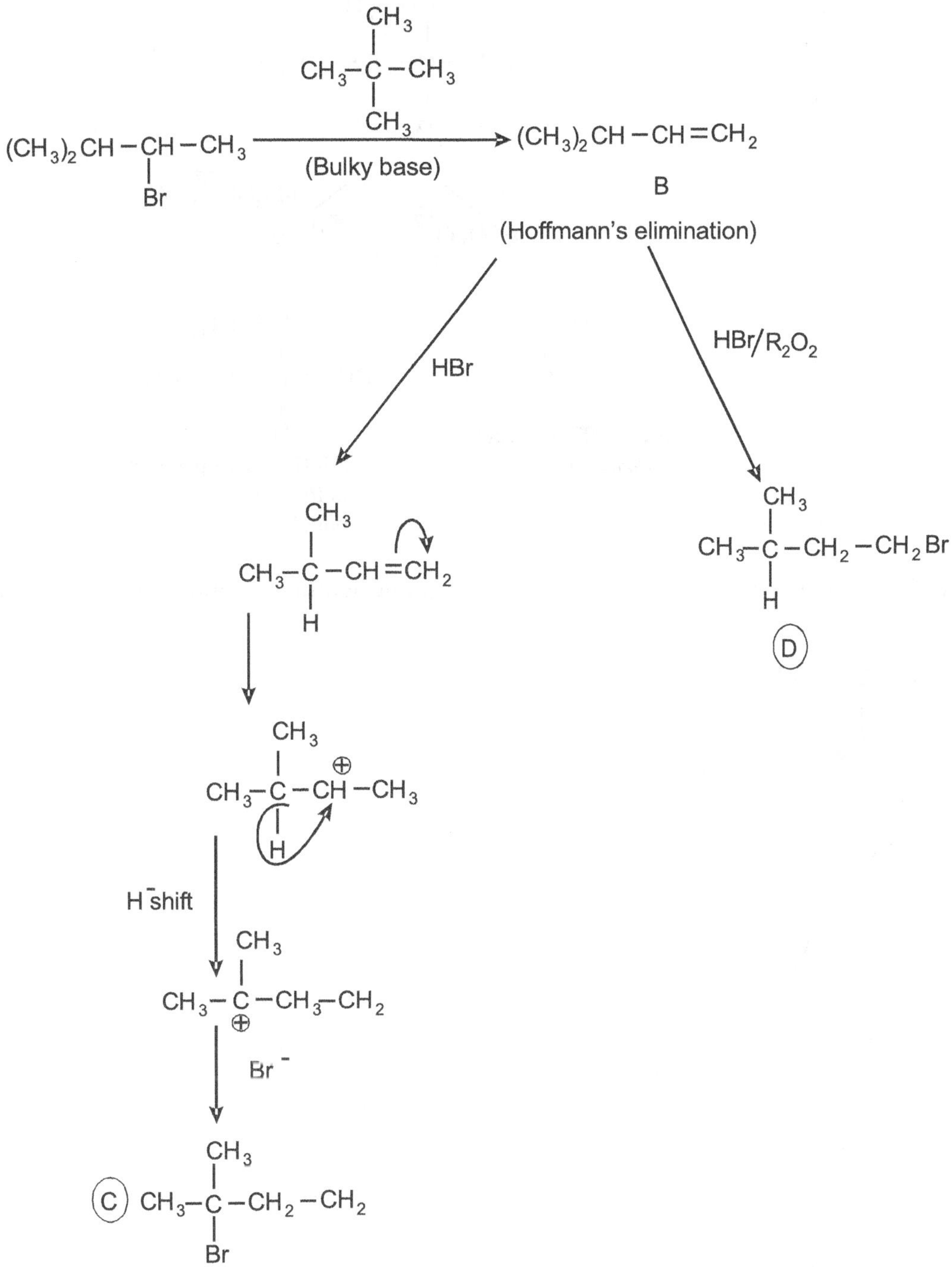

So, A is $(CH_3)_2\,CH-\underset{|}{\overset{Br}{CH}}-CH_3$ (2° halide) C is $CH_3-\overset{CH_3}{\underset{Br}{CH}}-CH_2-CH_3$ (3° halide)

D is $CH_3-\overset{CH_3}{\underset{H}{C}}-CH_2-CH_2-Br$ (1° halide)

49. (a)

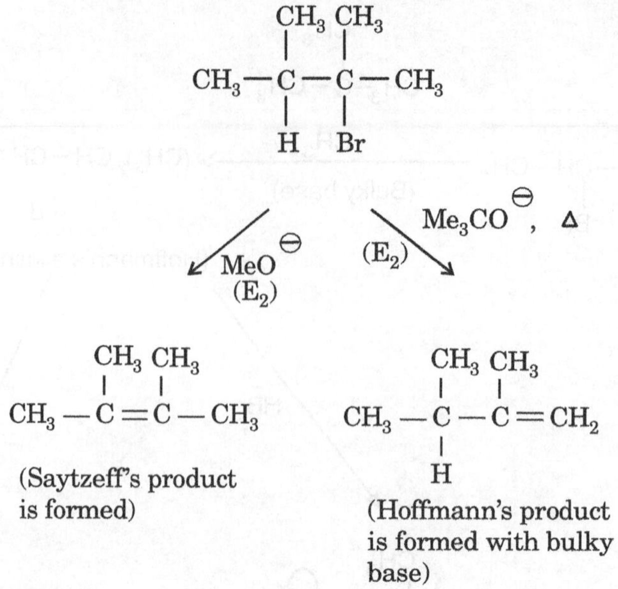

50. (a) Acid catalyzed hydration of alkene is an electrophilic addition reaction which occurs via carbocation formation.

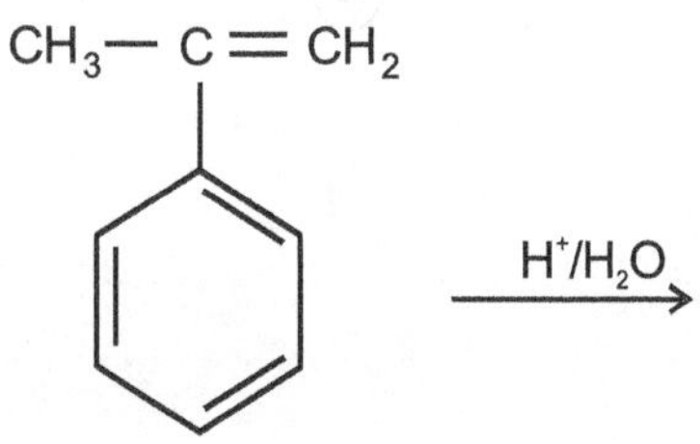

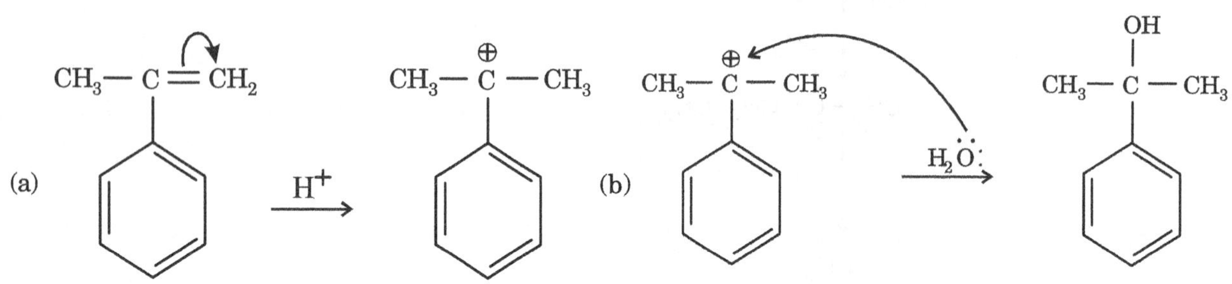

51. (d)

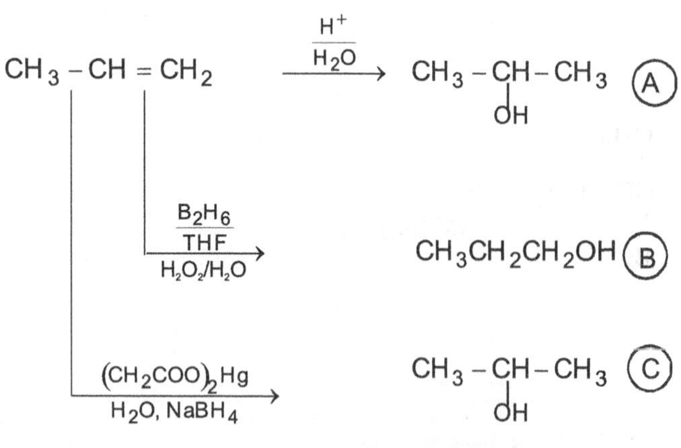

53. (d) The reaction occurs via formation of maximum stable intermediate.

54. (d) HCl does not show peroxide effect.

(a) 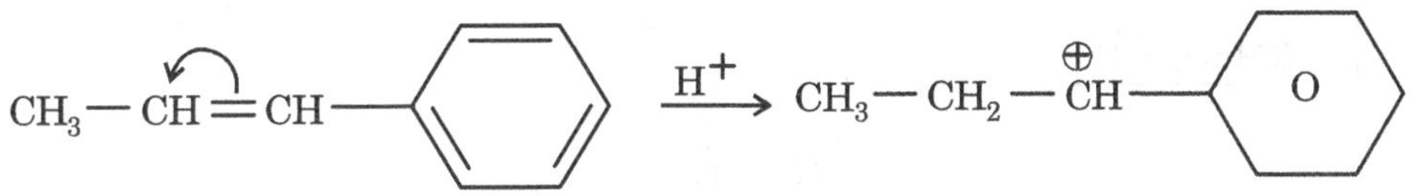

(b) 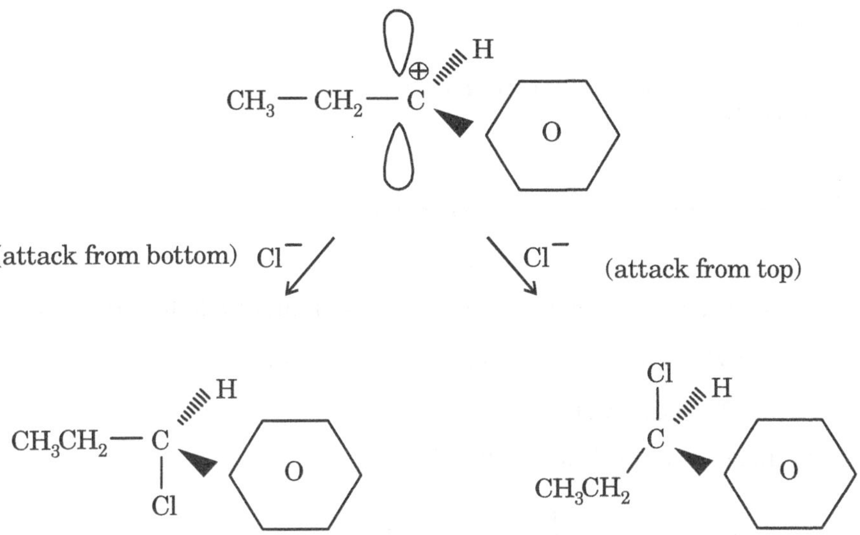

57. (c) Rate of acid catalyzed dehydration will depend on the stability of carbocation.

58. (b) 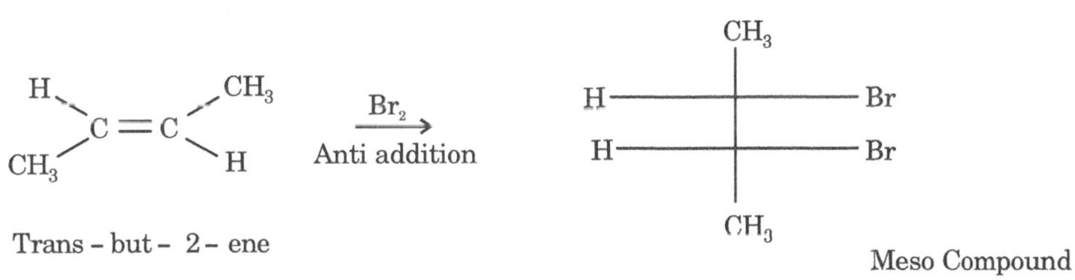

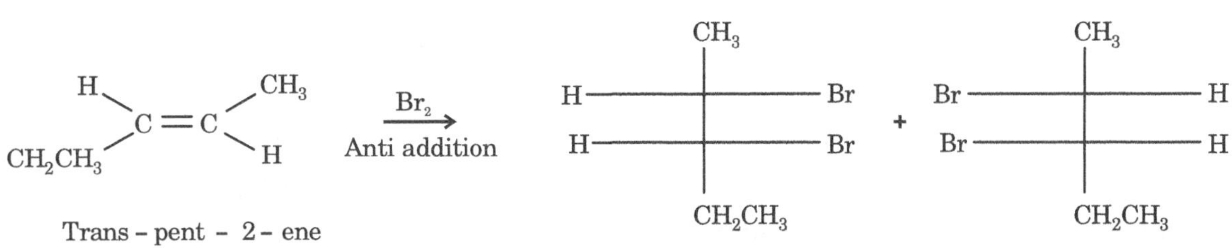

60. (c)

$$CH_3-CH=CH_2 \xrightarrow[H^+]{KMnO_4} CH_3COOH + CO_2 \uparrow$$

$$CH_3-CH=CH_2 \xrightarrow{KMnO_4,\ \overline{O}H} CH_3-\underset{OH}{\underset{|}{C}H}-\underset{OH}{\underset{|}{C}H_2}$$

$$CH_3-CH=CH_2 \xrightarrow[H_2O]{O_3,\ Zn} CH_3CHO + HCHO$$

$$CH_3-CH=CH_2 \xrightarrow[NaOH, H_2O_2]{B_2H_6} CH_3CH_2CH_2OH$$

$$CH_3-CH=CH_2 \xrightarrow{H^+/H_2O} CH_3-\underset{OH}{\underset{|}{C}H}-CH_3$$

61. (c) Reactivity towards electrophilic addition depends on the stability of intermediate carbocation.

$$CH_3-CH=CH_2,\ CH_3-O-CH=CH_2,$$

$$Ph-CH=CH_2$$

are, more reactivity than ethene towards electrophilic addition.

62. (a)

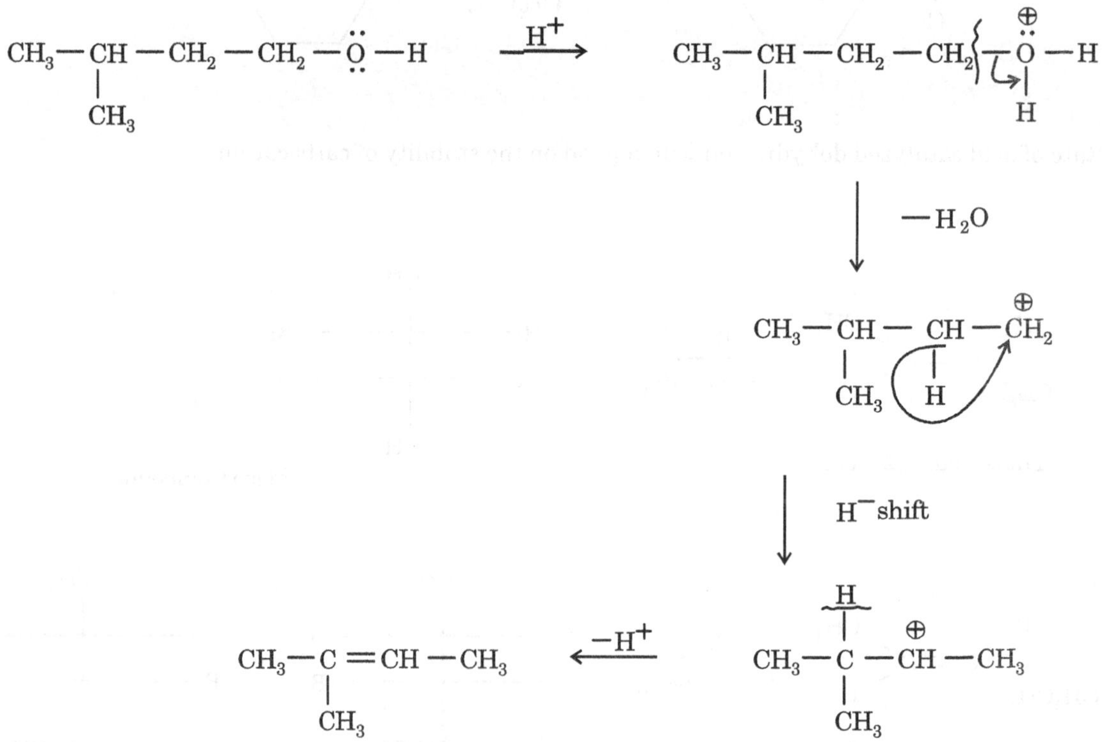

63. (b) Peroxide effect does not operate with HCl.

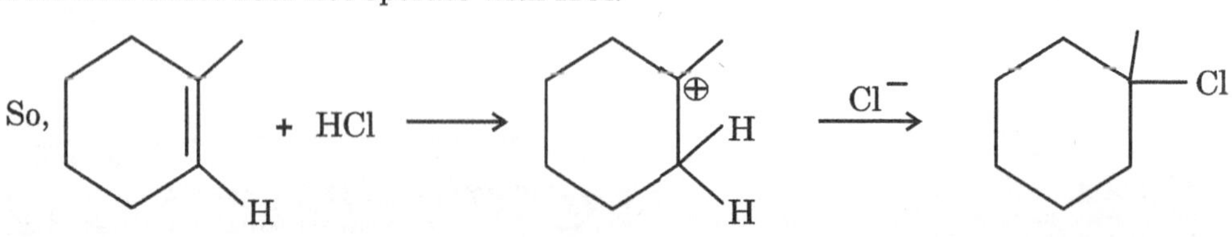

64. (d) (a) 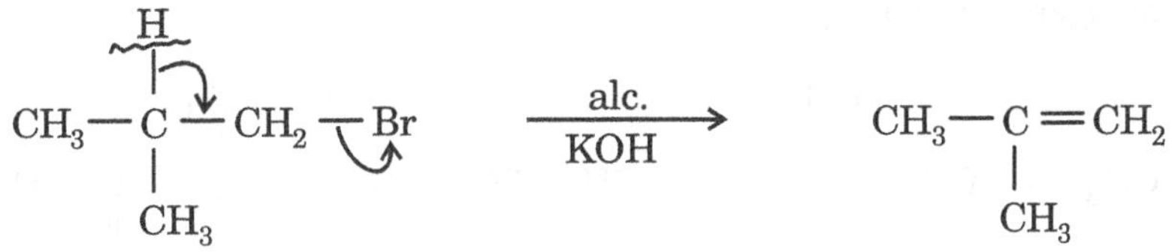

(6 ∝ - H)

(b) $CH_3 - CH_2 - Br \xrightarrow[KOH]{alc.} CH_2 = CH_2$

(c) 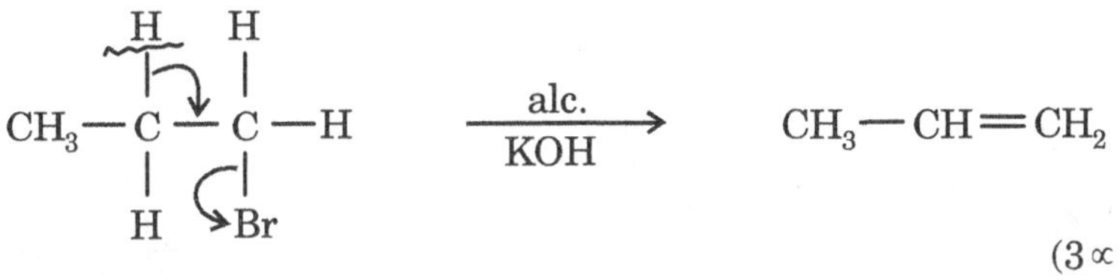

(3 ∝ - H)

Alkene formed is maximum Stable in Case 1.

So, $CH_3 - \underset{\displaystyle CH_3}{\underset{|}{CH}} - CH_2 - Br$ is max. reactive towards β - elimination.

65. (a) Anti addition occurs.

66. (a) → Since there is already a pre - existing chiral C which is having R configuration. On reaction with HBr, another chiral C is created. Because a planar carbocation is the intermediate, attack will occur from top & bottom giving a combination of RR & RS (Diastereomers)

67. (a) 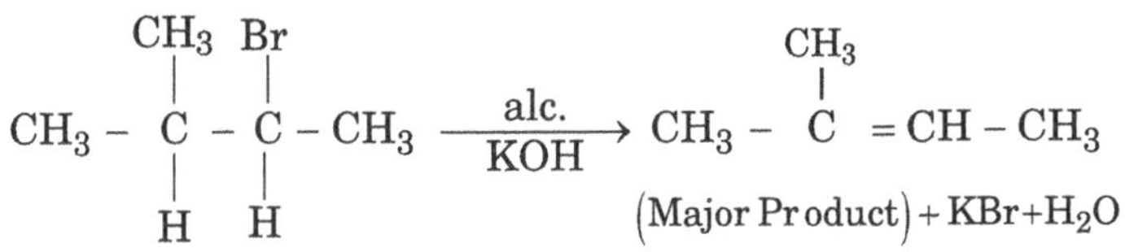

68. (b) 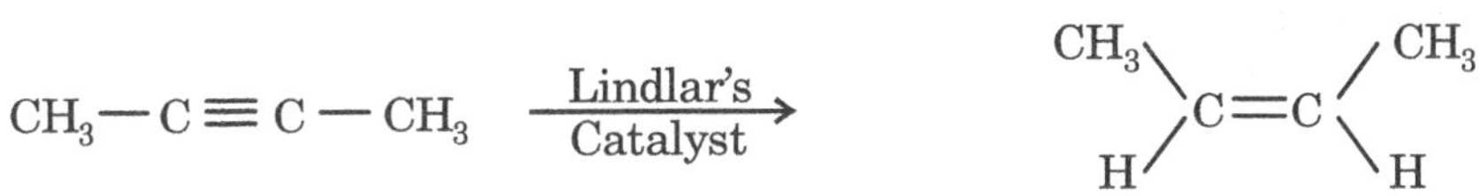

(Cis but-2-ene)

69. (a) 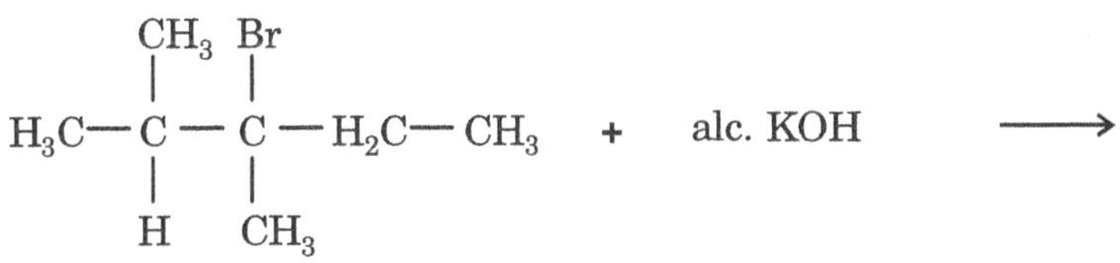

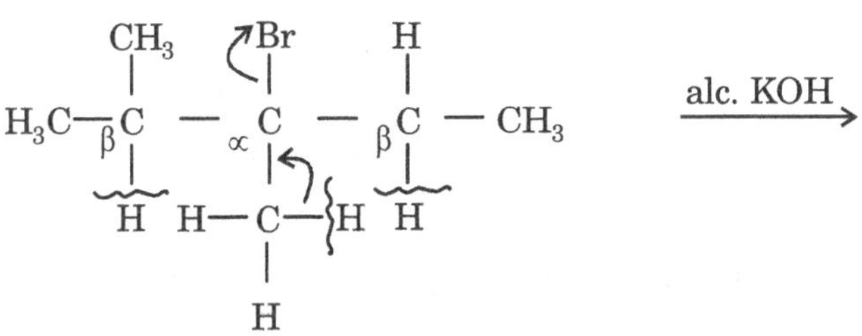

(Alc. KOH caused dehydrohologenation via E_2.)

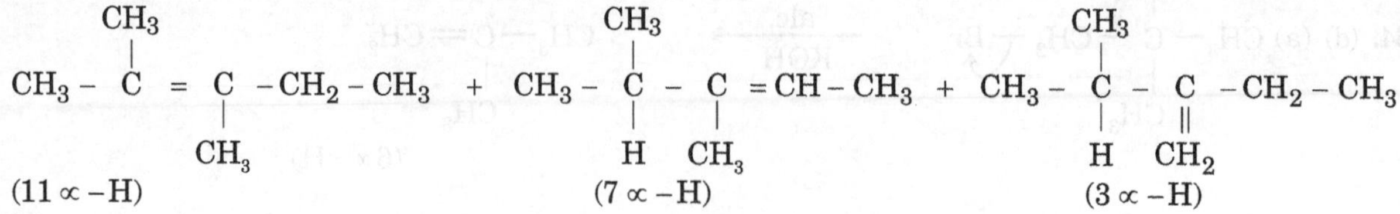

70. (b) $H_3PO_4 \rightleftharpoons H^+ + H_2PO_4^-$

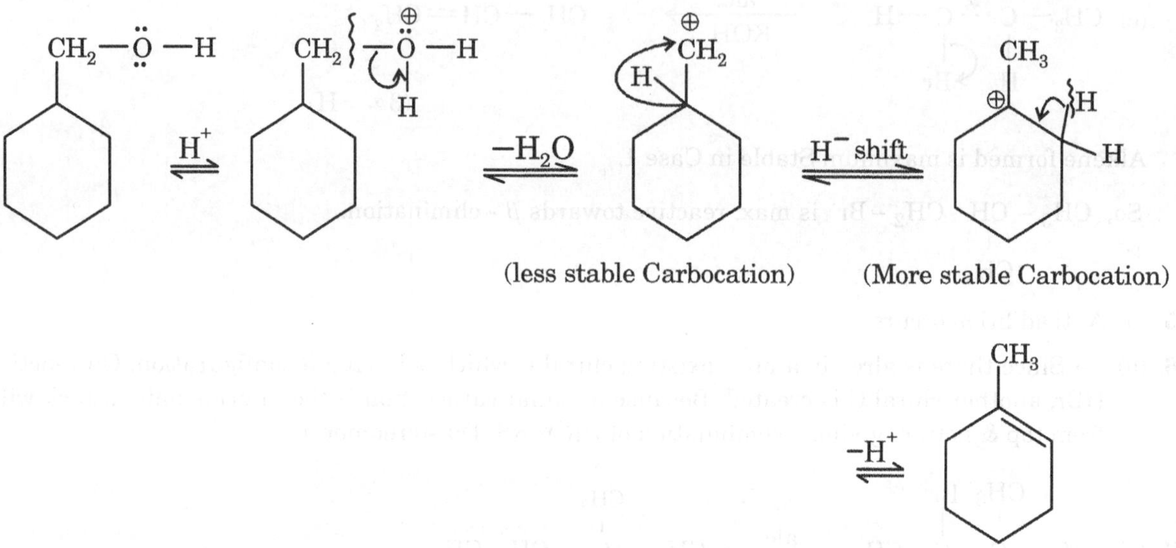

72. (b) It has no significance for symmetrical alkenes.

73. (b) $H_3C-H_2C-C\equiv C-H \xrightarrow[\text{reduction}]{\text{Controlled}} CH_3-CH_2-CH=CH_2$

$\downarrow$ HCl in peroxide

$CH_3-CH_2-CH(Cl)-CH_3$ (M. addition)

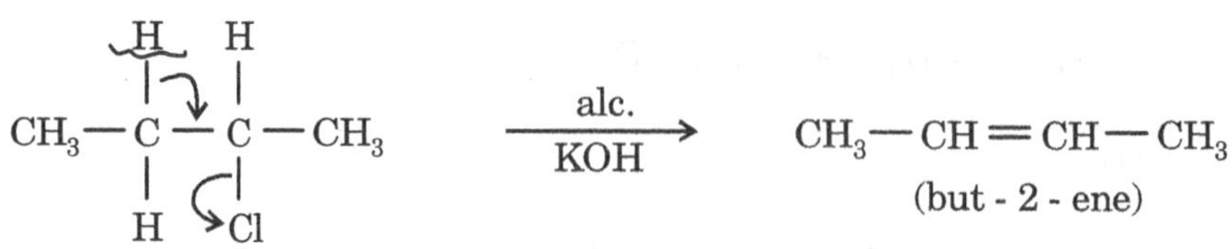

74. (c) $CH_3-CH=CH_2+H-Br \xrightarrow{\text{Peroxide}} CH_3-CH_2-CH_2Br$

(Anti M. addition) (Major product)

75. (d) It's a symmetrical alkene & addition of HBr (with or without M. Rulc) will give the same product.

76. (d) Addition of halogen proceeds via cyclic halonium ion.

77. (b) It is an example of electrophilic addition reaction and depends on stability of carbocations.

78. (b) To prepare propyl alcohol, indirect hydration has to be performed on propene

$$CH_3-CH=CH_2+\underset{\text{(Cold \& conc.)}}{H_2SO_4}\longrightarrow CH_3-\underset{\underset{HSO_4}{|}}{CH}-CH_3\xrightarrow{H_2O}CH_3-\underset{\underset{OH}{|}}{CH}-CH_3$$

(propan - 2 - ol is formed rather than propan - 1 - ol as major project)

To prepare C_2H_5OH, ethene is starting material

$$H_2C=CH_2\xrightarrow{H_2SO_4}CH_3-\underset{\underset{HSO_4}{|}}{CH_2}\xrightarrow{H_2O}CH_3CH_2OH$$

80. (c) 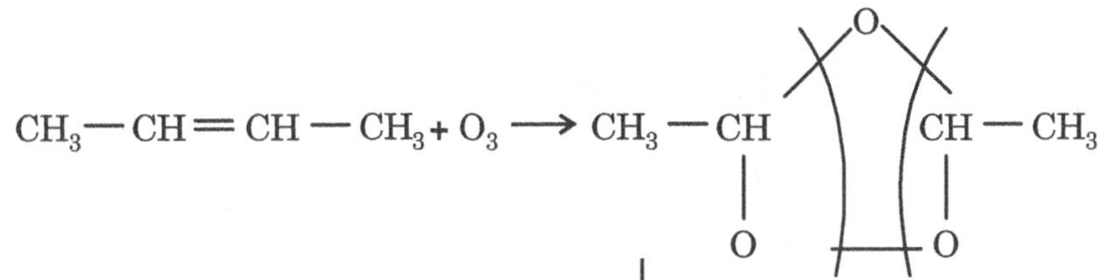

$$\downarrow H_2O$$

$$CH_3CHO+CH_3CHO+H_2O_2$$

$$CH_3CHO\xrightarrow{H_2O_2}CH_3COOH+H_2O$$

81. (b) 2 - methyl but - 2 - ene

$$H_3C-\overset{\overset{CH_3}{|}}{C}=\underset{\underset{H}{|}}{C}-CH_3\xrightarrow[(2)\,Zn,\,H_2O]{(1)\,O_3}CH_3-\overset{\overset{O}{||}}{C}-CH_3+CH_3-\overset{\overset{O}{||}}{C}-H$$

83. (d) Sodium succinate will give ethene (alkene) on Kolbe's electrolysis.

84. (b) Hex - 3 - ene will add one mole of H_2.

$$CH_3-CH_2-CH=CH-CH_2-CH_3\xrightarrow[KMnO_4]{hot}\underset{\text{(Propanoic acid)}}{CH_3CH_2COOH+CH_3CH_2COOH}$$

85. (b) $CH_3-CH=CH_2\xrightleftharpoons{H^+/H_2O}$

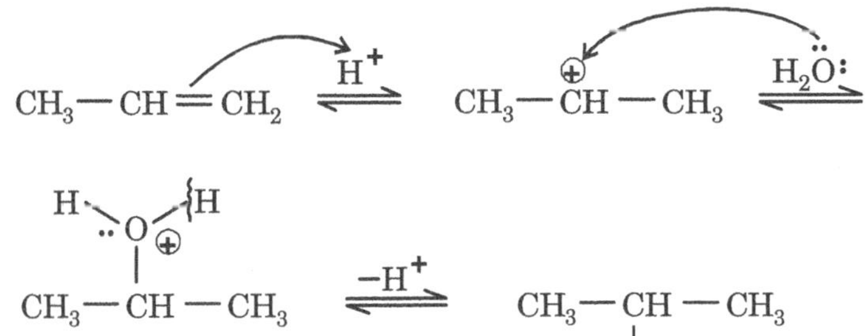

$$CH_3-\underset{\underset{OH}{|}}{CH}-CH_3$$

86. (c)

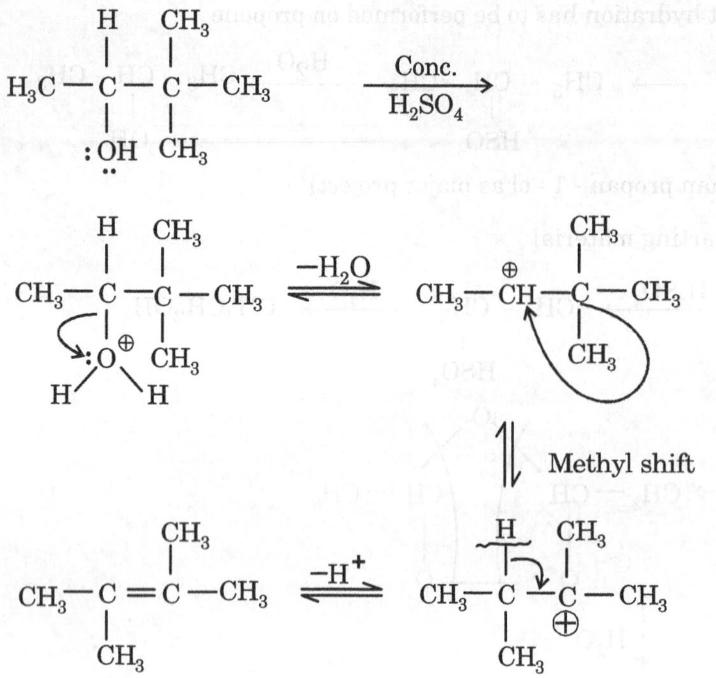

87. (c)

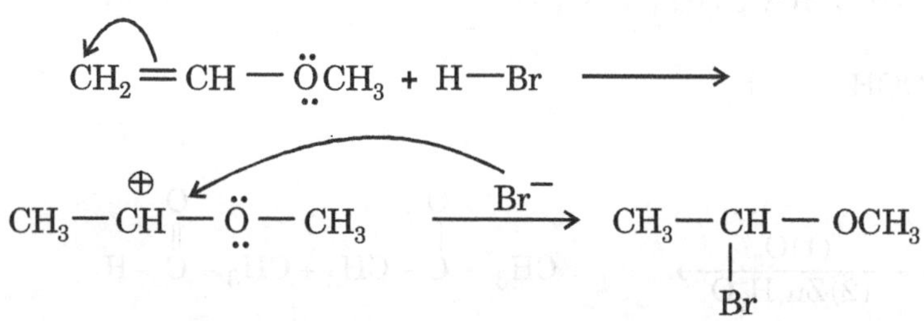

Cleavage of ether does not occur at room temperature

88. (a) Without light, Markownikoff's rule operates (electrophilic addition via carbocation formation) & in the presence of light, anti - Markowinkoff's rule operates.

89. (b)

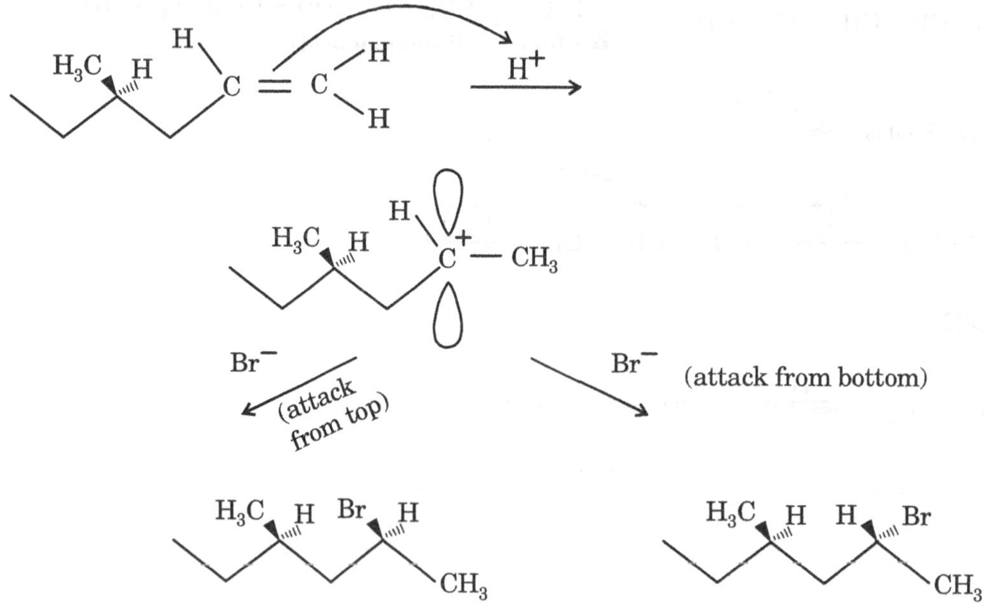

(2S, 4R) (2R, 4R)

90. (d) $CH_3-CH_2-CH=CH_2 + H-Br \longrightarrow$

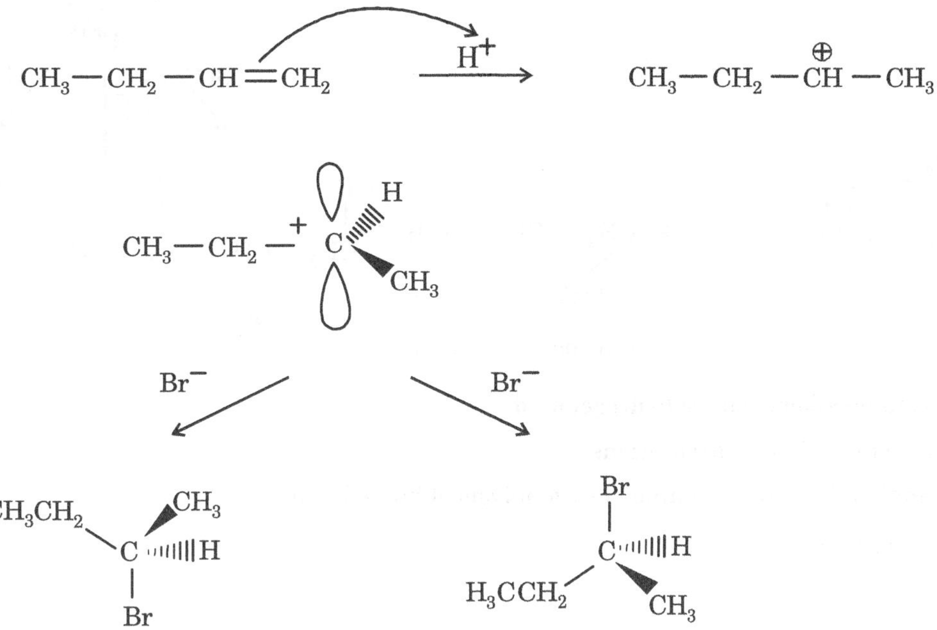

(R) – 2 – Bromobutane (50%) (S) –2 –Bromodutane (50%)

91. (c) Addition of HBr to alkene is an electrophilic addition reaction which occurs via carbocation formation. Stability of carbocation is the major concern.

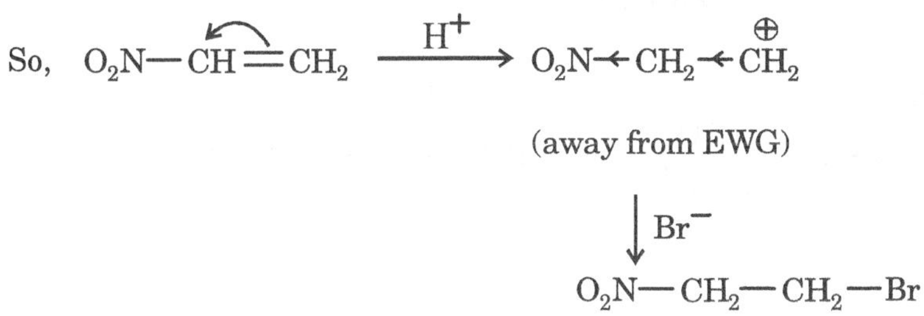

92. (a) The order depends on the stabilities of respective carbocations formed.

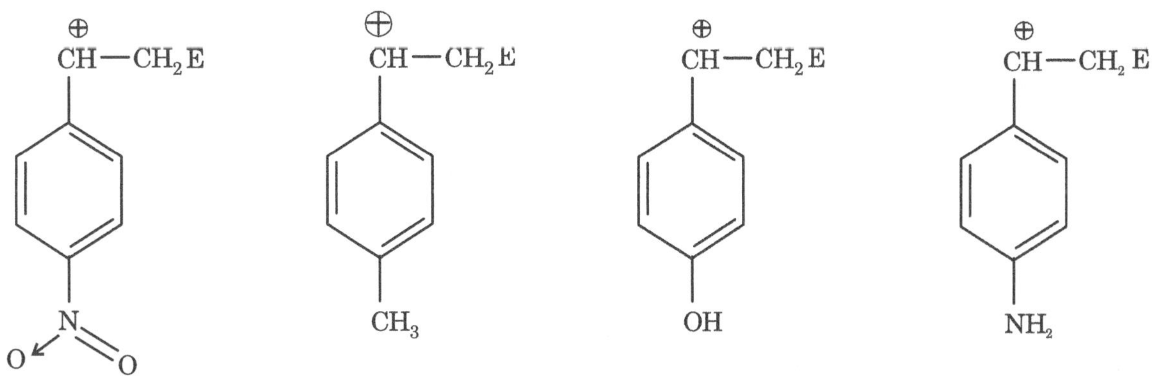

($-$M & $-$I by NO_2) highly unstable

(+ I & hyper conjugation of —CH_3)

(+M & – I of – OH) +M > – I

(+M & – I of NH_2) +M > – I

Where 'E' is electrophile. + M of $-NH_2$ > +M of – OH

93. (d)

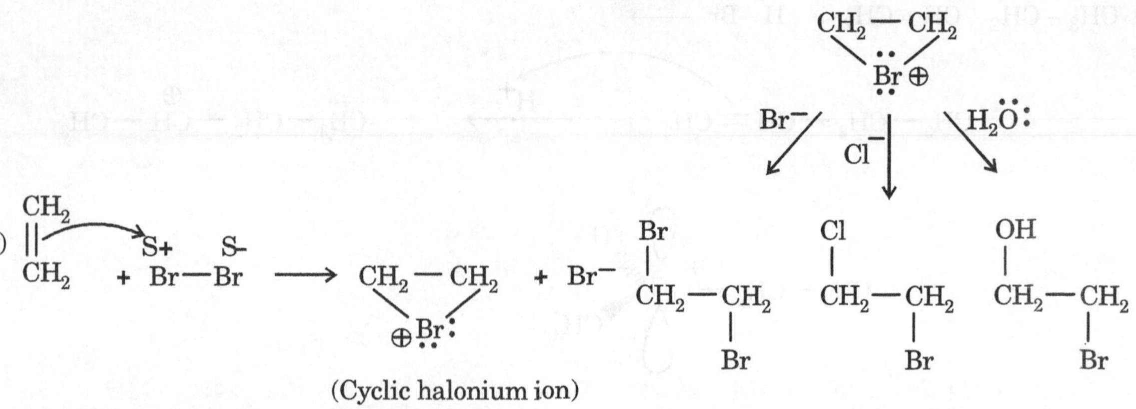

94. (c) Less stable alkene will be hydrogenated.

95. (a) Based on stability of carbocations.

96. (a) Ethanal (only) will be obtained by ozonolysis of but - 2 - ene.

$CH_3-CH=CH=CH_3$

Now, its isomers are – $CH_3-CH_2-CH=CH_2$ (But – 1 – ene) & $CH_3-\underset{}{\overset{CH_3}{\overset{|}{C}}}=CH_2$ (2 – methylpropene)

So, $$CH_3-CH_2-CH=CH_2 \xrightarrow[(2)\,Zn,\,H_2O]{(1)\,O_3} CH_3-CH_2CHO+HCHO$$

$$CH_3-\overset{CH_3}{\overset{|}{C}}=CH_2 \xrightarrow[(2)\,Zn,\,H_2O]{(1)\,O_3} CH_3-\overset{O}{\overset{||}{C}}-CH_3+HCHO$$

97. (d) (a)

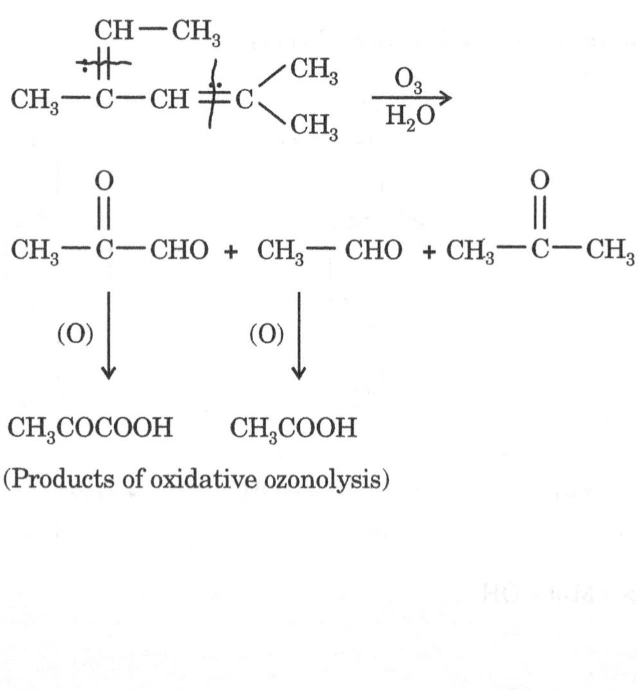

(Products of oxidative ozonolysis)

$$CH_3-\underset{}{\overset{C(CH_3)_2}{\overset{\|}{C}}}-CH=CH-CH_3 \xrightarrow[H_2O]{O_3}$$

(b) $CH_3-\overset{O}{\overset{\|}{C}}-CHO + CH_3-\overset{O}{\overset{\|}{C}}-CH_3 + CH_3-CHO$

$$CH_3-\overset{O}{\overset{\|}{C}}-CHO \xrightarrow{(O)} CH_3COCOOH \qquad CH_3-CHO \xrightarrow{(O)} CH_3COOH$$

98. (d) Its corresponding alkene i.e.,

$$CH_3-\underset{CH_3}{\overset{CH_3}{\overset{|}{\underset{|}{C}}}}=CH_2, \text{ is not possible}$$

99. (d) Acid catalyzed hydration of an alkene is an electrophilic addition reaction & occurs via carbocation formation. Stability of carbocations defines the order of reactivity.

100. (c) (a) $H-\underset{H}{\overset{H}{C}}-\underset{Cl}{\overset{H}{C}}-\underset{H}{\overset{H}{C}}-CH_2-CH_3 \xrightarrow[KOH]{alc.}$

$$\underset{\text{(Minor product)}}{CH_2=CH-CH_2-CH_2-CH_3} + \underset{\text{(Major product)}}{CH_3-CH=CH-CH_2-CH_3}$$

(b) $CH_3-CH_2-\underset{Cl}{\overset{H}{C}}-CH_2-CH_3 \xrightarrow[KOH]{alc.} CH_3-CH=CH-CH_2-CH_3$, but geometrical isomerism (Pent - 2 - ene) is possible in this , so more than 1 alkene with be possible.

(c) $CH_3-CH_2-CH_2-Cl \xrightarrow[KOH]{alc.} CH_3-CH=CH_2$

(Only 1 alkene is possible)

101. (a) $CH_3=CH-\underset{OCH_3}{\underset{|}{CH}}-CH_3 \xrightarrow{HBr}$

(a) $CH_2=CH-\underset{OCH_3}{\underset{|}{CH}}-CH_3 \xrightarrow{H^+} CH_3-\overset{\oplus}{CH}-\underset{OCH_3}{\overset{H}{C}}-CH_3$

$$\xrightarrow{H^- \text{ shift}} CH_3-CH_2-\underset{OCH_3}{\overset{\oplus}{C}}-CH_3 \xrightarrow{Br^-} CH_3-CH_2-\underset{OCH_3}{\overset{Br}{C}}-CH_3$$

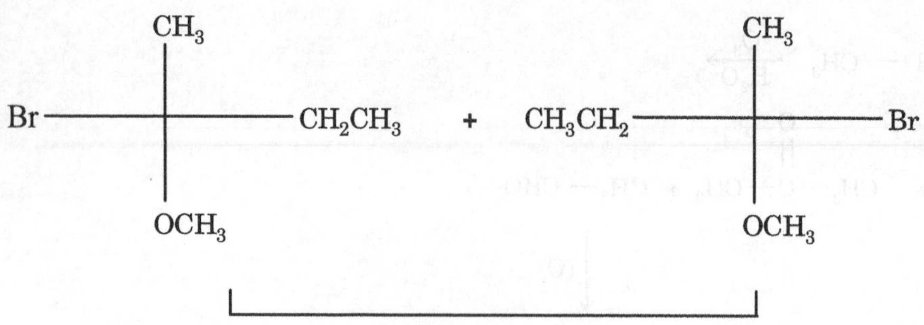

(A)

Racemic Mixture

$$\text{(B)}\quad CH_2{=}CH-\overset{*}{\underset{\underset{OCH_3}{|}}{CH}}-CH_3 \xrightarrow[R_2O_2]{HBr} \underset{\underset{Br}{|}}{CH_2}-CH_2-\overset{*}{\underset{\underset{OCH_3}{|}}{CH}}-CH_3\quad\text{(B)}$$

102. (b) (a) 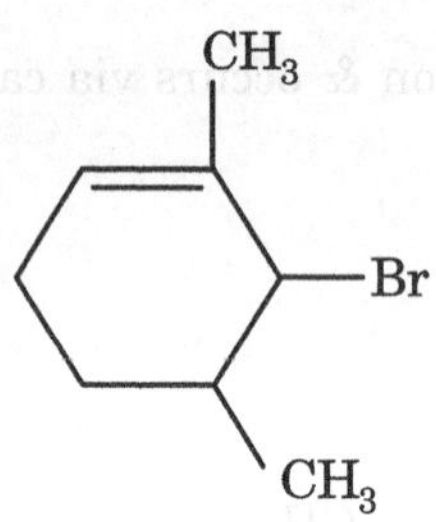 ⟶ Can undergo E_2 elimination.

(b) 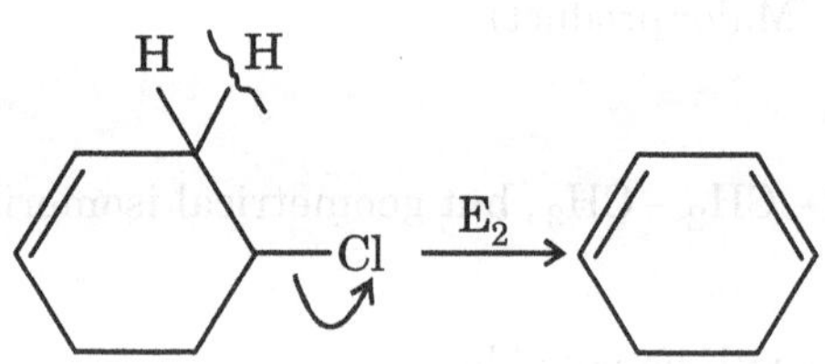

(Conjugated diene)
Chirality lost

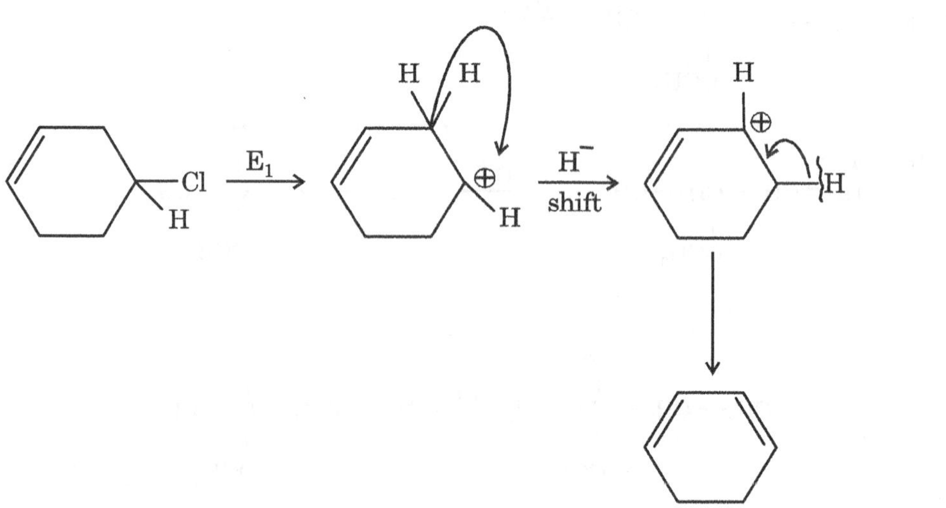

Conjugated diene (Chirality lost)

(c)

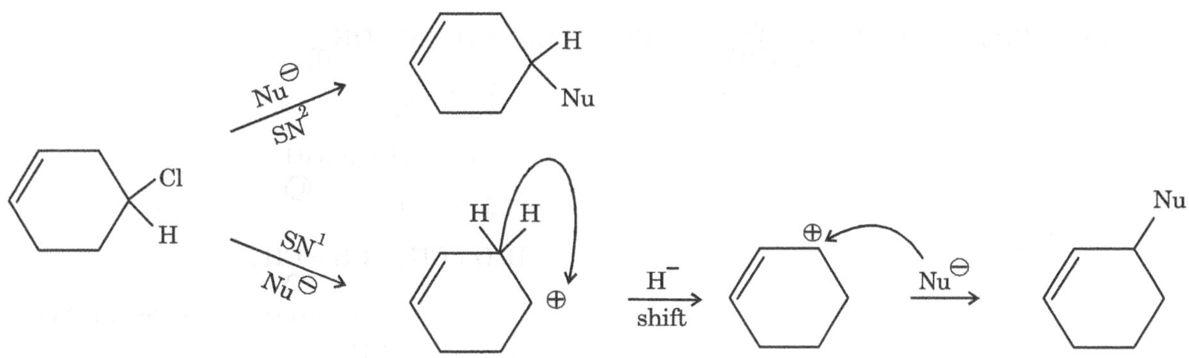

103. (d) Ammoniacal Cu_2Cl_2 will react with terminal alkyne i.e. 1 - butyne.

104. (a)

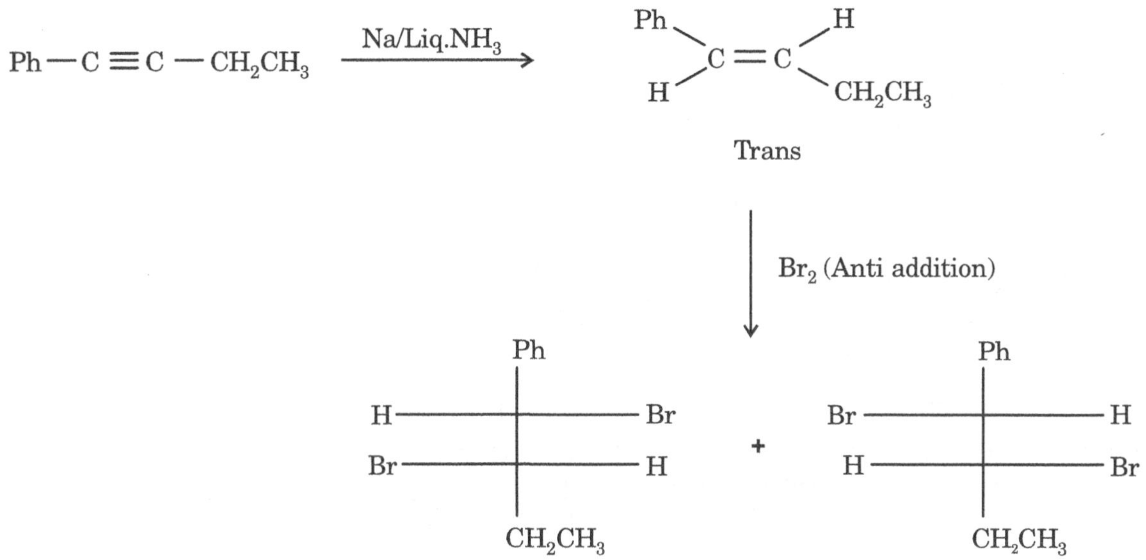

105. (b)

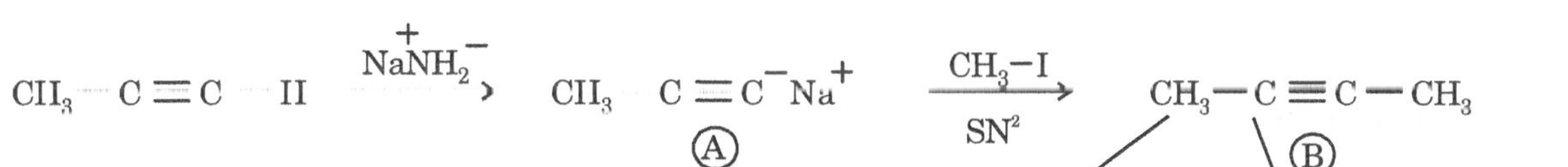

Cis & Trans forms of an alkene are also diastereomers.

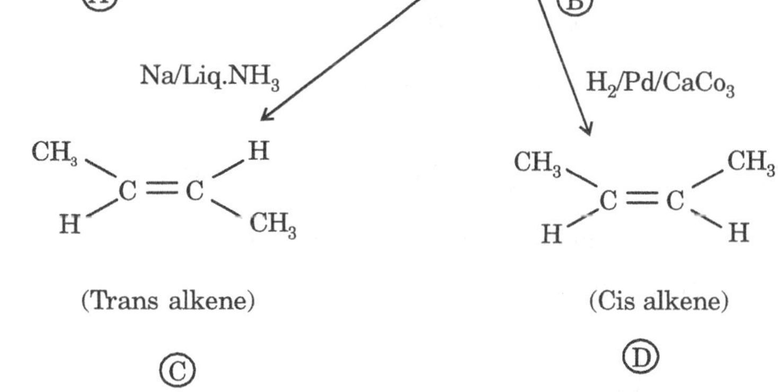

106. (a) But - 1 - yne is a terminal alkyne. So, it will give tollens test.

$$CH_3-CH_2-C\equiv C-H \xrightarrow[\text{test}]{\text{Tollen's}} CH_3-CH_2-C\equiv C^-Ag^+$$

$$CH_3-CH_2-C\equiv C-H \xrightarrow[\text{KMnO}_4]{\text{hot}} CO_2 + \underset{Ⓑ}{CH_3-CH_2-C-OH}$$

$$\downarrow LiAlH_4$$

$$\underset{Ⓒ}{CH_3-CH_2CH_2OH}$$

$$\downarrow H^+/\Delta$$

$$H_2O + \underset{Ⓓ}{CH_3-CH=CH_2}$$

$$\downarrow \text{oxymercuration} - \text{demercuration}$$

$$CH_3-\underset{\displaystyle OH}{\underset{|}{CH}}-CH_3$$

107. (b)

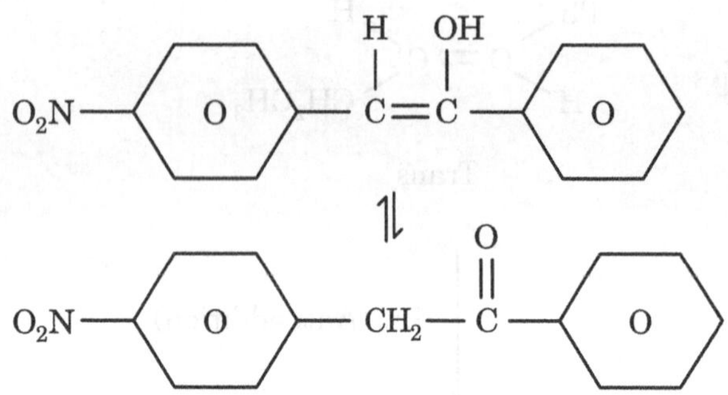

108. (d) (a)

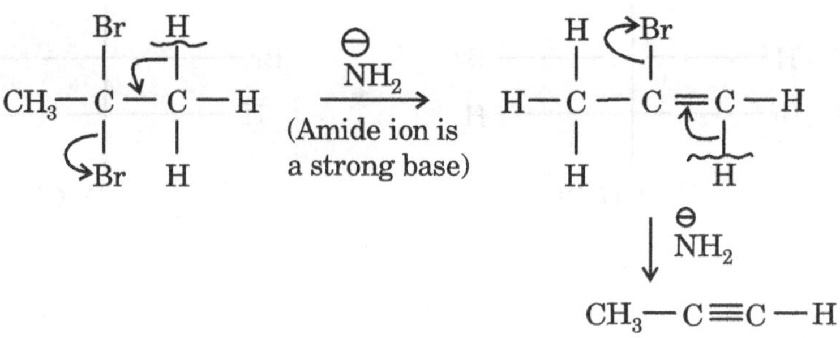

(b) $Cl_2CH-CHCl_2 + 2Zn \longrightarrow 2ZnCl_2 + H-C\equiv C-H$

(dehalogenation occurs)

(c) $CaC_2 + 2H_2O \longrightarrow Ca(OH)_2 + C_2H_2\uparrow$

109. (a) $H-C\equiv C-H + Na^+\overset{-}{O}H \rightleftharpoons H-C\equiv \overset{\ominus}{\ddot{C}}Na^+ + H_2O$

Water is more acidic than acetylene, so majority equilibrium shifts in the backward direction

110. (b) Acidic strength follows the order

$$R-\overset{\overset{\displaystyle O}{\|}}{C}-OH > H_2O > ROH > H-C\equiv C-H > NH_3 > CH_2=CH_2 > CH_3-CH_3$$

In 2nd reaction

$$H-C\equiv C-H + RO^{\ominus} \rightleftharpoons H-C\equiv C^{\ominus} + ROH$$

Alcohol is more acidic than acetylene. So, majority reaction shifts in backward direction.

111. (d) Terminal alkynes (with highly acidic H) react with tollins reagent.

But - 2 - yne I a non- terminal alkyne.

112. (c) 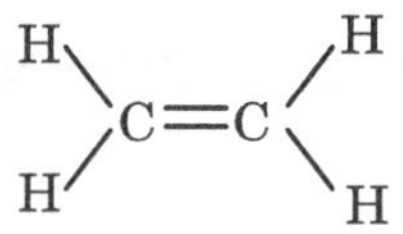

It is an unsaturated compound, so it will decolorize $KMnO_4$ but 'lacks sufficiently acidic H, so does not give precipitate with $AgNO_3$.

113. (a) alc. KOH converts terminal alkyne to non-terminal alkyne & $Na^+NH_2^-$ converts non-terminal to terminal alkyne. So, both alc. KOH & $NaNH_2$ are involved in this isomerization reaction.

114. (b) $CaC_2 + H_2O \longrightarrow Ca(OH)_2 + \underset{\text{(acetylene)}}{C_2H_2}$

$$H-C\equiv C-H \xrightarrow[H_2SO_4]{HgSO_4} H-\overset{OH}{\overset{|}{C}}=CH_2 \xrightleftharpoons{\text{TAUTO.}} CH_3-\overset{O}{\overset{||}{C}}-H$$

(hydration of alkyne)

115. (a) 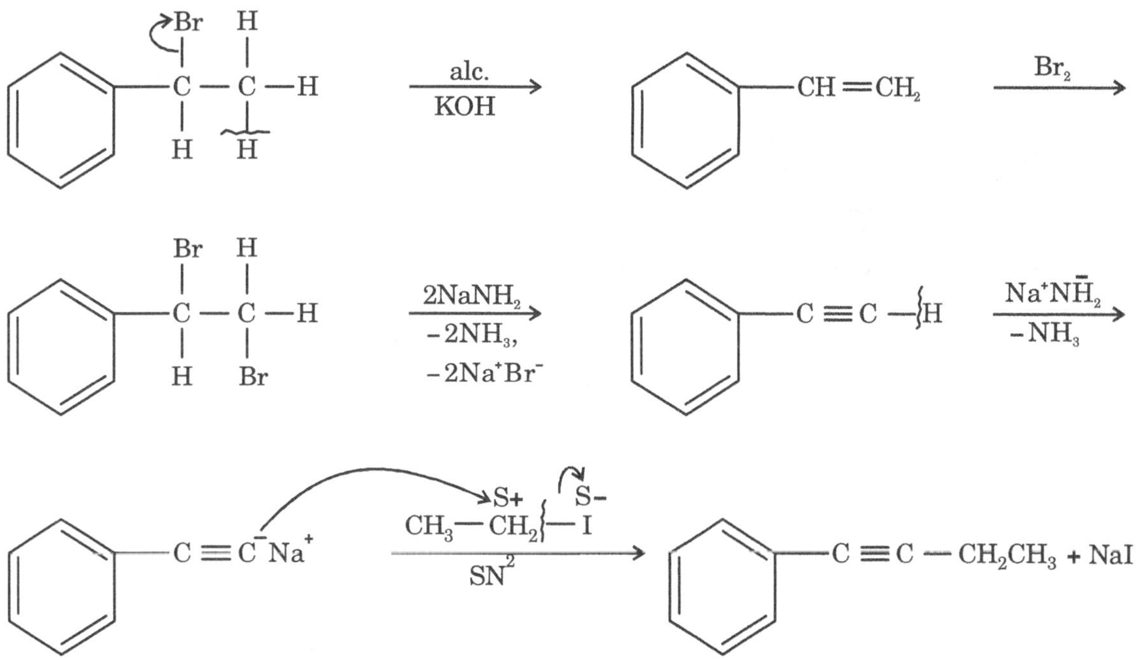

116. (b) $H-C\equiv C-H \xrightarrow{H-Br} CH_2=CH-Br \xrightarrow{HBr}$

$$CH_3-\underset{\underset{Br}{|}}{CH}-Br \xrightarrow[KOH]{alc.} CH_2=CH-Br \xrightarrow{NaNH_2}$$

(dehydrohalogenation)

$H-C\equiv C-H$

117. (b) (a) $CH_3-CH_2-C\equiv C-H \xrightarrow{Na^+NH_2^-} NH_3 + CH_3-CH_2-C\equiv C^-Na^+$

(b) $CH_3-CH_2-C\equiv C\ Na^+ + CH_3-Cl \xrightarrow{SN^2} CH_3-CH_2-C\equiv C-CH_3 + NaCl$

118. (d) Reaction I

(a)

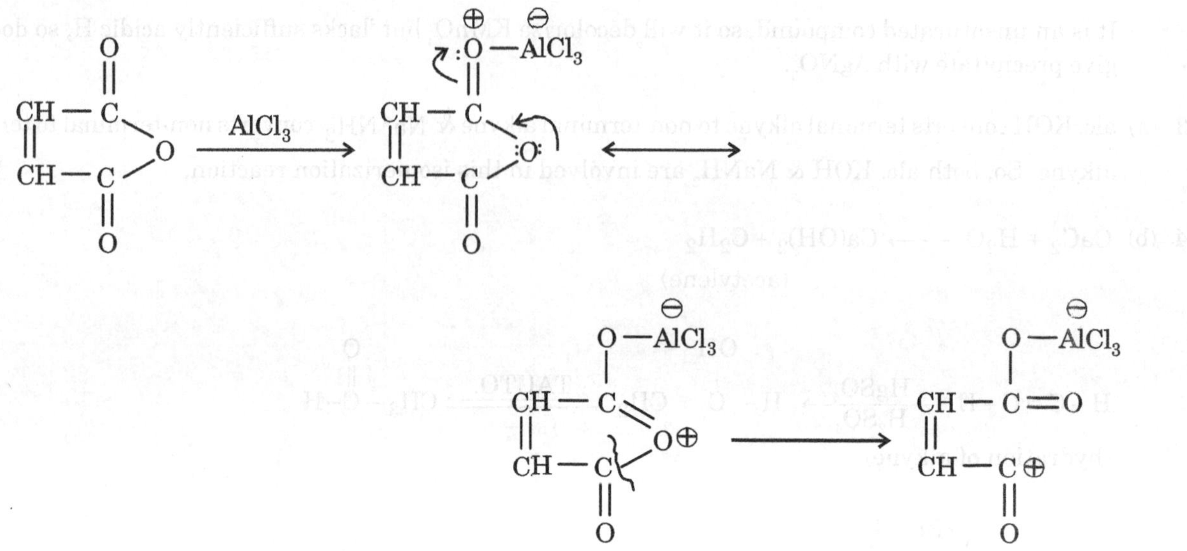

(electrophile generated)

(b)

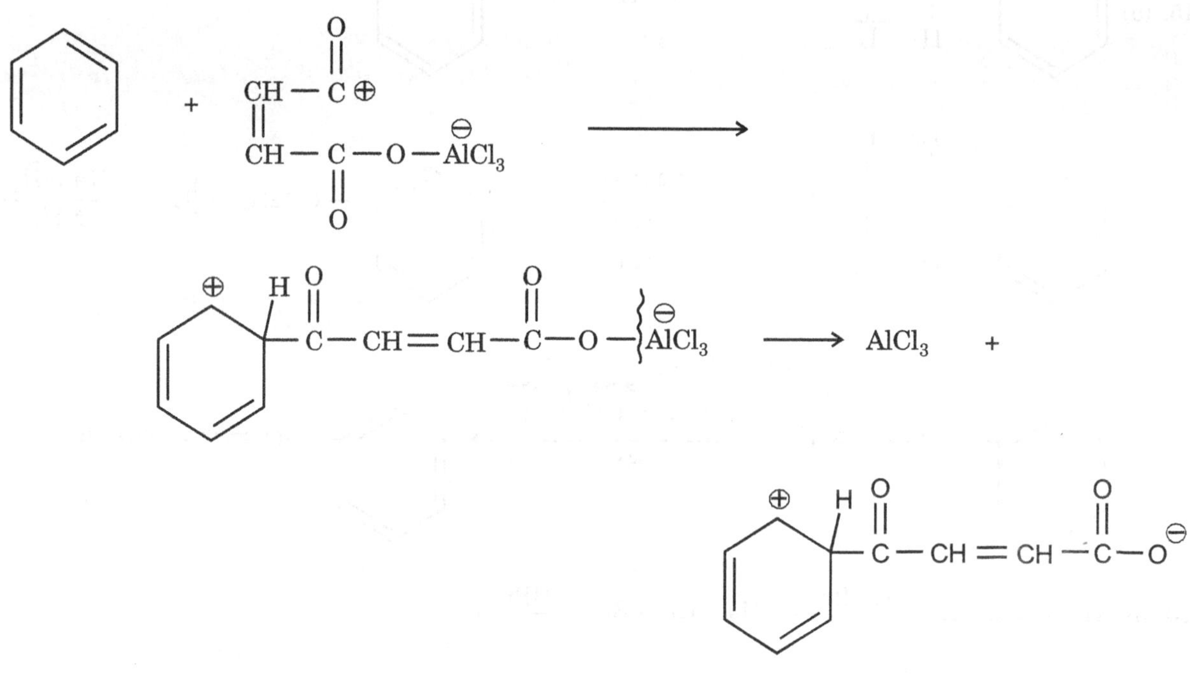

(c) Deprotonation

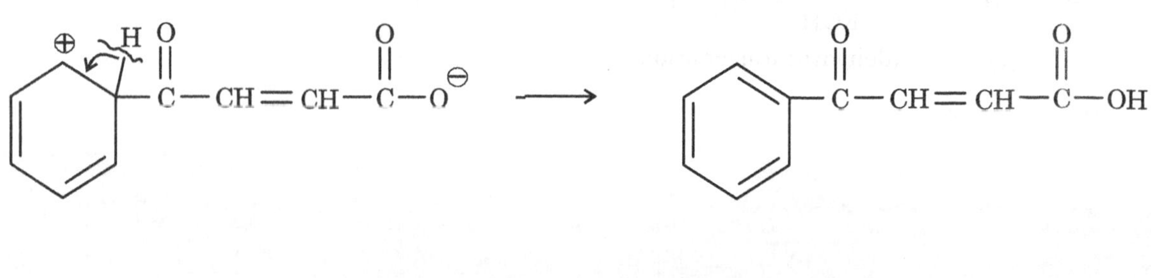

Reaction 2.

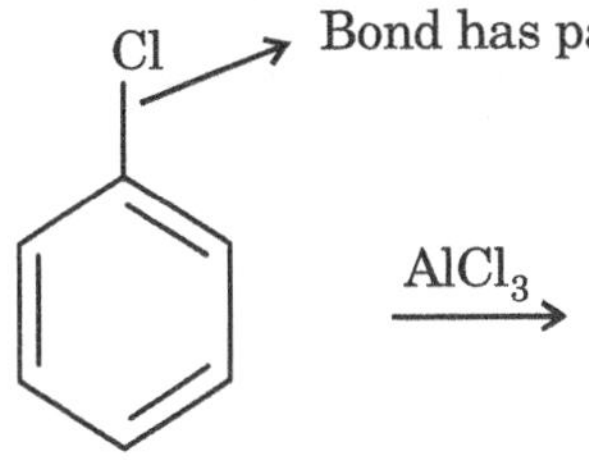

Also, phenyl Carbocation 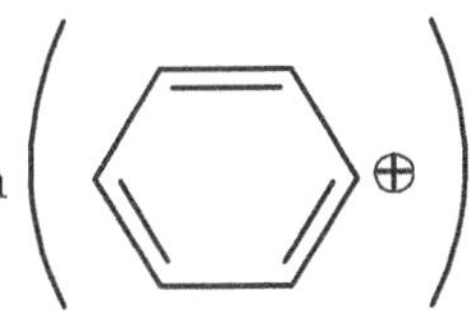 is unstable. So, reaction will not occur.

Reaction 3

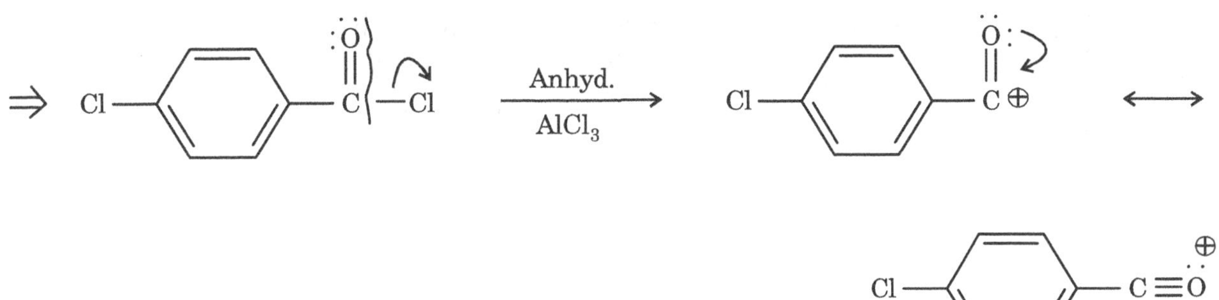

(electrophile generated)

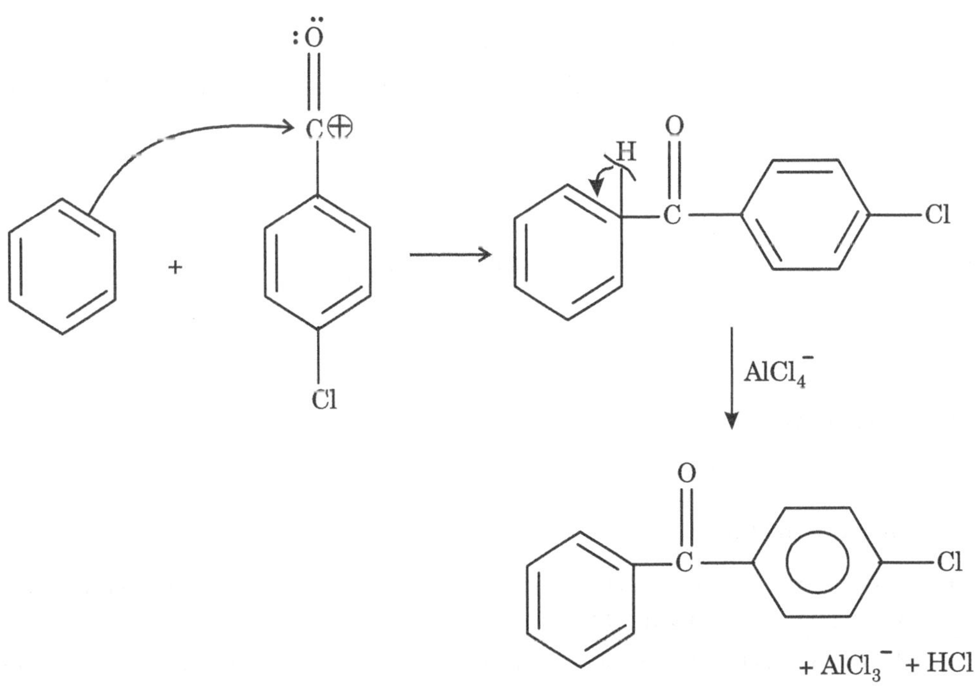

121. (d) (a) $R-Cl + AlCl_3 \rightleftharpoons R^+ + AlCl_4^-$

(b) $R-\ddot{O}-H + BF_3 \rightleftharpoons R-\overset{+}{O}(H)-\overset{-}{B}F_3 \rightleftharpoons R^+ + HOBF_3^-$

(c) $R-\ddot{O}-R + BF_3 \rightleftharpoons R-\overset{\oplus}{O}(R)-\overset{-}{B}F_3 \rightleftharpoons R^+ + ROBF_3^-$

(d) $CH_3-CH=CH_2 + H^+ \rightleftharpoons CH_3-\overset{\oplus}{C}H-CH_3$

122. (b)

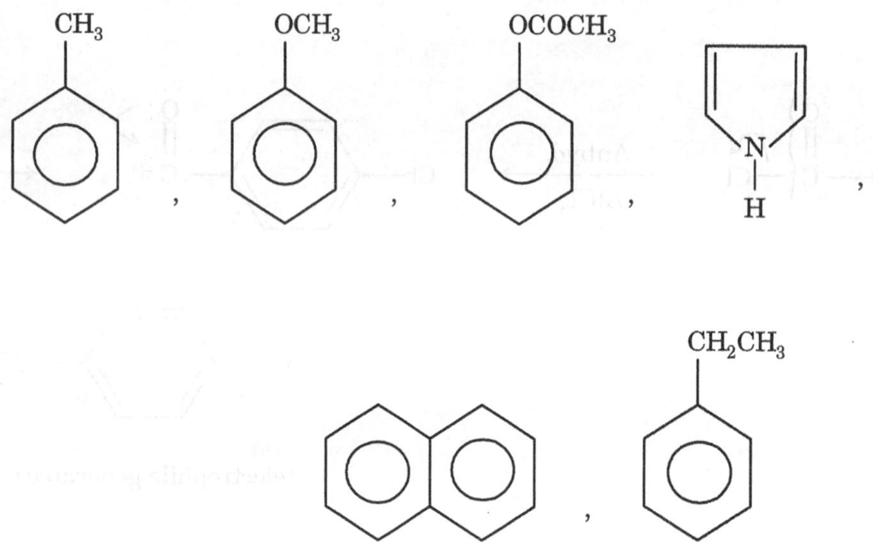

are more reactive than benzene towards nitration.

123. (a) It is an electrophilic substitution reaction. Groups that increase the electron density on benzene will make it more reactive towards nitration.

$-\ddot{O}CH_3$ is strongly activating.

$-NH-COCH_3$ & $-N(CH_3)(N=O)$ are moderatlely activating.

$-\overset{O}{\overset{\|}{C}}-CH_3$ is deactivating.

$-\overset{O}{\overset{\|}{C}}-H$ is also deactivating.

124. (c) In fused rings, if 1 ring is having activating group & other deactivating group then substitution takes place on the ring having activating group.

Now, $-OCH_3$ is o/p directing but para position wrt $-OCH_3$ is blocked. So, attack of electrophile occurs at orthowrt $-OCH_3$.

125. (d) In option 1,

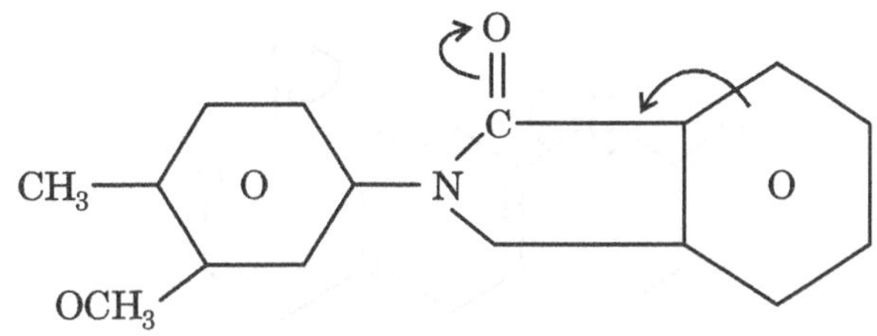

(activated ring) (deactivated ring)

On the left ring, $-OCH_3$ has a very strong + M. So, electrophile follows its direction.

(b) In the compound, 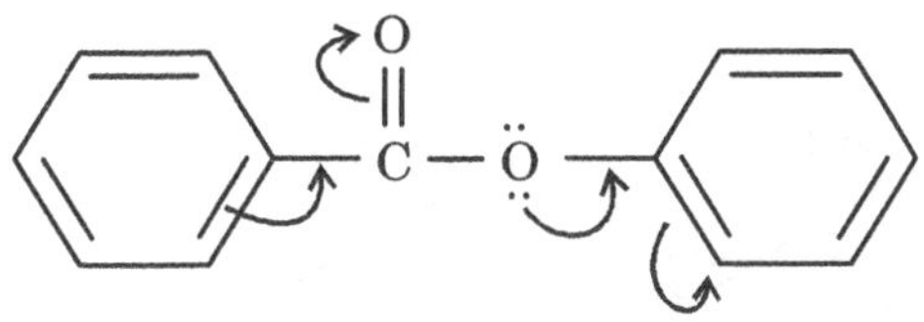 attack occurs at ring II.

(I) (II)

(deactivated ring) (activated ring)

(c) Toluene is more reactive than C_6H_6 because of hyper conjugation & + I of $-CH_3$ group.

126. (b) $-NHCOCH_3$ is more activatinggp. (o/p directing) than $-CH_3$ group. So, electrophile attack occurs at p - position wrt $-NHCOCH_3$.

127. (a)

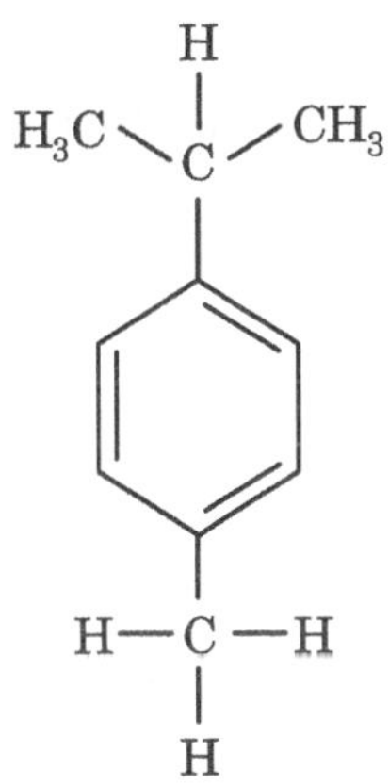

More activating, electrophile will attack at O, p wrt $-CH_3$ gp. But para position is blocked by isopropyl gp. $\left(-CH\begin{matrix} CH_3 \\ CH_3 \end{matrix}\right)$ so, attack occurs at orthowrt $-CH_3$.

128. (c) In III molecule,

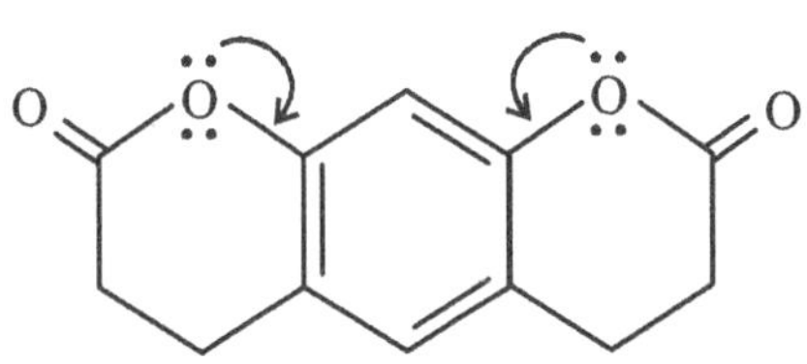

(+ M from both ends activate the central ring)

In II molecule,

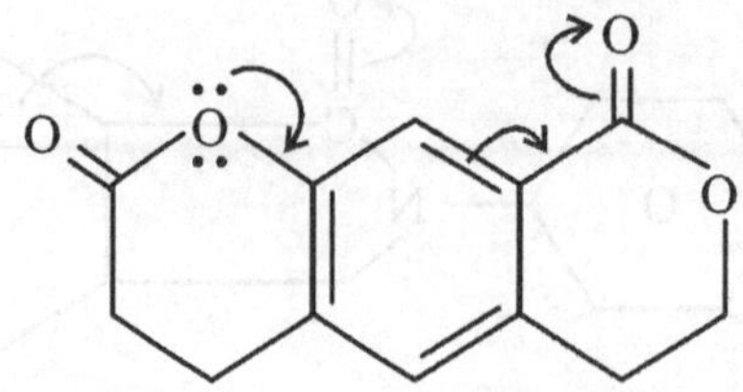

(+ M from here) (– M from here)

In molecule,

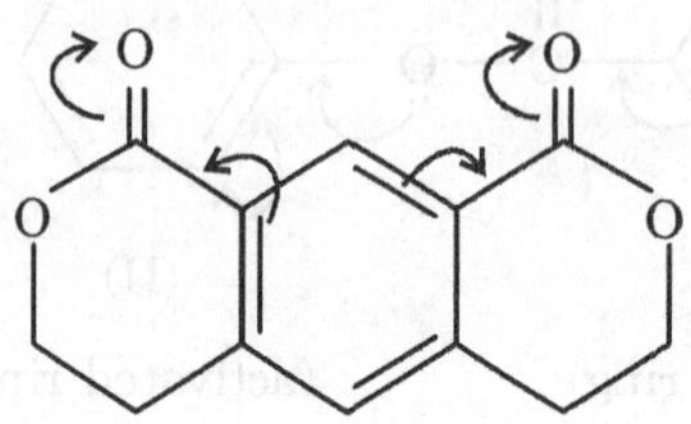

(- M from both sides, so central ring is highly deactivated)

129. (a) The ring on the right is more activated. Ortho wrt – $\ddot{\underset{\cdot\cdot}{O}}$ – is free while para is blocked by methyl.

130. (d) In option 1, Max.hyper conjugation is observed in 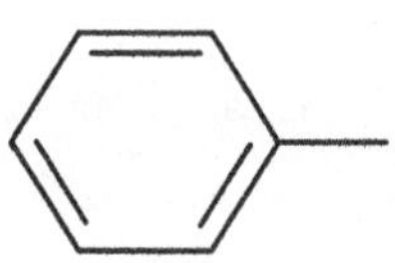 CH_3 on account of low BDE of C – H bond. So, ring is Max. activated.

In option 2, 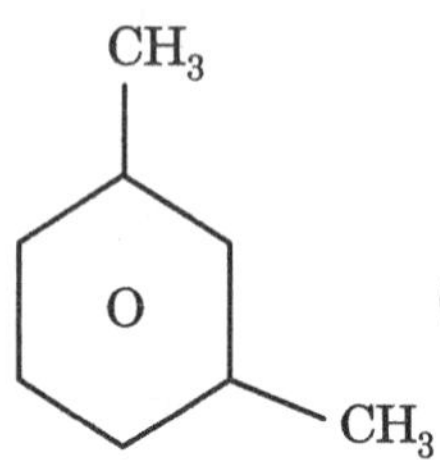 (Meta - xylene), both the methyl groups reinforce the effect of each other by hyper conjugation thus bringing - ve charge on same C's.

131. (b) 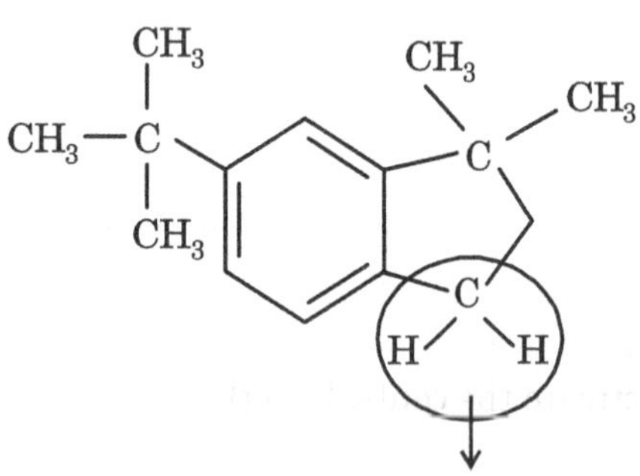

hyper conjugation activates the ring

132. (d)

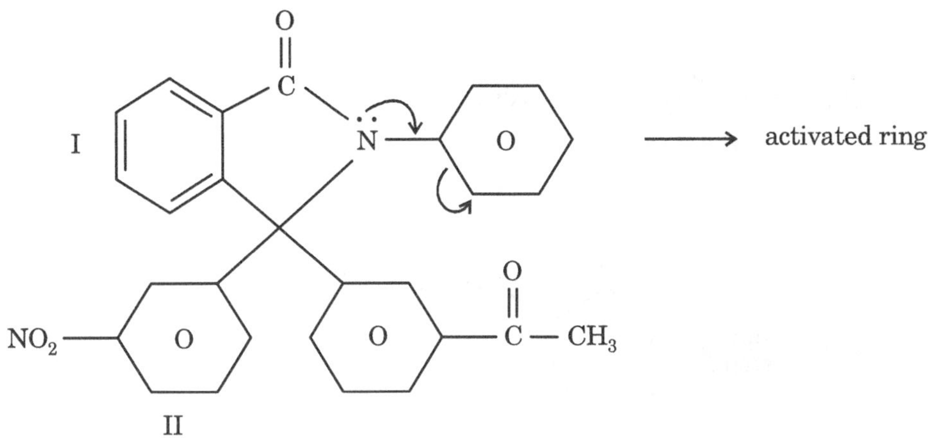

133. (c)

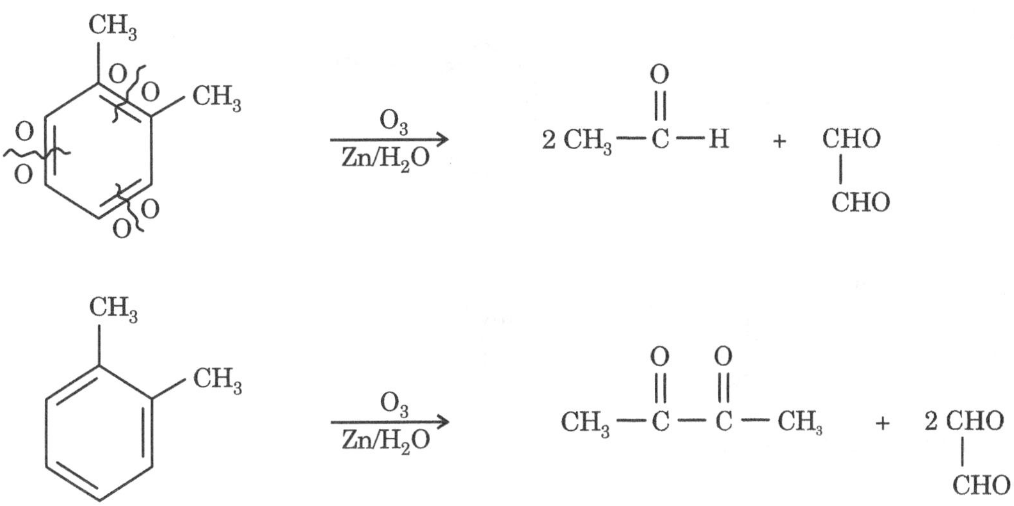

134. (a)

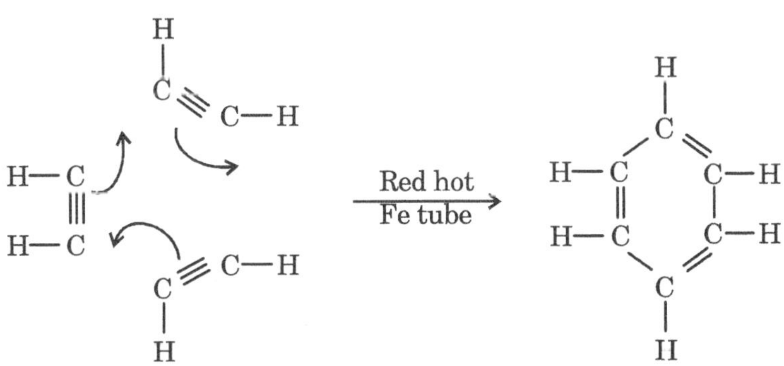

136. (d)

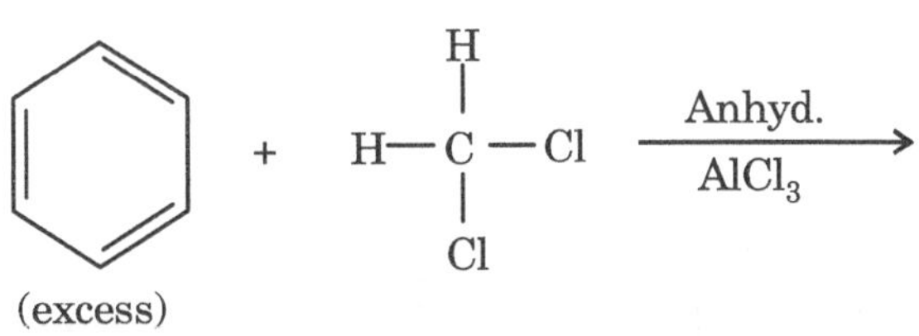

Step 1: Generation of electrophile

(a) 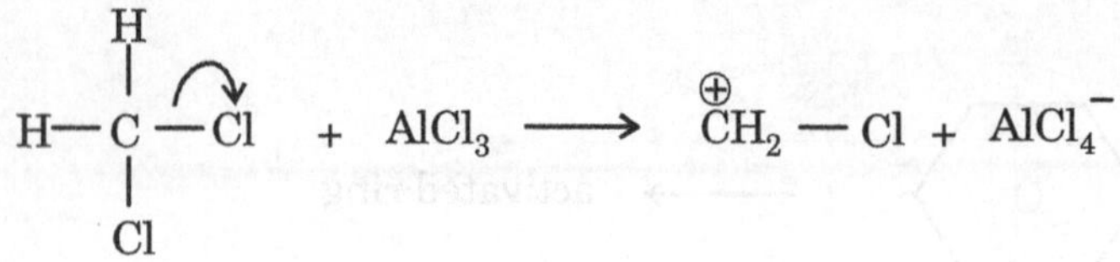

(b) Attack on Ring

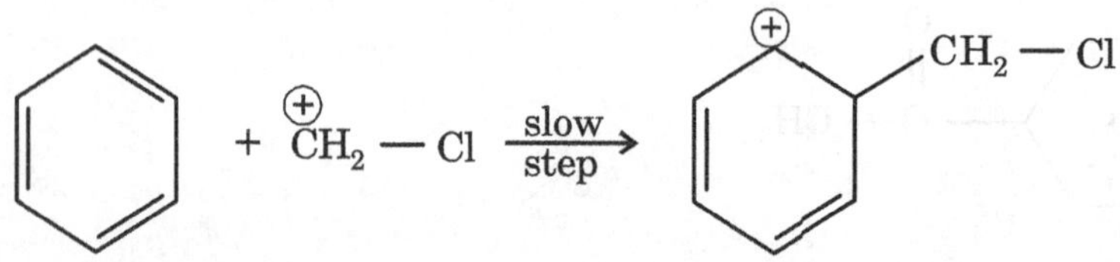

(c) Deprotonation

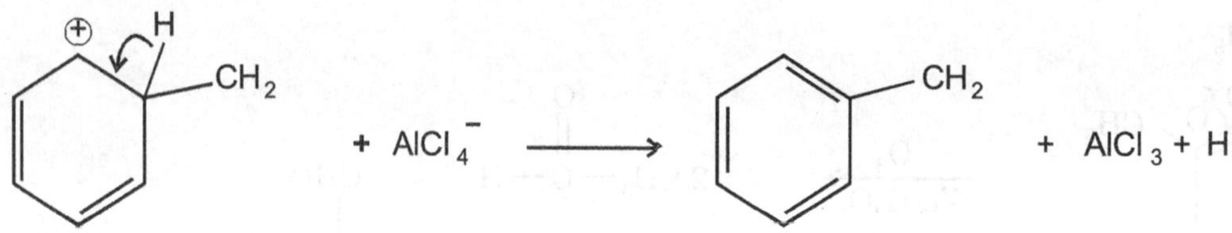

Again,

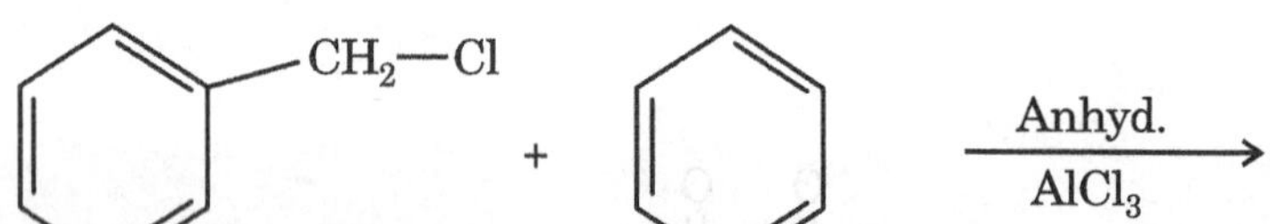

(a) Step 1 : Generation of electrophile

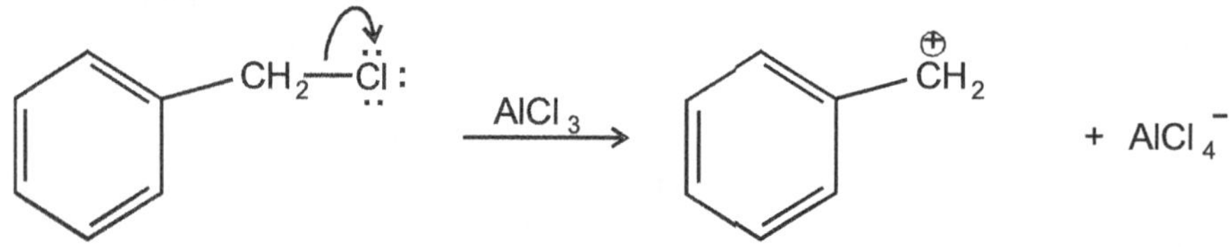

(b) Step 2 : Attack on Ring

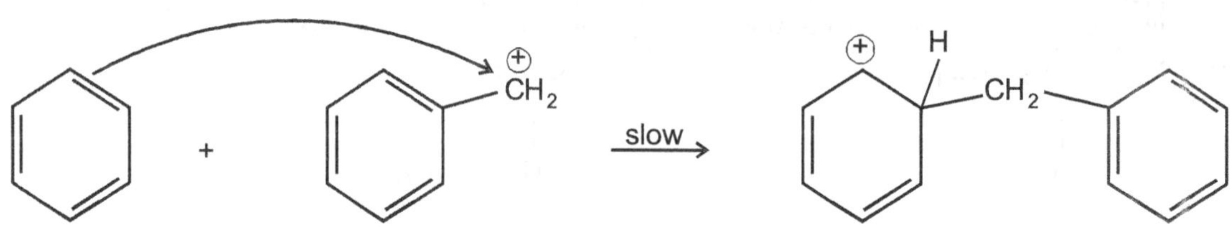

(c) Deprotonation

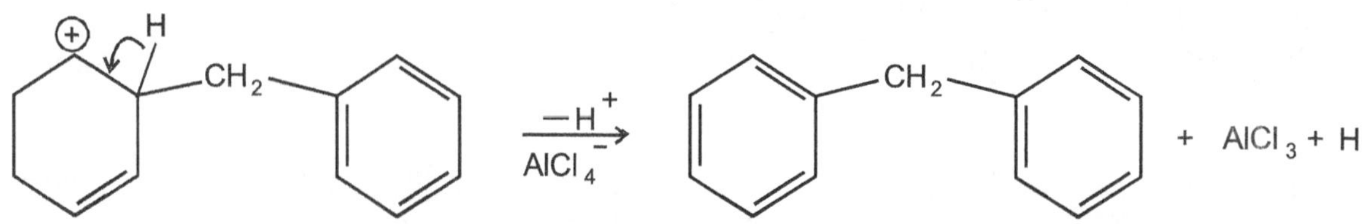

137. (d) Option (a) is an example of electrophilic substitution. Once, a chlorine is introduced on the ring, further halogenation occurs at o & p positions wrt halogen.

(Halogens are o, p directors but deactivators)

Option (b) & (c) are free radical addition reactions.

138. (b) (a) 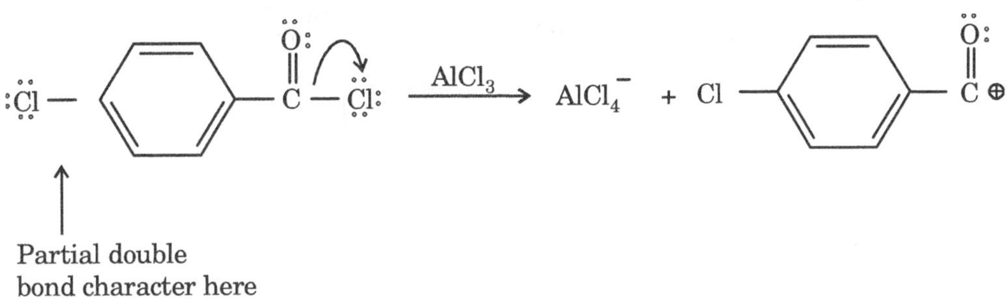

(b) 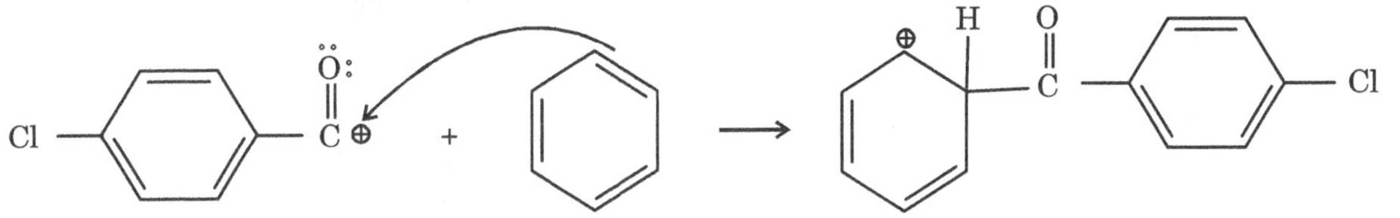

(c) 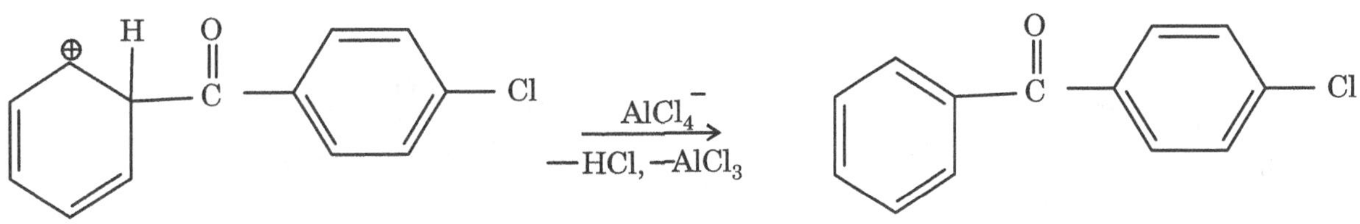

139. (d) For option (a)

(a) 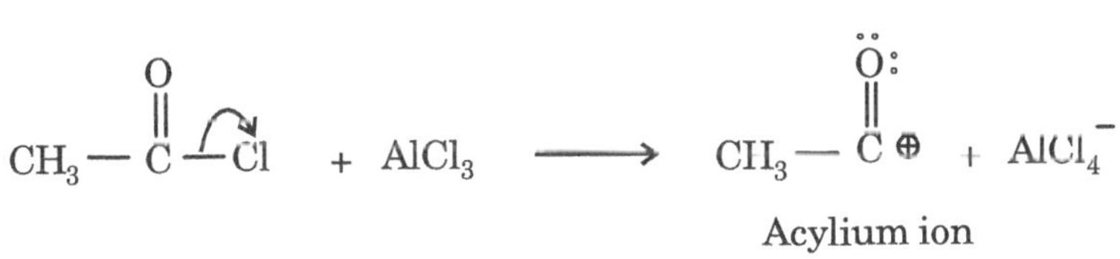

(b) 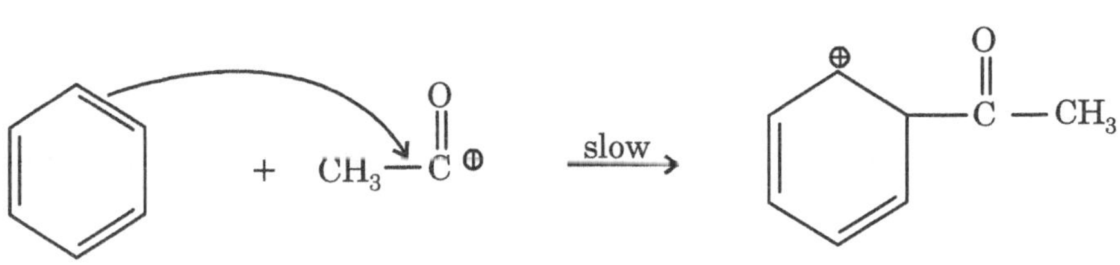

(c) 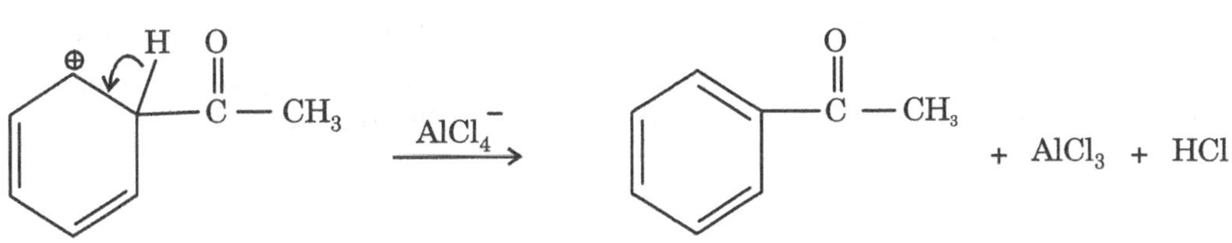

For option (b)

(a)

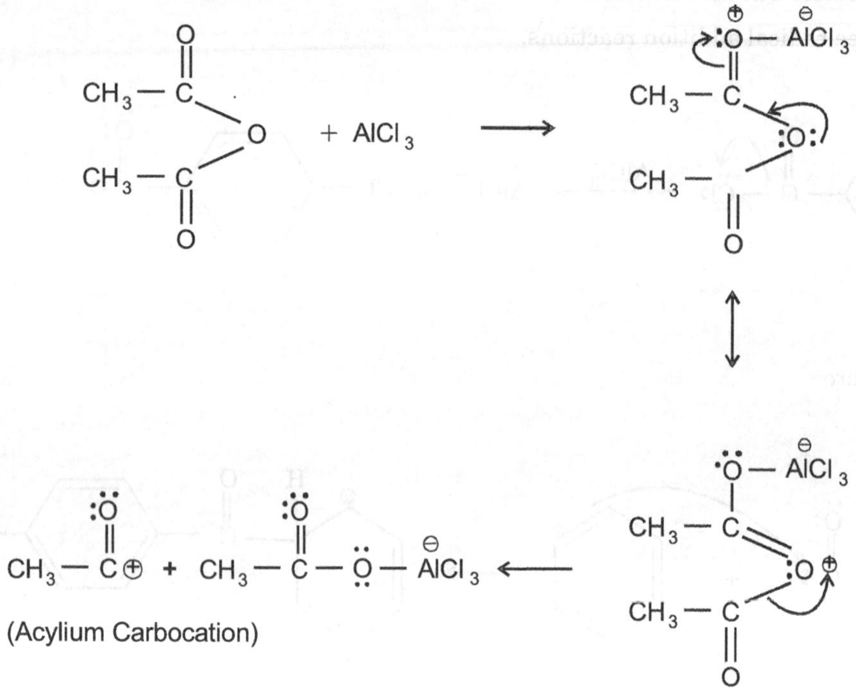

(b

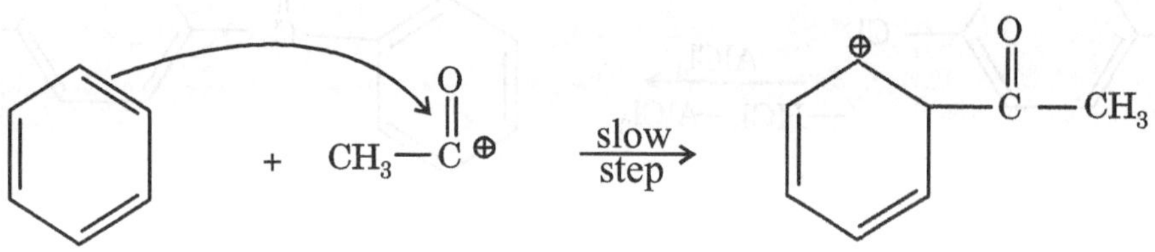

(c)

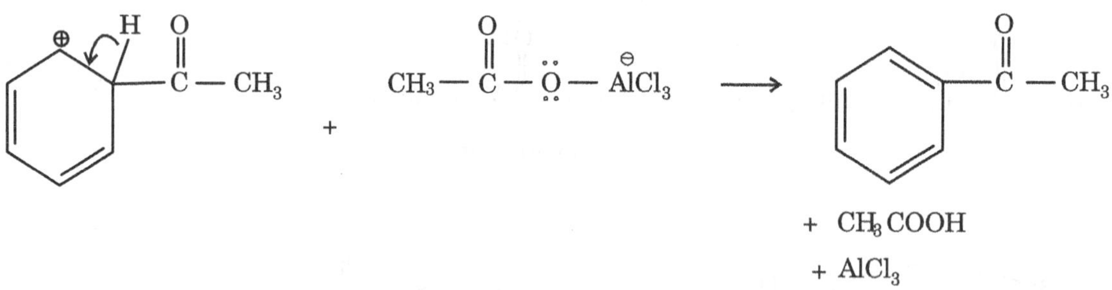

141. (b) In friedal crafts alkylation using an alkyl chloride, the lewis acid is used in catalytic quantities. In an acylation, however, the lewis acid can also complex to any O atoms present, in the $-\overset{O}{\overset{||}{C}}-$ group in the acetylchloride/acetic anhydride. As a result, in acylation reactions more amount oflewis acid is required.

142. (d) Sulphonation of benzene is an electrophilic substitution reaction. Among, the given compounds, toluene is highly activated because of hyper conjugation & +I effect of $-CH_3$ groups.

143. (d) 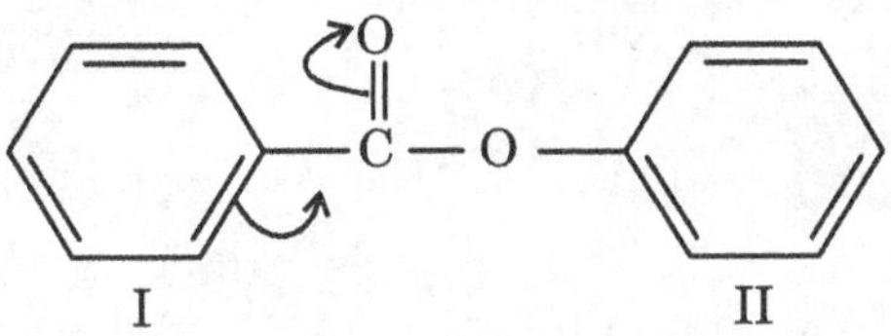

In phenyl benzoate, the ring labeled I is deactivated while the ring labeled to is activated.

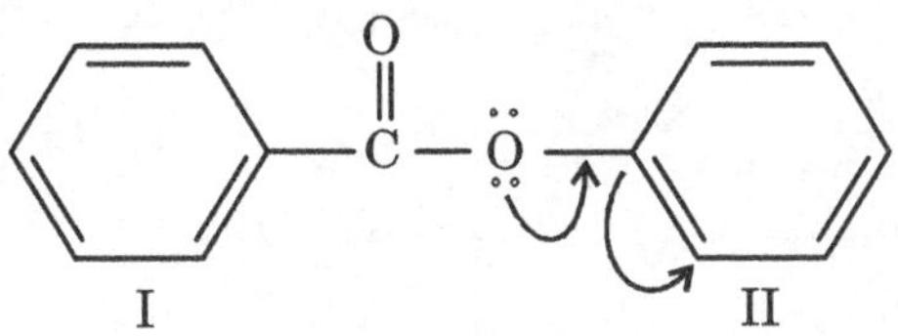

So, nitronium ion ($\overset{\oplus}{N}O_2$) attacks at O & p wrt 'O' atom & major product is obtained at p - position.

144. (b)

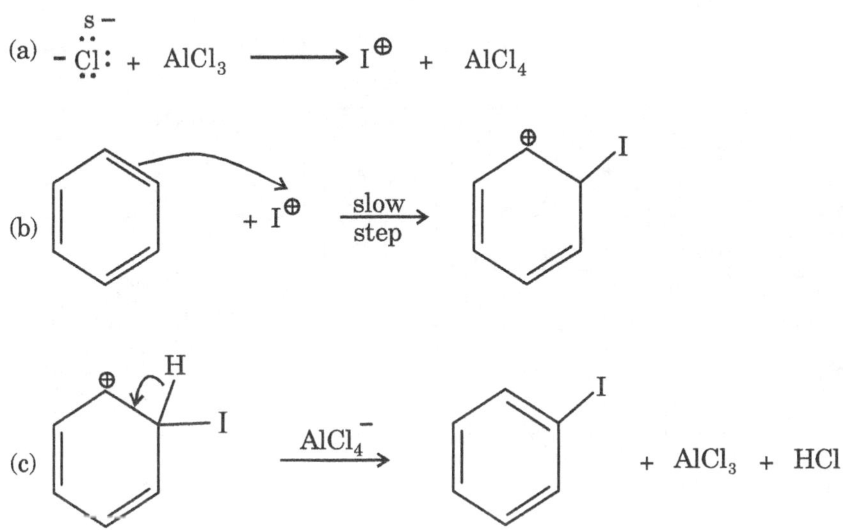

Smart Notes

16

CHAPTER

Environmental Chemistry

Air Pollution

1. Which of the following is not the consequence of global warming?
   (a) Increase in average temperature of earth
   (b) Melting of Himalayan Glaciers
   (c) Rise in sea level
   (d) Eutrophication
2. Which of the following conditions shows polluted environment?
   (a) pH of rain water is 5.6
   (b) Amount of CO_2 in the atmosphere is 0.03%
   (c) BOD is 4 ppm
   (d) Eutrophication
3. Methane gas is responsible for green house effect and global warming. It is produced naturally from
   (a) paddy fields (b) rotten garbage
   (c) coal mines (d) all the above.
4. The incorrect match is
   (a) oxide of S - chlorosis
   (b) oxide of N - lung irritant
   (c) particulate matter - shedding of leaves
   (d) oxide of C - asphyxiation
5. Which of the following statements is incorrect?
   (a) Oxides of nitrogen in atmosphere can cause depletion of ozone layer
   (b) Ozone absorbs infra red rays
   (c) Depletion of O_3 is due to its chemical reaction with halo alkanes
   (d) None of these
6. The fuel gas, whose presence in the exhaust fumes shows incomplete combustion of fuel, is
   (a) SO_2 (b) CO
   (c) NO_2 (d) none of these
7. Which of the following are the hazardous pollutants present in automobile exhaust gases?
   i. N_2 ii. CO
   iii. CH_4 iv. oxides of nitrogen
   (a) ii and iii (b) i and ii
   (c) ii and iv (d) i and iii
8. Which of the following gas causes green house effect to maximum extent?
   (a) CH_4
   (b) Water vapour
   (c) N_2O
   (d) CO_2

9. Match the entries in column-1 with their corresponding entries in

Column-I	Column-II
a. Troposphere	i. prevents UV rays coming to earth
b Stratosphere	ii. region of lower pressure
c. Mesosphere	iii. lowest region of atmosphere
d. Thermosphere	iv. region lies above 50 km above sea level

(a) a-i, b-ii, c-iii, d-iv (b) a-iii, b-i, c-iv, d-ii
(c) a-ii, b-iii, c-i, d-iv (d) a-iv, b-iii, c-ii, d-i

10. Which of the following is a secondary pollutant?
(a) NO (b) CO_2
(c) CO (d) Ketone

11. An example of major and most serious pollutant is
(a) O_2 (b) CO_2
(c) CO (d) N_2

12. A metallic pollutant which interferes with development and maturation of red blood cells. It is
(a) Pb (b) CO_2
(c) Mg (d) Dust

13. Which of the following are sources of dissolved oxygen in water?
a. Atmospheric oxygen
b. Photosynthesis
c. Organic matter
(a) both a and b (b) both b & c
(c) both a and c (d) a only

14. The lowest region of atmosphere is
(a) stratosphere (b) troposphere
(c) mesosphere (d) hydrosphere

15. Which one of the following statements is incorrect regarding green house effect?
(a) It occurs due to high concentration of in atmosphere
(b) It results in warming up of earth
(c) Gases likeCH_4, CFC's also contribute to green house effect
(d) It would result in lowering the level of oceans due to evaporation of water

16. Which of the following harmful effect is not caused by SO_2?
(a) Asthma (b) Asphyxiation
(c) Bronchitis (d) Chlorosis

17. The consequences of global warming may be
(a) increase in average temperature of the earth
(b) melting of Himalayan Glaciers
(c) increased biochemical oxygen demand
(d) both (a) and (b)

18. Which of the following are secondary pollutants?
(a) CO, CO_2, SO_2, hydrocarbon
(b) SO_3, NO_2, HNO_3, H_2SO_4
(c) Both (a) and (b)
(d) None of the above

Smog and Ozone Hole

19. In Antarctica, ozone depletion is due to hydrolysis of
(a) acrolein (b) PAN
(c) SO_2 and SO_3 (d) chlorine nitrate

20. The main constituents present in photochemical smog are
a. PAN
b. unsaturated hydrocarbons
c. acrolein
d. O_3
e. SO_2
(a) a, b, c & d (b) a, c, d & e
(c) a, d & e (d) a, b, c, d & e

21. The basic component of smog is
(a) PAN
(b) PBN
(c) NO_2
(d) all of these

22. Which is incorrect about classical smog?
(a) it is formed by reaction of particulate and SO_2
(b) it is oxidising in nature
(c) it contains SO_2 and carbon
(d) it may cause respiratory problems

23. Match the entries in column I with their corresponding entries in column-ll

Column-I	Column II
a. Photochemical smog	i. Reducing
b Classical smog	ii. SO_2
c. Primary pollutants	iii. Oxidising
d. Chlorosis	iv. NO_2

(a) a-iii,iv ; b-i,ii; c-ii,iv ; d-ii

(b) a-ii,iv ; b-ijii; c-iijv ; d-ii

(c) a-iii,i; b-iijv ; c-i,iv ; d-iv

(d) a-ii,iv ; b-irii; c-iijv ; d-i

24. Classical smog occurs in places of

(a) excess of NO_2 (b) warm dry climate

(c) cool humid climate (d) all of these

25. Which of the following statements about polar stratospheric clouds (PSCs) is not correct?

(a) PSCs are formed over antartica in summer season

(b) PSCs contain ice crystals and play significant role in ozone depletion'

(c) During winters when PSCs are formed over antarctica, a tight whirlpool of wind called polar vortex is formed which surrounds antarctica

(d) These provide surface on which reactions occur to form Cl radicals initiating ozone depletion

26. Which of the following methods are effective in controlling photochemical smog?

a. Use of catalytic converters in automobiles

b. Controlling primary precursors so that generation of secondary precursors is reduced

c. Planting trees like Pinus, Vitis, Quercus etc.

(a) a, b and c (b) a and b only

(c) b and c only (d) only a

27. CFC's is responsible for the decomposition of ozone to oxygen. Which of the following reacts with ozone to form oxygen?

(a) Cl_2 (b) Cl^-

(c) $\dot{Cl}$ (d) F-

28. Classical smog in nature is –

(a) oxidizing

(b) reducing

(c) both oxidizing and reducing

(d) neither oxidizing nor reducing

29. Among the following substances the one which catalyses conversion of SO_2 to SO_3 and causes classical smog is

(a) plants (b) I R radiations

(c) particulates (d) UV radiations

30. Regarding photochemical smog, which of the following statements is wrong?

(a) Its main components are produced by action of sunlight on emssions of automobiles and factories

(b) It has high concentration of oxidizing agents

(c) It contains smoke, fog and SO_2

(d) It is produced in warm, dry and sunny climate

31. Which of the following statements is wrong?

(a) Ozone is not responsible for greenhouse effect

(b) Ozone can oxidise sulphur dioxide present in the atmosphere to sulphur trioxide

(c) Ozone hole is thinning of ozone layer present in stratosphere

(d) Ozone is produced in upper stratosphere by the action of UV rays on oxygen.

32. Which of the following is responsible for depletion of ozone layer in the upper strata of the atmosphere?

(a) Polyhalogens (b) Ferrocene

(c) Fullerenes (d) Freons

Water pollution, Acid Rain

33. Marble acts as a sink for

(a) metallic pollutants (b) pollutants

(c) acidic pollutants (d) none of these

34. Which of the following statements is incorrect?

(a) High concentration of flouride ions causes brown mottling of teeth

(b) Blue baby syndrome is caused due to excess nitrate in drinking water

(c) Killing animal life by depriving them of oxygen by dense plant population in water body is called eutrophication

(d) Sulphate concentration above 500 ppm is safe in drinking water

35. Acid rain is produced by excess

(a) release of CO by incomplete combustion

(b) formation of CO_2 by combustion & respiration

(c) production of NH_3 by industries

(d) release of NO_2 and SO_2 from burning of fossil fuels

36. When huge amount of sewage is dumped in a river, the BOD

(a) will increase

(b) will remain unchanged

(c) will decrease

(d) may increase or decrease

37. Which of the following practices will not come under green chemistry?

(a) Use of CO_2 as solvent instead of Cl_2

(b) Use of H_2O_2 instead of Cl_2 for bleaching

(c) Synthesis of ethanal from ethene in one step

(d) Use of tetrachloroethene as a solvent for dry cleaning

38. Eutrophication causes reduction in

(a) dissolved salts

(b) dissolved oxygen

(c) nutrients

(d) all of these

39. Which of the following statement is NOT true?

(a) Non-polluted rain water should have pH of about 5.6

(b) Clean water should have BOD value of less than 5 ppm

(c) Oxides of S, N and C are the most wide spread air pollutants

(d) The main components of classical smog are produced by the action of sunlight on emissions of automobiles and factories

40. Which among the following is bio-degradable pollutant?

(a) Lead compounds

(b) Mercuric salts

(c) Pesticides

(d) Domestic wastes

41. Now a days, tetra chloroethene $(Cl_2C = CCl_2)$ solvent has been replaced by liquefied CO_2 for the dry cleaning purposes. This is because

(a) of less time consumption

(b) liquid CO_2 will result in less harm to ground water

(c) liquid CO_2 is suspected carcinogen

(d) liquid CO_2 is a better solvent

42. How many of the following statements are correct?

a. Excessive phytoplankton growth with in water causes water pollution

b. $NaClO_3$ and Na_3AsO_3 are herbicides and are used to kill weeds

c. Poly chlorinated biphenyls are used as fluids in transformers

d. Relative contribution of green house gases towards green house effect is

$$CO_2 > CH_4 > CFC's > O_3 > N_2O > H_2O$$

(a) 2 (b) 3

(c) 4 (d) 1

43. Excess of phosphate is water causes growth of

(a) fungi (b) algae

(c) bacteria (d) all of these

44. Flouride ions make the enamel of teeth harder by converting hydroxy apatite into

(a) chlorapatite (b) flourapatite

(c) calcium phosphate (d) none of the above

45. Which of the following industrial chemical pollutant is carcinogenic?

(a) CFC (b) PAN

(c) PCB (d) DDT

46. Clean water should have BOD value of less than_______ ppm whereas water having BOD value of ppm _______ or more is classified as highly polluted

(a) 6,8 (b) 4,10

(c) 5,12 (d) 5,17

47. SO_2 and NO_2 cause pollution by increasing

(a) Acidity

(b) Alkalinity

(c) Buffer action

(d) Both (a) and (c)

48. Match the entries in column I with their corresponding entries in column II

Column-I	Column-II
a. SO_2	i. Secondary pollutant
b. PAN	ii. Biodegradable
c. Cow dung	iii. Non-biodegradable
d. DDT	iv. Primary pollutant

(a) a-ii, b-iii, c-i, d-iv

(b) a-iii, b-i, c-ii, d-iv

(c) a-iv, b-i, c-ii, d-iii

(d) a-iii, b-ii, c-i, d-iv

49. Match the entries in column I with their corresponding entries in column II

Column-I	Column-II
a. SO_2	i. Sink of
b. Rain water	ii. Ammonium salts
c. Plants	iii. 5.6 pH
d. Atmospheric haze	iv. Green house gas

(a) a-ii, b-iii, c-i, d-iv

(b) a-iii, b-i, c-ii, d-iv

(c) a-iii, b-iv, c-ii, d-i

(d) a-iv, b-iii, c-i, d-ii

50. Which of the following statements is wrong?

(a) Greenhouse effect is responsible for global warming

(b) Ozone layer does not permit infracted radiation from the sun to reach the earth

(c) Acid rain is mostly because of oxides of nitrogen and sulphur

(d) Chlorofluorocarbons are responsible for ozone layer depletion

Answers

1. (d)	**2.** (d)	**3.** (d)	**4.** (c)	**5.** (b)	**6.** (b)	**7.** (c)	**8.** (d)	**9.** (b)	**10.** (d)
11. (c)	**12.** (a)	**13.** (a)	**14.** (b)	**15.** (d)	**16.** (b)	**17.** (d)	**18.** (b)	**19.** (d)	**20.** (a)
21. (c)	**22.** (b)	**23.** (a)	**24.** (c)	**25.** (a)	**26.** (a)	**27.** (c)	**28.** (b)	**29.** (c)	**30.** (d)
31. (a)	**32.** (d)	**33.** (c)	**34.** (d)	**35.** (d)	**36.** (a)	**37.** (d)	**38.** (b)	**39.** (d)	**40.** (d)
40. (c)	**41.** (b)	**42.** (c)	**43.** (b)	**44.** (b)	**45.** (c)	**46.** (d)	**47.** (a)	**48.** (c)	**49.** (d)
50. (b)									

Smart Notes